Chemistry of Arsenic, Antimony and Bismuth

Edited by

N.C. NORMAN
School of Chemistry
University of Bristol
Bristol, UK

BLACKIE ACADEMIC & PROFESSIONAL
An Imprint of Chapman & Hall
London · Weinheim · New York · Tokyo · Melbourne · Madras

Published by Blackie Academic & Professional, an imprint of Thomson Science, 2–6 Boundary Row, London SE1 8HN, UK

Thomson Science, 2–6 Boundary Row, London SE1 8HN, UK

Thomson Science, 115 Fifth Avenue, New York, NY 10003, USA

Thomson Science, Suite 750, 400 Market Street, Philadelphia, PA 19106, USA

Thomson Science, Pappelallee 3, 69469 Weinheim, Germany

First edition 1998

Thomson Science is a division of International Thomson Publishing I(T)P

Typset in 10/12 pt Times by Florencetype Ltd., Stoodleigh, Devon

Printed in Great Britain by St Edmundsbury Press Ltd, Bury St Edmunds, Suffolk

ISBN 0 7514 0389 X

A catalogue record for this book is available from the British Library

Library of Congress Catalog Card Number: 97–77012

Printed on acid-free text paper, manufactured in accordance with ANSI/NISO Z39.48-1992 (Permanence of Paper)

Contents

Contributors

C.J. Carmalt	Department of Chemistry and Biochemistry, The University of Texas at Austin, Austin, TX 78712, USA
W.I. Cross	Department of Chemistry, University of Manchester Institute of Science and Technology, Manchester M60 1QD, UK
S.M. Godfrey	Department of Chemistry, University of Manchester Institute of Science and Technology, Manchester M60 1QD, UK
C.A. McAuliffe	Department of Chemistry, University of Manchester Institute of Science and Technology, Manchester M60 1QD, UK
A.G. Mackie	Department of Chemistry, University of Manchester Institute of Science and Technology, Manchester M60 1QD, UK
Y. Matano	Department of Chemistry, Graduate School of Science, Kyoto University, Sakyo-ku, Kyoto 606-01, Japan
N.C. Norman	School of Chemistry, University of Bristol, Bristol BS8 1TS, UK
H. Onishi	Kasuga 3-chome 15-5, Tsukuba-shi, Ibaraki-ken 305, Japan
I.J. Polmear	Department of Materials Engineering, Monash University, Clayton, Vic 3168, Australia
R.G. Pritchard	Department of Chemistry, University of Manchester Institute of Science and Technology, Manchester M60 1QD, UK
J. Reglinski	Department of Pure and Applied Chemistry, University of Strathclyde, Thomas Graham Building, 295 Cathedral Street, Glasgow G1 1XL, UK
H. Suzuki	Department of Chemistry, Graduate School of Science, Kyoto University, Sakyo-ku, Kyoto 606-01, Japan
K.H. Whitmire	Department of Chemistry, Rice University, 6100 Main Street, Houston, TX 77005-1892, USA

Preface

This book deals with the chemistry of the elements arsenic, antimony and bismuth, although in a book of manageable size there is no possibility of providing a comprehensive treatment of all aspects of their chemistry, and some topics are inevitably covered in more detail than others. Nevertheless, the aim is to provide as broad a coverage as possible in the hope that there will be something for everybody with any interest in the chemistry of these three elements. Occasional duplication of subject matter in two or more chapters is also inevitable to a certain extent although this has been kept to a minimum as much as possible.

Chapter 1 is a brief introductory chapter which addresses some general properties and features of the heavier group 15 elements and their compounds (often including nitrogen and phosphorus for the sake of completeness) concentrating particularly on trends in physical and chemical properties. In terms of physical properties there is some overlap here in tabulated material with chapters 2 and 8, and many of the compounds used to illustrate various trends are also covered in subsequent relevant chapters.

Chapter 2 deals with the metallurgy of the elements and begins with a historical survey of their discovery and use. Subsequent sections deal with the occurrence of the elements in nature (also chapter 8), extraction and refining methods, a brief summary of some element characteristics and physical properties (some of which are also covered in chapter 1, Table 1.1) and a final section on uses of the elements which includes subsections on alloys, semiconductors and superconductors.

Chapter 3 focuses on the synthesis, physical properties and reactivity of the inorganic derivatives of the elements, dealing first with the hydrides followed by a detailed account of the chemistry of the halides, oxides and heavier chalcides, and oxoanions. Many of the latter two classes of compound occur naturally as minerals so there is some overlap here with chapters 2 and 8. As a final section, a range of binary and ternary intermetallics is surveyed.

Chapter 4 deals specifically with aspects of coordination chemistry and solution chemistry. The first section focuses on the Lewis basic properties of the tervalent compounds of arsenic, antimony and bismuth, and the following three sections cover Lewis acid chemistry, concentrating in turn on halogenoanions, complexes with group 15 donors and complexes with group 16 donor ligands. A final section describes aspects of the aqueous chemistry of these elements, there being some overlap here with chapter 8.

Chapters 5 and 6 cover organocompounds of arsenic, antimony and bismuth, chapter 5 dealing with the former two elements and chapter 6 with the latter element. In both chapters the emphasis is on synthesis, physical properties and reactivity, including the uses of these compounds in organic transformations. Both chapters are broadly subdivided into sections which deal separately with compounds in the (III) and (V) oxidation states and a brief treatment of compounds containing element–element multiple bonds is also provided.

Chapter 7 focuses on a topic which has developed into a very significant field in recent years, namely organotransition metal compounds with element to transition metal bonds, and is organized according to a classification of compound type. Although this chapter necessarily concentrates only on the elements arsenic, antimony and bismuth, it serves to illustrate a much broader field which represents a fusion between more classical organotransition metal and main group element chemistry.

Chapter 8 covers the rôle of the three elements in the field of medicinal and environmental chemistry. The first part of this chapter addresses the occurrence of the elements in the environment both from natural (also chapter 2) and man-made sources, and is followed by sections on such topics as toxicology, medicinal chemistry and immunochemistry and radiochemistry.

Finally, chapter 9 deals with analytical methods and covers all the various methods of analysis for the three elements, including analysis of biological samples.

1 Arsenic, antimony and bismuth: some general properties and aspects of periodicity

C.J. CARMALT and N.C. NORMAN

1.1 Some general characteristics of the group 15 elements

1.1.1 Physical properties of the elements

Group 15 constitutes the third column of the p-block of the periodic table and comprises the elements nitrogen, phosphorus, arsenic, antimony and bismuth (often referred to as the pnictogens or pnictides) (refs [1–8] are general references relevant to the material covered in this chapter; also relevant are the annual surveys published in *J. Organomet. Chem.* and *Coord. Chem. Rev.*, and the element surveys in the Gmelin series). For all elements, the ground state electronic configuration is ns^2np^3 (Table 1.1) which results in common oxidation states of +3 (III) and +5 (V), although the relative stabilities of these differ for the various elements, as discussed in Section 1.2.1 and in subsequent chapters. For the heavier elements, other oxidation states are only rarely encountered, at least in molecular compounds, but in the case of nitrogen, NO and NO_2 represent genuine examples of the +2 and +4 oxidation states respectively.

Despite being in the same group, however, the properties of the elements vary to a considerable degree as the group is descended. Thus whereas nitrogen and phosphorus are typical non-metals, arsenic and antimony are more usually described as metalloids or semi-metals, and bismuth has many properties associated with metallic behaviour. Indeed, group 15 (perhaps together with group 16) is probably second only to group 14 in exhibiting the full range of element chemistries from non-metallic through semi-metallic to metallic in a single group, part of the reason for which can be traced to the substantial decrease in element electronegativity as the group is descended.

In fact, much of the chemistry of the group 15 elements (indeed all elements) can be appreciated or rationalized through consideration of certain atomic properties, whether of isolated or bound atoms. Thus, the relationship between non-metallic, semi-metallic and metallic properties and electronegativity has been discussed in some detail by Allen in particular [9] who has pointed out that, on the standard Pauling scale of

Table 1.1 Some properties of the group 15 elements nitrogen, phosphorus, arsenic, antimony and bismuth. Note: some of the information presented in this table is also given in Table 2.2 and in Tables 9.1–9.3.

Property	N	P	As	Sb	Bi
(a) Properties of the isolated atom[a–c,e,f]					
Atomic number	7	15	33	51	83
Naturally occurring isotopes (percentage abundance is given in parentheses)[a,b]	^{14}N (99.63) ^{15}N (0.37)	^{31}P (100)	^{75}As (100)	^{121}Sb (57.4) ^{123}Sb (42.6)	^{209}Bi (100)
Relative atomic mass ($^{12}C = 12.0000$)[a]	14.00674	30.973762	74.9216	121.75	208.9804
Ground-state electron configuration (the term is given in parentheses)[a]	$[He]2s^22p^3(^4S_{3/2})$	$[Ne]3s^23p^3(^4S_{3/2})$	$[Ar]3d^{10}4s^24p^3(^4S_{3/2})$	$[Kr]4d^{10}5s^25p^3(^4S_{3/2})$	$[Xe]4f^{14}5d^{10}6s^26p^3(^4S_{3/2})$
Ionization energies (kJ mol^{-1})[a,c]					
$M \rightarrow M^+$	1402.3	1011.7	947.0	833.7	703.2
$M^+ \rightarrow M^{2+}$	2856.1	1903.2	1798	1794	1610
$M^{2+} \rightarrow M^{3+}$	4578.0	2912	2735	2443	2466
$M^{3+} \rightarrow M^{4+}$	7474.9	4956	4837	4260	4372
$M^{4+} \rightarrow M^{5+}$	9444.9	6274	6043	5400	5400
Main lines in atomic spectrum: λ (nm) (species in parentheses)[a,b,k]	**399.500** (II) 463.054 500.515 (II) 567.956 (II) 746.831 (I) 1246.962 (I)	213.617 (I) (AA) **952.573** (I) **956.344** (I) 979.685 (I) 1648.292 (I)	**193.759** (I) (AA) 419.008 (II) 445.847 (II) 446.635 (II) 449.423 (II) 450.766 (II) 454.348 (II)	206.833 (I) (AA) 217.581 (I) **231.147** (I) 252.852 (I) 259.805 (I)	202.121 (I) **206.170** (I) 211.026 (I) 223.061 (I) (AA) 289.798 (I) 306.772 (I)
Electron affinity [$M(g) \rightarrow M^-(g)$] (kJ mol^{-1})[a,e,h]	−7	72.0	78	101	91.3
Effective nuclear charge, Z^*					
Slater value[a]	3.90	4.80	6.30	6.30	6.30
Clementi value[a]	3.83	4.89	7.45	9.99	13.34
Froese–Fischer value[a]	3.46	5.28	8.98	12.37	16.90
Static average electric dipole polarizability for ground-state atom(Å^3)[b]	1.10	3.63	4.31	6.6	7.4

Thermal neutron capture cross-section (barns)[a]	1.91	0.180	4.30	5.4	0.034
Nuclei accessible to NMR (nuclear spin, I, in parentheses) I[a]	$^{14}N(+1)$ $^{15}N(-\frac{1}{2})$	$^{31}P(+\frac{1}{2})$	$^{75}As(-\frac{3}{2})$	$^{121}Sb(+\frac{5}{2})$ $^{123}Sb(+\frac{7}{2})$	$^{209}Bi(-\frac{9}{2})$
(b) Properties of the bound atom					
Electronegativity, χ					
Pauling scale[a]	3.04	2.19	2.18	2.05	2.02
Allred–Rochow scale[a]	3.07	2.06	2.20	1.82	1.67
Allen scale[d]	3.066	2.589	2.211	1.984	—
Atomic radius(Å)[a,f]	0.71	0.93 (white)	1.25	1.82	1.55
Single-bond covalent radius (Å)[a,f]	0.70	1.10	1.21	1.41	1.52
van der Waals radius (Å)[a,f]	1.54	1.90	2.00	2.20	2.40
X-ray diffraction: mass absorption coefficient, μ/ρ (cm^2) g^{-1})[a]					
CuK_α	7.52	74.1	83.4	270	240
MoK_α	0.916	7.89	697	33.1	120
(c) Properties of the element					
Crystal structure [cell dimensions (Å) in parentheses] and space group[a,g,i]	α-N_2: cubic ($a = 5.644$), $P2_13$; β-N_2: hexagonal close-packed ($a = 4.042$, $c = 6.601$), $P6_3/mmc$	α-P_4 white: cubic ($a = 18.51$), $I\bar{4}3m$; β-P_4: rhombohedral ($a = 3.377$, $c = 880.1$), $R\bar{3}m$; γ-P_4 white: cubic ($a = 2.377$), $Pm3m$; Red P: cubic ($a = 11.31$), $Pm3m$ or $P\bar{4}3$;	α-As: rhombohedral ($a = 4.1318$; $\alpha = 54° 10'$) $R\bar{3}m$; β-As, hexagonal ($a = 3.760$, $c = 10.548$)	α-Sb grey: rhombohedral ($a = 4.3084$; $c = 1124.7$), $R\bar{3}m$; cubic ($a = 2.986$), $Pm3m$; Metal: hexagonal close-packed ($a = 3.369$, $c = 5.33$) $P6_3/mmc$	α-Bi: rhombohedral ($a = 4.5495$, $c = 11.862$), $R\bar{3}m$

Table 1.1 *continued*

Property	N	P	As	Sb	Bi
		Hittorf's P: monoclinic ($a = 9.21$, $b = 9.15$, $c = 22.60$; $\beta = 1.061°$), $P2/c$; Black P: orthorhombic ($a = 3.316$, $b = 10.478$, $c = 4.3763$), $Cmca$			
Melting point (K)[a,b,f]	63.29	317.3 (P_4) 683 (red)[l]	1090 (α)[l]	903.89	544.5
Boiling point (K)[a,b,f]	77.4	553 (P_4)	889 (sublimes)	1908	1833
ΔH_{fusion} at melting point (kJ mol^{-1})[a,b,f]	0.720	2.51 (P_4)	27.7	20.9	10.48
ΔH_{vap} at 298.15 K (kJ mol^{-1})[a,b,f]	5.577	51.9 (P_4)	31.9	67.91	179.1
Standard entropy, S, at 298.15 K (J K^{-1} mol^{-1})[a]	191.61 gas (N_2) 153.298 (atoms)	41.09 solid (P_4) 22.80 solid (red) 163.193 gas	25.1 solid (α) 174.21 gas	45.69 solid 180.27 gas	56.74 solid 187.005 gas
Density (kg m^{-3})[a]	1026 (solid, 21 K)	1820 (P_4, 293 K)	5780 (α), 4700 (β) (293 K)	6691 (293 K)	9747 (293 K)
Electrical resistivity (Ω m)[a]	—	1×10^9 (P_4, 293 K)	26×10^{-8} (273 K)	39×10^{-8} (273 K)	106.8×10^{-8} (293 K)
Thermal conductivity at 300 K (W m^{-1} K^{-1})[a]	0.02598	0.235 (P_4)	50.0 (α) (300 K)	24.3 (300 K)	7.87 (300 K)
(d) Properties of the cations, and redox behaviour					
Ionic radius (Å)[a,b,e]					
M^{5+}	0.13[m]	0.35	0.46	0.62	0.74
M^{3+}	0.16[m]	0.44	0.58	0.76	0.96
ΔH_f [M^+(g)] (K cal mol^{-1})[b,c,e]	449.589	329.88	300.12	263.46	219.07
ΔH_f [M^{3+}(g)] (K cal mol^{-1})[b,c,e]	2229.41	1481.75	1422.14	1231.6	1196.4

	N	P	As	Sb	Bi
Standard reduction potentials, E (V)[a,b]					
$M^{III}_{(aq)} + 3e^- \rightarrow M_{(s)}$, $a_{H^+} = 1$	—	–0.502 [H_3PO_3]	0.240 [$HAsO_2$]	0.204 [$SbO]^+$	0.317 (neutral)
$M^{III}_{(aq)} + 3e^- \rightarrow M_{(s)}$, $a_{OH^+} = 1$	—	–1.73 [$HPO_3]^{2-}$	–0.68 [$AsO_2]^-$	–0.639 [$Sb(OH)_4]^-$	—
$M^{V}_{(aq)} + 2e^- \rightarrow M^{III}_{(aq)}$, $a_{H^+} = 1$	0.01 [$NO_3]^-$	–0.276 [H_3PO_4]	–0.560 [H_3AsO_4]	0.605 [Sb_2O_5]	*c.* 2 (neutral)
$M^{V}_{(aq)} + 2e^- \rightarrow M^{III}_{(aq)}$, $a_{OH^+} = 1$	0.94 [$NO_3]^-$	–1.12 [$PO_4]^{3-}$	–0.67 [$AsO_4]^{3-}$	–0.465 [$Sb(OH)_6]^-$	—
$M_{(s)} + 3e^- \rightarrow M^{-III}_{(aq)}$, $a_{H^+} = 1$	—	–0.063	–0.225	—	–0.97 (neutral)
$M_{(s)} + 3e^- \rightarrow M^{-III}_{(aq)}$, $a_{OH^+} = 1$	—	–0.89	1.37	–1.338	—
(e) Environmental properties[aj]					
Abundance					
Sun (relative to H = 1×10^{12})	8.71×10^7	3.16×10^5	n.a.	10	<80
Earth (ppm)	25	1000	1.5	0.2	0.048
Seawater (ppm)					
Atlantic surface	0.00008	0.0015	1.45×10^{-3}	*c.* 0.3×10^{-3}	5.1×10^{-8}
Atlantic deep	0.27	0.042	1.53×10^{-3}		n.a.
Pacific surface	0.00008	0.0015	1.45×10^{-3}		4×10^{-8}
Pacific deep	0.54	0.084	1.75×10^{-3}		0.4×10^{-8}
Residence time in ocean (in years) (oxidation state in parentheses)	6000 (V)	100 000 (V)	90 000 (V)	*c.* 3.5×10^5 (III)	n.a. (III)
Geological data[a,b,f,j]					
Chief ores and sources	Liquid air	Fluorapatite, [$Ca_5(PO_4)_3F$], occurs as vast deposits	realgar [As_4S_4], orpiment [As_2S_3], loellingite [$FeAs_2$]	stibnite [Sb_2S_3], ullmanite [NiSbS],	bismite [α-Bi_2O_3], bismuthinite [Bi_2S_3], bismutite [$(BiO)_2CO_3$], main source is by-product from Pb and Cu smelters
World production (tonnes per year)	44×10^6	153×10^{-6} (phosphate rock); $1.2 \times 10^6(P_4)$	47 000 (As_2O_3)	53 000	3000
Reserves (tonnes)	3.9×10^{15}	5.7×10^9	n.a.	44×10^{-6}	n.a.

Table 1.1 *continued*

Property	N	P	As	Sb	Bi
Biological role[a,j]	Constituent element of DNA; nitrogen cycle in nature	Constituent of DNA; phosphate cycle in nature; P_4 and many phosphorus compounds highly toxic	Essential to some species including humans; toxic; stimulatory carcinogenic	none; toxic; stimulatory	none; non-toxic
Human body (dry mass, g)	1800	780	0.018	n.a., but low	n.a., but low
Daily dietary intake (mg)	n.a., but high	900–1900	0.04–1.4	0.002–1.3	0.005–0.02
Toxic intake (mg)	some nitrogen compounds are toxic	Phosphates non-toxic	5–50	100	n.a.
Lethal intake (mg)	n.a.	60 (P_4)	50–340	n.a.	n.a.

n.a. – not available.

[a]Source: Emsley, J. (1991) *The Elements*, 2nd edn, Clarendon Press, Oxford.

[b]Source: Lide, D.R. (ed.) (1992–1993) *Handbook of Chemistry*, 73rd edn, CRC Press, Boca Raton, FL.

[c]Source: Johnson, D.A. (1982) *Some Thermodynamic Aspects of Inorganic Chemistry*, 2nd edn, Cambridge University Press, Cambridge.

[d]Source: Allen, L.C. (1989) *J. Am. Chem, Soc.*, **111**, 9003.

[e]Source: Dasent, W.E. (1982) *Inorganic Energetics: An Introduction*, 2nd edn, Cambridge University Press, Cambridge.

[f]Source: Dean, J.A. (1985) *Lange's Handbook of Chemistry*, 14th edn, McGraw-Hill, New York.

[g]Source: Cotton, F.A. and Wilkinson, G. (1988) *Advanced Inorganic Chemistry*, 5th edn, John Wiley, New York.

[h]Source: Greenwood, N.N. and Earnshaw, A. (1984) *Chemistry of the Elements*, Pergamon, Oxford.

[i]Source: Wells, A.F. (1984) *Structural Inorganic Chemistry*, 5th edn, Clarendon Press, Oxford.

[j]Source: Cox, P.A. (1989) *The Elements: Their Origin, Abundance and Distribution*, Oxford University Press, Oxford.

[k]I denotes lines arising from the neutral atom. Stronger lines only are listed, with the strongest shown in bold type. Lines with applications in atomic absorption spectroscopy are indicated by AA in parentheses.

[l]Under pressure.

[m]Calculated value.

electronegativities (0–4), a so-called metalloid band exists from about 1.9 to 2.2. Thus an element with an electronegativity within this range is a metalloid whereas elements with lower values are metals, and those with higher values are non-metals, the origin of this trend arising from the fact that increasing element electronegativity favours localized covalent bonding as opposed to delocalized metallic bonding. An examination of the Allen electronegativities shown in Table 1.1(b) reveals that, on this basis, nitrogen and phosphorus are clearly non-metals, whereas arsenic is borderline non-metal/metalloid and antimony is within the metalloid band (an Allen electronegativity value is not currently available for bismuth and, strictly speaking, this scale should be in Table 1.1(a) since Allen electronegativities are calculated for isolated atoms whereas those of the Pauling scale derive from thermochemical data for atoms in molecules).

Using values from either the Pauling or Allred–Rochow scales (amongst the two most widely used), it can be seen that whereas nitrogen clearly still lies in the non-metallic region, phosphorus lies within the metalloid range and should therefore be expected to have some metalloid properties; consistent with this is the fact that the black phosphorus allotrope is a semiconductor. Bismuth has the lowest electronegativity of all the group 15 elements and on the basis of the Allred–Rochow scale should be considered, together with antimony, as metallic, although the Pauling scale value would place both in the metalloid band. In fact, bismuth in particular has properties in common with metals and metalloids, the latter properties exemplified by its elemental structure (see later) and high resistivity (resistivities for all the group 15 elements except nitrogen are given in Table 1.1(c)). Thus, depending upon the electronegativity scale used, it should be expected that all the elements except nitrogen will have properties associated with metalloids, although non-metallic features should also be anticipated for phosphorus and arsenic, and bismuth and antimony will exhibit some properties in common with metals.

As a final point regarding electronegativity, it should be noted that the electronegativities of phosphorus and arsenic are quite similar as are those of antimony and bismuth; indeed for the former pair of elements the Allred–Rochow scale actually assigns a higher electronegativity to arsenic. In general, electronegativities are expected to decrease down a group [decreasing electronegativity down a group mirrors decreasing ionization energies (see below) and shows that the increasing size of the atoms, and hence increasing average distance of the valence electrons from the nucleus, more than offsets the increasing effective nuclear charges found on descending the group (see Table 1.1 for values of effective nuclear charge calculated by three different methods)] so we may restate the above features in terms of arsenic and bismuth both having higher than expected values. This apparent anomaly derives from the filling of the 3d row prior to arsenic and the 4f row prior to bismuth, both of which lead to higher

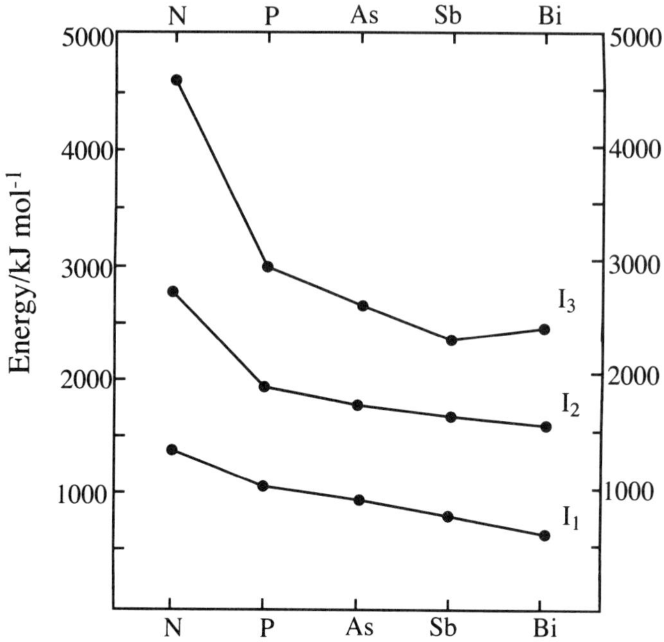

Figure 1.1 A graph of the first, second and third ionization energies for the group 15 elements.

effective nuclear charges as a result of inefficient screening of the nuclear charge by the 3d and 4f electrons, respectively.

Atomic properties closely related to electronegativity are ionization energy and electron affinity. Table 1.1(a) lists values for the first five ionization energies of all the group 15 elements together with a value for the electron affinity of the neutral atom. These values are mostly unremarkable in that they follow expected trends [10–12] but one particular point to note is that although the ionization energies decrease as expected on descending the group some of those for arsenic are quite similar to the corresponding values for phosphorus, as is also the case for antimony and bismuth; indeed for antimony and bismuth the third and fourth ionization energies of bismuth are actually slightly greater than those of antimony {an interesting and related point to note for phosphorus and arsenic is that the ionization energy of $AsCl_3$ (11.7 eV) is greater than that for PCl_3 (10.5 eV) [13]}. Both observations reflect the presence of the 3d and 4f rows which precede arsenic and bismuth in the periodic table; this is not only relevant to electronegativity values as mentioned above but also, in the latter case, to relativistic effects (Section 1.2.5). A graphical representation of the first three ionization energies for all the elements of group 15 is given in Fig. 1.1. With regard to the electron affinities, all

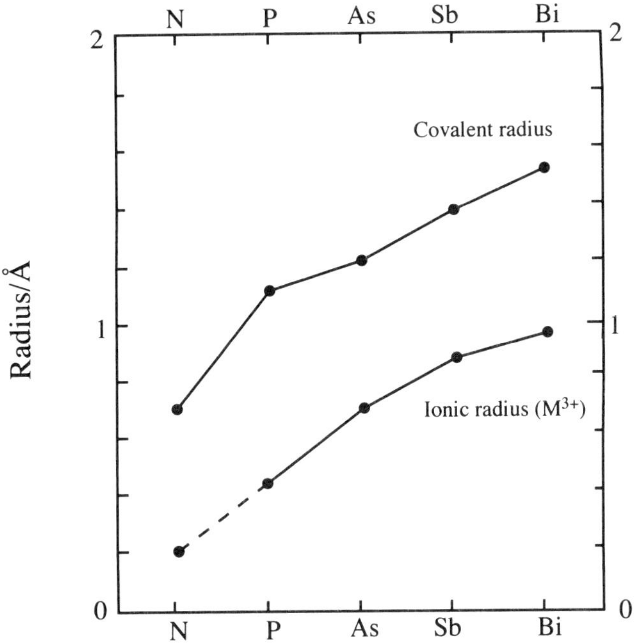

Figure 1.2 A graph of covalent and ionic radii for the +3 oxidation state for the group 15 elements.

are exothermic and of the order of 75–100 kJ mol^{-1}, with the obvious exception of the small endothermic value for nitrogen which can be traced to the disruption of a half-filled shell in a small atom.

Another trend to note is that relating to the various atomic radii (elemental or atomic, covalent, ionic and van der Waals). Some values are listed in Table 1.1(b), and a graphical illustration of covalent and ionic radii for the +3 oxidation state is presented in Fig. 1.2. From both sets of data, the relatively smooth increase in the size of the atoms as the group is descended is apparent.

A list of the properties of the better characterized isotopes of the group 15 elements (including nitrogen and phosphorus) is given in Table 1.2 (also in Table 2.1, chapter 2) which contains data on half-life and decay mode for radioactive isotopes together with information on nuclear spin and magnetic moment. In fact, the number of stable isotopes in group 15 is quite small compared with some groups [Table 1.1(a)], and naturally occurring phosphorus, arsenic and bismuth are all monoisotopic – ^{31}P, ^{75}As and ^{209}Bi (an interesting point to note is that ^{209}Bi is the heaviest stable isotope of any element). Antimony has two stable isotopes of roughly equal abundance, ^{121}Sb (57.4%) and ^{123}Sb (42.6%), whereas

Table 1.2 Properties of the better characterized isotopes of the group 15 atoms.

Nuclide	Atomic mass[a]	Half-life	Decay mode	Nuclear spin, I[b]	Nuclear magnetic moment, μ[c]
^{12}N	12.018613	11.00 ms	β^+, $\beta^+\alpha$	1+	0.457
^{13}N	13.005738	9.97 months	β^+	1/2–	0.3222
^{14}N	14.003074	stable	—	1+	0.40376
^{15}N	15.000108	stable	—	1/2+	–0.28319
^{16}N	16.006099	7.13 s	$\beta^-\alpha$	2–	—
^{17}N	17.008450	4.17 s	β^-, β^-n	1/2–	—
^{18}N	18.014081	0.63 s	β^-	1–	—
^{28}P	27.992313	270 ms	β^+	3+	—
^{29}P	28.981803	4.14 s	β^+	1/2+	1.2349
^{30}P	29.978307	2.50 months	β^+	1+	—
^{31}P	30.973762	stable	—	1/2+	1.13160
^{32}P	31.973907	14.28 days	β^-	1+	–0.2524
^{33}P	32.971725	25.3 days	β^-	1/2	—
^{34}P	33.973636	12.4 s	β^-	1+	—
^{35}P	34.973232	47 s	β^-	1/2+	—
^{67}As	66.939190	43 s	β^+, EC	5/2–	—
^{68}As	67.936790	2.53 months	β^+	3+	—
^{69}As	68.932280	15.1 months	β^+, EC	5/2–	1.2
^{70}As	69.930929	52.6 months	β^+, EC	4+	2.1061
^{71}As	70.927114	62 h	β^+, EC	5/2–	1.6735
^{72}As	71.926755	26.0 h	β^+	2–	–2.1566
^{73}As	72.923827	80.3 days	EC	3/2–	—
^{74}As	73.923827	17.78 days	β^+, EC	2–	–1.597
^{75}As	74.921594	stable	—	3/2–	1.43947
^{76}As	75.922393	26.3 h	β^-	2–	–0.906
^{77}As	76.920646	38.8 h	β^-	3/2–	—
^{78}As	77.921830	1.515 h	β^-	2–	—
^{79}As	78.920946	9.0 months	β^-	3/2–	—
^{80}As	79.922528	16 s	β^-	1+	—
^{81}As	80.922131	33 s	β^-	3/2–	—
^{82m}As	—	14 s	β^-	5–	—
^{82}As	81.924769	19 s	β^-	1+	—
^{83}As	82.924980	13 s	β^-	–	—
^{84}As	83.929060	5.5 s	β^-, n	1–	—
^{85}As	84.931820	2.03 s	β^-, n	3/2–	—
^{109}Sb	108.918143	18.3 s	β^+, EC	5/2+	—
^{110}Sb	109.916770	23.5 s	β^+, EC	3+	—
^{111}Sb	110.913220	75 s	β^+, EC	5/2+	—
^{112}Sb	111.912411	51 s	β^+, EC	3+	—
^{113}Sb	112.909372	6.7 months	β^+, EC	5/2+	—
^{114}Sb	113.909110	3.5 months	β^+, EC	3+	—
^{115}Sb	114.906601	32.1 months	β^+, EC	5/2+	3.46
^{116m}Sb	—	60 months	β^+, EC	8–	—
^{116}Sb	115.906800	16 months	β^+, EC	3+	2.72
^{117}Sb	116.904841	2.80 h	β^+, EC	5/2+	3.4
^{118m}Sb	—	5.00 h	EC	8–	2.3
^{118}Sb	117.905534	3.6 months	β^+, EC	1+	2.5
^{119}Sb	118.903948	36.1 h	EC	5/2+	3.45
^{120m}Sb	—	5.76 days	EC	8–	2.34
^{120}Sb	119.903821	15.9 months	β^+, EC	1+	2.3
^{121}Sb	120.903821	stable	—	5/2+	3.363
^{122m}Sb	—	4.21 months	IT	8–	—
^{122}Sb	121.905179	2.71 days	β^-	2–	–1.90

Table 1.2 *continued*

Nuclide	Atomic mass[a]	Half-life	Decay mode	Nuclear spin, I^b	Nuclear magnetic moment, μ^c
^{123}Sb	122.904216	stable	—	7/2+	2.550
^{124m2}Sb	—	20.3 months	IT	8–	—
^{124m1}Sb	—	96 s	IT	5+	—
^{124}Sb	123.905038	60.20 days	β^-	3–	1.2
^{125}Sb	124.905252	2.76 years	β^-	7/2+	2.63
^{201}Bi	200.976930	1.80 h	EC	9/2–	—
^{202}Bi	201.977660	1.72 h	β^+, EC	5+	—
^{203}Bi	202.976830	11.76 h	β^+, EC	9/2–	4.62
^{204}Bi	203.977740	11.2 h	EC	6+	4.28
^{205}Bi	204.977365	15.31 days	EC	9/2–	4.16
^{206}Bi	205.978478	6.243 days	EC	6+	4.60
^{207}Bi	206.978446	32.2 years	EC	9/2–	4.08
^{208}Bi	207.979717	3.86×10^5 years	EC	5+	—
^{209}Bi	208.980374	stable	—	9/2–	4.111
^{210m}Bi	—	3.0×10^6 years	α	9–	—
^{210}Bi	209.984095	5.01 days	β^-	1–	–0.0445
^{211}Bi	210.987255	2.14 months	α	9/2–	—

Sources: Emsley, J., *The Elements*, 2nd edn, Clarendon Press, Oxford, 1991; Lederer, C.M. and Shirley, V.S. *Table of Isotopes*, 7th edn, John Wiley, New York, 1978; Lide, D.R., *Handbook of Chemistry*, 73rd edn, CRC Press, Boca Raton, FL, 1992–1993.

[a]Relative to ^{12}C (12.0000).
[b]Units: $h/2\pi$.
[c]Units: nuclear magnetons.
Note: superscript m in isotope nomenclature indicates the nuclide is metastable.

nitrogen is found mostly as ^{14}N (99.63%) with only small amounts of ^{15}N (0.37%) occurring naturally. Of the many known radioactive isotopes, some have found use in radiochemical therapy as discussed in more detail in chapter 8.

A widely used spectroscopic method which relies on nuclear properties is nuclear magnetic resonance (NMR) spectroscopy and a list of important parameters for the naturally occurring nuclei is given in Table 1.3. For the three elements of primary concern here, however, the absence of a spin ½ nucleus and the resulting quadrupolar nature of the nuclei means that arsenic, antimony and bismuth NMR has been little used in contrast to the much more widely used ^{31}P and ^{14}N or ^{15}N NMR since signal line widths tend to be very broad in most compounds [14]. Another nuclear technique which has found some application however, ^{121}Sb Mössbauer spectroscopy [15, 16].

1.1.2 Occurrence of the elements

Some information on the occurrence of arsenic, antimony and bismuth in nature and their primary mineral sources is given in Table 1.1(e)–(g)

Table 1.3 NMR properties of nuclei of the group 15 atoms.[a]

Property	^{14}N	^{15}N	^{31}P	[^{75}As]	[^{121}Sb]	[^{123}Sb]	[^{209}Bi]
Natural abundance(%)	99.63	0.37	100	100	57.3	42.7	100
Nuclear spin ($h/2\pi$)	1+		½+	3/2–	5/2+	7/2+	9/2–
Magnetic moment (μ)	0.403 76	–0.283 19	1.13160	1.439 47	3.3592	2.5466	4.110
Relative sensitivity (^1H = 1.00)	1.01×10^{-3}	1.04×10^{-3}	6.63×10^{-2}	2.51×10^{-2}	0.16	4.57×10^{-2}	0.13
Receptivity (^{13}C = 1.00)	5.69	0.0219	377	143	520	111	777
Magnetogyric ration (10^7 rad T^{-1} s^{-1})	1.9331	–2.7116	10.8289	4.5804	6.4016	3.4668	4.2986
Quadrupole moment (m^2)	1.61×10^{-30}	—	—	0.30×10^{-28}	-0.53×10^{-28}	-0.68×10^{-28}	-0.4×10^{-28}
Frequency (^1H = 100 MHz; 2.3488 T) (MHz)	7.224	10.133	40.481	17.126	23.930	12.959	16.069
Normal reference	$MeNO_2$ or NO_3^-	—	85% H_3PO_4	$KAsF_6$	$Et_4N^+SbCl_6^-$		$KBiF_6$

Sources: Emsley, J. *The Elements*, 2nd edn, Clarendon Press, Oxford, 1991. Lide, D.R., *Handbook of Chemistry*, 73rd edn, CRC Press, Boca Raton, FL, 1992–1993.

[a]Nuclei in square brackets are seldom used for NMR studies.

but this topic is discussed much more fully in chapters 2 and 8 and will not be duplicated here (cf. Table 8.1 in chapter 8, and a discussion of arsenate minerals in chapter 3).

1.1.3 Structures of the elements

The differing properties of the group 15 elements were mentioned above in terms of electronegativity and its relationship to metallic, semi-metallic and non-metallic properties. A further, although clearly related, illustration of the differences in the chemistry of these elements is apparent from a consideration of the structures and properties of the elements themselves.

Under normal conditions, nitrogen exists as a colourless gas consisting of diatomic N_2 molecules held together by a strong N≡N triple bond; no other allotropic forms are known. In contrast, phosphorus has an extensive allotropy [2, 17] although not including the P_2 molecule which is only encountered in the gas phase at low pressures and high temperatures. White phosphorus (sometimes called yellow phosphorus) consists of tetrahedral P_4 molecules for which three crystalline modifications are known (Table 1.1). There are no appreciable intermolecular interactions between the P_4 molecules, and this is the least thermodynamically favourable and most reactive form of the element. The other commonly encountered form of phosphorus is red phosphorus, which is a polymeric, amorphous and much less reactive material. Additional, more crystalline modifications of red phosphorus have also been prepared, but all the red forms have only poorly defined structures [2]. Crystalline allotropes with well-defined polymeric structures are known, however, one of which is orthorhombic black phosphorus, the thermodynamically most stable form of this element at room temperature and pressure [18]. Two diagrammatic representations of orthorhombic black phosphorus are illustrated in Fig. 1.3(a) which shows that the phosphorus atoms have a trigonal pyramidal geometry and are arranged in puckered hexagonal sheets. Orthorhombic black phosphorus is transformed under pressure to a rhombohedral phase, which also consists of puckered hexagonal sheets [Fig. 1.3(b)], and, at higher pressures, a simple cubic phase is obtained [2]. The cubic phase, in which each phosphorus is octahedrally coordinated by six others, is illustrated in Fig. 1.3(c), and the lower representations in Figs 1.3(a) and 1.3(b) reveal how the idealized structures of the orthorhombic and rhombohedral forms, respectively, can be derived from this symmetrical cubic form; these illustrations have more than just graphical significance in that they illustrate how the structures of the orthorhombic and rhombohedral phases can be understood in terms of Peierls distortions of the symmetrical cubic form, a structure and bonding approach discussed by Burdett [19, 20].

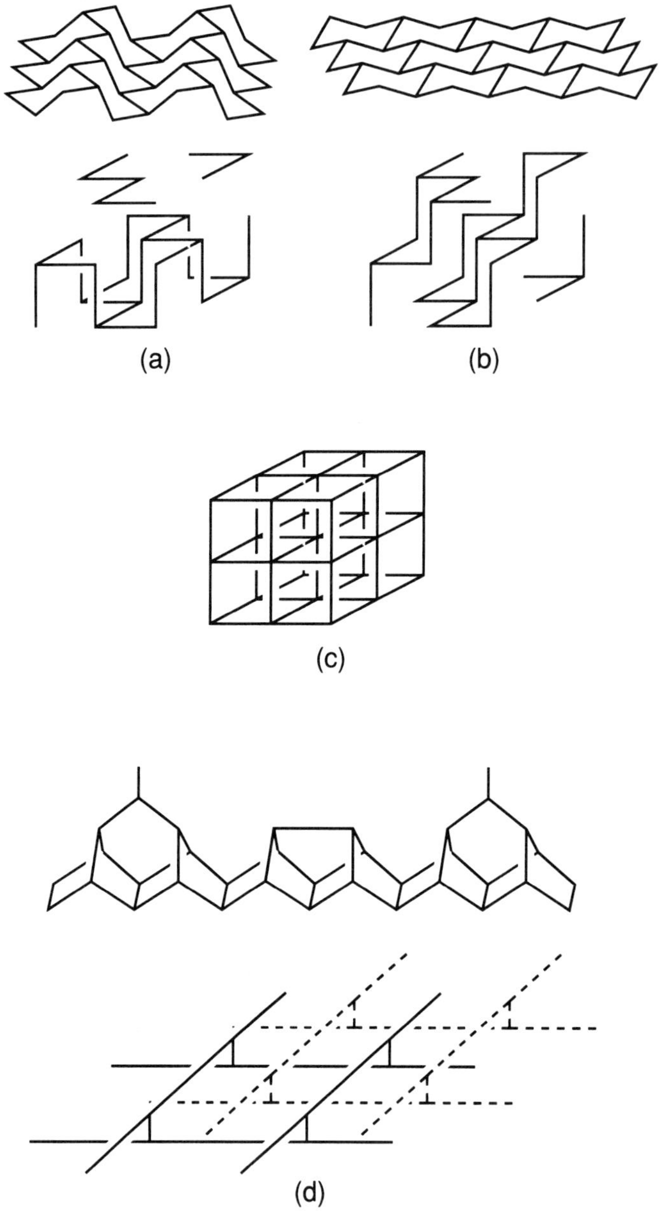

Figure 1.3 (a) Two views of the structure of orthorhombic black phosphorus; (b) two views of the structure of rhomobohedral black phosphorus; (c) cubic phosphorus; (d) two views of the structure of Hittorf's violet phosphorus.

As a final example, the allotrope known as Hittorf's violet phosphorus has also been structurally characterized (obtained by crystallizing phosphorus from molten lead!) and comprises a very complex structure, as illustrated in Fig. 1.3(d) [21]. All phosphorus centres are trigonal pyramidal in this allotrope and the structure can be described as comprising one-dimensional, linked norbornane-type units in which the bridgehead phosphorus atoms are either directly bonded or linked via a single phosphorus centre in the alternating manner shown in the upper diagram of Fig. 1.3(d). The three-dimensional structure is then constructed by cross-linking the one-dimensional units, as shown simplistically in the lower part of Fig. 1.3(d).

In contrast to phosphorus, the elements arsenic, antimony and bismuth have a less extensive allotropy. There are three crystalline forms of arsenic, six forms of antimony and 'several' polymorphs of bismuth [2], but of the various known forms the α-form is the most stable in each case, these being isostructural with rhombohedral black phosphorus [Fig. 1.3(b)]. For arsenic, the other crystalline forms include a yellow cubic modification obtained when the element is sublimed, which probably contains As_4 molecules analogous to white phosphorus (arsenic does exist as As_4 molecules in the vapour phase, As–As 2.435 Å), and ε-As (the mineral arsenolamprite), which is thought to be isostructural with orthorhombic phosphorus. For antimony, in addition to the α-form, the other crystalline allotropes include a very unstable yellow form (decomposition above –90°C), a black form obtained by cooling antimony vapour and an apparently unstable form produced electrolytically. The final two crystalline forms are both high pressure modifications, one of which (form I) is cubic wherein each antimony is surrounded octahedrally by six others (analogous to cubic phosphorus), and the other of which (form II) has a metallic hexagonal close-packed structure in which each antimony has twelve nearest neighbours (3.28 Å). Of the various known forms of bismuth, the only other well-characterized modification besides α-Bi is ζ-Bi, which has a body centred cubic structure in which each bismuth has eight nearest neighbours (3.291 Å) [2].

A more detailed comparison of the structures of the rhombohedral forms of the elements phosphorus to bismuth reveals an informative and general trend as indicated in Fig. 1.4 and Table 1.4. Figure 1.4 shows one hexagonal layer of the rhombohedral structure together with three atoms from an adjacent layer above one particular atom, the overall coordination for which is octahedral with three short bonds (length r_1) and three long bonds (length r_2). In Table 1.4 we list the r_1 and r_2 distances for the rhombohedral forms of the four elements together with the r_1/r_2 ratio from which it is apparent that as the group is descended the short primary bonds and the longer so-called secondary interlayer bonds come nearer to being the same length; in the limit, if all bond lengths were to become

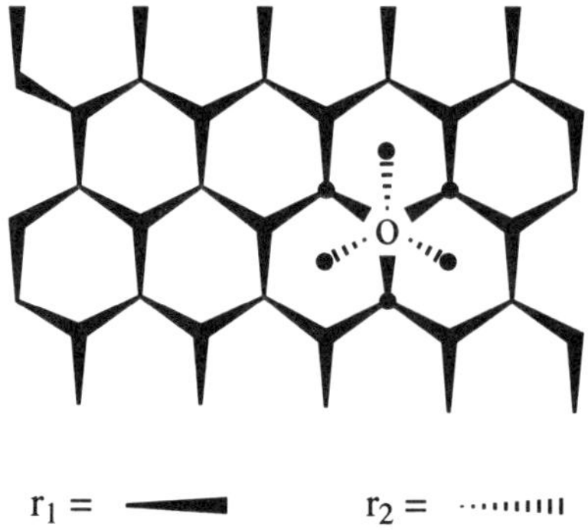

Figure 1.4 A representation of one puckered hexagonal layer of the rhombohedral black phosphorus structure. The atom indicated O (not oxygen) is shown linked to its three nearest neighbours (bonds of length r_1) and its next nearest neighbours (bonds of length r_2).

Table 1.4 A comparison of the interatomic distances, r_1 and r_2, in rhombohedral black phosphorus and the rhombohedral forms of arsenic, antimony and bismuth.

	r_1(Å)	r_2(Å)	r_2/r_1
Black P	2.231[a]	3.324[a]	1.490
α-As	2.517	3.120	1.240
α-Sb	2.908	3.355	1.153
α-Bi	3.072	3.529	1.149

[a]Average.

equal and all angles 90°, the simple cubic modification would be obtained. This structural trend well illustrates the tendency for coordination numbers to increase as the group is descended (resulting from increasing atomic size) together with the structural origin of the onset of increasing metallic character. As will be apparent later, secondary bonding interactions are an important feature of the structural chemistry of the heavier group 15 elements, particularly bismuth [22].

1.2 Some general characteristics of group 15 element compounds

1.2.1 Stability of oxidation states

As mentioned in Section 1.1.1, the common oxidation states for arsenic, antimony and bismuth are +3 (III) and +5 (V) although it should not be forgotten that the elements themselves have an oxidation state of zero,

Table 1.5 Standard reduction potential data (E^0, in V) for the elements arsenic, antimony and bismuth.

	(V)	E^0(V) →	(III)	E^0(V) →	0	E^0(V) →	–(III)
Arsenic							
Acid solution	H_3AsO_4	0.560 →	$HAsO_2$	0.240 →	As	–0.225 →	AsH_3
Alkaline solution	$[AsO_4]^{3-}$	–0.67 →	$[AsO_2]^-$	–0.68 →	As	1.37 →	AsH_3
Antimony							
Acid solution	Sb_2O_5	0.605 →	$[SbO]^+$	0.204 →	Sb		
Neutral solution	Sb_2O_5	0.699 →	Sb_4O_6	0.150 →	Sb	–0.510 →	
Alkaline solution	$[Sb(OH)_6]^-$	–0.465 →	$[Sb(OH)_4]^-$	–0.639 →	Sb	–1.338 →	SbH_3
Bismuth	Bi^{5+}	*c*.2 →	Bi^{3+}	0.317 →	Bi	–0.97 →	BiH_3

and the trihydrides, EH_3, are formally examples of the –3 (–III) oxidation state since hydrogen is more electronegative (slightly) than all of these three elements. (Note: E is used here as the common symbol to represent all group 15 elements.)

In terms of the stability of the oxidation states, some numerical data are given in Tables 1.1 and 1.5, and a graphical oxidation state diagram derived from ref. [2] (which includes some data for nitrogen and phosphorus) is shown in Fig. 1.5.

Particular points to note are that in the case of arsenic, one important difference in comparison with phosphorus is that, owing to the effects of the preceding 3d row (discussed in Section 1.1.1 in connection with electronegativity and ionization energies), there is a reluctance to achieve the +5 oxidation state. As a result, the oxide As_4O_{10} and the oxoacid H_3AsO_4 are oxidizing agents in contrast to the corresponding phosphorus species which are not. Examination of Fig. 1.5 reveals that antimony resembles arsenic reasonably closely but bismuth(V) is much more strongly oxidizing [Bi(V) will oxidize water] than is either As(V) or Sb(V), this latter feature being, at least qualitatively, predictable on the basis of the higher sum of the fourth and fifth ionization energies for the elements antimony and bismuth: 9.636 MJ mol^{-1} and 9.776 MJ mol^{-1}, respectively.

Another feature to note is that none of the +3 oxidation states for these three elements disproportionates in solution, nor do the elements themselves, as is found in the case of the reaction between P_4 and alkali, which yields PH_3 and hypophosphite. Furthermore, there is no particular stability for the +1 oxidation state for bismuth, a $d^{10}s^2p^2$ species, a feature discussed in Section 1.2.5 in conjunction with relativistic effects.

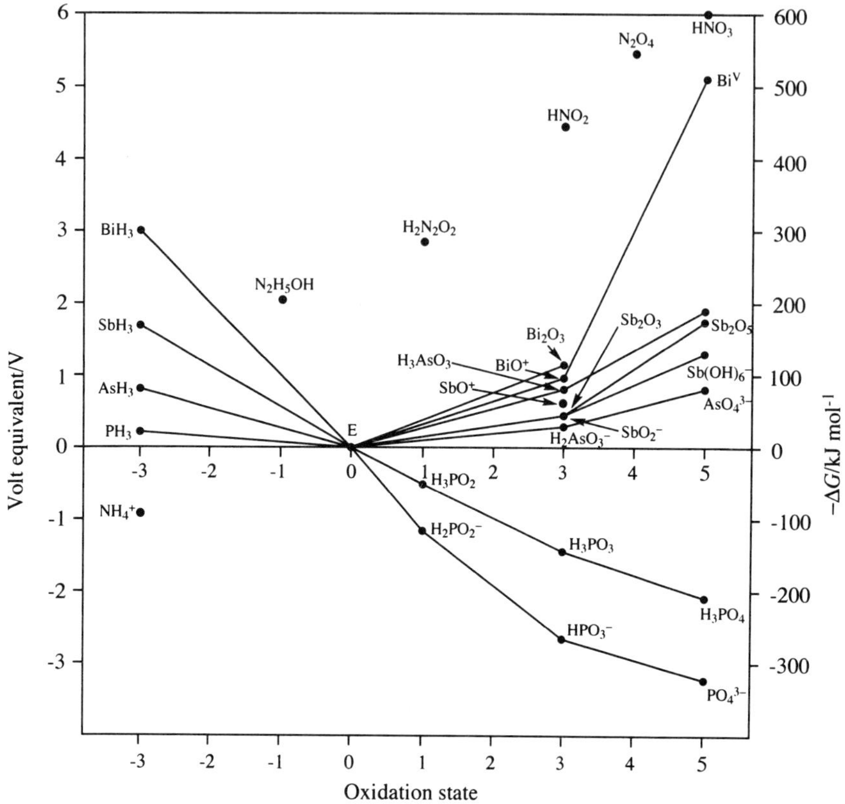

Figure 1.5 An oxidation state diagram for arsenic, antimony and bismuth in acid and alkaline solutions. Selected data for nitrogen and phosphorus species are also presented for comparison.

In seeking an explanation for the strongly oxidizing nature of Bi(V) in comparison with Sb(V), the higher fourth and fifth ionization energies for bismuth have already been mentioned. In fact the reluctance of the heavier p-block elements to achieve the group oxidation state (in this case +5) is a general feature in the chemistry of the p-block elements and is usually referred to as the 'inert pair effect', with the clear implication that there is an inert pair of electrons (the s pair) which are not easily involved in bonding. The large fourth and fifth ionization energies for bismuth provide some support for this assertion, the origins of which can be traced both to presence of the preceding 4f row discussed in Section 1.1.1, and also to a relativistic contraction and energetic stabilization of the valence 6s orbital described in Section 1.2.5 [similar arguments can be advanced regarding the oxidizing nature of As(V) and presence of the 3d row, but in the case of arsenic, relativistic effects are largely unimportant]. In fact this energetic stabilization of the s pair is only part of the picture and in some ways the term 'inert pair effect' is a misnomer. Also important is the fact that bonds get weaker as the group is descended (Section 1.2.2) such that for the heavier elements the energetic cost of involving the s pair is not compensated by the formation of two extra but rather weak bonds [10–12].

1.2.2 *Covalent bond energies*

A selection of covalent bond energies involving the elements arsenic, antimony and bismuth is given in Table 1.6 from which a number of trends are apparent (associated average covalent bond lengths are also given although the trends here mirror those described above for the various atomic radii; that is, bond lengths increase down the group). As with all bond energy data, the precise value is specific to a particular compound for which it was measured or deduced, and values may vary to a considerable extent from one compound to another; a good example is the energies of the N–N single bond in the compounds N_2H_4, N_2F_4 and N_2O_4, which are 167 kJ mol^{-1}, 88 kJ mol^{-1} and 57 kJ mol^{-1}, respectively. (Some definition of the term 'bond energy' is appropriate here. In general, the bond dissociation energy, ΔU_{el}, is greater than the true bond dissociation energy, ΔU_0, by an amount corresponding to the zero point energy even at 0 K. The value at room temperature (298 K), ΔU_{298}, is slightly different again since translational, rotational and vibrational energy must be taken into account, and the dissociation enthalpy, ΔH_{298}, includes a term relating to the work done to maintain constant pressure. ΔU_0, ΔU_{298} and ΔH_{298} are usually quite similar in magnitude, and ΔH_{298} is available by direct experimental measurement; the values used in Table 1.6 are mostly ΔU_0 values taken from the sources named therein.)

With regard to the E–E single bonds, the expected trend is for the bond energy to decrease as the group is descended because of the increasing size of the atoms and the correspondingly poorer overlap between the larger orbitals. This trend is observed for the elements phosphorus to bismuth, but the well-known weakness of the N–N single bond is immediately apparent, this bond being significantly weaker than the P–P single bond. The origin of this apparent anomaly can be traced to the smallness of the nitrogen atom, such that significant lone pair–lone pair repulsions arise across the N–N bond, and to the fact that relatively high energy antibonding orbitals are also populated.

The situation alters dramatically (with regard to nitrogen and phosphorus) when multiple bonds are considered wherein the N=N double and N≡N triple bonds are significantly stronger than the P=P double and P≡P triple bonds, respectively. In multiply bonded systems, lone pair–lone pair repulsions generally become less significant (especially in the E≡E species

Table 1.6 Selected bond energies (kJ mol^{-1}) and bond lengths (in parentheses in Å) for the group 15 elements (E).

	N	P	As	Sb	Bi
E–E	167 (1.45)	201 (2.21)	146 (2.43)	121 (–)	—
E=E	418 (1.25)	310 (2.00)	—	—	—
E≡E	942 (1.10)	481 (1.89)	380 (–)	295 (–)	192 (–)
E–H	386 (1.01)	322 (1.44)	247 (1.52)	—	—
E–C	305 (1.47)	264 (1.84)	200 (1.98)	215 (2.20)	143 (2.30)
E^{III}–F	283 (1.36)	490 (1.54)	484 (1.71)	440 (2.03)	393 (2.35)
E^{V}–F	—	—	406 (–)	402 (–)	297 (–)
E^{III}–Cl	313 (1.75)	326 (2.03)	322 (2.16)	315 (2.32)	274 (2.48)
E^{V}–Cl	—	—	—	248 (–)	—
E^{III}–Br	—	264 (–)	458 (2.33)	260 (2.51)	232 (2.63)
E^{V}–Br	—	—	—	184 (–)	—
E^{III}–I	—	184 (–)	200 (2.54)	195 (–)	168 (–)
E–O	201 (1.40)	335 (1.63)	301 (1.78)	314 (2.00)	339 (2.32)
E=O	607 (1.21)	544 (1.50)	389 (–)	—	—
E=S	—	335 (1.86)	—	—	—

where the lone pairs lie at 180° to the bond), but of more importance is the much better π overlap which occurs between 2p orbitals when compared with the π overlap between p orbitals of the subsequent periods. Thus, when considering single and multiple bonds together, it can be seen that whereas phosphorus (and the heavier elements) prefers to form two single bonds as opposed to one double bond since 2(P–P) > P=P, the reverse is true for nitrogen, where N=N > 2(N–N). Nitrogen therefore has a substantial chemistry involving multiple bonds (N≡N itself together with compounds containing N=N, N=O, N=C and N≡C bonds, for example) whereas analogous compounds involving phosphorus and the heavier elements are much less common and are generally obtained only in the presence of sterically demanding groups (as discussed in chapter 5 for arsenic and antimony and in chapter 6 for bismuth; also, see Section 1.2.3).

An alternative view of the relative strengths of single compared with multiple bonds can be had from a consideration of so-called bond increments discussed by Kutzelnigg [23]. In the particular case of a comparison of single with double bonds (i.e. E–E compared with E=E) for nitrogen, phosphorus and arsenic, the values (in kJ mol^{-1}) are 160:395, 200:145 and 175:120, respectively, these being defined in such a way that the first value is the energy of a single, or σ bond, and the second value the difference in energy between a double, or ($\sigma + \pi$) bond, and that of the single σ bond given by the first value, which can be taken as a measure of the strength of the π bond. Although values must be considered as approximate, it can be seen for nitrogen that the π bond is considerably stronger than the σ bond, whereas for phosphorus and arsenic the opposite is true, with the σ bond being significantly stronger than the π bond; this observation is quite general for the p-block, not just for group 15.

Further consideration of the values in Table 1.6 reveals that, as a general rule, bond energies decrease as the group is descended from nitrogen to bismuth (note, for example, the stengths of the E≡E triple bonds). The only notable exceptions to this rule occur when the attached atom has one or more lone pairs. Thus the N–F, N–Cl and N–O single bonds are weaker than the P–F, P–Cl and P–O single bonds, respectively, for reasons that are analogous to the explanation given for the relative energies of the N–N and P–P single bonds. Where adjacent lone pairs are not involved, as with E–C bonds, the energy of the bond to nitrogen is the largest of the group. Notice that when adjacent lone pairs are present but where multiple bonds are involved, such as in a comparison of N=O with P=O, the bond to nitrogen is stronger as for the homonuclear E=E bonds. Some discussion of the relative strengths of arsenic, antimony and bismuth bonds to carbon and transition metals is given in chapter 7.

As a final point, it should be noted that where bond energies are available for the same pair of elements but with the group 15 element in different oxidation states, the lower oxidation state bond has the higher energy; this is particularly noticeable for antimony for which data are available for fluorides, chlorides and bromides in both the +3 and the +5 oxidation state. In many ways this observation is counterintuitive since the lower oxidation state compounds usually have the longer bond lengths as a result of the presence of a lone pair at the group 15 element centre. The reason for this observation can be traced to the poor hybridization between valence s and p orbitals for the heavier elements. For compounds in the +3 oxidation state, bonding can occur using only the higher energy valence p orbitals, whereas for the +5 oxidation state it is necessary to involve the s orbital. As discussed by Kutzelnigg [23] the quite different radial extensions of the valence s and p orbitals for all but the 2p elements results in poor s–p hybridization and consequently weaker bonds.

1.2.3 Multiple bonds

In the previous section it was shown, on the basis of bond energy considerations, why single bonding is expected to be much more common than pπ–pπ multiple bonding for the heavier p block elements and in particular for nitrogen compared with phosphorus and arsenic. However, an analysis which deals only with strengths of particular bonds does not encompass other important features. The synthesis of stable compounds containing isolated multiple bonds has been achieved by employing sterically bulky substituents, typical examples of which are $C_6H_2R_3$–2,4,6 [R = Me, Et, Pr^i (isopropyl), Bu^t (tert-butyl), CF_3], $(Me_3Si)_3C$, $(Me_3Si)_2CH$ and Ph_3Si (Scheme 1.1), which are discussed in more detail in chapters 5 and 6 [24–31].

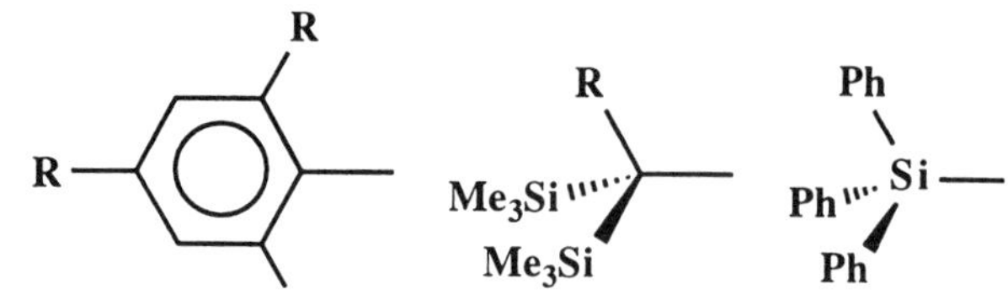

R = Me, Et, Pr^i, Bu^t, CF_3 R = Me_3Si, H

Scheme 1.1

The efficacy of such groups in this regard has often been stated to be kinetic in origin, with the implication that they protect an unstable functional group from oligomerization by means of their size. However, although there is some truth in this assertion, it is not really kinetic stabilization. If we take as an example some phosphorus compounds, then for the RP or phosphinidene unit, when R is phenyl, cyclic pentameric and hexameric species are formed whereas when R is the much more bulky aryl substituent $C_6H_2Bu^t_3$–2,4,6, it is the doubly bonded diphosphene (RP=PR) which is obtained. If we now consider various possible ring sizes and also that a diphosphene can be thought of, albeit somewhat perversely, as a two-membered ring, it is clear from the diagrams in Scheme 1.2 that the more bulky substituents will favour smaller ring sizes since the R–P–P angle(s), and hence the R···R separation, increase as the ring size decreases. This argument assumes that the P_n rings are planar and that the bonded atoms of the R groups are coplanar with the P_n ring, which will not generally be the case, but the basic point of the argument remains in that smaller rings will more readily accommodate larger R groups (Scheme 1.2).

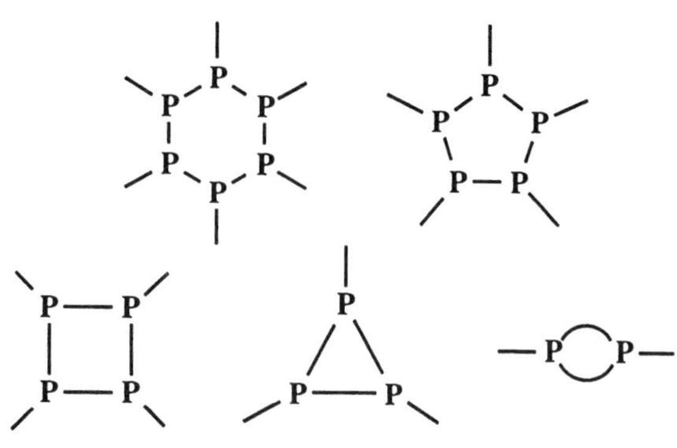

Scheme 1.2

On this basis, it is apparent that the stabilization of a diphosphene by sterically demanding substituents is not really kinetic stabilization but is a form of thermodynamic stabilization since it is certainly not the case that they are kinetic products which are somehow prevented (by a large activation energy) from forming a larger, and thermodynamically more stable, P_n ring. Thus π bonded compounds (two-membered rings) can be rendered quite stable if the σ bonded alternatives (larger rings) are made less stable by using large and bulky side groups; in effect, these large groups exert thermodynamic control over oligomerization and kinetic control over other reactions which may occur at the double bond.

1.2.4 Trends in acidity and basicity

Lewis acidity amongst the group 15 elements is generally associated with the +5 oxidation state wherein expansion of the coordination number (often from five to six) commonly occurs either through bridging interactions, in the case of the pentahalides for example, or coordination of neutral or anionic atoms or ligands, for example, $SbF_5 + F^- \rightarrow SbF_6^-$ (further examples are given in chapters 3 and 4). As such, compounds such as SbF_5 may be thought of as both coordinatively unsaturated, in that they can readily accommodate additional atoms or groups (usually one more), as well as electronically unsaturated. In the latter case, the ability to form extra bonds has traditionally been associated with the presence of vacant d orbitals, but more recent work [32] has cast doubt on this premise, the origin of the electronic unsaturation deriving from the high formal positive charge at the element centre resulting from its high oxidation state; the bonding in both five-coordinate and six-coordinate species is now more usually discussed in terms of delocalized three-centre, four-electron interactions involving only the valence s and p orbitals.

Perhaps less well appreciated is the Lewis acidity of the elements in the +3 oxidation state, since the presence of a lone pair of electrons generally leads to the expectation of Lewis basic behaviour. That group 15 compounds in the +3 oxidation state can act as Lewis bases is well documented and is exemplified by the ability of triorganyl derivatives, for example EPh_3, to act as ligands to transition metals. This topic is addressed in chapter 7, but the point is made that in contrast to the very large number of metal phosphine complexes, far fewer such complexes are known for the heavier elements, with the number of well characterized examples dropping dramatically as the group is descended; only a few triorganobismuthine compounds have been prepared. This observation illustrates a more general point relating to the decrease in basicity as the group is descended. The reason for this observation is sometimes discussed in terms of the low energy and hence unavailability of the E(III) lone pair as the group is descended (which probably is not correct), but a better explanation is more likely associated with the fact that, as discussed in Section 1.2.2, the increasingly poor hybridization of the valence s and p orbitals resulting from their increasingly disparate radial extensions results in the bonding occurring primarily through p orbitals, with the lone pair residing in an orbital of largely s character. Since s orbitals have poorer directional properties compared with s–p type hybrids, it is partly for this reason that bonds become progressively weaker as the group is descended (in addition to the general trend towards weaker bonds for larger atoms) thereby accounting for the reduced basicity. Alternatively, and in a complementary sense, it may be argued that on formation of a fourth bond, some degree of s–p type hybridization becomes necessary, but since the resulting bonds are weaker there is less of a thermodynamic driving force for such a process to occur. A structural property associated with the decrease in E(III) basicity is the increasingly pyramidal nature of the three-coordinate geometry (interbond angles approach 90°) and associated increase in the pyramidal inversion energy barriers, both of which are addressed further in Section 1.2.5.

Returning to the Lewis acidity of arsenic, antimony and bismuth in the +3 oxidation state, it is apparent that whatever basicity such compounds may exhibit, a considerable degree of Lewis acidity is also evident especially where the group 15 element is bonded to significantly more electronegative atoms or groups. Thus, antimony and bismuth trihalides, SbX_3 and BiX_3, and phenyl dihalides, $PhSbX_2$ and $PhBiX_2$, for example, are sufficiently Lewis acidic to exhibit quite an extensive coordination chemistry with ligands such as phosphines (PR_3), phosphine oxides (OPR_3), ethers (OR_2), and halide ions (X^-) (chapter 4) [33–35]. The origin of this Lewis acidity together with many of the trends in structural features observed for these complexes, such as bond lengths and coordination geometry, can be rationalized on the basis of a bonding model in which Sb−X or Bi−X σ* orbitals are the primary acceptor orbitals through which the ligands bond (as opposed to a more conventional description which employs vacant d orbitals as noted above in relation to the +5 oxidation state) as is evident from some of the examples discussed below.

Thus, the structures of the compounds $[BiPhX_2(thf)]$ (X = Cl, Br, I; thf = tetrahydrofuran) are very similar [36, 37] and comprise monomeric units, $BiPhX_2(thf)$, linked by halide bridges to form a one-dimensional polymer such that the overall coordination number of the bismuth centre is five, with a regular, square-based pyramidal coordination geometry, the phenyl group occupying the apical site and the basal plane containing three halides (one from an adjacent monomeric unit) and the oxygen atom of a coordinated thf ligand (structure A, Scheme 1.3).

For each bismuth centre, this stereochemical arrangement may be considered as a coordination complex of a trigonal pyramidal $BiPhX_2$ unit to which a thf ligand and a halide from a neighbouring unit are attached

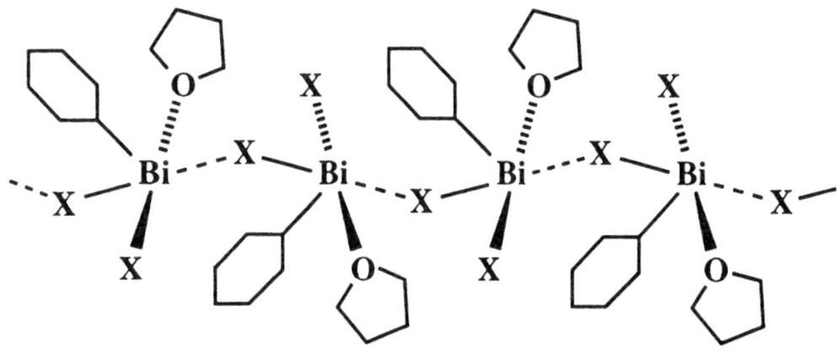

X = Cl, Br, I

(A)

Scheme 1.3

such that they lie opposite the original Bi–X bonds, thereby resulting in the observed square-based pyramidal geometry. Furthermore, for the two *trans* related halides, there is a shorter and a longer bond which may be defined respectively, as primary and secondary bonds, after Alcock [22]; the remaining halide (*trans* to the thf) can be considered as terminal. Relevant bond distances for the structures where X = Cl, Br and I are listed in Table 1.7 from which a number of trends are apparent.

First, it is clear that on going from chloride to iodide, the difference between the primary and secondary Bi–X bond lengths (Δ) decreases (were the distances to become equal, i.e. Δ = 0, the distinction between primary and secondary would be lost). A complementary description, in this case, is that the Bi–X–Bi bridges become more symmetric. Moreover, the comparison is particularly appropriate for chloride and bromide since the crystals are isomorphous; the iodide has a slightly different crystal structure but the configuration and conformation of the one-dimensional polymer is very similar to that found in the other two. This feature of the

Table 1.7 Bond lengths (Å) for the compounds $[BiPhX_2(thf)]$ (X = Cl, Br, I).

	X			
Bond	Cl[a]	Br[a]	I[b]	
Bi–X[c]				
terminal	2.543(3)	2.684(1)	2.882(2)	2.892(2)
primary	2.654(3)	2.825(1)	3.056(2)	3.079(2)
secondary	2.934(3)	3.038(1)	3.227(2)	3.211(2)
Δ	0.280	0.213	0.171	0.132
δ	0.111	0.141	0.174	0.187

[a]Isomorphous structures. [b]Two crystallographically independent monomeric units. Δ = difference between Bi–X primary and Bi–X secondary bonds; δ = difference between Bi–X primary and Bi–X terminal bonds.

primary and secondary E–X bonds becoming more equal in length on progressing to the heavier halides is, in fact, quite general [38]. Second, the structures of $SbCl_3$ and the isomorphous α-$SbBr_3$ can also be considered wherein the difference between the three primary Sb–X bonds and a set of three *trans* secondary Sb–X bonds is about 1.25 Å and 1.10 Å respectively (chapter 3).

In both the above examples, this trend towards smaller values of Δ can be taken as an indication of the increasing importance of secondary bonding (involving halides) where the heavier halides are concerned. Furthermore, it is observed when comparing a range of structures (for given E and X) that primary and secondary E–X bond distances are correlated such that as the secondary distance gets shorter the primary distance gets longer [22, 38]. An additional measure of the increasing length of the primary Bi–X bond in the aryl halide structures (resulting from a decrease in Δ) can be obtained by comparing the difference (δ) between the so-called primary and terminal Bi–X bonds, this difference increasing from the chloride to the iodide as secondary bonding becomes more pronounced.

The above examples focus on the effect on primary and secondary bond distances of changing the halide for a given group 15 element E. An example of the opposite situation where the same halide is involved but the group 15 element is changed is found in the isomorphous, centrosymmetric diphosphine antimony and bismuth bromide complexes $[E_2Br_6(dmpe)_2]$ (dmpe = $Me_2PCH_2CH_2PMe_2$; E = Sb, Bi) (structure B, Scheme 1.4) [39, 40]. The important feature here is the asymmetry of the E–Br–E bridge; for the antimony compound the relevant Sb–Br distances are 2.828 Å and 3.595 Å (Δ = 0.767 Å) whereas for the bismuth derivative the corresponding Bi–Br distances are 2.887 Å and 3.345 Å (Δ = 0.458 Å). Clearly, the bridge asymmetry is considerably smaller for bismuth than it is for antimony, consistent with secondary bonding interactions being more pronounced for bismuth.

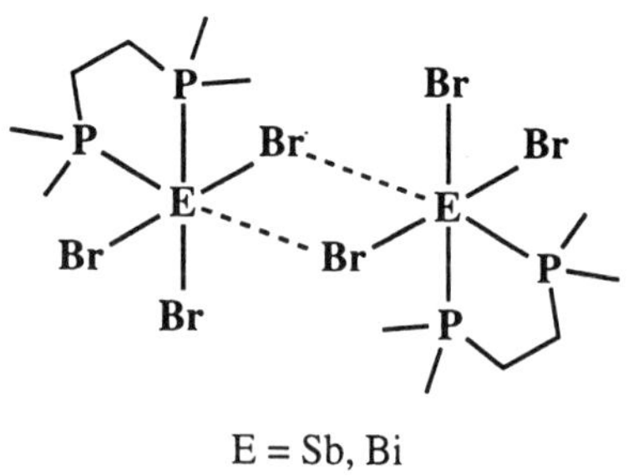

E = Sb, Bi

(B)

Scheme 1.4

This second observation is also quite general, and a further example is found in the isomorphous element triiodide [EI_3 (E = As, Sb, Bi)] structures (chapter 3) [2, 17]. These structures are based on a hexagonal close-packed array of iodines in which ⅔ of the octahedral holes in adjacent layers (⅓ of the total number of octahedral holes) are occupied by the E atoms. In AsI_3, the arsenic atom is displaced towards three mutually *cis* iodines, giving three shorter As–I bonds, and three *trans* bonds, which are longer, whereas in BiI_3, the bismuth is in a regular octahedral environment with all six Bi–I distances equal. In SbI_3 the situation is intermediate, the relevant bond lengths and bond length differences for the three compounds being shown in Table 1.8 (note also the increasing importance of secondary bonding in the structures of the elements themselves as discussed in Section 1.1.3).

Furthermore, some quantitative calculated reaction energy data shown in Table 1.9 is consistent with an increasing affinity of the heavier group 15 element trifluorides for the fluoride ion to form the EF_4^- anions [41].

Thus, overall, there are two important trends, at least as far as the element–halide interactions are concerned. First, there is a tendency for secondary bonding to be more pronounced for the heavier halides, and, second, secondary bonding is also seen to be more significant for the heavier group 15 elements; in the case of bismuth triiodide, the distinction between primary and secondary bonding has been lost. Many other examples of these structural trends are given in the chapters that follow.

In accounting for the above observations, the Lewis acidity and resulting expansion of the coordination sphere of the heavier group 15 elements (heavier p-block elements in general) has often been attributed to the

Table 1.8 Bond lengths (Å) for the compounds EI_3 (E = As, Sb, Bi).

Compound	primary E–I	secondary E–I	Δ (E–I)
AsI_3	2.591	3.467	0.876
SbI_3	2.868	3.32	0.45
BiI_3	3.1	3.1	0

Table 1.9 Calculated reaction energies, *E*, for $EF_3 + F^- \rightarrow EF_4$ (E = P, As, Sb, Bi).

Reaction	*E* (kJ mol^{-1})
$PF_3 + F^- \rightarrow PF_4^-$	–204
$AsF_3 + F^- \rightarrow AsF_4^-$	–247
$SbF_3 + F^- \rightarrow SbF_4^-$	–293
$BiF_3 + F^- \rightarrow BiF_4^-$	–296

Source: Nagase, S., in *The Chemistry of Organic Arsenic, Antimony and Bismuth Compounds* (ed. S. Patai), John Wiley, Chichester, Sussex, Ch. 1.

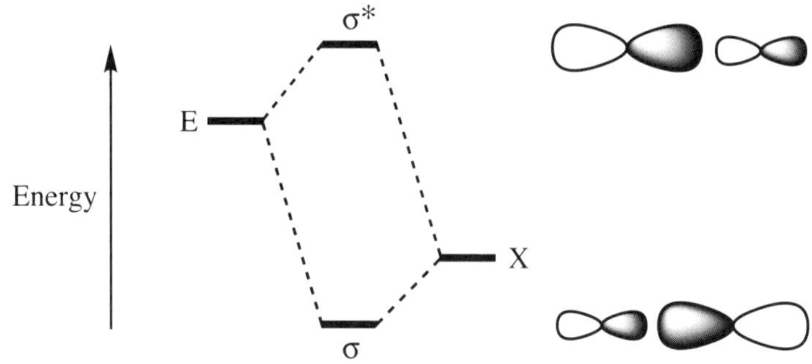

Figure 1.6 A simple two-orbital interaction diagram for an E−X bond (E = group 15 element; X = halogen), showing the nature of the σ*-orbital.

presence of vacant d orbitals, as discussed above. More recently, this approach has been questioned as a result of calculations which indicate that the necessary d orbitals are probably too high in energy, with the result that any d orbital participation is minor at best [32]. An alternative model centres on the availability of vacant E−X σ* orbitals, and although there is as yet little calculational support for this approach the model is at least qualitatively useful. In considering a two-orbital σ interaction between atoms E and X, where X is the more electronegative, the first point to note is that the bonding σ orbital will be polarized towards X, since this is the lower energy atomic orbital, whereas the antibonding σ* orbital will be polarized more towards the less electronegative element E (Fig. 1.6). If this σ* orbital is sufficiently low in energy, it can act as an acceptor orbital on E through which a Lewis base, L (either a neutral or anionic ligand or a bridging atom from an adjacent monomer unit), can bond. Moreover, the polarization of this orbital towards the E atom will tend to ensure that the Lewis base bonds to E (rather than X) and also that the geometry of the L−E−X interaction should be close to linear. Furthermore, as the L−E interaction becomes stronger (and the L−E bond shorter) the population of the σ* orbital is increased, which will lead to a lengthening of the E−X bond, precisely the correlation which is observed.

In considering why heavier element E complexes should show more pronounced secondary bonding, two factors become important. First, on descending the group, the electronegativity of E will decrease, which will be reflected in a raising of the energy of the valence E orbital(s). This will result in a greater difference between the energies of the atomic orbitals of E and X (providing X is more electronegative than E) leading to poorer overlap and a correspondingly lower energy σ* orbital (and higher energy σ orbital). Second, the increasing size of the heavier element

E means larger and more diffuse orbitals, which will also result in poorer overlap, with, importantly, the same consequences for the energies of the σ orbital and σ^* orbital. The lower the energy of the σ^* orbital, the more available for bonding this orbital becomes, resulting in a corresponding increase in the likely extent of secondary bonding interactions.

In the case where X is varied for a given element E, it is also observed that secondary bonding is apparently more significant, where the heavier, and less electronegative, halides are concerned, but the situation here is a little more complicated as not only must the acceptor properties of E be considered but also the donor properties of X

With regard to the E centre, increasing the size of X will lead to poorer orbital overlap, just as it did for E, resulting in a decrease in the energy of the σ^* orbital. However, raising the energy of the X atomic orbitals (as a result of decreasing electronegativity) will reduce the energy separation between E and X, which should tend to raise the energy of the σ^* orbital and reduce the extent of polarization of this orbital on E. Clearly, these two factors work in opposition rather than in concert and care should therefore be exercised in the application of these qualitative ideas with regard to secondary bonding at the E centre. In addition, an important factor in understanding the observed trend towards increased secondary bonding for the halides is that since the heavier halides are less electronegative they will be better electron donors and should therefore tend to bridge more readily. Thus it may well be the donor properties of X rather than the acceptor properties of E that are important here.

Clearly, a detailed and more quantitative understanding of secondary bonding is required since the qualitative ideas outlined here, although useful, are not without some ambiguity in certain cases. Preliminary calculations indicate that secondary bonding for the heavier p-block elements in general is difficult to model satisfactorily since the potential well associated with the secondary bond is very shallow and only calculations of a high order are likely to be of any use. Some relevant work in this area has been perfomed by Pyykö [42], who has suggested that secondary bonding has much in common with van der Waals type interactions, although such a mechanism less readily accounts for the observed directional properties of secondary bonds, that is that they are formed *trans* to primary bonds to (usually) the most electronegative element.

1.2.5 Structural trends

In addition to many of the structural trends discussed above, some mention of the coordination geometries associated with the group 15 element centres is warranted.

For compounds in the +5 (V) oxidation state, the coordination numbers encountered are most commonly four, five and six for which the observed

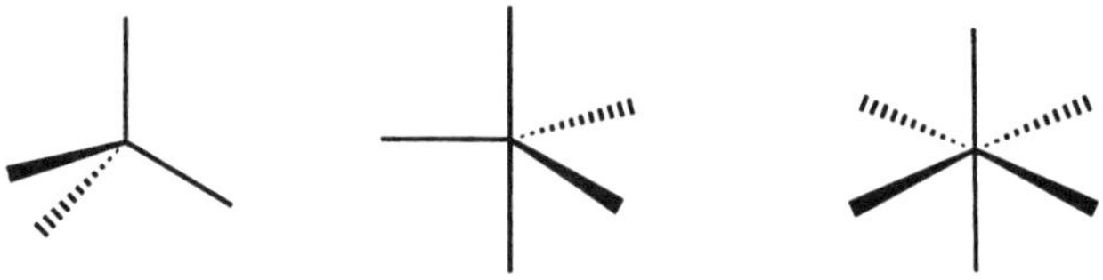

C D E

Scheme 1.5

geometries are tetrahedral (structure **C**, Scheme 1.5), trigonal bipyramidal (structure **D**) and octahedral (structure **E**), respectively, in line with the predictions made by valence shell electron pair repulsion (VSEPR) theory [43]. An interesting exception is found for the structure of $BiPh_5$ and some related penta-aryls of bismuth in that these compounds are found to have a square-based pyramidal geometry rather than the expected trigonal bipyramidal structure. This topic has been discussed in some detail by Seppelt [44] and is addressed more fully in chapter 6, the origin of the unusual structures having much to do with the importance of relativistic effects (see Section 1.2.6) for bismuth.

Much more structurally diverse are compounds of the elements in the +3 (III) oxidation state as a result of the presence of a non-bonded or lone pairs of electrons. For three, four and five coordination, the structures are generally trigonal pyramidal (structure **F**, Scheme 1.6), equatorially vacant, trigonal bipyramidal (or disphenoidal) (structure **G**) and square-based pyramidal (structure **H**) as expected for these AB_3E, AB_4E and AB_5E systems on the basis of VSEPR (using Gillespie's nomenclature where, in this case, A is the group 15 element, B is the groups to which it is bonded and E is the lone pair. Somewhat more problematical are six-coordinate compounds, AB_6E, which have seven electron pairs since there are a number of possible structures which are quite close in energy. In general, for seven coordination, the three structures encountered are pentagonal bipyramidal, monocapped octahedral and monocapped trigonal prismatic. For a AB_6E system, any one of these might be adopted wherein one vertex is occupied by the lone pair but, because of the closeness in energy of these various forms, any prediction as to which is observed in any particular case is problematical.

However, it is often found that AB_6E structures adopt regular (or nearly regular) octahedral geometries wherein the lone pair is stereochemically inactive and, in such cases, the lone pair is often said to reside in a pure s orbital and hence exert no stereochemical influence. It is not appropriate here to discuss this matter in detail since a number of texts deal more fully with this matter [43], but one approach involving a consideration of σ^* orbitals is outlined below.

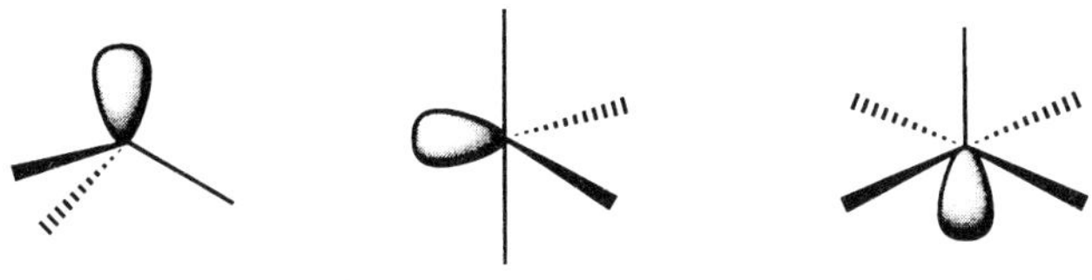

Scheme 1.6

Thus, if we consider a trigonal pyramidal unit EX_3 we can readily account for the geometry according to VSEPR rules. If we now consider a coordination complex with further ligands (for example X^-), and if the acceptor orbitals which give rise to the Lewis acidity are indeed the E−X σ* orbitals, we may represent the structure as shown in Scheme 1.7. Addition of one ligand, X_1, then results in a four-coordinate structure (EX_4^-) with the expected equatorially vacant, trigonal bipyramidal structure. Addition of two ligands, X_1 and X_2, leads to a square-based pyramidal geometry (EX_5^{2-}) whereas three ligands, X_1–X_3, result directly in an octahedral geometry (EX_6^{3-}). Moreover, this latter situation will result in a regular octahedral geometry when the lengths of the primary and secondary bonds are equal, as is found, for example, in the BiI_3 structure described above or in many of the EX_6^{3-} halogenoanions discussed in chapter 4.

If we wish now to consider the effect of the lone pair and any stereochemical activity which is sometimes observed, we can start from the regular octahedral geometry and recognize that the lone pair can become localized either towards a vertex of the octahedron, towards an edge or towards a face along the lines expected for a T_{1u} distortion of a regular (O_h) octahedron according to a second order Jahn–Teller distortion (Scheme 1.8) [45, 46]. In practice, most distortions of six-coordinate

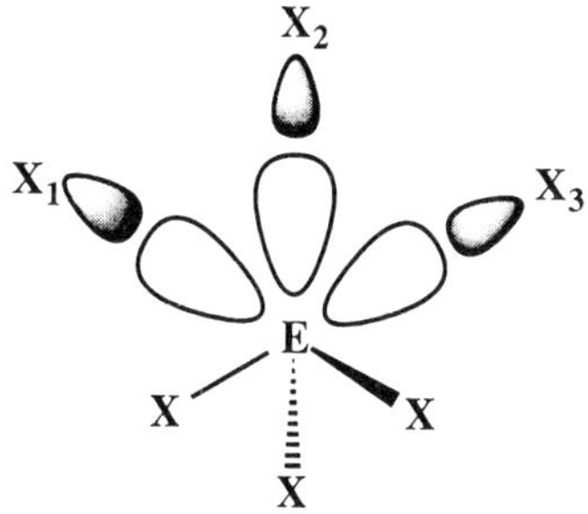

Scheme 1.7 An EX_3 unit showing the three E−X σ* orbitals and the three ligands X_1–X_3, where E is the group 15 element.

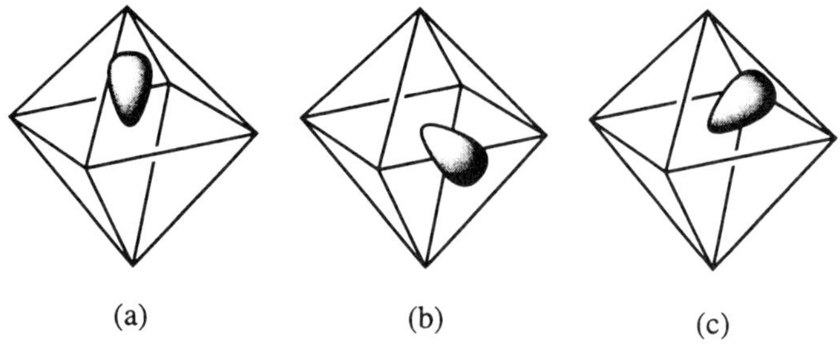

Scheme 1.8 Lone pair stereochemical activity through (a) a vertex; (b) an edge; (c) a face of an octahedron.

structures away from a regular octahedron fall into one of these three categories.

As a final point relating to structural trends for compounds in the trivalent state, it is usually observed that for three-coordinate compounds, EX_3, the X–E–X angles decrease as the group is descended. This is shown for the trihydrides and trifluorides of the group 15 elements in Table 1.10 which also lists the associated pyramidal inversion energy barriers.

The decrease in bond angle, which is most marked between nitrogen and phosphorus, is often discussed in terms of the inert pair effect. Thus as the group is descended, the s pair of electrons becomes less available such that bonding occurs primarily through the p orbitals, with the consequence that the interbond angles approach 90°. As with many discussions of the inert pair effect, the contention that the lone pair is unavailable for bonding because it is too low in energy is probably not really correct. More likely as an explanation in this context is the fact that, as pointed

Table 1.10 H–E–H and F–E–F bond angles (in degrees) for the group 15 trihydrides (EH_3) and trifluorides (EF_3)and approximate calculated pyramidal inversion energy barriers (kJ mol^{-1}) for EH_3 and EF_3.[a]

Compound	X–E–X angle	EX_3 inversion barrier
NH_3	107.5	25
PH_3	93.2	147
AsH_3	92.1	165
SbH_3	91.6	184
BiH_3	—	265
NF_3	102.3	—
PF_3	97.8	220
AsF_3	95.8	192
SbF_3	87.3	158
BiF_3	—	141

Source: Nagase, S., in *The Chemistry of Organic Arsenic, Antimony and Bismuth Compounds* (ed. S. Patai), John Wiley, Chichester, Sussex, Ch. 1, and references therein.

[a]Vertex inversion for EH_3 and edge inversion for EF_3.

out by Kutzelnigg [23], s–p hybridization becomes much less efficient for the 3p to 6p elements as a result of their increasingly disparate radial extensions, this being the main reason why bonding occurs through unhybridized p orbitals.

An alternative explanation is provided by thinking of a trigonal pyramidal structure as a second-order Jahn–Teller distortion arising from a more symmetric trigonal planar reference structure. Thus, according to the qualitative molecular orbital diagram shown in Fig. 1.7, mixing of the $2a_1'$ LUMO and the a_2'' HOMO in D_{3h} symmetry can lead, according to rules governing a second-order Jahn–Teller distortion [45, 46], to a trigonal pyramidal geometry in which these orbitals have the same symmetry ($3a_1$ and $2a_1$, respectively), the distortion being driven by the stabilization of the HOMO (a_2'' in D_{3h} becoming $2a_1$ in C_{3v}). The magnitude of the distortion (for which $\Delta E'$ in Fig. 1.7 is a measure) is then a function of

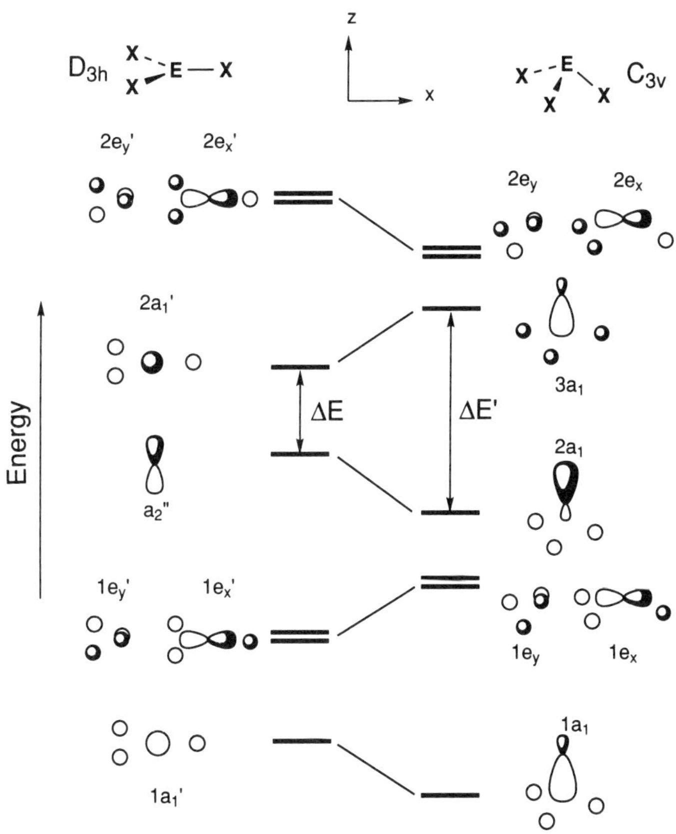

Figure 1.7 A qualitative molecular orbital diagram showing the relationship between the frontier orbitals of trigonal planar (D_{3h}) and trigonal pyramidal (C_{3v}) geometries for EX_3 (E = group 15 element; X = ligand).

the closeness in energy of the $2a_1'$ and a_2'' orbitals in the D_{3h} geometry (ΔE in Fig. 1.7). In general, as the group is descended, this energy separation diminishes for two reasons. First, the a_2'' HOMO is raised in energy as the elements become less electronegative and, second, the $2a_1'$ LUMO is lowered in energy as overlap becomes poorer (for larger atoms) and the interaction becomes less antibonding. A fuller description of this type of argument is given in refs [45, 46] but it should be noted that hybridization arguments are not necessary in this type of approach.

In relation to the trihydrides, the increasing pyramidal inversion energy barriers can also be understood using either of the two main arguments outlined above. Thus, in terms of hybridization, the trigonal planar (D_{3h}) transition state through which inversion occurs [Scheme 1.9(a)] necessarily involves some degree of s–p hybridization (specifically sp^2) which becomes less favoured for the heavier elements thereby resulting in a higher barrier to inversion. In terms of the second-order Jahn–Teller argument, the increased stabilization of the trigonal pyramidal geometry (C_{3v}), resulting from a smaller ΔE in the D_{3h} structure, also results in a higher activation energy for inversion (a crude measure of the inversion barrier would be given by $\frac{1}{2}\Delta E' - \frac{1}{2}\Delta E$ if other orbitals are ignored, this value increasing as the group is descended).

Table 1.10 reveals, however, that in contrast to the inversion barriers for the trihydrides, which increase as the group is descended, the inversion barriers for the trifluorides show the opposite trend. This matter was resolved by Dixon and Arduengo [47–50] who showed that for the fluorides the trigonal planar geometry is not the transition state for inversion but rather that inversion occurs through a T-shaped (C_{2v}) transition state [Scheme 1.9(b)]. Dixon and Arduengo showed that the T-shaped transition state is favoured for electron-withdrawing groups such as fluoride and that the barrier to inversion through this C_{2v} geometry decreases in energy as the group is descended as a result of the increasing electronegativity difference between the atoms. A more detailed discussion of this matter is given in chapter 1 of ref. 3.

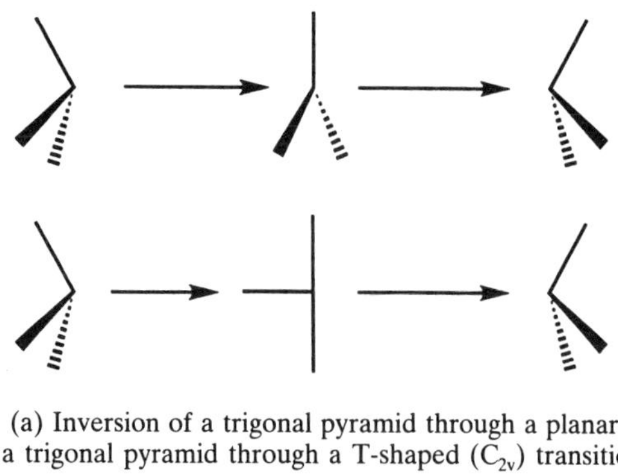

(a)

(b)

Scheme 1.9 (a) Inversion of a trigonal pyramid through a planar (D_{3h}) transition state: (b) inversion of a trigonal pyramid through a T-shaped (C_{2v}) transition state.

1.2.6 Relativistic effects

There are a number of reviews of so-called relativistic effects and their influence on chemical properties [51]. Briefly stated, appreciable relativistic effects arise when the velocity (v) of an electron in an atom approaches the speed of light (c) since, according to the special theory of relativity, the mass (m) of an electron increases with speed:

$$m = m_0\,[1 - (v/c)^2]^{-1/2} \tag{1.1}$$

where m_0 is the rest mass of the electron.

The effect is most apparent in the case of s orbital electrons for the heavier elements of the periodic table. Thus, since s orbitals have no angular momentum they can approach the nucleus quite closely which, for the heavier elements having a large positive nuclear charge, results in electron speeds which are an appreciable fraction of the speed of light. In the case of the 1s electron in bismuth, for example, the electron reaches a speed of around 60% c which results in a mass of $1.26m_0$, leading to a reduction in the radial extension (Bohr radius) of the 1s orbital of about 20% (the Bohr radius is inversely proportional to m). Similar radial contractions (direct relativistic orbital contractions) occur for the s orbitals of higher principal quantum number and, to a much smaller extent, for p orbitals as well, but any direct effect is relatively unimportant for d and f orbitals, although these latter two types of orbital are affected indirectly by the contraction of the s and p orbitals such that they are less well shielded and hence more diffuse than they would otherwise be (indirect relativistic orbital expansion).

One of the consequences of the relativistic contraction of valence s orbitals and the associated lowering in energy is that, for the heavier elements (in this case, 6s for bismuth), the s orbitals in particular become less available for bonding, with the result that s-p hybridization is less efficient and bonding tends to occur primarily through the valence p orbitals. This accounts, at least in part, for some of the observed structural trends (increasing pyramidality for EX_3) and oxidation state trends (inert pair effect) discussed in previous sections.

Another important relativistic effect is a strong coupling of the electron spin and any orbital angular momentum such that a distinction between the quantum numbers l and m_s ceases to be valid; a better description is in terms of a new quantum number j where $j = l + s$. One result of this spin-orbit coupling is that the normally degenerate p orbitals (usually denoted as p_x, p_y and p_z) now split into one $p_{1/2}$ and two $p_{3/2}$ orbitals, the former being spherically symmetric and the latter two, doughnut-shaped. Moreover, the $p_{1/2}$ orbital is stabilized by relativistic effects in much the same way as for the s orbitals described above such that $p_{1/2}$ is lower in energy than $p_{3/2}$. This lifting of the degeneracy of the p orbitals is a purely relativistic effect and might be expected to lead to a double inert pair for bismuth resulting in a stable +1 oxidation state [Bi(0) $6s^2 6p_{1/2}{}^2 6p_{3/2}{}^1 \rightarrow$ Bi(I) $6s^2 6p_{1/2}{}^2 6p_{3/2}{}^0$]. However, since the effect is relatively small for bismuth it is not obvious in the chemistry of this element that a stable +1 oxidation state exists. For element 115, however, which would be below bismuth in group 15, the effect is expected to be much more pronounced and a stable +1 oxidation state would be a likely feature of the chemistry of this element.

1.2.7 Thermodyanamic stability of the halides

The halides of the heavier group 15 elements are treated fully in chapter 3 and will not be discussed in any detail here. It is interesting to note, however, the variation in thermodynamic stability of the element trichlorides of the group 15 elements as the group is descended. This information is presented in Table 1.11 from which it is apparent that there is little difference in the stability of the trichlorides for phosphorus, arsenic, antimony and bismuth which are all exothermic compounds, but that NCl_3 is strongly endothermic. This can be traced both to the stabilty of N_2 and to the weakness of the N–Cl bonds.

As a final point concerning the group 15 halides it is notable that for the +5 (V) oxidation state, PCl_5 and $SbCl_5$ are stable compounds at room temperature whereas $AsCl_5$ decomposes above –50°C [52]. This provides a striking example of the difference in stability of the +5 oxidation state for arsenic in comparison with this oxidation state for phosphorus and antimony, reflecting the importance of the preceding 3d row for arsenic.

Table 1.11 A thermodynamic cycle for the formation of ECl_3 (E = group 15 element):

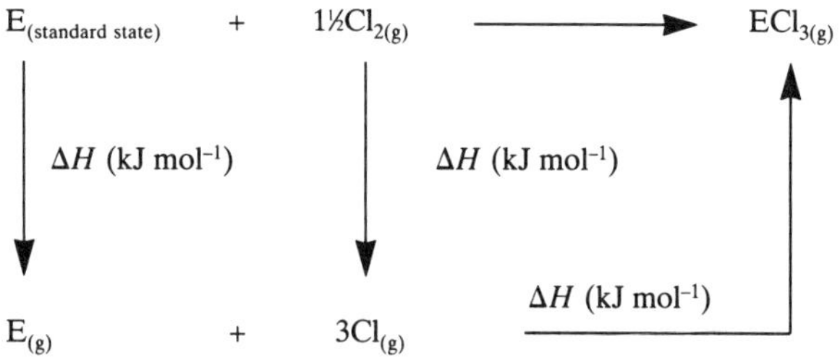

	ΔH (kJ mol^{-1}) N	P	As	Sb	Bi
$E_{(standard\ state)} \rightarrow E_{(g)}$	473	315	302	262	207
$0.5\ Cl_{2(g)} \rightarrow 3Cl_{(g)}$	363	363	363	363	363
$E_{(g)} + 3Cl_{(g)} \rightarrow ECl_3$	–578	–965	–927	–941	–836
$E_{(standard\ state)} + 1.5\ Cl_{2(g)} \rightarrow ECl_3$	258	–287	–262	–316	–266

References

1. Cotton, F. A. and Wilkinson, G. (1988) *Advanced Inorganic Chemistry*, 5th edn, John Wiley, New York.
2. Greenwood, N. N. and Earnshaw, A. (1984) *Chemistry of the Elements*, Pergamon, Oxford.
3. Patai, S. (ed.) (1994) *The Chemistry of Organic Arsenic, Antimony and Bismuth Compounds*, John Wiley, Chichester, Sussex.
4. Wilkinson, G., Abel, E. W. and Stone, F. G. A. (eds) (1982) *Comprehensive Organometallic Chemistry*, Pergamon, Oxford.
5. Abel, E. W., Stone, F. G. A. and Wilkinson, G. (eds) (1995) *Comprehensive Organometallic Chemistry (II)*, Pergamon, Oxford.
6. Wilkinson, G., Gillard, R. D. and McCleverty, J. A. (eds) (1987) *Comprehensive Coordination Chemistry*, Pergamon, Oxford.
7. Bailar, J. C., Emeleus, H. J., Nyholm, R. and Trotman-Dickenson, A. F. (eds) (1973) *Comprehensive Inorganic Chemistry*, Pergamon, Oxford.
8. King, R. B. (ed.) (1994) *Encyclopedia of Inorganic Chemistry*, John Wiley, Chichester, Sussex.
9. Allen, L. C. (1989) *J. Am. Chem. Soc.*, **111**, 9003.
10. Norman, N. C. (1997) *Periodicity and the s- and p-Block Elements*, Oxford University Press, Oxford.
11. Owen, S. M. and Brooker, A. T. (1991) *A Guide to Modern Inorganic Chemistry*, Longman, Harlow, Essex.
12. Barrett, J. (1991) *Understanding Inorganic Chemistry*, Ellis Horwood, Chichester, Sussex.
13. Lide, D. R. (ed.) (1992–1993) *Handbook of Chemistry*, 73rd edn, CRC Press, Boca Raton, FL.
14. Harris, R. K. and Mann, B. E. (eds) (1978) *NMR and the Periodic Table*, Academic Press, London.
15. Long, G. J. (ed.) (1985) *Mössbauer Spectroscopy Applied to Inorganic Chemistry*, vol. 1, Plenum Press, New York.
16. Berry, F. J. and Dickson, D. P. E. (1985) *Mössbauer Spectroscopy in Perspective*, Cambridge University Press, Cambridge.
17. Wells, A. F. (1984) *Structural Inorganic Chemistry*, 5th edn, Clarendon Press, Oxford.
18. Brown, A. and Rundqvist, S. (1965) *Acta Cryst.*, **19**, 684.
19. Burdett, J. K. and Lee, S. (1983) *J. Am. Chem. Soc.*, **105**, 1079.
20. Burdett, J. K. (1994) *Chem. Soc. Rev.*, 299.
21. von Thurm, H. and Krebs, H. (1969) *Acta Cryst.*, **B25**, 125.
22. Alcock, N. W. (1972) *Adv. Inorg. Chem. Radiochem.*, **15**, 1.
23. Kutzelnigg, W. (1984) *Angew. Chem., Int. Ed. Engl.*, **23**, 272.
24. Regitz, M. and Scherer, O. J. (1990) *Multiple Bonds and Low Coordination in Phosphorus Chemistry*, Georg Thieme, Stuttgart.
25. Norman, N. C. (1993) *Polyhedron*, **12**, 2431.
26. Appel, R. and Knoll, F. (1989) *Adv. Inorg. Chem.*, **33**, 259.
27. Cowley, A. H. and Norman, N. C. (1986) *Prog. Inorg. Chem.*, **34**, 1.
28. Cowley, A. H. (1984) *Acc. Chem. Res.*, **17**, 386.
29. Cowley, A. H. (1984) *Polyhedron*, **3**, 389.
30. Cowley, A. H. (1990) *J. Organomet. Chem.*, **400**, 71.
31. Weber, L. (1992) *Chem. Rev.*, **92**, 1839.
32. Reed, A. E. and Schleyer, P. V. R. (1990) *J. Am. Chem. Soc.*, **112**, 1434.
33. Carmalt, C. J., Clegg, W., Elsegood, M. R. J. *et al.* (1996) *Inorg. Chem.*, **35**, 3709.
34. Carmalt, C. J., Cowley, A. H., Decken, A. and Norman, N. C. (1995) *J. Organomet. Chem.*, **496**, 59.
35. Norman, N. C. and Pickett, N. L. (1995) *Coord. Chem. Rev.*, **145**, 27.
36. Clegg, W., Errington, R. J., Fisher, G. A. *et al.* (1992) *J. Chem. Soc., Dalton Trans.*, 1967.
37. Clegg, W., Errington, R. J., Fisher, G. A. *et al.* (1993) *J. Chem. Soc., Dalton Trans.*, 637.

38. Norman, N. C. (1994) *Phosphorus Sulfur*, **87**, 167.
39. Clegg, W., Elsegood, M. R. J., Graham, V. *et al.* (1994) *J. Chem. Soc., Dalton Trans.*, 1743.
40. Clegg, W., Elsegood, M. R. J., Norman, N. C. and Pickett, N. L. (1994) *J. Chem. Soc., Dalton Trans.*, 1753.
41. Nagase, S. (1994) in *The Chemistry of Organic Arsenic, Antimony and Bismuth Compounds*, (ed. S. Patai), John Wiley, Chichester, Sussex, Ch. 1.
42. Pyykkö, P. (1997) *Chem. Rev.* **97**, 597.
43. Gillespie, R. J. and Hargittai, I. (1991) *The VSEPR Model of Molecular Geometry*, Allyn and Bacon, Boston, MA.
44. Seppelt, K. (1992) *Adv. Organomet. Chem.*, **34**, 207.
45. Albright, T. A., Burdett, J. K. and Whangbo, M. H. (1985) *Orbital Interactions in Chemistry*, John Wiley, New York.
46. Gimarc, B. M. (1979) *Molecular Structure and Bonding*, Academic Press, New York.
47. Dixon, D. A. and Arduengo, A. J. (1987) *J. Am. Chem. Soc.*, **109**, 338.
48. Dixon, D. A., Arduengo, A. J. and Fukunaga, T. (1986) *J. Am. Chem. Soc.*, **108**, 2461.
49. Arduengo, A. J., Dixon, D. A. and Roe, D. C. (1987) *J. Am. Chem. Soc.*, **109**, 6821.
50. Dixon, D. A. and Arduengo, A. J. (1987) *J. Chem. Soc., Chem. Comm.*, 498.
51. Pyykkö, P. (1988) *Chem. Rev.*, **88**, 563.
52. Seppelt, K. (1976) *Angew. Chem., Int. Ed. Engl.*, **15**, 377.

2 Metallurgy of the elements

I.J. POLMEAR

2.1 Early history

Small quantities of each of the elements arsenic, antimony and bismuth have been found in the native form but all mostly occur as simple or complex sulphides or oxides. Of the three, arsenic has perhaps the most interesting historical record.

2.1.1 Arsenic

Arsenic compounds such as the sulphides orpiment (As_2S_3) and realgar (AsS) were known to the Greeks and Romans. Orpiment was also called 'arsenic', and Disoscorides noted that calcination of orpiment produced a white powder (As_2O_3) [1]. It is probable that arsenic metal was first produced by the German alchemist Albertus Magnus in AD 1250. Later the famous Swiss-born, German physician/alchemist Paracelsus (1493–1541), who is attributed with establishing the role of chemistry in medicine, described a method of preparing arsenic by heating orpiment with eggshells and organic matter. The probable reaction was $As_2S_3 + 3CaCO_3 + 6C \rightarrow 9CO_{(g)} + 3CaS + 2As_{(g)}$.

Much earlier, arsenical materials contributed to the metallurgical development of bronzes at the beginning of the Bronze Age (*c.* 3500 BC) [2–4]. Minerals containing arsenic, such as enargite (Cu_3AsS_3), are often found in association with deposits of sulphides rich in copper, for example chalcopyrite ($Cu_2S.Fe_2S_3$). Because of this, arsenic would no doubt have been introduced by chance during some of the early melting and refining of copper in the primitive furnaces of the time. The smiths of those days would have observed that some copper castings could be hammered to yield weapons and tools of greater hardness and strength than others (Fig. 2.1) [5]. Moreover, it may be presumed that a conscious progressive selection and utilization of particular minerals led to the development of copper–arsenic alloys [2–6].

Such melting operations would have generated copious fumes of toxic, garlic-smelling arsenious oxide (As_2O_3) and doubtless had an adverse effect on the health and life expectancy of the smiths of the time. Degenerative changes would have occurred to them as arsenic was absorbed into their bodies, which may have led to medical conditions such

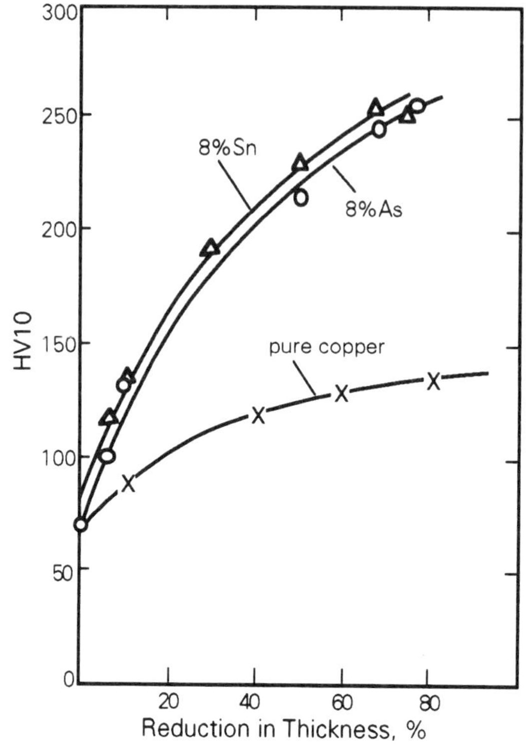

Figure 2.1 Work hardening of copper–arsenic and copper–tin bronzes as compared with copper. Redrawn from Maréchal, S.R., *Hétaux-Corrosion-Ind.*, number 397, 377, 1958.

as polyneuritis and muscular atrophy in limbs. One consequence of this could have been lameness, and it may be relevant to note that, of the Greek gods on Mount Olympus mentioned by Homer, only Hephaestus (the god of fire and forge) was described as being physically imperfect (lame) [7]. Accordingly, Charles [6] has speculated that Hephaestus may reflect the view at the time that the smiths' craft was an unhealthy one. Perhaps this early problem with occupational health and safety was one factor that encouraged the smiths to experiment with the smelting of other minerals to harden copper, leading to the wide use of cassiterite (SnO_2) and the general acceptance of the tin-bronzes. Thus the period over which copper–arsenic alloys were exploited exclusively may have been relatively short, although it seems that both bronzes may have been used together for some 500 years in widely scattered metallurgical centres [6].

Another deliberate use of arsenic as an alloying element occurred much later when craftsmen in the eighteenth century found that it lowered the melting point of platina (naturally occurring platinum containing other elements such as palladium, rhodium and iridium). Hitherto, platina could not be melted because of the high melting points of the elements concerned (e.g. 1769°C for platinum). The technique of adding arsenic was used, for example, by Marc Etienne Janety, a Paris goldsmith during the reign of Louis XVI, and an ornamental sugar bowl he created still exists on display in the New York Metropolitan Museum of Art [8].

The toxicity of arsenic to humans was formally recognized by James Marsh, an English chemist whose forensic investigations suggested that 65 mg of As_2O_3 was sufficient to be fatal [1]. Nevertheless, it is interesting to note that an organic arsenic-containing drug known as Salvarsan 606 was once widely used to treat syphilis [9]. This drug was sufficiently virulent that occasionally it was reputed to terminate the disease by eliminating the sufferer!

2.1.2 Antimony

The natural sulphide of antimony was used as an eyebrow paint by the Babylonians, and a vase found at Tello, Chaldea, was reported to be made from cast antimony [10]. Thus the metal, or at least its sulphide, was known as far back as 4000 BC. Copper articles covered with a thin coating of metallic antimony, found in Egypt and identified with the period 2500 to 2200 BC, indicate that the early Egyptians were also using the metal [10]. Later, Pliny (AD 50) referred to the metal as stibium, from which the chemical symbol Sb is derived, although the final name antimony is attributed to the Greek *anti* and *monos*, which means 'enemy of solitude' or 'metal not found alone'.

The melting of ores containing antimony was mentioned by the German metallurgist Georgius Agricola in the famous book *De Re Metallica* [11] written in the middle of the sixteenth century. At the same time, the Italian Biringuccio described the use of antimony as an alloying element to increase the tone of bell metal, as well as for the production of pewter and metal mirrors [10]. Antimony compounds were also used as a yellow pigment for painting earthenware as well as for tinting enamels and glass. Such compounds were also prescribed as medication for stomach ulcers, although it is now realized that they may cause harmful effects on body tissues, resembling arsenic poisoning [12, 13].

2.1.3 Bismuth

The name bismuth is variously attributed to the Arabic *wiss majaht*, which refers to its ease of melting; the Persian *sipedak* or *isfidaz*, which means white; or the German terms *blei-weiss*, meaning white lead, or *weisse masse* (*wismut*), which is an old mining term for white mass [9]. Apart from the Arabic term, all translated derivatives incorporate a reference to the colour of the metal which is consistent with the fact that bismuth was commonly found in its native or elemental state. Because of its similarity in appearance to lead and tin it was probably not identified until the fifteenth century when it was described by Basil Valentine, a German monk [14]. Agricola [11] referred to it as *bisemutium*, but credit for showing bismuth to be distinct from lead is actually attributed to Claude Geoffroy the Younger in France [1].

In contrast to arsenic and antimony, bismuth has proved benign in its effects on humans, and compounds such as bismuth subsalicylate have been used for many years as a palliative for upset stomachs and intestinal distress, as well as for outlining the alimentary tract during X-ray examination [9, 14]. Moreover, the substitution of bismuth for arsenic in a modified form of the drug Salvarsan 606 maintained its potency in treating syphilis while largely eliminating the risk to the patient [9] (chapter 8).

2.2 Occurrence

Although found native and widely distributed throughout the earth's crust in a variety of minerals, arsenic, antimony and bismuth are comparatively rare elements. Arsenic is the most abundant and estimates vary from $1.8\ g\ t^{-1}$ [15] to $5\ g\ t^{-1}$ [10], which is similar in extent to tin and molybdenum. The crustal abundance of antimony is $0.2\ g\ t^{-1}$, which is similar to cadmium and indium, whereas bismuth (0.001–$0.002\ g\ t^{-1}$) is the least common of the three elements [15, 16]. It may be compared in abundance to the precious metals silver ($0.075\ g\ t^{-1}$) and platinum ($0.005\ g\ t^{-1}$).

The amount of arsenic contained in copper and lead ores is thought to be over 1×10^6 t, half of which is located in Chile (260 000 t), Mexico (60 000 t), the USA and Canada (50 000 t each), Peru and the Philippines (40 000 t each) [17]. China, Kazakhstan and Russia have large but unspecified reserves. Total world resources of arsenic are considered to be 11×10^6 t. Identified world reserves of antimony amount to 5.1×10^6 t, some 52% of which is in China [17]. Other important sources are Bolivia (7.3%), Thailand and the Commonwealth of Independent States (6.4% each) and South Africa (5.6%). World reserves of bismuth are much less. Some 107 000 t are estimated to be present in association with lead and copper ores, with world resources totalling some 250 000 t [17]. Coal ash and manganese nodules in the oceans are other potential sources.

Over 150 arsenic-bearing minerals have been identified which occur mainly as sulphides and which are usually associated with ores containing copper, lead, zinc, gold and silver [10]. The most important is arsenopyrite or mispickel (FeAsS). Other naturally occurring minerals include löllingite ($FeAs_2$), realgar (AsS), orpiment (As_2S_3), niccolite (NiAs), gersdorffite (NiAsS), cobaltite (CoAsS), enargite (Cu_3AsS_4) and sperrylite

($PtAs_2$). Arsenious oxide As_2O_3 is found as the oxidation product of various arsenical minerals.

Antimony occurs mainly in the grey sulphide mineral stibnite (Sb_2S_3). It is also found as various antimonides of the heavy metals such as copper, silver and mercury. Examples are breithauptite (NiSb), dyscrasite (Ag_3Sb) and pyrargyrite ($3Ag_2S.Sb_2S_3$) [10]. Another antimonide, stibiopalladinide (Pd_5Sb_2) is the principal source of palladium [13]. Where stibnite has been weathered by exposure to water and oxygen, it is converted to oxides such as valentenite (Sb_2O_3), cevantite ($Sb_2O_3.Sb_2O_5$) and kermesite ($2Sb_2S_3.Sb_2O_3$).

Bismuth minerals are commonly associated with ores containing copper, lead or tin [10]. The most important are bismuthite (Bi_2S_3), bismite (Bi_2O_3) and bismutite $(BiO)_2CO_3.H_2O$.

2.3 Extraction and refining

Arsenic, antimony and bismuth are recovered mainly as by-products during the processing of complex ores that are mined to obtain copper, lead, zinc, tin, gold and silver. A basic motivation for their removal is the quality specified for these various base metals which require the application of effective refining processes in order to remove these three elements along with other impurities.

World production figures and annual values for 1992 are estimated to be [17]:

- arsenic as As_2O_3 (76% arsenic): 49 300 t, value $US32 million;
- antimony: 75 000 t, value $US125 million (note that an approximately equal amount is also present in lead alloy scrap and is recycled);
- bismuth: 3740 t, value $US18 million.

2.3.1 Arsenic

Arsenic can be obtained directly by sublimation during smelting of the ores arsenopyrite or löllingite at 650°C–700°C in the absence of air, or by the reduction of As_2O_3 with charcoal in a steel retort heated to 700°C–800°C [10]. In both cases, arsenic vapour is collected in a condenser. However, as mentioned above, most arsenic is produced as a by-product during the processing of complex ores that are mined to extract copper, lead, zinc, tin, gold or silver. In recent years, the demand for these base metals has generally been high, with the result that the amount of arsenic produced has far exceeded demand. For this reason, some ores high in arsenical minerals have in fact faced a cost penalty at the smelter, or are deliberately avoided during mining [18]. Surplus arsenic has been stock-

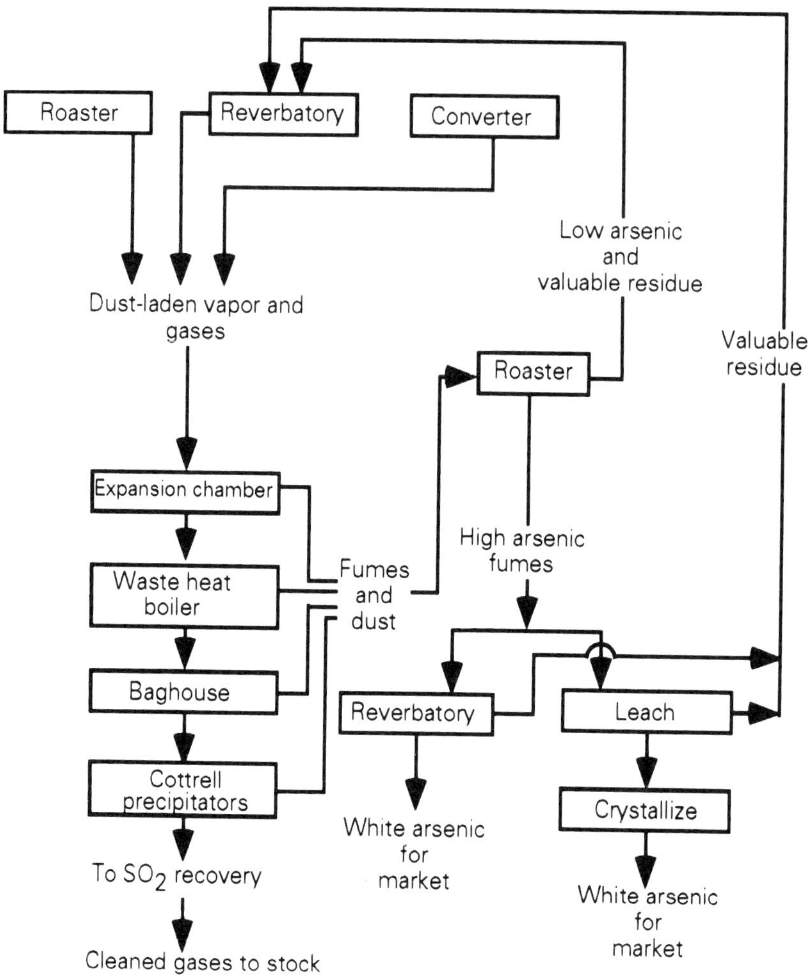

Figure 2.2 Arsenic trioxide flowsheet in a typical copper smelter. Redrawn from Grayson, M. (ed.) *Kirk-Othmer Encyclopedia of Chemical Technology*, 3rd edn, John Wiley, New York, 1983.

piled, mainly as the impure oxide As_2O_3 although, more recently, environmental regulations in some countries have required that arsenic be stored in a chemical compound that is insoluble in water. Examples are calcium and iron arsenates. Production of arsenic doubled from around 30 000 t to 60 000 t during the decade 1980–90 [18]. Leading producers are the USA, Sweden, France, Mexico and countries of the former Soviet Union.

The first stage of copper production involves heating the sulphide concentrates under oxidizing conditions at a high temperature to eliminate much of the sulphur. This operation also provides the opportunity to remove arsenic as the highly volatile oxide As_2O_3 which is collected as flue dust in an electrostatic precipitator or baghouse [10]. Crude flue dust may contain up to 30% As_2O_3, with the balance being oxides of copper, lead and perhaps other metals such as antimony and zinc. The content of As_2O_3 in the flue gas is then upgraded to around 95% by further roasting, after which it can be resublimed to give a product of 99.0%–99.9% purity [10, 19]. The As_2O_3 flowsheet for a typical copper smelter is shown in Fig. 2.2.

In the processing of lead concentrates in a lead blast furnace, most of the arsenic collects in the molten lead bullion together with other impurities, including antimony, copper, bismuth, silver and gold [18]. For example, the bullion treated at the world's largest lead refinery at Port Pirie in South Australia may contain around 0.5% each of arsenic and antimony [20, 21]. Sulphur is added first to remove copper as a slag or dross after which air is blown on to the molten lead, that is maintained at 700°C to 730°C, to remove arsenic and antimony into a low melting point $PbO–As_2O_3–Sb_2O_3$ slag. This part of the process is called softening and the levels of arsenic and antimony in the lead bullion are reduced typically to 0.003% and 0.03%, respectively. The slag, which is removed at regular intervals, may contain several per cent each of arsenic and antimony together with about 75% lead as the oxide. This oxide product is treated separately in a furnace where it is reduced with charcoal to recover arsenic, by evaporation, and antimony, which remains dissolved in lead in what is a saleable lead–antimony alloy known as antimonial lead [22].

2.3.2 Antimony

The main antimony-containing mineral stibnite has a relatively low melting point (546°C) which provides a simple method of recovery from other minerals in a suitable ore body [10]. Heating to 550°C–600°C allows selective melting, or liquation, of the stibnite which can be collected as a crude concentrate. In such a case, the next step is to roast the concentrate in air to produce the volatile oxide Sb_2O_3 that can be retrieved as relatively pure dust in a collection system such as a baghouse. The oxide can then be reduced readily with charcoal or coke in a blast or reverberatory furnace making use of the slag to remove many of the impurity elements. Loss of antimony by volatilization can be high (12%–20%) and must be collected as flue gas and recycled [10]. One variant of the reverberatory furnace is to charge the stibnite concentrate directly with the furnace with soda ash and scrap iron. The soda ash forms a liquid slag whereas the iron displaces antimony from the molten sulphide thereby forming an iron sulphide matte and allowing the antimony to be liberated.

The antimony metal produced by these reduction processes still contains impurities, the most important being lead, arsenic, sulphur, iron and copper. The iron and copper may be removed by treating the metal with an application of fresh stibnite, resulting in formation of an iron–copper sulphide matte which is skimmed off from the antimony surface. Arsenic and sulphur are then removed with an oxidizing flux of sodium nitrate with caustic soda or sodium carbonate. Lead is more difficult to remove but the remaining material can form the basis of lead–antimony alloys, as mentioned earlier [22].

More complex ores can be treated by leaching followed by electrolysis to recover antimony, although this practice is not as economical as the pyrometallurgical processes [10]. The most successful leaching processes use an alkali hydroxide or sulphide as the solvent for stibnite. The filtered leach solution containing sodium thioantimonate, Na_3SbS_4, is electrolysed in a diaphragm cell using an iron or lead anode and an iron or mild steel cathode. Antimony (93%–99% pure) is deposited at the cathode, although problems can arise because the electrolyte becomes fouled with thiosulphate, sulphates and other compounds. Impure antimony metal may also be refined by electrolysis by using an electrolyte containing antimony fluoride and sulphuric acid [18]. The impure antimony is used as the anode but, if the concentrations of copper and arsenic are high, they will co-deposit with the antimony at the cathode. In such a case, it may be necessary to remelt using an oxidizing flux which lowers the arsenic content in particular. Under favourable conditions, the purity of the antimony may exceed 99.9%.

Antimony-containing minerals are nearly always associated with lead concentrates, the smelting and refining of which releases some antimony by volatilization which is collected in electrostatic precipitators or in a baghouse [10]. Antimony remaining in the molten lead can be removed as a low melting point $PbO–Sb_2O_3–As_2O_3$ slag by the softening process described earlier. Reduction of the slag with carbon produces the antimonical lead alloy. If it is necessary to remove arsenic as an impurity, this alloy can be treated with caustic soda or the arsenic can be removed by vacuum distillation.

As will be discussed later (Section 2.5.1), the dominant use of antimony is as an alloying element to harden lead, notably for grids in motor car batteries. This provides a major source of so-called secondary antimony and a substantial part of the requirements for antimony are met simply by the recycling of battery plates and other lead products [10, 18].

2.3.3 Bismuth

Some native bismuth occurs in veins associated with silver, lead, zinc and tin areas of Bolivia, Canada and Germany [16]. However, once again, the major source of this element is as a by-product from the treatment of lead, copper, tin, zinc and silver ores.

As with antimony, bismuth accompanies molten lead through the smelting and refining operations. Two important methods for removing bismuth are the Betterton–Kroll and Betts processes [10, 16, 18]. The former process, which is responsible for the recovery of over 90% of bismuth metal and is widely used throughout the world, [16, 18] relies on the formation of compounds such as Ca_2Bi_2 and Mg_3Bi_2 that have low densities and high melting points. During refining in the so-called furnace kettle these compounds float to the surface of the molten lead and can be skimmed off as dross. The dross is then heated in other small kettles, any entrapped lead removed by liquation, and the dross chlorinated to remove calcium and magnesium. A final treatment with caustic soda produces bismuth which may be 99.95% pure.

In the Betts process the impure lead bullion is cast into anode plates, a number of which are placed in an electrolytic cell along with pure lead cathodes. The electrolyte is a solution of lead fluorosilicate, $PbSiF_6$, and fluorosilicic acid, H_2SiF_6 to which is added a small amount of glue or other suitable agent. During electrolysis the lead from the anodes dissolves and is redeposited on the cathodes, whereas the impurities, including bismuth, remain on the face of the anode as a porous slime. This slime is selectively oxidized to remove arsenic, antimony and lead, whereas the bismuth remains in a slag. This bismuth-rich slag is fluxed and reduced with carbon to form a copper matte together with crude bismuth metal which is removed and refined by the same procedure as used in the Betterton–Kroll process.

Lead is available in different grades (or purities), all of which contain some bismuth as an impurity [9]. Maximum allowable levels of bismuth may vary from 15 ppm for the low bismuth grade, to 250 ppm for refined pure lead, and 500 ppm for what is defined as pure lead. Levels of bismuth that are tolerated in copper are much less and have to be as low as 1 ppm for copper that is to be fabricated into rod and wire [23]. This stringent requirement arises because bismuth is apt to cause 'hot shortness' in copper castings which is a form of embrittlement that may arise when a low melting point element segregates to grain boundaries during solidification as a consequence of its low solid solubility in a parent metal. If such a cast alloy is then hot worked, cracking may occur because melting of the grain boundary element greatly reduces the cohesion between grains. This situation occurs with bismuth (melting point 271°C) which has a low solubility in copper (melting point 1083°C).

Bismuth present in copper concentrates distributes itself between the matte, slag and flue dusts during smelting in the following approximate proportions: 10% in the matte, 80% in the flue dust and 10% in the slag [9]. Unlike most impurity base metals present in copper concentrates, bismuth demonstrates an increasing preference for the matte phase which mainly comprises copper sulphide. For example, if the concentration of bismuth in the matte is increased tenfold, say from 75 ppm to 750 ppm, the distribution in the matte, flue dust and slag may change to 18%, 80% and 2%, respectively.

The fact that 80% of the bismuth remains in the flue dust is not helpful to its removal from the molten copper because modern smelting processes generate sufficient volumes of flue dust to compel its return to the primary smelting circuits in order to recover the copper it also contains [9]. Thus it is necessary either to treat the flue dust to remove bismuth, to recover the copper from the flue dust by a separate process or to pass on part of the problem of bismuth removal to the later electrorefining operation [9].

The complexity of bismuth removal during copper smelting is illustrated in Fig. 2.3 [24]. The next stage in copper production is conversion of the molten copper sulphide matte through oxidation to an impure form of copper known as blister copper. The behaviour of bismuth at this stage is similar to its behaviour during smelting, with about 20% partitioning to the converter slag, 70% to the flue dust and 10% to the blister copper which itself is cast into large anodes for subsequent electrorefining.

Electrorefining of copper involves the dissolution of the blister copper in the electrolyte and its transfer to cathodes which are in the form of copper starter sheets. Conditions are adjusted so that minimal amounts of impurity elements, including bismuth, are transferred to the cathodes. In practice, bismuth may collect on the anode surfaces as insoluble compounds such as bismuth arsenate in a slime which sloughs off and collects at the bottom of the tank. Bismuth metal can then be extracted by treating these slimes.

Bismuth can also be recovered from tin concentrates by leaching with hydrochloric acid [10]. The acid leach liquor is clarified by settling or filtration and the bismuth is precipitated as bismuth oxychloride when the liquors are diluted with larger volumes of water. The bismuth oxychloride is impure and is usually redissolved in hydrochloric acid and reprecipitated several times. It is then dried, mixed with soda ash and carbon, and reduced to the metal. Additional refining must be carried out if high-purity bismuth is required.

2.4 Characteristics and physical properties

Arsenic, antimony and bismuth belong to group 15 of the periodic table. The stable forms of all are crystalline, have a grey, metallic lustre and, although they are comparatively soft, are brittle. Arsenic lies between germanium and selenium in the first long period of the table, and, like these two elements, is classified chemically as a non-metal or metalloid.

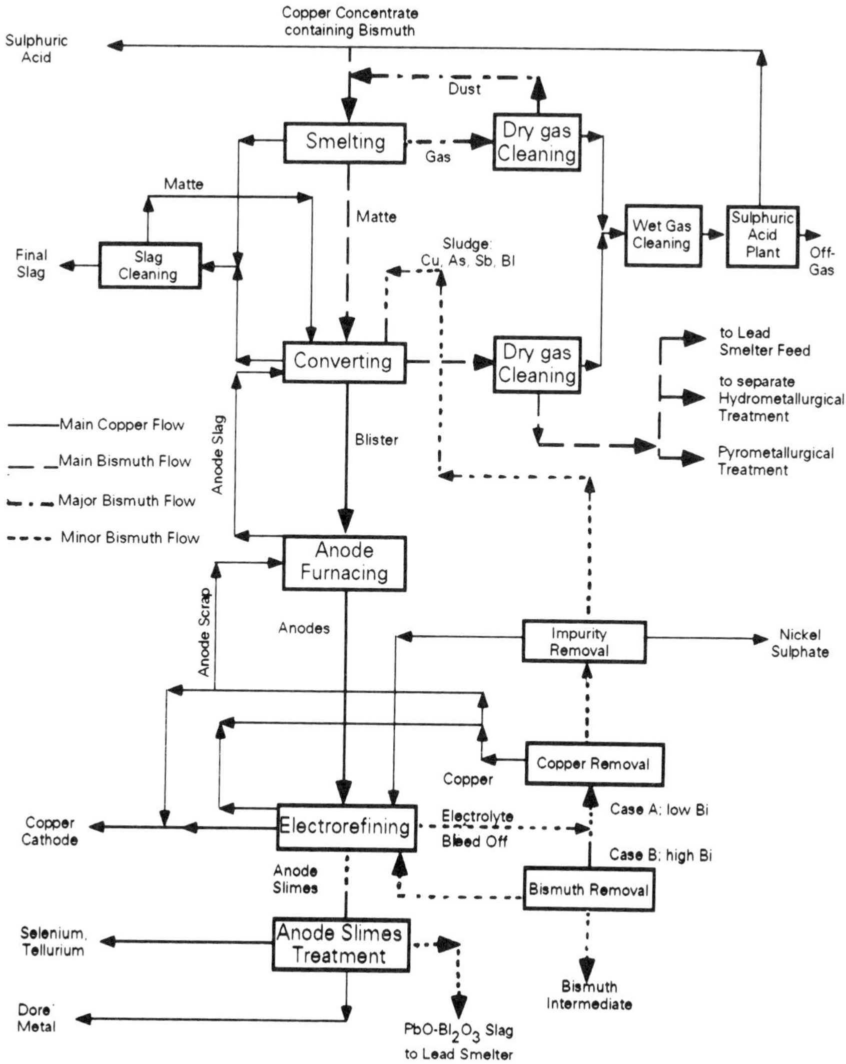

Figure 2.3 Flowsheet showing bismuth removal during copper smelting and refining. Redrawn from Piret, N.L. (1994) *J. Metals*, **46** (10), 15.

On the other hand, antimony and bismuth are clearly more metallic in their characteristics. Arsenic and bismuth exist entirely as the single isotopes ^{75}As and ^{209}Bi, whereas antimony is composed of two stable isotopes ^{121}Sb (57.25%) and ^{123}Sb (42.75%). Each has several radioactive nuclides, all of which have comparatively short half-lives (Table 2.1) [15].

Table 2.1 Key isotopes of the elements.

Element, by nuclide	Atomic mass	Natural abundance	Half-life
Arsenic			
^{73}As	—	0	80.3 days
^{74}As	73.924	0	17.9 days
^{75}As	74.9216	100	stable
^{76}As	75.922	0	26.5 hours
Antimony			
^{121}Sb	120.9038	57.3	stable
^{122}Sb	—	0	2.8 days
^{123}Sb	122.9041	42.7	stable
^{124}Sb	—	0	60.4 days
^{125}Sb	—	0	2.71 years
Bismuth			
^{206}Bi	—	0	6.3 days
^{207}Bi	—	0	30.2 years
^{209}Bi	208.9804	100	stable
^{210m}Bi	0	0	3×10^6 years
^{210}Bi	—	trace	5.0 days
^{212}Bi	—	trace	60.6 months
^{214}Bi	—	trace	19.7 months

Source: Lide, D.R. (ed.), *Handbook of Chemistry and Physics*, 76th edn, CRC Press, Boca Raton, FL, 1995–1996

Arsenic normally exists as the α-crystalline form (density 5.73 g cm^{-3}) although several other allotropes have been reported [10, 16]. A yellow cubic form (density 1.97 g cm^{-3}) may be obtained by rapid cooling of arsenic vapour. However, this allotrope readily reverts to the stable grey form on warming [25]. Another metastable form, β (density 4.6–5.2 g cm^{-3}), has also been reported [26, 27] which is black, has an amorphous structure and is produced by the thermal decomposition of arsine (AsH_3). Antimony has also been reported to occur as several allotropes, although evidence is only available for the existence of two or perhaps three forms. The stable allotrope (density 6.79 g cm^{-3}) has a lustrous blue-white colour and a flaky crystalline texture [10, 15, 27]. It is easily pulverized [26]. A black, amorphous form (density 5.3 g cm^{-3}) can be prepared by rapidly cooling antimony vapour [25]. A yellow form, equivalent to yellow arsenic, is obtained at low temperatures by oxidizing stibine (SbH_3). However, this form is impure and may not be a distinct allotrope [15, 25]. Moreover, it changes readily to the metastable black form [26]. Another possible allotrope, known as explosive antimony, has been reported which is produced by electrolysis of antimony chloride, iodide or bromide and is believed to be in a strained amorphous state. When the electrodeposit is scratched, bent or rapidly heated, it explodes and becomes crystalline [26].

Bismuth exists as a single allotrope and has a silver-white colour uniquely tinged with red. It is the most metallic of the three metals and may show some ductility.

The physical properties of the three elements are compared in Table 2.2. Each has the same rhombohedral structure and all are relatively poor conductors of heat and electricity. Densities vary from 5.78 g cm^{-3} for arsenic to 9.79 g cm^{-3} for bismuth. Although arsenic is shown as having the highest melting point (817°C) it does in fact sublime above 616°C. Antimony differs from normal metals in having a lower electrical conductivity as a solid than as a liquid, a feature it shares with bismuth. Bismuth also has several other special characteristics:

- it is the most diamagnetic of all metals, that is, it exhibits the greatest opposition to being magnetized;
- it has a high electrical resistivity and displays the highest Hall effect of any metal (that is, the greatest increase in electrical resistance when placed in a magnetic field);
- it shares with gallium (and water) the property of expanding slightly

Table 2.2 Physical properties of stable allotropes of the elements.

Property	Arsenic	Antimony	Bismuth
Atomic number	33	51	83
Relative atomic mass (^{12}C = 12.0000)	74.9216	121.750	208.9804
Boiling point (°C)	616[a]	1587	1564
Melting point (°C)	817	630.7	271.4
Crystal structure	rhombohedral	rhombohedral	rhombohedral
a (nm)	0.3760	0.4307	0.4546
c (nm)	1.0548	1.1273	1.1860
Valence	−3, +3, +5	−3, +3, +5	+3, +5
Density at 20°C (g cm^{-3})	5.778	6.697	9.79
Electrical resistivity at 0°C ($\mu\Omega$ cm)	33.3	39.0	107
Heat of fusion (kJ kg^{-1})	370.3	163.2	53.98
Heat of evaporation at boiling point (kJ kg^{-1})	426.77[a]	1602	854.78
Specific heat at 25°C (J K^{-1} kg^{-1})	328	207	122
Magnetic susceptibility (volumetric) (m.k.s.)	-3.9×10^{-6}	-10.2×10^{-6}	-1.68×10^{-5}
Thermal conductivity at 27°C (W m^{-1} K^{-1})	50	24.3	7.87
Thermal expansion, linear coefficient at 25°C (K^{-1})	5.6×10^{-6}	11.0×10^{-6}	13.4×10^{-6}
Volume change on freezing (%)		−0.8	+3.32
Thermal neutron capture cross-section (barns)	4.30 ± 0.10	5 ± 1	0.034

Sources: ASM International, *Metals Handbook*, vol. 2, 10th edn, Materials Park, OH, 1990; Grayson, M. (ed.), *Kirk–Othmer Encyclopedia of Chemical Technology*, 3rd edn, John Wiley, New York, 1983; Lide, D.R. (ed.), *Handbook of Chemistry and Physics*, 76th edn, CRC Press, Boca Raton, FL, 1995–1996.

[a]Sublimation occurs.

on solidification rather than the more usual contraction (it has also been claimed that antimony expands on solidification, although values reported vary from +1% to a contraction of –1.4% and probably depend upon the purity of the metal [10]; more recently Kirshenbaum and Cahill [27] clearly showed that pure antimony in fact contracts 0.79 ± 0.14 vol% on solidification);

- it has the lowest thermal conductivity of all metals except mercury.

2.5 Uses of the elements

With only minor exceptions arsenic, antimony and bismuth are not used as base metals. Each forms intermetallic compounds having unique properties which has led to their use for specialized purposes. However, for most applications, the three elements are used as major or minor alloying additions to other metals. The different areas in which the elements are used will now be considered.

2.5.1 Antimonial lead

Pb–Sb alloys are normally used in the cast state and account for the greatest single application of lead [28]. These two metals form a simple eutectic system with a eutectic point at 11.1% Sb and a melting temperature of 252°C (Fig. 2.4) [29]. It will also be noted that the solubility of antimony in lead decreases from a maximum of 3.5% at 252°C to as little as 0.25% at 25°C whereas lead has little solubility in antimony. This provides the opportunity to obtain supersaturated Pb–Sb alloys by fast cooling after casting or by a separate process involving solution treatment at around 250°C followed by quenching to room temperature and then ageing. Ageing, for example at ambient temperatures, promotes precipitation of Sb particles on the $\{1\,1\,1\}_\alpha$ planes of the lead matrix [30] which increases strength and hardness [28]. For example, the yield strength of the alloy Pb–3% Sb can increase from 34 MPa to 55 MPa after ageing for 30 days at 25°C. A further response to age hardening is possible by the addition of traces of elements such as arsenic [31] and copper [32], which refine the size of the Sb precipitates.

Pb–Sb alloys have been used for more than 100 years [28] as electrode plates, connectors and terminals for lead-acid storage batteries [28] and this application commonly accounts for 80%–85% of all antimony used in metal products [10]. The alloys are creep-resistant, corrosion-resistant in H_2SO_4 and resist structural changes caused by charge–discharge cycles in the battery. However, the use of Pb–Sb alloys for the positive grids of storage batteries does have one disadvantage because some corrosion occurs during charging which releases antimony into the electrolyte. This

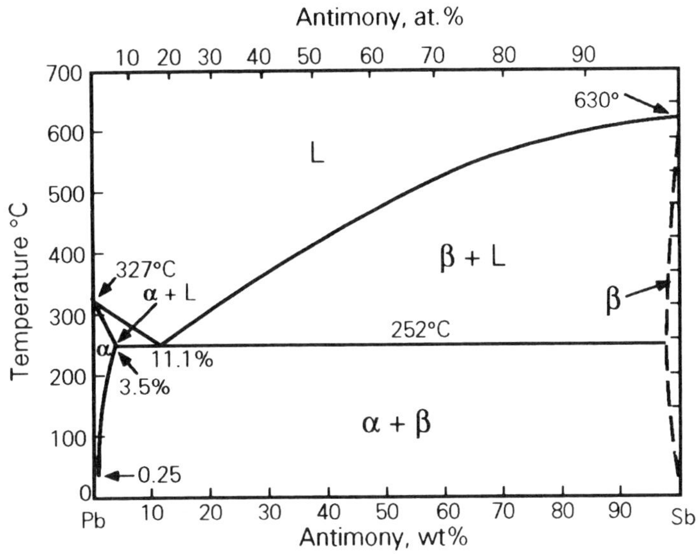

Figure 2.4 Lead–antimony phase diagram. Redrawn from Massalski, T.B. (ed.), *Binary Alloy Phase Diagrams*, 2nd edn, ASM International, Materials Park, OH (1990).

antimony plates out on the cathode and has the effect of reducing the voltage at which water in the electrolyte decomposes to H_2 and O_2 during charging. Battery water is thus consumed by this gassing and must be replaced [33]. The amount of antimony transferred increases with service life and the antimony content of the positive grid. For this reason the trend has been to lower the antimony content of the alloy to levels of 2.5% to 3%, although the casting quality is reduced as a consequence. In this regard, trace additions of arsenic tend to compensate for the loss of strength associated with the lower antimony content [10].

Pb–Sb alloys are also used for cable sheathing, as anodes for the electrowinning and plating of other metals such as nickel, as collapsible tubes and as anodes for the cathodic protection (by means of an applied electrical current) of pipelines, ocean-going vessels and oil drilling platforms [28]. The alloys are also used for ammunition in the form of shot, which is produced by dropping the molten alloy through small holes drilled in an iron pan and allowing it to freefall in air into water at the bottom of a tower. Such alloys also contain small amounts of arsenic since this element raises the surface tension of the alloy thereby enabling the molten droplets to attain a round shape during freefall [28]. Pb–Sb alloys containing larger amounts of Sb (10%–16%) as well as 1%–7% Sn are used for bearings and for moveable type in printing presses [28].

2.5.2 *Fusible alloys*

The main use of bismuth is in the manufacture of low melting point alloys, many of which will melt at temperatures below 100°C. These so-called fusible alloys include a group of binary, ternary, quaternary and occasionally quinary alloys, many of which form eutectics (Table 2.3). A well-known example is Woods metal (50% Bi–26.7% Pb–13.3% Sn–10% Cd) which melts at 70°C, although there is another composition that melts at only 47°C.

Fusible alloys have several industrial applications, one of which is in automatic safety devices such as fire sprinklers, boiler plugs and controls for furnaces. For these purposes a eutectic alloy is selected which has sufficient strength at ambient temperatures to hold parts together yet which, at a specific elevated temperature, will melt thereby disconnecting the parts. In fire sprinklers, melting of the link allows water to be released from the associated piping systems so as to extinguish the fire. Boiler plugs and furnace controls react similarly in that an increase in temperature beyond some safe limit will melt the plug thereby allowing release of pressure in a boiler on cessation of the fuel supply to a furnace which is in danger of overheating.

As well as eutectic alloys which melt at a particular temperature there are numerous compositions containing bismuth which melt over a range of temperatures (Table 2.3). Such materials are useful in staking or positioning rods and tubing in assemblies, because the alloy is distributed to the various surfaces while it is molten and then provides a firm anchorage after it solidifies.

Table 2.3 Compositions and melting temperatures of selected fusible alloys containing bismuth.

Composition(%)					Melting temperature (°C)
Bi	Pb	Sn	Cd	In	
44.7	22.6	8.3	5.3	19.1	47
49.0	18.0	12.0	—	21.0	58
50.0	26.7	13.3	10.0	—	70
51.6	40.2	—	8.2	—	91.5
52.5	32.0	—	15.5	—	95
54.0	—	26.0	20.0	—	102.5
55.5	44.5	—	—	—	124
58.0	—	42.0	—	—	138.5
60.0	—	—	40.0	—	144
50.5	27.8	12.4	9.3	—	70–73
50.0	34.5	9.3	6.2	—	70–79
35.1	36.4	19.05	9.45	—	70–101
56.0	22.0	22.0	—	—	95–104
33.3[a]	33.3[a]	33.3[a]	—	—	101–143
40.0	—	60.0	—	—	138–170

Source: ASM International, *Metal Handbook*, vol. 6, 10th edn, Materials Park, OH (1990).

[a]Figures do not add up to 100% owing to rounding errors.

Fusible alloys based on bismuth are also used to assist in grinding and polishing glass or transparent plastic optical lenses. In this case, the low melting point allows the alloy to be cast directly on to the optical blank to provide support during these operations.

2.5.3 Solders

The American Welding Society defines soldering as metal coalescence below 800°F (426.5°C); as a method of joining it is relatively cheap and involves minimal heat changes to the parts concerned. The main industrial solders are based on the lead–tin system in which the compositions are adjusted to meet the particular requirements of the assembly [28, 34]. The presence of arsenic as an impurity causes the progressive deterioration of the quality of lead–tin solders, and as little as 0.005% can adversely affect melting. Low levels of bismuth generally do not cause problems although some discoloration of soldered surfaces occurs at levels above 0.5%. As an impurity, antimony tends to reduce the spread of high-lead solders and specifications normally require a limit of 0.05% of this element. However, antimony (e.g. 3%) is deliberately added to strengthen and improve the creep resistance of other lead–tin solders. These properties are, in fact, highest in the binary tin–antimony alloy (95% Sn–5% Sb) which has a relatively higher melting range (233°C–240°C). This particular solder is often used when joining stainless steels, in order to avoid the possibility of lead contamination. Pb–Sb–Sn alloys are also commonly used for automobile body solder [28].

The fusible alloys based on bismuth (Table 2.3) may also serve as solders where the temperatures must not exceed 180°C. However, it is then usually necessary to clean the surfaces being joined with rather corrosive fluxes or to include a pre-timing operation, because these solders do not have good wetting characteristics.

Recently interest in lead-free solders has been stimulated by the introduction in 1990 of a bill in the US Congress to propose the banning of lead from a wide variety of uses, including solders for electronic circuits. Although an immediate ban on lead in these solders seems unlikely, major electronic companies and their suppliers of solder are now investigating lead-free alternatives. Tin-base solders are receiving particular attention and the tin–bismuth system is currently the best understood [35]. Tin and bismuth form a simple eutectic which has the composition 42% Sn–58% Bi and melts at 139°C (Fig. 2.5) [31]. The equilibrium phases are bismuth and tin, with the tin component dissolving 21 wt% bismuth at the eutectic temperature. As the alloy cools, the solubility of bismuth falls and this element precipitates in the tin phase. The volume ratio of tin to bismuth in this eutectic is 49:51, and a classical lamellar microstructure is obtained (Fig. 2.6) [36]. Tin–bismuth alloys are an exception to the general rule

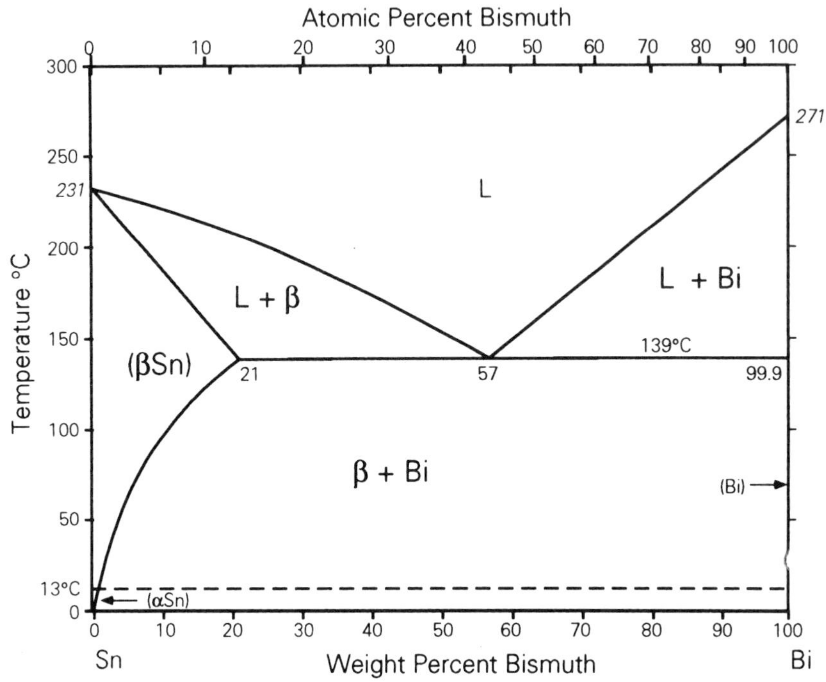

Figure 2.5 Tin–bismuth phase diagram. Redrawn from Massalski, T.B. (ed.), *Binary Alloy Phase Diagrams*, 2nd edn, ASM International, Materials Park, OH (1990).

that solders, as with most other metals and alloys, contract during solidification. As mentioned earlier (Section 2.4), bismuth expands during solidification, whereas tin contracts, but to a lesser amount, with the overall effect that tin–bismuth solders containing more than 47% bismuth expand as they solidify [37]. Tin–bismuth solder shows reasonable strength even up to 100°C despite its low melting point, and other properties compare well with the lead–tin solders it may replace [35].

2.5.4 *Bearing alloys*

In a lubricated journal bearing, the hydrodynamic pressure that builds up in the oil film between the moving surfaces will, under ideal conditions, prevent metallic contact. However, departures from these conditions do occur, leading to oil breakdown, and it is necessary to surround the journal with a bearing alloy to which the oil film will adhere, thereby promoting lubrication [38, 39].

Bearing alloys have conflicting property requirements. On the one hand they must be relatively soft so that they can both adapt to stresses arising from elastic and plastic deformations under which they must operate and

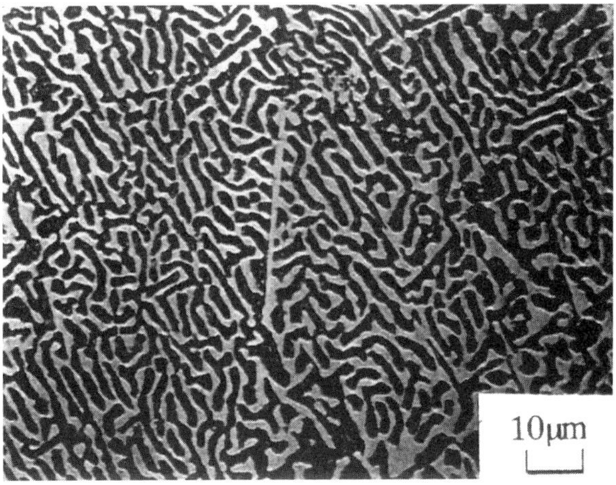

Figure 2.6 Lamellar microstructure in eutectic tin–bismuth alloy. Reproduced from Morris, J.W. Jr, Freer Goldstein, J.L. and Mei, Z., *The Mechanics of Solder Alloy Interconnects* (eds D.R. Frear *et al.*), Van Nostrand Reinhold, New York, (1994), p. 7.

Figure 2.7 Microstructure of white metal bearing alloy, showing cuboids of the hard SbSn phase. (Magnification ×250). Reproduced from Lloyd, K., *Bearing Metallography*, Metallurgical Services, Betchwith, UK (1971).

have the ability to absorb or imbed hard particles that may be generated as a result of wear. They must also be significantly softer than the journal they support, otherwise cutting or scoring will occur. On the other hand the bearing alloy must be hard enough to support the operating load imposed upon it and to minimize its own wear. One way of meeting these various requirements is for the bearing alloy to have a duplex microstructure in which hard particles are distributed in a soft matrix (Fig. 2.7). This is the situation for the so-called white metal bearings, commonly known as babbits, which are named after Isaac Babbit who was granted a patent in 1839 when knowledge of microstructures did not in fact exist.

The white metal bearings are based on tin or lead and usually contain antimony and copper as major alloying additions. As an example, Fig. 2.8 shows the tin-rich corner of the Sn–Sb–Cu phase diagram [40]. The matrix is a solid solution of antimony in tin and it will be noted that the compound SbSn is formed when the antimony content exceeds 7%–9%. This phase, which is easily recognized by its geometric, cuboidal shape (Fig. 2.7) is significantly harder than the softer, grey matrix [41]. The needle-like phase Cu_6Sn_5 also makes a useful contribution to the bearing qualities of the alloy. Similar microstructures are observed in lead-base alloys and in each case the bearings are prepared as either a thin shell bonded to a stronger steel strip or, less commonly, as solid bearings without a supportive backing. Typical compositions are Sn–7.5% Sb–3% Cu and Pb–10% Sb–6% Sn. White metal alloys have been replaced by other

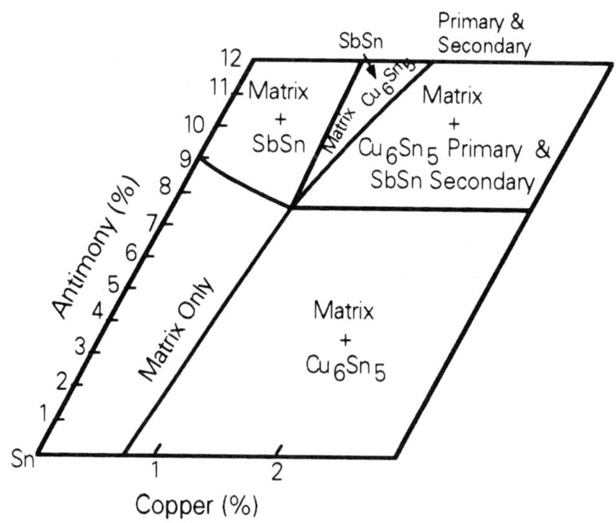

Figure 2.8 Section of the ternary tin–antimony–copper phase diagram. Redrawn from Barry, B.T.K. and Thwaites, C.J., *Tin and its Alloys and Compounds*, Ellis Horwood, Chichester, Sussex (1983).

bearing materials for many applications such as high compression automotive engines, but they still account for the second greatest use of antimony [10].

2.5.5 *Semiconductors*

Semiconductors were first developed with use of the group 14 element germanium but now they are based largely on silicon because this element is plentiful and cheaper to produce. Group 14 elements do not, however, possess the most outstanding electronic properties and for some applications compounds made by combining elements from groups 13 and 15 offer superior performance. Of these, gallium arsenide (GaAs) is attracting particular attention because it has the following advantages over silicon [42–44]

- There is a narrower gap between the electron conduction and valence bands so that less energy is required to induce conduction. This allows more compact integrated circuits to be assembled without the danger of overheating. Moreover, gallium arsenide semiconductors can operate at higher temperatures than those made from silicon.
- Electron mobility is some six times greater, a feature which has special significance for fast electronic switching in applications such as computing.
- As with most other compound semiconductors, the energy evolved when electrons move out of the conduction band is usually in the form of light waves having frequencies in the visible and infra-red ranges. As a consequence, devices can be constructed to exploit this two-way conversion between light waves and electronic currents and include light-emitting diodes (LEDs) that were first used for displays on calculators and digital watches, laser diodes and photodetectors. The opportunity to link electronic components to optical communications systems has revolutionized telecommunications and has led to the development of the optoelectronic industry.
- They are less susceptible to radiation damage and for this reason find application in space probes and satellites.

Gallium arsenide is more difficult than silicon to produce as bulk crystals that are pure and free of defects. Accordingly it is used more commonly for thin-film devices that are produced by techniques such as molecular beam epitaxy [45]. These techniques also enable layered devices to be prepared which may contain other arsenides including InGaAs and AlGaAs, which are used for various lasers. All these arsenide semiconductors are more expensive than those made from silicon and as a consequence they are used in niche markets rather than for general circuitry. Nevertheless they already constitute a multimillion-pound industry.

High purity antimony (> 99.999%) has also found a much more limited market in semiconductor devices. Compounds AlSb, GaSb and InSb have been used for infra-red detectors, diodes and Hall-effect devices [16].

2.5.6 Superconductors

Since the achievement of superconductivity in ceramic materials of the general formula $La_{2-x}Sr_xCuO_4$ in 1986 [46] a large number of other compounds have been developed with a range of superconducting transition temperatures (T_c) [15]. Included are three compounds containing bismuth: $Bi_2Sr_2CuO_x$ (2201 phase) [47], $Bi_2Sr_2CaCu_2O_x$ (2212 phase) [48] and $Bi_2Sr_2Ca_2Cu_3O_x$ (2223 phase) [48] with T_c values of < 20 K, 80 K and 110 K, respectively. These compounds are attracting particular attention because they have so far proved amenable to the powder-in-tube (PIT) process for fabricating superconducting wires and tapes that can carry relatively high electrical currents [49]. They offer the advantage that they can be produced as glasses [50] from which filamentary crystals can be formed which in turn can be aligned in a desirable texture by well-known thermomechanical processes [51]. Cross-sections of a flat tape and round wire in which the superconducting compounds are sheathed in silver alloys are shown in Fig. 2.9 [51]. Manufacturing plants have been established in

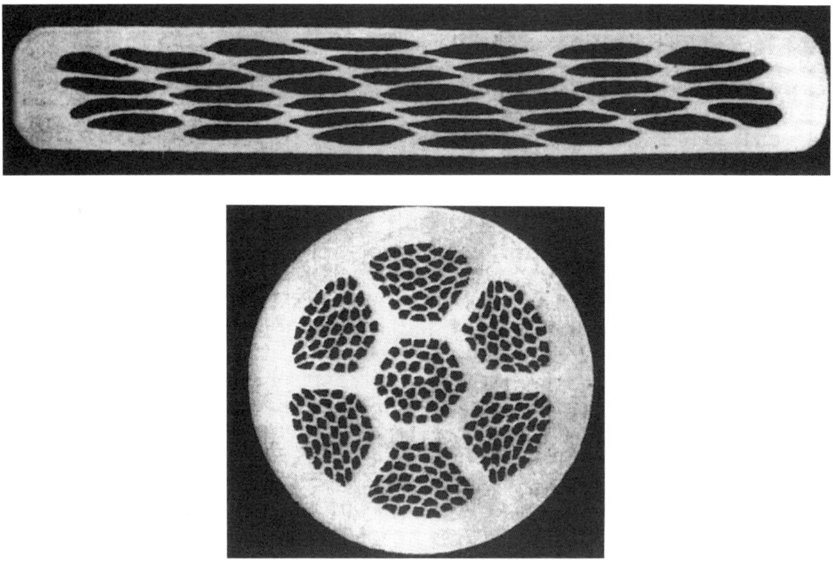

Figure 2.9 Sections of (a) tape; (b) wire cable, containing bismuth compounds (dark areas) sheathed in silver (white areas). Reproduced from Sokolowski, R.S., Selamanickam, V., Hazelton, D.W. *et al.* (1995) *J. Metals*, **47**, (7), 61.

the USA and Japan which can produce tapes and wires kilometres long. Many prototype applications are being tested and the first commercial applications may be only a few years away [49, 52, 53].

2.5.7 Superplasticity

A number of alloys, when deformed slowly in tension at temperatures close to half their melting points in Kelvin, are capable of exhibiting neck-free plastic extensions that may exceed 1000% [54–56]. Such extreme elongations compare with values of 50% or less for conventional metals and alloys. Whereas the phenomenon remained a laboratory curiosity for many years it now has important practical implications for hot-forming complex shapes in certain steels and in alloys based on a variety of metals including aluminium, titanium, copper, zinc and nickel.

In addition to the prerequisites mentioned above, it is usually necessary for these alloys to have stable, fine-grained microstructures comprising two distinct phases. This situation exists with many eutectics such as that shown for the Sn–Bi alloy in Fig. 2.6, and it is interesting to note that it was the historic photograph published by Pearson [57] in 1934 (Fig. 2.10) that first stimulated an interest in superplasticity. In this case, the specimen of the Sn–Bi eutectic alloy (Fig. 2.5) was extended 1950%. Currently the largest extension reported has been 8000% for a commercial aluminium bronze [58].

It has been proposed that ancient arsenical bronzes used in Turkey *c.* 2200 BC may have inadvertently exhibited superplasticity during hand

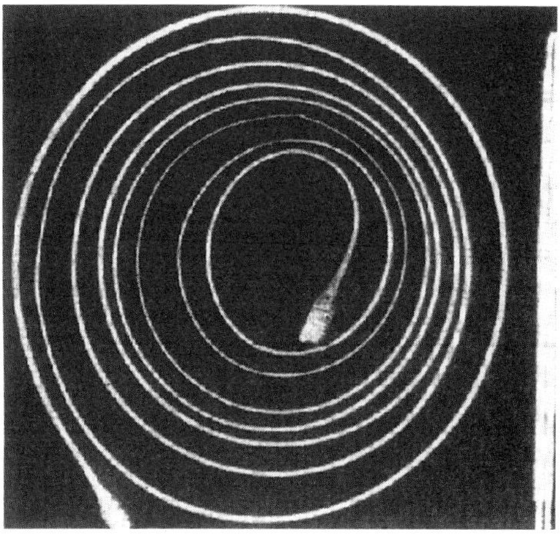

Figure 2.10 Historic photograph by C.P. Pearson of superplastic extension of 1950% in an extruded tin–bismuth alloy eutectic. Reproduced from Pearson, C.P. (1934) *J. Inst. Met.*, **54**, 111.

forging to intricate shapes [59]. Furthermore, it has been noted that the ancient steels of Damascus in use from 300 BC to the late nineteenth century are very similar in composition to some modern steels that display superplastic characteristics [56]. The Sn–Bi alloy mentioned above has no commercial importance. However, it has special interest because, as a consequence of its low melting point, the alloy is one of only a few that do exhibit superplasticity when deformed slowly at ambient temperatures [60].

2.5.8 *Sundry applications*

Thermoelectric alloys. The intermetallic compound Bi_2Te_3 is used for Peltiér cooling devices and for low-temperature thermoelectric power production [61]. The compound also exhibits excellent properties as a magnetometer in the temperature range 1.2 to 300 K.

Magnets. A permanent magnetic alloy (bismanol) made from the compound MnBi is used for specialized applications requiring a high coercive force and maximum energy product [10].

Rectifiers. Eutectic alloys of Bi–Sn and Bi–Cu have been used as counterelectrode materials in the manufacture of selenium rectifiers [10].

Mouldings. An alloy Sn–40% Bi exhibits zero contraction or expansion after castings and is sometimes used for producing dimensionally accurate plastic mouldings [10].

Pewter. The general class of alloys known as pewter have been used for domestic utensils and decorative ware for some 2000 years. All are tin-based, although the earlier alloys contained substantial amounts of lead (e.g. 35%) which is now recognized as a health hazard. The best of modern pewters are lead-free and usually contain antimony to increase hardness. A typical composition is Sn–7.5% Sb–1.5% Cu.

2.5.9 *Minor alloying additions*

Reference has already been made to the embrittling effect which trace amounts (e.g. 1 ppm) of bismuth may have on copper alloys (Section 2.5.2). Arsenic is also an undesirable impurity in some alloys, an example being the significant decrease it causes in the conductivity of copper used for electrical conductors. There are, however, a number of other examples where small amounts of arsenic, antimony or bismuth can improve the properties of specific alloys:

- antimony is sometimes added to copper to increase the recrystallization temperature of that metal [16];
- aluminium–magnesium alloys are susceptible to cracking during hot working, for example rolling, if contaminated by small quantities of sodium; alloying with traces of bismuth (0.01%–0.02%) restores ductility by the removal of sodium into the innocuous compound NaBi [62];
- larger amounts of bismuth (0.2%–0.6%) together with similar amounts of lead are added to certain wrought aluminium alloys to improve machining characteristics; these elements form discrete particles of an insoluble compound which is dispersed throughout the microstructure and assists in promoting chip formation [16];
- additions of 0.2% antimony are sometimes made to cast Al–Si alloys in order to refine the size of the silicon phase of the eutectic and thereby improve the mechanical properties [63];
- bismuth in quantities of 0.004%–0.02% is sometimes added to malleable cast irons to stabilize carbides and to provide uniform graphitization during subsequent annealing; additions of around 50 ppm of antimony, preferably together with cerium, to the class of spheroidal graphite cast irons also has the capacity to make graphite fully nodular in the centre of thick castings;
- minor additions of arsenic in the range 0.02–0.1% are made to several brasses, for example Muntz alloy (Cu–40% Zn) and aluminium brass (Cu–23% Zn–2.1% Al) to improve heat resistance [10].

Acknowledgement

The assistance given by Professor F. Lawson in preparation of Section 2.3 is acknowledged and appreciated.

References

1. Trifonov, D. N. and Trifonov, V. D. (1980,1982) *Chemical Elements: How They Were Discovered*, MIR Publishers; Russian edn, 1980, English translation 1982.
2. Garland, H. and Bannister, C. O. (1927) *Ancient Egyptian Metallurgy*, Charles Griffen, London.
3. Tylecote, R. F. (1976) *A History of Metallurgy*, The Metals Society, London.
4. Aitcheson, L. (1960) *A History of Metals*, vols 1 and 2, Macdonald and Evans, Harlow, Essex.
5. Maréchal, S. R. (1958) *Métaux-Corrosion-Ind.*, no. 397, 377.
6. Charles, J. A. (1979) *J. Metals*, **31**, (7) 8.
7. Grimal, P. (1973) *Larousse World Mythology*, Hamlyn, London.
8. Young, G. (1983) *National Geographic*, **164**, 686.
9. Hoffman, J. E. (1994) *J. Metals*, **46**, (10) 10.
10. Grayson, M. (1983) *Kirk–Othmer Encyclopedia of Chemical Technology*, 3rd edn, John Wiley, New York.

11. Agricola, G. (1912) *De Re Metallica* (trans H. C. Hoover and L. H. Hoover), The Mining Magazine, London.
12. Sax, N. I. and Lewis, R. J. (1989) *Dangerous Properties of Industrial Materials*, Van Nostrand Reinhold, New York.
13. *Encyclopedia Britannica* (1987) 15th edn, **1**, 457.
14. *Encyclopedia Britannica* (1987) 15th edn, **2**, 242.
15. Lide, D. R. (ed) (1995–1996) *Handbook of Chemistry and Physics*, 76th edn. CRC Press, Boca Raton, FL.
16. *Metals Handbook*, vol. 2, 10th edn, ASM International, Materials Park, OH.
17. Crowson, P. (1995) *Minerals Handbook 1994–5: Statistics and Analysis of the World's Mineral Industry*, Stockton, New York.
18. Cahn, R. W. (ed.) (1993) *Encyclopedia of Materials Science in Engineering, Supplement Volume 3*, Pergamon Press, Oxford.
19. Edelstein, D. L. (1985) *Arsenic: Mineral Facts and Problems*, Bureau of Mines, US Department of Interior, Washington, DC.
20. Saint, G., Legg, J. E. and Pizzey, M. (1993) *Australasian Mining and Metallurgy*, vol. 1, 2nd edn (eds J. T. Woodcock and J. K. Hamilton), Monograph and Report Series No. 19, The Australasian Institute of Mining and Metallurgy, Parkville, Vic., p. 599.
21. Moffat, R. J. (1993) in *Australasian Mining and Metallurgy*, vol. 2, 2nd edn (eds J. T. Woodcock and J. K. Hamilton), Monograph and Report series No. 19, The Australasian Institute of Mining and Metallurgy, Parkville Vic., p. 1215.
22. Plunket, P. A. (1985) in *Antimony: Mineral Facts and Problems*, Bureau of Mines, US Department of Interior, Washington, DC, p. 33.
23. Serrano, J., Berger, D. and Bridges, W. Jr, (1994) *J. Metals*, **46**, (10), 12.
24. Piret, N. L. (1994) *J. Metals*, **46** (10), 15.
25. Considine, D. M. (ed.) (1976) *Van Nostrand's Scientific Encyclopedia*, 5th edn, Van Nostrand Reinhold, New York.
26. Lynch, C. T. (ed.) (1994) *Practical Handbook of Materials Science*, CRC Press, Boca Raton, FL.
27. Kirshenbaum, A. D. and Cahill, J. A. (1962) *Trans ASM*, **55**, 849.
28. Prengamon, R. D. (1966) *Lead Alloys, Materials Science and Technology, Vol. 8: Structure and Properties of Non Ferrous Alloys* (ed. K. H. Matucha), Pergamon Press, Oxford, 71.
29. Massalski, T. B. (ed.) (1990) *Binary Alloy Phase Diagrams*, 2nd edn, ASM International, Materials Park, OH.
30. Aballe, M., Redigor, J. T., Sistraga, J. M. and Torralba, M. (1972) *Z. Metallk.*, **63**, 564.
31. Hopkin, L. M. T. and Thwaites, C. J. (1952–1953) *J. Inst. Met.*, **81**, 255.
32. Mao, G., Larson, J. G. and Rao, P. (1969) *J. Inst. Met.*, **97**, 343.
33. Ruetschi, P. and Cahan, B. D. (1975) *J. Electrochem. Soc.*, **104**, 406.
34. *Metals Handbook*, vol. 6, 10th edn., ASM International, Materials Park, OH, p. 1069.
35. Glazer, J. (1995) *Inter. Mater. Rev.*, **40**, 65.
36. Morris, J. W. Jr, Freer Goldstein, J. L. and Mei, Z. (1994) in *The Mechanics of Solder Alloy Interconnects*, (ed. D. R. Frear *et al.*), Van Nostrand Reinhold, New York, p. 7.
37. Manko, H. H. (964) *Solders and Soldering*, McGraw-Hill, New York.
38. Pratt, G. C. (1973) *Inter. Met. Rev.*, **18**, 63.
39. Suri, A. K. and Bannerjee, S. (1966) in *Lead Alloys, Materials Science and Technology, Vol. 8: Structure and Properties of Non Ferrous Alloys* (ed. K. H. Matucha), Pergamon Press, Oxford, p. 23.
40. Barry, B. T. K. and Thwaites, C. J. (1983) *Tin and its Alloys and Compounds*, Ellis Horwood, Chichester, Sussex.
41. Lloyd, K. (1971) *Bearing Metallography*, Metallurgical Services, Betchwith.
42. *Metals Handbook*, vol. 2, 10th edn, ASM International, Materials Park, OH, p. 733.
43. US Bureau of Mines (1988) *Gallium and Gallium Arsenide: Supply, Technology and Uses*, Report IC 9208.
44. Jones, S., Thomas, G. B. and Wiltshire, B. (1986) *Met. Mater.*, **2**, 353.
45. Tu, C. W. (1995) *J. Metals*, **47**, (12), 34.
46. Bednorz, J. G. and Müller, K. A. (1986) *Z. Phys.*, **B64**, 189.
47. Michel, C., Hervieu, M., Bolel, M. M. *et al.* (1987) *Z. Phys.*, **B68**, 421.

48. Maeda, H., Yanaka, Y., Fukutomi, M. and Asano, T. (1989) *Jap. J. Appl. Phys.*, **28**, L1121.
49. Goyal, A. (1995) *J. Metals*, **47**, (7), 55.
50. Dimesso, L., Matsubara, I., Ogura, T. *et al.* (1994) *Mater. Res.*, **9**, 2501.
51. Sokolowski, R. S., Selamanickam, V., Hazelton, D. W. *et al.* (1995) *J. Metals*, **47**, (7), 61.
52. Brokenborough, W. and Malozemoff, A. P. (1995) *J. Metals*, **47**, (7), 59.
53. Sato, K. (1995) *J. Metals*, **47**, (7), 65.
54. Underwood, E. E. (1962) *J. Metals*, **14**, 914.
55. Edington, J. W., Melton, K. N. and Cutler, C. P. (1976) *Prog. in Materials Science*, **21**, (2), 63.
56. Sherby, O. D. and Wadsworth, J. (1989) *Prog. in Materials Science*, **33**, (3), 169.
57. Pearson, C. E. (1934) *J. Inst. Met.*, **54**, 111.
58. Higashi, K., Ohnishi, T. and Nakatani, Y. (1985) *Scripta Metall.*, **19**, 821.
59. Geckinli, E. (1987) *Prog. in Materials Science*, private communication from Technical University of Istanbul, Istanbul.
60. Chaudhari, P. (1968) *Science and Technology*, 81, 42.
61. *McGraw-Hill Encyclopedia of Science and Technology*, vol. 18, McGraw-Hill, New York, p. 295.
62. Talbot, D. E. J. and Granger, D. A. (1995) *J. Metals*, **47** (2), 44.
63. Khan, S. and Elliot, R. (1994) *J. Mater. Sci.*, **29**, 736.

3 Inorganic derivatives of the elements

S.M. GODFREY, C.A. McAULIFFE, A.G. MACKIE and R.G. PRITCHARD

3.1 Introduction

In this chapter the hydrides, halides, chalcides, oxoanion and intermetallic compounds of the pnictogens are described. The chemistry is confined to the homoleptic species, that is, essentially EX_n species where E = As, Sb, Bi and X = the element(s) in question.

3.2 The hydrides of arsenic, antimony and bismuth

Only arsenic has stable hydrides. The hydrides of antimony and bismuth are endothermic compounds which decompose spontaneously at room temperature. Thus this section is necessarily dominated by the chemistry of arsenic and hydrogen.

3.2.1 Arsenic–hydrogen compounds

This treatment of binary arsenic hydrogen compounds is divided into two parts. First, the transient AsH_x ($x = 1, 2$) species observed spectroscopically are described, some of which are important commercially for their role in metallo-organic chemical vapour deposition (MOCVD) processes. This is followed by an account of arsine and its homologues.

Arsenic–hydrogen diatomic (AsH) and triatomic (AsH_2) species and their ions. Interest in the arsenic hydride transients formed in the thermolysis of arsine has developed not only because of their significance in the MOCVD production of gallium arsenide and related materials but also because of their importance in modern instrumentation for the monitoring of arsenic in the environment (chapters 8 and 9).

Arsenic monohydride and its ion. Arsenic monohydride is an unstable transient molecule. It may be generated in sufficient concentration for spectroscopic examination by the flash photolysis of arsine at 20 torr [1], by the reaction of arsenic powder with dihydrogen in the microwave discharge [2] or by the reaction of hydrogen atoms with AsH_3 [3].

Table 3.1 Physical parameters of diatomic AsH: experimental (exp.) and calculated (calc.) values.

State	Re (A)		M_e (cm^{-1})		T_e (cm^{-1})		D_e (eV)		μ_e (D)
	calc.	exp.	calc.	exp.	calc.	exp.	calc.	exp.	
$x^3\Sigma^-$	1.528	1.534	2194	2130	0		2.60	2.80	0.178
$^1\Delta$	1.525		2190		9 968		3.03		0.205
$^1\Sigma^+$	1.522		2197		17 114		0.26		0.185
$A^3\Pi$	1.577	1.58	1398	1208	32 244	30 000			0.779
1Π	2.467		441		36 608				0.211
$^3\Pi$(II)	2.187		1386		41 461				–1.315
$^1\Pi$(II)	2.561		710		44 319				–0.225
$^1\Pi$(III)	1.603		3606		63 288				–1.872
$^3\Pi$(III)	1.875		3624		67 848				0.717
$^1\Delta$(II)	1.553		2042		72 881				0.525
$^1\Sigma^+$(II)	1.552		2059		75 312				0.523

Sources: experimental values from Berkaritz, J., *J. Chem. Phys.* (1988) **89**, 7065; Dixon, R.N. and Lamberton, H.M., *J. Mol. Spectrosc.* (1968) **25**, 12; calculated values from Balasubramanian, K. and Nannegari, V., *J. Mol. Spectrosc.* (1989) **138**, 482.

Note: positive polarity means polarity of As^-H^+.

The radical cation AsH^+ is formed in the mass spectrometer, from neutral AsH [3] and AsH_3 [4] enabling the measurement of ionization potentials [3]. AsH^+ has also been studied spectroscopically [1, 5], and has been the subject of quantum mechanical calculations [6, 7]. Balasubramanian and Nannegari [7] computed the spectroscopic constants and potential energy curves for 16 electronic states of AsH. The physical parameters, measured and calculated, are given in Table 3.1.

Arsenic dihydride and its ions. Arsenic dihydride is an unstable fugitive species. The neutral AsH_2 radical is formed by the reaction [3.1] between atomic hydrogen and arsine [3]:

$$AsH_3 + H \rightarrow AsH_2 + H_2 \qquad [3.1]$$

The geometry of the $^2B_1$ ground state of neutral AsH_2 was determined by Dixon, Duxbury and Lambeston [8]. They also found that the H–As–H bond angle opened up from 90.7° to 123° in the $^2A_1$ excited state with no change in the As–H bond length.

The *ab initio* calculations of Balasubramanian [9] agree with the experimental results [8]. However, the compound geometries indicate As–H bond shortening in the excited states of both neutral AsH_2 and its cation AsH_2^+ (Table 3.2).

Arsine and the hydroarsenics. Arsine, AsH_3, commands interest as the parent compound of two series of derivatives. First, it is the initial member of the hydroarsenic homologous series As_nH_{n+2}. Second, and more impor-

Table 3.2 The calculated molecular parameters of AsH_2 and AsH_2^+, including bond length d (As–H) and bond angle θ (H–As–H).

	State	d (As–H) (Å)	θ (H–As–H) (°)
AsH_2	$^2B_1$	1.521	90.7
AsH_2	$^2A_1$	1.490	122.9
AsH_2^+	$^1A_1$	1.520	91.4
AsH_2^+	$^3B_1$	1.50	121.8

Source: Hellwege, K.H. and Landolt-Börnstein, vol. 4, Springer, Berlin, 1967

Table 3.3 Arsine physical properties.

Melting point (°C)	–117
latent heat of diffusion (kcal kg^{-1})	3.67
Boiling point (°C)	–62.48
latent heat of vaporization (kcal kg^{-1})	51.2
Triple point (at 0.030 bar) (°C)	–116.9
Critical point (at 66 bar) (°C)	–99.9
Density of liquid (g ml^{-1})	1.634

Sources: *Gas Encyclopaedia*, Elsevier, Amsterdam, 1976; Hellwege, K.H. and Landolt-Börnstein, vol. 4, Springer, Berlin, 1967.

Note: 1 cal ≃ 4.187 J.

tantly, it is the parent of the large family of general formulae $H_{3-n}AsR_n$ ($n = 0, \ldots, 3$) known collectively as the arsines. These compounds are of great scientific interest and commercial value and are dealt with primarily in chapter 5.

Arsine is a colourless gas at ambient temperature with a characteristic garlic-like odour. Some physical data are listed in Table 3.3. AsH_3 is highly toxic, with the safe working threshold of 0.2 mg m^{-3} being revised downward; its lethal concentration (vol./vol.) for one hour is 10 ppm.

The toxicity of arsine has, if anything, been underestimated and it has been shown in the case of rats [10] that, although a single exposure to ten times the (1990) threshold limit value (TCV) had no detectable effects, repeated exposure to half the TCV results in significant anaemia.

Arsine is slightly soluble in water (Bunsen coefficient 0.20 at 20°C), the solution being neutral in reaction.

The preparation of arsine. Arsine is readily prepared on the laboratory scale and many methods have been developed. Arsine is sometimes referred to as arsenurated or arseniurated hydrogen in the early literature. Classically it has been prepared by the zinc-acid reduction of arsenic(III) compounds according to reaction [3.2]:

$$As^{3+} + 3Zn + 3H^+ \rightarrow 3Zn^{2+} + AsH_3 \qquad [3.2]$$

This venerable route was developed in a 1988 patent [11] where As_3O_3 is dissolved in aqueous NaOH and then acidified with H_2SO_4 whereupon the reduction is effected with zinc dust. In general, the route chosen by workers will of course depend on the resources available to them. Below are listed some of the more popular protocols. Jolly and Drake [12] reduced arsenic(III) oxide with KBH_4 in an aqueous medium:

$$As_2O_3 + BH_4^- + H^+ \rightarrow AsH_3 + H_3BO_4 + H_2O \qquad [3.3]$$

The tetrahydroborate reduction of arsenic compounds to arsine is widely used in analytical chemistry, although catalysts and adulterants often have an effect on the conversion [13].

Drake and Riddle [14] decomposed powdered Na_3As at 25°C by the slow and cautious addition of water, resulting in up to 90% AsH_3:

$$Na_3As + 3H_2O \rightarrow 3NaOH + AsH_3 \qquad [3.4]$$

Tzschach, Heinicke and Schoen [15] have disclosed details of a convenient laboratory method to prepare up to 2 mol AsH_3 from zinc arsenide:

$$Zn_3As_2 + 6HCl_{(aq)} \xrightarrow{70°C\text{–}95°C} 3ZnCl_{2(aq)} + 2AsH_3 \qquad [3.5]$$

Full details of a somewhat more sophisticated route via a liquid ammonia solution of alkali metal (M) arsenides has been given by Johnson and Pechukas [16]:

$$M_3As + 3NH_4Br \xrightarrow{NH_{3(l)}} 3MBr + NH_3 + AsH_3 \qquad [3.6]$$

For small quantities of pure arsine, the solid state reduction of arsenic(III) oxide by lithium aluminium hydride at 148°C is useful despite the low yield (10%) since AsH_3 is the only volatile product [17] and the reagents are inexpensive.

The foregoing protocols are for the relatively small-scale batch production of AsH_3 for synthetic uses, for example the preparation of synthons such as $KAsH_2$.

MOCVD processes in microelectronics require a continuous flow of arsine. Concentrations of 2%–50% vol./vol. usually with dihydrogen as the carrier, at flow rates up to 1 l min^{-1} are typical for MOCVD or vapour phase epitaxy (VPE) processes. Higher concentrations, 80%–100% vol./vol., albeit at a lower flow rate of 20 ml min^{-1} or less, are required for molecular beam epitaxy. For arsenic doping of, for example, silicon wafers, quite low concentrations of arsine in dihydrogen suffice, typically 1–100 ppm. The obvious solution of using commercial bottled gas is potentially lethal in the event of a system failure and the extreme toxicity of arsine gives rise to rational fears regarding the transportation, storage and end use of this material. Various on-demand point-of-use arsine generators have been proposed as alternatives to the bottled gas. They fall into three basic types:

- chemical flow reactors;
- electrochemical generators;
- sorption–desorption storage systems.

First, with regard to *flow reactors*, Salzman and Davis [18] passed a hydrogen carrier gas through arsenic trichloride. The $H_2/AsCl_3$ gas stream was then passed into $LiAlH_4$ dissolved in a non-volatile ether [tetraglyme, $Me(OCH_2CH_2)_4OMe$, vapour pressure 0.01 T at 20°C] producing AsH_3. The H_2/AsH_3 stream was purified by treatment with a liquid Ga/In/Al alloy. Jensen, Omstead and Annapragada [19] are developing an arsine generator based on the direct reaction between arsenic and hydrogen in a microwave plasma reactor. Argon has been found to aid the reaction efficiency ($Ar:H_2 = 1:1$ for maximum yields), with metastable argon species thought to act as energy transfer intermediates in the overall reactions. Since the arsenic and hydrogen can be obtained in extremely high purity, deleterious elements are excluded from the system. The arsine is not pure, however, containing appreciable amounts of diarsine although in microelectronic applications this may be advantageous.

Second, with regard to *electrochemical generators*, Valdes, Cadet and Mitchell [20] used a high-purity (99.9999%) solid arsenic electrode in an electrolyte of 1 M NaOH (semiconductor grade 99.99%) and 18 M Ω cm resistivity ultrapure water. The current efficiency for the production of arsine was shown to be 95%–97%. The generator was used in the preparation of an InGaAs film [21].

Third, with regard to *atmospheric storage systems*, there are two types of sorption storage – the physical systems such as charcoal, silica or zeolites and the chemical based systems. Sorption of AsH_3 on zeolites has been advocated by, among others, Freitas and Sillmon [22]. They found that 6 l of CaA-type zeolite (bulk density *c.* 0.65 kg l^{-1}) would store 1 kg AsH_3. The arsine was readily desorbed by heating to a maximum of 190°C to afford a 100% recovery at a flow rate of 50 ml min^{-1}. The formation of thermally labile complexes achieves a similar end. Zheng and Bradley [23] have proposed complexing arsine with trisperfluorophenylborane. The complex $[(C_6F_5)_3BAsH_3]$ is a low-hazard adduct from which pure AsH_3 is recovered by heating to over 55°C. An added advantage to this system is that any moisture in the original AsH_3 is adsorbed by the borane.

Purification of arsine. In normal laboratory practice where 99.9% purity is acceptable arsine is purified by washing with aqueous acids to remove basic impurities (often NH_3 from metal nitride impurities in arsenides), with aqueous alkali to remove acidic adulterants and finally with water to remove any carryover from the previous treatment. The water vapour is removed by successive treatment with anhydrous $CaCl_2$ and P_2O_5; alternatively zeolites may be used. A final low-temperature trap to capture distillation will afford material suitable for most purposes. Arsine may be stored in glass vessels for several months with less than 5% decomposition, although hydrocarbon grease is blackened by prolonged exposure to AsH_3 [14]. For the greater purity required by microelectronics laboratories special care is necessary first in the purification of the starting materials for the AsH_3 preparation, thereby removing potential problems at source, and second, in the removal of final traces of dioxygen and moisture by means of several special procedures which have been developed.

Shealy *et al.* [24] recommend bubbling AsH_3 through a solution of aluminium (1.8%) in a gallium–indium (10:1) alloy (melting point 24°C) as a getter for O_2 and H_2O in the preparation of high-purity arsine for MOCVD processes.

A US patent [25] discloses a method of purifying AsH_3 based on the exchange of impurities from the gas phase to a resin-bound AsH_2 moiety (ResAs):

$$AsH_3(O_2/H_2O) + ResAsH_2 \rightarrow ResAsH_2(O_2/H_2O) + AsH_3 \qquad [3.7]$$

The structure of arsine. Arsine has been shown by spectroscopic analysis to be a trigonal pyramidal molecule. The molecular parameters determined by various groups are given in Table 3.4(a). *Ab initio* calcu-

Table 3.4 Molecular parameters of arsine [bond length, d (As–H) and bond angle, θ (H–As–H)]: (a) observed; (b) from *ab initio* calculations.

d (As–H)(Å)	θ (H–As–H)(°)	Reference
(a)		
1.53 ± 0.02	90	Stevenson, D.P. (1940) *J. Chem. Phys.*, **8**, 285
1.519	91.85	Loomis, C.C. and Strandberg, M.W.P. (1955) *Phys. Rev.*, **81**, 798
1.515	91.52	Blevins, G., Jache, A.W. and Gordy, W. (1954) *Phys. Rev.*, **95**, 229
1.5275	91.93	Oka, T. and Chu, F.Y. (1974) *J. Chem. Phys.*, **60**, 4612
1.511	92.1	Dunning, T.H. Jr (1977) *J. Chem. Phys.*, **66**, 1382
1.513	92.08	McRae, G.A., Gerry, M.C.C., Wong, M. *et al.* (1987) *J. Mol. Spectrosc.*, **123**, 321
(b)		
1.517	92.2	Balasubramanian, K. and Dai, D. (1990) *J. Chem. Phys.*, **93**, 1837
1.517	93.7	Dobbs, K.D., Tractman, M., Bock, C.W. and Cowley, A.H. (1990) *J. Phys. Chem.*, **94**, 5210
1.510	92.4	Graves, R.M. and Scusaria, G.E. (1992) *J. Chem. Phys.*, **96**,
1.509	92.4	Pyykkö, P., Laakkonen, L.J. and Schwerdtfeger, P. (1992) *J. Chem. Phys.*, **96**, 6807

lations indicate the ground state to be $^1A_1$ symmetry with a trigonal pyramidal (C_{3v}) geometry. The values calculated for the AsH_3 parameters by various research groups are given in Table 3.4(b).

Since arsine has a trigonal pyramidal structure (C_{3v}) inversion is possible, and the process is reasonably thought to occur via a trigonal planar transition state (D_{3h}). This process can theoretically be observed or inferred from spectroscopic data. *Ab initio* calculations by Pyykkö, Laakkonen and Schwerdtfeger [26] indicate the inversion barrier potential to be 164 kcal mol^{-1}, indicating a stable configuration (cf. NH_3 with a value of 24.2 kcal mol^{-1} and PH_3 at 132 kcal mol^{-1}). Consequently, the tunnelling frequencies are too small to be detected experimentally even with ultra-high-resolution spectroscopy of AsH_3, for example as carried out by Di Lonardo and Fusina [27] at 0.006 cm^{-1} resolution. For the bulk material, inversion has no significance since there are no isomers to resolve. However, in principle it is possible to synthesize arsine tagged with the three hydrogen isotopes, that is, AsHDT, which in principle could be resolved.

The detection and analysis of arsine. In this section we describe some of the more important aspects of the detection and analysis of arsine *per se*, not the general analysis of arsenicals (chapter 9).

Environmental monitoring has become progressively more important as industrialization has increased. Arsenic is a notorious poison and there is therefore a public demand for a strict 'arsenic audit' by industry. Various on-line analysers reduce arsenic compounds to arsine which is then analysed by atomic emission spectrometry [28]. In the case of low concentrations, pre-separation and/or pre-concentration techniques have to be used [29]. These aspects of environmental monitoring are covered more fully in chapter 9.

Classical wet methods of analysis such as the oxidation of As(–III) to As(III) by $AgNO_3$ [30], shown in reaction [3.8], are not now widely used:

$$AsH_3 + 6AgNO_3 + 3H_2O \rightarrow H_3AsO_3 + 6HNO_3 + 6Ag \quad [3.8]$$

The most widely used wet method for the detection of arsine is based on its reaction with mercury(II) chloride. The well-known commercial Dräger tubes [31] base their arsine detectors on the reaction

$$AsH_3 + Au^{3+} \rightarrow Au_{(colloidal)} \quad [3.9]$$

The colour change is white to grey-violet, with a detection limit of 0.05–3 ppm.

Arsine is formed adventitiously in many industrial processes, for example electrowinning and/or refining of metals, or the action of acids or water on residues from processes in which arsenic traces may have formed metal arsenides, such as in ore extraction. Arsine has itself become an important material in the microelectronics industry. These factors have led to the development of an industrial market for automated arsine detecting and monitoring equipment.

Electrochemical sensors for arsine are of two basic types – those based on a liquid-phase electrolyte and those based on a solid-phase electrolyte. Systems using liquid-phase electrolytes are interfaced to the environment by means of using gas permeable membranes. Yamashita *et al.* [32] used a catalytic permeable membrane incorporating gold particles as one of the working electrodes. The working electrolyte was 6 M sulphuric acid, the reaction detected between gold electrodes being

$$AsH_3 + 2O_2 \rightarrow H_3AsO_4 \qquad [3.10]$$

A lower limit of detection of 0.01 ppm is claimed and a 90% response time of 30 s for 0.05 ppm AsH_3. Bale, Kirchnerova and Skeaff [33] used a solid electrolyte of composition $(Ag_2O)_{1.34}\|Al_2O_3$ based on β-alumina, the reaction controlling the electromotive force of the sensor being

$$AsH_3 + Ag/\beta\text{-}Al_2O_2 \xrightarrow{O_2} Ag_3AsO_4/\beta\text{-}Al_2O_3 \qquad [3.11]$$

The system created the galvanic cell: $Ag(s) \mid Ag^+ - \beta Al_2O_3 \mid Ag_3AsO_4 \mid$ air, (AsH_3), Pt. The detector tollerates 1% CO_2, 10 ppm SO_2 and 0.1% H_2. Other conducting ceramics such as sodium-β-alumina, $NaCaAsO_4$ and $AgZr_2(AsO_4)_3$ have been investigated as arsine detectors [33]. Solid-state sensors based on microelectronic technology are becoming increasingly important in all branches of chemical analysis. There are several arsine detectors based on metal oxides; for example, the conductance of an SnO_2 film is increased in proportion to the arsine concentration [34].

The relative performance of zinc(II) oxide and tin(IV) oxide were compared by Yakimov *et al.* [35]. The SnO_2 system has a greater sensitivity but response time is a function of temperature and this is optimized at *c.* 400°C. The ZnO system optimizes at 440°C but at a lower absolute sensitivity than achieved by SnO_2 at 320°C. The trade off between sensitivity and temperature-dependent response time is the subject of further development.

Chemiluminescence detectors based on the reaction between arsine and ozone are capable of detecting concentrations of 0.002 ppm vol./vol. AsH_3 with a response time of around 10 s.

The power of infra-red (IR) spectroscopy for the detection and estimation of arsine is shown by its detection in the atmospheres of the major planets, initially by Noll [36] and by Bezard *et al.* [37]. Unlike the case with the oxidizing terrestrial atmosphere, the larger planets provide a stable reducing ambient for arsine. It is, however, degraded by ultraviolet (UV) radiation in the tropsphere [37]. The mole fraction of AsH_3 in the atmosphere of Saturn is *c.* 3×10^{-9}, which is about 10 times higher than that of Jupiter [38].

The reactions of arsine. Arsine burns in air with a bluish-white flame (reaction [3.12]) which deposits a black stain of metallic arsenic on a cold surface. The black stain is dissolved by $NaClO_{(aq)}$ (the Marsh test for arsine).

$$AsH_3 + {}^3/_2\,O_2 \rightarrow {}^1/_4\,As_4O_6 + {}^3/_2H_2O, \quad \Delta H = -822\ \text{kJ mol}^{-1} \qquad [3.12]$$

The flammability limit for arsine in air is 5.1%–78% vol./vol. [39].

The structure of arsine may be interpreted in terms of VSEPR theory as a trigonal pyramidal species with the lone pair occupying the apical pyramidal site. This has two consequences. First, the arsine lone pair is made available for donor–acceptor bond formation in which arsine acts as a Lewis base. Second, the availability of the lone pair enables arsine to act as a nucleophile in many reactions, for example with halogens. Some characteristic reactions of arsine are illustrated in Scheme 3.1.

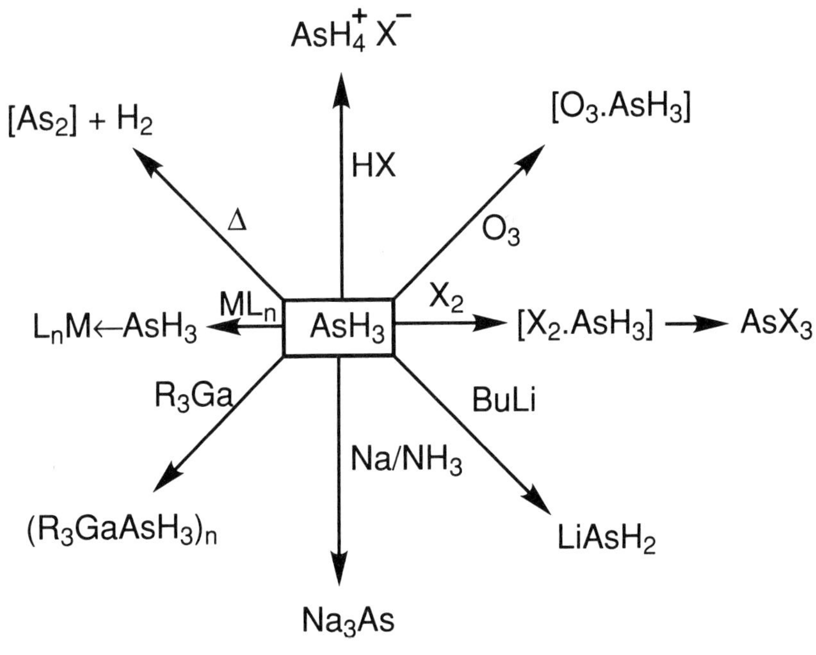

Scheme 3.1 Some characteristic reactions of arsine.

Examples of each of these modes of behaviour will be found in the following subsections, in which some of the reactions of arsine with s-, p- and d-block elements and their compounds are reviewed.

s-Block elements. In its reactions with group 1 metals, arsine behaves as a typical Brønsted acid forming salt-like hydroarsenides, for example in liquid ammonia [40]:

$$AsH_3 + Na \xrightarrow{NH_{3(l)}} Na[AsH_2] \quad [3.13]$$

Since arsine has three available hydrogens it is a tribasic acid, and excess alkali metal affords arsenides (As^{3-}, Zintl phases). It should be noted that this is one of the few arsine reactions that is not an oxidation, arsenic being in the formal –III oxidation state throughout.

s-Block arsenides are also prepared by reactions with other strong bases such as butyllithium [41]:

$$AsH_3 + BuLi \xrightarrow{DME} LiAsH_2DME + BuH \quad [3.14]$$

Where DME is dimethyloxyethane.

p-Block elements.

Group 13. The Lewis base behaviour of arsine is probably best displayed in its reactions with the group 13 Lewis acids R_3M where R = H, halogen or organo moiety, and M = group 13 metal:

$$R_3M + AsH_3 \rightarrow [R_3M.AsH_3]_n \quad [3.15]$$

The compounds are often oligomeric, but not all the possible reactions have yet been demonstrated.

The reaction of excess AsH_3 with Bu^t_3Al (neat or as a C_6H_6 solution) at 0°C followed by heating to 50°C affords a cyclotrimer, for which it is not unreasonable to postulate a monomeric donor–acceptor complex as a reaction intermediate [42]:

$$Bu^t_3Al + AsH_3 \rightarrow [Bu^t_3Al \leftarrow AsH_3] \rightarrow [Bu^t_2AlAsH_2]_3 \quad [3.16]$$

The pyrolysis of these compounds is of great importance in the microelectronic industries for the production of III–V (or 13–15) semiconductor materials MAs, where M = Al, Ga or In, of which gallium arsenide is pre-eminent. The ideal MOCVD single-source precursor for GaAs would be $H_3Ga{\cdot}AsH_3$, but this has proved elusive, although calculations by Trachtman *et al.* [43] indicate that it should be observable in the gas phase. The adduct $[Me_3Ga{\cdot}AsH_3]$ has been detected by Ault and Piocos [44] by IR spectroscopy in an argon matrix.

Group 14. Arsine does not react with simple olefins but the π-bond is made vulnerable to the nucleophilic attack of the arsine lone pair by electron-withdrawing substituents. In these cases AsH_3 will add across an olefenic bond, forming a secondary arsine which may further react with olefins to form a secondary arsine and finally a tertiary arsine [45]:

$$R_2C{=}CR_2' \xrightarrow{AsH_3} R_2CHC(R_2')AsH_2 \xrightarrow{R_2C=CR_2'} (R_2CHCR_2')_2AsH$$
$$\xrightarrow{R_2C=CR_2'} (R_2CHCR_2')_3As \quad [3.17]$$

The American Cyanamid Co. [46] disclosed a process for the production of tertiary butyl arsine from butylene with a modest (37%) yield:

$$CH_2{=}CMe_2 + AsH_3 \xrightarrow[12\ atm,\ 80°C–95°C,\ 4\ h]{n\text{-octane/MeSO}_3H_{(aq)}} Me_3CAsH_2 \quad [3.18]$$

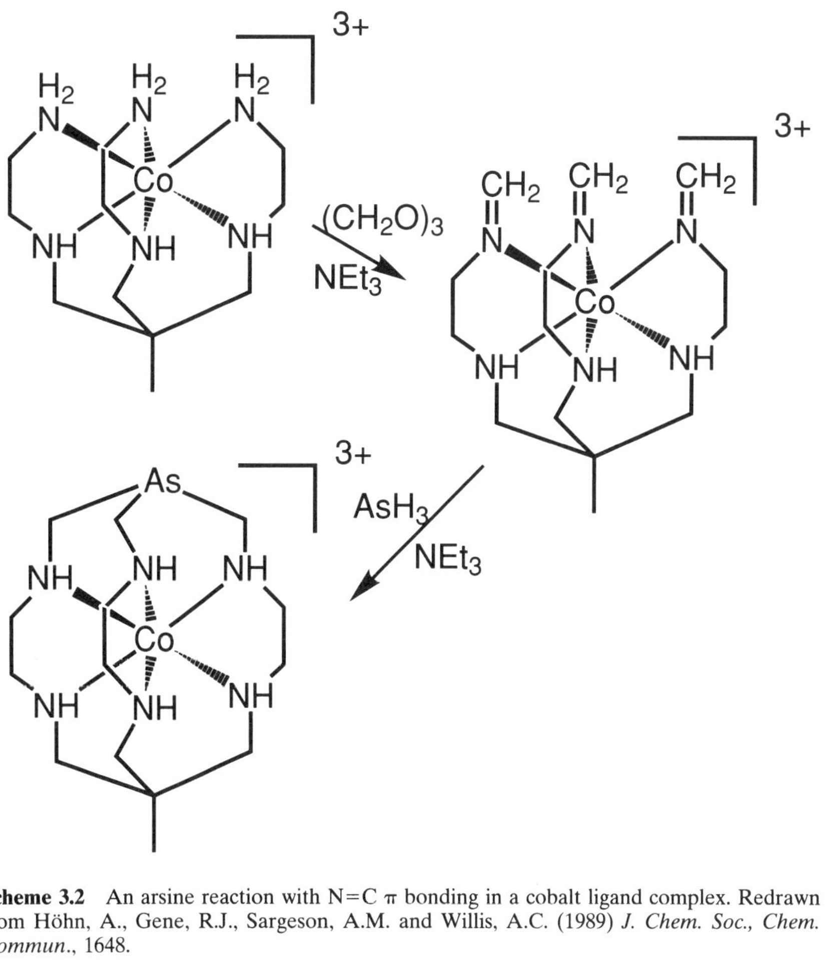

Scheme 3.2 An arsine reaction with N=C π bonding in a cobalt ligand complex. Redrawn from Höhn, A., Gene, R.J., Sargeson, A.M. and Willis, A.C. (1989) *J. Chem. Soc., Chem. Commun.*, 1648.

Lithium tetraarsinoaluminate is a useful synthon in the preparation of substituted arsines, for example silyl and germyl arsines. It is formed from arsine and $LiAlH_4$ in ether-type solvents [47]:

$$LiAlH_4 + AsH_3 \xrightarrow{BrEH_3} Li[Al(AsH_2)_4] \longrightarrow H_3EAsH_2, \quad E = Si, Ge \qquad [3.19]$$

Group 15 elements. Ang and Lee [48] found that $(CF_3)_2NONO$ formed an iminoarsine with AsH_3, the first such compound prepared:

$$(CF_3)_2NONO + AsH_3 \rightarrow (CF_3)_2NON = AsH + H_2O \qquad [3.20]$$

Arsine will also add across the N=C π bond as shown by Höhn *et al.* [49] in a spectacular synthesis involving a cobalt ligand sphere (Scheme 3.2).

Group 16 elements. The reaction with ozone, chemiluminescence and the destructive oxidation of arsine are discussed as separate topics. Ang and Koh [50] showed that AsH_3 reacts with $(CF_3)_2NO$ to form the As−O bonded $[(CF_3)_2NO]_3As$. They postulated the initial formation of $(CF_3)_2NOAsH_2$, which then decays via two routes:

$$AsH_3 + (CF_3)_2NO \rightarrow \qquad [3.21]$$

$$[(CF_3)_2NOAsH_2] \begin{cases} \rightarrow [(CF_3)_2NO]_3As + AsH_3 \\ \rightarrow [(CF_3)_2NOH + AsH \rightarrow As_4 + H_2 \end{cases}$$

Group 17 elements. The charge-transfer complexes between the arsines and halogens and their further reaction to give dihalo arsoranes are fully discussed later. Remarkably, the 'parent complex' $H_3As \rightarrow F_2$ has been observed by Andrews and McInnes [51], the complex being observed in a solid argon matrix, together with the partial substitution product AsH_2F. Similarly, the complexes $H_3As–Cl–X$ (X = Cl, F) have been inferred by Ault and Machara [52] from the co-deposition of AsH_3/Cl_2 and AsH_3/ClF in argon matrices.

Under 'normal' laboratory conditions, halogens react violently with arsine to afford the corresponding trihalide:

$$AsH_3 + X_2 \rightarrow AsX_3 + HX, \; X = \text{halogen} \qquad [3.22]$$

d-Block elements. In this section we will briefly describe some of the reactions of arsine with transition metal complexes. The various transition metal–arsine structures are shown in Scheme 3.3. This is, of course, only the 'base set' since where there are two or more metal atoms shown these may be different or may be part of larger clusters.

In three-coordinate arsine species a normal σ bond is implied in $H_2As−M$ moieties, that is, essentially, the metal has replaced hydrogen

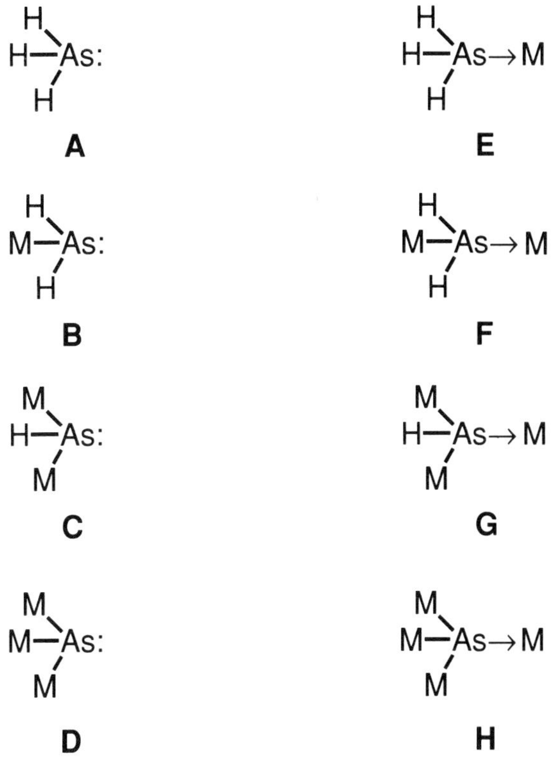

Scheme 3.3 Three-coordinate (A–D) and four-coordinate (E–H) arsenic species.

in the arsenic ligand sphere, leaving the lone pair available for further donor bond formation. No hierarchical order of substitution is implied by the scheme. In a bimetallic species the formation of species C or F (Scheme 3.3) will be influenced by the nature of M (or M′ not illustrated) with its associated ligand sphere (L_n), and the outcome may be kinetically or thermodynamically determined. Further, although we have written the structures, for example species F with two types of M–As bonds, this is only for illlustrative purposes. In real structures, of course, we would predict or expect a resonance hybrid resulting in indistinguishable (degenerate) M–As bonds.

Arsine forms simple donor complexes (species E) with transition metal compounds, for example [$TiCl_4AsH_3$] was shown by Drake and Riddle [53] to be a trigonal bipyramidal molecular complex with the arsine occupying an equatorial site.

Table 3.5 Geometrical distortion of AsH_3 on complex formation: changes to bond length [d (As–H)] and bond angle [θ (H–As–H)]

AsH_3 species	d (As–H)(Å)	θ (H–As–H) (°)
Free AsH_3		
calculated	1.519	91.8
experimental	1.511	92.1
$[(CO)_5CrAsH_3]$	1.530	95.2

When the complex $[CpMn(CO)_2L]$ (L = cyclohexene) in benzene solution is stirred under an atmosphere of arsine at 40°C the cyclohexane ligand is exchanged with the formation of $[CpMn(CO)_2AsH_3]$. This species is also a E type complex.

In contrast a complex of type C (Scheme 3.3) is formed by the reaction between arsine and $[Cp^*Mn(CO)_2thf]$ (thf = tetrahydrofuran). Ziegler and co-workers [54] calculated, using density functional theory (DFT), the distortion of the AsH_3 donor ligand on complexing with chromium carbonyl to form $[(CO)_5CrAsH_3]$ (Table 3.5). The As–H bond length increases by approximately 1% on complex formation, and the *trans* C–Cr bond length increases by approximately 1.5%. These differences, although small, are nonetheless spectroscopically significant

The square planar iridium complex *trans*-$[Ir(CO)Cl(PEt_3)_2]$ reacts with arsine to form a complex of type B (Scheme 3.3). The square planar iridium complex *trans*-$[Ir(CO)Cl(PEt_3)_2]$ and arsine affords an oxidative addition product, $[Ir(CO)H(Cl)(PEt_3)_2(AsH_2)]$, the σ-bonded AsH_2 moiety still having the lone pair available for further reactions, which Ebsworth and co-workers [55] demonstrated by reaction with B_2H_6 forming a B–As donor complex $[L_nIr{-}AsH_2 \rightarrow BH_3]$. The arsine lone pair coordination was also used in a bridging interaction forming an Ir–As–Ru cluster of type F (Scheme 3.3).

The protonation of $[Ir(CO)H(Cl)(PEt_3)_2(AsH_2)]$ by HCl in toluene has been postulated on the strength of NMR evidence to be a reversible reaction initially; however, the slow chloride sustitution reaction results in the decay of the cation.

The hydroarsenics

The arsine homologues, As_nH_{n+2}. The arsine homologues As_nH_{n+2} are a limited series with only diarsine (As_2H_4) being well characterized. Jolly and Shriver [56] have described a reliable protocol.

Diarsine is fairly stable in the gas phase but is very unstable in the liquid phase, above –100°C. It is often produced as a minor or trace by-product in the production of arsine, for example in the decomposition of

Al–Mg–As alloys by acids [57] and in the microwave plasma synthesis of arsine from the elements [19].

The independent existence of triarsine, $AsH_2AsHAsH_2$, has yet to be confirmed, although its transient existence in the mass spectrometer has been demonstrated [4, 58].

Other hydroarsenics. There are many catanated arsenic chains, heterocycles and cages that are stabilized by bulky substituents or coordination to transition metals (chapter 5). The corresponding hydrogen parent compounds have so far proved elusive, thus As_7R_3 is known but As_7H_3 is not.

Jolly, Anderson and Beltrami's reports of a yellow-brown solid species AsH_x have not been confirmed or further investigated [59]:

$$As_2H_4 \rightarrow AsH_{0.57} + {}_{0.07} \qquad [3.23]$$

Pentavalent arsenic hydrogen compounds. The parent pentavalent arsenic hydride AsH_5 is unknown, although theoretical calculations show that it would be thermochemically stable with a trigonal bipyramidal geometry [60]. The only known species of pentavalent hydrides are the salts $[AsH_4]^+X^-$ (X = Br, I) which are extremely unstable and exist only at low temperatures (110 K) [61].

3.2.2 Antimony–hydrogen compounds

The relatively extensive chemistry of arsenic and hydrogen is not continued by the heavier group 15 metals. The antimony–hydrogen compounds are divided into two groups: the transient SbH_x ($x = 1, 2$) species and the relatively stable SbH_3.

Antimony–hydrogen transient species (SbH_x, $x = 1,2$). The transient diatomic species SbH (or SbD) has been generated by the flash photolysis of $SbH_3(SbD_3)$ by Bollmark and Lindgren [62]. *Ab initio* calculations by Balasubramanian and Dai [9, 63] on the triatomic SbH_2 indicate a bent structure; the molecular parameters are given in Table 3.6.

Table 3.6 Molecular parameters of antimony–hydrogen transient species: bond length [*d* (Sb–H)] bond angle [*θ* (H–Sb–H)] and ionization potential

Species	*d* (Sb–H)(Å)		*θ (H–Sb–H)(°)*	Ionization potential (eV)
	Calculated	Experimental		
SbH_2	1.726	—	89.8	8.38
SbH	1.731	1.72	—	8.66

Table 3.7 The molecular parameters of stibine: bond length [*d* (Sb–H)] and bond angle [θ (H–Sb–H)]

d (Sb–H) (Å)	θ (H–Sb–H) (°)	Reference
1.7102	91.7	Helminger, P., Beeson, E.C. and Gordy, W. (1971) *Phys. Rev. A. Gen. Phys.*, **3**, 122
1.707	91.3	Jache, A.W., Blevins, G.S. and Gordy, W. (1955) *Phys. Rev.*, **97**, 680
1.700	91.54	Lummila, J., Lukka, T., Halonen, L. *et al.* (1996) *J. Phys. Chem.*, **104**, 488

Stibine (SbH_3)

Preparation. Stibine, like arsine, gives rise to three series of organic substituted compounds: the primary stibines, $RSbH_2$, the secondary stibines, R_2SbH, and the stibines R_3Sb which are described elsewhere. They are all more stable than the parent stibine. Stibine, SbH_3, is the only relatively stable antimony–hydrogen binary compound.

Antimony metal has been shown to react with hydrogen radicals (H·) generated by thermal, electric discharge or photochemical methods to give low yields of stibine [57].

Classically, stibine was prepared by a dissolving metal reduction of an antimony alloy in low yield; for example Durrant, Pearson and Robinson [64] obtained 10 g SbH_3 from 250 g of a magnesium–antimony alloy approximating to Mg_3Sb_2. Full practial details are given by Schlenk [65]. For practical amounts of SbH_3, the reduction of $SbCl_3$ by $NaBH_4$ is the preferred route. The reaction was first studied by Jolly and co-workers [66] who assumed the stoichiometry shown in reaction [3.24]:

$$3BH_4^- + 4SbCl_4^- + 9H_2O \rightarrow 3H_3BO_3 + 4SbH_3 + 9H^+ + 16Cl^- \qquad [3.24]$$

The yields were modest (*c.* 25% based on BH_4^-), but were much improved in a later Russian study. Zorin *et al.* [67] showed that the yield was a maximum at a $NaBH_4/SbCl_3$ ratio of 10:1 and with the pH in the range 7–11. They claimed these conditions increased the yields to *c.* 80%. They purified the stibine by low-temperature distillation at *c.* 400 torr.

Point-of-use SbH_3 electrochemical generators have been developed by Reismann *et al.* [68] and Burk [69]. A pure (99.998%) antimony cathode reacts with electrochemically generated hydrogen atoms ($H_{(ads)}$) to give the hydride; the electrolyte is 22% KOH at –15°C:

$$\left.\begin{aligned} &H^+ + e^- \rightarrow H_{(ads)} \\ &Sb + 3H_{(ads)} \rightarrow SbH_3 \end{aligned}\right\} \qquad [3.25]$$

Overall, the reaction may be written as

$$Sb + 3H^+ + 3e^- \rightarrow SbH_3 \qquad [3.26]$$

Structure. Stibine, like arsine, is a trigonal pyramidal molecule with the lone pair occupying the apical site. The Sb–H (or Sb–D) bond length is 1.7102 Å (1.7075 Å) and the H–Sb–H bond angles are 91.7°, as shown by microwave spectra (Table 3.7) [70].

Properties. Stibine is a malodorous, toxic colourless gas at ambient temperatures; some physical data are listed in Tables 3.8 and 3.9.

SbH_3 is slightly soluble in water and freely soluble in organic solvents such as carbon disulphide and ethanol. It is decomposed by air with the formation of Sb_2O_3.

Stibine decomposes slowly into the elements slightly above the boiling point and decomposes rapidly on heating over 200°C. Berka [71] describes the decomposition, initiated by heating SbH_3 at *c.* 380 torr in a tube with a bunsen flame, as taking place within 10 s, a bright yellow-orange flash travelling along the decomposition tube resulting in the deposition of Sb

Table 3.8 Observed vapour pressures of stibine comapred with the calculated vapour pressures with use of equation (3.1) in text.

Temperature (K)	Vapour pressure (torr)	
	observed	calculated
250.18	621.4	627.8
227.66	223.4	222.7
221.73	164.2	163.2
209.21	80.2	79.4
194.70	30.4	30.3
192.49	25.4	25.8
187.83	18.0	18.1

Sources: Jolly, W.L., Berka, L.H., Briggs, T. and Millard, M. (1960) *J. Inorg. Nucl. Chem.*, **14**, 190

Note: the vapour pressure over solid SbH_3 at 178 K is 8.2 torr.

Table 3.9 Selected physical data of stibine.

Melting point (°C)	–88[a] or –94.3[b]
Boiling point (°C)	–18.4
Critical temperature (°C)	167.2
Density of liquid at –18°C (g ml^{-1})	2.204
ΔH^0 (kJ mol^{-1})	145.1
Critical pressure (atm)	72.1

Sources: Berka, L. Lawrence Radiation Laboratory Report UCRL-878, 1959; Durrant, A.A., Pearson, T.G. and Robinson, P.L. (1934) *J. Am. Chem. Soc.*, **56**, 730; Jolly, W.L., Berka, L.H., Briggs, T. and Millard, M. (1960) *J. Inorg. Nucl. Chem.*, **14**, 190; Yaws, C.L., Setty, H.S.N., Hopper, J.R. and Swinderman, E.M. (1974) *Solid State Techn.*, **17**, 47.

[a]According to Durrant, Pearson and Robinson.
[b]According to Jolly *et al.*

metal. This thermal lability of stibine is used in MOCVD processes, but because of its instability it has to be generated on-site; for example, Burk [69] used SbH_3 to prepare a film of $InP_{0.88}Sb_{0.12}$.

Berka [71] showed that values obtained from equation (3.1) below were in agreement with the experimental measurements of stibine vapour pressure (P_v) variations with temperature (T):

$$\log (P_{atm}/\text{torr}) = 0.011290\,(P_v/\text{torr}) - 1446.34/(T/K) - 3.11997 \log (T/K) + 13.17136 \quad (3.1)$$

Where P_{atm} is atmospheric pressure. The vapour pressure at the melting point of chloroform (–63.5°C) is 81 torr, and at the melting point of chlorobenzene (–45.2°C) 224 torr [72].

Toxicity. Stibine is extremely toxic [LD_{50}(mice) 10 ppm]. It has been associated with sudden infant death syndrome (SIDS); it is thought that SbH_3 may be generated from antimony fire retardants used in the manufacture of cot mattreses by the action of urine and/or microbiota [73,74]. Recent research challenges this as a cause of sudden infant death (chapter 8).

Pentavalent antimony hydrides. Although Heinemann [61] demonstrated the existence of AsH_4^+ (final subsection of Section 3.2.1) the analogous SbH_4^+ cations were not formed under the same conditions.

3.2.3 Bismuth–hydrogen compounds

The binary bismuth–hydrogen compounds are divided into two sections. First, the transient BiH_x species ($x < 3$) are discussed, followed by a description of the only 'stable' hydride, BiH_3.

Transient bismuth–hydrogen species. Early investigations on BiH were carried out by Hulthen and co-workers [75, 76] and by Lindgren and Nilsson [77]. The diatomic species was generated by passing dihydrogen over heated bismuth metal at *c.* 1000°C at 25 torr by Bernath, Bopegedera

Table 3.10 Calculated and experimental parameters of the bismuth bydrides, BiH_x ($x = 1, 2$): bond length [d (Bi–H)] and bond angle [θ (H–Bi–H)].

Molecule	d (Bi–H) (Å)		θ (H–Bi–H) (°)
	calculated	experimental	
BiH	1.869	1.809	—
BiH_2	1.876	—	88.9

Source: Balasubramanian, K. and Dai, D. (1990) *J. Chem. Phys.*, **93**, 1837.

and Brazier [78] who found a Bi–H ground state bond length of 1.80867 Å and a Bi–D ground state bond length of 1.80671 Å.

The molecular parameters calculated for BiH and BiH_2 by Balasubramanian and Dai [63] are given in Table 3.10.

Bismuthine (BiH_3). Bismuthine is a colourless volatile liquid which decomposes at ambient temperature. Synthetic routes that afford AsH_3 and SbH_3 give only trace yields of BiH_3. Amberger [79] used the low-temperature disproportionation reaction of $MeBiH_2$, which gives good yields of BiH_3:

$$\left.\begin{array}{l} MeBiCl_2 + LiAlH_4 \xrightarrow{Me_2O \text{ at } -50°C} MeBiH_2 \\ 3MeBiH_2 \xrightarrow{-45°C} 2BiH_3 + Me_3Bi \end{array}\right\} \qquad [3.27]$$

Bismuthine is stable only at low temperatures (< –60°C). The vapour pressure–temperature data are given in Table 3.11 together with the calculated values obtained by means of equation (3.2) below, valid between –116°C and –43°C:

$$\log (P_v/\text{torr}) = -1314.2/(T/K) + 7.4146 \qquad (3.2)$$

The extrapolated boiling point is 16.8°C.

Bismuthane is a pyramidal molecule with a calculated [63] B–H bond length of 1.865 Å and H–Bi–H bond angle of 90.3°.

Owing to the difficulty of preparation and handling BiH_3, its chemistry has not been developed. It decomposes at ambient temperatures to give dihydrogen quantitatively:

$$BiH_3 \rightarrow Bi + {}^3/_2H_2 \qquad [3.28]$$

Table 3.11 Observed vapour pressures of bismuthine compared with calculated vapour pressures with use of equation (3.2) in text.

Temperature (K)	Vapour pressure (torr)	
	observed	calculated
157.9	0.10	0.11
168.2	0.39	0.40
184.7	1.97	2.00
203.0	8.42	8.79
219.6	27.02	27.02
224.0	36.21	35.41
229.7	48.73	49.50
289.9[a]		760

[a]Extrapolated by rearranging equation (3.2) in text to make T/K the subject.

3.3 The halides of arsenic, antimony and bismuth

In this section the halides of arsenic, antimony and bismuth are treated in descending order down group 17, that is, fluorine, chlorine, bromine and iodine, and in order of increasing oxidation state of the pnictogens. First, the lower valent and transient compounds are briefly discussed.

3.3.1 Low valent transient halogen pnictogen species

The twelve transient vapour phase diatomic species EX (E = As, Sb, Bi; X = F, Cl, Br, I) have all been observed spectroscopically. They are usually prepared by the flash photolysis of the corresponding EX_3 vapour at low pressure, and usually decay into the elements as follows:

$$2AsCl \rightarrow As_2 + Cl_2 \qquad [3.29]$$

The triatomic species EX_2 have similarly been observed, for example AsI_2 from the photodissociation of AsI_3 [80]. These triatomics are bent, not linear, molecules. The spectra and molecular parameters of these species are available in physical chemistry reference works.

3.3.2 Low valent stable halogen pnictogen species

Arsenic forms only one stable low valent halide, the diarsenic(II) tetra-iodide, As_2I_4. This material is obtained by heating stoichiometric amounts of the elements at 260°C in a sealed tube [81]. The compound is stable in an inert atmosphere but disproportionates at higher temperatures (300°C–350°C) [82]:

$$6As_2I_4 \rightleftharpoons 8AsI_3 + As_4 \qquad [3.30]$$

It is readily oxidized and hydrolysed in moist air. The antimony analogue is unstable, only being observed spectroscopically or by other techniques for studying transients.

In contrast to the other members of the triad, bismuth forms several low valent compounds with halogens which have remarkable structures. A chloride ‘BiCl’ was first prepared in the 1900s by reducing $BiCl_3$ with Bi metal at 325°C. The mixture is cooled slowly to 270°C over 2 weeks after which excess $BiCl_3$ is removed by sublimation. It has since been shown that the stoichiometry is $[Bi_{24}Cl_{28}] \equiv Bi_6Cl_7$ or $BiCl_{1.167}$. Corbett and Hershaft [83] showed the crystal structure to consist of Bi_9^{5+} clusters and two different bismuth choroanions, $[BiCl_5]^{2-}$ and the chlorine-bridge $[Bi_2Cl_8]^{2-}$ anion, in the ratios of 2:4:1 – $[(Bi_9^{5+})_2(BiCl_5^{2-})_4\ (Bi_2Cl_8^{2-})]$. The Bi_9^{5+} cluster is a trigonal prism with the three square faces capped by Bi. The bromo analogue $BiBr_{1.167}$ was shown by von Benda, Simon and Anhoffar [84] to have a similar structure.

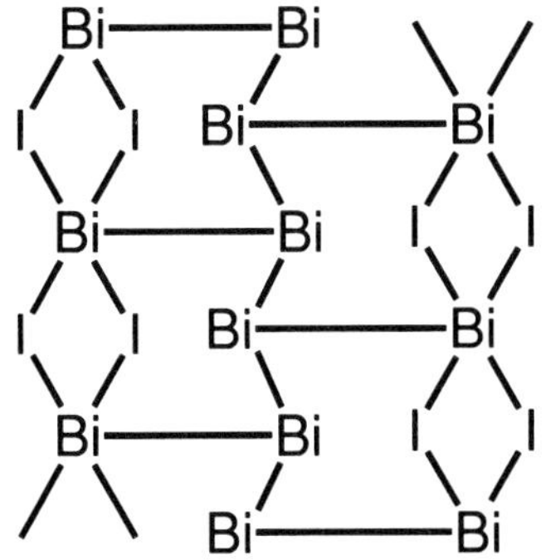

Figure 3.1 The structure of BiI, showing $[Bi(Bi)_3]$ and $[Bi(I_4Bi)]$ units that make up the ${}_{\infty}^{1}[Bi_4I_4]$ chains.

Bismuth also forms a true monoiodide, BiI [85], when stoichiometric amounts of the elements are heated in a sealed tube. It has three crystal modifications but ultimately disproportionate, to BiI_3 and Bi:

$$\alpha\text{-BiI} \xrightarrow{>97°C} \beta\text{-BiI} \xrightarrow{>291°C} \gamma\text{-BiI} \xrightarrow{>312°C} BiI_3 + Bi \qquad [3.31]$$

The formation of β-BiI by a martenistic transition (lattice shear, no diffusion) is followed by peritectic decomposition of the high-temperature γ modification to the more stable BiI_3. BiI has a remarkable mixed valent structure: $[Bi(0)\{(Bi(II)I_2\}]$. There are two bismuth centres, a bismuth chain of $BiBi_3$ trigonal pyramids and iodine bridged BiI_4Bi units (Fig. 3.1). These form one-dimensional infinite chains ${}_{\infty}{}^{1}[Bi_4I_4]$. The stacking of these chains is complex and varies with temperature. There is an analogous BiBr compound [86].

3.3.3 Tervalent arsenic, antimony and bismuth halides

All the possible 12 compounds EX_3 (E = As, Sb, Bi; X = F, Cl, Br, I) are known, stable compounds and are all readily available by straightforward syntheses.

Fluorides of the tervalent pnictogens

Preparation. The fluorides, EF_3 are usually prepared by the action of HF on the oxide. Antimony and bismuth fluorides can be prepared by using concentrated aqueous HF, but in the case of arsenic a CaF_2/concentrated H_2SO_4 mixture is used. The sulphuric acid sequesters the water, preventing the hydrolysis of the AsF_3 formed [87]:

$$As_4O_6 + 12HF \rightarrow 4AsF_3 + 6H_2O \qquad [3.32]$$

Table 3.12 The tervalent pnictogen (E) fluorides: selected physical properties and molecular parameters.

	AsF_3	SbF_3	BiF_3
Melting point (°C)	–6.0	290	649
Boiling point (°C)	62.8	*c.* 345	900
Density (g ml^{-1})	2.67 (at 0°C)	4.38 (at 25°C)	*c.* 5.3
Colour and phase	Colourless liquid	Colourless crystals	Light-grey powder
ΔH_f^0(kJ mol^{-1})	–956.5	–915.5	–900
Bond length (E–F) (Å)	1.708[a]	1.92[b,c] 2.61[b,c]	
Bond angle (F–E–F) (°)	95.97[a]	87.3[b]	

[a]Gas phase.
[b]Crystal structure.
[c]All three E–F bonds.

Details of the laboratory scale preparations of all three fluorides are given by Kwasnik [88]. Some selected physical data for the EF_3 compounds are given in Table 3.12.

BiF_3 is prepared by dissolving Bi_2O_3 in HF with the formation of $BiF_3{\cdot}3HF$ (H_3BiF_6) which yields BiF_3 on heating. Although the addition compound H_3BiF_6 is readily hydrolysed by H_2O to BiOF, BiF_3 is unaffected by water and is almost insoluble (0.1 g l^{-1}). BiF_3 does not readily form complexes, but $BiF_3{\cdot}3HF$ and NH_4 [BiF_4] are known.

Structures. AsF_3 is a molecular compound and has been shown by electron diffraction [89, 90] and microwave spectroscopy [91] to have a pyramidal structure. In common with other arsenic pyramidal species AsF_3 is capable of inversion. Dixon and Arduengo [92] have shown by *ab initio* calculations that the electronegative fluorine ligands change the 'normal' vertex inversion via a planar D_{3h} intermediate to an edge inversion via a T-shaped intermediate. The calculated parameters of the T-shaped intermediate reveal As–F bond lengths of 1.774 Å [axial (ax)] and 1.676 Å [equatorial (eql)] and angles for F_{ax}–As–F_{ax} of 169.5° and F_{eql}–As–F_{ax} of 84.8°.

SbF_3 has a solid-state structure with three short Sb–F distances and three longer Sb–F bonds, giving an overall distorted SbF_6 octahedral geometry [93].

The case of bismuth is more complex, β-BiF_3 having a YF_3-type structure [94], that is, a nine-coordinate BiF_9 tricapped trigonal prism with eight approximately equal Bi–F bond lengths and one longer Bi–F interaction.

Reactions. Arsenic trifluoride has some advantages as a fluorinating agent: its ease of preparation, availability and ease of handling. The disadvantage is of course that the toxicity of arsenic is added to that of fluorine, resulting in increased disposal problems. An example of such a reaction is the monofluorination of hexachloroethane by AsF_3, catalysed by SbF_5 [95]:

$$CCl_3{\cdot}CCl_3 + AsF_3 \xrightarrow{SbF_3\ \text{catalyst/reflux}} \underset{(35\%)}{CCl_2F{\cdot}CCl_3} + \underset{(5\%)}{CCl_2F{\cdot}CCl_2F} \qquad [3.33]$$

The fluorination of certain compounds with use of antimony fluorides is known as the Swarts reaction, a few examples of which are given below in reactions [3.34]:

$$SbF_3 \begin{cases} \xrightarrow{R_3PS} R_3PF_2 \\ \xrightarrow{CF_3PCl_2} CF_3PF_2 \\ \xrightarrow{PhPCl_2} PhPF_4 \\ \xrightarrow{C_2Cl_6} CFCl_2{\cdot}CFCl_2 \\ \xrightarrow{SiCl_4} SiFCl_3,\ SiF_2Cl_2,\ SiF_3Cl \\ \xrightarrow{Me_2P(S)P(S)Me} Me_2PF_3 \end{cases} \qquad [3.34]$$

Antimony trifluoride is also used in a very convenient preparation of PF_3 [96]:

$$SbF_3 + PCl_3 \xrightarrow{SbCl_5} PF_3 + SbCl_3 \qquad [3.35]$$

SbF_3 is less reactive than the higher members of group 15 and is very soluble in methanol (1540 g l^{-1}) from which it is recovered unchanged [97].

Chlorides of the tervalent pnictogens

Preparation. Although the heavy pnictogen(III) chlorides are readily available commercially, their laboratory scale preparation may be useful in remote locations. The accounts in Brauer are recommended for $AsCl_3$, $SbCl_3$ and $BiCl_3$ (pages 596, 608 and 621 in ref. [65], respectively).

$BiCl_3$ can be prepared by dissolving bismuth metal in aqua regia, giving the dihydrate $BiCl_3{\cdot}2H_2O$ on evaporation. Distillation, which decomposes the hydrate, gives $BiCl_3$ as colourless deliquescent crystals. The vapour density indicates the presence of monomeric $BiCl_3$ and similarly in dilute solution (0.1 M in Et_2O), but association occurs at greater concentrations. The molten material exhibits a specific conductance of the same order of magnitude as that of ionic compounds (0.58 S cm^{-1} at 427°C). Selected physical data are presented in Table 3.13.

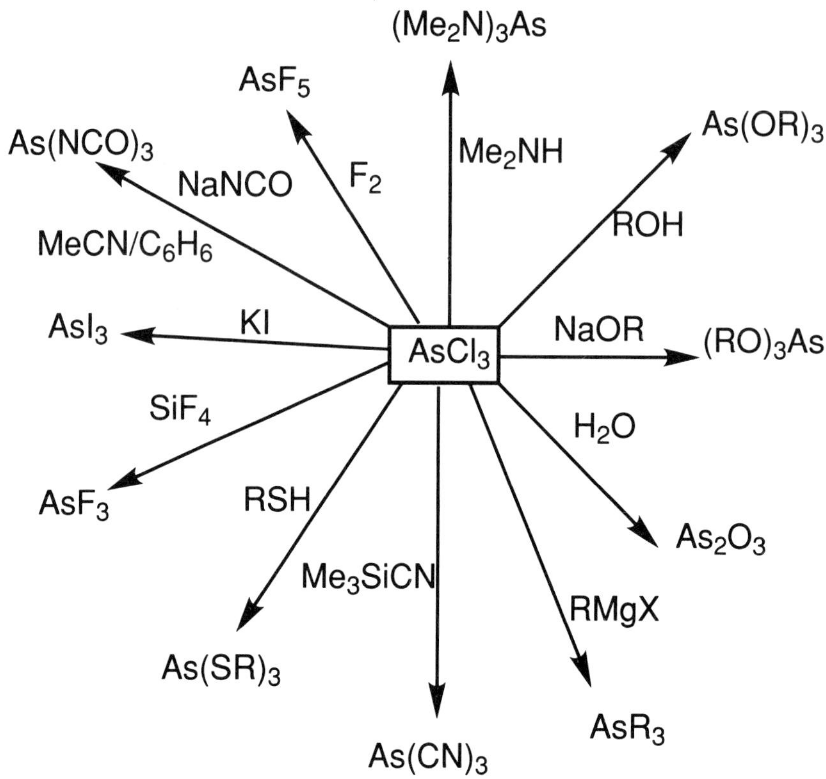

Scheme 3.4 Some reactions of arsenic trichloride. R = alkyl or aryl groups; X = halogen.

Table 3.13 The tervalent pnictogen (E) chlorides: selected physical properties and molecular parameters.

	$AsCl_3$	$SbCl_3$	$BiCl_3$
Melting point (°C)	–16.2	73.4	233.5
Boiling point (°C)	130.2	223	441
Density (g ml^{-1})	2.205 (at 0°C)	3.14	4.75
Colour and phase	Colourless liquid	Colourless cystals	Colourless cystals
ΔH^0_f(kJ mol^{-1})	–305.0	–382.2	–379
Bond length (E–Cl) (Å)	2.1621[a]	2.33[a]	2.423[a]
	2.17[b]	2.36[b,c]	2.50[b]
Bond angle (Cl–E–Cl) (°)	98.63[a]	97.2[a]	97.3[a]
	97.7[b]	97 or 96[b]	84 or 94[b]

[a]Determined by electron diffraction; gas phase.
[b]Determined by X-ray crystallography.
[c]Average.

Structure. Arsenic(III) trichloride has a pyramidal structure in the vapour phase and in the solid state [98], the pyramidal 'nest' affording $AsCl_6$ distorted octahedra, effectively three short molecular As–Cl bonds and three longer intermolecular bonds.

Although $SbCl_3$ [99] and $BiCl_3$ [100] have pyramidal molecular structures in the vapour phase, their solid-state structures are quite distinct from that of $AsCl_3$. Both antimony [101] and bismuth [102] have three short E–Cl bonds and five longer-distance interactions. However, although both have bicapped trigonal prismatic geometries (ECl_8 units), the ratio of the three short Sb–Cl bonds to the five longer Sb–Cl bonds is greater than the corresponding ratio in the case of bismuth showing the greater tendency for bismuth to adopt higher coordinating numbers.

Raman spectra [103] of $AsCl_3$ and $BiCl_3$ indicate the presence of ethane-like dimers, that is, $Cl_3E{-}ECl_3$.

Lunelli, Cazzoli and Lattarzi [104] pointed out that chlorine has an isomer ratio $^{35}Cl{:}^{37}Cl$ of *c.* 75:25. Thus one mole of ECl_3 is composed of four isotopomers in the proportions shown in Table 3.14. This simple analysis applies only to arsenic (100% ^{75}As) and bismuth (100% ^{209}Bi) because antimony itself has two nuclear isotopes ($^{121}Sb{:}^{123}Sb$, 57:43). However, the isotopic mass of Sb has no effect on the symmetry arguments, therefore ^{121}Sb and ^{123}Sb are equivalent as far as the structure is concerned, giving the same mole fractions for each structural type as for As and Bi, with each fraction split into two; that is, for the C_{3v} symmetrical species involving the ^{35}Cl isomer:

$$\text{mole fraction } (^{121}Sb^{35}Cl_3) + \text{mole fraction } (^{123}Sb^{35}Cl_3) = 0.4309. \qquad (3.3)$$

These isotopomers are detectable spectroscopically.

Reactions. $AsCl_3$ is not as reactive to nucleophilic attack as PCl_3 is. A reaction scheme for $AsCl_3$ (Scheme 3.4) is presented on two grounds. First, $AsCl_3$ is one of the major commercial arsenic compounds and, second, it exemplifies the EX_3 reactions in general. Of course, there are variations in the chemistry of the twelve compounds, but these are the 'expected' reactions. In Scheme 3.4 $AsCl_3$ is shown as yielding trisubstituted products. These reactions all proceed stepwise and it is much easier

Table 3.14 The isotopomers of ECl_3 (E = As, Bi).

Isotopomer	Mole fraction[a]	Symmetry
$E^{35}Cl_3$	0.4309	C_{3v}
$E^{35}Cl_2{}^{37}Cl$	0.4188	C_s
$E^{35}Cl^{37}Cl_2$	0.1357	C_s
$E^{37}Cl_3$	0.0147	C_{3v}

[a]Owing to rounding errors, the mole fractions in this column do not sum to 1.0000.

to stop at a lower level of substitution with $AsCl_3$ than is the case with PCl_3.

This lowered reactivity is shown dramatically in the behaviour of $AsCl_3$ and PCl_3 with water. PCl_3 is violently reactive towards water whereas 1 mol of $AsCl_3$ (*c.* 84 ml) can be dissolved in 9 mol of water (162 ml), to give a specific gravity $\simeq$ 1.53, without decomposition. This solution may be further diluted to give a solution of specific gravity $\simeq$ 1.35. Further dilution precipitates As_2O_3.

Antimony(III) chloride forms a clear solution with water, and there is no evidence for Sb^{3+} ions; dilution results in precipitation of insoluble oxychlorides of various composition, for example SbOCl, $Sb_4O_5Cl_2$ and $Sb_8O_{11}Cl_2$. Some reactions of $SbCl_3$ are shown in Scheme 3.5.

Antimony(III) chloride has been widely used as a solvent, especially for chloride ion transfer reactions and solvo-acid/solvo-base reactions.

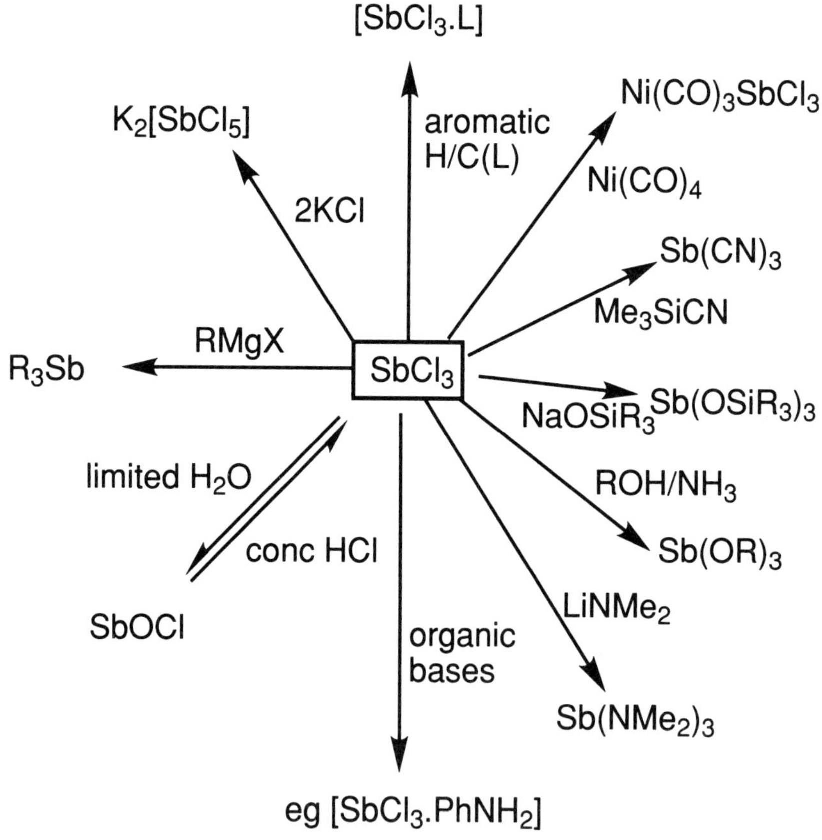

Scheme 3.5 Representative reactions of antimony (III) chloride. R = akyl or aryl groups; L = aromatic hydrocarbon.

It is a weak electron pair donor but it is an excellent acceptor, especially for halide ions and amines. The nominal self-ionization equation is as follows:

$$2SbCl_3 \rightleftharpoons SbCl_2^+ + SbCl_4^- \qquad [3.36]$$

but the low conductivity impels the equilibrium almost wholly to the left. Some physical properties and solubilities in $SbCl_3$ are given in Table 3.15. It may be noted that molecular solids, for example I_2, S_8, SbOCl, are soluble and that ionic solids are generally insoluble, the exception being large monovalent ions. This area has been excellently reviewed by Baughan [105].

Antimony trichloride, like the other EX_3 compounds is a powerful halide acceptor and this aspect of its chemistry and Lewis acid–base behaviour is discussed in chapter 4.

Reactions with metal halides are dealt with in the next chapter (chapter 4) but the following reactions of antimony and bismuth chlorides are given as examples not included in Scheme 3.5.

Antimony(III) trichloride can remove halide from s-block metals, for example $MgCl_2$ [106]:

$$MgCl_2 + 2SbCl_3 \xrightarrow{MeCN} [Mg(MeCN)_6][Sb_2Cl_8] \qquad [3.37]$$

Bismuth trichloride is used to build clusters with transition metals; for example with $Na[Mn(CO)_5]$, $BiCl_3$ affords the pyramidal bismuth cluster

Table 3.15 Solubilities in $SbCl_3$.

Easily soluble (sometimes with evolution of heat)
Elements: Cl, I, S
Monovalent halides: chlorides and bromides of K, Rb, Tl, NH_4, and substituted ammonium ions, KI (with colour formation)
Polyvalent halides: $HgCl_2$, $HgBr_2$, $AlCl_3$, SbF_3, $SeCl_4$ and $TeCl_4$
Oxides and sulphides: only Sb_2O_3, Sb_2S_3 and As_2O_3 (see below)
Oxyacid salts: sulphate and perchlorate of $N(CH_3)_4$; acetates of K, Hg(II) and Sb(III)

Slightly soluble:
Chlorides: LiCl, NaCl, $FeCl_3$ and $BiCl_3$

Insoluble:
Halides: KF, CuCl and AgCl
Divalent chlorides: Ca, Sr, Ba, Zn, Cd, Pb, Mn, Co, Ni and Hg_2Cl_2 (and $CrCl_3$)
Divalent oxides: Mg, Ca, Zn, Mn and Pb (and Al_2O_3)
Divalent sulphides: Zn, Cu, Hg and Pb
Oxyacid salts: sulphates of Na, K, NH_4, Mg, Ba, Zn and Sb(III), $KClO_4$; $AgClO_4$; K_2CrO_4; $KMnO_4$ and KCN

Soluble with evolution of gas:
Nitrates and carbonates: KNO_3 (NO_x released); K_2CO_3 (CO_2 released)

Slightly soluble or insoluble, with gas evolution:
Nitrates: Na and Ag
Carbonates: Li, Na, Mg, Ba, Zn, Mn and Pb

$[Bi\{Mn(CO)_5\}_3]$ with a bond length (Bi–Mn) of 2.90 Å (average) and Mn–Bi–Mn bond angles of 108.2° (average).

$AsCl_3$ has found some use as a solvent in preparative chemistry in, for example, fluorine–chlorine replacement reactions when AsF_3 is used [107, 108]. The AsF_3 is converted into the solvent thus giving fewer impurities and by-products:

$$NbCl_5{\cdot}PCl_5 + 2AsF_3 \xrightarrow{AsCl_3} NbCl_4F + PF_5 + 2AsCl_3 \quad [3.38]$$

However, $AsCl_3$ (dielectric constant 12.8 at 25°C) has not been as extensively studied as $SbCl_3$ (dielectric constant 33.2 at 75°C), as it is an inferior solvent. Through the equilibrium reaction

$$AsCl_3 + AsCl_3 \rightleftharpoons AsCl_2^+ + AsCl_4^- \quad [3.39]$$

typical bases (chloride ion released) are pyridine, $[R_4N]Cl$ and PCl_5, and typical acids are $AlCl_3$, $FeCl_3$, $SnCl_4$, $TlCl_4$, VCl_4 and $SbCl_5$; $TeCl_4$ is amphoteric [107, 108]

Bromides of the tervalent pnictogens.

Preparation. The bromides, EBr_3 are prepared by similar routes to the chlorides, with full preparative details given for $AsBr_3$, $SbBr_3$ and $BiBr_3$ on pages 597, 613 and 623 of ref. [65], respectively. Selected physical data are given in Table 3.16.

Structure. Arsenic tribromide is a pyramidal molecule in the gas phase [89, 109, 110] condensing to a crystalline solid with distorted octahedral $AsBr_6$ units containing three short bonds and three longer interactions [111]. The molecular parameters are shown in Table 3.16.

Table 3.16 The tervalent pnictogen (E) bromides: selected physical properties and molecular parameters.

	$AsBr_3$	$SbBr_3$	$BiBr_3$
Melting point (°)	31.1	96.0	219
Boiling point (°C)	221	288	462
	89 at 11 torr		
Density (g ml^{-1})	3.28 (at 50°C)	4.15	5.72
Colour and phase	Yellow crystals	White crystals	Yellow crystals
ΔH_f^0 (kJ mol^{-1})	–197.0	–259.4	–276
Bond length (E–Br) (Å)	2.33[a]	2.49[a]	2.63[a]
	2.36[b]	2.50[b]	2.66[b]
Bond angle (Br–E–Br) (°)	99.7[a]	98[a]	100[a]
	97.7[b]	95[b]	

[a]Determined by electron diffraction; gas phase.
[b]Determined by X-ray diffraction; solid phase.

Antimony tribromide was also shown to be pyramidal in the gas phase by an electron diffraction study [99]. However, the solid phase exists as both α–$SbBr_3$ [112] and β–$SbBr_3$ [113].

Bismuth tribromide was also shown to be a pyramidal molecule in the vapour phase [114]. In the solid phase $BiBr_3$ also exists in two forms [115], with a transition temperature of 158°C:

$$\alpha\text{-}BiBr_3 \rightleftharpoons \beta\text{-}BiBr_3 \qquad [3.40]$$

The low-temperature α-$BiBr_3$ has a distorted $BiBr_6$ octahedral structure, with three short Bi–Br bond lengths (2.66 Å mean) and three longer interactions (3.32 Å mean). The high-temperature β-$BiBr_3$ has an $AlCl_3$-type structure.

Arsenic tribromide fumes in moist air and is decomposed by water giving As_2O_3 and HBr. It is soluble in organic solvents such as Et_2O, benzene, hydrocarbons and chlorinated solvents.

Iodides of the tervalent pnictogens.

Preparation. Arsenic triiodide can be prepared in aqueous solution [116], for example:

$$As_2O_3 + HCl(conc)_{(aq)} \rightarrow AsCl_3 \xrightarrow{KI_{(aq)}} AsI_3 \qquad [3.41]$$

Useful laboratory procedures for the preparation of AsI_3, SbI_3 and BiI_3 are available in refs [116], [117] and [118], respectively. Selected physical data are given in Table 3.17.

Structure. Arsenic triiodide, as with the other halides of arsenic, is a pyramidal molecule in the vapour phase, as shown by microwave spec-

Table 3.17 The tervalent pnictogen (E) iodides: selected physical properties and molecular parameters.

	AsI_3	SbI_3	BiI_3
Melting point (°C)	140.4	170.5	408.6
Boiling point (°C)	*c.* 400	401	*c.* 540 extrap.
Density (g ml^{-1})	4.39	4.93	5.64
Colour and phase	Red crystals	Red crystals (trigonal) greenish yellow crystals (monoclinic)	Greenish black crystals
ΔH^0_f(kJ mol^{-1})	–58.2	–100.4	–150
Bond length (E–I) (Å)	2.55[a] 2.59[b]	2.719[a] 2.87[b]	3.07[b]
Bond angle (I–E–I) (°)	100.2[a] 99.7[b]	99.1[a] 96[b]	90[b]

[a]Determined by electron diffraction; gas phase.
[b]Crystal structure.

Table 3.18 Distortion of the EI_6 octahedron in EI_3 (E = As, Sb, Bi).

	AsI_3	SbI_3	BiI_3
Short E–I bonds (Å)	2.59[a]	2.87[a]	3.07[b]
Lone E–I bonds (Å)	3.47[a]	3.32[a]	3.07[b]
Bond angle (I–E–I) (°)	100	96	90

[a]Three bonds.
[b]Six bonds.

troscopy and electron diffraction [110]. Similarly antimony triiodide was shown by electron diffraction to be pyramidal in the gas phase [119].

The solid-phase structures of arsenic [120], antimony [121] and bismuth [121] triiodides show a progressive change from distorted octahedral AsI_6 to symmetrical octahedral BiI_6, that is, the pnictogen atom (As, Sb, Bi) progressively moves to the centre of the I_6 octahedron as group 15 is descended (Table 3.18).

Antimony triiodide has two crystal modifications, the red trigonal crystalline form just described and a greenish-yellow monoclinic phase. The structure of monoclinic SbI_3 [122] is intermediate between the trigonal (close-packed iodine layers) and a layered molecular structure with discrete SbI_3 units forming strongly distorted SbI_6 octahedra: Sb–I short bond lengths average 2.765 Å; longer interactions (I′) average 3.675 Å; bond angles (I–Sb–I) of 95.1°, 96.6° and 100.2°, and with bond angles (I′–Sb–I′) of 110.7°, 75.9° and 114.8°.

Aqueous solutions of AsI_3 are strongly acidic, a 0.1 M solution having a pH of about 1. Iodine is liberated from AsI_3 by moist air. Arsenic(III) iodide was used in the past in the treatment of syphilis in the form of an aqueous solution of HgI_2 and AsI_3 (Donovan's solution).

Tervalent pnictogen mixed halides. The mixed halides $EX_nX'_{3-n}$ are not usually stable with respect to the re-formation of the EX_3 compounds. However, Hass, Bechstein and Seefield [86] have shown that mixtures of $AsCl_3$ and AsI_3 favour the formation $AsICl_2$:

$$2AsCl_3 + AsI_3 \rightleftharpoons 3AsICl_2 \qquad [3.42]$$

It has been shown by ^{19}F NMR that small amounts of AsF_2Cl and $AsFCl_2$ are formed in a mixture of AsF_3 and $AlCl_3$. Similarly, the presence of $AsCl_2Br$ and $AsBr_2Cl$ have been detected in mixtures of $AsCl_3$ and $AsBr_3$.

3.3.4 Pentavalent pnictogen halides

The most notable feature of the pnictogen halides, EX_5, is the large number of missing members. Of the twelve possible compounds for As, Sb and Bi, only five are known, the pentafluorides being the only complete set. No pentabromides or iodides have been characterized to date. This instability of the higher group valency in heavier p-block elements is general and is ascribed to the 'd-block contraction'. This reduces the availability of the ns^2 electrons necessary for the achievement of the group valency. In the case of group 15 there is a tendency for the senior members to revert from E(V) to E(III). This may result in the oxidation of the halogen EX_5 to X_2:

$$AsCl_5 \rightarrow AsCl_3 + Cl_2 \quad [3.43]$$

In extreme cases this tendency may prevent the formation of EX_5, for example no BiX_5 (X = Cl, Br, I) are known. In the case of bismuth, not only is there the analogous 'f-block contraction', but also there is the 'relativistic contraction' [123–125], reinforcing the non-bonding tendency of the $6s^2$ electrons, that is, the inert pair effect.

The main use of EX_5 compounds is as powerful halide acceptors, forming EX_6^- anions. The large anions are a versatile synthetic tool, this aspect of their chemistry being discussed in chapter 4.

Fluorides of the pentavalent pnictogens

Preparation. The action of fluorine on arsenic gives AsF_5 [88], a colourless gas soluble in organic solvents (Et_2O, C_6H_6), but instantly hydrolysed by water. Dry AsF_5 is inert to glass, but a trace of H_2O or HF catalyses a rapid etching reaction with potentially catastrophic laboratory consequences. Convenient laboratory preparations for SbF_5 and BiF_5 are described in ref. [88].

SbF_5 is a syrupy viscous liquid (*c.* 850 centipoise at 20°C which is a similar value to glycerol) probably as a result of the polymeric chains of *cis*-bridged SbF_6 octahedra. The viscosity is greatly reduced by the addition of small amounts of $SbCl_5$ which breaks the Sb–F–Sb bridges. This is of industrial importance in processes such as the Swarts reaction.

Structure. In the vapour phase AsF_5 has the expected trigonal bipyramidal structure with axial bond lengths of 1.711 Å and equatorial bond lengths of 1.656 Å, as revealed by electron diffraction [91].

Both antimony and bismuth would be expected to have similar vapour phase structures. In the crystalline state SbF_5 exists as a tetramer $[SbF_5]_4$.

Antimony(V) fluoride is the most powerful Lewis acid known and, in conjunction with HSO_3F, forms 'magic acid', the most powerful of the superacids. Antimony pentafluoride is also a powerful Friedel–Crafts catalyst because of its ability to generate cations; among the first structural determinations of an oxocarbonium ion was that of $[CH_3CO]^+[SbF_6]^-$. The only stable bismuth(V) halide is BiF_5, a white crystalline material.

Table 3.19 The pentavalent pnictogen fluorides: selected physical properties and molecular parameters.

	AsF_5	SbF_5	BiF_5
Melting point (°C)	–79.8	8.3	154.4
Boiling point (°C)	–53.2	141	230
Density (g ml^{-1})	2.33 (–53)	3.11 (25)	5.40 (25)
Colour and phase	Colourless gas Yellow liquid	Colourless viscous liquid	White needles
ΔH^0_f(kJ mol^{-1})	–948.9	–908.8	–887.0
Bond length (E–F) (Å)	1.711 (ax)[a] 1.656 (eq)[a]	1.82 (term)[b,c] 2.02 (br)[b,c]	1.90[b] (term) torr 2.11 (br)
Bond angle (°):			
F_{ax}–E–F_{ax}	180	141[d]	180
F_{ax}–E–F_{eq}	90	170[d]	90

[a]Gas phase determined by electron diffraction; gas phase.
[b]Determined by X-ray crystallography.
[c]Average.
[d]Sb–F–Sb.

Note: ax = axial; br = bridging atom; term = terminal atom.

The crystal structure consists of BiF_6 octahedra in infinite *trans*- chains in the manner of α-UF_5 [126].

BiF_5 reacts explosively with water to form OF_2, O_3 and a brown solid [Bi(V) oxyfluoride]. BiF_5 is a powerful fluorinating agent, for example it will fluorinate paraffin oil to fluorocarbons at moderate temperatures (>50°C).

Selected physical data of the pentavalent pnictogen fluorides are given in Table 3.19.

Chlorides of the pentavalent pnictogens. Although antimony pentachloride is a well-known compound [88] until the 1970s it remained the only known member of this group. $AsCl_5$ was finally synthesized by Seppelt [127] although it decomposes above –50°C. The gas-phase structure of $SbCl_5$ was shown by electron diffraction to be the expected trigonal bipyramid [128]. This geometry was also found for the crystal structure [129]. Selected physical data are given in Table 3.20.

Pentavalent pnictogen mixed halides. The mixed halides $EX_nX'_{5-n}$, like the tervalent compounds EX_nX_{3-n}, are essentially unstable species. However, the series $AsF_{5-n}Cl_n$ ($n = 0, \ldots, 5$) have all been observed, albeit some of them only in an argon matrix. AsF_5 and $AsCl_5$ (n = 0 and 5) have already been described.

The attempted oxidation of $AsCl_3$ by F_2 or of AsF_3 by Cl_2 gives a product of empirical formula AsF_3Cl_2 [130]. This product is in fact ionic, with the formula $[AsCl_4]^+[AsF_6]^-$. These ionic compounds are discussed in chapter 4.

Table 3.20 The pentavalent pnictogen chlorides: selected physical properties and molecular parameters.

	$AsCl_5$	$SbCl_5$
Melting point (°C)	–50 (decomp)	2.8
Boiling point (°C)	—	92°/30 torr
Density (g ml^{-1})	—	2.346
Colour and phase	Pale yellow solid	Pale yellow liquid
ΔH^0_f (kJ mol^{-1})	NA	–438.5
Bond length (E–Cl) (Å)	2.43 [a] 1.67(ax)	2.43 (ax) 2.31 (eq)[b] 2.34 (ax)[c] 2.29 (eq)[c]
Bond angle (°):		
Cl_{ax}–E–Cl_{ax}	*c.* 180[a]	*c.* 180
Cl_{eq}–E–Cl_{eq}	*c.* 120	*c.* 120

[a]Determined by Raman spectroscopy
[b]Determined by X-ray crystallography.

Note: ax = axial; eq = equatorial; decomp = decomposed; NA = not applicable.

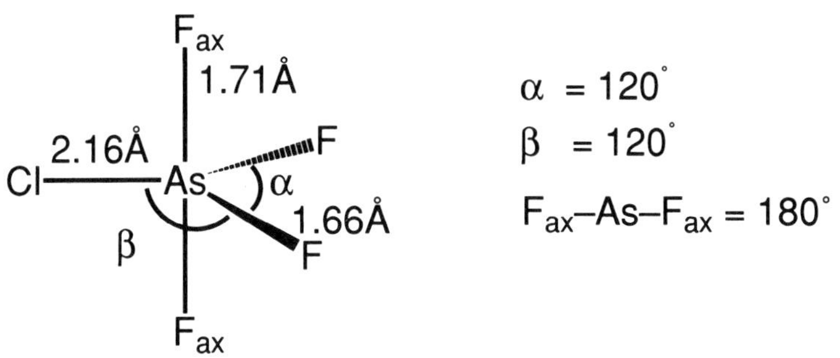

Figure 3.2 The proposed structure of AsF_4Cl from IR–Raman spectroscopy. Redrawn from Minkwitz, R., Clans, F., Glaser, M. and Wölfel, V. (1984) *Z. Anorg. Allg. Chem.*, **516**, 207.

The remaining series numbers $n = 1$–4 are as follows. For n = 1, AsF_4Cl is prepared by the pyrolysis of $[AsCl_4]^+[AsF_6]^-$ at 100°C, the product being condensed at –196°C [131]. It decomposes above –90°C in the liquid phase. A spectroscopic analysis (IR–Raman) pointed to a structure consistent with the structure shown in Fig. 3.2

For $n = 2$, AsF_3Cl_2 is also isolated from the pyrolysis of $[AsCl_4]^+[AsF_6]^-$ by condensation at –196°C. The compound AsF_3Cl_2 decomposes in the liquid phase above –75°C [132]. It was shown by electron diffraction [133] to have the expected trigonal bipyramidal structure (Fig. 3.3).

For $n = 3$, AsF_2Cl_3 is prepared by treating AsF_3Cl_2 with $CaCl_2$ [134]. The product was characterized by vibrational spectroscopy below –196°C showing D_{3h} symmetry.

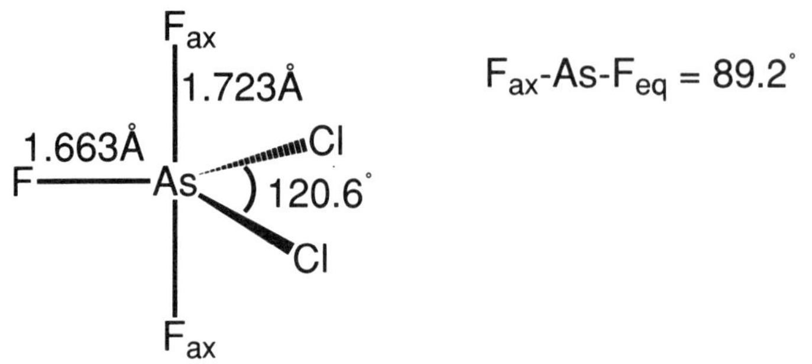

Figure 3.3 The structure of AsF_3Cl_2, determined by electron diffraction. Redrawn from Minkwitz, R., Prenzel, A., Schardy, A. and Overhammer, H. (1987) *Inorg. Chem.*, **26**, 2730.

For $n = 4$, $AsFCl_4$ was also isolated by Minkwitz and Clans [135] at –196°C. On warming, it decomposes:

$$AsFCl_4 \rightarrow AsCl_3 + FCl \qquad [3.44]$$

A variety of mixed antimony(V) chlorides and fluorides have been described, in the main arising from the fluorination of $SbCl_5$ although the product of fluorination is strongly dependent on the fluorinating agent. Bridging fluorides exist in the cyclic tetramer $SbCl_4F$, similar to $(SbF_5)_4$. One of the products from the HF reaction, $SbCl_3F_2$ is also a tetramer with bridging fluorine, bond length (Sb–F) 2.07 Å [136, 137]. The other HF fluorination product, $[SbCl_4][Sb_2Cl_2F_9]$, contains the unit $[ClF_4Sb-F-SbF_4Cl]$, with a distorted octahedral geometry about antimony, with a bond angle (Sb–F–Sb) of 163°C.

Ballard, Birchall and Slim [136, 137] found that the fluorination of $SbCl_4F$ gave a product best described as a tetramer of $SbCl_3F_2$ with partial replacement of the fluorine by chlorine and corresponding to the general formula $Sb_4Cl_{12+x}F_{8-x}$ ($x = 0, \ldots, 4$). In the case where $x = 1$,

$$SbCl_4F \xrightarrow[SO_{2(l)}]{SbF_3} Sb_4Cl_{13}F_7 \quad \text{melting point 49°C–54°C} \qquad [3.45]$$

the substitution results in a compound with a disordered crystal structure.

3.4 The binary pnictogen chalcogen compounds

The pnictogen (E = As, Sb, Bi) chalcides [Ch = O, S, Se, Te (Po)] are the most important group of pnictogen compounds inasmuch as they are found in the minerals and source ores that feed their industrial uses.

Transient species such as the diatomics are omitted from this discussion although most if not all the possible ECh combinations have been described. The account of each chalcogen is divided into three main sections corresponding to the group 15 valencies: the tervalent compounds are dealt with first, then the mixed valent compounds are examined; finally, the pentavalent compounds are discussed.

3.4.1 *Oxides of the pnictogens*

Oxides of tervalent pnictogens. The tervalent pnictogen oxides E_2O_3 are dealt with in descending order of group 15. Some selected data and molecular parameters are given in Table 3.21.

Arsenic. Arsenic(III) oxide, As_4O_6, is usually referred to as arsenic trioxide and is the most important arsenic compound of commerce produced on a large scale. Chemically, As_4O_6 is important since it is the most stable of the oxides, and many arsenic compounds are converted to As_4O_6 when ignited in air (Scheme 3.6). Above 250°C arsenic glows when heated in dioxygen at subatmospheric pressure (lower limit of O_2 pressure 4–10 torr, upper limit 200 torr at 260°C, 300 torr at 300°C) [138].

Arsenic(III) oxide is the basal source of commercial arsenic compounds and is important in analytical and preparative laboratories (Scheme 3.6). The structure [139] of As_4O_6 may be thought of as an oxygen-bridged As_4 tetrahedron. The resulting adamantine-like structure is extremely stable. The As_4O_6 unit is present in solid, liquid and gaseous phases, dissociation only beginning at 800°C. It crystallizes in cubic (arsenolite) and monoclinic (claudetite) forms. The latter exists in two forms, claudetite I and II, each of which has a layered structure of arsenic atoms joined though oxygen atoms [140]. The O–As–O angle, 95.5°, results in these rings being puckered. Taking the plane of the oxygen atoms as a reference plane, the arsenic atoms may be above (or below) the plane. Various combinations are theoretically possible, but only two have so far been described. Some reactions of As_4O_6 are shown in Scheme 3.7.

Table 3.21 The tervalent pnictogen oxides, E_2O_3.

	As_2O_3	Sb_2O_3	Bi_2O_3
Melting point (°C)	193 (sublimes)	655	820
Boiling point (°C)	—	1425 (sublimes)	—
Colour phase	White	White	Yellow
Density (g ml^{-1})	3.71	5.51	8.5
ΔH^0_f (kJ mol^{-1})	–656.8	–689.9	—573.9
Bond length (E–O) (Å)	1.79	1.98	2.196[a]
Bond angle (O–E–O) (°)	95.5	96	—

[a]Average.

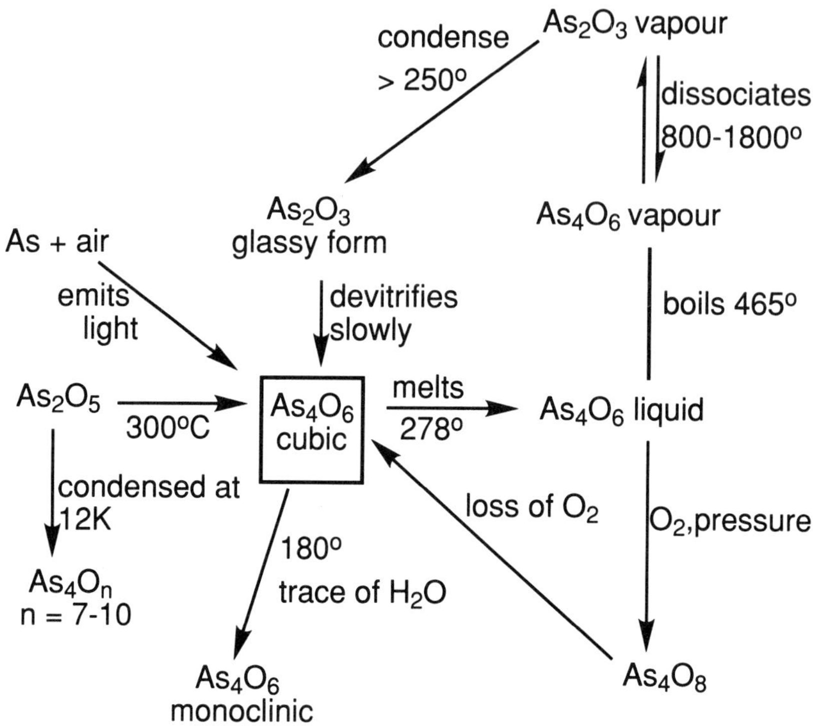

Scheme 3.6 Relationships between the oxides of arsenic.

Antimony. Antimony(III) oxide occurs naturally as the minerals serarmontite and valentinite. It is insoluble in water, dilute HNO_3 or dilute H_2SO_4 but is soluble in hydrochloric acid and in organic acids. It is an amphoteric oxide shown by its solubility in bases to give antimonates(III). The aqueous chemistry of antimony is discussed in more detail in chapter 4.

Laboratory prepared antimony(III) oxide by, for example, the hydrolysis of antimony(III) chloride is a mixed-phase material composed of varying proportions of cubic Sb_4O_6 and orthorhombic Sb_2O_3. Single-phase pure material is produced by calcining at 500°C (cubic, senarmontite) or 600°C (orthorhombic, valentinite), as follows:

$$\underset{\text{(cubic)}}{Sb_4O_6} \xrightleftharpoons{570^\circ C} \underset{\text{(orthorhombic)}}{[Sb_2O_3]_n} \rightleftharpoons \underset{\text{(liquid phase)}}{Sb_4O_6} \rightleftharpoons \underset{\text{(vapour phase)}}{Sb_4O_6} \qquad [3.46]$$

The structure of antimony(III) oxide in the vapour phase has been shown [141] to be discrete Sb_4O_6 molecules of Sb_4O_6 which dissociate into SbO_3

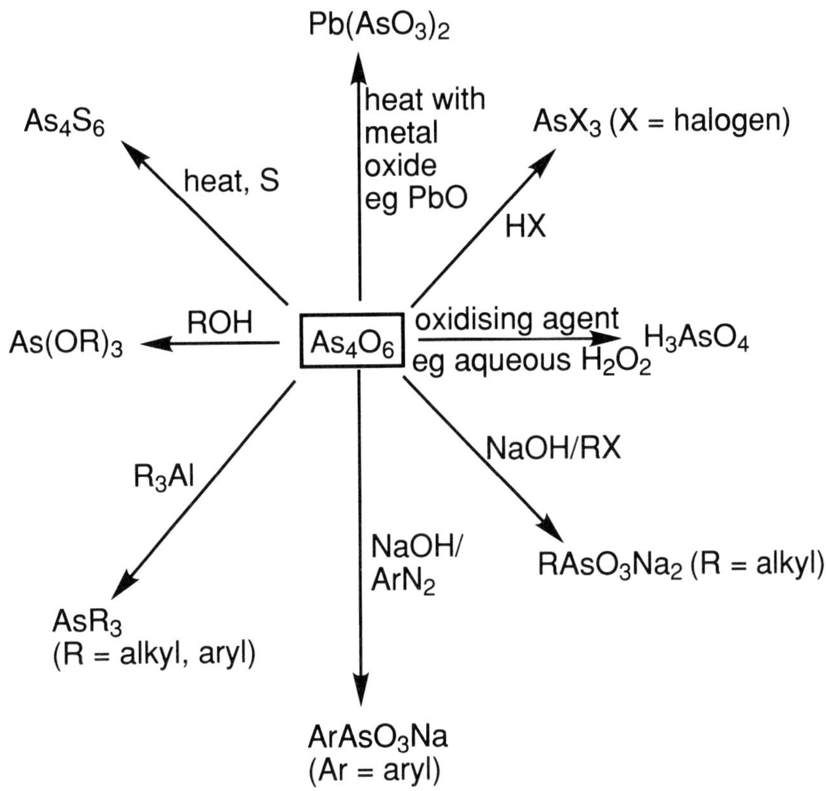

Scheme 3.7 Reactions of As_4O_6.

trigonal pyramidal units at high temperatures. The solid cubic phase consists of Sb_4O_6 units with each Sb having three Sb–O intramolecular bridging bonds (length, 1.98 Å; O–Sb–O angle, 96°) and three intermolecular Sb ··· O distances of *c.* 3.8 Å. Upon heating, the cubic phase changes to a denser orthorhombic form (density 5.20–5.79 g ml^{-1}) and a corresponding structural change to a polymeric form, $[Sb_2O_3]_n$.

Bismuth. Bismuth trioxide Bi_2O_3 is the only Bi(III) oxide. It occurs naturally as the mineral bismite. Bismuth trioxide is the most basic of the heavy pnictogen oxides showing typical p-block trends. Bi_2O_3 dissolves in mineral acids, forming salts; these and the aqueous chemistry of bismuth(III) are discussed in chapter 4.

There are several polymorphs, the normal pale yellow phase, stable at room temperature, being termed α-Bi_2O_3. The complex relationships between the polymorphs are shown in Scheme 3.8 [142, 143]. The crystal structure of bismuth trioxide was determined by Malmros [144] and

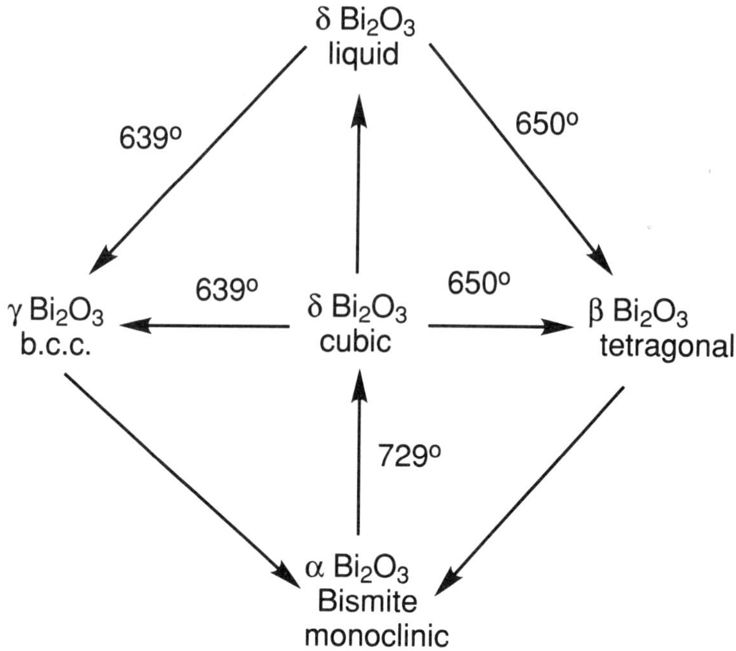

Scheme 3.8 Relationships between the Bi_2O_3 oxides. b.c.c. = body centred cubic.

refined by Harwig [142]. There are two distinguishable bismuth coordination spheres: Bi(1), BiO_6E and Bi(2), BiO_5E (E = lone pair). The oxygen polyhedron around Bi(1) is a distorted octahedron caused by the lone pair; the geometry around Bi(2) is a distorted octahedron with the lone pair occupying one of the sites. These polyhedra are linked together by shared corners and edges forming tunnels (minimum diameter 2.5 Å), and the bismuth $6s^2$ lone pairs are directed towards these tunnels.

Oxides of the mixed valent pnictogens. The mixed valent pnictogen oxides E_2O_4 are a small group of oxides if only binary compounds are considered. They can be considered a special case of 'mixed' metal oxides, for example phases such as $SbNbO_4$ and $SbTaO_4$ are known which form a much larger class. However, we shall confine ourselves here to the narrow definition.

Arsenic. Arsenic(III, V) oxide is commonly called arsenic dioxide because of its empirical formula AsO_2 which implies an As(IV) oxidation state. However it is more usefully written as As(III) As(V) O_4.

It is formed by heating As_4O_6 in an autoclave for 14 days at 260°C under dioxygen pressure (5–50 MPa) [145]. The hygroscopic colourless crystals have been shown to consist of infinite layers of arsenic atoms coordinated to three or four oxygen atoms. From the As–O bond lengths and the coordination numbers, the arsenic, which is trigonally coordinated to three oxygens at right angles, is considered to be As(III), and the arsenic coordinated to four oxygens in a distorted tetrahedron is considered to be As(V); the short As(V)–O bond length (1.611 Å) suggests some double-bond character.

Either of the two types of arsenic may be replaced by other group 5 elements, the arsenic being replaced depending on the electronegativity of the substituent. For example, when an equimolar mixture of Sb_2O_3 and

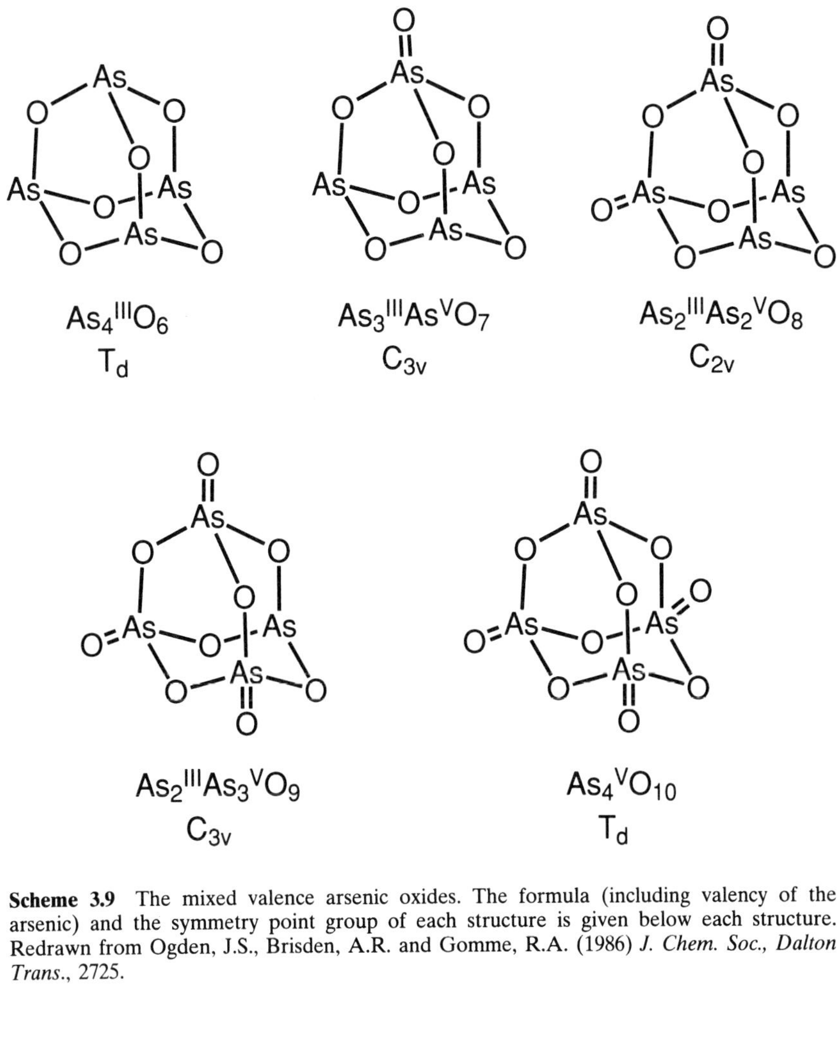

Scheme 3.9 The mixed valence arsenic oxides. The formula (including valency of the arsenic) and the symmetry point group of each structure is given below each structure. Redrawn from Ogden, J.S., Brisden, A.R. and Gomme, R.A. (1986) *J. Chem. Soc., Dalton Trans.*, 2725.

$As_2O_5.\frac{5}{3}H_2O$ are heated for a week in a closed vessel, a white crystalline mass of composition $AsSbO_2$ is obtained. X-ray analysis revealed the structure as Sb(III)As(V)O_4 [146]. The oxide has a layered structure as has As(III)As(V) oxide, but there are no terminal oxygen atoms – Sb and As are both four-coordinate.

The complete range of possible oxides between As(III)$_4O_6$ and As_4(V)O_{10} has been observed by means of matrix isolation techniques [147]: arsenic(V) oxide is heated in a platinum boat to 730°C and the vapour swept out in a stream of nitrogen and condensed at 12 K into the CsI window of an IR spectrophotometer. Arsenic(V) oxide enriched with ^{18}O has been similarly treated. The structures proposed on the basis of the IR analysis and a previous mass spectral study of As_2O_5 are shown in Scheme 3.9.

Antimony. The corresponding antimony mixed valent oxide Sb(III)-Sb(V)O_4 occurs in two crystal forms. The yellow mineral cervantite is orthorhombic (α) Sb_2O_4 and is converted to the monoclinic (β) form by heating at *c.* 1100°C in oxygen. Their structures are similar, one being sheared to give the other. Sb(V)O_6 octahedra form wrinkled layers by sharing the four equatorial vertices, with the Sb(III) between the layers forming distorted pyramidal Sb(III)O_4 units [148, 149]. Bond lengths from antimony to oxygens 1–6 (O^1–O^6): Sb$-O^1$ 2.09 Å, Sb$-O^2$ 1.86 Å, Sb$-O^3$ 2.37 Å, Sb$-O^4$ 2.17 Å. Bond angles: O^1-Sb-O^3 146°, O^2-Sb-O^4 90°.)

Bismuth. Bismuth(III) may be accommodated into such a structure, where it also resides in a distorted tetrahedral geometry [150].

An orange-red compound of empirical formula BiO_2 is formed as an insoluble precipitate when potassium bismuthate $KBiO_3$ is refluxed with 10% $HClO_4$ for 10 h [151]:

$$KBiO_3 + HClO_4 \rightarrow KClO_4 + BiO_2 + \tfrac{1}{4}O_2 \qquad [3.47]$$

The BiO_2 is hydrated and the structure has not been determined, but is not unreasonably thought to be a mixed valent species analogous to the other pnictogen dioxides. Gattow and Klippel [152] prepared an orange phase identified as BiO_2 by heating the presumed Bi(V) pentoxide at 150°C–180°C (described in the next subsection).

Oxides of the pentavalent pnictogens. Only two pentavalent pnictogen oxides E_2O_5 are known – As_2O_5 and the less stable Sb_2O_5 – although species approximating to a Bi_2O_5 stoichiometry are known for bismuth. Thus, even with highly electronegative oxygen, the lower members of group 15 have difficulty in exhibiting the maximum group valency, which is typical for the heavy p-block elements. Selected physical data are presented in Table 3.22.

Table 3.22 The pentavalent heavy pnictogen oxides, E_2O_5.

	As_2O_5	Sb_2O_5	Bi_2O_5
Melting point (°C)	315[a]	300[a]	150[a]
Boiling point (°C)	—	—	—
Colour and phase	White	Pale yellow	Dark brown
Density (g ml^{-1})	4.32	3.78	5.10

[a]Decomposes.

Arsenic. Arsenic(V) oxide is prepared by the oxidation of arsenic with use of nitric acid, followed by careful dehydration of the $As_2O_5.xH_2O$. As_2O_5 cannot be prepared by direct oxidation with dioxygen owing to the back dissociation reaction:

$$As_2O_3 + O_2 \rightleftharpoons As_2O_5 \qquad [3.48]$$

Arsenic(V) oxide is thermally unstable, easily losing dioxygen at 300°C to form As_4O_6 (Scheme 3.6). The pentoxide is deliquescent and highly soluble in water (2.30 g ml^{-1} at 20°C). The ease of oxygen loss has made it difficult to obtain the pure material for X-ray analysis, but this was finally achieved by Jansen [153]. The arsenic is in two different coordination states, half as four-coordinate tetrahedra and half as six-coordinate octahedra. The polyhedra are linked by their corners, resulting in cavities.

Other pnictides may occupy some of the arsenic sites in As_2O_5, for example antimony in $AsSbO_5$ which was shown by IR and Raman spectroscopy to be isostructural with As_2O_5 [154], with Sb occupying the octahedral sites. The compound with phosphorus substitution $AsPO_5$ has the phosphorus in the tetrahedral sites.

Antimony. An impure form of Sb_2O_5 is formed as a yellow powder on careful drying of the hydrated hydrolysis product of $SbCl_5$:

$$SbCl_5 \xrightarrow{NH_4Cl/H_2O} Sb_2O_5{\cdot}xH_2O \xrightarrow{275°C} Sb_2O_5 \qquad [3.49]$$

A pure antimony(V) oxide is produced in the high-pressure, high-temperature oxidation of Sb_2O_3 [155]. The structure is based on SbO_6 octahedra in sheets sharing vertices and edges [156].

Bismuth. Bismuth(V) oxide, Bi_2O_5, has not been prepared in a pure state. Compounds having approximate Bi_2O_5 stoichiometry have been claimed to have been prepared by oxidizing alkaline aqueous suspensions of Bi_2O_3 with H_2O_2 or chlorine. In a systematic study, Gattow and Klippel [152] attempted to prepare Bi_2O_5 by a variety of routes. Although they did not prepare any pure compounds they did present strong evidence for

Table 3.23 Proposed bismuth(V) oxides.

Empirical formula of bismuth–oxygen phase	Proposed composition of pure phase	Colour
$BiO_{1.98}$	BiO_2	Orange
$BiO_{2.33}$	α-Bi_2O_5	Light brown to grey brown
$BiO_{2.43}$	β-B_2O_5	Dark brown
$BiO_{2.48}$	γ-Bi_2O_5	Dark brown
$BiO_{2.24}$	δ-B_2O_5	Light red

Source: Gattow, G. and Klippel, W., *Z. Anorg. Allg. Chem.*, **470**, 25, 1980.

the existence of several Bi_2O_5 polymorphs and for a presumed mixed valent BiO_2 phase based on thermal analysis and powder X-ray analyses (Table 3.23). They attribute the stoichiometry in reaction [3.50] to the observed thermal transitions:

$$Bi_2O_5{\cdot}xH_2O \xrightarrow{80°C–100°C} Bi_2O_5 \xrightarrow{150°C–180°C} BiO_2 \xrightarrow{350°C} \alpha\text{-}Bi_2O_3 \quad [3.50]$$

The complexity of bismuth(V)–oxygen phases resembles that of Nb_2O_5 [157], but without achieving true stoichiometry.

3.4.2 Sulphides of the pnictogens

The organization of this section echoes that of the previous section on oxides; that is, three sections, the first on the tervalent compounds E_2S_3, the second on mixed valent species and the final section on the pentavalent phases E_2S_5, each split into a further three subsections, on arsenic, antimony and bismuth, respectively.

Sulphides of the tervalent pnictogens. The tervalent pnictogen sulphides E_2S_3, corresponding to the lower of the group 15 valencies, occur naturally

Table 3.24 The tervalent pnictogen sulphides, E_2S_3.

	As_2S_3	Sb_2S_3	Bi_2S_3
Melting point (°C)	320	546	850
Boiling point (°C)	707	—	—
Colour	Yellow	Black	Dark brown to grey black
Density (g ml^{-1})	3.49	4.64	6.78
ΔH^0_f (kJ mol^{-1})	–169.0	–160.0	–143.1
	2.25[a]		2.682[c]
	2.24[b]	2.53[b]	
Bond angle (S–E–S) (°)	114[a]		
	99[b]	—	—

[a]Determined by electron diffraction; vapour phase.
[b]Determined by X-ray crystallography.
[c]Average.

as the minerals yellow orpiment, (As_2S_3); black stibnite, (Sb_2S_3); and brown to black bismuthinite, (Bi_2S_3), selected data for which are presented in Table 3.24.

Arsenic. A yellow arsenic(III) sulphide is prepared by heating the elements in stoichiometric amounts or by heating As_2O_3 with sulphur. It is also formed by passing H_2S into acidified aqueous solutions containing As(III) compounds. In the solid phase the structure is similar to that of As_2O_3, that is layered trigonal pyramidal AsS_3 units [158]. It sublimes above its melting point (320°C) and the vapour has been shown by electron diffraction to be As_4S_6 [159] with a cage structure similar to As_4O_6.

The substitution of half the arsenic in As_2S_3 by antimony is found in the dark-red mineral getchellite, $AsSbS_3$. The mineral is unusually soft (1.5–2 Mohs scale) and sectile, with a perfect {0 0 1} cleavage. The crystal structure [160] of a synthetic sample revealed an asymmetric structural unit containing four pnictogen and six sulphur atoms. The pnictogen coordination is pyramidal ES_3, but the site occupancy is disordered. Although getchellite has a sheet structure it is unrelated to that of orpiment (As_2S_3) or stibnite (Sb_2S_3). It is possible the native mineral, formed on a geological timescale, has a more ordered structure.

Antimony. Antimony(III) sulphide is the most important sulphur compound of antimony. It occurs naturally as the mineral stibnite. It is usually a black crystalline mass of needles, although there is also an amorphous red-bronze form, and if both occur together the colour is carmine. It may be prepared in the laboratory by passing H_2S into a solution of $SbCl_3$. It is soluble in concentrated HCl and alkalis. The crystal structure shows it to consist of infinite Sb_4S_6 chains parallel to the *b* axis (the needle axis) [161]. The chains are linked to form crumpled sheets at right angles to the *a* axis. The sheets are held together by weaker (layer) Sb···S interactions in the range 3.37–3.64 Å (cf. the sum of the van der Waals radii, 4.05 Å).

When co-precipitated Sb_2S_3 and Sb_2O_3 were heated in a saturated Na_2CO_3 solution, bronze-coloured prismatic crystals were found in the reaction products. Single-crystal X-ray analysis showed this to be $Sb_{10}S_{15}$ [162] isostructural with fülöppite, $Pb_3Sb_8S_{15}$. There are vacancies in three of the Sb sites preserving charge neutrality for this unusual Sb(III) species. $Sb_{10}S_{15}$ is metastable at room temperature. When the reaction product is cooled slowly (12 h from 250°C to 125°C) the only product is Sb_2O_2S.

Bismuth. Bismuth(III) sulphide, Bi_2S_3, occurs naturally as the mineral bismuthinite in the form of a dark-brown to grey-black crystalline solid isostructural to stibnite. The laboratory preparation of Bi_2S_3 is similar to that of Sb_2S_3.

The coordination of bismuth in Bi_2S_3 is complex [163]. Owing to its large size and large effective nuclear charge there is a tendency for bismuth to have a potentially large ligand environment, accommodating more than the usual group 15 upper coordination number of 6. There are two crystallographically distinct bismuth atoms (Bi^1 and Bi^2). The three shortest bonds in each case form a trigonal pyramid with mean bond lengths of 2.678 Å and 2.686 Å for $Bi^1{-}S$ and $Bi^2{-}S$, respectively. In the case of Bi^1 the next three $Bi^1{-}S$ bonds are at a mean distance of 3.060 Å, which form a triangular distorted antiprism with a seventh $Bi^1{-}S$ distance of 3.329 Å capping one of the prism faces. The other bismuth atom has four $Bi^2{-}S$ bonds of an average length of 2.865 Å as a square pyramidal base with a $Bi^2{-}S$ apical distance of 2.542 Å. There are two longer $Bi^2{-}S$ interactions at 3.276 Å under the square pyramid base.

A bismuth–sulphur phase, composition BiS_2, is formed as grey needles by direct synthesis from the elements at 1250°C and 50 kbar pressure. The structure $Bi^{2+}[BiS_4]^{3-}$ has been proposed.

Sulphides of the mixed valent pnictogens. Arsenic is the only element in the heavy pnictogen triad to have a well-developed mixed valent chemistry with sulphur. In part this is probably because of the occurrence of several of these species as arsenic minerals. It is anticipated that this area of antimony sulphur chemistry will develop further.

Tervalent arsenic sulphur cage compounds. The arsenic sulphides discussed in this section are more correctly termed tervalent arsenic rather than arsenic(III) since the oxidation number varies from (I) to (III), reflecting the arsenic–arsenic connectivity in some of the species. The relationships may be visualized as the sequential insertion of sulphur bridges into a tetra-arsenic cluster:

$$As_4 \xrightarrow{3S} As_4S_3 \xrightarrow{S} As_4S_4 \xrightarrow{S} As_4S_5 \xrightarrow{S} As_4S_6 \quad [3.51]$$

In fact the various compounds can be prepared by heating mixtures of the elements with the appropriate stoichiometry in sealed vessels. Substitution compounds are known where arsenic or sulphur are partly replaced by other pnictogen or chalogens.

Tetra-arsenic trisulphide, As_4S_3, occurs as the orange-yellow mineral dimorphite which is found in α and β forms. The transition temperature to the α form is 130°C. Dimorphite is converted to realgar by heating with a carbon disulphide solution of sulphur:

$$As_4S_3 + S \xrightarrow{CS_2} As_4S_4 \quad [3.52]$$

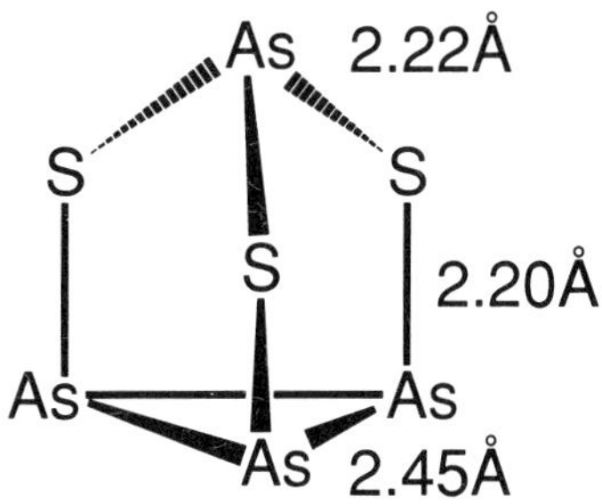

Scheme 3.10 The crystal structure of α-dimorphite, α-As_4S_3. Redrawn from Whitfield, H.J. (1970) *J. Chem. Soc. A*, 1800.

The structures of α-dimorphite (Scheme 3.10) and β-dimorphite were elucidated by Whitfield [164, 165], both exhibiting C_{3v} symmetry. Their differences are a result of crystal packing effects.

The substitution of one of the arsenic atoms by phosphorus is not a random process:

$$P_4S_3 + 3As_4S_3 \rightleftharpoons 4PAs_3S_3 \qquad [3.53]$$

The product was rationalized by Gimarc and Joseph [166] in terms of 'topological charge stabilization'. The electronegative sulphur should accept bridging positions in the cage structure and the less electronegative phosphorus should occupy the apical site [166].

Pale yellow crystals of $P_2As_2S_3$ were grown by vacuum sublimation from a melt of composition P:As:S of 2:2:3. The crystals were shown to be isomorphous with β-dimorphite crystals. All the arsenic sites were partially replaced by phosphorus, although the apical site was preferentially so occupied. Gillespie, Christian and Sawyer were able to rationalize the observed bond lengths by weighted averaging of the P–S and As–S bond lengths [167].

Tetra-arsenic tetrasulphide, As_4S_4, occurs as the orange-red to orange-yellow mineral realgar (from the arabic *rahjal ghár*, 'ore powder'). Sunlight changes the colour to yellow when the ore becomes friable. The synthetic mineral has two forms as a result of crystal packing differences. The α form is stable at ambient temperature and is identical to the mineral realgar. The β form is a high-temperature modification. These differences are reflected in their Raman spectra, two strong lines in the α form (at 177 cm^{-1} and 191 cm^{-1}) being replaced in the β form by a single intense line (at 187 cm^{-1}) [168]. The commercial synthetic material is known as 'red-glass', 'ruby arsenic' or 'arsenic blend' and is used in tanneries as a depilatory in the production of fine leather. Some reactions of realgar are shown in Scheme 3.11. Although realgar is a molecular compound it does

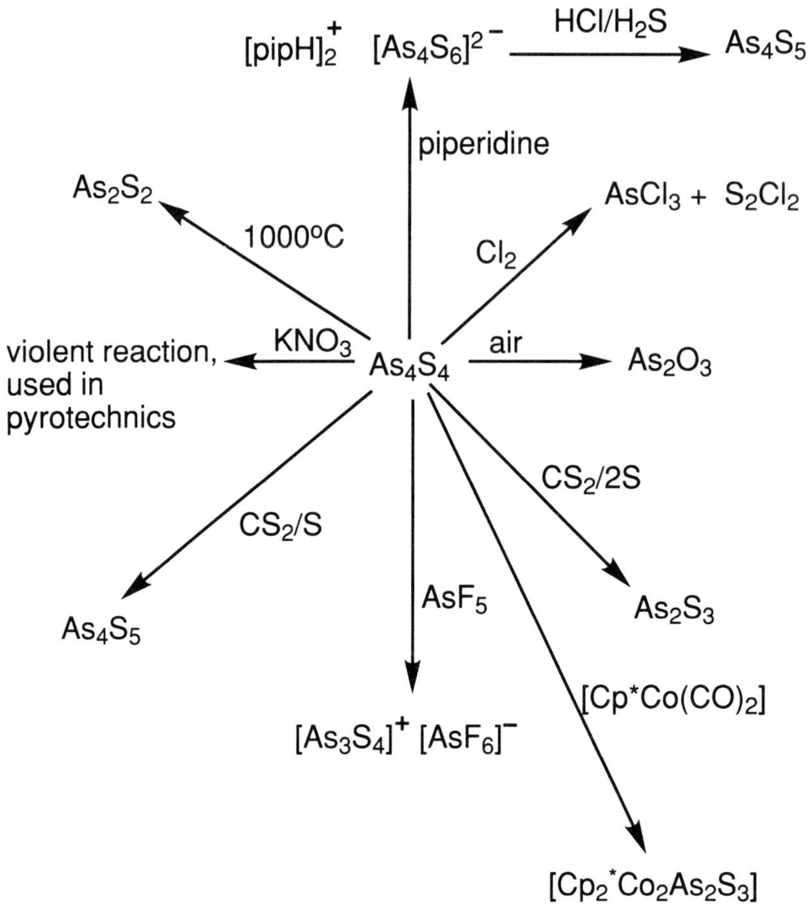

Scheme 3.11 Some reactions of realgar, As_4S_4.

not form simple coordination compounds with transition metals. The As_4S_4 structure is fragmented on reacting with $[CpCo(CO)_2]$ [169] in thf irradiated with UV:

$$[Cp^*Co(CO)_2] + As_4S_4 \rightarrow [Cp_2{}^*Co_2(As_2S_3)] \qquad [3.54]$$

As_4S_4 boils at 565°C to give a vapour the colour of chlorine; this vapour dissociates into As_2S_2 completely at 1000°C:

$$\alpha\text{-}As_4S_{4(s)} \xrightarrow{130°C} \beta\text{-}As_4S_{4(s)} \xrightarrow{307°C} As_4S_{4(l)} \xrightarrow{565°C}$$
$$As_4S_{4(g)} + As_4S_{2(g)} \xrightarrow{1000°C} As_2S_{2(g)} \qquad [3.55]$$

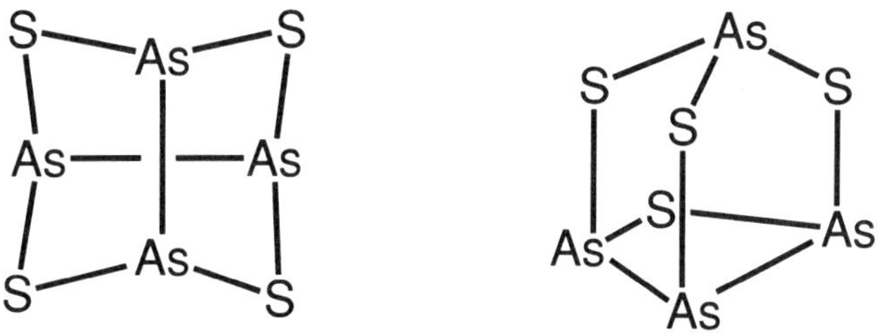

α and β -As_4S_4 γ -As_4S_4

Scheme 3.12 The crystal structure of α, β and γ realgar. Redrawn from Kutoglu, A. (1976) *Z. Anorg. Allg. Chem.*, **419**, 176.

The structures of the As_4S_4 units in the α form [168, 170], the β form [168] and in the vapour phase [139] are, within experimental limits, identical (Scheme 3.12).

There is a third structural isomer of realgar, the so-called γ-As_4S_4 (Scheme 3.12). In γ-realgar each arsenic coordination sphere is composed of one arsenic and two sulphur neighbours.

Although reaction [3.55] shows the overall reaction, the mechanism is not simply a case of building sulphur bridges since if the melt at 550°C is quenched rapidly the γ-As_4S_4 structure is frozen out, containing AsS_3, $AsAsS_2$ and $AsAs_2S$ pyramidal subunits as shown in the crystal structure [171, 172]. This structure requires new As–As bond formation to reach the α-As_4S_4 structure.

The formal oxidation state of arsenic in the As_4S_4 forms is two, although in the γ-isomer As(I), As(II) and As(III) sites may be identified.

Although As_4S_7 is not known the mixed pnitide $As_2P_2S_7$ has been characterized. It is synthesized from the elements in a sealed ampoule at 350°C. IR and Raman spectra indicate C_{2v} symmetry (Scheme 3.13), with arsenic the (II) oxidation state [173].

$As_2P_2S_7$ may be further oxidized by the addition of sulphur, to afford the polymeric $[As_2P_2S_8]_n$. This material also results from the reaction between arsenic(III) and phosphorus(V) sulphides:

$$n\mathrm{As_4S_6} + n\mathrm{P_4S_{10}} \xrightarrow{400^\circ\mathrm{C},\ 3\ \mathrm{days}} 2(\mathrm{As_2P_2S_8})_n \qquad [3.56]$$

The $[As_2P_2S_8]$ units also have C_{2v} symmetry (as determined by IR spectroscopy), supporting the proposed structure shown in Scheme 3.14.

The complexity of the arsenic–sulphur phases has not yet been observed in antimony or bismuth sulphur chemistry.

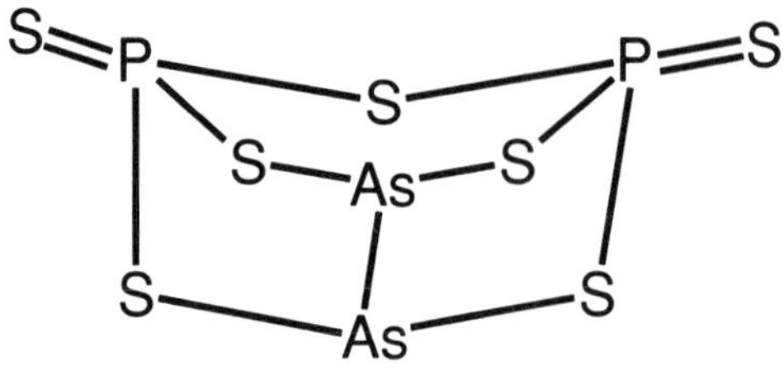

Scheme 3.13 The proposed structure of $As_2P_2S_7$ (peritectic melting point *c.* 400°C, density 2.61 g cm^{-3}).

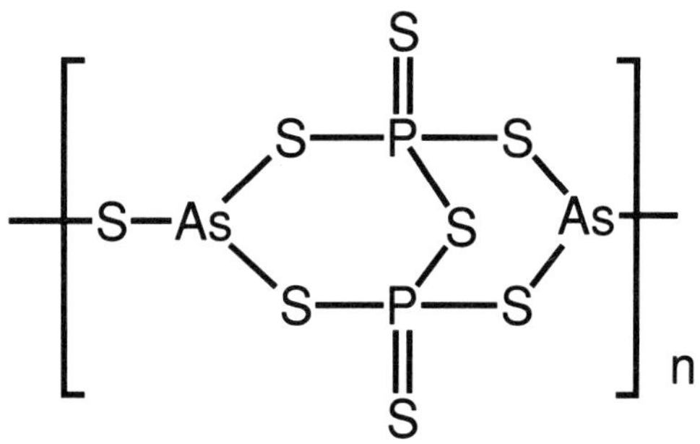

Scheme 3.14 The proposed structure of the polymer $[As_2P_2S_7]_n$ (peritectic melting point *c.* 440°C, density 2.45 g cm^{-3}).

Sulphides of the pentavalent pnictogens. The pentavalent heavy pnictogen sulphides are either ill-defined or of doubtful existence. Only As_2S_5 is a 'respectable' compound, being prepared from H_2S and a solution of an As(V) compound [174]:

$$2H_3AsO_4 + 5H_2S \xrightarrow{H^+} As_2S_5 + 8H_2O \qquad [3.57]$$

or by the acidification of a thioarsenate salt [174]:

$$2Na_3AsS_4 + 6HCl \rightarrow As_2S_5 + 6NaCl + 3H_2S \qquad [3.58]$$

On heating, As_2S_5 loses sulphur to form As_2S_3.

The antimony analogue is a non-stoichiometric compound, the composition varying with the method of preparation. Mössbauer studies [175, 176] showed only the presence of Sb(III). However, the theoretical status of 'antimony pentasulphide' does not worry its industrial users in firework manufacture or in pigmentation applicants.

Bi_2S_5 does not seem to exist; at any rate, it has not been prepared to date. Selected physical data relating to the pentavalent pnictogen sulphides are given in Table 3.25.

Table 3.25 The pentavalent pnictogen sulphides, Pn_2S_5.

	As_2S_5	$(Sb_2S_5)^a$	$(Bi_2S_5)^a$
Melting point (°C)	d. 500	d. 135°	—
Colour	Yellow	Golden	—
Density (g ml^{-1})	—	4.12	—

[a]Existence doubtful.

3.4.3 *Selenides of the pnictogens*

Selenium forms the expected binary compounds E_2Se_3 with the heavy pnictogens, but with higher valencies the limitations mirror that of sulphur, that is, only arsenic forms an identifiable phase. However, selenium forms many more phases with the heavy pnictogens which tend in properties to metallic alloys, reflecting the increasing metallic character of the p-block elements on descending a group.

Selenides of the tervalent pnictogens. The heavy pnictogen triselenides E_2Se_3 correspond to the lower group 15 valency of 3. All three compounds are known and can be synthesized from the elements by fusion in stoichiometric amounts.

Arsenic(III) selenide is synthesized by fusion of the elements at high temperature. It is isomorphous with As_2S_3 with the molecular parameters given in Table 3.26.

Antimony(III) selenide is a grey material formed by fusion of the elements (selected physical data and molecular parameters are given in Table 3.26). It has a chain structure [161, 177] involving short Sb−Se bonds (2.58–2.77 Å, the chains being held in pairs by weaker bonding (2.9 Å); these pairs are in turn bonded into sheets with Sb−Se interactions of about 3.26 Å, and the sheets held by Sb ··· Se interactions of about 3.46 Å (cf. van der Waals radii for Sb−Se of 4.20 Å).

Bismuth(III) selenide is synthesized by heating (at 475°C) the elements in stoichiometric amounts in an evacuated quartz tube. It occurs naturally as the mineral guanajuatite. It decomposes when heated in air, is insoluble in water and is attacked by nitric acid. Mixed selenium–tellurium compounds can be prepared, for example Bi_2Te_2Se [178].

Table 3.26 The tervalent pnictogen selenides, E_2Se_3.

	As_2Se_3	Sb_2Se_3	Bi_2Se_3
Melting point (°C)	380	612	706
Colour	Brown	Grey	Black
Density (g ml^{-1})	4.80	5.81	7.50
Bond length (E−Se) (Å)	2.40^a	2.68^a	—
Bond angle (Se−E−Se) (°)	91–105	—	—

[a]Average.

Table 3.27 Binary arsenic selenium compositions that have been given registry numbers in *Chemical Abstracts*.

Composition	Number	Composition	Number	Composition	Number
AsSe	12006-05-2	$AsSe_{10}$	66106-59-0	As_2Se_5	1303-37-3
$AsSe_2$	22626-92-2	$AsSe_{11}$	66106-60-3	As_2Se_2	53800-83-2
$AsSe_3$	22739-14-6	$AsSe_{20}$	12006-07-4	As_3Se_7	12044-57-4
$AsSe_4$	12006-06-3	As_2Se_3	1303-36-2	As_4Se_3	63897-50-7
$AsSe_8$	66106-58-9	As_2Se_4	37190-00-4	As_4Se_4	52126-83-7
$AsSe_9$	12044-31-4				

Selenides of mixed valent pnictogens. Although there are selenium pnictogen compounds analogous to those of sulphur (Section 3.4.2), there are complexities that are difficult to fit into our earlier system for oxygen and sulphur ligands. These miscellaneous compounds and phases will be touched on here, for example arsenic and selenium form a large number of glasses of varying composition that are of interest to the electronics industry. Various compositions are listed in *Chemical Abstracts* (Table 3.27), but in the main these are not compounds with established structures, in fact many are solid solutions.

As_4Se_4 has been shown to be isomorphous with realgar As_4S_4 [179, 180]. Although As_4Se_3 [181] is isomorphous with As_4S_3, the bond energies are different and this results in different chemical behaviour. As_4S_3 undergoes an exchange reaction with P_4S_3, resulting in apical-PAs_3S_3. In the case of As_4Se_3 the distribution of phosphorus in the molecule was found to be statistically random, showing that energy differences between As(apical) and As(base) are smaller in As_4Se_3 than in As_4S_3.

The bismuth–selenium system affords three identifiable phases (by differential thermal analysis and powder X-ray) [182]. Bi_2Se_3 (melts congruently at 606°C), BiSe and Bi_4Se_3 decompose peritectically at 610°C and 467°C, respectively.

Selenides of the pentavalent pnictogens. In common with the other higher members of group 16 the group 15 selenides are only represented by one pentavalent species, namely As_2Se_5, which has a black metallic lustre. The bismuth analogue cannot, as yet, be prepared, and the antimony analogue is doubtful.

As_2Se_5, produced in a manner similar to As_2S_5, is similarly ill-defined, but in the glasses $AsSe_x$, when the mol% equals or exceeds that of As_2Se_5, then in the formal sense the arsenic may be thought of as pentavalent. These compounds have not, however, had their structures elucidated.

Table 3.28 The tervalent pnictogen tellurides, E_2Te_3.

	As_2Te_3	Sb_2Te_3	Bi_2Te_3
Melting point (°C)	360	620	585
Colour	—	Grey	Grey
Density (g ml^{-1})	6.25	6.5	7.64
Bond length (E–Te) (Å)	—	2.98–3.17	—
Bond angle (Te–E–Te) (°)	—	—	—

3.4.4 Tellurides of the pnictogens

The tervalent pnictogens form tellurides but the pentavalent species are, again, much less stable, Bi_2Te_5 being unknown. The binary compounds and their doped derivatives have been studied mainly for their semiconductor and allied electronic properties rather than for their chemistry *per se.* The earlier literature to about 1970 is reviewed by Dutton [183]. Selected physical data are given in Table 3.28.

Arsenic(III) telluride, As_2Te_3, is prepared by the fusion of the elements in a sealed tube [184].

Antimony(III) telluride is a grey material (melting point 620°C) formed by direct synthesis from the elements at elevated temperatures. Structurally, the tellurium atoms are in two different layers, Te^1 and Te^2. The Te^2 layer has an Sb layer on each side, whereas the Te^1 layer has an Sb layer on one side and a Te^1 layer on the other. Selenium can be substituted for some of the tellurium in the synthesis, resulting in materials $Sb_2Te_{3-x}Se_x$. Anderson and Krause [185] showed that the substitution is not random: the Te^2 layer is replaced first, and only when this is complete does Te^1 substitution take place.

Bismuth(III) telluride occurs naturally as the mineral tellurobismuthite, Bi_2Te_3, and with partial sulphur substitution in tetradymite, Bi_2Te_2S. Rhombohedral Bi_2Te_3 (unit cell length $a = 10.267$ Å, $\alpha = 25.5°$, $Z = 1$) has a space group $R3m$ [186].

Various mixed valent phases have been described, for example AsTe, As_8Te_3, SbTe, Sb_2Te, BiTe, Bi_2Te, Bi_7Te_3 but these, like the selenium–pnictogen phases, may be solid solutions [183].

3.5 Oxoanion compounds

The IUPAC recommended systematic nomenclature for oxoacids and oxoanions has not yet passed into common usage. The recommended name for example of H_3AsO_4 is tetraoxoarsenic acid and of H_3AsO_3 is trioxoarsenic acid. Also recommended is the 'hydrogen nomenclature' whereby the acidic hydrogen atoms are treated as if they were the cations of a salt which gives the names trihydrogen tetraoxoarsenate and trihydrogen trioxoarsenate (3–) for H_3AsO_4 and H_3AsO_3, respectively. The recommended stems for As, Sb and Bi are arsen-, antimon- and bismuth-, respectively.

We shall use the traditional nomenclature in conjunction with the recommended stems, for instance we will refer to H_3AsO_4 as arsenic acid, which gives rise to arsenates, and to H_3AsO_3 as arsenous acid, affording the arsenites. Similarly, antimonic and antimonate and bismuthous and bismuthite, and so on will be used.

The principal valence states of group 15 give rise to two series of oxoanions derived from the ill-defined or hypothetical fully hydroxylated species $E(OH)_3$ and $E(OH)_5$. We will first describe the simple E(III) oxoanions and then their more complex derivatives. Similarly, the discussion of the simple E(V) oxoanions will be followed by a discussion of their more complex derivatives. Finally, we will deal with some miscellaneous topics that do not fall easily into the above categories.

3.5.1 Oxoanion derivatives of the pnictogen(III) Acids

Although the free acids $E(OH)_3$ *per se* or in solution are ill-defined species, in the case of arsenic the salts are well characterized and many are commercially important. Arsenic(III) species are numerically and commercially the most significant, thus discussion of these compounds will form the larger part of this section. The formation and properties of the oxoanion derivatives of the heavy pnictogens is a clear-cut example of the similarity of arsenic to phosphorus and its divergence from antimony and bismuth as a result of the 3d and 4f contractions and, especially in the case of bismuth, as a result of relativistic effects.

Oxoanion derivatives of arsenous acid. As a tribasic acid, arsenous acid forms a series of salts with the alkali metals (M): MH_2AsO_3, M_2HAsO_3 and M_3AsO_3, all of which are soluble in water. The group 2 metal salts are slightly soluble in water and the heavy metal arsenites are insoluble. The salts of arsenous acid may be successively dehydrated to two other series of salts, the pyroarsenites, $As_2O_5^{4-}$, and the meta-arsenites, AsO_2^-, a polymeric anion. $Na_3AsO_3{\cdot}6H_2O$ is prepared from NaOH and As_4O_6 in aqueous methanol [187].

Alkali metal meta-arsenites have been obtained as isolated $MAsO_2$ molecules at low temperatures (12 K) by using matrix-isolation techniques [188], and IR analysis indicates a C_{2v} symmetry of ring structures for $CsAsO_2$ as shown in Fig. 3.4. X-ray analyses have revealed that many salts have complex structures in the solid state. Sodium meta-arsenite has been shown [189] to be a polymeta-arsenite, $[NaAsO_2]_n$, consisting of pyramidal AsO_3 units which are linked together by shared oxygen atoms into linear chains (Scheme 3.15). Four chains run through the unit cell, held

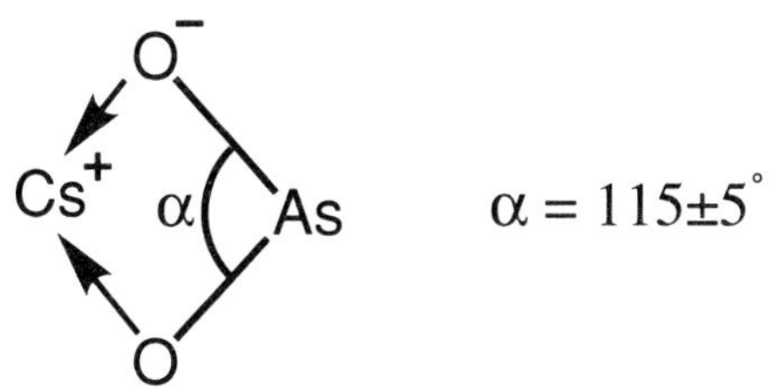

Figure 3.4 Proposed structure of matrix isolated caesium meta-arsenite.

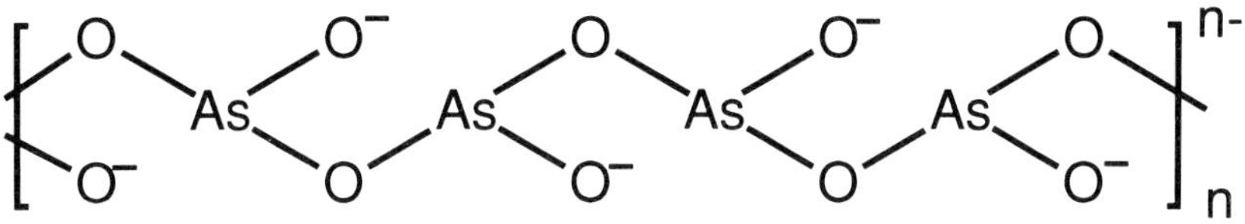

Scheme 3.15 The structure of the anion in the polymer $[NaAsO_2]_n$. Each arsenic has a lone pair which results in a trigonal pyramidal geometry.

together in two sets by the electrostatic attraction of sheets of sodium cations. This structure may be likened to the monoclinic modification of As_2O_3, and it is interesting to note that monoclinic As_2O_3 crystallizes from KOH solutions of As_2O_3 [189].

The synthetic $Ba_2As_6O_{11}$ contains the highly condensed polyarsenite anion $[As_6O_{11}]^{4-}{}_n$ [190] and is prepared as follows:

$$2BaO + 3As_2O_3 \xrightarrow[\text{Hydrothermal reaction}]{200°C,\ 7\ days} Ba_2As_6O_{11} \quad [3.59]$$

X-ray analysis reveals a double chain structure in which the chains are bridged by two μ-AsO_3 units forming $As_{10}O_{10}$ rings. The double chains are linked into double sheets through relatively strong As···O bonds with an average length of 2.68 Å.

AsO_3 pyramids are corner connected into As_4O_8 rings in the mineral stenhuggarite, $CaFeSbAs_2O_7$ [191] and in $Pb(II)(AsO_3)_2$ [192] although the ring conformations are different. The only As_3O_6 ring revealed to date is that in Schweinfurter Grün, $Cu_2(MeCOO)As_3O_6$ [193]. The arsenite minerals were briefly reviewed in 1985 by Hawthorne [194]. Selected structural data are presented in Table 3.29.

Arsenite, AsO^{3-}, forms lacunary heteropoly anions (inorganic cryptands) with M_4^{2-} (M = group 6 metal, especially W); for example when sodium tungstate and sodium arsenite are dissolved in boiling water and the pH adjusted to 4 by $HCl_{(aq)}$, subsequent reduction of the volume and

Table 3.29 Metal arsenites and related compounds.

Compound	Bond length (As–O) (Å)	Bond Angle (°) O–As–O	As–O–As	Reference
$NaAsO_2 \cdot 4H_2O$	}			
$Na_2(HAsO_3) \cdot 5H_2O$	}			Sheldrick, E.S. and Häusler, H.J. (1987) *Z. Anorg. Allg. Chem.*, **549**, 177
$Na_5(HAsO_3)(AsO_3) \cdot 12H_2O$	}			
$Na_2(H_2As_4O_8)$	}			
$KHAs_2O_4 \cdot 5H_2O$	1.68–1.82	93–98	125	Fuchs, J. (1980) *Z. Naturforsch.*, **35B**, 1344
$K_3(HAs_2O_4)(As_2O_4) \cdot 1.5H_2O$				Sheldrick, W.S. (1987) *Z. Anorg. Allg. Chem.*, **551**, 101
$K_3HAs_2O_5 \cdot 6H_2O$	1.72–1.78, 1.83[a]	89–103	126	Sheldrick, W.S. and Häusler, H.J. (1985) *Z. Naturforsch.*, **40B**, 1622
$CaAs_2O_6$	1.86–1.91			Nord, A.G., Stefandis, T. and Kierkrgaard, P. (1985) *Z. Krist.*, **173**, 313
$K_2Cu_3As_2O_5$	1.764, 1.838[a]		123	Effenberger, H. and Pertlik, F. (1987) *J. Solid State Chem.*, **70**, 219
$Cu_2(AcO)As_3O_6$	1.72–1.86, 1.76[b]	95.6–103.1		Pertlik, F. (1977) *Z. Krist.*, **145**, 35
$Pb_6Cu(AsO_3)_2Cl_7$	1.76			Pertlik, F. (1986) *Monatsch. Chem.*, **117**, 1257
$Ba_2As_6O_{11}$	1.787[b]	89.9[c]		Sheldrick, W.S. and Häusler, H.J. (1986) *Z. Anorg. Allg. Chem.*, **538**, 45
μ_3-AsO_3	1.877[b], 1.689[c]			
μ_2-AsO_3		93.8[d], 97.9[d]		
$BaAs_2O_4H_2O$	1.67–1.85	90.3–99.5		Sheldrick, W.S. (1987) *Z. Anorg. Allg. Chem.*, **554**, 137
$PbAs_2O_4$	1.79–1.83[c]			Pertlik, F., Effenberger, H., Kugler, A. *et al.* (1988) *Acta Cryst.*, **44**, 2043
	1.73, 1.70[e]			
$Zn_3(AsO_3)_2$ (reinerite)	1.753–1.772	95.2–100		Ghose, S. (1977) *Am. Min.*, **62**, 1129
$Pb_5Cl(AsO_3)_3$ (finnemanite)	1.74–1.87	97–100		Effenberger, H. (1979) *Tscher. Min. Petl. Mitt.*, **26**, 95
$Pb_2As_2O_5$ (paulmooreite)	1.73–1.77[e]	95–101		
	1.83–1.84[a]			Moore, P. (1980) *Am. Min.*, **65**, 340
$CaFeSbAs_2O_7$ (stenhuggarite)				Coda, A., Dal Negro, A., Sabelli, G. and Tazzoli, V. (1977) *Acta Cryst.*, **B33**, 1807
$Mn_5(OH(AsO_3)_3$ (magnussonite)	cubic			Gabrielson, O. (1957) *Ark. Min. Geol.*, **2**, 133
$CuAs_2O_4$ (trippkeite)				Pertlik, F. (1975) *Tscher. Min. Petr. Mitt.*, **22**, 211
$Pb_3Mn(AsO_3)_2(HAsO_3)$ (trigonite)	1.75–1.89			Pertlik, F. (1978) *Tscher. Min. Petr. Mitt.*, **25**, 95

[a]Bridge atoms. [b]Average. [c]Terminal atoms [d]$O_{(ring)}-As-O_{(oxo)}$. [e]Ring atoms.

cooling causes the tungstoarsenite to crystallize according to the following [195]:

$$40Na_2WO_4 + 4Na_3AsO_3 + (32 + n)H_2O \rightarrow Na_{27}(NaAs_4W_{40}O_{140})\cdot nH_2O + 64NaOH \quad [3.60]$$

The ionic weight of these anions (*c.* 10^4) is among the highest recorded for heteropolyanions. The Na^+ may be exchanged for other cations and, indeed, unless it is replaced by a larger cation, usually K^+, the polyanion is metastable. There are also four surface sites available for complex formation. The structure of the complex $(NH_4)_{23}[NH_4As_4W_{40}O_{140}(CoOH_2)_2]\cdot$*c.* $20H_2O$ has been determined by X-ray diffraction [196].

Four bridging WO_6 octahedra act as hinges in the flexible $As_4W_{40}O_{140}$ structure enabling two of the four ligating sites to accommodate the bulk of the transition metal cation. These structural adjustments are claimed to be analogous to the allosteric effects of enzyme reactions.

Müller *et al.* [197] have succeeded in obtaining a crystal structure of a blue As(III) heteropoly anion, $[H_6As_{10}{}^{III}Mo_8{}^{V}Mo_{16}{}^{VI}O_{90}]^{8-}$. It is, in a formal sense, composed of two $As_4Mo_{12}O_{42}$ spherical moieties of approximate C_{3v} symmetry with a central AsO_3 group. Their publication contains an excellent stereo view of this complex anion.

In a class of compounds which can be described as arsenic(III) complex spiroanions arsenic is four coordinate with respect to the oxygen ligands, that is, it is an AsO_4 species. In a formal sense these species may be thought of as being formed by the addition of OR^- to $As(OR)_3$:

$$As(OR)_3 + M[OR] \rightarrow M^+[As(OR)_4]^- \quad [3.61]$$

where M is an alkali metal.

The arsenic in these compounds is in a four coordinate, ten-electron state, 10-As-4 in the notation of Arduengo. These complex anions are stabilized by ring formation, although they are partly dissociated in solution, according to ^{13}C NMR spectroscopy [198]. These compounds are well illustrated in the dihydroxycyclohexane dicarboxylic complexes. Typically, the acid is added to a cooled solution of sodium arsenite when an exothermic reaction occurs forming the dimeric products.

X-ray analysis shows [198] the geometry about arsenic to be the expected equatorially vacant, trigonal bipyramid with a stereochemically active lone pair. In fact, potassium di-*o*-phenylenedioxyarsenate, $K[As(C_6H_4O_2)]$, was the first arsenic compound to be shown to have an AsO_4 pseudotrigonal bipyramidal structure for which the axial OAsO angle is 166.2° [199]. If these compounds are compared with the structure of arsenic(V) spirocycloarsonates, the lone pair in the former occupies the site of the organo (R) group in the As(V) compound. The bis μ-(tartarato) bis arsenate has a similar structure [200].

Oxoanion derivatives of antimonous acid. Antimonous acid H_3SbO_3 or $Sb(OH)_3$ is not well-characterized and its salts are also ill-defined. By fusing the oxides of sodium and antimony(III) one obtains the cubic, colourless Na_3SbO_3 [201] shown by single-crystal X-ray diffraction to be a tetramar, $[Na_3SbO_3]_4$.

Transition metal 'antimonites' are better considered as mixed metal oxides; for example, the mineral schafarzikite $Fe^{II}Sb_2{}^{III}O_4$ [202] consists of FeO_6 octahedra and pyramidal SbO_3 with a stereochemically active lone pair.

Antimony(III) oxide forms a large group of complex oxoanions with organic acids, and many of these compounds are of clinical significance (chapter 8). The best known example in this class is tartar emetic, dipotassium bis[μ-D-tartaratodiantimonate]trihydrate. The structure was first elucidated as the ±D,L salt by X-ray crystallography in China by Mu and Hsueh Ting [203] and further refined by Jacobson and Grass [204] as the + isomer by three dimensional X-ray and white radiation neutron diffraction. The structure of $[Sb_2(D\text{-}C_4H_2O_6)_2]^{2-}$ is shown in Scheme 3.16; the coordination about antimony is that of a highly distorted trigonal bipyramid. The axial oxygen atoms are from the carboxylate moiety and the equatorial oxygens are hydroxyl derived.

The bond lengthening between the longer axial Sb–O bond and the equatorial Sb–O bond is, on average, 0.17 Å. Between the dimers there is a weak Sb–O interaction involving the uncoordinated carboxylate oxygen atom. The dimers are bound loosely to clusters of three water molecules, forming sheets. Potassium ions lie between the sheets holding them in place in the crystal.

Oxoanion derivatives of bismuthous acid. Bismuth(III) oxide is the most basic of the pnitogen oxides, thus the ill-defined hydrated Bi_2O_3 has few acid properties. However, at elevated temperature group 1 oxides and

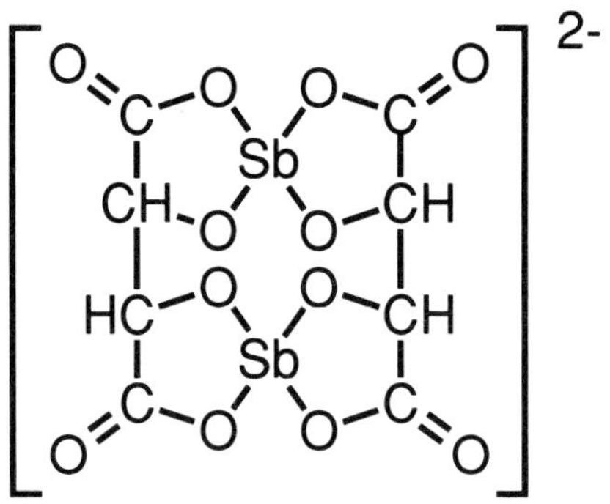

Scheme 3.16 The structural connectivity found in tartar emetic.

Bi_2O_3 afford two series of compounds: first, the colourless $MBiO_2$ (M = Li [205], Na [206], K, Rb, Cs [207]) in which BiO_2 moieties are twisted through 180° forming infinite chains, synthesized as follows:

$$KO_{0.64} + Bi_2O_3 \xrightarrow[\text{silver bomb}]{700°C,\ 14\ h} KBi^{(III)}O_2 \qquad [3.62]$$

and, second, M_3BiO_3 {M = Li (triclinic) [208], Na (cubic) [203]} show an interesting variation in structure, and which are synthesized as follows:

$$Li_2O + Bi_2O_3 \xrightarrow[\text{silver bomb}]{750°C,\ 20\ d} Li_3BiO_3 \qquad [3.63]$$

Na_3BiO_3 has a similar structure to Na_3SbO_3 described in the previous subsection but whereas Na_3BiO_3 only fills half the sites of the cube, Li_3BiO_3 shows BiO_3 pyramids at each cubic site (Scheme 3.17). Hoppe and Hübenthal [208] consider the bismuth lone pairs all to be directed into the cube, although the steric crowding results in a slightly distorted cube.

The metastable β-Bi_2O_3 structure is stabilized by the addition of transition metal oxides. These phases show anti-glass disorder [209] and are termed β-bismuthates although they are Bi(III) compounds, for example yttrium β-bismuthate of composition $YBi_{15}O_{24}$. The yttrium has been shown by neutron diffraction to have replaced bismuth randomly in the fluorite-type super structure [210].

Although not strictly oxoanionic compounds of bismuth, the mixed oxide superconducting phases, for example $Sr_2Ca_2Bi_2Cu_3O_{10}$ [211] are also noted here.

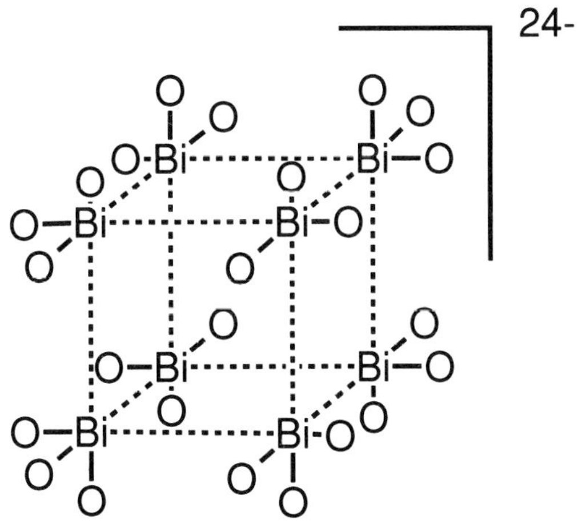

Scheme 3.17 The structure of the anion of the compound Li_3BiO_3, showing the BiO_3 pyramids at each cubic site.

3.5.2 Oxoanion derivatives of the pnictogen(V) acids

As with the pnictogen(III) acids, the pentavalent acid derivatives are heavily biased towards arsenic both in terms of stability and in terms of numbers of compounds. Again this demonstrates the similarity between arsenic and phosphorus and emphasizes the changes in character on descending group 15.

Oxoanion derivatives of arsenic acid. This section is divided into six parts, the first four being a description of the arsenates of the s-, p-, d- and f-blocks. There is then a discussion on the hydrates of the arsenates. Finally, a selection of minerals is presented. General properties of the compounds – both synthetic and mineral – described throughout the following subsections are presented in Tables 3.31 and 3.32.

s-Block arsenates. The preparation of most metal arsenates is straightforward. In the case of soluble arsenates, simple neutralization of 'arsenic acid' solutions with the appropriate base will suffice, for example, neutralization with NaOH affords $NaH_2AsO_4{\cdot}H_2O$ after concentration and crystallization.

Calcium arsenate was used in agriculture but has been banned in most of the advanced economies. The acid calcium arsenates, $CaHAsO_4$ and $Ca(H_2AsO_4)_2$, are harmful to plant life, hence the agrochemical named 'calcium arsenate' in fact contains excess lime. Commercially, disodium arsenate, Na_2HAsO_4, is prepared in an oxidative neutralization process by roasting a mixture of As_2O_3, Na_2CO_3, and $NaNO_3$ in air. This material is then used in the manufacture of other arsenates. Fusion methods are often used to avoid the occluded contaminants inherent in precipitation methods. For example, $MgCO_3$ and As_2O_5 are fused and the melt cooled slowly to give $Mg_3(AsO_4)_2$, suitable for single-crystal X-ray analysis [212]. In those arsenates where the AsO_4 tetrahedra are isolated in the crystal lattice, the As–O bonds and the O–As–O angles are normal for tetrahedral geometry. The cation may impose constraints on crystal growth, for example the ethylenediammonium hydrogenarsenate, $[H_3NCH_2CH_2NH_3]^{2+}[HOAsO_3]^{2-}$, crystal contains hydrogen bonded pairs of $HOAsO_3$ tetrahedra with an As···As distance of 4.994 Å [213].

In a formal sense, condensed arsenic acids are formed when the acid H_3AsO_4 loses water, forming linear, cyclic or cage species. The acids are unknown in the free state, although the As_2O_5 hydrates may be written as acids; however, many of the derived anions have been characterized. The simplest such anions are the pyroarsenates, $As_2O_7^{4-}$, which are usually easily prepared by heating a metal acid arsenate:

$$2Na_2HAsO_4{\cdot}7H_2O \xrightarrow{900°C} Na_4As_2O_7 + 15H_2O \quad [3.64]$$

Alternatively, pyroarsenates can be formed when stoichiometric amounts of the appropriate oxides and/or oxide precursors are heated together, for example, the synthesis of $CaK_2As_2O_7$:

$$K_2CO_3 + CaO + As_2O_5 \xrightarrow[\text{Pt crucible}]{700^\circ C,\ 4\ h} CaK_2As_2O_7 + CO_2 \quad [3.65]$$

To obtain crystals suitable for X-ray diffraction it is often necessary to cool the molten salt very slowly, in this case from 700°C to 500°C at 3°C h^{-1}, that is, for *c.* 2½ days.

Sodium pyroarsenate, $Na_4As_2O_7$, has been shown to have a number of slowly reversible transitions between ambient temperatures and 227°C, with a further reversible transition at 680°C [214]. $Na_4As_2O_7$ is not isostructural with $Na_4P_2O_7$. $Li_4As_2O_7$ also shows many phase transitions [215]. The difference between the temperature required to prepare the pyroarsenate and its dissociation temperature is sometimes quite small. For example, for $Li_4As_2O_7$ [216]:

$$LiAsO_3 + Li_3AsO_4 \xrightarrow{720^\circ C} Li_4As_2O_7 \quad [3.66]$$

$$3Li_4As_2O_7 \xrightarrow{>735^\circ C} 4Li_3AsO_4 + As_2O_5 \quad [3.67]$$

in which all the phases have been characterized by X-ray powder attraction.

The formation of $Mg_2As_2O_7$ is used for the gravimetric quantitative estimation of arsenates precipitated as ammonium magnesium arsenate [217]:

$$AsO_4^{3-} + MgCl_2 + NH_4Cl + 6H_2O \longrightarrow Mg(NH_4)AsO_4{\cdot}6H_2O + 3Cl^- \xrightarrow{\text{ignition}} Mg_2As_2O_7 \quad [3.68]$$

Structurally, the pyroarsenates are corner-sharing AsO_4 tetrahedra with the As–O–As angle of the oxygen bridge varying with the crystal constraints arising from the metal cations. The AsO_3 moieties are usually in an eclipsed conformation. The vibrational spectra of $M_2As_2O_7$ (M = Ba, Sr) have indicated that the As–O bond order in the As–O–As bridge is 1.0. The bond order in the terminal As–O bonds is 1.45. The angle made by the bridging groups (α in Fig. 3.5) was *c.* 115°–120° [218]. Selected structures are shown in Fig. 3.5.

Sodium triarsenate $Na_5As_3O_{10}$ has been prepared by heating a mixture of the hydrogen and dihydrogen arsenates:

$$NaH_2AsO_4 + 2Na_2HAsO_4 \xrightarrow{300^\circ C\text{–}400^\circ C} Na_5As_3O_{10} \quad [3.69]$$

It is isostructural with the low-temperature form of $Na_5P_3O_{10}$ [219]. The

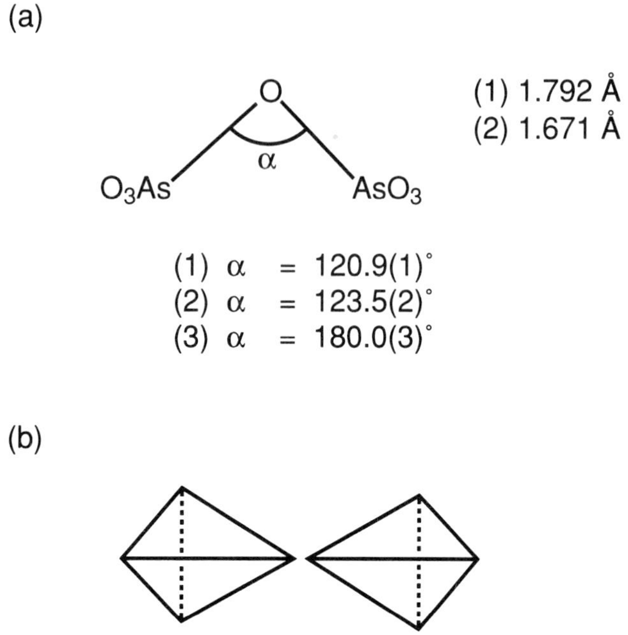

Figure 3.5 (a) The structure of $As_2O_7^{4-}$ in (1) $CaK_2As_2O_7$; (2) $Na_4As_2O_7$; (3) $Ca_2As_2O_7$, $Mg_2As_2O_7$. (b) AsO_3 moieties in eclipsed conformation [Fig. 3.5(a)(3) above]. Sources: part (a)(2) data from Calvo, C. and Leung, K.Y. (1975) *Can. J. Chem.*, **51**, 2080; part (a) (3) data for $Ca_2As_2O_7$ from Pertlik, F. (1980) *Monatsch. Chem.*, **111**, 399; part (a)(3) data for $Mg_2As_2O_7$ from Calvo, C. and Neelakantan, K. (1970) *Can. J. Chem.*, **48**, 890.

anion $As_3O_{10}^{5-}$ (as the potassium salt) has been shown to be three-corner-sharing AsO_4 tetrahedra, giving a chain that is highly bent, with an angle through the arsenic atoms of 105° [220]:

```
   O    O    O
   |    |    |
O–As–O–As–O–As–O
   |    |    |
   O    O    O
```

A tetrameric As_4O_{14} unit is formed in $NaHAs_2O_6$ and $BaH_6As_4O_{14}$. The anion is unusual in that it contains arsenic(V) in both tetrahedral and octahedral coordination.

In the barium salt [221] the As_4O_{14} units are isolated structures in the crystal lattice, whereas in the sodium salt [222] they are oxygen-bridged structures forming a two-dimensional network. The As_4O_{14} unit occurs in the more highly condensed species. In $NaH_2As_3O_9$ the As_4O_{14} rings are linked by two bidentate bridging AsO_4 tetrahedra via the ring AsO_6 octahedra to form infinite chains [223]. The structure of $Na_2As_4O_{11}$ also features the As_4O_{14} ring in this highly condensed arsenate, with the lowest O:As ratio 1:2.75 to date. All the oxygen atoms are shared, save one on each AsO_4 tetrahedron. One of the As–O–As bridges between the AsO_6 octahedra in the As_4O_{14} ring is opened up to form new As–O–As bridges, with AsO_4 tetrahedra connecting the new rings (As_4O_{15}). The links are increased by sharing of the edges of the AsO_6 octahedra, completing the three-dimensional structure [224].

Alkali meta-arsenates $MAsO_3$ (M = alkali metal) are prepared by heating the dihydrogen arsenates (MH_2AsO_4). When sodium dihydrogen arsenate hydrate is heated slowly, the di-, tri- and polyarsenates are formed successively [225]:

$$\begin{array}{ccccc} NaH_2AsO_4H_2O & \xrightarrow{50^\circ C} & NaH_2AsO_4 & \xrightarrow{90^\circ C} & Na_2H_2As_2O_7 \\ & & & & \downarrow{\scriptstyle 135^\circ C} \\ \text{melts} & \xleftarrow{615^\circ C} & (NaAsO_3)_n & \xleftarrow{>230^\circ C} & Na_3H_2As_3O_{10} \end{array} \qquad [3.70]$$

The K_2O–As_2O_5 phase diagram has been shown to contain four compounds: α-K_3AsO_4, $K_4As_2O_7$, $K_5As_3O_{10}$ and γ-$KAsO_3$ [220]. Potassium meta-arsenate has two high-temperature modifications [226], β-$KAsO_3$ is metastable at room temperature, whereas α-$KAsO_3$ is unstable:

$$\begin{array}{ccccc} \beta\text{-}KAsO_3 & \underset{\text{slow cooling}}{\overset{530^\circ C}{\rightleftharpoons}} & \alpha\text{-}KAsO_3 & \xrightarrow{600^\circ C} & \text{melts} \\ {\scriptstyle 430^\circ C}\uparrow\downarrow{\scriptstyle H_2O} & & \downarrow{\scriptstyle \text{rapid quenching}} & & \\ KH_2AsO_4 & \xleftarrow{H_2O} & \gamma KAsO_3 & & \end{array} \qquad [3.71]$$

Although all the alkali metal meta-arsenates have the same general formula, their structures are quite different. The fibrous crystals of sodium meta-arsenate have been shown [227] to be polymeric; AsO_4 tetrahedra in trimeric repeat units form parallel chains. Layers of these chains alternate with layers of sodium atoms which are enclosed in distorted octahedra of oxygen atoms. Lithium meta-arsenate has a similar structure [228], but with a dimeric repeating AsO_4 tetrahedral unit. In potassium meta-arsenate, $KAsO_3$, the anion has been shown by X-ray analysis to be a cyclic trimer.

Calcium meta-arsenate, $CaAs_2O_6$, is prepared by heating $CaCO_3$ with As_2O_3 in the presence of excess nitric acid and heating the evaporate at 500°C for four days. If the meta-arsenate is fused (at 1300°C) it loses arsenic(V) oxide:

$$3CaAs_2O_6 \xrightarrow{1300°C} Ca_3(AsO_4)_2 + 2As_2O_5 \qquad [3.72]$$

Calcium meta-arsenate has a structure of chains of arsenic octahedra AsO_6 and calcium octahedra [229]. The As–O bond lengths are longer in AsO_6 octahedra than in AsO_4 tetrahedra. Octahedral arsenic–oxygen structures usually occur in combination with arsenic tetrahedra, for example As_2O_5, $H_5As_3O_{10}$, $BaH_6As_4O_{16}$, $NaHAs_2O_6$. $CaAs_2O_6$ is unusual inasmuch as all the arsenic is in octahedral coordination.

The presence of monomeric ion pairs $M^+AsO_3^-$ (M = K, Rb, Cs) has been demonstrated in the vapour phase over $MAsO_3$ (platinum cell) at 1150–1200 K by the matrix isolation technique of the vapour at 12 K. The structure of the ion pair has been suggested on the basis of infrared spectroscopy (Scheme 3.18) [230].

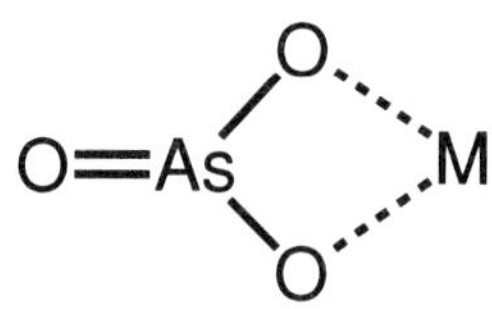

Scheme 3.18 The proposed structure of $M^+AsO_3^-$ in the vapour phase.

In a given pnictogen compound, substitution by another pnictogen is often possible, giving rise to mixed species. The tendency is particularly marked with the phosphorus and arsenic diad; attention has already been drawn to their mixed oxides in Section 3.4.1.

The arsenatophosphate anion $[AsPO_7]^{4-}$ is formed preferentially when equimolar arsenate and phosphate precursors are heated together rather than a random mixture of $[As_2O_7]^{4-}$, $[P_2O_7]^{4-}$ and $[AsPO_7]^{4-}$:

$Na_2HPO_4 + NaHAsO_4 \xrightarrow{Heat} Na_4$ 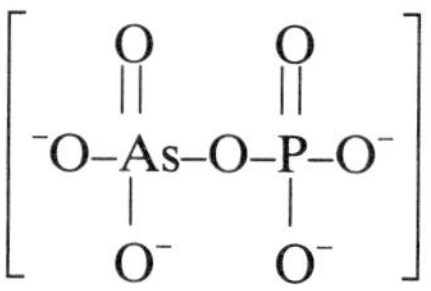 [3.73]

By heating NaH_2PO_4 and NaH_2AsO_4 (in a mixture of 1:2 mole ratio) one obtains the following species:

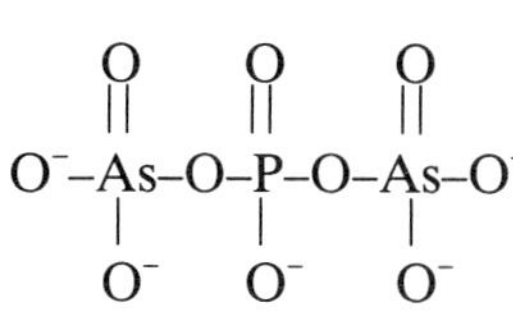

whereas in a mixture of these in a 2:1 mole ratio one obtains $[AsPO_7]^{4-}$ and $[P_2O_7]^{4-}$ and ultimately the random polyarsenatophosphate $[AsPO_3]_n^{n-}$ [231].

In the case of tetra-arsenates, systematic replacement by phosphorus is possible. This has been demonstrated by Grunze and Thilo [232] in an X-ray powder diffraction study of condensed magnesium arsenato-phosphates. When magnesium hydrogenarsenate monohydrate is heated the water of crystallization is lost, followed by the condensation of the arsenate to the polymeric meta-arsenate. Stronger heating results in the formation of the pyroarsenate:

$$Mg[HAsO_4]_2H_2O \xrightarrow{80°C-150°C} Mg[HAsO_4]_2 \xrightarrow{200°C-600°C} Mg[AsO_3]_2 \xrightarrow{>600°C} Mg_2As_2O_7] \quad [3.74]$$

If H_3PO_4 is substituted for a proportion of the As_2O_5 (that is, H_3AsO_4) used in the preparation of $Mg[H_2AsO_4]_2$, the compound $Mg[H_2(As,P)O_4]_2$ is obtained. On heating this product to 500°C either a tetrameric anion is obtained, $Mg_2[(As,P)_4O_{12}]$, or a polymeric species $Mg[(As,P)O_3]_2$ depending on the As:P ratio. High arsenic values result in the polymeric material (As:P of 1:1 or greater), whereas lower proportions of arsenic (As:P of 1:2–1:9) give tetrameric species. The structures were proposed on the basis of powder X-ray diffraction patterns. The tetrameric anion is shown in Scheme 3.19(a), and the simplified form (the ring atoms only) is illustrated in Scheme 3.19(b).

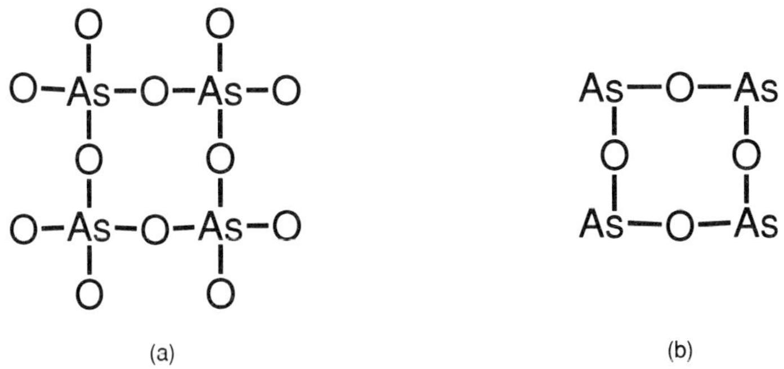

Scheme 3.19 (a) The structure of the tetrameric anion $[As_4O_{12}]^{4-}$; (b) the structure of the tetrameric anion $[As_4O_{12}]^{4-}$ in simplified form, showing only the ring atoms.

It has been proposed that the mixed anion $[(As,P)_4O_{12}]^{4-}$ undergoes stepwise substitution from $[As_4O_{12}]^{4-}$ through $[(As,P)_4O_{12}]^{4-}$ to $[P_4O_{12}]^{4-}$ as shown in Scheme 3.20. A similar scheme was proposed by Dostal and Kocman [233] for $Tl_4\,[(As,P)_4O_{12}]$. Mixtures of NaH_2PO_4 and NaH_2AsO_4 when fused together followed by quenching form glasses. The resulting

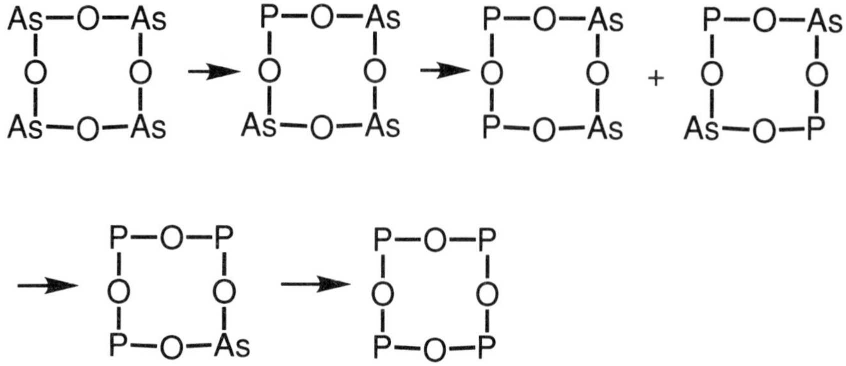

Scheme 3.20 The stepwise substitution from $[As_4O_{12}]^{4-}$ through $[(As,P)_4O_{12}]^{4-}$ to $[P_4O_{12}]^{4-}$.

polyarsenatophosphates consist of corner-sharing AsO_4 and PO_4 tetrahedra situated randomly in chains.

Dissolution in water results in hydrolysis to monomeric AsO_4^{3-} anions and a mixture of polyphosphate anions, graphically contrasting the vulnerability of the polyarsenate anions to hydrolysis [231], with the stability of the polyphosphate anions under identical conditions. In compounds where the arsenate and phosphates are isomorphous the AsO_4^{3-} may randomly replace PO_4^{3-} in a crystal. The inorganic component of human bone is phosphorus hydroxylapatite, $Ca_{10}(OH)_2(PO_4)_6$. In chronic arsenic poisoning random replacement of PO_4^{3-} by AsO_4^{3-} has been observed.

An interesting ordered mixed anionic species was prepared by Averbuch-Pouchot, Boubig and Durif [234], from a stoichiometric solution of ammonium sulphate and arsenic acid:

$$(NH_4)_2SO_4 + H_3AsO_4 \rightarrow (NH_4)_2H_3AsO_4SO_4 \qquad [3.75]$$

The SO_4 and AsO_4 tetrahedra are ordered in the crystal lattice, linked by hydrogen bonding, as $[H_3AsO_4SO_4]^{2-}$ groups.

p-Block arsenates. Many main group elements form ternary compounds with arsenic and oxygen. They often have formulae which result in the appellation 'arsenate,' whereas their real nature is that of a mixed oxide rather than a salt. Boron arsenate, $BAsO_4$, is taken as the archetypal element-arsenic mixed oxide.

Boron arsenate crystallizes in the tetragonal system in a β-cristobalite-type (SiO_2) structure, in which boron and arsenic alternate in the silicon sites. The AsO_4 and BO_4 tetrahedra are corner linked. A high-temperature and high-pressure modification has been prepared which remains stable on pressure release:

$$H_3BO_3 + H_3AsO_4 \rightarrow \underset{\text{(tetragonal)}}{BAsO_4} \xrightarrow[\text{(85 000 atm)}]{1000°C–1200°C} \underset{\text{(hexagonal)}}{BAsO_4} \qquad [3.76]$$

The hexagonal lattice is analogous to quartz [235]. The mineral terruggite, $Ca_4Mg[AsB_6O_{11}(OH)_6]_2$, contains an arsenic–boron heteropoly anion [236]. Aluminium arsenate, $AlAsO_4$, and gallium arsenate, $GaAsO_4$, are similar to $BAsO_4$, that is, they are quartz-type structures [237]. The high-pressure/high-density form of $AlAsO_4$ however has a rutile-type structure in which aluminium is six-coordinate (AlO_6 octahedra). Aluminium arsenate occurs naturally as the mineral mansfieldite.

An examination of the system B_2O_3–Al_2O_3–As_2O_5–Na_2O by Jouini and Driss [238] led to the isolation of $Na_2AlBAs_4O_{14}$ with a new aluminoboroarsenate anion $[Al_2B_2As_8O_{28}]^{4n-}$. The structure is built up from centrosymmetrical $B_2As_4O_{18}$ rings with pyroarsenate branches. The AlO_6 octahedra form a three-dimensional framework via the bridging oxygen atoms. Jouini and Driss [239] also separated a phase of composition $Na(Al_{1.5}As_{0.5})(As_2O_7)_2$ from the ternary system Na_2O–As_2O_5–Al_2O_3. The structure is composed of AlO_6 octahedra corner-sharing with pyroarsenate As_2O_7 moieties. The layers are linked by *trans* AlO_6 octahedral bridges through Al–O–As bonds. The tunnels formed in the three-dimensional framework are occupied by the sodium cations. One quarter of the aluminium sites is occupied by arsenic in the relatively rare AsO_6 coordination mode.

The heavier main group elements form more metallic type arsenates, although the dividing line between the mixed oxide type structure and the ionic metallic structures is not clear-cut.

The formula of tin(II) *ortho*arsenate $2SnHAsO_4{\cdot}H_2O$ has been challenged. Nabar and Dalvi [240] adduce evidence to support the formula $Sn_2O(H_2AsO_4)$, with an Sn–O–Sn moiety bridging $O_2As(OH)_2$ groups. The complete structure is still unknown.

The lead arsenates are of diminishing importance in agriculture owing to their double environmental negatives – both lead and arsenic either persist in the soil or leach into the water-table. Lead hydrogen arsenate, $PbHAsO_4$, was the most important of these compounds. It is insoluble in water but may be dissolved in nitric acid. Partial hydrolysis to the basic lead arsenate occurs in the presence of strong alkali or acids:

$$5PbHAsO_4 + H_2O \rightarrow Pb(OH)(H_2AsO_4){\cdot}Pb_3(AsO_4)_2{\cdot}Pb(H_2AsO_4)_2 \qquad [3.77]$$

Some other basic lead arsenates that have been reported include $5PbO{\cdot}Pb_3(AsO_4)_2$, $Pb(OH)_24Pb_3(AsO_4)_2$, $PbO{\cdot}Pb_3(AsO_4)_2$ and $Pb(OH)_2$ $9Pb_3(AsO_4)_2$. Mimetestite, georgiadesite and heliophyllite are among the minerals containing lead arsenate. Bismuth arsenate, the mineral rooseveltite, has a monazite-type ($CePO_4$) structure with distortions attributed to the influence of the Bi(III) lone pair [241].

When heated above 280°C the white lead hydrogen arsenate loses water and forms the yellow condensed pyroarsenate:

$$2PbHAsO_4 \xrightarrow{>280°C} Pb_2As_2O_7 + H_2O \qquad [3.78]$$

Lead dihydrogen arsenate forms an acid pyroarsenate:

$$Pb(H_2AsO_4)_2 \xrightarrow{140°C} Pb_2H_2As_2O_7 + H_2O \qquad [3.79]$$

Although condensed arsenate species are usually hydrolysed in aqueous environments, it is not always the case. In the system Tl_2O–As_2O_5–H_2O two complex phases separate besides TlH_2AsO_4; namely, $Tl_2O{\cdot}2As_2O_5{\cdot}3H_2O$ and $2Tl_2O{\cdot}As_2O_5{\cdot}H_2O$ or $Tl_2H_4As_4O_{13}{\cdot}H_2O$ and $Tl_4As_2O_7{\cdot}H_2O$ [242, 243].

d-Block arsenates. A large number of transition metal arsenates, pyroarsenates and related condensed anionic species has been prepared and a significant proportion of these has had an X-ray crystal structure elucidated. The variety is illustrated here by the cobalt arsenates. When cobalt(II) meta-arsenate pentahydrate is heated in air, after the dehydration stage, As_2O_5 is lost in stages to give finally cobalt(II) oxide [244]. Equation [3.80] shows part of the CoO–As_2O_5 system, the conventional (to chemists) notation is used together with the old oxide notation (given underneath in braces) which emphasizes the arsenic loss:

$$\underset{\{CoO{\cdot}As_2O_5{\cdot}5H_2O\}}{Co(AsO_3)_2{\cdot}5H_2O} \xrightarrow[\text{in air}]{\text{Heat to 1000°C}} \underset{\substack{\{3CoO{\cdot}As_2O_5\}\\ \text{(deep pink in colour)}}}{Co_3(AsO_4)_2} + \underset{\substack{\{(c.\ 4CoO)As_2O_5\}\\ \text{(purple crystals)}}}{Co_{6.95}As_{3.62}O_{16}} \xrightarrow[\textit{in vacuo}]{850°C}$$

$$\underset{\{\text{'}6CoO{\cdot}As_2O_5\text{'}\}}{\text{'}Co_6As_2O_{11}\text{'}} \xrightarrow[\textit{in vacuo}]{1000°C} \underset{\{CoO\}}{CoO} \qquad [3.80]$$

Note: the phase '$6CoO{\cdot}As_2O_5$' was shown to be non-stoichiometric and is more nearly 5.44 $CoO{\cdot}As_2O_5$.

Two polymorphs of cobalt arsenate have been described, a tetragonal form prepared by Gopal, Rutherford and Robertson [245]:

$$Co_2As_2O_7 + As_2O_5 \xrightarrow[\text{sealed vessel}]{\substack{1250°C,\ N_2\\ \text{(atmospheric pressure)}}} Co_3(AsO_4)_2\ \text{(tetragonal)} \qquad [3.81]$$

and a high-temperature monoclinic modification prepared by Jasper-Tönnes and Müller-Buschbaum [246] from cobalt oxalate and arsenic(V) oxide:

$$CoC_2O_4 + As_2O_5{\cdot}\tfrac{5}{3}H_2O \xrightarrow[\text{(b) laser heating to 1800°C}]{\text{(a) 1 h at 550°C}} Co_3(AsO_4)_2 \text{ (monoclinic)} \quad [3.82]$$

Arsenic and a transition metal may form a complex anionic array as in the series $MTiOAsO_4$ (M = K, Rb, Tl) [247]. The members of this series are prepared by prolonged heating of the mixed solids; first for 12 h at 400°C and finally for 48 h at 700°C:

$$\tfrac{1}{2}M_2CO_3 + NH_4HAsO_4 + TiO_2 \xrightarrow[\text{(b) 48 h at 700°C}]{\text{(a) 12 h at 400°C}} MTiOAsO_4 + 2NH_3 + \tfrac{1}{2}CO_2 + \tfrac{3}{2}H_2O \quad [3.83]$$

The product is cooled slowly to afford a crystalline material of potential use in laser technology. The structure was revealed by X-ray crystallography to consist of parallel chains of AsO_4–TiO_6–AsO_4 polyhedra bridged by AsO_4 tetrahedra.

Vanadyl arsenate ($VOAsO_4$) has a layered structure and forms intercalation compounds [248]. When it is heated with pyridine at 150°C a greenish-yellow adduct is formed:

$$VOAsO_4 + C_5H_5N \xrightarrow[\text{sealed vessel}]{\text{10 d at 150°C}} VOAsO_4Py_{0.9} \quad [3.84]$$

If the vanadyl arsenate hydrate is used a greater uptake is observed and the product is orange:

$$VOAsO_4{\cdot}2H_2O + C_5H_5N \xrightarrow{\text{Reflux for 6 d}} VOAsO_4Py_{2.5} \quad [3.85]$$

The majority of transition metal arsenates are very insoluble and so precipitation techniques may be used, for example a nitric acid solution of As_2O_5 when treated with $Fe(NO_3)_3$ solution precipitates the light-brown $FeAsO_4$. To give a product suitable for powder neutron diffraction, the dried filter cake is calcined at 800°C for a week [249]. More complex arsenates can be synthesized by these precipitation methods, for example $AgCu_3Cu(AsO_4)_3$. The stoichiometric reaction mixture is heated with a little water in an autoclave for a week at 300°C [250].

$$As_2O_5 + AgNO_3 + CuSO_4{\cdot}5H_2O \xrightarrow[\text{autoclave}]{\text{7 d at 300°C}} AgCu_3Cu(AsO_4)_3 \quad [3.86]$$

The copper was found by X-ray analysis to be $1 \times CuO_4$ square planar and $3 \times CuO_6$ Jahn–Teller distorted octahedra. The silver is in its relatively rare eight coordinate form (AgO_8), but the arsenate is in the normal tetrahedral geometry (AsO_4).

Tantalum arsenate has been recommended as an ion exchange material [251].

In qualitative analysis the formation of the chocolate-brown precipitate of Ag_3AsO_4 has been used to distinguish between AsO_4^{3-} and PO_4^{3-}:

$$AsO_4^{3-}{}_{(aq)} + AgNO_{3(aq)} \longrightarrow Ag_3AsO_4 \text{ (chocolate brown)} \quad [3.87]$$

$$PO_4^{3-}{}_{(aq)} + AgNO_{3(aq)} \longrightarrow Ag_3PO_4 \text{ (yellow)} \quad [3.88]$$

Silver arsenate has found applications in electronics owing to the high mobility of Ag^+ in arsenate glasses. The Ag^+ mobility has been found to be especially high in the mixed iodo salt $Ag_4(I_4)AsO_4$. The structure of these non-crystalline compounds is unknown [252].

Manganese(III) arsenate hydrate ($MnAsO_4{\cdot}1.2H_2O$) is of interest because of the unusual water stoichiometry which varies only between 1.1–1.2 mol mol^{-1} of arsenate. This water is lost only on strong heating (>270°C) showing that it is powerfully held by the structure. A further unusual feature of this salt is that on stronger heating (580°C), forming the pyroarsenate, the manganese is reduced [253]. The overall reaction is

$$MnAsO_4{\cdot}1.2H_2O \rightarrow 0.5Mn_2As_2O_7 + 1.2H_2O + 0.25O_2 \quad [3.89]$$

A chromium arsenate $[Cr_2H_2(As_2O_7)(As_4O_{12})]$ contains both the pyroarsenate ion and a cyclic tetramer [254].

These arsenochromates are hydrolysed in aqueous solution and unlike the case with arsenoheteropolyanions of group 5 or 6 (e.g. Mo) CrO_4 tetrahedra may replace the AsO_4 in, for example, $BaHCr_2AsO_{10}$ [255] and $K_2HCr_2AsO_2$ [256].

Arsenate- or pyroarsenate-type structures are obtained depending on the stoichiometry when K_2CO_3, $(NH_4)(H_2AsO_4)$ and molybdic acid (85%) are heated to 350°C, eliminating NH_3, H_2O and CO_2. The products are calcined at 600°C–650°C giving $KMoO_2AsO_4$ [257] or $K_2MoO_2As_2O_7$. The structures consist of AsO_4 tetrahedra and MoO_6 octahedra.

There is a large class of heteropolyanions formed between the pnictogens and group 6 and, to a lesser extent, group 5 transition metals. Although the phosphomolybdates and phosphotungstates predominate there is a considerable number of the arsenic compounds. The pnictogen moiety, usually an EO_4 tetrahedron, is either encapsulated by MO_6 octahedra or forms part of a surface feature of the polyanion. In general these heteropolyanions form air-stable aqueous solutions. The anions are large (up to 25 Å) and are of high ionic weight, typically 1000–10 000. Pope has written an excellent monograph on these materials [258]. The formation of arsenic heteropolyanions is used in the analysis of arsenic; for example $[AsMo_{12}O_{40}]^{3-}$ is used in the indirect atomic adsorption spectroscopy [259] of arsenic.

The monoarsenic encapsulated species, Keggin compounds, for example α-$[AsW_{12}O_{40}]^{3-}$ are much less resistant to hydrolysis than are the phosphorus analogues. The Keggin structures are compact and highly symmetrical (T_d) consisting of four groups of three edge-sharing octahedra (W_3O_{13}) surrounding the central AsO_4 tetrahedron. The W_3O_{13} are connected to each other and the AsO_4 by corner-sharing. Theoretically, each of the W_3O_{13} groups may be independently rotated by $\pi/3$ affording five possible isomers.

Another common class of arsenic heteropolyanions is based on the Dawson structure $[As_2W_{18}O_{62}]^{6-}$. These are more stable than the Keggin compounds, being formed, for example, when aqueous sodium tungstate and excess H_3AsO_4 are refluxed. The two independent AsO_4 octahedra are encapsulated by WO_6 octahedra in two isomeric forms. The more stable α-isomer has virtual D_{3h} symmetry.

Apart from their formation, the buried AsO_4 tetrahedra do not exhibit much arsenic chemistry. There are arsenic heteropolyanions with exposed arsenic moieties, for example $[(HOAsO_3)_4Mo_4O_{10}]^{4-}$ [260] (Scheme 3.21) and $[(HOAsO_3)_6V_4O_6]^{4-}$ [261]. They are easily prepared, for example the arsenopolyvanadate is forming by refluxing sodium metavanade and 50% aqueous arsenic acid for 10 minutes:

$$4NaVO_3 + 6H_3AsO_4 \xrightarrow[100°C\ \text{for 10 min}]{H_2O} Na_4[V_4O_6(HOAsO_3)_6] + 6H_2O \quad [3.90]$$

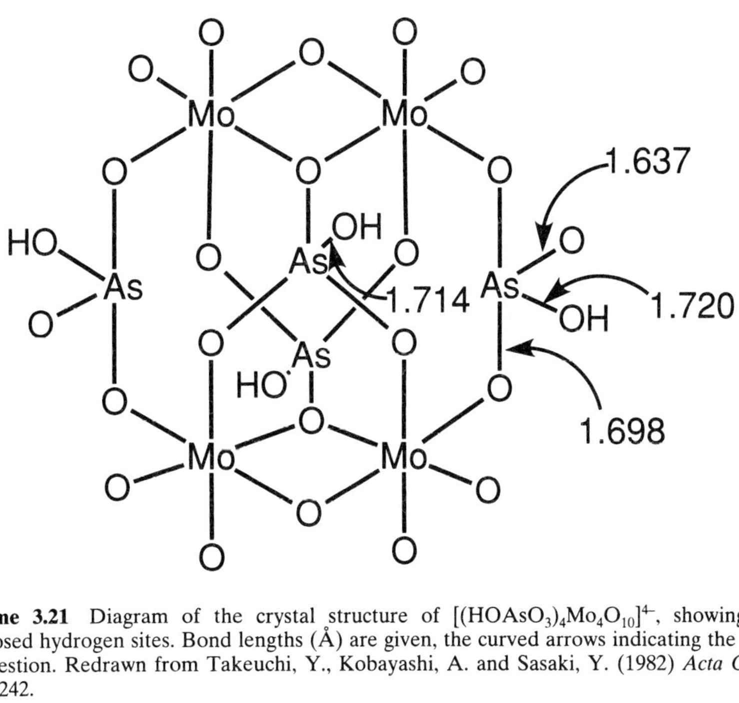

Scheme 3.21 Diagram of the crystal structure of $[(HOAsO_3)_4Mo_4O_{10}]^{4-}$, showing the proposed hydrogen sites. Bond lengths (Å) are given, the curved arrows indicating the bond in question. Redrawn from Takeuchi, Y., Kobayashi, A. and Sasaki, Y. (1982) *Acta Cryst.*, **B38**, 242.

After adding NH_4CO_3 the orange prisms separate on cooling.

Finally, mention must be made of the 'heteropoly blues' since they are used in the analytical chemistry of arsenic. The intense blue colour produced in reductive processes of heteropolyanions is a function of the MO_6 octahedra, more specifically of those anions which have MO_6 groups with one terminal (non-bridging) oxygen. Pope has discussed the underlying processes [258]. The analytical use is based on the strong absorption maximum at *c.* 840 nm of the heteropolyarsenomolybdate, the so-called molybdenum blue complex [217].

f-Block arsenates. The lanthanide arsenates $LnAsO_4$ (Ln = lanthanide metal) exist in two crystalline types, a monoclinic monazite-type in the case of the larger lanthanides and a tetragonal zircon-type for the earlier members [262]. In the corresponding phosphate series the change in structural type occurs at terbium phosphate. However, with the larger arsenate the change occurs at promethium:

$$\underset{\text{(black-violet)}}{PmAsO_4{\cdot}xH_2O} \xrightarrow{800°C} \underset{\text{(red)}}{PmAsO_4} \qquad [3.91]$$

$PmAsO_4$ has the zircon-type ($ZrSiO_4$) structure (tetragonal, space group $I4_1/amd$) [263].

Several uranium(VI) arsenates occur as minerals and several have been prepared in the laboratory [264]:

$$UO_2(OAc)_2 + H_3AsO_4/H_2O \rightarrow UO_2HAsO_4{\cdot}nH_2O \quad n = 3, 4, 12 \qquad [3.92]$$

(the mineral trogerite is formed when $n = 12$);

$$UO_2(OAc)_2 + Me_3AsO_4/H_2O \rightarrow MUO_2AsO_4{\cdot}nH_2O \quad M = Na, K, NH_4 \qquad [3.93]$$

$$UO_2(OAc)_2 + M_3(AsO_4)_2/H_2O \rightarrow M\,[UO_2AsO_4]_2{\cdot}nH_2O \quad M = Ca, Cu \qquad [3.94]$$

(the mineral uranospinite is formed when M = Ca).

Plutonium(VI) forms an analogous series of arsenates, $PuO_2HAsO_4{\cdot}nH_2O$ ($n = 2, 3$), as greyish-white precipitates:

$$PuO_3{\cdot}xH_2O + H_3AsO_{4(aq)} \xrightarrow{pH\ 1.5} PuO_2HAsO_4{\cdot}nH_2O \qquad [3.95]$$

The H of PuO_2HAsO_4 is exchangeable with monovalent (H^+, K^+, Rb^+, NH_4^+) and divalent (Ca^{2+}, Sr^{2+}) cations to give pale-green to olive-green compounds (except for the beige calcium salt) of formulae $MPuO_2AsO_4{\cdot}yH_2O$ and $M(PuO_2AsO_4)_2{\cdot}zH_2O$ [265]. The various phases have been characterized by X-ray diffraction. The pale-blue $NpO_2HAsO_4{\cdot}4H_2O$ is similarly formed [266]. The light-green $PuAsO_4$ is precipitated from acid solutions of Pu^{3+} by arsenates [267]:

$$Pu^{3+} + (NH_4)_2HAsO_4 \xrightarrow{0.1\ \text{M HCl}} PuAsO_4 \qquad [3.96]$$

The pale-rose $AmAsO_4$ was formed in a solid-state reaction [267]:

$$Am(NO_3)_3 + (NH_4)_2HAsO_4 \xrightarrow{1000°C} AmAsO_4 \qquad [3.97]$$

The lemon-yellow americium(VI) arsenates have been prepared as shown in reaction [3.98] below, the pH in the final state being critical [268]:

$$Am(OH)_3 \xrightarrow{M_2S_2O_8} Am(VI) \xrightarrow{H_3AsO_{4(aq)}} HAmO_2AsO_{4(aq)} \xrightarrow[\text{pH 3.5}]{M_2CO_3} MAmO_2AsO_4{\cdot}yH_2O \qquad [3.98]$$

Arsenate hydrates. Metal arsenates commonly form hydrates, for example five hydrates have been characterized in the $MgO–As_2O_5–H_2O$ system $Mg_3(AsO_4)_2{\cdot}nH_2O$ ($n = 1$, 2.5, 7, 8, 10) [269]. In the $CaO–As_2O_5–H_2O$ system there are several mineral phases. Of the neutral salts, $Ca_3(AsO_4)_2{\cdot}nH_2O$, the synthetic phases $n = 0$, 2, 3, 4, 5, 8 have been prepared [270]. The secondary minerals phaunouxite ($n = 11$) and rauenthalite ($n = 10$) form an interesting pair. Phaunouxite readily loses water in dry air to form rauenthalite:

$$\underset{\text{(phaunouxite)}}{Ca_3(AsO_4)_2{\cdot}11H_2O} \xrightarrow{\text{dry air}} \underset{\text{(rauenthalite)}}{Ca_3(AsO)_2{\cdot}10H_2O} \qquad [3.99]$$

indeed samples of the former always contain the latter owing to *in situ* dehydration [271]. In this solid-state reaction the two-dimensional elements of the structure (layers) are conserved.

The mineral machatschkite was originally thought to be a member of this series ($n = 9$) [272], but it has been shown that the variable phosphorus–sulphur content is an essential feature of the structure [273]. There are many mineral species of the acidic calcium arsenates, including weilite ($CaHAsO_4$), with the related dihydrate haidingerite ($CaHAsO_4{\cdot}2H_2O$), and the hydrates of $Ca_5(HAsO_4)_2(AsO_4)_2{\cdot}nH_2O$, inter alia sainfeldrite ($n = 4$), vladimirite ($n = 5$), and the dimorphous ferrarisite and guerinite ($n = 9$). There is a partial replacement of calcium by magnesium in picropharmacolite, $Ca_4Mg(HAsO_4)_2(AsO_4)_2{\cdot}11H_2O$.

$NaH_2AsO_4{\cdot}H_2O$ has been shown [274] to contain $AsO_2(OH)_2^-$ anions linked into sheets by O–H···O hydrogen bonds. The H_2O is also involved in sheet formation; $NaO_4(H_2O)_2$ octahedra join the sheets by sharing edges.

Table 3.30 Comparison of lammerite and synthetic copper(II) arsenate.

	Lammerite	Copper(II) arsenate
Formula	$Cu_3(AsO_4)_2$	$Cu_3(AsO_4)_2$
Crystal form	monoclinic	monoclinic
Lattice parameters		
a (Å)	5.079	6.327
b (Å)	11.611	8.642
c (Å)	5.394	11.313
β (°)	111.72	92.04
Z	2	4
Space group	$P2_1/a$	$P2_1/c$
Calculated density (g cm^{-3})	5.263	5.032

Magnesium hydrogen arsenate heptahydrate has been shown by single-crystal X-ray analysis [275] to have a struvite-type structure, that is, an interpenetrating layer-type structure with open channels at the centre. Owing to the larger AsO_4 tetrahedron these channels are not as prominent in $MgHAsO_4{\cdot}7H_2O$ as in $MgNaPO_4{\cdot}7H_2O$.

When an aqueous solution of K_2HAsO_4 and $MgAsO_4{\cdot}7H_2O$ is adjusted to pH 7.0 with KOH and cooled to 5°C, thin plates of a water-rich hydrate, $Mg_2KH(AsO_4)_2{\cdot}15H_2O$, are obtained. The structure consists of $Mg(H_2O)_6$ octahedra hydrogen bonded to $H(AsO_4)_2$ units with K^+/H_2O occupying channels in the lattice [276]. In gallium arsenate dihydrate IR and Raman spectroscopy indicate that the lattice water is held as H_3O^+ and OH^- [277].

$HUO_2AsO_4{\cdot}4H_2O$ [268], one of the class of $M^{2+}{}_{1/z}\,(UO_2)(AsO_4){\cdot}nH_2O$, has been shown to have a very high proton conductivity of 0.6 s m^{-1} at 310 K [278]. When the acid proton is replaced by lithium ($LiUO_2AsO_4{\cdot}4D_2O$), the lithium is tetrahedrally coordinated to four water molecules as shown by powder neutron diffraction [279]. This adjustment of the water molecules in the structure results in loss of proton conductivity. The larger cations in $KUO_2AsO_4{\cdot}3H_2O$ [280] and $NH_4UO_2AsO_4{\cdot}3H_2O$ [280] replace H_3O^+ in the $HUO_2AsO_4{\cdot}4H_2O$ water network, also reducing the proton conductivities. These structures are illustrative of the divergence that can occur between the solid-state properties of compounds and their formal chemical classification.

The hydrates of pyroarsenates are rare since the $[As_2O_7]^{4-}$ anion is usually prepared by dehydrating $[AsO_4]^{3-}$ salts. However, pyroarsenates can be prepared under pressurized hydrothermal conditions; for example $Mg(OH)_2$ and aqueous As_2O_5 when heated at 350°C in an autoclave for 4 d afford $Mg_2As_2O_7{\cdot}2H_2O$ [281]. Transition metal derivatives have also been prepared, for example $Co_2As_2O_7{\cdot}2H_2O$ [282].

Arsenate-containing minerals. Arsenate minerals are often termed secondary minerals because they are formed in the upper parts of a

Table 3.31 Arsenate (AsO_4) containing minerals characterized by X-ray analysis.

Mineral	Formula	(As–O) bond length (Å)	O–As–O bond angle (°)	Reference
Abernathyite	$K(UO_2(AsO_4)\cdot 3H_2O$	1.68[a]	108.5	Ross, M. and Evans, H.T. (1964) *Amer. Min.*, **46**, 1578
Adamite	$Zn_2(OH)AsO_4$	1.68[a]		Hill, R.J. (1976) *Amer. Min.*, **61**, 979
Agardite[b]	$(Ca, Y)Cu_6(OH)_6(AsO_4)_3\cdot 3H_2O$	1.66–1.71		Aruga, A. and Nakai, I. (1985) *Acta Cryst.*, **C41**, 161
Allactite	$Mn_7(OH)_8(AsO_4)_2$	1.65–1.72		Moore, P.B. (1968) *Amer. Min.*, **53**, 733
Angelellite	$Fe_4O_3(AsO_4)_2$	1.67–1.69		Moore, P.B. and Araki, T. (1978) *Neues. Jb. Miner.*, **132**, 91
Arsenbrackebushite	$Pb_2(Fe, Zn)(AsO_4)_2(OH, H_2)$	1.67–1.73		Hofmeister, W. and Tillmanns, E. (1978) *Tsch. Min. Petr. Mitt.*, **25**, 153
Arsenostruvite	$(NH_4)MgAsO_4\cdot 6H_2O$	1.69–1.70		Ferraris, G. and Frankchini-Angela, M. (1973) *Acta. Cryst.*, **B29**, 859
Arthurite	$CuFe_2(OH)_2(AsO_4)_2\cdot 4H_2O$	1.670–1.721		Keller, P. and Hess, H. (1978) *Neues, Jb. Miner. Mh.*, **133**, 291
Bayldonite	$(Cu, Zn)_3Pb(OH)_2(AsO_4)_2$	1.67–1.72		Ghose, S. and Wan, C. (1979) *Acta Cryst.*, **B35**, 819
Bergslagite	$CaBe(OH)AsO_4$	1.664–1.698		Hansen, S., Fälth, L. and Johnsen, O. (1984) *Z. Krist.*, **166**, 73
Brandtite	$MnCa_2(AsO_4)_2\cdot 2H_2O$	1.61–1.76		Dahlman, B. (1952) *Ark. Min. Geol.*, **1**, 339
Brassite	$MgHAsO_4\cdot 4H_2O$	1.67; 1.74[c]		Protas, J. and Gindt, R. (1976) *Acta Cryst.*, **B32**, 1460
Caberite	$(Ni, Mg)_3(AsO_4)_2\cdot 8H_2O$	1.68–1.70		Giuseppettei, G. (1982) *Bull. Miner.*, **105**, 333
Cahnite	$Ca_2B(OH)_4AsO_4$	1.68[a]		Prewitt, C.T. and Buerger, M.J. (1961) *Amer. Min.*, **46**, 1077
Chalcophyllite	$Cu_9Al(OH)_{12}(SO_4)_{1.5}(AsO_4)_2\cdot 18H_2O$	1.62–1.68		Sabelli, C. (1980) *Z. Krist.*, **151**, 129
Chlorotite	$LnCu_6(OH)_6(AsO_4)_3\cdot 3H_2O$	1.67–1.71		Hess, H. (1983) *Neues. Jb. Miner., Mh.*, part 9, 385
Conichalcite	$CaCu(OH)AsO_4$	1.60–1.74		Qurashi, M.M. and Barnes, W.H. (1963) *Can. Min.*, **7**, 561
Euchorite	$Cu_2(OH)AsO_4\cdot 3H_2O$	1.688[a]	109.4[a]	Hawthorne, F.C. and Eby, R.K. (1989) *Acta Cryst.*, **C45**, 1479
Eveite	$Mn_2(OH)AsO_4$	1.65–1.73		Moore, P.B. (1972) *Amer. Min.*, **57**, 593
Ferrarsite	$Ca_5H_2(AsO_4)_4\cdot 9H_2O$	1.665–1.739		Ferraris, G., Catti, M. and Chiari, G. (1980) *Bull. Miner.*, **103**, 541
Flinkite	$Mn_3(OH)_4AsO_4$	1.64–1.76		Moore, P.B. (1967) *Amer. Min.*, **52**, 1603

Table 3.31 *continued*

Mineral	Formula	(As–O) Bond Length (Å)	O–As–O bond angle (°)	Reference
Fluckite	$CaMn(HAsO_4)_2{\cdot}2H_2O$	1.65–1.67; 1.74[c]		Bari, H., Cesbron, F., Permingeat, F. and Pillard, F. (1980) *Bull. Miner.*, **103**, 122
Fornacite	$Pb_2(CuFe)[CrO_4(As, P)O_4(OH)]$	1.57–1.65		Cocco, G., Fanfarni, L. and Zanazzi, P.F. (1967) *Z. Krist.*, **124**, 385
Haidingerite	$CaHAsO_2{\cdot}2H_2O$	1.66–1.73	104.4–113.3	Ferraris, G., Jones, D.W. and Yerkess, J. (1972) *Acta Cryst.*, **B28**, 209
Hedyphane	$Ca_2Pb_3(Cl)(AsO_4)_3$	1.66–1.73		Rouse, R.C., Dunn, P.J. and Peacor, D.R. (1984) *Amer. Miner.*, **69**, 920
Hematolite	$(Mn, Mg, Al)_{15}(OH)_{23}(AsO_3)(AsO_4)_2$			Moore, P.B. and Araki, T. (1978) *Amer. Min.*, **63**, 150
Koritnigite	$ZnHAsO_4{\cdot}H_2O$	1.65–1.69; 1.73–1.74[c]		Keller, P., Hess, H. and Riffel, H. (1980) *Neues. Jb. Miner.*, **138**, 316
Kottigite	$(Zn, Co, Ni)_3AsO_4{\cdot}8H_2O$	1.669–1.697		Hill, R.J. (1979) *Amer. Min.*, **64**, 376
Krautite	$Mn(HAsO_4){\cdot}H_2O$	1.65–1.68; 1.73[c], 1.77[c]		Catti, M. and Franchini-Angela, M. (1979) *Amer. Miner.*, **64**, 1248
Lammerite	$Cu_3(AsO_4)_2$	1.67–1.71; 1.690[a]	109.3[a]	Hawthorne, F.C. (1986) *Amer. Miner.*, **71**, 206
Legrandite	$Zn_2(OH)AsO_4{\cdot}H_2O$	1.67–1.69		McLean, W.J., Anthony, J.W., Finney, J.J. and Laughon, R.B. (1971) *Amer. Min.*, **56**, 1147
Liroconite	$Cu_2Al(OH)_4(As, P)O_4{\cdot}4H_2O$	*c.* 1.60		Kolesova, R.V. (1967) *Sov. Phys. Doklady*, **12**, 538 (in English)
Mapimite	$Zn_2Fe_3(OH)_4(AsO_4)_3{\cdot}10H_2O$	1.66–1.70		Ginderow, D. and Cesbron, F. (1981) *Acta Cryst.*, **B37**, 1040
Metazeunerite	$Cu(UO_2)_2AsO_4{\cdot}8H_2O$	1.77		Hanic, F. (1960) *Czech. J. Phys.*, **10**, 169
Nickelaustinite	$Ca(Ni, Zn)(OH)AsO_4$	1.68[a]		Cesbron, F.P., Ginerow, D., Girand, R. *et al.* (1987) *Canad. Miner.*, **25**, 401
Olivente	$Cu_2(OH)AsO_4$	1.62–1.73		Toman, K. (1977) *Acta Cryst.*, **B33**, 2638
Paradamite	$Zn_2(OH)AsO_4$	1.68–1.70		Hawthorne, F.C. (1979) *Acta Cryst.*, **B35**, 720
Pharmacolite	$CaHAsO_4{\cdot}2H_2O$	1.661–1.678; 1.729[c]		Ferraris, G., Jones, D.W. and Yerkess, J. (1971) *Acta Cryst.*, **B27**, 349
Preisingerite	$Bi_3O(OH)(AsO_4)_2$	1.66–1.71		Mereiter, K. and Bedlivy, D. (1982) *Amer. Miner.*, **67**, 833

Phaunouxite	$Ca_3(AsO_4)_2 \cdot 11H_2O$	1.65–1.70		Catti, M. and Ivaldi, G. (1983) *Acta Cryst.*, **B39**, 4
Picropharmacolite	$Ca_4Mg(HAsO_4)_2(AsO_4)_2 \cdot 11H_2O$	1.60–1.77		Catti, M., Ferraris, G. and Ivaldi, G. (1981) *Amer. Miner.*, **67**, 833
Properite	$Ca_2Zn_2(AsO_4)_4 \cdot H_2O$	1.677–1.696		Keller, P., Riffel, H. and Hess, H. (1982) *Z. Krist.*, **158**, 33
Rauenthalite	$Ca_3(AsO_4)_2 \cdot 10H_2O$	1.65–1.70		Catti, M. and Ivaldi, G. (1983) *Acta Cryst.*, **B39**, 4
Retzian	$Mn_2Y(OH)_4AsO_4$	1.60[a]		Moore, P.B. (1967) *Amer. Min.*, **52**, 1603
Richelsdorfite	$Ca_2Cu_5Sb[Cl(OH)_6(AsO_4)_4] \cdot 6H_2O$	1.66–1.72		Süsse, P. and Tillmann, B. (1987) *Z. Krist.*, **179**, 323
Rooseveltite	α-$BiAsO_4$	1.67–1.70		Mereiter, K. and Bedlivy, D. (1982) *Acta Cryst.*, **B38**, 1559
Roselite	$Ca_2(Co, Mg)(AsO_4)_2 \cdot 2H_2O$	1.67–1.70		Hawthorne, F.C. and Ferguson, R.B. (1977) *Canad. Miner.*, **15**, 36
Rosslerite	$MgHAsO_4 \cdot 7H_2O$	1.67–1.71		Whitaker, A. (1973) *Z. Krist.*, **137**, 194
Sarkinite	$Mn_2(OH)AsO_4$	1.66–1.70		Dal Negro, A., Giuseppettei, G. and Pozas, J.M.M. (1974) *Tsch. Miner., Petr. Mitt.*, **21**, 246
Schultenite	$PbHAsO_4$	1.68–1.70		Effenberger, H. (1986) *Tsch. Min. Petr. Mitt.*, **35**, 157
Scorodite	$FeAsO_4 \cdot 2H_2O$	1.670–1.685		Hawthorne, F.C. (1986) *Acta Cryst.*, **B32**, 2891
Stranskiite	$CuZn_2(AsO_4)_2$	1.670–1.712		Keller, P., Hess, H. and Dunn, J.P. (1979) *Tsch. Min. Petr. Mitt.*, **26**, 167
Symplesite	$Fe_3(AsO_4)_2 \cdot 8H_2O$	1.60–1.69		Mori, H. and Ito, T. (1950) *Acta Cryst.* **3**, 1
Synadelphite	$Mn_9(OH)_9(AsO_3)(AsO_4)_2 \cdot 2H_2O$	1.68, 1.70; 1.76[d]		Moore, P.B. (1970) *Amer. Min.*, **55**, 2023
Talmessite	$Ca_2(Mg, Co)(AsO_4)_2 \cdot 2H_2O$	1.66–1.70		Catti, M., Ferraris, G. and Ivaldi, G. (1977) *Bull. Soc. Fr. Miner., Cryst.*, **100**, 230
Tilasite	$CaMg(F)AsO_4$	1.65–1.72	104–114	Bladh, K.W., Corbett, R.K., McLean, W.J. and Laughon, R.B. (1972) *Amer. Miner.*, **57**, 1880
Tsumcorite	$Pb(Fe, Zn)_2(OH, H_2O)(AsO_4)_2$	1.66–1.72		Tillmanns, E. and Gebert, W. (1973) *Acta Cryst.*, **B29**, 2789
Warkahnite	$Zn_3(AsO_4)_2 \cdot 2H_2O$	1.667–1.723		Keller, P., Riffel, H. and Hess, H. (1980) *Tsch. Min. Petr. Mitt.*, **27**, 187
Weilite	$CaHAsO_4$	1.66–1.70; 1.72[c]		Ferraris, G. and Chiari, G. (1970) *Acta Cryst.*, **B36**, 403

[a]Average. [b]Calcium-rich. [c]As–OH [d]As^{III}–O.

Table 3.32 Arsenic–oxygen bond lengths, *d*, and angles, *θ*.

Compound	*d* (As–O) (Å)	*d* (As–OH) (Å)	*θ* (As–O–O) (°)	Reference
$Na_3AsO_4 \cdot 12H_2O$	1.669		109.5[a]	Bauer, W.H. and Tillmans, E. (1971) *Acta. Cryst.*, **B27**, 2124
$Ca(H_2AsO_4)_2$	1.650	1.723	100–110, 109.2[a]	Ferraris, G., Jones, D.W. and Yerkes, J. (1972) *Acta. Cryst.*, **B28**, 2430
$CaHAsO_4 \cdot 3H_2O$	1.68[a]	1.734	115[b], 105.6–113.8[c]	Catti, M. and Ferraris, G. (1973) *Acta. Cryst.*, **B29**, 90
$NaD_2AsO_4 \cdot D_2O$	1.655, 1.696	1.731[e]	113.3[c] 107[d] 107.8[f], 114.7[f] 97.2[g]	Rendle, D.F. and Trotter, J. (1977) *Acta. Cryst.*, **B33**, 44
$BaH(AsCr_2O_{10})$		1.69[a]	121.4[h], 132.7[h] 106.7[c], 115.8[c]	Averbuch-Pouchot, M.T., Blum, D. and Guitel, J.C. (1979) *Acta Cryst.*, **B35**, 726
$K_5As_3O_{10}$		1.78[a,i] 1.65[k]	131.4[j], 124.9[j]	Hornstra, J. and Verweij, H. (1980) *Acta Cryst.*, **B36**, 1634
$Ba(OH)_6As_4O_9$	Octahedra	1.887[a,i] 1.839[i,l] 1.690[a,m]		Blum, D., Durif, A. and Guitel, J.C. (1977) *Acta Cryst.*, **B33**, 3222
$Ba(OH)_6As_4O_8$	Tetrahedra	1.767[l,m] 1.721[i] 1.690[a,m]		
$NaHAs_2O_6$	Octahedra	1.87[i,n] 1.80[i,l] 1.77[m]		Dung, N-H. and Tahar, J. (1978) *Acta Cryst.*, **B34**, 3727
$MgNH_4 \cdot AsO_46H_2O$	1.688[a]		108.7–111.3	Giuseppettei, G. (1982) *Bull. Miner.*, **105**, 333
$Na_2HAsO_4 \cdot 7H_2O$	1.662	1.728	103–113	Bauer, W.H. and Khan, A.A. (1973) *Acta Cryst.*, **B29**, 859
$Na_2HAsO_4 \cdot 7H_2O$	1.670	1.742	102–113	Ferraris, G. and Chiari, G. (1970) *Acta. Cryst.*, **B26**, 1574
$(NH_4)_2HAsO_4$	1.670	1.742	100–113, 109.4[a]	Khan, A.A., Straumanis, M.E. and James, W.J. (1970) *Acta Cryst.*, **B26**, 1889

[a]Average. [b]As–O–H. [c]O–As–O. [d]D–O–As. [e]As–OD. [f]DO–As–O. [g]DO–As–OD. [h]Cr–O–As. [i]Bridge bond. [j]As–O–As. [k]Non-bridge bond. [l]Equatorial bond. [m]Terminal atom. [n]Axial bond.

'mother lode.' The arsenate is formed by hydrothermal oxidation. For example, mimetite $Pb_5Cl(AsO_4)_3$ occurs in the upper oxidized parts of lead veins in, for instance, Cornwall, UK, and Eureka, Utah, USA. Similarly, scorodite, $FeAsO_4{\cdot}2H_2O$ occurs on a deposit of the hot springs in the Yellowstone National Park, USA. These secondary arsenate minerals are used as indicator minerals, that is they show the presence of the 'mother lode'; for example the economically unimportant mineral erythrite (cobalt bloom), $Co_3(AsO_4)_2{\cdot}8H_2O$, is used as a guide to valuable cobalt minerals, for instance cobalt arsenide, and the often associated native silver. It is not possible in a book of this nature to do other than indicate the great variety of minerals that have been characterized. This variety is born of the fact that all environmental and crustal arsenic must ultimately become arsenate.

Laboratory-produced material is often different from geologically-produced mineral, as is the case with the polymorphs of $Cu_3(AsO_4)_2$. The synthetic material [283] has a much more open structure than does the mineral [284] lammerite (Table 3.30). Selected structural data for arsenic species described in the preceding subsections are presented in Tables 3.31 and 3.32.

Oxoanion derivatives of antimonic acid. Antimonic acid is poorly characterized, and reported examples in the older literature proved to be hydrated Sb_2O_5 of unknown structure. It is known in its salts, however, which are of two main types – the 'hydrates' and the anhydrous materials. In neither case is the simple oxoanion SbO_4^{3-} found.

The 'hydrated' salts, often still written as simple hydrates, for example $LiSbO_3{\cdot}3H_2O$, were all found to contain the $[Sb(OH)_6]^-$ anion. Thus the aforementioned $LiSbO_3{\cdot}3H_2O$ should be written $Li[Sb(OH)_6]$. Some examples of the old classical formulae and their re-interpretation are given in Table 3.33.

Dehydration of these hydrates affords edge-sharing polymeric octahedra $[SbO_6]_n^{n-}$. A wide variety of compounds have been prepared with the SbO_6 octahedral unit, and these materials are usually described as mixed oxides rather than salts. These antimony transition metal mixed oxide phases find use industrially as catalysts [285].

Table 3.33 Revision of antimonate formulae.

Old formula	New formula
$Na_2H_2Sb_2O_7{\cdot}5H_2O$	$Na[Sb(OH)_6]$
$Mg(SbO_3)_2{\cdot}12H_2O$	$[Mg(H_2O)_6][Sb(OH)_6]_2$
$Cu(NH_3)_3(SbO_3)_2{\cdot}9H_2O$	$[Cu(NH_3)_3(H_2O)_3][Sb(OH)_6]_2$
$LiSbO_3{\cdot}3H_2O$	$Li[Sb(OH)_6]$
$NaF{\cdot}SbOF_3{\cdot}H_2O$	$Na[SbF_4(OH)_2]$

Oxoanion derivatives of bismuthic acid. The hydration of the ill-defined Bi_2O_5 does not yield $Bi(OH)_5$ but only the equally ill-defined $Bi_2O_5{\cdot}xH_2O$.

If a mixture of Bi_2O_3 and Na_2O is heated in the presence of dioxygen yellow-orange sodium metabismuthate $NaBiO_3$ is obtained. The metabismuthates are powerful oxidizing agents, for example they can oxidize pale-pink manganese(II) in nitric acid solution to purple permanganate. This reaction forms the basis of a qualitative test for manganese. Prior to the discovery of antibiotics sodium bismuthate found medical use in the treatment of syphilis.

3.6 Pnictogen intermetallic compounds

Intermetallic chemistry is a complex subject not helped by the taxonomic minefield set not only by the definition of 'intermetallic,' but also by the definitions of alloys, interstitial compounds, clatherates and non-stoichiometric compounds. We will try and tread the path of chemical common sense rather than picking our way along the narrow track of pedantic definition.

The outstanding feature of the intermetallic chemistry of the heavy pnictogens is the increase in metallic character down the group. The metalloids arsenic and antimony show some aspects of p-block non-metallic chemistry especially with the s-block metals. However, even in this area there are many examples of compounds that shade into genuine intermetallic compounds, for example LiAs.

Most metals form pnictides with wide variations in stoichiometry. These will be described in the following sections on s-, p-, d-, and f-block pnictides, followed by a brief description of some important pnictogen-containing alloys.

3.6.1 s-Block intermetallics of the pnictogens

The s-block metals of groups 1 and 2 form the expected compounds with the pnictogens (E), of stoichiometry M_3E (M = group 1) and M_3E_2 (M = group 2).

The vast majority of s-block pnictides are Zintl compounds, that is the s-block metals act as electron donors and the pnictogen moiety forms bonds conforming to the octet rule. Before this elucidation, stoichiometries such as M_3E_7 (M = group 1 metal) were difficult to understand.

Group 1 metals form a variety of antimonides, for example Li_3Sb, NaSb, KSb_2, Rb_5Sb_4, Cs_3Sb_7 [286]. These examples are chosen to show the wide range of composition (Zintl phases). For many years their nature and non-aqueous solution chemistry was a matter of controversy with very little firm structural data. This situation was transformed by the isolation of the heptaantimonide (3–) anion as the Na(crypt)$^+$ salt by Adolphson, Corbett and Merrymard [287]. Such anions are termed (polyatomic) Zintl ions.

K_5Bi_4 and K_3Bi_2 have also been found to react with solutions of 2,2,2-crypt to give a crystalline compound (crypt-K)$_2{}^+Bi_4{}^{2-}$ which was investigated by X-ray spectroscopy. The $Bi_4{}^{2-}$ anion is almost a square, with sides (Bi–Bi) of 2.936 Å and 2.941 Å and angles 89.9° and 90.2°.

When the alkali metal alloy $Na_2Sn_2Bi_2$ is dissolved in ethylenediamine the anion $[Sn_2Bi_2]^{2-}$ is formed, which was confirmed by Sn NMR spectroscopy [288].

Group 1 pnictides of the type M_3E are the valency intermetallics to be expected, but their structures are not simple $(M^+)_3E^{3-}$ ions. The Li_3As phase may be viewed as the lithium salt of the amphoteric AsH_3. It is formed on the fusion of the elements in stoichiometric quantities, yielding a brown-black powder [289]. A solution of lithium in liquid ammonia reacts with a suspension of powdered arsenic in the same solvent to yield Li_3As as a brown powder [290]. However, for most synthetic purposes it is most conveniently prepared from arsine [291].

$$3Bu^nLi + AsH_3 \xrightarrow{Et_2O} Li_3As + Bu^nH \qquad [3.100]$$

The intermediates LiH_2As and Li_2HAs may also be isolated by using the appropriate stoichiometry; Li_3As, a useful synthon in organoarsenic chemistry, is not normally isolated *per se* (cf. Grignard reagents).

Na_3As has a hexagonal anti-LaF_3-type structure in which each As has five near Na (three at 294 pm and two at 299 pm) and six, more distant, at 330 pm giving $AsNa_{11}$ units. The Na atoms are in two environments, one with three As at 294 pm and six Na at 330 pm, the other has one near As at 299 pm and three As at 330 pm forming a distorted tetrahedron together with six Na at 328–330 pm. These high coordination numbers are typical of metallic structures. These M_3E-type compounds exhibit metallic conduction or are semiconductors.

Compounds of the type M_3E_7, for example Na_3As_7 [292] and Cs_3As_7 [293] have cage-like $As_7{}^{3-}$ anions. They are highly coloured compounds, Cs_3As_7 being ruby-red, melting point 788°C. It undergoes a first order transition at 547°C.

Finally, the group 1 pnictide ternary and higher phases with other metals are noted, for example the phases obtained by heating caesium arsenic and a transition metal, typically at 500°C–700°C for 2 d affording $CsFe_2As_2$ [294]. Selected examples are listed in Table 3.34.

The technique of using crypt to form a cation that prevents recombination with the anion has been applied to other systems. The ternary alloy KGeSb reacts with crypt in ethylenediamine solution to give $K_2{}^+Sb_4{}^{2-}$ as hard black crystals [295]. A bond order of 1.25 for the $Sb_4{}^{2-}$ anion is

Table 3.34 A selection of ternary group 1 pnictides, by structure type.

Compound	Reference
Structure-type: tetragonal $CaAl_2Si_2$	
Li_3NdAs_2	[A]
Li_3YSb_2	[A]
Structure-type: tetragonal $ThCr_2Si_2$	
$RbFe_2As_2$	[B]
KRu_2As_2	[B]
$RbRu_2As_2$	[B]
$RbCo_2As_2$	[B]
$RbRh_2As_2$	[B]
KIr_2As_2	[B]
$RbIrAs_2$	[B]
$CsFe_2As_2$	[C]
$CsCo_2As_2$	[C]
$CsIr_2As_2$	[C]
Structure-type: orthorhombic K_2PdP_2	
K_2NiAs_2	[B]
Rb_2NiAs_2	[B]
Rb_2PdAs_2	[B]
Rb_2PtAs_2	[B]
$CsNiAs_2$	[C]
$CsPtAs_2$	[C]

References

A. Grund, I., Müller, P. and Schuster, H.U. (1984) *Z. Anorg. Allg. Chem.*, **515**, 151.

B. Schuster, H.U. and Wenz, P. (1984) *Z. Naturforsch.*, **39B**, 1816.

C. Czybulka, A., Noack, M. and Schuster, H.U. (1992) *Z. Anorg. Allg. Chem.*, **609**, 122.

implied by the molecular orbital diagram. The Sb_7^{3-} anion has similarly been isolated.

Group 2 pnictides also show the expected M_2E_3 stoichiometry. The pnictides in these compounds form catenated anions; Ca_2As_3 [296] forms As_4 and As_8 chains whereas Sr_2Sb_2 [297] and Ba_2Sb_3 [298] form Sb_6^{8-} chains. Mg_3Sb_2 [296] has a structure of the La_2O_3-type. The structure is described in terms of $SbMg_7$ units. Other stoichiometries occur, for example Ca_2Sb, Ca_2Bi [299] in which the Sb and Bi atoms are coordinated to nine Ca atoms to form a distorted tetragonal anti-prism with the ninth Ca atom located above the plane. Ca_5Sb_3 and Ca_5Bi_3 [300] (with interatomic distances for Ca–Sb of 3.249–3.371 Å) exhibit partial ionic character.

Sr_2Sb_3 has been shown by Eisenmann and Deller [301] to have Sb_6 chains and Ba_2Sb_3 has been shown to be isostructural [302].

$Ca_{11}Sb_{10}$ has been shown to have Sb_4 rings and Sb_2 units as well as isolated Sb atoms. There is also a bismuth analogue. In contrast $SrSb_2$ has twisted chains of antimony atoms (with Sb–Sb bond lengths of 2.895–2.919 Å).

A phase diagram of the magnesium–arsenic system has been given by Clark and Nayeb-Hashemi [303]. Other group 2 pnictides include Ba_3As_4 [304] and SrAs [305] [Sr–As bond length 3.13 Å].

There are also many ternary phases, for example $SrPd_2As_2$ [306] with a $ThCr_2Si$-type structure (cf. Table 3.34) and $SrPd_2Sb_2$ and $SrPd_2Bi_2$ with $CaBe_2Ge_2$-type structures. Hofmann, Alemany and Alvarez [307] synthesized the Zintl phase $Ba_7Ga_4Sb_9$, a compound they claim probes the limits of the Zintl concept. There is a two-dimensional net of $Ga_4Sb_9^{14-}$ anion units. It is made up of one-dimensional alternating four- and eight-membered Ga–Sb rings linked together into a two-dimensional structure by long but bonding Sb–Sb contacts.

3.6.2 *p-Block intermetallics of the pnictogens*

These compounds are dominated by the III–V semiconductors (now more properly called 13–15 semiconductors) and especially by gallium arsenide.

In the case of indium bismuthides the (In–Bi) distance decreases with the increase in the proportion of Bi, for example in In_5Bi_3 and In_2Bi it is 3.17 Å, shortening to 3.13 Å in InBi [308].

Group 14 affords the examples GeAs (melting point 737°C) and $GeAs_2$ (melting point 732°C). Tin arsenide, SnAs (NaCl-type structure), is superconducting at less than 3.5 K, and Sn_4As_3 at below 1.2 K. The many important alloys formed with p-block metals and the heavy pnictogens are described in Section 3.6.5.

Critchlow and Corbett [295] have also produced the heteroatomic Zintl ion $Pb_2Sb_2^{2-}$ as the K crypt$^+$ salt. The structure is tetrahedral, but since it is also disordered it is not possible to assign Pb or Sb to the various apices in a regular manner. Positions A and B are 66% and C is 27% occupied by lead while D is largely antimony. By averaging, the Sb–Pb bond length is assessed at 2.953 Å.

Badawi [309] has proposed the formation of the compound Pb_3Bi_2 in liquid lead–bismuth systems.

3.6.3 *d-Block intermetallics of the pnictogens*

The d-block intermetallic compounds of the heavy pnictogens are the archetypal intermetallics. We can no more than touch on this vast body of compounds and phases. They include many minerals, for example $NiAs_2$ (rammelsbergite) and Ag_3Sb (dyscrasite). Binary compounds with metals are made by heating together stoichiometric amounts of the elements. There have been reports of a large number of these compounds, (for example $CrSb_2$, $FeSb_2$, $RuSb_2$ and $OsSb_2$) with Loellingite-type structures; with pyrite type structures ($PdSb_2$, $PtSb_2$ and $AuSb_2$); and with the marcasite structure ($NiSb_2$). Investigations have been made into transition metal antimonides [310]; $FeAs_2$ (loellingite) [311] $CoAs_3$ (skutterudite) and NiAs are important structural types that have been found to occur widely. The NiAs structure is a very common ME geometry, evident in, for example, TiAs, TiSb, VSb, CrSb, MnAs, MnSb, MnBi, FeSb, CoSb, CoAs, NiSb, NiBi, RhBi, IrSb, IrBi, PdSb, PtSb, PtBi. In NiAs the arsenic forms an hexagonal close packed lattice with Ni in interstitial sites.

The metallic compound Rh_2As has low temperature (α) cubic and high-temperature (β) forms, with a transition temperature 680°C [312]. This material has found use in GaAs technology owing to its compatible structure [313]. Manganese arsenide, MnAs, exhibits a unique first-order magnetic phase transition at $T_c = 313$ K which destroys an ordered ferromagnetic state [313]. This property finds use in microelectronics. Ta_3As [314] has a structure [315] with $AsTa_{10}$ units with a bi-capped square antiprism geometry average length Ta–As 2.77 Å. Niobium arsenide, Nb_3As, has a similar structure.

An all d-block ternary phase $Y_3Au_3Sb_4$ [316] has four Y–Sb distances at 3.349 Å and four at 3.448 Å, averaging 3.399 Å, and four Au–Sb distances at 2.766 Å.

Zr_5Sb_3 [317] is of interest since it has been shown [318] to be an intermetallic host forming Zr_5Sb_3E (E = Al, As, Co, Ni, Cu, Zn, Ag, Ru and the non-metallic elements C, O, Si, P, S, Se). Excess antimony can be added to the compound to form the non-stoichiometric $Zr_5Sb_{3.12}$. The arsenide Mo_5As_4 [319] has also been used as a host to prepare metal insertion compounds [320] $M_xMo_5As_4$. $Cu_4Mo_5As_4$ is an electrical conductor whereas $Al_2Mo_5As_4$ and $Ga_2Mo_5As_4$ are semiconductors. The transition from conductor to semiconductor was associated with the attainment of a 24-electron count for the cluster.

3.6.4 f-Block intermetallics of the pnictogens

The f-block metals form pnictides, for example uranium arsenide [321] U_4As_3, U_5As_3, U_3As_2 [322], lanthanide antimonides [323,324] $LaSb_2$, $CeSb_2$, $NdSb_2$, $SmSb_2$ (all of the $LaSb_2$ type). The phase diagram for the ytterbium–antimony system indicates seven stable phases, YbSb, $YbSb_2$, α- and β-Yb_5Sb_4, Yb_4Sb_3, Yb_5Sb_3 and Yb_2Sb_2 [325]. Actinide antimonides [326] have also been prepared.

Bismuth forms alloys with the rare earths and yttrium and two series of phases with nominal composition Ln_5Bi_3 (Ln = Y, rare earth metal). The elements La, Ce, Pr, Nd and Gd form compounds with an Mn_5Si_3-type structure, and Y, Dy, Ho, Er and Tm form compounds with the Y_5Bi_3 structure.

In the La_5Bi_3 series, the Mn_5Si_3-type structure is stable with a radius ratio Ln:Bi $\geqslant$ 1.05. At this junction the Ln ion core becomes no longer compressible and the Y_5Bi_3-type becomes the more stable type without the short Ln–Ln single bond distance.

Binary compounds are also formed with the actinides, for example americium heated with bismuth in a sealed tube yields material with NaCl cubic structure with unit cell $a_0 = 6.338$ Å:

$$Am + Bi \xrightarrow[48\ h]{975°C} AmBi \qquad [3.101]$$

Plutonium and uranium give material with $a_0 = 6.350$ Å and 6.364 Å, respectively.

A lanthanide ternary series, $M_3Au_3Sb_4$ (M = Nd, Sm, Gd, Tb, Dy, Ho, Er, Tm and Lu), has been prepared [316]. Their structure is isotypic with $Y_3Au_3Sb_4$. The phases MPd_2As_2 (M = La, Ce, Pr, Nd, Sm, Eu) have the $ThCr_2Si_2$-type structure [306]. The phases $M_2Ni_{12}As_7$ (M = Pr to Lu) are isostructural with $Zr_2Fe_{12}P_7$ [327].

A novel intermetallic compound $CeCu_{1-x}Bi_2$ ($x = 0.3$) [328] was synthesized by the bismuth self-flux method. It has been shown to have the $ZrCuSi_2$-type structure.

The ternary compound $Eu_3Pd_4As_4$ was prepared by heating stoichiometric quantities of the elements in a sealed quartz ampoule at 800°C for 15 h [329]. The arsenic atoms occur as isolated pairs [As–As length 2.505(4) Å] in the structure, which is composed of a combination of $CeMg_2Si_2$-type and AlB_2-type units. The As–Eu distances are between 3.207(1) Å and 3.273(3) Å and those of As–Pd are between 2.445(2) Å and 3.010(1) Å.

3.6.5 Pnictogen alloys

The ability of antimony to combine with metals has been known since its discovery by the medieval alchemists. One of the alchemists' symbols for antimony was the wolf, 'because in the fire it devours the greater part of metals'.

An extremely large number of antimony alloys have been described. It is not the purpose of a chapter of this nature to comment on the complexities of metallurgy and the properties of alloys (but see chapter 2), beyond mentioning that various dilute alloys of antimony are well-known and widely-used materials. Pewter contains about 10% antimony and 1%–2% copper, the bulk metal being tin. Type metal is lead with 5%–10% antimony and tin. Battery lead, for lead accumulators, contains a few per cent antimony, which can give rise to traces of SbH_3 in unfavourable circumstances. Tungsten oxide bronzes containing small quantities of antimony, corresponding to $Sb_{0.07}WO_3$ and $Sb_{0.15}WO_3$, have

Table 3.35 Vapour pressure of arsenic, $P(As_4)$, over arsenic–lead alloys at 703°C.

Arsenic (wt%)	$P(As_4)$ (torr)
0.52	0.0078
1.00	0.0147
1.41	0.0351
1.97	0.0821
2.61	0.238
3.28	0.563

Source: McClincy, R.J. and Larson, A.H. (1969) *Trans. Metall. Soc.*, **621**, 57.

been prepared. Non-stoichiometry is widespread, but many alloys are made with a chemical formula as a basis of composition, but such a composition does not necessarily mean that a compound in the chemical sense exists.

The difficulty of removing arsenic from arsenic–lead alloys is shown by the low vapour pressure of As_4 over these alloys, shown in Table 3.35 [330].

Bismuth is the most metallic of the heavy pnictogens. It is the most diamagnetic of the metals (specific susceptibility $-16 \times 10^{-9}\, m^3\, kg^{-1}$ at 20°C) and exhibits the largest Hall effect (increase in resistivity in a magnetic field).

Bismuth forms a large number of alloys from high melting point intermetallic compounds (Zintl phases) such as Li_3Bi (melting point 1145°C) to the low melting point so-called fusible alloys, for example In–Cd–Pb–Sn–Bi alloy, melting point 47°C. The fusible alloys are industrially important and have some interesting properties. Some exhibit no volume change on solidification (useful in precision moulding) such as the alloy Bi 40%–Sn 60%. Other alloys expand on cooling and then further expand in the solid state, the alloy Bi 56%–Sn 20%–Pb 24% has the largest such expansion. Some examples of these alloys and their eutectic temperatures are given in Table 3.36. Not all metals form alloys with bismuth, for example iron is insoluble in bismuth and so iron vessels may be used for manipulating liquid bismuth. It is also rare for there to be complete miscibility between all compositions of liquid and solid phases in binary bismuth systems, but such is the case in the bismuth–antimony system, which has been widely studied. A mass spectra study of this system detected only nine molecular species: Sb_4, $BiSb_3$, Bi_2Sb_2, Bi_3Sb, Bi_4, Sb_2, BiSb, Bi_2, and Bi.

Finally, a group of alloys, which are important in various safety devices, must be noted. The non-eutectic bismuth alloys under tensile load lose their mechanical cohesion and yield, breaking an electrical circuit, for example. Some examples are given in Table 3.37.

Table 3.36 Bismuth low melting point alloys: melting temperature, T_m, and composition.

T_m (°C)	Composition (%)						
	Bi	Pb	Sn	Cd	In	Tl	Other
47	44.7	22.6	8.3	5.3	19.1	0.0	0.0
58	49.4	18.0	11.6	0.0	21.0	0.0	0.0
61	33.0	0.0	16.0	0.0	51.0	0.0	0.0
62	30.0	0.0	0.0	8.0	62.0	0.0	0.0
64	48.0	25.6	12.8	9.6	4.0	0.0	0.0
70	50.0	26.7	13.3	10.0	0.0	0.0	0.0
72	33.7	0.0	0.0	0.0	66.3	0.0	0.0
79	58.0	0.0	26.0	0.0	16.0	0.0	0.0
91	55.2	33.3	0.0	0.0	0.0	11.5	0.0
92	52.0	40.0	0.0	8.0	0.0	0.0	0.0
94.6	49.1	0.0	23.5	18.2	0.0	9.2	0.0
96	52.0	32.0	16.0	0.0	0.0	0.0	0.0
103	54.0	0.0	26.0	20.0	0.0	0.0	0.0
109	68.0	0.0	0.0	0.0	32.0	0.0	0.0
124	50.0	0.0	35.7	0.0	0.0	14.3	0.0
125	56.5	43.5	0.0	0.0	0.0	0.0	0.0
130	56.0	0.0	40.0	0.0	0.0	0.0	4 (Zn)
138	57.0	43.0	0.0	0.0	0.0	0.0	0.0
144	60.0	0.0	0.0	40.0	0.0	0.0	0.0
167	44.0	0.0	31.0	0.0	0.0	25.0	0.0
186	42.2	9.8	0.0	0.0	0.0	48.0	0.0
188	47.5	0.0	0.0	0.0	0.0	52.5	0.0
198	76.5	0.0	0.0	0.0	0.0	23.5	0.0
218	97.0	0.0	0.0	0.0	0.0	0.0	3 (Na)
240	82.0	0.0	0.0	0.0	0.0	0.0	18 (Au)
243	99.5	0.0	0.0	0.0	0.0	0.0	0.5 (Li)
254	97.3	0.0	0.0	0.0	0.0	0.0	2.7 (Zn)
262	97.5	0.0	0.0	0.0	0.0	0.0	2.5 (Ag)
551	59.0	0.0	0.0	0.00	0.0	0.0	41 (Mg)

Table 3.37 Non-eutectic bismuth alloys: melting range, T_m^{range}, and yield temperature, T_y.

Composition (%)				Sb	T_m^{range} (°C)	T_y (°C)
Bi	Pb	Sn	Cd			
14.0	43.0	43.0	0.0	0.0	143–163	154
5.0	32.0	45.0	18.0	0.0	132–139	135
48.0	28.5	14.5	0.0	9.0	103–227	116
59.4	14.8	25.8	0.0	0.0	95–114	100
52.0	31.7	15.3	1.0	0.0	83–92	90
42.5	37.7	11.3	8.5	0.0	70–90	72

References

1. Dixon, R.N., Duxbury, G. and Lamborton, H.M. (1966) *J. Chem. Soc., Chem. Commun.*, 460.
2. Kawaguchi, K. and Hirota, E. (1984) *J. Mol. Spectrosc.*, **106,** 423.
3. Berkowitz, J. (1988) *J. Chem. Phys.*, **89,** 7065.
4. Svec, H.J. and Saalfeld, F.E. (1963) *Inorg. Chem.*, **2,** 46, 50.
5. Dixon, R.N. and Lamberton, H.M. (1968) *J. Mol. Spectrosc.*, **25,** 12.
6. Lewerenz, M., Bruna, P.S., Peyerimhoff, S.D. and Bueuker, R.J. (1983) *Mol. Phys.*, **49,** 1.
7. Balasubramanian, K. and V. Nannegari, (1989) *J. Mol. Spectrosc.*, **138** 482.
8. Dixon, R.N., Duxbury, G. and Lamberton, H.M. (1968) *Proc. Roy. Soc., Ser. A*, **305**, 275.
9. Balasubramanian, K. (1989) *J. Chem. Phys.*, **91**, 2443.
10. Fowler, B.A., Blair, P.C., Tompson, B.C. *et al.* (1988) *Fundam. Appl. Toxicol.*, **14**, 776.
11. Piffard, J.F. and Rade, J.Y. (1988) *Fr. Demande,* FR 2609978.
12. Jolly, W.L. and Drake, J.E. (1968) *Inorg. Syn.*, **7** 34.
13. Brindle, I.D. and Chen, H. (1991) *Talanta*, **38**, 1137.
14. Drake, J.E. and Riddle, C. (1987) *Inorg. Syn.*, **13**, 14.
15. Tzschach, A., Heinicke, J. and Schoen, H. (1985) *Ger (East) Patent* DD 225 684; *Chem. Abs.*, **104**, 529, 74ff.
16. Johnson, W.C. and Pechukas, A. (1937) *J. Amer. Chem. Soc.,* **59**, 2065.
17. MacDiarmid, A.G. and Bellama, J.M. (1983) *Inorg. Chem.*, **7**, 2070.
18. Salzman, K.A. and Davies, G.A. (1989) *J. Appl. Phys.*, **66**, 1185.
19. Jensen, K.F., Omstead, T.R. and Annapragada, A.V. (1990) *Appl. Phys. Lett.*, **57**, 2543.
20. Valdes, J.L., Cadet, G. and Mitchell, J.W. (1991) *J. Electrochem. Soc.*, **138**, 1654.
21. Buckley, D.N., Bylsma, R.B., Cadet, G. *et al.* (1990) *Appl. Phys. Lett.*, **57**, 1684.
22. Freitas J.A. Jr and Sillmon, R.S. (1990) *Appl. Phys. Lett.*, **56**, 174.
23. Zheng, D.H. and Bradley, D.C. (1991) *Euro. Pat. Appl.*, 443 739.
24. Shealy, J.R., Kreismanis, V.G., Wagner, D.K. and Woodall, J.M. (1983) *Appl. Phys. Lett.*, **42**, 83.
25. Tom, G.M. (1987) *US Patent* 4 659 552 (Hercules Inc.).
26. Pyykkö, P., Laakkonen, L.J. and Schwerdtfeger, P. (1992) *J. Chem. Phys.,* **96**, 6807.
27. Di Lonardo, G. and Fusina, L. (1984) *J. Mol. Spectrosc.*, **104**, 282.
28. Brindle, I.D., Alarabi, H., Karshman, S. *et al.* (1992) *Analyst*, **117**, 407.
29. Apte, S.C., Howard, A.G. and Campbell, A.T. (1989) in *Environmental Analysis Using Chromatography Interfaced with Atomic Spectroscopy* (eds R.M. Harrison and S. Rapsomanikis), Ellis Horwood, Chichester, Sussex, Ch. 9.
30. Cristau, B. (1958) *Bull. Soc. Chim. Fr.*, **5**, 717.
31. Drägerwerk Aktiengesellschaft (1994) *Dräger-Tube Handbook*, 9th edn, Lübeck.
32. Yamashita, S., Kume, S., Hemmi, A. *et al.* (1993) *Sens. Actuat. B.*, **13–14**, 466.
33. Bale, C.W., Kirchnerova, J. and Skeaff, J.M. (1990) *Sens. Actuat. B.*, **2**, 7.
34. Heiland, G., Kohl, D. and Mokwa, M. (1985) *Sens. Actuat.*, **8**, 101.
35. Yakimov, S.S., Varfolomeer, A.E., Volkov, A.I. *et al.* (1992) *Sens. Actuat. B.*, **7**, 727.
36. Noll, K.S. (1988) *Icarus,* **75**, 409.
37. Bezard, B., Drossart, P., Lellouch, E. *et al.* (1988) *Bull. Amer. Astron. Soc.,* **20**, 879.
38. Noll, K.S. and Larson, H.P. (1990) *Icarus,* **89**, 168.
39. Ohtani, H., Horiguchi, S., Urano, Y. *et al.* (1989) *Comb. Flame*, **76**, 307.
40. Johnson, W.C. and Pechukas, A. (1937) *J. Am. Chem. Soc.*, **59**, 2068.
41. Becker, G., Schmidt, M. and Westerhausen, M. (1992) *Z. Anorg. Allg. Chem.*, **607**, 101.
42. Cowley, A.H., Atwood, D.A., Atwood, J.L. *et al.* (1993) *Organometallics*, **12**, 24.
43. Trachtman, M., Bock, C.W., Dobbs, K.D. and Cowley, A.H. (1990) *J. Phys. Chem.,* **94**, 5210.
44. Ault, B.S. and Piocos, E.A. (1991) *J. Phys. Chem.,* **95**, 6827.
45. Wolfsberger, (1985) *Chem. Z.,* **109**, 53.
46. Valentine, D. Jr (1989) *Euro. Pat. Appl.*, EP 324 958 (American Cyanamid Co.).

47. Drake, J.E. and Anderson, J.W. (1970) *J. Chem. Soc. (A)*, 1131.
48. Ang, H.G. and Lee, L.K. (1989) *Polyhedron*, **8**, 1461.
49. Höhn, A., Gene, R.J., Sargeson, A.M. and Willis, A.C. (1989) *J. Chem. Soc., Chem. Comm.*, 1648.
50. Ang, H.G. and Koh, C.H. (1992) *J. Fluor. Chem.*, **58**, 9.
51. Andrews, L. and McInnes, T.C. (1991) *Inorg. Chem.*, **30**, 2990.
52. Ault, B.S. and Machara, N.P. (1988) *Inorg. Chem.*, **27**, 2383.
53. Drake, J.E and Riddle, C. (1969) *Inorg. Nucl. Chem. Lett.*, **5**, 665.
54. Ziegler, T., Kraatz, H.B., Jacobsen, H. and Boorman, P.M. (1993) *Organometallics*, **12**, 76.
55. Ebsworth, E.A.V., Gould, R.O., Mayo, R.A. and Walkinshaw, M. (1987) *J. Chem. Soc., Dalton Trans.*, 2831.
56. Jolly, W.D. and Shriver, D.F. (1958) Report UCRL-5148, Lawrence Radiation Laboratory, Livermore, CA.
57. Jolly, W.L. and Norman, A.D. (1968) *Prep. Inorg. React.*, **4**, 1.
58. Royen, P. and Rocktaeschel, C. (1966) *Z. Anorg. Allg. Chem.*, **346**, 290.
59. Jolly, W.L., Anderson, L.B. and Beltrami, R.T. (1957) *J. Am. Chem. Soc.*, **79**, 2443.
60. Breidung, J., Thiel, W. and Komornicki, A. (1991) *Inorg. Chem.*, **30**, 1067.
61. Heinemann, A. (1961) *Naturwiss.*, **48**, 568.
62. Bollmark, P. and Lindgren, B. (1967) *Chem. Phys. Lett.*, **1**, 480.
63. Balasubramanian, K. and Dai, D. (1990) *J. Chem. Phys.*, **93**, 1837.
64. Durrant, A.A., Pearson, T.G. and Robinson, P.L. (1934) *J. Am. Chem. Soc.*, **56**, 730.
65. Schenk, P.W. (1963) in *Handbook of Preparative Inorganic Chemistry*, vol. 1 (ed. G. Brauer), Academic Press, London, Ch. 10.
66. Jolly, W.L., Berka, L.H., Briggs, T. and Millard, M. (1960) *J. Inorg. Nucl. Chem.*, **14**, 190.
67. Zorin, A.D., Frolav, I.A., Zabrdyaev, V.S. and Nosyrev, S.A. (1974) *J. Appl. Chem., USSR*, **47**, 1233.
68. Reismann, A., Berkenblit, M., Haas, E.C. and Gaimes, A. (1954) *J. Electrochem. Soc.*, **101**, 387.
69. Burk, A.A. (1992) *J. Cryst. Chem.*, **124**, 292.
70. Helminger, P., Beeson, E.C. Jr and Gordy, W. (1971) *Phys. Rev. A. Gen. Phys.*, **3**, 122.
71. Berka, L. (1959) Lawrence Radiation Laboratory report UCRL-8781.
72. Jolly, W.L. and Drake, J.E. (1963) *Inorg. Syn.*, **7**, 43.
73. Dewolff, F.A., (1995) *Brit. Med. J.*, **310**, 1216.
74. Thompson, M. and Faull, J.L. (1995) *Lancet*, **346**, 1557.
75. Heimer, A. and Hulthen, E. (1931) *Nature*, **127**, 557.
76. Hulthen, E. and Neuhaus, H. (1956) *Phys. Rev.*, **102**, 1415.
77. Lindgren, B. and Nilsson, C. (1985) *J. Mol. Struct.*, **55**, 407.
78. Bernath, P.F., Bopegedera, A.M.R.P. and Brazier, C.R. (1989) *Chem. Phys. Lett.*, **162**, 301.
79. Amberger, E. (1961) *C. Ber.*, **94**, 1447.
80. Kawasaki, M. and Bersohn, R. (1978) *J. Chem. Phys.*, **68**, 2105.
81. Bandler, M. and Stassen, H.J. (1966) *Z. Anorg. Allg. Chem.*, **343**, 244.
82. Hillel, R., Bouix, J. and Favre, (1975) *R. Bull. Soc. Chem. Fr.*, 2458.
83. Corttbett, J.D. and Hershaft, A. (1963) *Inorg. Chem.*, **2**, 979.
84. von Benda, H., Simon, A. and Anhoffar, A.B. (1978) *Z. Anorg. Allg. Chem.*, **438**, 53.
85. von Schnering, H.G., von Benda, H. and Kalveram, C. (1978) *Z. Anorg. Allg. Chem.*, **438**, 37.
86. Hass, D., Bechstein, O. and Seefield, V. (1984) *Z. Anorg. Allg. Chem.*, **509**, 161.
87. Greenwood, N.N. and Woolf, A.A. (1950) *J. Chem. Soc.*, 2200.
88. Kwasnik, W. (1963) in *Handbook of Preparative Inorganic Chemistry*, vol. 1 (ed. G. Brauer), Academic Press, London, §4.
89. Konaka, S. (1970) *Bull. Chem. Soc., Jpn.*, **43**, 3107.
90. Avre, R., Hillel R. and Bouix, R. (1975) *R. Bull. Soc. Chim. Fr.*, 2458.
91. Clippard, F.B. and Bartell, L.S. (1970) *Inorg. Chem.*, **9**, 805.
92. Dixon, D.A. and Arduengo, A.J. III (1987) *J. Am. Chem. Soc.*, **109**, 338.
93. Edwards, A.J. (1970) *J. Chem. Soc. (A)*, 2751.

94. Aurivillius, B. and Lundqrist, T. (1955) *Acta Chem. Scand.*, **9**, 1206, 1209.
95. Hudlicky, M. (1976) *Chemistry of Organic Fluorine Compounds*, 2nd edn, Chichester, Sussex.
96. Clark, R.J. and Belefant, H. (1990) *Inorg. Syn.*, **28**, 310.
97. Anderson, F., Bak, B. and Hillebent, A. (1953) *Acta Chem. Scand.*, **7**, 236.
98. Cazzoli, G., Forti, P. and Lunelli, A. (1978) *J. Mol. Spectroscop.*, **69**, 71.
99. Spridonov, V. and Gershikov, A.G. (1986) *J. Mol. Struct. (Theochem)*, **140**, 173.
100. Haaland, A., Hougen, J., Samdal, S. and Tremmel, J. (1988) *Acta Chem. Scand.*, **42A**, 409.
101. Lipka, A. (1979) *Acta Cryst.*, **B35**, 3020.
102. Nyburg, S.C., Ozin, G.A. and Szymanski, J.T. (1971) *Acta Cryst.*, **B27**, 2298.
103. Kozulin, A.T., Gogoler, A.V. and Shishkin, S.G. (1977) *Izv Vyssh Uchebn Zaved, Fiz*, **20**, 42, Chem Abst. **87**, 46107q.
104. Lunelli, B.L., Cazzoli, G. and Lattanzi, F. (1983) *J. Mol. Spectrosc.*, **100**, 174.
105. Baughan, E.C. (1976) in *The Chemistry of Non-aqueous Solvents*, vol. 4 (ed. J.J. Lagowski) Academic Press, New York, Ch. 5.
106. Claire, P.P.K., Drew, M.G.B. and Willey, G.R. (1988) *J. Chem. Soc., Dalton Trans.*, 215.
107. Gutmann, V. (1956) *Quart. Rev.*, **10**, 451.
108. Kolditz, L. and Furcht, G. (1961) *Z. Anorg. Allg. Chem.*, **312**, 11.
109. Robiette, A.G. (1976) *J. Mol. Struct.*, **35**, 81.
110. Morino, Y., Ukaji, T. and Ito, T. (1966) *Bull. Chem. Soc. Jpan.*, **39**, 71.
111. Trotter, J. (1965) *Z. Krist.*, **122**, 230.
112. Cushen, D.W. and Hulme, R. (1964) *J. Chem. Soc.*, 4162.
113. Cushen, D.W. and Hulme, R. (1962) *J. Chem. Soc.*, 2218.
114. Skinner, H.A. and Sutton, L.E. (1940) *Trans. Faraday Soc.*, **36**, 681.
115. von Benda, H. (1980) *Z. Krist.*, **151**, 271.
116. Bailer, J.C. Jr (1939) *Inorg Syn.*, **1**, 103.
117. Cundy, P.F. and Bailer, J.C. Jr (1939) *Inorg. Syn..*, **1**, 104.
118. Watt, G.W., Hakki, W.W. and Choppin, G.R. (1953) *Inorg. Syn.*, **4**, 114.
119. Almenningen, A. and Bjorvatten, T. (1963) *Acta Chem. Scand.*, **17**, 2573.
120. Galy, J. and Enjalbert, R. (1980) *Acta Cryst.*, **B36**, 914.
121. Trotter, J. and Zobel, T. (1966) *Z. Krist.*, **123**, 67.
122. Pohl, S. and Saak, W. (1984) *Z. Krist.*, **169**, 177.
123. Pitzer, K.S. (1979) *Acc. Chem. Res.*, **12**, 271.
124. Pyykkö, P. and Desclaux, J.P. (1979) *Acc. Chem. Res.*, **12**, 276.
125. Noorby, L.J. (1991) *J. Chem. Ed.*, **68**, 110.
126. Hebecker, C. (1971) *Z. Anorg. Allg. Chem.*, **384**, 111.
127. Seppelt, K. (1977) *Z. Anorg. Allg. Chem.*, **434**, 5.
128. Roualt, M. (1940) *Ann. Phys. Lpz.*, **14**, 78.
129. Ohlberg, S.M. (1950) *J. Am. Chem. Soc.*, **81**, 811.
130. Priess, H. (1971) *Z. Anorg. Allg. Chem.*, **380**, 45.
131. Minkwitz, R., Clans, F., Glaser, M. and Wölfel, V. (1984) *Z. Anorg. Allg. Chim.*, **517**, 207.
132. Minkwitz, F., Clans, R. and Glaser, M. (1983) *Z. Anorg. Allg. Chem.*, **506**, 178.
133. Minkwitz, R., Prenzel, A., Schardy, A. and Oberhammer, H. (1987) *Inorg. Chem.*, **26**, 2730.
134. Minkwitz, R. and Prenzel, H. (1987) *Z. Anorg. Allg. Chem.*, **548**, 103.
135. Minkwitz, R. and Clans, F. (1983) *Z. Anorg. Allg. Chem.*, **501**, 19.
136. Ballard, J.G., Birchall, T. and Slim, D.R. (1977) *J. Chem. Soc., Dalton Trans.*, 1469.
137. Ballard, J.G., Birchall, T. and Slim, D.R. (1979) *J. Chem. Soc., Dalton Trans.*, 62.
138. Emeléus, H.J. (1927) *J. Chem. Soc.*, 783.
139. Donohue, J. and Lu, C.S. (1944) *J. Am. Chem. Soc.*, **66**, 819.
140. Pertlick, F. (1975) *Monatsh. Chem.*, **106**, 755.
141. Akisin, P.A. and Spiridonor, V.P. (1961) *Zh. Struct. Khim.*, **2**, 542.
142. Harwig, H.A. (1978) *Z. Anorg. Allg. Chem.*, **444**, 151.
143. Harwig, H.A. and Weenk, H.W. (1978) *Z. Anorg. Allg. Chem.*, **444**, 167.
144. Malmros, G. (1970) *Acta Chem. Scand.*, **24**, 384.

145. Jones, P.J., Beesk, W., Sheldrick, G.M. and Schwarzmann, E. (1980) *Acta Cryst*, **B36**, 439.
146. Jones, P.G., Sheldrick, G.M. and Schwarzmann, E. (1980) *Acta Cryst.* **B36**, 1923.
147. Ogden, J.S., Brisden, A.R. and Gomme, R.A. (1986) *J. Chem. Soc., Dalton Trans.*, 2725.
148. Rogers, D. and Skapski, A.C. (1964) *Proc. Chem. Soc.*, 400.
149. Thornton, G. (1977) *Acta Cryst.*, **B33**, 1271.
150. Aurivillius, B. (1952) *Arkiv Kemi.*, **3**, 153.
151. Scholder, R. and Stobbe, H. (1941) *Z. Anorg. Allg. Chem.*, **247**, 392.
152. Gattow, G. and Klippel, W. (1980) *Z. Anorg. Allg. Chem.*, **470**, 25.
153. Jansen, M. (1978) *Z. Anorg. Allg. Chem.*, **441**, 5.
154. Jansen, M. and Ehrhardt, H. (1983) *Z. Anorg. Allg. Chem.*, **504**, 128.
155. Berndt, W., Rumel, H. and Schwarzmann, E. (1977) *Z. Naturforsch.*, **32B**, 617.
156. Jansen, M. (1978) *Angew. Chem. Int. Ed. Engl.*, **17**, 137.
157. Gruehn, R. and Hibst, H. (1978) *Z. Anorg. Allg. Chem.*, **440**, 137.
158. Nowacki, W. and Mullen, D.J.E. (1972) *Z. Krist.*, **136**, 48.
159. Hildenbrand, D.L., Lau, K.H. and Brittain, R.D. (1982) *J. Chem. Phys.*, **86**, 4429.
160. Gnillermo, T.R. and Wuensch, B.J. (1973) *Acta Cryst.*, **B29**, 2536.
161. Nowacki, W. and Bayliss, P. (1972) *Z. Krist.*, **135**, 308.
162. Swinnea, J.S., Tenorio, A.J. and Steinfink, H. (1985) *Amer. Miner.*, **70**, 1056.
163. Kupcik, V., Vesela-Novakova, L. and Tschermaks, (1970) *Miner. Petrog. Mitt.*, **14**, 55.
164. Whitfield, H.J. (1970) *J. Chem. Soc. (A)*, 1800.
165. Whitfield, J.H. (1973) *J. Chem. Soc., Dalton Trans.*, 1737.
166. Gimarc, B. and Joseph, P.J. (1984) *Angew. Chem. Int. Ed. Engl.*, **23**, 506.
167. Gillespie, R.J., Christian, B.H. and Sawyer, J.F. (1987) *Acta Cryst.*, **C43**, 187.
168. Sheldrick, G.M. and Porter, E.J. (1972) *J. Chem. Soc., Dalton Trans.*, 1347.
169. Brunner, H., Kauermann, H., Nuber, B. *et al.* (1986) *Angew. Chem. Int. Ed. Engl.*, **25**, 557.
170. Ito, T., Morimoto, N. and Sadamaya, R. (1952) *Acta. Cryst.*, **5**, 775.
171. Kutoglu, A. (1976) *Z. Anorg. Allg. Chem.*, **419**, 176.
172. Brockner, W. and Wibbelmann, C. (1981) *Z. Naturforsch.*, **36A**, 836.
173. Behrens, H. and Glasser, L. (1955) *Z. Anorg. Allg. Chem.*, **278**, 174.
174. Thilo, E., Hertzog, K. and Winkler, A. (1970) *Z. Anorg. Allg. Chem.*, **373**, 111.
175. Bowen, L.H., Long, G.G., Ruby, S.L. and Stevens, J.G. (1969) *Inorg. Nucl. Chem. Lett.*, **5**, 21.
176. Valle, B.D. and Birchall, T. (1970) *Chem. Soc. Chem. Commun.*, 675.
177. Tideswell, N.W., Kruse, N.H. and McCullough, J.D. (1957) *Acta. Cryst.*, **10**, 99.
178. Bland, J.A. and Basinski, S.J. (1961) *Can. J. Phys.*, **39**, 1040.
179. Baston, T.J. and Whitfield, H.J. (1973) *J. Chem. Soc., Dalton Trans.*, 1739.
180. Goldstein, P. and Paton, A. (1977) *Acta. Cryst.*, **B30**, 915.
181. Whitfield, H.J. and Baston, T.J. (1977) *J. Chem. Soc., Dalton Trans.*, 959.
182. Gardes, B., Brun, G. and Tedenac, J.C. (1989) *Eur. J. Solid State Inorg. Chem.*, **26**, 221.
183. Dutton, W.A. (1971) in *Tellurium* (ed W.C. Cooper), Van Nostrand Reinhold, New York, Ch. 4.
184. Eifert, J.R. and Peretti, E.A. (1968) *J. Mater. Sci.*, **3**, 293.
185. Anderson, T.L. and Krause, H.B. (1974) *Acta Cryst.*, **B30**, 1307.
186. Olsson, P.O. and Kihlborg, L. (1989) *Ultramicrosc.*, **27**, 387.
187. Pietsch, R. and Ludwig, P. (1964) *Z. Anorg. Allg. Chem.*, **331**, 344.
188. Bencivennie, L. and Gingerrich, K.A. (1983) *J. Mol. Struct.*, **99**, 23.
189. Menary, J.W. (1958) *Acta Cryst.*, **11**, 742.
190. Sheldrick, W.S. and Häusler, H.J. (1986) *Z. Anorg. Allg. Chem.*, **538**, 45.
191. Coda, A., Dal Negro, A., Sabelli, G. and Tazzoli, V. (1977) *Acta Cryst*, **B33**, 1807.
192. Pertlik, F., Effenberger, H., Kugler, A. *et al.* (1988) *Acta Cryst*, **44**, 2043.
193. Pertlik, F. (1977) *Z. Krist.*, **145**, 35.
194. Hawthorne, F.C. (1985) *Can. Miner.*, **23**, 675.
195. Leyrie, M. and Herve, G. (1978) *Nouv. J. Chim.*, **2**, 233.
196. Robert, F., Leyrie, M., Hervé, G. *et al.* (1980) *Inorg. Chem.*, **19**, 1746.
197. Müller, A., Krickemeyer, E., Penk, M. *et al.* (1990) *Angew. Chem. Int. Ed. Engl.*, **29**, 88.

198. Tapscott, R.E., Marcovich, D., Duesler, D.E. and Them, T.F. (1982) *Inorg. Chem.*, **21**, 3336.
199. Skapski, A.C. (1966) *J. Chem. Commun.*, 10.
200. Matsumoto, K., Kawaguchi, H., Kuroya, H. and Kawaguchi, S. (1974) *Bull. Chem. Soc. Jpn.*, **46**, 2424.
201. Hoppe, R. and Stöver, H.D. (1980) *Z. Anorg. Allg. Chem.*, **468**, 137.
202. Mellini, M. and Merlino, S. (1979) *Am. Min.*, **64**, 1235.
203. Mu, H.C. and Hsueh Ting, K.C. (1966) *K'o Hsueh T'ung*, **17**, 502.
204. Jacobson, R.A. and Grass, M.E. (1974) *Inorg. Chim. Acta.*, **8**, 209.
205. Schwedes, B. and Hoppe, R. (1971) *Rev. Chim. Miner.*, **8**, 583.
206. Schwedes, B. and Hoppe, R. (1972) *Z. Anorg. Allg. Chem.*, **391**, 313.
207. Schwedes, B. and Hoppe, R. (1972) *Z. Anorg. Allg. Chem.*, **392**, 97.
208. Hoppe, R. and Hübenthal, R. (1989) *Z. Anorg. Allg. Chem.*, **576**, 169.
209. Bruckhardt, H.G. and Trömel, (1983) *Acta Cryst.*, **C39**, 1322.
210. Ducke, J., Trömel, Holwein, M., Kizler, P. (1983) *Acta Cryst.*, **C39**, 1322.
211. Burns, G. (1992) *High Temperature Superconductivity, An Introduction*, Academic Press, London.
212. Calvo, C. and Drishnamachari, N. (1973) *Acta. Cryst.*, **B29**, 2611.
213. Averbuch-Pouchot, A.T. and Durif, A. (1987) *Acta. Cryst.*, **C43**, 1894.
214. Calvo, C. and Leung, K.Y. (1973) *Can. J. Chem.*, **51**, 2082.
215. Remy, F. (1969) *Bull. Soc. Chim. Fr.*, 3380.
216. Leyrie, M., Herve, G., Teze, A. and Jeannin, Y. (1980) *Inorg. Chem.*, **19**, 1746.
217. Vogel, A.I. (1978) *Vogel's Textbook of Quantitative Inorganic Chemistry*, 4th edn, Longman, Harlow, Essex.
218. Baran, E.J., Pedregosa, J.C. and Aymonino, P.J. (1974) *J. Mol. Struct.*, **22**, 377.
219. Thilo, E. and Winkler, A. (1966) *Monatsber. Deut. Akad. Wiss., (Berlin)*, **8**, 301; (1967) *Chem. Abs.* **67**, 15781.
220. Hornstra, J. and Verweij, H. (1980) *Acta. Cryst.*, **B36**, 1634.
221. Blum, D., Durif, A. and Guitel, J.C. (1977) *Acta. Cryst.*, **B33**, 3222.
222. Jones, P.G., Sheldrick, G.M. and Schwarzmann, E. (1980) *Acta. Cryst.*, **B36**, 1923.
223. Jouini, T., Driss, A.S., Durif, A. and Averbuch-Pouchot, M.T. (1988) *Acta. Cryst.*, **C44**, 1507.
224. Jouini, T., Driss, A. and Mezzine, M.O. (1988) *Acta. Cryst.*, **C44**, 788.
225. Thilo, E. and Plaetschuke, I. (1949) *Z. Anorg. Allg. Chem.*, **260**, 315.
226. Thilo, E. and Dostal, K. (1959) *Z. Anorg. Allg. Chem.*, **298**, 100.
227. Liebau, F. (1956) *Acta. Cryst.*, **9**, 811.
228. Hilmer, W. and Dornberger-Schiff, K. (1956) *Acta. Cryst.*, **9**, 87.
229. Nord, A.G., Stefandis, T. and Kierkegaard, P. (1985) *Z. Krist.*, **173**, 313.
230. Gingerich, K.A., Gencivenni, L. and Teghil, R. (1984) *Inorg. Chim. Acta.*, **85**, L11.
231. Thilo, E. (1963) *Adv. Inorg. Chem. Radiochem.*, **4**, 1.
232. Grunze, I. and Thilo, E. (1964) *Z. Anorg. Allg. Chem.*, **331**, 102.
233. Dostal, K. and Kocman, K. (1969) *Z. Anorg. Allg. Chem.*, **367**, 92.
234. Averbuch-Pouchot, M.T., Boubia, M. and Durif, A. (1985) *Acta. Cryst.*, **C41**, 1562.
235. Dachille, F. and Glasser, L.S.D. (1959) *Acta. Cryst.*, **12**, 820.
236. Kumbasar, L., Dal Negro, A. and Ungaretti, L. (1973) *Am. Min.*, **58**, 1034.
237. Philppot, E., Oiffon, A.G., Bayle, G. *et al.* (1983) *Rev. Chim. Miner.*, **20**, 338.
238. Jouini, T. and Driss, A. (1988) *Acta. Cryst.*, **C44**, 791.
239. Jouini, T. and Driss, A. (1988) *Acta Cryst.*, **C44**, 356.
240. Nabar, M.A. and Dalvi, A.P. (1978) *Z. Anorg. Allg. Chem.*, **440**, 265.
241. Mereiter, K. and Bedlivy, D. (1982) *Acta Cryst.*, **B38**, 1559.
242. Le But, J., Hagen Muller, P. and Baron, A. (1961) *Bull. Soc. Chim. Fr.*, 500.
243. Le But, J., Hagen Muller, P. and Grosvald, G. (1961) *R. Bull. Soc. Chim. Fr.*, 1655.
244. Calvo, C. and Krishnamachari, N. (1973) *Can. J. Chem.*, **51**, 46.
245. Gopal, R., Rutherford, J.S. and Robertson, B.E. (1980) *J. Solid State Chem.*, **32**, 29.
246. Jasper-Tönnes, B. and Müller-Buschbaum, H. *Z. Anorg. Allg.*
247. Durand, J. and El-Brahimi, M. (1986) *Rev. Chim. Miner.*, **23**, 146.
248. Johnson, J.W., Jacobson, A.J., Brody, J.F. and Rich, S.M. (1982) *Inorg. Chem.*, **21**, 3820.
249. Cheetham, A.K., David, W.F., Eddy, M.M. *et al.* (1986) *Nature*, **320**, 46.

250. Keller, P., Hess, H. and Riffel, H. (1985) *Z. Anorg. Allg. Chem.*, **530**, 60.
251. Prawat, J.P. and Mujtaba, S.Q. (1975) *Can. J. Chem.*, **53**, 2586.
252. West, A.R. (1984) *Solid State Chemistry and Its Applications*, John Wiley, Chichester, Sussex.
253. Aranda, M.A.G., Bruque, S. and Attfield, J.P. (1991) *Inorg. Chem.*, **30**, 2043.
254. Jansen, M. and Brachtel, G. (1980) *Monatsh. Chem.*, **111**, 377.
255. Averbuch-Pouchot, M.T., Blum, D. and Guitel, J.C. (1979) *Acta Cryst.*, **B35**, 726.
256. Ellerman, J., Veit, A., Lindner, E. and Hoehne, S. (1982) *J. Chem. Soc. Chem. Comm.*, 382.
257. Zid, M.F. and Jouini, T. (1996) *Acta Cryst.*, **C52** (3), 133.
258. Pope, M.T. (1983) *Heteropoly and Isopoly Oxometalates*, Springer, New York.
259. Simon, S.J. and Boltz, D.F. (1985) *Anal. Chem.*, **47**, 1758.
260. Takeuchi, Y., Kobayashi, A. and Sasaki, Y. (1982) *Acta. Cryst.*, **B38**, 242.
261. Durif, A. and Averbuch-Pouchot, M.T. (1979) *Acta. Cryst.*, **B35**, 1441.
262. Brown, R.G. and Ross, S.D. (1970) *Spectrochim. Acta.*, **26A**, 955.
263. Weigel, F. and Scherer, V. (1967) *Radiochem. Acta.*, **7**, 46.
264. Weiss, A., Hartl, K. and Hofmann, U. (1957) *Z. Naturforsch.*, **12B**, 669.
265. Fischer, R., Werner, G.D. (1981) *J. Less. Comm. Met.*, **80**, 121.
266. Markin, T.L. and Street, R.S. (1967) *J. Inorg. Nucl. Chem.*, **29**, 2265.
267. Keller, C. and Walter, K.H. (1965) *J. Inorg. Nucl. Chem.*, **27**, 1253.
268. Lawaldt, D., Marquart, R., Werner, G.D. and Weigel, F. (1982) *J. Less. Comm. Met.*, **85**, 37.
269. Mattrat, P. and Guerin, H. (1960) *R. Bull. Soc. Chim. Fr.*, 601.
270. Guerin, H., Stahl-Brasse, R. and Ariguib-Kbir, N. (1971) *R. Bull. Soc. Chim. Fr.*, 2828.
271. Catti, M. and Ivaldi, G. (1983) *Acta. Cryst.*, **B39**, 4.
272. Walenta, K. (1977) *Tsch. Min. Petre. Mitt.*, **24**, 125.
273. Effenberger, H., Zeman, J., Mereiter, K. and Pimminger, M. (1982) *Tsch. Min. Petr. Mitt.*, **30**, 145.
274. Rendle, D.F. and Trotter, J. (1977) *Acta. Cryst.*, **B33**, 2684.
275. Whitaker, A. (1973) *Z. Krist.*, **137**, 194.
276. Brown, W.E., Mathew, M. and Takagi, S. (1982) *Acta. Cryst.*, **B38**, 44.
277. Tarte, P. and Paques-LeDent, M.T. (1969) *Spectrochim. Acta.*, **25A**, 1115.
278. Shilton, M.G. and Howe, A.T. (1980) *J. Solid State Chem.*, **34**, 149.
279. Fitch, A.N., Fender, B.E.F. and Wright, A.F. (1982) *Acta Cryst.*, **B38**, 1108.
280. Appleman, D.E., Evans, H.T. and Ross, M. (1964) *Amer. Mineral.*, **49**, 1603.
281. Wu, C.H., Chen, T.C. and Wang, S.L. (1996) *Acta Cryst.*, **C52**, 1326.
282. Wang, S.L., Horng, J.C. and Lee, Y.H. (1994) *J. Chem. Soc., Dalton Trans.*, 1681.
283. Calvo, C. and Poulsen, S.J. (1968) *Can. J. Chem.*, **48**, 890.
284. Hawthorne, F.C. (1986) *Am. Min.*, **71**, 206.
285. Long, G.J. (ed.) (1984) *Mössbauer Spectroscopy in Heterogeneous Catalysis*, Plenum Press, New York.
286. Corbett, J.D., Adolphson, D.G., Corbett, J.G. and Merryman, D.J. (1975) *J. Am. Chem. Soc.*, **97**, 6267.
287. Adolphson, D.G., Corbett, J.G. and Merryman, D.J. (1976) *J. Am. Chem. Soc.*, **98**, 7234.
288. Wilson, W.L., Rudolph, R.W., Lohr, L.L. *et al.* (1986) *Inorg. Chem.*, **25**, 1535.
289. Dönges, E. (1963) in *Handbook of Preparative Inorganic Chemistry*, 2nd edn (ed. G. Brauer) Academic Press, London.
290. Nicholls, D. (1979) *Inorganic Chemistry in Liquid Ammonia*, Elsevier, Oxford.
291. Bürger, H. and Goetze, U. (1968) *J. Organomet. Chem.*, **12**, 451.
292. Hönle, W. and von Schnering, H.G. (1986) *Z. Krist.*, **174**, 82, 203.
293. Somer, M., Hönle, W. and von Schnering, H.G. (1989) *Z. Naturforsch.*, **44B**, 296.
294. Czybulka, A., Noack, M. and Schuster, H.U. (1992) *Z. Anorg. Allg. Chem.*, **609**, 122.
295. Critchlow, S.C. and Corbett, J.D. (1984) *Inorg. Chem.*, **23**, 770.
296. Brauer, G., Haase, A. and Martinez-Ripoll, M. (1974) *Z. Naturforsch.*, **29B**, 13.
297. Eisenmann, B. (1979) *Z. Naturforsch.*, **34B**, 1162.
298. Schäfer, H., Eisenmann, B. and Jordan, H. (1985) *Z. Naturforsch.*, **40B**, 1603.
299. Schäfer, H. and Eisenmann, B. (1974) *Z. Naturforsch.*, **29B**, 13.

300. Brauer, G. and Martinez-Ripoll, M. (1974) *Acta Cryst.,* **B30**, 1083.
301. Eisenmann, B. and Deller, K. (1979) *Z. Naturforsch.,* **34B**, 1162.
302. Eisenmann, B., Jordan, H. and Schäfer, H. (1985) *Z. Naturforsch.,* **40B**, 1603.
303. Clark, J.B. and Nayeb-Hashemi, A.A. (1985) *Bull. Alloy. Phase Diagrams,* **6**, 432.
304. von Schnering, H.G., Wittmann, M. and Sommer, D. (1984) *Z. Anorg. Allg. Chem.,* **510**, 61.
305. Franceschi, E. and Iandelli, A. (1983) *J. Less Comm. Met.,* **30**, 211.
306. Hoffmann, W.K. and Jeitschko, W. (1985) *Monatsch. Chem.,* **116**, 569.
307. Hoffmann, R., Alemany, P. and Alvarez, S. (1990) *Inorg. Chem.,* **29**, 3070.
308. Wang, R., Giessen, B.C. and Grant, N.J. (1969) *Z. Krist.,* **128**, 244.
309. Badawi, W.A. (1988) *Bull. Chem. Soc. Jpn.,* **61**, 2262.
310. Nataf, G. and Verie, C. (1981) *J. Cryst. Growth,* **55**, 87.
311. Tossell, J., Vaughan, D.J. and Burdett, J.K. (1981) *Phys. Chem. Miner.,* **7**, 177.
312. Secoué, M., Auvray, P., Toudic, Y. *et al.* (1986) *J. Cryst. Growth,* **76**, 135.
313. Gover, G.A. (1986) *Sov. Phys. Solid State,* **28**, 18.
314. Murray, J.J. (1976) *J. Less Comm. Met.,* **46**, 311.
315. Wang, Y. and Taylor, J.B. (1979) *Acta Cryst.,* **B35**, 1447.
316. Dwight, A.E. (1977) *Acta Cryst.,* **B33**, 1579.
317. Boller, H. and Parthém, E. (1963) *Monatsch. Chem.,* **94**, 225.
318. Corbett, J.D. and Garcia, E. (1990) *Inorg. Chem.,* **29**, 3274.
319. Jensen, P. and Kjekshus, A. (1966) *Acta Chem. Scand.,* **20**, 1309.
320. Nanjundaswamy, K.S. and Gopalakrishnan, J. (1988) *J. Chem. Soc., Dalton Trans.,* 1.
321. McWhan, D.B., Vettier, C., Isaacs, E.D. *et al.* (1990) *Phys. Rev. B Cond. Mater.,* **42**, 6007.
322. Gschneidner, K.A. and Calderwood, F.W. (1986) *Bull. Alloy Phase Diagr.,* **7**, 279, 291.
323. Brixner, L.H. (1960) *J. Inorg. Nucl. Chem.,* **15**, 199.
324. Wang, R. and Steinfink, H. (1967) *Inorg. Chem.,* **6**, 1685.
325. Bodner, R.E. and Steinfink, H. (1967) *Inorg. Chem.,* **6**, 327.
326. Dell, R.M. and Bridger, N.J. (1972) *Inter. Rev. Sci.,* **7**, 211.
327. Pivan, J.Y., Pena, O., Ghadraoui, E.H., Guerin, R. and Bonville, P. (1990) *Physica B.,* **163**, 185.
328. Matsumoto, T., Ye, J., Huang, Y.K. and Kadowaki, K. (1996) *Acta Cryst.,* **C52**, 1323.
329. Johrendt, D. and Mewis, A. (1995) *Z. Anorg. Allg. Chem.,* **621**, 57.
330. McClincy, R.J. and Larson, A.H. (1969) *Trans. Metall. Soc., A.I.M.E.,* **245**, 173.

4 Coordination chemistry and solution chemistry

S.M. GODFREY, C.A. McAULIFFE, A.G. MACKIE and R.G. PRITCHARD

4.1 Arsenic, antimony and bismuth compounds as Lewis bases

The ability of ER_3 (E = As, Sb, Bi) to act as Lewis bases has long been recognized and numerous structural examples have been described in the literature. All such compounds arise from the donation of the lone pair of electrons on the E atom to a suitable electrophile. The majority of such complexes may be considered as two electron donors. The basicity of ER_3 compounds does not vary smoothly down the group but rather AsR_3 compounds are recognized as being stronger complexing ligands than SbR_3 and especially BiR_3 ligands. The two main reasons for this have been proposed to be the post transition metal effect (d-orbited contraction) between phosphorus and arsenic, and the lanthanide contraction between antimony and bismuth. A further influence on the nature of these complexing ligands is attributable to relativistic effects (chapter 1). Detailed discussion of these phenomena are given in the literature [1–3]. An additional reason for the relatively weak Lewis basicity of BiR_3 compounds is that the lone pair of electrons resides in an essentially s-character orbital [4]. Although the Lewis basicity of a given ER_3 species is generally thought to be governed by electronic factors, that is, the nature of the R groups bound to the central E atom, there is also evidence that the cone angle of the ER_3 species is of importance. The cone angle concept, first expounded by Tolman [5] for tertiary phosphine complexes has been extended to cover the heavier group 15 atoms (As, Sb, Bi) by Imyanitov [6]. A table of relevant cone angles is displayed in Table 4.1 (for ER_2R' as well as for ER_3). It is interesting to note that the cone angles show a decrease as the group 15 atom becomes heavier, and, in general, the cone angles of tertiary arsines are slightly larger than those of the corresponding tertiary phosphines, whereas the cone angles become smaller for a given ER_3 species (compared with PR_3) for SbR_3, and BiR_3, respectively. Clearly, the cone angle and thus the steric bulk of a given ER_3 species will affect how strong it will be as a complex ligand. The Lewis base interaction of ER_3 with dihalogens has received considerable study and two different structural modifications have been identified,

Table 4.1 Cone angles for arsines, stibines and bismuthines.

	Cone angle (°)		
	E=As	E=Sb	E=Bi
ER_3			
R=alkyl:			
EMe_3	121	119	117
EEt_3	134	131	129
EPr^n_3, EBu^n_3	134	131	129
EBu^i_3	145	141	139
EPr^i_3	162	157	155
EBu^s_3	162	157	155
EBz_3	166	162	159
$E(CH_2CMe_3)_3$	181	176	173
EBu^t_3	183	178	175
R=aryl:			
EPh_3	147	143	141
E(*o*-tolyl)$_3$	195	189	185
E(mesityl)$_3$	212	205	202
ER_2R'			
EMe_2Ph	125	122	120
EMe_2CF_3	127	124	123
EPh_2Me	138	135	133
EPh_2Et	140	137	135
EPh_2Pr^i	152	148	146
EPh_2Bu^t	158	154	152
EBu^t_2Ph	171	166	164
$E(CH_2CH_2CN)_3$	134	131	129
$E(CF_3)_3$	139	136	134
EPh_3	185	179	176

several crystal structures being available including Ph_3AsX_2 (X = F [7], Br [8]), Me_3AsCl_2 [9], R_3SbX_2 (R = Ph [10], Me [12]; X = Cl, Br, I) and Ph_3BiCl_2 [13], which are all trigonal bipyramidal. However, the compounds Ph_3AsI_2 [14], Me_3AsX_2 (X = Br, I) [8] adopt a molecular four-coordinate structure in which the arsenic atom adopts tetrahedral geometry. In this latter case, the lone pair of electrons on the arsenic atom interacts with the LUMO σ^* orbital of X_2. The X_2 moiety may therefore be considered as being sp^3d hybridized and the complex would be expected to have linear As–X–X bond angles. This has been established by the near-to-linear bond angles for all crystallographically characterized compounds of this type. In addition, significant lengthened X–X bonds are also exhibited, as expected since electron density is being put into the antibonding orbital of the X_2 by the tertiary arsine donor. The reasons why these two structural modifications exist has been reasoned to be due to the relative basicities of the ER_3 species. A number of other adduct species have been described. The reaction of $BiPh_3$ with $BiBr_3$ and subsequent recrystallization from tetrahydrofuran (thf)/hexane produced the equatorially vacant trigonal bipyramidal or disphenoidal $BiPh_2Br(thf)$. The complex was described as a bismuth acceptor complex of thf and compared with the established $BiPhBr_2$(thf) [15]. A similar acceptor complex has been established for the dimethylsulphoxide complex [(1,2-O_2-3-OHC_6H_3) CpSb(Me_2SO)] [16]. The weak interaction of tetracyanoethylene with EPh_3 (E = As, Sb, Bi) has also been described [17].

A number of tetracoordinate complexes containing nitrogen, oxygen and/or nitrogen donor ligands have been described, including the polymer diphenylacetatostibine, which exhibits a disphenoidal geometry [18]. Also adopting this geometry at the antimony centre are the compounds $Ph_2Sb(O_2PPh_2)$ and $Ph_2Sb(OSPPh_2)$ [19]. All the three complexes have the stereochemically active lone pair in equatorial positions. The five coordinate antimony containing anions [20] $[SbCl_3(CN)_2]^{2-}$ and $[MeClSbBr_2]^-$ have also been described, the latter [21] being prepared from $MeSbBr_2$ and Ph_4SbCl. The bismuth complexes $BiPhX_2$(thf) (X = Br, I) form polymer chains containing long halogen–bismuth interactions, the bismuth atoms adopting a square pyramidal geometry [15]. An example of a tridentate ligand coordinating to an antimony(III) centre is provided by the complexes $[Ph_2SbS_2EPh_2]_2$ (E = P, As) [22], which contain intramolecular as well as intermolecular interactions forming weakly bound dimers. The ability of R_3E species to act as Lewis bases in transition metal and main group metal complexes as a result of their donor properties is well established, especially in transition metal chemistry. With regard to main group complexes, a number of gallium complexes have been formed from the reaction of organoarsenic(III) compounds with gallium(III) salts [23–26]. The resultant complexes are four-membered heterocycles $[R_2As{-}GaR']_2$ [R = CH_2SiMe_3; R′ = Cl, Ph, $As(CH_2SiMe_3)_2$]. In contrast, reaction of $GaCl_3$ with $Bu^t_2SbSiMe_3$ forms [27] the six-membered ring complex $[R_2SbGaCl_2]_3$ (R = Bu^t), the heterocyclic ring adopting a boat conformation. An example of an indium(III) complex containing a dialkylstibine ligand is $[R_2Sb\text{-}InClSbR_2]_2$ (R = CMe_3) prepared at −78°C from two equivalents of $Bu^t_2SbSiMe_3$ and one of $InCl_3$ [28]. A very interesting illustration of the differences in structures available for main group complexes and how this is influenced by Lewis acidity or basicity is a comparison of the complexes $As_2Cl_6(AsEt_3)_2$ [29], which is a discreet dimeric species, and $Sb_2I_6(PMe_3)_2$ thf [30], which is best described as a loosely bound polymer of strongly bound dimers. Presumably it is the slightly lower Lewis basicity of $AsEt_3$ compared with PMe_3 [or the higher Lewis acidity of the antimony(III) iodide compared with the arsenic(III) chloride centres] which means that, unlike in the antimony complex, in the arsenic complex no long-range As . . . Cl interactions prevail to complete octahedral coordination around the arsenic(III) centres and thus there is no polymeric structure.

Transition metal complexes containing ER_3 ligands are numerous, although less so than the corresponding tertiary phosphine complexes, and are rather limited to organometallic chemistry. This is presumably because of the reluctance of the relatively soft ER_3 (E = As, Sb, Bi) Lewis bases to coordinate to the higher metal oxidation states (and thus harder acid centres) associated with coordination chemistry [31, 32]. Consequently, it is mainly with the larger and later members of the transition elements that examples of complexes containing ER_3 ligands (E = As, Sb, Bi) have been clearly established. However a number of nickel(II) complexes have been described containing either monodentate or bidentate tertiary arsine ligands. The dinuclear nickel complex $[(PMe_3)Ni(\mu\text{-}AsBu^t_2)]_2$ was prepared [33] from $AsBu^t_2Li$ and $NiCl_2(PMe_3)_2$ and contains a nickel–nickel bond and bridging $AsBu^t_2$ ligands. The nickel(III) complex, $NiI_3(AsMe_3)_2$ has also been structurally characterized [34], being obtained unexpectedly from the reaction of $[\{Ni(\mu\text{-}Bu^t_2As)(PMe_3)_2\}_2]$ with diiodine. $NiI_3(AsMe_3)_2$ may also be prepared from the reaction of Me_3AsI_2 with nickel powder, the equation being balanced by the formation of diiodine [35]. Me_3AsI_2 also reacts with cobalt metal powder to produce the isostructural $CoI_3(AsMe_3)_2$ [36]. A number of bis-tertiary arsine nickel(II) complexes [37] which contain the metal in five-coordinate geometry are $[Ni(mtas)_2]^{2+}$, $[Ni(ptas)]^{2+}$ and $[Ni(diars)(mtas)]^{2+}$ (diars = 1,2-$(Me_2As)_2C_6H_4$); mtas = bis(2-dimethylarsinophenyl) methyl arsine; ptas = bis(2-dimethylarsinophenyl) phenyl arsine). A large series of complexes of ER_3 bond to silver to form species of the type $[Ag(ER_3)_n]^+ X^-$ (E = As; R = Ph, $n = 3$, X = NO_3; R = Ph, $n = 4$, X = $SnPh_2(NO_3)_2Cl$; R = Ph, $n = 3$, X = Cl; R = C_6F_5, $n = 2$, X = ClO_4]. Such complexes have been studied in detail to evaluate the Lewis base strengths of the ER_3 ligands [38–40]. $BiPh_3$ is also able to coordinate to $AgClO_4$ but considerably less strongly, again reflecting the poor Lewis basicity compared with ER_3 (E = As, Sb) [41].

4.2 Halogenoanion complexes of the pnictogens

4.2.1 *Tervalent halopnictogen complex anions*

A number of reviews on the halopnictogens are available [42–45]. In the later sections of this chapter the structures are described as reported by the authors because there is not yet a sufficiently compelling theoretical basis on which to reinterpret definitively the older data. In earlier papers the stereochemically active lone pair was more frequently cited to account for distortions. The more recent papers often attribute distortions to hydrogen bonding, cation–anion interactions or crystal forces.

The formation of halopnictogen anions is essentially a Lewis acid–base reaction:

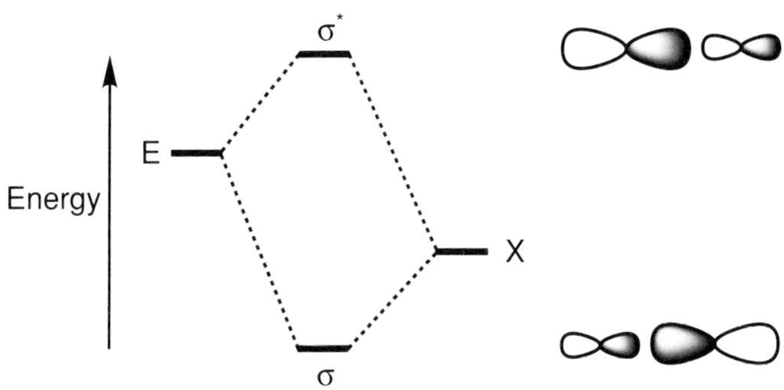

Figure 4.1 A σ two-orbital diagram for E–X.

$$\underset{\text{base}}{X^-} + \underset{\text{acid}}{EX_3} \rightleftharpoons [EX_4]^- \underset{-X^-}{\overset{+X^-}{\rightleftharpoons}} [EX_5]^{2-} \underset{-X^-}{\overset{+X^-}{\rightleftharpoons}} [EX_6]^{3-} \qquad [4.1]$$

The heavy pnictogens are powerful halide acceptors and this was rationalized by the classic model employing vacant d orbitals to form hybrid sets, for example dsp^3 and d^2sp^3 for five- and six-coordinate complexes respectively [46]. This model has been questioned for some time particularly by Schleyer and Reed [47] who reinterpret the data in terms of E-X σ* orbitals (chapter 1). In the case of two orbitals, σ-bond formation between E and the more electronegative halide X gives rise to the orbital interaction diagram illustrated in Fig. 4.1. Since the σ orbital will be polarized towards the more electronegative X, the antibonding σ* orbital will be polarized towards the pnictogen. Providing the σ* orbital is of sufficiently low energy, it can act as an acceptor orbital for Lewis base X^- to form a bond with the pnictogen atom as shown in Fig. 4.2.

The addition of X^- ligands to EX_3 uses the σ* orbitals *trans* to the E–X bond resulting sequentially in $[EX_4]^-$, $[EX_5]^{2-}$ and $[EX_6]^{3-}$. This model and valence shell electron pair repulsion (VSEPR) theory give

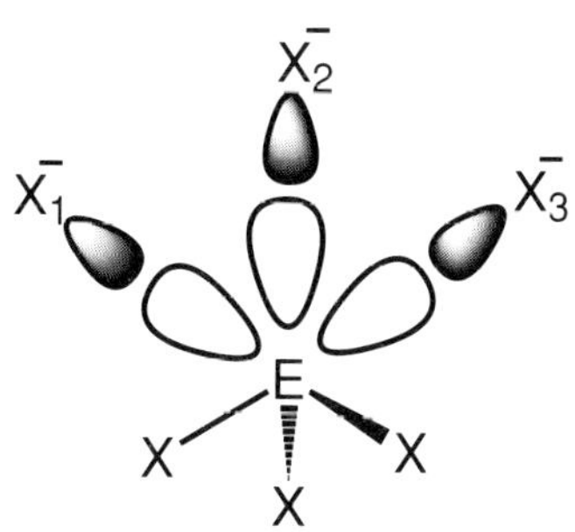

Figure 4.2 An interaction diagram involving the E–X σ* orbital.

similar structures for $[EX_4]^-$ and $[EX_5]^{2-}$. VSEPR theory always must predict a distorted structure for $[EX_6]^{3-}$ structures because of the necessity of accommodating the lone pair. However, the model using σ* orbitals will result in symmetrical $[EX_6]^{3-}$ octahedral geometry. This topic is discussed more fully in chapter 1.

The halopnictide complex anions show a great variety of structural types ranging from simple anions, $[EX_4]^-$ (E = As, Sb, Bi; X = halogen) to large anionic clusters, for example $[E_8X_{28}]^{4-}$. The influence of the cation is important, large cations promoting the growth of large anion complexes in crystal structures [48, 49]. Crystal structure data often show two broad groups of E−X bond distances in EX_6 units of three shorter bonds, which Alcock [50] termed primary bonds, and three longer bonds (secondary bonds). These secondary bonds may range from being equal in length to the primary bonds to being of the order of the sum of the van der Waals radii.

Empirical formula can be misleading, for example, at first sight Cd_3AsCl_3 may seem to be an As−Cl compound but in fact it has no such bonds, the arsenic coordination being $AsCd_4$ tetrahedra [51]. The chlorine atoms form layers with rhombic symmetry. In this rectangular lattice the centres of the hexagons correspond to the $AsCd_4$ tetrahedra; the stacking of the Cl layers is distorted hexagonal.

The success of the VSEPR model in predicting and explaining main group molecular geometry tends to obscure its less successful predictions. This is especially true of the heavier p-block elements such as antimony and bismuth. A compound which shows the VSEPR predicted structures is a confirmation of the theory, but one that does not conform is described as having an inert pair in a spherically symmetrical stereochemically inactive ns^2 orbital, by way of discounting the problem. This simple view is now being questioned by, for example, Wheeler and Kumar [52]. Their extended Hückel molecular orbital and crystal orbital calculations show an underlying tendency for a second-order Jahn–Teller distortion from octahedral to trigonal coordination of the central pnictogen (chapter 1).

Whether or not the distortion is structurally realized for a $[EX_6]^{3-}$ molecule depends on the balance between HOMO–LUMO mixing, the mixing between the HOMO and lower energy, $1a_1$ orbital with ligand lone pair and E p_z character and repulsions among ligand lone pairs.

Haloarsenic(III) anions. Arsenic(III) halides show good Lewis acid strength when acting as halide acceptors:

$$MX + AsX_3 \rightarrow M^+[AsX_4]^- \qquad [4.2]$$

where M = cationic species, X = halogen, although this simple reaction does not always occur. The stoichiometry may be $MAsX_4$ but the structure is not necessarily monomeric since in $[PyH][AsI_4]$ [53] the anion has a chain structure composed of AsI_6 octahedra. The anticipated reaction may not occur if the balance of forces, described in the previous, introductory, subsection is unfavourable; for example, Drew, Claire and Willey [54] intended to prepare transition metal cationic species by reactions such as

$$TiX_4 + AsX_3 \xrightarrow{\text{Solvent (solv.)}} [TiX_3\,(\text{solv.})_n]^+[AsX_4]^- \qquad [4.3]$$

which are successful in the case of $SbCl_5$. The products obtained, however, contained no arsenic where $POCl_3$ was used as a solvent:

$$TiCl_4 + AsCl_3 \xrightarrow{POCl_3} [Ti_2Cl_6(POCl_3)_2(POCl_2)_2] \qquad [4.4]$$

Changing the solvent to acetonitrile afforded $[TiCl_4(MeCN)_2]$. Similarly, when magnesium or zinc chloride were reacted with $AsCl_3$ in acetonitrile the product was $[MCl_2(MeCN)_2]$ (M = Zn, Mg).

The unpredictability of the products in the case of transition metals is well illustrated in the reaction of a nickel salt with AsI_3 in thf/butanol [55]:

$$[Ni(H_2O)_6][ClO_4]_2 + AsI_3 + \text{triphos} \xrightarrow[N_2]{\text{thf/butane}} 2[(\text{triphos})NiI][As_6I_8] \qquad [4.5]$$

The structure was shown to be of cubane-type, but with As–As bonding. The structure may be described as the chair conformer of cyclic $(AsI)_6$ capped by two I^- anions.

The influence of the cation is crucial in crystal formation. Sheldrick, Kaus and Häusler [53] reacted organoammonium iodides with AsI_3 in acetonitrile, which afforded a variety of products, some of which are shown in reaction [4.6]:

$$AsI_3 + [Q]I \xrightarrow{MeCN} \begin{cases} \xrightarrow{Q\,=\,PyH} [PyH][AsI_4] \\ \xrightarrow{Q\,=\,Bu_4N} [Bu_4N]_2[As_2I_8] \\ \xrightarrow{Q\,=\,Et_3NH} [Et_3NH]_4[As_8I_{28}] \end{cases} \qquad [4.6]$$

The structure of the iodoarsenate anion varied with the cation. The $[AsI_4]^-$ anion has already been mentioned. The $[As_2I_8]^{2-}$ anion consists of discrete dimeric units with C_i symmetry. The arsenic lone pair is stereochemically active, occupying an octahedral site, that is, the structure is effectively an edge-sharing square-based pyramid with *trans* apices as shown in Scheme 4.1. The $[As_8I_{28}]^{4-}$ anion consists if eight edge-bridged AsI_6 octahedra.

$AsBr_3$ and $[Et_3NH]Br$ in $MeNO_2$ afford $[Et_3NH]_3[As_3Br_{12}]$, the anion being shown by X-ray crystallography [56, 57] to consist of a central $AsBr_6$

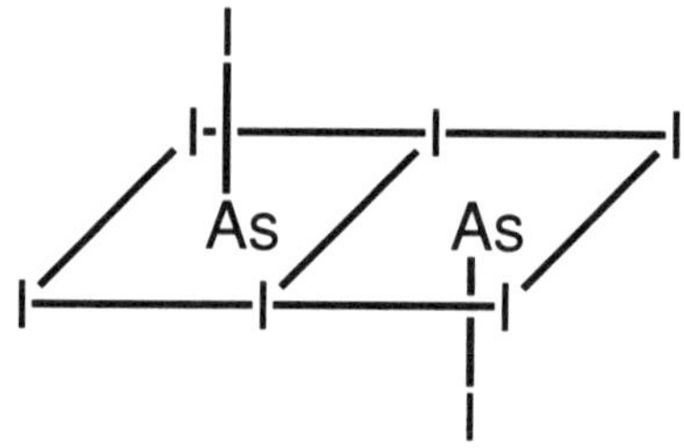

Scheme 4.1 The structure of the $[As_2I_8]^{2-}$ di-anion.

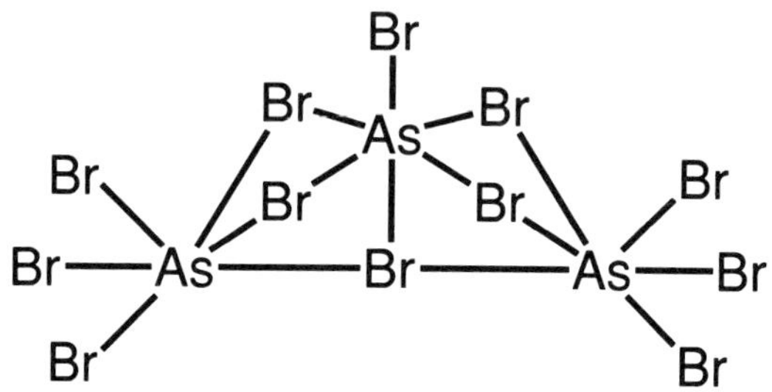

Scheme 4.2 The structure of the $[As_3Br_{12}]^{3-}$ tri-anion.

octahedron connected via two faces with a common corner to two other $AsBr_6$ octahedra, as shown in Scheme 4.2.

$AsCl_3$ forms the As_2Cl_8 anion with a structure [58] analogous to that of the iodide. However, the anion with stoichiometry $[As_8Cl_{28}]^{4-}$ is different from $[As_8I_{28}]^{4-}$. The reaction of trithiazyl chloride, $N_3S_3Cl_3$, with As_2O_3 affords the complex chloroarsenic anionic compound $[S_5N_5]_4[As_8Cl_{28}]\cdot 2S_4N_4$. A central $[As_4Cl_{16}]^{4-}$ unit with a cubane-type structure has four loosely bound $AsCl_3$ molecules [59], that is, the anion is better represented as $[As_4Cl_{16}(AsCl_3)_4]^{4-}$. $AsCl_3$ [60] and $AsBr_3$ [61] form haloanions of the type $[As_2X_9]^{3-}$ which consist of face sharing AsX_6 octahedra.

The s-block haloarsenic anions compounds $M_3[As_2X_9]$ are decomposed on heating, reforming the reactants:

$$M_3[As_2X_9] \xrightarrow{\Delta} 3MX + 2AsX_3 \qquad [4.7]$$

In a thermogravimetric study of the decomposition of $M_3[As_2X_9]$ (M = Rb, Cs; X = Cl, Br), the dissociation temperatures found by Zalewicz and Ptaszynski [62] are given in Table 4.2. Their attempts to prepare $M_3[As_2I_9]$ by the standard methods [63, 64] failed.

Haloantimony(III) anions. Monomeric haloantimonate(III) anions $[SbX_4]^-$, $[SbX_5]^{2-}$ and $[SbX_6]^{3-}$ are formed by the addition of MX

Table 4.2 Thermolysis of $M_3[As_2X_9]$ compounds.

Compound	Temperature range of decomposition(°C)	DTA[a] peak temperatures (°C)
$Rb_3[As_2Cl_9]$	162–320	242, 732
$Rb_3[As_2Br_9]$	200–292	285, 698
$Cs_3[As_2Cl_9]$	200–325	319, 475, 642
$Cs_3[As_2Br_9]$	245–380	370, 640

Source: Zalewicz, M. and Ptaszynski, B. (1979) *Polish J. Chem.* **53**, 1437.
[a]Different in thermal analysis.

(M = alkali metal or organocationic species, X = halogen) to SbX_3. The $[SbX_4]^-$ anions have a disphenoidal structure whereas the $[SbX_5]^{2-}$ anions have a square-based pyramidal structure resulting from the stereochemically active $5s^2$ lone pair on antimony. However, the $[SbX_6]^{3-}$ anions generally have a symmetrical octahedral geometry. Sowerby *et al.* [65] use an alternative model in which successive additions of X^- anions are made *trans* to X in pyramidal SbX_3 (X–Sb–X angle *c.* 90°). As the $Sb\cdots X^-$ bond forms the Sb–X bonds adjust until all the bonds are approximately equal in length and the charge is equally shared. This is equivalent to treating each $X\cdots Sb\cdots X$ moiety as a three-centre, four-electron system, which affords the observed $[SbX_4]^-$ and $[SbX_5]^{2-}$ structures and, of necessity, an undistorted octahedral geometry for the $[SbX_6]^{3-}$ anions (4.2.1 and Chapter 1).

Theoretical calculations [66] have, however, cast doubts on the validity of the assumption that the geometry of the $[SbF_5]^{2-}$ anion is influenced by the lone pair. The stereochemical aspects of antimony(III) halides and their anionic derivatives has been reviewed by Sawyer and Gillespie [44].

An interesting fluoroantimonate anion $[Sb_4F_{17}]^-$ is formed when SbF_5 is reacted with selenium in sulphur dioxide solution for a week at 100°C to yield a yellow-green crystalline product [67]:

$$Se + SbF_5 \xrightarrow{SO_2} [Se_4]_2[Sb_4F_{17}][SbF_6]_3 \qquad [4.8]$$

The complex anion $[Sb_4F_{17}]^-$ has also been prepared by Gillespie and co-workers [68] in $[Se_3S][Sb_4F_{17}][SbF_6]_3$. It also occurs in the complex binary antimony fluoride Sb_8F_{30}, or $[Sb_3F_7]^{2+}[Sb_4F_{17}]^-[SbF_6]_3^-$ [69].

The anion may be considered as composed of two SbF_3 moieties bridged by a fluoride ion and this unit coordinated to two SbF_5 ligands, that is, $F_5Sb\cdots F_3Sb\cdots F\cdots SbF_3\cdots SbF_5$.

Compounds of the type $M_3[Sb_2X_9]$ (M = alkali metal or organic cationic species) have been shown to have a polymeric structure for X = F or Cl [70]. However, in $[NMe_4]_3[Sb_2Br_9]$ [65] a discrete nonabromodiantimonate anion is formed with three bridging bromine atoms. Sowerby and co-workers [65] have also shown that partial substitution of the chlorine in the polymeric $[Sb_2Cl_9]^{3-}$ anion by bromine affords $[Sb_2Br_3Cl_6]^{3-}$, a discrete anion with the bromine selectively occupying bridging sites. The change in structure was attributed to the size of the halogen and the residual charges carried by the bridging atoms.

The polymeric structure of the $[Sb_2Cl_9]^{3-}$ anion can be broken up by partial hydrolysis. Thus $[PyH]_3[Sb_2Cl_9]$ when recrystallized from moist ethanol formed $[PyH]_2[Sb_2OCl_6]$. The crystal structure [71] showed discrete anions with two chlorine bridging atoms and one oxygen bridging atom.

The $[Sb_2Br_8]^{2-}$ anion was prepared by Müller and Mohammed [72], from Sb_2S_3 and HBr in the presence of a phosphonium bromide:

$$2PPh_4Br + Sb_2S_3 + 6HBr \rightarrow [PPh_4]_2[Sb_2Br_8] + 3H_2S \quad [4.9]$$

The anion was shown by X-ray analysis to consist of two square-based pyramidal $SbBr_5$ units sharing a common edge with the apical bromine atoms in a *trans* disposition, as shown for the $[As_2I_8]^{2-}$ anion in Scheme 4.1.

Interestingly, the iodine analogue $[PPh_4]_2[Sb_2I_8]$ is not isostructural, crystallizing as a di-solvate of MeCN. An X-ray investigation by Pohl, Saak and Haase [73] revealed each antimony atom to have an η^2 interaction with one of the phenyl rings of the PPh_4^+ action, giving a coordination sphere SbI_5C_2, an uncommon seven-coordinate species. Pohl, Saak and Haase [73] suggest that the iodine substituents both lower the Lewis acidity of antimony and favour the inert pair effect, that is, the lone pair become stereochemically inactive.

The $[Sb_2X_9]^{3-}$ anions show no distortions attributable to a lone pair effect. In contrast $[Co(NH_3)_6][Sb_2F_9]$ has been shown [74] to have only one bridging fluorine atom, that is, $[F_4Sb{-}F{-}SbF_4]^-$, strongly suggesting a stereoactive lone pair.

A representative selection of references for iodoantimonate crystal structures is given in Table 4.3. The wide range of (Sb−I) distances and the associated bond angles makes average values almost meaningless and hence they have been omitted.

Halobismuth(III) anions. Halobismuth anions are formed by the reaction of halide donors with bismuth(III) halides either in a hydrothermal reaction, for example in constant boiling HBr (47%) [75]:

$$RbBr + BiBr_3 + HBr_{(aq)} \xrightarrow{20\ d} Rb_3[BiBr]_6 \quad [4.10]$$

or by fusion methods, for example stoichiometric amounts heated in a sealed quartz tube [76]:

Table 4.3 A selection of available iodoantimonate crystal structures.

Cation	Anion	Reference
$[Me_3BzN]^+$	$[Sb_2I_7]^-$	[i]
$[CH_2C(4\text{-}C_6H_4NO_2)CH_2NMe_3]$	$[Sb_3I_{10}]^-$	[ii]
$[C(NMe_2)_3]^+$	$[Sb_3I_{10}]^-$	[iii]
$[Ph_4P]^+$	$[Sb_3I_{10}]^-$	[iv]
$[K(15\text{-crown-}5)]^+$	$[Sb_3I_{12}]^{3-}$	[v]
$[Me_4N]^+$	$[Sb_5I_{18}]^{3-}$	[vi]
$[H\{OPN(Me_2)_3\}_2]$	$[Sb_5I_{18}]^{3-}$	[vii]
$[FeCp_2]^+$	$[Sb_6I_{22}]^{4-}$	[vi]
$[Fe(1,10\text{-phen})_3]^+$	$[Sb_6I_{22}]^{4-}$	[viii]
$[PPh_4]^+$	$[Sb_8I_{28}]^{4-}$	[iv]
$[Me_4N]^+$	$[Sb_8I_{28}]^{4-}$	[i]
$[Me_3S]^+$	$[Sb_8I_{28}]^{4-}$	[i]

References

i Pohl, S., Lotz, R., Saak, W. and Haase, D. (1988) *Z. Naturforsch*, **43B**, 1144.
ii Norman, N. C., Carmalt, C. J. and Farrugia, L. J. (1994) *Polyhedron*, **13**, 1655.
iii Pohl, S., Saak, W., Mayer, P. and Schmidpeter, A. (1986) *Angew. Chem. Int. Ed. Engl.*, **25**, 825.
iv Saak, W., Haase, D. and Pohl, S. (1987) *Z. Naturforsch.*, **42B**, 1493.
v Dehicke, K., Weller, F. and Borgsen, B. (1991) *Z. Anorg. Allg. Chem.*, **596**, 55.
vi Saak, W., Lotz, R., Haase, D. and Pohl, S. (1989) *Angew. Chem. Int. Ed. Engl.*, **28**, 344.
vii Norman, N. C., Carmalt, C. J. and Farrugia, L. J. (1993) *Polyhedron*, **12**, 2081.
viii Lotz, R., Haase, D., Pohl, S. and Saak, W. (1988) *Z. Naturforsch*, **43B**, 1033.

$$2NaI + BiI_3 \xrightarrow{600°C} Na_2[BiI]_5 \qquad [4.11]$$

Some typical examples of these metal cation halobismuthate compounds are given in Table 4.4.

Alkali metal fluorobismuth compounds are prepared by fusion techniques, for example, a series of compounds $M_2M'[BiF]_6$ (M = Cs, Rb; M' = Na, K, Rb, Tl) was prepared by Cousson, Vedrine and Consseins [77].

$K_3[BiF]_6$ was prepared by Hobecker [78] from $KBiS_2$. The product was the thermally metastable β (tetragonal) isomer, which spontaneously rearranges to the α (cubic) form at room temperature over several days:

$$K[BiS]_2 \xrightarrow{KF/370°C} \beta\text{-}K_3[BiF]_6 \xrightarrow{slowly} \alpha\text{-}K_3[BiF]_6 \qquad [4.12]$$

The organocation halobismuthates are normally prepared in solvent systems, usually aqueous based, for example Bi_2O_3 is dissolved in constant boiling hydrogen halide and added to a hot aqueous solution of the hydrohalide salt. In the case of 4,6-dimethylpyrimidine-2-(1H)-thione (L), the

Table 4.4 Selected metal-halobismathate compounds.

Compound	Synthetic method	Colour of crystals	Density (g cm^{-3})	Reference
$Rb_3[BiCl_3]$	Hydrothermal	White	3.24	[i]
$Rb_7[Bi_3Cl_{16}]$	Hydrothermal	White	3.53	[i]
$Rb_3[BiBr_6]$	Hydrothermal	Pale yellow	4.19	[ii]
$Rb_3[Bi_2Br_9]$	Hydrothermal	Yellow	4.51	[ii]
$Rb_7[Bi_2Br_{16}]$	Hydrothermal	Pale yellow	4.25	[ii]
$Ag[BiI_4]$	Fusion	Black opaque	4.82	[iii]
$Na_2[BiI_5]$	Fusion	Ruby	3.14	[iv]
$Ba[BiI_5]$	Fusion	Bright red prismatic	5.03	[v]
$Ag_3[BiI_6]$	Fusion	Black opaque	5.15	[iii]
$Rb_3[Bi_2I_9]$	Fusion	Orange	—	[vi]

References

i Vlasova, I. V., Stepina, S. B., Sokolova, N. P. and Plyushchev, V. E. (1968) *Russ. J. Inorg. Chem.*, **13**, 1326.
ii Stancheva, L. I., Vlasova, I. V., Stepina, S. B. and Plyushchev, V. E. (1966) *Russ. J. Inorg. Chem.*, **11**, 761.
iii Dzervanova, K. B., Bukhalova, G. A. and Kaloev, N. I. (1985) *Russ. J. Inorg. Chem.*, **30**, 2983.
iv Dzervanova, K. B., Bukhalova, G. A., Kaloev, N. I. and Gazdanov, I. V. (1986) *Russ. J. Inorg. Chem.*, **31**, 756.
v Dzervanova, K. B., Bukhalova, G. A. and Kaloev, N. I. (1985) *Russ. J. Inorg. Chem.*, **30**, 1240.
vi Dzervanova, K. B., Bukhalova, G. A. and Kaloev, N. I. (1985) *Russ. J. Inorg. Chem.*, **30**, 1397.

red iodide $[HL]_3$ $[Bi_2I_9]$ was precipitated instantaneously in contrast to the yellow $[HL]_3[BiX_6]$ (X = Cl, Br) which only separated after three days on cooling to 0°C [79].

Phase diagram studies have been made of BiI_3 with NaI ($Na_2[BiI_5]$ melts incongruently at 525°C) [76], RbI ($Rb_3[Bi_2I_9]$ melts congruently at 503°C) [80], BaI_2 ($Ba[BiI]_5$ melts incongruently at 406°C) and AgI ($Ag_3[BiI_6]$, polymorphic transition over 170°C–180°C) [81]. Metal halides do not always afford identifiable products with bismuth halides; for example, in the system $PdCl_2$–$BiCl_3$ [82] a eutectic (3% $PdCl_2$), melting point 228°C, is observed but no compound was identified.

A thermoanalytical study [62] of $Rb_3[BiX_6]$ (X = Cl, Br) and $Cs_3[Bi_2X_9]$ (X = Cl, Br, I) showed that the decomposition of the rubidium salts was a one-stage process

$$\beta\text{-}Rb_3[BiCl_6] \xrightarrow{441°C} Rb_3[BiCl_6] \xrightarrow{605°C} 3RbCl + BiCl_3 \qquad [4.13]$$

and that of the caesium compounds a two-stage reaction:

$$Cs_3[Bi_2Cl_9] \xrightarrow{460°C} Cs_3[BiCl_6] \xrightarrow{803°C} 3CsCl + BiCl_3 \qquad [4.14]$$

$Rb_3[BiCl_6]$ showed a phase change at 441°C. The decomposition and DTA peak temperatures are given in Table 4.5.

Table 4.5 Thermoanalytical data of the decomposition of rubidium and caesium halobismuthates.

Compound	Temperature range of decomposition (°C)	DTA peak temperatures (°C)
$Rb_3[BiCl_6]$	580–920	433, 441, 605, 650, 950
$Rb_3[Bi\ Br_6]$	435–822	532, 587, 946
$Cs_3[Bi_2Cl_9]$	460–803	612, 695, 925
	803–950	
$Cs_3[Bi_2Br_9]$	520–680	620, *c.*920
	680–*c.* 920	
$Cs_3[Bi_2I_9]$	380–480	480, 633, 882, 966
	480–940	

Source: Zalewicz, M. and Ptaszynski, B. (1979) *Polish J. Chem.*, **53**, 1437.

The crystal structures of the halobismuthate anions are based mainly on the octahedral BiX_6 unit. The anions are mononuclear or multinuclear and monomeric or polymeric. The negative charge per bismuth atom can vary from one to three, although where neutral BiX_3 moieties are coordinated to the cluster averaging the charge gives fractional charge per Bi.

The multinuclear and polymeric structures are built up via a bridging halogen, that is, Bi–X–Bi moieties, which often exhibit unequal bond lengths. We will discuss the various halobismuthate structures as derivatives of $[BiX_4]^-$, $[BiX_5]^{2-}$ and $[BiX_6]^{3-}$.

Compounds of the type $MBiX_4$ (M = metal or organo cation, X = halogen) are expected to be polymeric because the BiX_4 moiety is highly coordinatively unsaturated with a strong tendency to achieve octahedral geometry. $NH_4[BiF_4]$ [83], $[Et_2NH_2]$ $[BiCl_4]$ [84], $[FeCp_2][BiCl_4]$ [85] and [2-picolinium]$[BiX_4]$ [86] (X = Br, I) form polymeric structures in which the non-bridging halogens are always *cis* as shown in Scheme 4.3.

The structure of the $[BiF_4]$ anion [87] is somewhat anomolous, as is often the case in fluorine chemistry. The bismuth atoms are coordinated by nine fluorine atoms with distances in the range 2.19(3)–2.86(3) Å. The coordination polyhedra are linked by sharing edges and faces to infinite layers of composition $[BiF_4]^-$ parallel to thc *yz* plane.

Unlike the arsenic and antimony analogues, tetrabismuthates do not form stable dimers because of the powerful tendency to form octahedral

Scheme 4.3 Structures of polymeric $[BiX_4]^-$ anions.

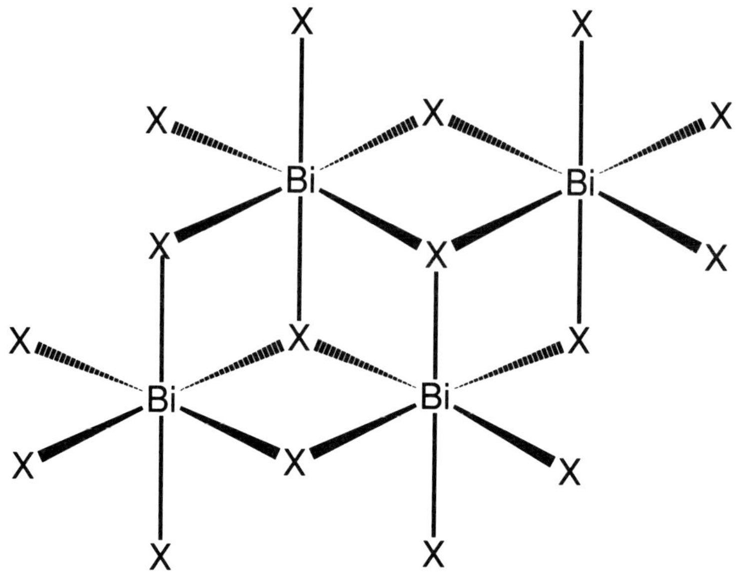

Scheme 4.4 The structure of the $[Bi_4X_{16}]^{4-}$ anions (X=Cl, Br).

BiX_6 structures. The only exception to date is $[BiCl_2$(18-crown-6)]$[Bi_2Cl_8]$ which, owing to crystal constraints has an edge- sharing bi-square pyramidal structure [88]. However, tetramers $[Bi_4X_{16}]^{4-}$ are often formed which may be considered to be 'dimeric dimers'. Examples are $[Mg(MeCN)_6]$ $[Bi_4Cl_{16}]$ [89] and $[FeCp_2]_4[Bi_4Br_{16}]$ [90] as shown in Scheme 4.4.

Compounds of the type $M_2[BiX_5]$ are not found as isolated BiX_5 units for the same reasons cited for $[BiX_4]^-$ above. The coordinative unsaturation is relieved by polymerization or dimerization. Thus in [piperidinium]$_2$ $[BiBr_5]$ [91], a polymeric structure (Scheme 4.5) results in which the Br–Bi–Br bridging is nearly symmetrical (3.02 Å and 3.13 Å) in contrast to the highly asymmetric bonding in the antimony analogue: examples of the dimers are $K_4[Bi_2X_{10}]$ [92], $[NH_4]_4[Bi_2Br_{10}]$ [92], $[Sr(H_2O)_8][Bi_2Br_{10}]$ [93] and $[HL]_4[Bi_2Br_{10}]$ (L = 2,5-diamino-1,3,4-thiazdiazole) [94]. The structure of the $[Bi_2Br_{10}]^{4-}$ anion is shown in Scheme 4.6.

The $[Bi_2Br_{10}]^{4-}$ anion also occurs as a structural element in $Na_7[Bi_3Br_{16}]$ [95] described later. In the triply charged species $M_3[BiX_6]$, the bismuth atom is sufficiently stabilized by six halogen ligands for the $[BiX_6]^{3-}$ anion to exist in isolation, that is, at greater than van der Waals radii contacts with the other components of the crystal. Examples characterized by single crystal X-ray structural analysis include $Cs_2Na[BiCl_6]$ [96], $[Me_2NH_2]_4[BiCl_6]Cl$ [97], $[HL]_3[BiCl_6]\cdot 2H_2O$ [79] (L = 4,6-dimethylpyrimidine-2(1*H*)thione), $Rb_3[BiBr_6]$ [98], $[Et_2NH_2]_3$ $[BiBr_6]$ [99] $[Me_2NH_2]_3$

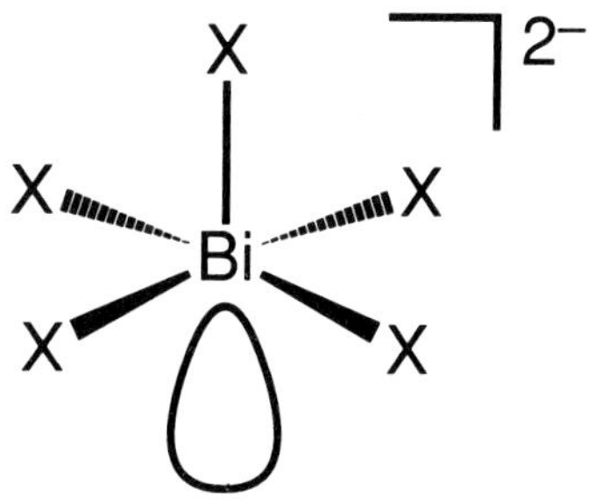

Scheme 4.5 The structure of the $[BiBr_5]^{2-}$ anions in [piperidinium]$_2$ $[BiBr_5]$.

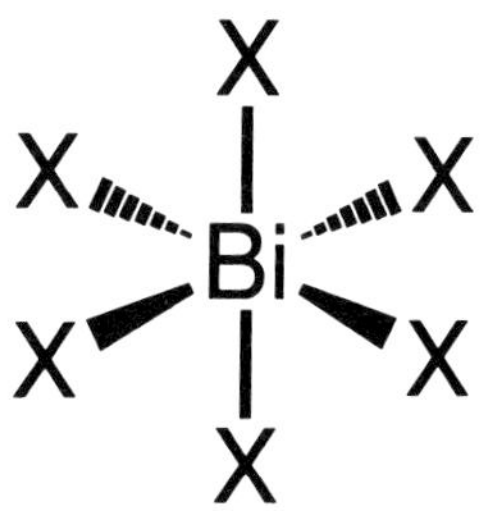

Scheme 4.6 The structure of the $[Bi_2BR_{10}]^{4-}$ anion.

$[BiBr_6]$ [100], $Na_7[BiBr_6][Bi_2Br_{10}]\cdot18H_2O$ [95], $Rb_5[BiI_6][I_3][I]\cdot2H_2O$ [101] and a range of alkali metal $[BiF_6]^{3-}$ salts [77]. The structures all approach symmetrical octahedral geometry with only minor distortions.

The $[Bi_2X_9]^{3-}$ (X = Cl, Br, I) anions are very stable being present in many crystal structures with large cations. Slow evaporation at room temperature from a saturated solution of Et_2NH and BiI_3 (3:2 mol ratio) in *n*-propanol-acetone-40%HI (90:9:1) afforded well-developed crystals of $[Et_2NH_2]_3[Bi_2I_9]$ [112]. The structure is that of face-sharing BiI_6 octahedra, as shown in Scheme 4.7. Similar structures were found in $[Ph_4P]_3[Bi_2Br_9]$ [87] and $[Et_2NH_2]_3$ $[Bi_2Br_9]$ [103].

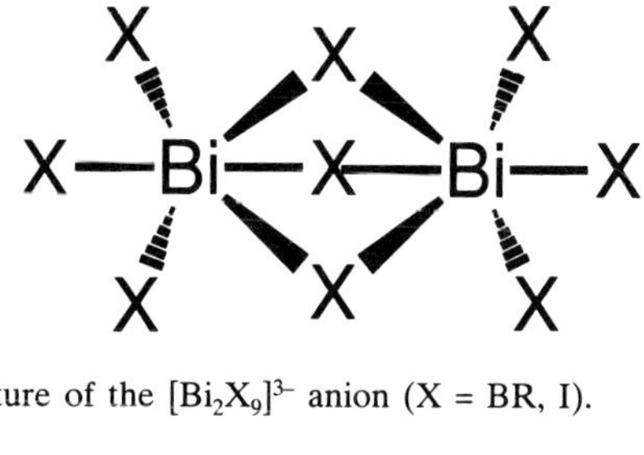

Scheme 4.7 The structure of the $[Bi_2X_9]^{3-}$ anion (X = BR, I).

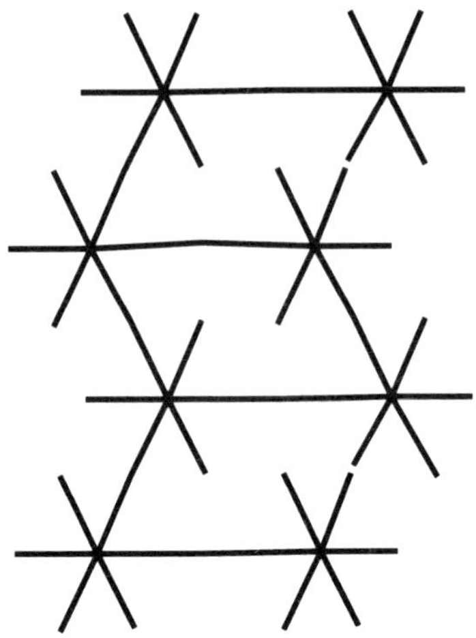

Scheme 4.8 The structure of the $[Bi_2Cl_9]^{3-}$ anion in $Cs_3[Bi_2Cl_9]$.

The BiX_6 octahedra may bridge via apices, rather than the face bridging of the dimers, in which case a polymeric structure is formed as found in α- and β-$Cs_3[Bi_2Cl_9]$ which forms an infinite double chain structure (Scheme 4.8). The two forms differ in their crystal stacking with a phase transition temperature of 373°C.

These double chain polymeric structures also occur in $Cs_3[Bi_2Br_9]$ [105] and $Cs_3[Bi_2I_9]$ [106]. The anion $[Bi_3X_{12}]^{3-}$ may be considered as the second member of the series $[BiX_6{\cdot}(BiX_3)_n]^{3-}$, $n = 1, \ldots, 4$, that is a $[BiX_6]^{3-}$ anion bi-capped by BiX_3. Examples of this type include $[Bu_4N]_3$ $[Bi_3I_{12}]$ [107] and $[Bi(dmpu)_6][Bi_3I_{12}]$ [43] (dmpu = dimethylpropylene urea) (Scheme 4.9).

The series is completed to date with $[Ph_4P]_3[Bi_5I_{18}]$ [43] with an extended linear structure, as shown in Scheme 4.10.

In $[MeNH_3]_5[Bi_2Cl_{11}]$ [108] and the isomorphous $[MeNH_3]_5[Bi_2I_{11}]$ [109] the two BiX_6 octahedra are connected apically, that is by one linear Bi–Br–Bi bridge (Scheme 4.11). The tetrabismuth anion $[Bi_4X_{18}]^{6-}$ may

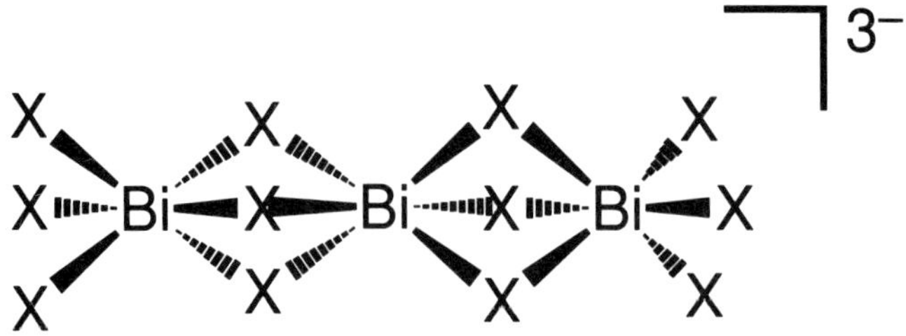

Scheme 4.9 The structure of the $[Bi_3X_{12}]^{3-}$ anion (X = I).

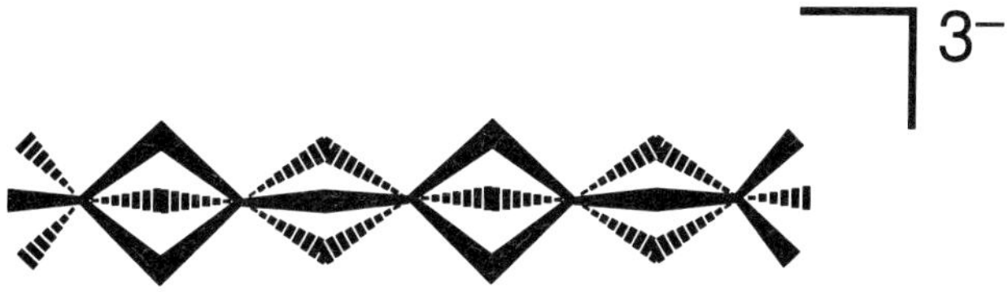

Scheme 4.10 The structure of the $[Bi_5X_{18}]^{3-}$ anion (X = I).

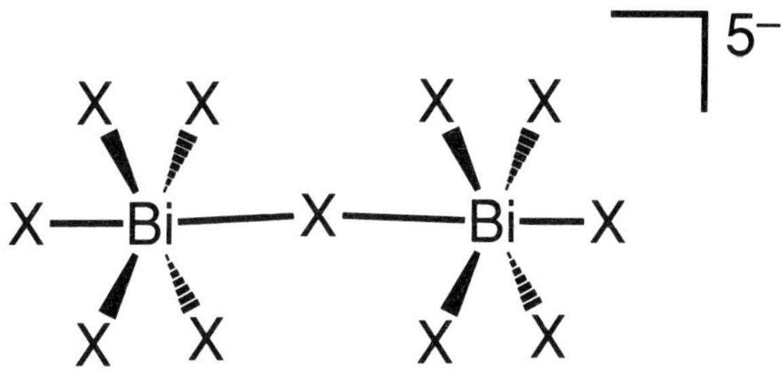

Scheme 4.11 The structure of the $[Bi_2X_{11}]^{5-}$ anion (X = Cl, I).

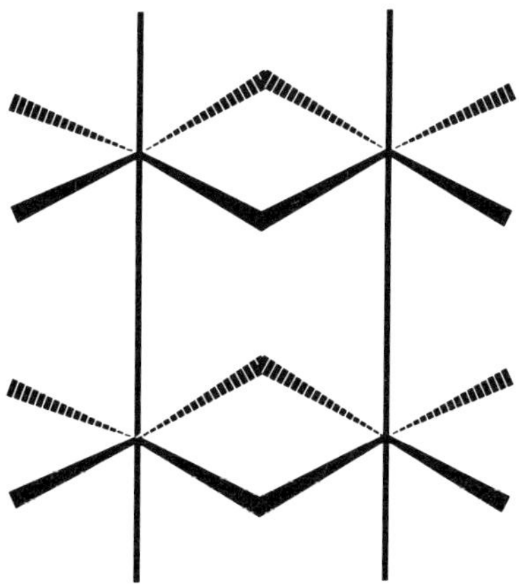

Scheme 4.12 The structure of the $[Bi_4X_{18}]^{6-}$ anion (X = Cl).

be considered to be a dimer of $[Bi_2X_9]^{3-}$ formed by mutual breaking one of the Bi−X−Bi bridges and reforming the bridges between the two anions. The structure of the anion in $[PyH]_6[Bi_4Cl_{18}]$ [110] is shown in Scheme 4.12.

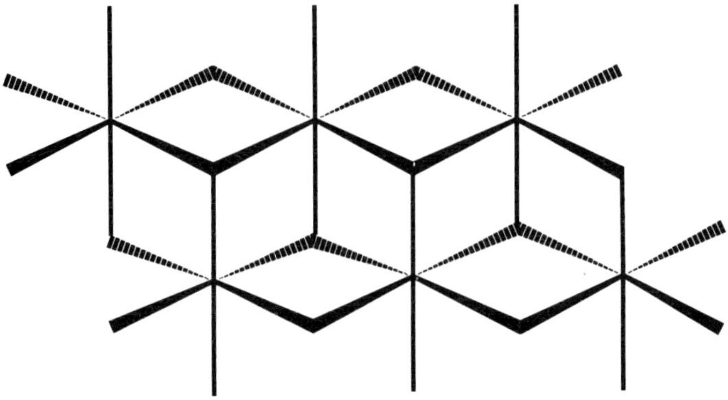

Scheme 4.13 The structure of the $[Bi_6X_{22}]^{4-}$ anion (X = I).

The hexanuclear anions $[E_6X_{22}]^{4-}$ may be considered members of a series that starts with the basic structure of $[E_2X_{10}]^{4-}$ and of general formula $[E_{2n}X_{6n+4}]^{4-}$. The structures are built up by capping two faces with neutralBiI_3 molecules. The structure of the anion $[Bi_6I_{22}]^{4-}$ is shown in Scheme 4.13 [111].

The largest discrete halobismuthate anion characterized to date is $[Bi_8Cl_{30}]^{6-}$ as the tetraethylammonium salt [112]. This structure may be envisaged as being built on a base unit of $[E_4X_{16}]^{4-}$ by the addition of two $[E_2X_7]^{-}$ anions. The structure of the anion is shown in Scheme 4.14.

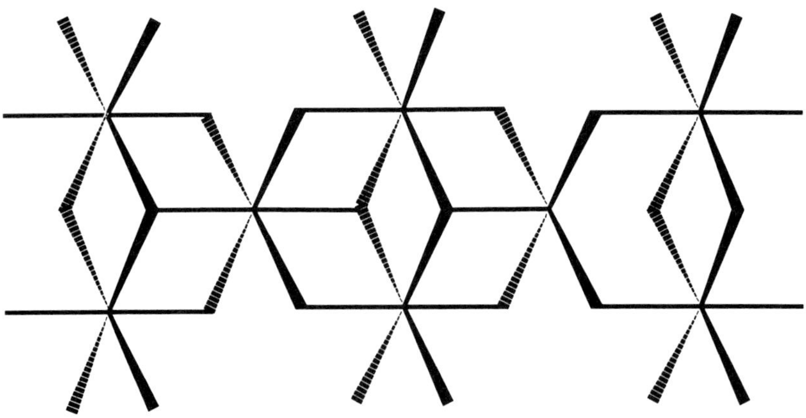

Scheme 4.14 The structure of the $[Bi_8X_{30}]^{6-}$ anion (X = Cl).

4.2.2 Pentavalent halopnictogen complex anions

The pentavalent halopnictogens, like their tervalent cogeners are also powerful Lewis acid halide acceptors. The basics of the bonding theory controversies for group 15 halocomplexes are similar for the pentavalent species and will not be repeated.

We have omitted from this discussion some interesting systems because the materials are not well characterized, notably the 'super acids' [113] and the graphite intercalation compounds of AsF_5 [114] and SbF_5 [115]

Haloarsenic(V) anions. Arsenic pentahalides, AsX_5, are powerful Lewis acids which are widely used to stabilize cationic species or even other arsenic species. Although $AsBr_5$ does not exist the $[AsBr_4]^+$ cation has been prepared as the hexafluoroarsenate:

$$3AsF_5 + 2AsBr_3 + Br_2 \rightarrow 2[AsBr_4]^+ [AsF_6]^- + AsF_3 \qquad [4.15]$$

It is a pure white compound decomposing slowly at –5°C into AsF_3 and Br_2 [116]. As(III) acts as the halide acceptor when $[As_2Cl_8]^{2-}$ is oxidized by chlorine at –40°C to give an intense yellow product containing the $[AsCl_6]^-$ anion [117]:

$$[PPh_4]_2[As_2Cl_8] \xrightarrow[\text{5 h at } -40°C]{Cl_2/CH_2Cl_2} [PPh_4][AsCl_6] \text{ (yield 86\%)} \qquad [4.16]$$

The hexafluoroarsenate anion in $K[AsF_6]$ was shown by Gafuer and Kruger [118] to be an almost perfect octahedron, F–As–F angle 89.3° and 90.7° and As–F length 1.72 Å. The high degree of symmetry of the $[AsF_6]^-$ anion generates relatively simple IR–Raman spectra. An instructive analysis is given in some detail by Fadini and Schnepel [119]. The mixed halides $[AsCl_nF_{6-n}]^-$ as the $[NEt_4]^+$ salts were characterized by Dove and co-workers [120] by multinuclear NMR.

There are many examples of formal pnictogen(V) halogen compounds, mainly of the R_3EX_2 type reacting with metals (chapter 5). In contrast AsF_5 in liquid SO_2 reacts with Mn, Fe, Ni [121], Cu [122] and Ag [123] to give $Mn[AsF_6]_2$, $FeF[AsF_6]$, $NiF[AsF_6]$, $Cu[AsF_6]$ and $Ag[AsF_6]$, respectively. AsF_5 reacts with monel in the presence of SO_2 to afford $Cu[AsF_6]$ and $Ni[AsF_6]_2{\cdot}2SO_2$, which can be an expensive reaction! Mercury under these conditions gives the remarkable silvery-golden compound $Hg_3[AsF_6]$, often reported as non-stoichiometric $Hg_{2.86}[AsF_6]$. The crystal structure consists of two orthogonal and non-intersecting linear chains of $Hg^{0.334+}$ cations skewering a lattice of octahedral $[AsF_6]^-$ anions [124]. The conflict between the elemental analysis data ($Hg_3[AsF_6]$ indicated) and structural results which indicate only 2.86 mercury atoms per chain in a unit cell is resolved by the existence of a defect structure with *c.* 1 $[AsF_6]^-$ unit in 20 missing from the structure. Thus this mercury-chain material is a stoichiometric compound in an incommensurate structure stabilized by anion vacancies [124].

An unusual pentagonal bipyramidal geometry is shown in silver hexafluoroarsentate compound $[AgF][AsF_6]$ in which five As–F distances involving bridging fluorines are in the range 1.714–1.729 Å in contrast to the non-bridging fluorine distance of 1.677 Å [125].

The power and utility of $[AsF_6]^-$ as an anion for stabilizing and isolating unusual cations for crystallographic studies is further shown in the case of $[H_3O][AsF_6]$ [126].

Haloantimony(V) anions. Pentahaloantimony compounds are powerful halide acceptors, usually forming the $[SbX_6]^-$ anion but occasionally forming the more complex $[Sb_2F_{11}]^-$ or $[Sb_3F_{16}]^-$ anions in the case of the fluorides. Antimony penatahalides are the most powerful of the pnictogen halide Lewis acids. $SbCl_5$ acts as a halide acceptor in the presence of (i.e. in competition with) $AsCl_3$ and chlorine [127], although the As(III) is oxidized to As(V):

$$SbCl_5 + AsCl_3 + Cl_2 \xrightarrow{80°C \text{ for } 3\text{ h}} [AsCl_4][SbCl_6]\cdot AsCl_3 \qquad [4.17]$$

The mixed halides $[NEt_4][SbX_nF_{6-n}]$ (X = Cl, Br) were characterized by Dove and co-workers [120] by ^{121}Sb and ^{19}F NMR. Clark and Trumble [128] prepared a caesium hexafluoroantimonate of unusual stoichiometry, $Cs_2[SbCl_6]$. The dark-blue solid proved to be the mixed valent compound $Cs_4[SbCl_6]^-[SbCl_6]^{3-}$. A mixed pnictogen analogue has been described, $Cs_4[SbCl_6]^-[BiCl_6]^{3-}$.

A number of unusual and highly electrophilic cations have been stabilized by $[SbF_6]^-$, $[Sb_2F_{11}]^-$ or $[Sb_3F_{16}]^-$, anions which include I_2^+ and Br_2^+ [129], Me_2Sn^{2+} [130], $[Au(CO)_2]^+$ [131] and $[Pd(CO)_4]^{2+}$ [132].

Although the hexafluoroantimonate anion $[SbF_6]^-$ is thought of as non-coordinating, this is not always the case as found for the product of the reaction between $[(Me_2PhP)W(CO)_5]$ and $[NO_2]$ $[SbF_6]$, which was assumed to be the hexafluorantimonate salt, until a crystal structure determination revealed a bridging fluorine interaction [133].

Halobismuth(V) anions. Although BiF_5 is a weaker fluorine acceptor than SbF_5 it is still widely used to stabilize unusual cationic species. BiF_5 was shown to be a weaker Lewis acid than SbF_5 by the fact that it dissolves in HF without showing Raman lines of $[BiF_6]^-$, unlike the antimony case where evidence of $[SbF_6]^-$ was observed [134]. However, on the addition of caesium fluoride, $Cs[BiF_6]$ was formed. Preparative details and vibrational spectra are given for $M[BiF_6]_2$ (M = Mg, Ca, Sr, Ba) by Popov *et al.* [135]. The hexafluorobismuthate anion has been widely used to isolate

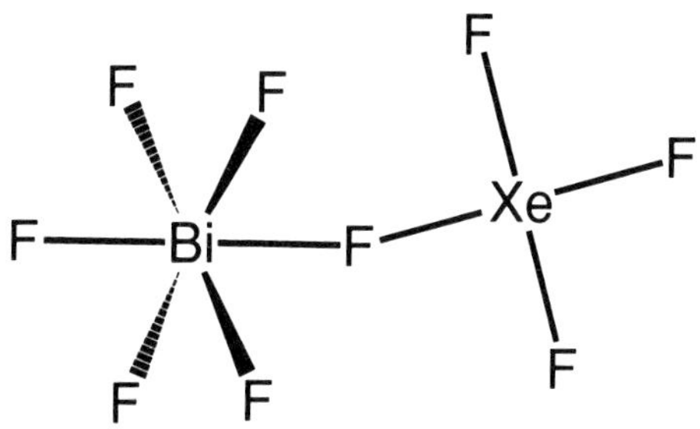

Scheme 4.15 The structure of the adduct between XeF_4 and BiF_5.

cationic species for further investigations such as $[O_2]^+$ $[BiF_6]^-$ [136], $[NO]^+$ $[BiF_6]^-$ [137], $[NF_4]^+$ $[BiF_6]^-$ [138], $[H_3O]^+$ $[BiF_6]^-$ [139], and $[ClOF_2]^-[BiF_6]^+$ [140].

That BiF_5 is a considerably weaker Lewis acid than SbF_5 is further shown by the structure of the adduct with xenon tetrafluoride, the X-ray structure of which was determined by Gillespie *et al.* [141], as shown in Scheme 4.15.

The antimony compound is described as the ionic $[XeF_3]^+[SbF_6]^-$ on the basis of the bridging fluorine ligand bond lengths to Xe of 2.49 Å [143]. In $[XeF_3][BiF_6]$ the μ-F has a shorter Xe−F bond length of 2.25 Å and has clearly much greater covalent character than that of the Sb anologue and is thus equally well described as the adduct $XeF_4{\cdot}BiF_5$.

All the products from the reactions of BiF_5 with group 18 halides (reactions [4.18]–[4.20] below) are ionic to a first approximation, but all have strong bridging fluorine anion–cation interactions [143]:

$$BiF_5 \xrightarrow{KrF_2} KrF_2{\cdot}BiF_5 \quad [4.18]$$

$$BiF_5 \xrightarrow{XeF_2} XeF_2{\cdot}BiF_5 + XeF_2{\cdot}2BiF_5 + 2XeF_2{\cdot}BiF_5 \quad [4.19]$$

$$BiF_5 \xrightarrow{XeF_4} XeF_4{\cdot}BiF_5 + XeF_4{\cdot}2BiF_5 \quad [4.20]$$

4.3 Complexes containing group 15 donor ligands

4.3.1 Arsenic(V) complexes

Several neutral arsenic(V) fluoride complexes containing nitrogen donor ligands have been crystallographically characterized [144–147], all are simple monomeric octahedral species with As−N bond lengths varying between 1.962(4)–2.116(3) Å.

4.3.2 Antimony(V) complexes

Only one neutral antimony(V) fluoride complex has been crystallographically characterized, the dinitrile complex [148] $[(NC-CN)SbF_5]$ which exhibits an Sb–N bond length of 2.232(2) Å. In contrast to arsenic, a number of neutral antimony(V) chloride complexes have also been reported; the first of these was the acetonitrile complex $[(CH_3CN)SbCl_5]$, prepared by Binas [149] in 1967. Further examples subsequently characterized by other workers include a diiodoisopropylcyanide [150] complex, $[\{(CH_3)_2CH\}_2NCNSbCl_5]$, and a diisopropylcarbodiimide [151] complex, $[\{(CH_3)_2CH\}\{(CH_3)_2CHNC\}CSbCl_5]$. All three complexes have Sb–N bond lengths varying between 2.145–2.232(2) Å, and near to perfect octahedral geometry for the antimony atoms.

4.3.3 Arsenic(III) complexes

The 1:1 addition compound $[AsCl_3(NMe_3)]$ has been described by Webster and Keats [152], and was prepared by vacuum condensation of trimethylamine onto arsenic trichloride. The structure consists of discreet $[AlCl_3(NMe_3)]$ molecules, the geometry of the arsenic(III) centre being disphenoidal with the stereochemically active lone pair lying in the equatorial plane.

Synthesis of the novel arsenic(III) complex $[As\{PMe_2C(SiMe_3)_2\}_3]$, containing phosphorus–arsenic bonds has been described recently. An X-ray crystallographic study of this compound revealed that the arsenic atom is in a pyramidal environment surrounded by three phosphorus atoms (sum over all angles 316.8°) [153].

4.3.4 Antimony(III) complexes

The structure of $[SbCl_3(PhNH_2)]$ has been determined [154], and, like $[AsCl_3(NMe_3)]$ [152], adopts a disphenoidal geometry with the stereochemically active lone pair on the antimony lying in the equatorial plane. In contrast, the 1:1 complex formed between antimony(III) chloride and bipyridine, $[SbCl_3(N_2C_{10}H_8)]$, exhibits a weakly associated dimeric structure [155] in which the antimony(III) centres adopt a distorted octahedral geometry. The weak Sb–Cl bonds are 3.34(2) Å (van der Waals radius sum for the atoms is 4.0 Å).

More recently, the 1:1 adduct of antimony(III) fluoride and terpyridine (2,2′:6′2″-terpyridine) has been prepared and crystallographically characterized [156]. Although two molecules are present in the asymmetric unit, there are no long Sb–Cl intermolecular contacts. Each independent molecule consists of an antimony(III) centre coordinated by a single terpyridine, three fluorine atoms (two nearly accommodated in a plane and the third at the apical site of a pseudo-pentagonal-bipyramidal arrangement). The seventh coordination site is occupied by the antimony lone pair.

Until recently complexes of antimony(III) containing tertiary phosphine ligands were virtually unknown, although their possible existence was indicated by early workers principally from NMR studies [157, 158]. No X-ray crystallographic information was available until 1993 when Norman and co-workers [159] reported the ionic complex $[H(py)_2]$ $[SbI_4(dmpe)]$ [py-pyridine, dmpe = 1,2-bis(dimethylphosphino)ethane] which was prepared serendipitously from the dissolution of a compound of formula $[SbI_3(dmpe)]$ in pyridine. It was reasoned that the ionic complex resulted from partial hydrolysis of $SbI_3(dmpe)$ by traces of moisture in the pyridine. The complex is described as ionic, the cation, $[H(py)_2]^+$, consisting of a protonated pyridine which is hydrogen bonded to a second pyridine in a nearly linear fashion ($N^1 \ldots H^2{-}N^2$ angle 160°). The $[SbI_4(dmpe)]^-$ anion adopts a structure in which the antimony centre is approximately octahedrally coordinated by four iodines and two ligand phosphorus atoms. The unusual distortions from ideal octahedral geometry cannot be accounted for on steric grounds and Clegg *et al.* postulate the structure is evidence for an arrested double S_N2 substitution. A further example [160] of an ionic complex containing antimony(III) tertiary phosphine bonds is $[HPEt_3][Sb_2Br_7(PEt_3)_2]$ which was isolated from the reaction of triethylphosphine with antimony(III) bromide in tetrahydrofuran (thf) solution. The unintentional hydrolysis of the complex again underlines the acute moisture sensitivity of these complexes. The cation of the complex, $[HPEt_3]^+$, is unremarkable, the anion, $[Sb_2Br_7(PEt_3)_2]^-$, consists of discreet dinuclear units. The geometry of the antimony(III) centres is best described as in between face-shared bioctahedral and two square-based pyramids sharing a vertex, from a consideration of the bond lengths and angles.

Further studies by the same workers [161] have established a trimethylphosphine complex of antimony(III) iodide, $[SbI_3(PMe_3)]_n$, prepared by the direct reaction of SbI_3 with PMe_3 in thf solution. The structure is best described as a polymer of dimers, the dimeric units consisting of two antimony atoms with a square-based pyramidal geometry coordinated by a PMe_3 ligand in the apical site and four iodine atoms in the basal plane. The basal planes of the pyramids are then linked through an edge. Two coordinated iodine atoms form weak bridging interactions to adjacent dimer units [3.547(2) Å and 3.698(2) Å] to form the polymeric array. The bond angles in the complex are all close to idealized values, implying that the antimony(III) lone pair has no appreciable stereochemical activity.

In addition to the (mono)tertiaryphosphine derivatives described above, two isomeric forms of the ditertiaryphosphine antimony(III) bromide complex of empirical formula $[SbBr_3(dmpe)]$ have been described [161].

The first of these, $[Sb_2Br_6(dmpe)_2]$, has a dimeric edge-shared bioctahedral structure, whereas the second, $[Sb_4Br_{12}(dmpe)_4]$, may be described as a centrosymmetric tetramer. A particularly noteworthy phenomenon regarding this complex is the great asymmetry of the Sb–Br bridges [2.787(2) Å and 3.493(2) Å], which is reasoned to be a result of the *trans* effect of the coordinated phosphines on one halogen atom in the halide bridge. The antimony(III) atoms again adopt an essentially octahedral geometry.

4.3.5 Bismuth (III) complexes

As with antimony, although bismuth(III) complexes containing tertiary phosphine ligands were virtually unknown until recently, there now exists a variety of complexes, reflecting the stereochemical variety associated with a bismuth(III) centre which has previously been observed with other donor ligands. The first neutral bismuth(III) complex containing a tertiaryphosphine ligand, $[Bi_2Br_6(PMe_3)_4]$, was synthesized and crystallographically characterized by Norman and co-workers [161]. The complex was prepared by stirring a suspension of $BiBr_3$ in neat PMe_3 and recrystallizing the resultant yellow solid from dichloromethane solution. $[Bi_2Br_6(PMe_3)_4]$ adopts a centrosymmetric edge-shared bioctahedral structure, with the bismuth(III) atoms in a near to ideal octahedral geometry. In the analogous reaction of dimethylphenylphosphine with bismuth(III) bromide, the mixed phosphine/phosphine oxide bismuth(III) complex, $[Bi_2Br_6(PPhMe_2)_2(OPPhMe_2)_2]$, was isolated, the oxidation of two of the dimethylphenylphosphine ligands occurring accidentally during the recrystallization process. In common with $[Bi_2Br_6(PMe_3)_4]$, this complex is also an example of an edge-shared bioctahedral structure with the bismuth(III) centres having almost perfect octahedral geometry.

In contrast, the reaction of bismuth(III) bromide with two equivalents of triethylphosphine in thf solution produced the tetrameric bismuth(III) complex $[Bi_4Br_{12}(PEt_3)_4]$ [161] which possesses C_i symmetry with two unique bismuth atoms each bonded to one phosphine atom and five bromine atoms. The structure may be viewed as two linked edge-shared bioctahedral units, again with the bismuth(III) atoms having almost perfect octahedral geometry. A number of complexes of bismuth(III) containing chelating tertiary phosphine ligands have also been reported recently. The complex $[Bi_2Br_6(dmpe)_2]$ has been described and is isomorphous with the corresponding antimony complex $[Sb_2Br_6(dmpe)_2]$ [161].

The reaction of bismuth(III) chloride with dppm and dppe [dppe = 1,2-bis(diphenylphosphino)ethane, dppm = 1,2-bis(diphenylphosphino)methane] in acetonitrile solution has been investigated by Willey *et al.* [162]. The resulting complexes, $[Bi_2Cl_6(dppm)_2]$ and $[Bi_4Cl_{12}(dppe)_5]$, respectively, were studied by single crystal X-ray diffraction. $[Bi_2Cl_6(dppm)_2]$ has a centrosymmetric structure consisting of coplanar halide-bridged Bi_2Cl_6 units with each of the two dppm ligands in a bidentate bridging mode between the two metal centres. The overall coordination geometry at each bismuth(III) atom is approximately octahedral. In contrast the structure of $[Bi_4Cl_{12}(dppe)_5]$ has two crystallographically independent molecules in the asymmetric unit which may be identified as $[Bi_2Cl_6(dppe)_2]$ and $[Bi_2Cl_6(dppe)_3]$. For the former species, the centrosymmetric structure is made up of halide bridged Bi_2Cl_6 units with bidentate chelate attachment of a dppe ligand to each bismuth atom giving rise to an overall edge-shared bioctahedral geometry. The $[Bi_2Cl_6(dppe)_3]$ species is a centrosymmetric dimer containing two $BiCl_3(dppe)$ moieties connected by a bridging dppe ligand. Thus both chelating and bridging dppe ligands are observed. The geometry of the bismuth(III) atoms is mer-octahedral, complexed by three chloride and three phosphorus (ligand) atoms.

As well as neutral bismuth(III) phosphine complexes, a few anionic bismuth(III) phosphine complexes have also been described. Of these, the complex $[PMe_3H][Bi_2Br_7(PMe_3)_2]\cdot 2MeCN$ was the first to be crystallographically characterized [163]. The structure consists of a polymeric anion, the monomeric unit of which contains a planar Bi_2Br_6 core, each bismuth being further bonded to a PMe_3 ligand and an additional bromide such that the bismuth resides in an octahedral environment. The Bi–P distance is 2.762(3) Å. The corresponding reaction between BiI_3 and PEt_3 yielded crystals of $[PEt_4]_4[Bi_6I_{22}]$. A low-precision X-ray crystallographic study of $[H(py)_2][BiI_4(dmpe)]$ indicated, from the position of the heavy atoms, that this complex is isostructural with the analogous antimony complex, $[H(py)_2][SbI_4(dmpe)]$ described in Section 4.3.4 [159].

A further bismuth(III) anionic complex containing a tertiary phosphine ligand is $[PPh_4][BiI_4(PMe_2Ph)_2]$ prepared from the reaction of BiI_3, $[PPh_4]I$ and an excess of PMe_2Ph in nitromethane solution. This complex has also been studied by X-ray crystallography. The anion, $[BiI_4(PMe_2Ph)_2]^-$, comprises an octahedral bismuth(III) centre bonded to four iodines and two *cis*-phosphine ligands, although significant deviations from regular octahedral geometry are noted, the reasons for which, partly steric, are discussed in detail elsewhere [160].

4.4 Complexes containing group 16 donor ligands

Complexes containing the heavier group 15 atoms in the +5 oxidation state are confined to simple octahedral or distorted octahedral molecular complexes usually containing five chloride or fluoride ligands with the sixth coordination site occupied by a neutral donor ligand. The hardness of the E(V) centre means that examples of such complexes are limited to the lighter donors of groups 15 and 16.

4.4.1 Arsenic(V) complexes

The ionic complexes $[PPh_4][AsX_4(thf)_2]$ (X = Br, Cl) have been prepared [164] from the reaction of $W(CO)_6$ and $[PPh_4]_2[As_2Cl_8]$ and of $Cr(CO)_6$ and $[PPh_4][As_2SBr_5]$ in thf, respectively. Both complexes have octahedral coordination with *trans* configuration. The arsenic(V) catecholate complex $[H_7O_3]^+[As(C_6H_4O_2)_3]$ has also been prepared [165] and again exhibits octahedral geometry with the anionic arsenic-containing fragment having D_3 symmetry. There are no arsenic(V) complexes containing heavier group 16 donor atoms.

4.4.2 Antimony(V) complexes

Antimony(V) complexes containing oxygen donor ligands have long been recognized. The first crystallographically characterized example was $[SbCl_5(OPCl_3)]$ reported in 1959 by Branden and Lindquist [166], and the same workers also reported the related compound, $[SbCl_5(OPMe_3)]$ [167]. Both complexes exhibited regular octahedral geometry. Several antimony(V) complexes containing phosphate donor ligands have been synthesized and crystallographically characterized [168]. Both monomeric and dimeric species have been identified, for example, bis(dimethylamino)phosphate monohydrate reacts with $SbCl_5$ to yield the 1:1 adduct $[Me_2N(Me_2NH)PO_2SbCl_5]$ whereas trimethylphosphate reacts with $SbCl_5$ to yield the dimeric $[Cl_4SbO_2PMe_2]_2$ [169]. All characterized complexes have octahedral geometry for the antimony atoms. A comprehensive study of a variety of antimony(V) complexes containing chelating oxygen donor ligands has been reported and studied by means of nuclear quadruple resonance (NQR) [170]. The crystal structures are reported of the acetylacetanato [171] derivatives, $SbCl_4(O_2C_5H_7)$ and the benzoylbenzophenato [172] derivative, $SbCl_4(O_2C_{15}H_{11})$. Both complexes have *cis* octahedral geometry. Also octahedral is the monodentate oxygen donor complex $SbCl_5(OCClC_6H_5)$ prepared from the direct reaction of $SbCl_5$ with benzoyl chloride in CCl_4. The 1:1 complex of $SbCl_5$ and DMSO has also attracted interest [173] because of its ability to oxidize alcohols to aldehydes in a simple one-step reaction. An X-ray crystallographic study of this adduct exhibited distorted octahedral geometry with disorder of the coordinated DMSO molecule [173]. Reaction of simple tetrachloroantimony(V) carboxylates and tetrachloroantimony alkoxides with carboxylic acids, alcohols or water produces novel cyclic antimony(V) species such as $Cl_3Sb(CH_3CO_2)OH(O)SbCl_3$ [174].

4.4.3 Arsenic(III) complexes

The crystal structure of 2-chloro-4,4,6,6-tetramethyl-1,3,2-dioxarsenane has been described [175], the complex exhibiting As–O bond lengths of

1.738(3) Å and 1.743(3) Å. There are also a number of arsenic(III) complexes containing crown ether ligands which are essentially complexes of arsenic(III) halides multiply coordinated by the crown ether. Examples include trichloro- and tribromo-(15-crown-5)-arsenic [176, 177] and trichloro-(12-crown-4)-arsenic [176]. These compounds have also been studied by means of synchrotron X-ray absorption studies [178].

There are two reports of the crystal structure of tris(diethyldithiocarbamato)arsenic(III) [179, 180]. The complex has an octahedral geometry with each ligand acting in a bidentate manner with delocalization of electrons across the NCS_2As unit, as determined from a consideration of bond lengths and angles in the complex. A series of arsenic(III) complexes containing sulphur donor ligands have also been described, and X-ray crystallographic data [181] are available for the complex $AsCl_3$(1,3-dimethyl-2(3H)imidazolethione). This species contains trigonal bi-pyramidal arsenic(III) centres weakly linked into dimers by long [3.12(1) Å] As−Cl bonds. IR and Raman spectroscopic data on a series of related complexes suggest that weak dimeric association of these species is a common feature. The toluene-3,4-dithiol (tdt) complex $AsCl_3$(tdt) has been synthesized from the direct reaction of $AsCl_3$ with tdt in chloroform solution [182]. The arsenic atom is three coordinate with a trigonal pyramidal geometry, the lone pair occupying an apical site. A number of related complexes of general formula $AsCl_3$(L) (L = 5-chloro-1,4,6,5-oxadithiarsoane; 2-chloro-1,3,6,2-trithiarsocane) have also been crystallographically characterized [183, 184]. All have a similar stereochemistry illustrating the spatial requirements of the lone pair on the arsenic atom.

4.4.4 *Antimony(III) complexes*

The X-ray crystal structure of the anion $[Sb(C_2O_4)_3]^{3-}$ has been determined [185] to illustrate a sterochemically active lone pair in six coordination. The antimony is approximately pentagonal bipyramidal with one vacant site. The triphenylphosphine [186] and triphenylarsine oxide [187] complexes $[SbCl_3(OEPh_3)_2]$ (E = P, As) have been described. Both contain the antimony atom in a nearly regular square pyramidal geometry. The structures of tris(diethyldithiocarbamato)antimony(III) [188] and tris(*o*-methyldithiocarbamato)antimony(III) [189] have been described as distorted dodecahedral, which in some cases exhibited weak intermolecular S–S and Sb–Sb contacts, reflecting the well-known property of antimony to increase its coordination number through intermetallic and/or Sb–S interactions. The reaction of tris(diethyldithiocarbamato)antimony(III) with diiodine in $CHCl_3$ yields $SbI(S_2CNEt_2)_2$ which consists of an infinite one-dimensional polymeric array of $Sb(S_2CNEt_2)_2$ units linked by iodide bridges [190]. The related complex, $SbI(S_2NC_5H_8)_2$, adopts a similar structure [191]. In contrast, reaction of $SbCl_3$ with tetramethylthiourea produces a 1:1 adduct containing the antimony atoms in a disphenoidal geometry (including the stereochemically active lone pair) [192]. The discreet molecules are linked into a polymeric sheet by intermolecular Sb–Cl contacts (3.123(2) Å). Further evidence for this structure is provided from ^{121}Sb Mössbauer data. A further example of an antimony complex utilizing intermolecular contacts to increase its coordination number is provided by [$SbCl_3$(dmit)] (dmit = 1,3-dimethyl-2-(3H)-imidazolethione) [193]. If only the bonded atoms are considered, the antimony atom appears to adopt a disphenoidal geometry; however, there are two intermolecular contacts from neighbouring molecules that are within the van der Waals radii of the atoms involved. Thus the total coordination geometry is more accurately described as a distorted octahedron. The Sb–Cl contact of 2.98(1) Å indicates a relatively strong interaction compared with the van der Waals radius for these atoms (4.0 Å). Distorted octahedral geometry [194] is also exhibited by $SbCl_3$ (di-isopropyldithiooxamide)$_{1.5}$ which is a three-dimensional polymer, the Sb atom lying on a threefold axis, being strongly coordinated to three chlorine atoms [2.413(2) Å] and weakly bound to three sulphur atoms from different ligands.

Although the coordination number of antimony is usually increased by secondary bonding to sulphur or, more commonly, halogen atoms on neighbouring molecules, an interesting study by Pohl *et al.* [195] illustrated that weak interactions between antimony(III) centres and phenyl groups may also be observed, which also serve to increase the coordination number of the antimony(III) centre. The compounds [$Sb_2Br_6(SPPh_3)_2$] and [$Sb_2I_6(SePPh_3)_2$]·$2CH_2Cl_2$, prepared from $SbBr_3$ and SbI_3 and the corresponding thio- or selenophosphorane, respectively, in dichloromethane solution contain centrosymmetric dimeric units with bridging halogen atoms. In both compounds weak interactions between antimony(III) atoms and phenyl groups complete the sixfold coordination of Sb(III) [Sb−C bond length 3.39(2)–3.65(2) Å]. Similar long-range Sb–C contacts are observed in the ionic complex $[PPh_4]_2[Sb_2Br_8]$, where one phenyl ring of the $[PPh_4]^+$ moieties completes the sixfold coordination sphere of the dimeric anion. No such Sb–C contacts are observed in the related tetrameric complex $[SbBr_3(SPMe_2Ph)]_4$, which contains bridging halogen and sulphur atoms [195].

4.4.5 *Bismuth(III) complexes*

The polymeric complex, bismuth(III) tris(methoxyethanoate), $Bi(OC_2H_4OMe)_3$, has been prepared by a metathesis reaction between bismuth(III) chloride or bromide and sodium methoxyethanolate in thf [196]. The complex has a stereochemically active lone pair, adopting tetragonal pyramidal geometry at the bismuth(III) atoms; the extended structure may thus be viewed as a polymerization of $Bi(OC_2H_4OMe)_3$ units to a chain of $Bi(OC_2H_4OMe)_5$ pyramids transfused at a basal edge. Three bismuth tris-carboxylates have also been structurally characterized, namely the formate, $Bi(O_2CH)_3$ [197], acetate $Bi(O_2CMe)_3$ [198] and the pivalate (2,2-dimethylpropanate) [199], the latter complex being prepared via a ligand exchange reaction between bismuth(III) acetate and 2,2-dimethylpropanoic acid. The crystal structure of this complex is of particular interest, consisting of isolated tetrameric units, $[Bi(O_2CCMe_3)_3]_4$, the bismuth atoms being weakly linked through long Bi–O bonds [2.87(2)–3.16(2) Å]. The great volatility of this compound, and relative stability towards hydrolysis, led Troyanov and Pisarevsky [199] to conclude it may find some utility as a bismuth oxide precursor. A novel seven coordinate bismuth complex has recently been crystallographically characterized, $[Bi\{N(CH_2CH_2N{=}CHC_6H_4OH)_3\}]$, which contains a Schiff base ligand prepared from the condensation of 2,2′,2″-triaminotriethylamine with salicaldehyde [200].

A number of tertiaryphosphine and ditertiaryphosphine and arsine oxide complexes have also been described. In contrast to the antimony(III) halide complexes of tertiaryphosphine oxide and tertiaryarsine oxide, which are monomeric square pyramidal species [186, 187], further aggregation occurs in the bismuth complexes. Two structural modifications have been identified: the triphenylphosphine oxide complex, $[BiI_3(OPPh_3)_3]_2$ which is dimeric, being built of two octahedra sharing an edge through I atoms and with triphenylphosphine oxide ligands in *cis* positions [201], and the triphenylarsine oxide complex, $[Bi_2I_6(OAsPh_3)_3]$, which consists of two distorted octahedra sharing a face through I atoms and with all of the triphenylarsine oxide ligands linked to one of the Bi atoms [202]. A further example [203] of a bismuth(III) iodide complex containing tertiary phosphine oxide ligands is $[Bi_2I_6\{OP(NMe_2)\}_3]_n$, a complex in which even further aggregation occurs so that individual dimeric repeat units are linked to form a one-dimensional polymer with the $OP(NMe_2)_3$ ligands, one per bismuth, alternating on either side of the $(Bi_2I_6)_n$ chain. Variation in the experimental conditions in the reaction of BiI_3 with $OP(NMe_2)_3$ allowed Clegg *et al.* [203] to isolate and crystallographically characterize the ionic complex $[BiI_2\{OP(NMe_2)_3\}_4][I_3]$. The cationic fragment of this complex contains a mononuclear bismuth atom coordinated to two iodine atoms and four $OP(NMe_2)_3$ ligands with a *trans* octahedral geometry. The existence of a complex of this formula was extensively discussed and investigated by means of EXAFS (extended X-ray absorption fine structure) measurements [203].

The attempted synthesis of some ditertiaryphosphine and ditertiaryarsine complexes of bismuth(III) chloride has been described by Willey *et al.* [204]; however, during recrystallization adventitious oxidation of the group 15 atoms occurred to produce the corresponding ditertiaryphosphine and ditertiaryarsine oxide ligands. Two such structures were crystallographically characterized: $[BiCl_3\{Ph_2P(O)CH_2P(O)P_2\}_2]$, which exists as an edge-shared bioctahedral species, the bismuth centres being bonded to two terminal chlorine atoms, two bridging chlorine atoms and two oxygen (ligand) atoms; and $[BiCl_3\{As(O)MePh_2\}\{Ph_2As(O)CH_2CH_2As(O)Ph_2\}]_n$, consisting of one-dimensional polymeric chains where each bismuth(III) atom is bonded in an approximately *mer*-octahedral arrangement to three terminal chlorine atoms and three oxygen (ligand) atoms [204].

A large number of complexes have been reported containing bismuth(III) and sulphur donor ligands, amongst the first examples to be structurally characterized [168] being those containing the tris(diethyldithiocarbamato) ligand and associated derivatives. Thus the parent complex bismuth(III) tris(diethyldithiocarbamate) was shown to consist of six-coordinate bismuth atoms in distorted dodecahedral geometry (taking into account the stereochemically active lone pair) weakly linked into dimers (Bi–S length, 3.210(5) Å, compared with the van der Waals radius of 4.05 Å for the atoms). The crystal structure of tris(*o*-methyldithiocarbamato)bismuth(III) complex has also been reported and is isostructural [189] with the analogous antimony(III) complex. A large variety of halogen-substituted complexes of bismuth(III) tris(diethyldithiocarbamate) have been described [205–210], for example $Bi(S_2CNEt_2)_2X$ (X = Br, I) and $Bi(S_2CNEt_2)X_2$ (X = Cl, Br, I). Recrystallization of these complexes from N-donor solvents, such as pyridine, results in isolation of complexes of the type $[Bi(S_2CNEt_2)X_2(py)_4]$ (X = Cl, Br, I). Analogous complexes have been prepared with bipyridyl and terpyridyl donor ligands. The structures adopted by these complexes are very complicated, illustrating varied and irregular stereochemistries, including the implied presence of lone pairs on the metal atom. This, coupled with the copious potential for the formation of polymeric and oligomeric systems by bridging halide and dithiocarbamate groups as well as long range intermolecular contacts, means that a concise overview of the work is impossible and interested parties are referred to individual papers.

Two adducts of bismuth(III) chloride and thiourea (tu) have been reported, a yellow form, $[Bi_2Cl_2(\mu\text{-}Cl)_2(tu)_6]^{2+}[BiCl_5(tu)]^{2-}$ [213] and a red form, consisting of a complicated array of octahedral bismuth atoms coordinated by chloride and/or sulphur (ligand) atoms of nominal formula $[Bi(tu)_6][Bi\{(tu)_{1.5}Cl_{1.5}\}Cl_3]_2$ [212]. Two bismuth(III) thiourea derivatives have also been described, tetra[1-allyl-3-(2pyridyl)thiourea–*S*]bismuth(III) and hexa[1-allyl-3-(2-pyridyl)thiourea-*S*]bismuth(III) nitrate [213]. The former structure consists of a binuclear centrosymmetric neutral complex in which two sulphur atoms from two ligand molecules and four chlorine atoms coordinate to each bismuth atom, resulting in distorted octahedral geometry. The latter structure is ionic, the bismuth(III) atom being coordinated by three chelating thiourea ligands, producing distorted octahedral geometry, the charge being balanced by three discreet nitrate anions. Another example of a bismuth(III) complex illustrating distorted octahedral geometry and complexed by three chelating sulphur donor ligands is tris[*P*,*P*,*P*′,*P*′-tetraphenylamidobis(phosphine sulphido)-*S*,*S*′]bismuth(III), prepared by Williams, Quicksall and Barkigia [214]. Further geometri-cal diversity for bismuth(III) complexes is illustrated by two 1-pyrrolidinecarbodithioate complexes prepared from bismuth(III) chloride, thiourea and 1-pyrrolidinecarbodithioate [215, 216]. The first complex, μ-[1′-pyrrolidine–carbodithioato-*S*($Bi^{1,2}$) *S*′(Bi^2)]–1,1,1,2,2-pentakis(1′-pyrrolidinecarbo-dithioato-*S*,*S*′)dibismuth(III), contains the two bismuth atoms in a different geometry. Bi^1 is trigonal prismatic with an extra position on a rectangular face, whereas Bi^2 is pentagonal pyramidal. A sulphur atom joins the two polyhedra with an asymmetric bridge. The second complex, di-μ-chloro-bis[bis(1–pyrolidinecarbodithioato-*SS*′)(thiourea-*S*)bismuth(III) contains a pentagonal bipyramidal bismuth(III) centre with one apical and one equatorial chlorine atom, both bridging, so that two adjacent polyhedra share an edge.

Other sulphur donor ligands to be complexed to bismuth(III) metal centres include thiophosphinate, thiolate and sulphide containing moieties. The complex $Bi(S_2PMe_2)_3$, containing the ligand dimethyl-dithiophosphinate, has been described [217]. The compound is a centrosymmetric dimer in which pentagonal-bipyramidal monomeric units are held together by secondary bismuth–sulphur interactions. The existence of a stereochemically active lone pair is considered but no definite conclusions are drawn. A comprehensive study of coordination complexes of the bismuth(III) thiolate, $Bi(SC_6F_5)_3$, has recently been described by Farrugia, Lawlor and Norman [217, 218]. Reaction of $Bi(SC_6F_5)_3$ and $SPPh_3$ produces the complex $[Bi(SC_6F_5)_3(SPPh_3)]$, which has a disphenoidal geometry with the $SPPh_3$ ligand *trans* to one thiolate group. An additional intermolecular interaction is also present owing to a weakly bridging thiolate sulphur which results in a centrosymmetric dimer with each bismuth(III) centre having square-based pyramidal five coordination. Similar bismuth geometry was noted for the anion in the ionic complex. [K(18-crown-6)] $[Bi(SC_6F_5)_3(NCS)]$ (18-crown-6 = 1,4,7,10,13,16-hexaoxacyclooctadecane) derived from the reaction of $Bi(SC_6F_5)_3$ and [K(18-crown-6)]NCS. Reaction of $Bi(SC_6F_5)_3$ with $OPPh_3$, $OP(NMe_2)_3$ or *N*,*N*′-dimethylpropyleneurea gave the bis(ligand) complexes $[Bi(SC_6F_5)_3L_2]$. All three complexes are monomeric with a five-coordinate square-based pyramidal geometry around the bismuth(III) centre. Reaction of $Bi(SC_6F_5)_3$ with $S{=}C(NHMe)_2$ yielded the tris(ligand) complex $[Bi(SC_6F_5)_3\{S{=}C(NHMe)_2\}_3]$ which has almost regular octahedral geometry for the bismuth(III) atom [218].

Concerning the reports of bismuth(III) containing sulphide donor ligands, two are of particular interest, the first of which, $[SMe_3]_2[Bi_2I_8(SMe_2)_2]$, was prepared in low yield from a mixture of BiI_3 and SMe_2 in hexane [219]. The anionic portion of the complexes may be viewed as a complex of the dimeric dianion $[Bi_2I_8]^{2-}$ and two SMe_2 ligands with an overall edge-shared crystallographically centrosymmetric, bioctahedral geometry. Clegg, Norman and Pickett [219] conclude that the formation of this complex may be the result of a small degree of hydrolysis resulting from adventitious water, although it seems more likely that its formation may be explained by methyl migration between dimethylsulphide ligands, a phenomenon previously observed in reactions involving, for example, triphenylantimony [220]. The ditertiaryphosphine sulphide complex $[BiCl_3\{Pr_2{}^nP(S)P(S)Pr_2{}^n\}]$ has been reported, prepared from the direct reaction of bismuth(III) chloride and the ligand in CH_3CN. The crystal structure of the complex reveals a halide-bridged dimeric structure in which each bismuth atom is surrounded by an essentially octahedral arrangement of four chlorine (two terminal and two bridging) and two sulphur atoms. Each of two ligand molecules is involved in bidentate (*S,S'*) chelation to a bismuth atom to form a five-membered ring [221].

There are no examples of a bismuth(III) complex containing a coordinated tertiary phosphine selenide. Attempts were made to prepare such a complex from the reaction of $SeP(NMe_2)_3$ and bismuth(III) chloride in acetonitrile. Crystallographic characterization of the product isolated, however, revealed the ionic complex $[(Me_2N)_3PSeSeP(NMe_2)_3]^{2+}[(BiCl_4)_2]_n{}^{2n-}$, which was presumed to be formed as a result of contamination from adventitious water. The bismuth-containing anion is best described as dimeric $[BiCl_4]_2$ units linked by further halide bridging to form an infinite chain structure of edge-shared octahedra [221].

4.5 The aqueous chemistry of the pnictogens

Much of the contemporary heavy pnictogen chemistry, for example high T_c superconductors, metallo-organic chemical vapour deposition (MOCVD) processes, transition metal complexes, and E=E multiple bonds to name but four topics, is either high-temperature or non-aqueous chemistry. This was not always the case and much classical pnictogen chemistry was aqueous chemistry. This is the chemistry of our environment both on our timescale (pollution problems) and on the geological timescale (e.g. weathering of rocks). Although the concentration of arsenic in sea water is low (*c.* 0.003 mg l^{-1}) this is equivalent to *c.* 14 tons per cubic mile or *c.* 5×10^9 tons in the oceans as a whole [223].

This section is divided into a treatment of tervalent and pentavalent species.

4.5.1 The aqueous chemistry of tervalent pnictogen compounds

The tervalent oxides E_2O_3 show a progression from the essentially acidic As_2O_3 to the basic Bi_2O_3. In the following, the more important features of their aqueous chemistry are given, in the main, sequentially, that is arsenic through antimony to bismuth. The standard aqueous reduction potentials for both E^{III} and E^{V} species are grouped together in Table 4.6 (also in Table 1.5) [223].

Arsenous acid, $As(OH)_3$. Arsenic does not form any well-characterized hydroxides and As_4O_6 is only sparingly soluble in water (1.7 g per 100 cm^3 at 25°C) to give a weakly acid solution (about as strong as boric acid):

$$H_2O + As(OH)_3 \rightleftharpoons [AsO(OH)_2]^- + H_3O^+, \quad K = 6 \times 10^{-10} \qquad [4.21]$$

When aqueous solutions of As_4O_6 are evaporated it is As_4O_6 that crystallizes out, not $As(OH)_3$, although the main species in acid solution is the pyramidal $As(OH)_3$. In basic solutions $[AsO(OH)_2]^-$, $[AsO_2(OH)]^{2-}$

Table 4.6 Standard aqueous reduction potentials at 25°C of the pnictogens arsenic, antimony and bismuth.

Reaction	E^0 v(V)
Arsenic	
$As + 3H^+ + 3e^- \rightleftharpoons AsH_3$	–0.608
$As_2O_3 + 6H^+ + 6e^- \rightleftharpoons 2As + 3H_2O$	0.234
$HAsO_2 + 3H^+ + 3e \rightleftharpoons As + 2H_2O$	–0.248
$AsO_2^- + 2H_2O + 3e^- \rightleftharpoons As + 4OH^-$	–0.68
$H_3AsO_4 + 2H^+ + 2e^- \rightleftharpoons HAsO_2 + 2H_2O$	0.56
$AsO_4^{3-} + 2H_2O + 2e^- \rightleftharpoons AsO_2^- + 4OH^-$	–0.71
Antimony	
$Sb + 3H^+ + 3e^- \rightleftharpoons SbH_3$	–0.51
$Sb_2O_3 + 6H^+ + 6e^- \rightleftharpoons 2Sb + 3H_2O$	–0.152
$Sb_2O_5 + 4H^+ + 4e^- \rightleftharpoons Sb_2O_3 + 2H_2O$	0.649[a]
$Sb_2O_{5(s)} + 6H^+ + 4e^- \rightleftharpoons 2SbO^+ + 3H_2O$	0.581
$SbO^+ + 2H^+ + 3e^- \rightleftharpoons Sb + 2H_2O$	0.212
$SbO_2^- + 2H_2O + 3e^- \rightleftharpoons Sb + 4OH^-$	–0.66
$SbO_3^- \rightleftharpoons H_2O + 2e^- \rightleftharpoons SbO_2^- + 2OH^-$	–0.59
Bismuth	
$Bi^+ + e \rightleftharpoons Bi$	0.5
$BiCl_4^- + 3e^- \rightleftharpoons Bi + 4Cl^-$	0.16
$Bi_2O_3 + 3H_2O + 6e^- \rightleftharpoons 2Bi + 6OH$	–0.46
$Bi_2O_4 + 4H^+ + 2e^- \rightleftharpoons 2BiO^+ + 2H_2O$	1.59
$BiO^+ + 2H^+ + 3e^- \rightleftharpoons Bi + H_2O$	0.32
$BiOCl + 2H^+ + 3e^- \rightleftharpoons Bi + Cl^- + H_2O$	0.1583
$Bi^{3+} + 3e^- \rightleftharpoons Bi$	0.308
$Bi^{3+} + 2e^- \rightleftharpoons Bi^+$	0.2
$Bi + 3H^+ + 3e^- \rightleftharpoons Bi\ H_3$	–0.8

[a]Valentinite (senarmontite 0.671 V).

and $[AsO_3]^{3-}$ have been identified by Raman spectroscopy [224]. Both $As(OH)_3$ and $[AsO_3]^{3-}$ spectra are consistent with C_{3v} symmetry. The other anionic species, $AsO(OH)_2$ and $[AsO_2(OH)]^{2-}$, exhibit C_s symmetry but in both cases, the three-coordinate arsenic centre resides in a trigonal pyramidal coordination environment.

Arsenic trihalides are hydrolysed by water on dilution, but 1 mol of $AsCl_3$ (*c.* 84 ml) for example, can be dissolved in 9 mol (*c.* 162 ml) of water unchanged. Dilution results in the precipitation of As_2O_3. This is in marked contrast to the violent reaction of PCl_3 with water.

An aerated aqueous solution of Na_3AsO_3 is oxidized to Na_3AsO_4 by ionizing radiation; for example, Daniels and Weiss [225] using 200 kV X-rays found the oxidation occurs over a wide range of pH.

Oscillation is observed in the iodate-arsenite reaction system in a stirred tank reactor [226]. The visual oscillations of the reactions are accompanied by oscillations in the iodide concentration and the redox potential. The oscillations are inhibited by the addition of ClO_3^-, Cl^- and $NaHCO_3$.

As_2S_3 readily forms a colloidal solution in water by adding a 1% solution of As_2O_3 to a saturated solution of H_2S while passing a stream of H_2S through the solution. Excess H_2S is removed from the intense yellow solution by a stream of dihydrogen. A fairly stable Sb_2S_3 orange solution may be made by the same technique, substituting tartar emetic for the As_2O_3. The colloidal particles are negatively charged by surface adsoprtion of sulphide ions.

The oxidation of AsH_3 by H_2O_2 in aqueous media, although thermodynamically allowed, is very slow or does not take place without halide catalysts. Dorfman *et al.* [227] have proposed on the basis of detailed kinetics studies that the first stage is the protonation of H_2O_2 followed by its reaction with the halide catalyst:

$$H_2O_{2(aq)} + H_3O^+ \rightleftharpoons H_3O_2^+{}_{(aq)} + H_2O \qquad [4.22]$$

$$H_3O_2^+{}_{(aq)} + X^- \rightleftharpoons [H_3O_2X_{(aq)}] \qquad [4.23]$$

It is this H_3O_2X intermediate which is the active oxidizing species and the rate limiting step is:

$$H_3O_2X_{(aq)} + AsH_3 \rightarrow 2H_2O + HX + AsH \qquad [4.24]$$

The subsequent oxidation of AsH to H_3AsO_4 has no effect on the kinetics of the process.

Arsine is also oxidized by aqueous potassium permanganate in the presence of halide. In this case Dorfman *et al.* [228] postulate the formation of an intermediate arsine manganese complex, $[MnO_3(AsH_3)(OH)X]^-$ (X = Cl, Br, I). The oxidation of the AsH_3 is thought to take place within this manganese ligand set. Similarly, the rate limiting step in the oxidation of arsine by $Na_2Cr_2O_7$ in aqueous solution was found by the same group [229] to be the intramolecular electron transfer in the complex $[CrO_3(AsH_3)(H_2O)X]$ (X = Cl, Br, I).

Water-soluble arsenic species, particularly As(III), are toxic. The most effective antidotes depend on the reaction between $As(III)_{(aq)}$ and mercaptides in which less toxic species are formed that are eliminated by the body. Chelating ligands such as 2,3-dimercaptosuccinic acid are proving to be more efficacious in the treatment of acute arsenic poisoning (chapter 8) [230, 231].

The basic behaviour of As_4O_6. The basic dissociation of $As(OH)_3$ in aqueous solution is as follows:

$$As(OH)_3 \rightleftharpoons [As(OH)_2]^+ + OH^- \qquad [4.25]$$

showing it to be an extremely weak base. It is unable to form stable salts with oxyacids, but species such as $[As(OH)(HSO_4)_2]$ have been suggested for solutions of As_4O_6. The solubility of As(III) oxide is higher in concentrated hydrochloric acid than it is in water, owing to the equilibrium:

$$As(OH)_3 + 3HCl \rightleftharpoons AsCl_3 + 3H_2O \qquad [4.26]$$

Although $AsCl_3$ is not to be thought of as a salt, nevertheless this reaction with HCl is indicative of the growing basicity of group 15 elements on descending the group. A more formally convincing salt is arsenic(III) tris(trifluoroacetate); however, its low melting point (45°C–46°C) together with IR and mass spectral evidence show that the compound is composed of discrete pyramidal molecular units [232].

Antimony. The lower acid of antimony is known only as the hydrated oxide, and the few Sb antimony(III) salts, the metaantimonates (e.g. $NaSbO_2$) are described in chapter 3. Sb_2O_3 is an amphoteric oxide showing typically increased basic properties associated with descending the p-block groups. Sb_2O_3 will dissolve in aqueous HCl but is precipitated when the solution is sufficiently diluted.

Antimony is not attacked by dilute aqueous acids but is attacked by hot concentrated sulphuric acid to give antimony(III) sulphate.

In contrast, hot concentrated nitric acid gives a mixture of oxides (III and V) rather than the nitrate. Aqua regia dissolves antimony to afford a mixture of antimony(III) and antimony(V) chlorides. The amphoteric nature of the metal is shown by its reaction with fused sodium hydroxide to produce sodium antimonate(III) with the evolution of dihydrogen together with some SbH_3 formed via reduction with the nascent hydrogen.

Antimony, in contact with aluminium and an electrolyte, can act as a metal and form a galvanic pile. This is what causes the corrosion observed when there is contact between antimony or antimony-containing materials and aluminium [233].

The reaction of stibnite, Sb_2S_3, with aqueous HCl to give $SbCl_3$ and after purification hydrolysis to very pure Sb_2O_3 is an important industrial process. The use of antimony in producing fire-resistant materials accounted for 57% of total antimony consumption in the USA in 1985 [234]. Copur, Colak and Yapici [234] made a detailed study of the $Sb_2S_3/HCl/H_2O$ system and developed a kinetic model to represent the dissolution process:

$$-\ln(1-x) = 9.79 \times 10^{-10}\ (D/\mathrm{mm})^{-0.908}\ (C/\mathrm{wt}\%)^{10.600} \left(\frac{S/\mathrm{g}}{L/\mathrm{ml}}\right)^{-0.32} \exp\left(\frac{-6244}{T/\mathrm{K}}\right) (t/\mathrm{min}), \tag{4.1}$$

where x is the conversion fraction of the mineral, D is the particle size, C the acid concentration, S/L is the solid-to-liquid ratio, T is temperature and t is time. The activation energy for the dissolution process was calculated as 52 kJ mol^{-1}.

Bismuth. Bi_2O_3 will dissolve in concentrated aqueous solutions of strong oxyacids to afford the bismuth(III) salt. On dilution, partial hydrolysis gives the basic salts; for example Bi_2O_3 dissolves in 70% $HClO_4$, the solution yielding $Bi(ClO_4)_3{\cdot}5H_2O$, which may be partly dehydrated to $Bi(ClO_4)_3{\cdot}H_2O$ and finally to an anhydrous $Bi(ClO_4)_3$ by heating to 80°C–100°C.

Metallic bismuth dissolves in nitric acid, forming a solution of bismuth(III) nitrate from which a pentahydrate, $Bi(NO_3)_3{\cdot}5H_2O_4$, crystallizes from the acid solution. Reduction of the pH results in partial hydrolysis to a compound of approximate composition $BiO(NO_3){\cdot}H_2O$, which has pharmaceutical applications.

The classical methods of analysis of bismuth, both qualitative and quantitative, are, in the main, aqueous-based methods (chapter 9). Although now superseded by instrumental methods in modern laboratories, the older procedures are still of use at remote, less-well-equipped sites. Bismuth phosphate, $BiPO_4$, is precipitated quantitatively from very dilute HNO_3 solutions. In the absence of transition metals EDTA (ethylenediaminetetraacetic acid) titrations are selective and quantitative.

Stability constants for all the various combinations of $BiCl_n(NO_3)_m^{3-n-m}$ bismuth(III) anions ($n + m = 2, \ldots, 4$) have been estimated via Sillèn's method [235]. At high ionic strengths, species containing five groups occur, although no mixed hexacoordinated species have been identified.

Bismuth trioxide is a basic oxide and dissolves in mineral acids to give bismuth salts such as $Bi_2(SO_4)_3$, $Bi(NO_3)_3{\cdot}5H_2O$ and $BiPO_4$. The hydroxide $Bi(OH)_3$ may be precipitated by the addition of a base. The composition of $Bi(OH)_3$ is variable and compounds such as BiO(OH) have been described, but no structures have yet been confirmed. Before precipitation takes place complex cations are formed. At one time it was thought [236] that these cations were $[Bi_6(OH)_{12}]^{6+}$, but it has been shown [237] that the cation is most commonly $[Bi_6O_4(OH)_4]^{6+}$. The X-ray structure of $Bi_6O_4(OH)_4(NO_3)_6{\cdot}H_2O$ showed the six bismuth atoms to be at the apices of an octahedron at non-bonding distances. The octahedron is face-capped by OH and O groups forming Bi–O–Bi bridges. A similar structure was demonstrated for $Bi_6O_4(OH)_4(ClO_4)_6{\cdot}7H_2O$ [238]. The lone pairs on the bismuth atoms are directed away from the cage. Each bismuth atom is bonded to two equatorial oxide oxygen atoms and two axial hydroxide oxygen atoms resulting in a disphenoidal structure about each bismuth atom. The bismuth atoms are also weakly bonded to anions in the crystal lattice. The structures are thought to exist in solution and are consistent with Raman spectra previously interpreted as $[Bi_6(OH)_{12}]^{6+}$ species.

Bismuth trifluoride is the only bismuth halide not hydrolysed by pure water even at the boiling point. However, the addition of HF results in the formation of BiOF.

The other bismuth trihalides are partly hydrolysed by water, forming oxyhalides (BiOX). Bismuth oxychloride exhibits a nacreous (mother-of-pearl) effect, giving it some commercial interest. It imparts a pearl-like effect to manufactured goods and a frosted effect in cosmetics.

$[Bi(OH)]^+$ and $[Bi_6(OH)_{12}]^{6+}$ (now thought to be $[Bi_6O_4(OH)_4]^{6+}$) are major species in the hydrolysis of Bi^{3+}, as shown by enthalpy titrations [239] with a minor amount of $[Bi_3(OH)_4]^{5+}$.

$BiBr_3$ is also hydrolysed, forming white BiOBr. Bismuth triiodide is insoluble in and stable to cold water, but is hydrolysed by hot water to the brick-red BiOI. Interestingly, it can be dissolved in aqueous HI from which the hydrate $HBiI_4{\cdot}4H_2O$ can be crystallized.

4.5.2 *The aqueous chemistry of pentavalent pnictogen compounds*

Arsenic(V) acid. Unlike arsenous acid, arsenic acid can be isolated in the solid state, albeit only in hydrated forms. The crystalline compounds are often formulated as hydrates of As_2O_5 and their relationship is expressed as follows:

$$As_2O_5{\cdot}7H_2O \xrightarrow{-30^\circ C} As_2O_5{\cdot}4H_2O \xrightarrow{36^\circ C} As_2O_5{\cdot}\tfrac{5}{3}H_2O \xrightarrow{170^\circ C} As_2O_5 \qquad [4.27]$$

The solution chemistry of $[AsO_4]^{3-}$ is summarized in Scheme 4.16 and Table 4.7. H_3AsO_4 is a stronger acid than is H_3AsO_3 and it is about as strong an acid as chloroacetic acid.

A saturated aqueous solution of H_3AsO_4 when saturated with gaseous ammonia precipitates $(NH_4)_3AsO_4.3H_2O$ crystals. The crystals lose water and ammonia in air to give $(NH_4)_2HAsO_4$. By heating the original solution of $(NH_4)_3AsO_4$ one obtains $(NH_4)H_2AsO_4$, after loss of NH_3.

Table 4.7 Dissociation constants of arsenic(V) acid, H_3AsO_4 at 18°C.

	K_a	pK_a
1st	5.62×10^{-3}	2.25
2nd	1.70×10^{-7}	6.77
3rd	3.95×10^{-12}	11.60

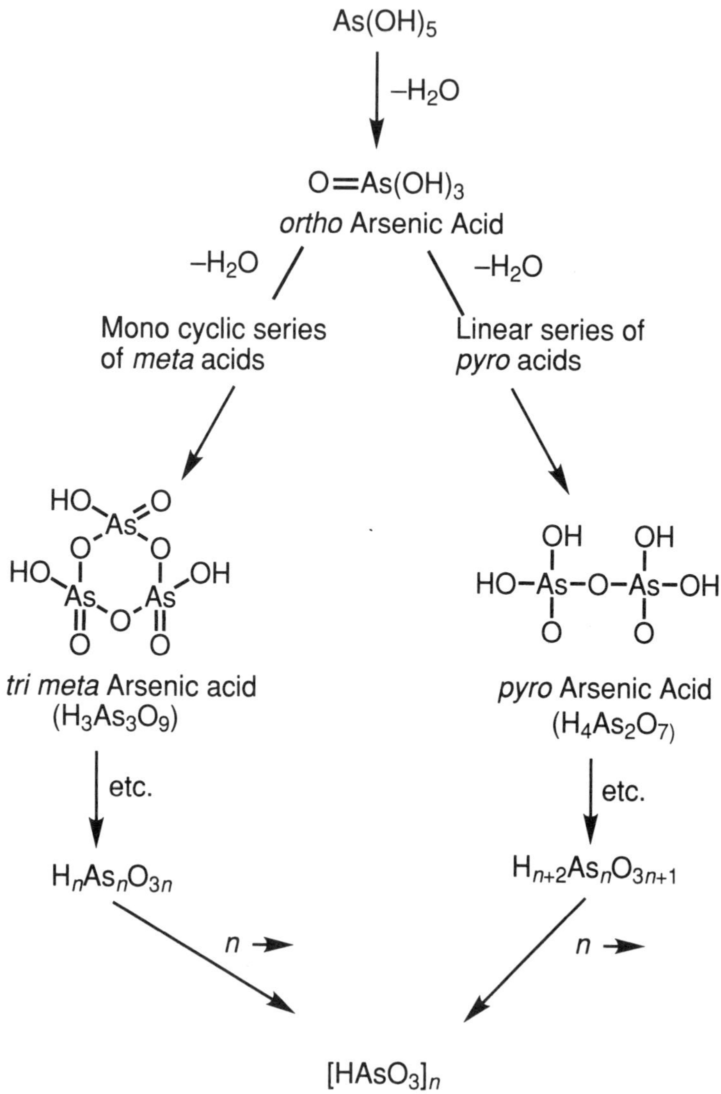

Scheme 4.16 The formal relationships of the various arsenic(V) acids.

Hydrothermal synthesis. Hydrothermal synthesis is the technique of carrying out reactions in aqueous media at temperatures over 100°C by use of high-pressure sealed apparatus. The techniques have been reviewed by Laudise [240]. Although the temperatures (*c.* 500°C) used are high for aqueous chemistry they are usually considerably below those used in solid state synthesis (over 1000°C). This has advantages in terms of thermal efficiency and the reduction of crystal defects. For example, the non-linear optical (NLO) material potassium titanyl arsenate, $KTiOAsO_4$ (KTA), is of contemporary interest. The flux growth method involves heating TiO_2, KH_2AsO_4 and K_2CO_3 at *c.* 1000°C. The isomorphous $KTiOPO_4$ prepared at high temperatures has been shown to have K_2O vacancy defects associated with temperature. Hydrothermal synthesis offers a relatively low-temperature alternative [241, 242]. Full details of this preparation are given by Phillips and Stucky [243]. The reaction is as follows:

$$TiO_2 + KH_2AsO_4 \xrightarrow[2700\ \text{bar}]{H_2O/700°C} KTiOAsO_4 + H_2O \qquad [4.28]$$

The alkali metal and ammonium arsenates are soluble in water, but the alkali earth metals together with the d-block and f-block metals are insoluble. Table 4.8 lists the solubility products of the more common arsenates. Insoluble arsenates are of industrial interest since they are used to stabilize (i.e. immobilize) arsenic, commonly in the form of iron(III), calcium

Table 4.8 Solubility products of selected arsenates

Compound	pK_{sp}	K_{sp}
$AlAsO_4$	15.8	1.6×10^{-16}
$NH_4\,(UO_2)AsO_4$	23.77	1.7×10^{-24}
$Ba_3(AsO_4)_2$	50.11	8.0×10^{-51}
$BiAsO_4$	9.36	4.4×10^{-10}
$Cd_3(AsO_4)_2$	32.66	2.2×10^{-33}
$Ca_3(AsO_4)_2$	18.7	6.8×10^{-19}
$CrAsO_4$	20.11	7.7×10^{-21}
$Co_3(AsO_4)_2$	28.12	7.6×10^{-29}
$Cu_3(AsO_4)_2$	35.12	7.6×10^{-36}
$Fe^{III}AsO_4$	20.24	5.7×10^{-21}
$Pb_3(AsO_4)_2$	35.39	4.0×10^{-36}
$Li(UO_2)AsO_4$	18.82	1.5×10^{-19}
$Mn_3(AsO_4)_2$	28.72	1.9×10^{-29}
$Mg_3(AsO_4)_2$	19.68	2.1×10^{-20}
$Ni_3(AsO_4)_2$	25.51	3.1×10^{-26}
$K(UO_2)AsO_4$	22.60	2.5×10^{-23}
Ag_3AsO_4	22.00	1.0×10^{-22}
$Na(UO_2)AsO_4$	21.87	1.3×10^{-22}
$Sr_3(AsO_4)_2$	18.09	$8 \times 1 \times 10^{-19}$
$(UO_2)HAsO_4$	10.50	3.2×10^{-11}
$Zn_3(AsO_4)_2$	27.89	1.3×10^{-28}

or magnesium arsenate. The solubility data in the older reference books, for example Seidell [244, 245] are now thought to be doubtful, for arsenates at any rate. Robins [246] has shown that the solubilities of $FeAsO_4$, $Ca_3(AsO_4)_2$ and $Mg_3(ASO_4)_2$ are significantly higher in the regions of minimum solubility than previously assumed. Arsenic in groundwater is in a much more complex situation [247, 248] and the stability of 'insoluble' arsenates should be established and not assumed.

Stannic arsenate has been utilized in paper chromatography, electrochromatography and thin-layer chromatrography (TLC). Sharma and Sethi [249] have suggested a correlation between R_f values on stannic arsenate and the K_{sp} of the metal cations (Table 4.9).

Chlorotin arsenate, $[SnCl(H_2O)AsO_4]\cdot 2H_2O$, has been shown to have useful intercalation properties [250]. One of the protons is exchangeable with metal ions, which indicates that the formulation $H[SnCl(OH)AsO_4]\cdot 2H_2O$ is a better indication of its properties. A wide range of nitrogen base guest molecules has been studied, and the changes in interlayer spacing have been followed by X-ray powder diffraction. Trimethylamine *N*-oxide expands the chlorotin arsenate layered spacing to 12.3 Å (from 7.09 Å in water or 6.30 Å dehydrated).

Zirconium(IV) arsenate vanadate [251] has been suggested as an ion-exchange material with good exchange capacity and fair stability to dilute mineral acid.

Antimony. Antimony pentoxide, Sb_2O_5, is not as well characterized as its arsenic congener and is virtually insoluble in water. The salts of the hypothetical antimonic acid, for example sodium antimonate, have been shown to contain the anion $[Sb(OH)_6]^-$, indicating a parent acid $HSb(OH)_6$ (chapter 3). The alkali metal salts have remarkably low solubilities, this fact being used as a qualitative wet test for sodium. The solubility was re-determined by Burgess, Blandammer and Peacock [252] (Table 4.10), the enthalpy of solution being 33 kJ mol^{-1}.

Table 4.9 K_{sp} values for metal arsenates derived from R_f values.

Metal arsenate	K_{sp}
$Hg_3(AsO_4)_2$	2.5×10^{-39}
$Pd_3(AsO_4)_2$	4.6×10^{-30}
$IrAsO_4$	2.8×10^{-21}
$Be_3(AsO_4)_2$	6.9×10^{-28}
$GaAsO_4$	2.2×10^{-17}
$Fe_3(AsO_4)_2$	5.2×10^{-26}
$YAsO_4$	1.7×10^{-22}
$LaAsO_4$	2.1×10^{-23}
$CeAsO_4$	2.1×10^{-23}
$Ce_3(AsO_4)_4$	1.9×10^{-25}

Table 4.10 The solubility of sodium hexahydroxoantimonate in aqueous solution.

Temperature (K)	298.2	308.2	323.2	343.2	353.2
Solubility (mol dm^{-3})	3.3×10^{-3}	4.4×10^{-3}	6.0×10^{-3}	8.4×10^{-3}	9.3×10^{-3}

Antimony pentoxide will dissolve in concentrated hydrochloric acid, although the dissolved species have not been identified. $SbCl_5$ prepared from $SbCl_3$ with chlorine fumes in air, and with small amounts of water forms the hydrates $SbCl_5 \cdot H_2O$ and $SbCl_5 \cdot 4H_2O$. If the solution in concentrated HCl is diluted, hydrated antimony(V) oxide is precipitated. H_2S precipitates the orange-red Sb_2S_5 from the HCl solution. The Sb_2S_5 is soluble in aqueous ammonium sulphide:

$$2SbCl_5 + 5H_2S \rightarrow Sb_2S_5 + 10HCl \qquad [4.29]$$

$$Sb_2S_5 + 3(NH_4)_2S \rightarrow 2(NH_4)_3SbS_4 \qquad [4.30]$$

Analysis for the heavy pnictogens of samples from environmental water usually require some form of preconcentration, often reduction of the E(III) and/or E(V) species to EH_3 as volatiles (chapter 9) [253], the analysis being completed by using chromatography interfaced with some other technique such as mass spectroscopic analysis [254]. The direct separation of As(V) and As(III) in solutions containing iron and sulphuric acid has been achieved using ion-exclusion chromatography [255]. This is of practical importance for industries producing arsenic wastewaters.

Bismuth. Pentavalent bismuth species are not common and the aqueous chemistry is limited, sodium and potassium bismuthate ($MBiO_3$) being used as powerful oxidizing agents. $NaBiO_3$ is decomposed by hot water, forming Bi_2O_3, NaOH and liberating dioxygen. It is used for the determination of manganese in alloys, the Mn being oxidized to $[MnO_4]^-$ in hot acidic solutions.

References

1. Pyykkö, P. (1988) *Chem Rev.*, **88**, 563.
2. Pitzter, K. S. (1979) *Acc. Chem. Res.*, **12**, 271.
3. Huheey, J. E. (1988) *Anorganische Chemie*, W. de Gruyter, Berlin, New York.
4. Norman, N. C. (1988) *Chem. Soc. Rev.*, **17**, 269.
5. Tolman, C. A. (1970) *J. Am. Chem. Soc.*, **92**, 2953.
6. Imyanitov, N. S. (1985) *Koord. Khim*, **11**, 1041, 1181.
7. Augustine, A. Ferguson, G. and March, F. C. (1975) *Can. J. Chem.* **53**, 1647.
8. Bricklebank, N., Godfrey, S. M., Lane, H.P. *et al.* (1995) *J. Chem. Soc., Dalton Trans.*, 3873.
9. Hursthouse, M. B. and Steer, I. A. (1971) *J. Organomet. Chem.*, **27**, C11.
10. Begley, M. J. and Sowerby, D. B. (1993) *Acta Crystallogr. Sect. C.*, **49**, 1044.
11. Bricklebank, N. S., Godfrey, M., Lane, H. P. *et al.* (1995) *J. Chem. Soc., Dalton Trans.*, 2421.

12. Wells, A. F. (1938) *Z. Kristallogr.*, **99**, 367.
13. Hawley, D. M. and Ferguson, G. (1968) *J. Chem. Soc. (A)*, 2539.
14. McAuliffe, C. A., Beagley, B., Gott, G. A. *et al.* (1987) *Angew. Chem. Int. Ed. Engl.*, **26**, 264.
15. Clegg, W., Errington, R. J., Fisher, G. A. *et al.* (1962) *J. Chem. Soc., Dalton Trans.*, 1967.
16. Rastogi, M. K. (1984) *Synth. React. Inorg. Met. Org. Chem.*, **14**, 799.
17. Frey, J. E., Cole, R. D., Kitchen, E.-C. *et al.*(1985) *J. Am. Chem. Soc.*, **107**, 748.
18. Bone, S. P. and Sowerby, D. B. (1980) *J. Organomet. Chem.*, **184**, 181.
19. Begley, M. J., Sowerby, D. B., Weseolek, D. *et al.* (1986) *J. Organomet. Chem.*, **316**, 281.
20. Rastogi, M. K. (1987) *Synth. React. Inorg. Met.-Org. Chem.*, **17**, 525.
21. Weiber, M. and Walz, J. (1990) *Z. Anorg. Allg. Chem.* **583**, 102.
22. Hall, M., Sowerby, D. B. and Felshaw, C. P. (1986) *J. Organomet. Chem.*, **315**, 321.
23. Arif, A. M., Benac, B. L., Cowley, A. H. *et al.*(1986) *J. Chem. Soc., Chem. Commun.*, 1543.
24. Wells, R. L., Purdy, A. P., Higa, K. T. *et al.* (1986) *J. Organomet. Chem.*, **325**, C7.
25. Pitt, C. G., Purdy, A. P. and Higa, K. T. (1986) *Organometallics*, **5**, 1266.
26. Wells, R. L., Purdry, A. P., McPhail A. T. and Pitt, C. G. (1986) *J. Organomet. Chem.*, **308**, 281.
27. Cowley, A. H., Jones, R. A., Kidd, K. B. *et al.* (1988) J. *Organomet. Chem.*, **341**, C1.
28. Barron, A. R., Cowley, A.-H., Jones, A. R. *et al.* (1988) *Polyhedron*, **7**, 77.
29. Baum, G., Greiling, A., Massa, W. *et al.* (1989) *Z. Naturforsch, Teil B*, **44**, 560.
30. Clegg, W., Errington, R. J., Flynn, R. J. *et al.* (1992) *J. Chem. Soc., Dalton Trans.*, 1973.
31. Pearson, R. G. (1988) *J. Am. Chem. Soc.*, **110**, 7684.
32. Jensen, W. B. (1978) *Chem. Rev.*, **78**, 1.
33. Jones, R. A. and Whittlesey, B. R. (1986) *Inorg. Chem.*, **25**, 825.
34. Chandler, D. J., Jones, R. A. and Whittlesey, B. R. (1987) *J. Coord. Chem.*, **16**, 19.
35. Godfrey, S. M., Kelly, D. G., Mackie, A. G. *et al.* (1991) *J. Chem. Soc., Chem. Commun.*, 1447.
36. Godfrey, S. M., Mackie, A. G., McAuliffe, C. A. and Pritchard, R. G. (1992) *Angew. Chem. Int. Ed. Engl.* **32**, 919.
37. Downard, A. J., Hanton, L. R. and Paul, R. L. (1992) *J. Chem. Soc., Chem. Commun.*, 2354.
38. Navdelli, M., Pelizzi, C., Pelizzi, G. and Tarasconi, P. (1985) *J. Chem. Soc., Dalton Trans.*, 321.
39. Pelizzi, C., Pelizzi, G. and Tarasconi, P. (1985) *J. Organomet. Chem.*, **281**, 403.
40. Baudler, M., Koch, D. and Carlson, B. (1978) *Chem. Ber.*, **11**, 1217.
41. Hulten, F. and Persson, I. (1987) *Inorg. Chim. Acta*, **43**, 128.
42. Norman, N. C. (1994) *Phos. Sulf. Si.*, **87** 167.
43. Norman, N. C. and Fisher, G. A. (1994) *Adv. Inorg. Chem.*, **41**, 233.
44. Sawyer, J. F. and Gillespie, R. J. (1968) *Prog. Inorg. Chem.*, **34**, 65.
45. Buslaev, Y. A. and Klyagina, A. P. (1993) *Coord. Chem. Rev.*, **126**, 149.
46. Corbridge, D. E. C. (1980) *Phosphorus*, 2nd edn, Elsevier, Oxford.
47. Schleyer, P. V. R. and Reed, A. E. (1990) *J. Amer. Chem. Soc.*, **112**, 1434.
48. Basolo, F. (1968) *Coord. Chem. Rev*,. **3**, 213.
49. Martinson, A. and Songstad, J. (1977) *Acta Chem. Scand.*, **31A**, 645.
50. Alcock, N. W. (1972) *Adv. Inorg. Chem. and Radiochem.*, **15**, 1.
51. Rebbah, A., Yazbeck, J. and Deschanvres, A. (1980) *Acta Cryst.*, **B36**, 1744.
52. Wheeler, R. A. and Kumar, P. N. V. P. (1992) *J. Amer. Chem. Soc.*, **114**, 4776.
53. Sheldrick, W. S., Kaub, J. and Haüsler, H. J. (1986) *Z. Naturforsch.*, **43B**, 789.
54. Drew, M. G. B., Claire, P. P. K. and Willey, G. R. (1989) *J. Chem. Soc. Dalton Trans.*, 57.
55. Midollini, S., Ghilardi, C. A., Moneti, S. and Orlandini, A. (1988) *J. Chem. Soc. Chem. Commun.*, 1241.
56. Sheldrick, W. S. and Kiefer, J. (1992) *Z Naturforsch*, **B47**, 1079.
57. Haüsler, H. J. and Sheldrick, W. S. (1987) Angew. *Chem. Int. Ed. Eng.*, **26**, 1172.
58. Müller, U. and Mohammed, A. T. (1985) *Acta Cryst.*, **C41**, 329.

59. Willing, W., Müller, U., Eicher, J. and Dehnicke, K. (1986) *Z. Anorg. Allg. Chem.*, **537**, 145.
60. Sheldrick, W. S. and Kaub, J. (1984) *Z. Naturforsch*, **39B**, 1252.
61. Kaub, J. and Sheldrick, W. S. (1984) *Z. Naturforsch*, **39B**, 1257.
62. Zalewicz, M. and Ptaszynski, B. (979)*Polish J. Chem.*, **53**, 1437.
63. Cavalco, L., Nardellini, M. and Braibanti, M. (1955) *Gazz. Chim. Ital.*, **85**, 1239.
64. Stepina, S. B., Vlasova, I. V., Sokolova, N. P. and Plyushchev, V. E. (1968) *Z. Neorg. Knim.*, **13**, 2570, 2865.
65. Sowerby, D. R., Hall, M., Nunn, M. and Begley, M. J. (1986) *J. Chem. Soc. Dalton Trans.*, 1231.
66. Gutsev, G. L., Kostin, V. I., Sergienko, V. I. and Ignateva, L. N. (1986) *J. Fluor. Chem.*, **32**, 367.
67. Nowicki, J., Borrmann, H. and Minkwitz, R. (1991) *Z. Naturforsch.*, **46B**, 629.
68. Gillespie, R. J., Sawyer, J. F., Collins, M. J. and Schrobilgen, G. J. (1986) *Inorg. Chem.*, **25**, 2053.
69. Borrmann, H., Nowicki, J. and Minkwitz, R. (1991) *Z. Anorg. Allg. Chem.*, **605**, 109.
70. Kihara, K. and Sudo, T. (1974) *Acta Cryst.*, **30B**, 1088.
71. Sowerby, D. R., Nunn, M. and Hall, M. (1986) *J. Chem. Soc. Dalton Trans.*, 1239.
72. Müller, U. and Mohammed, T. A. (1985) *Z. Naturforsch.*, **40B**, 562.
73. Pohl, S., Saak, W. and Haase, D. (1987) *Angew. Chem. Int. Ed. Engl.*, **26**, 467.
74. Jacobsen, R. A. and Schröder, D. R. (1973) *Inorg. Chem.*, **12**, 515.
75. Stancheva, L. I., Vlasova, I. V., Stepina, S. B. and Plyuschev, V. E. (1966) *Russ. J. Inorg. Chem.*, **11**, 761.
76. Dzeranova, K. B., Bukhalova, G. A., Kaloev, N. I. and Gazdanov, I. V. (1986) *Russ. J. Inorg. Chem.*, **31**, 756.
77. Cousson, A., Vedrine, A. and Consseins, J. C. (1972) *Compt. Rend.*, **274C**, 864.
78. Hobecker, C. (1971) *Z. Anorg. Allg. Chem.*, **384**, 12.
79. Battaglia, L. P., Corradi, A. B., Battistuzzi, R. and Manfredini, T. (1986) *Inorg. Chim. Acta*, **121**, 131.
80. Dzeranova, K. B., Bukhalova, G. A. and Kaloev, N. I. (1985) *Russ. J. Inorg. Chem.*, **30**, 1397.
81. Dzervanova, K. B., Bukhalova, G. A. and Kaloev, N. I. (1985) *Russ. J. Inorg. Chem.*, **30**, 2983.
82. Safonov, V. V. and Mireev, V. A. (1985) *Russ. J. Inorg. Chem.*, **30**, 1396.
83. Aurivillius, B. and Lindblom, C. I. (1964) *Acta Chem. Scan.*, **18**, 1554.
84. Lazarini, F. and Blazic, B. (1985) *Acta Cryst.*, **C41**, 1619.
85. Rheingold, A. L., Mammano, N. J., Zalkin, A. and Landers, A. (1977) *Inorg. Chem.*, **16**, 297.
86. Robertson, B. K., McPherson, W. G. and Meyers, E. A. (1967) *J. Phys. Chem.*, **71**, 3531.
87. Lazarini, F. (1977) *Acta Cryst.*, **B33**, 2686.
88. Alcock, N. W., Ravindran, M. and Willey, G. R. (1989) *J. Chem. Soc., Chem. Comm.*, 1063.
89. Willey, G. R., Collins, H. and Drew, M. G. B. (1991) *J. Chem. Soc., Dalton Trans.*, 961.
90. Rheingold, A. L., Uhler, A. D. and Landers, A. G. (1983) *Inorg. Chem.*, **22**, 3255.
91. Meyers, E. A. and McPherson, W. G. (1968) *J. Phys. Chem.*, **72**, 532.
92. Lazarini, F. (1977) *Acta Cryst.*, **B33**, 1954.
93. Lazarini, F. and Leban, I. (1980) *Acta Cryst.*, **B36**, 2745.
94. Benedetti, A., Fabretti, A. C. and Malavasi, W. (1972) *J. Cryst. Spectrosc. Res.*, **22**, 145.
95. Lazarini, F. (1980) *Acta Cryst.*, **B36**, 2748.
96. Robinson, W. R. and Morss, L. R. (1972) *Acta Cryst.*, **B28**, 653.
97. Herdtweck, E. and Kreusel, U. (1993) *Acta Cryst.*, **C49**, 318.
98. Lazarini, F. (1978) *Acta Cryst.*, **B34**, 2288.
99. Lazarini, F. (1985) *Acta Cryst.*, **C41**, 1617.
100. McPherson, W. G. and Meyer, E. A. (1968) *J. Phys. Chem.*, **72**, 3117.
101. Lazarini, F. (1977) *Acta Cryst.*, **B33**, 1957.
102. Lazarini, F. (1987) *Acta Cryst.*, **C43**, 875.

103. Lazarini, F. (1987) *Acta Cryst.*, **C43**, 637.
104. Meyer, G. and Schönemund, A. (1980) *Z. Anorg. Allg. Chem.*, **468**, 185.
105. Lazarini, F. (1977) *Acta Cryst.*, **B33**, 2961.
106. Chalbot, B. and Parthe, E. (1978) *Acta Cryst.*, **B34**, 645.
107. Williams, J. M., Geiser, U., Wade, E. and Wang, H. H. (1990) *Acta Cyrst.*, **C46**, 1547.
108. Jakubas, R., Lefebvre, J. and Carpenter, P. (1991) *Acta Cryst.*, **B47**, 228.
109. Jakubas, R., Matuszewski, J., Sobczyk, L. and Glowiak, T. (1990) *Acta Cryst.*, **C46**, 1385.
110. Aurivillius, B. and Stålhandske, C. (1978) *Acta. Chem. Scand.*, **32A**, 715.
111. Norman, N. C., Clegg, W., Fisher, G. A. *et al.* (1991) *Chem. Ber.*, **124**, 2457.
112. Jakubas, R., Zaleski, J., Glowiak, T. and Sobczyk, L. (1989) *J. Phys. Chem. Solids*, **50**, 1265.
113. O'Donnel, T. A. (1987) *Chem. Soc. Rev.*, **16**, 1.
114. Aubke, F. and Karunanithy, S. (1984) *J. Fluor. Chem.*, **25**, 339.
115. Aubke, F. and Karunanithy, S. (1983) *J. Fluor. Chem.*, **23**, 541.
116. Passmore, J. and Klapötke, T. (1990) *J. Chem. Soc., Dalton Trans.*, 3815.
117. Müller, U, and Bebendorf, J. (1990) *Z. Naturforsch.*, **45B**, 927.
118. Gafner, G. and Kruger, G. J. (1974) *Acta Cryst.*, **30B**, 250.
119. Fadini, A. and Schnepel, F. M. (1989) *Vibrational Spectroscopy*, Ellis Horwood, Chichester, Sussex.
120. Dove, M. F. A., Sanders, J. C. P., Jones, E. L. and Parkin, M. J. (1984) *J. Chem. Soc., Chem. Comm.*, 1578.
121. Dean, P. A. W. (1975) *J. Fluor. Chem.*, **5**, 499.
122. Passmore, J. and Desjardins, C. D. (1975) *J. Fluor. Chem.*, **6**, 379.
123. Roesky, H. W. and Witt, M. (1986) *Inorg. Synth.*, **24**, 72.
124. Williams, J. M., Schultz, A. J., Miro N. D. *et al.* (1978) *Inorg. Chem.*, **17**, 646.
125. Russell, D. R., Gantar, D., Frlec, B. and Holloway, J. H. (1987) *Acta Cryst.*, **C43**, 618.
126. Mootz, D. and Wiebroke, M. (1986) *Inorg. Chem.*, **25**, 3095.
127. Preiss, H. (1971) *Z. Anorg. Allg. Chem.*, **380**, 71.
128. Clark, R. J. H. and Trumble, W. R. (1975) *J. Chem. Soc., Chem. Comm.*, 318.
129. Wilson, W. W., Thompson, R. C. and Aubke, F. (1980) *Inorg. Chem.*, **19**, 1484.
130. Mallela, S. P., Yap, S., Sams, J. R. and Aubke, F. (1986) *Rev. Chim. Miner.*, **23**, 572.
131. Aubke, F., Willner, H., Schaebs, J. *et al.* (1992) *J. Amer. Chem. Soc.*, **114**, 8972.
132. Aubke, F., Hwang, G., Wang, C. *et al.* (1993) *Can. J. Chem.*, **71**, 1532.
133. Hersh, W. H. (1985) *J. Amer. Chem. Soc.*, **107**, 4599.
134. Surles, T., Quarterman, L. A. and Hyman, H. H. (1973) *J. Inorg. Nucl. Chem.*, **35**, 670.
135. Popov, A. I., Scharabarin, A. V., Sukhorerkhov, V. F. and Tchumaevsky, N. A. (1989) *Z. Anorg. Allg. Chem.*, **576**, 242.
136. Griffiths, J. E., Sunder, W. A. and Falconer, W. E. (1975) *Spectrochim. Acta*, **31A**, 1207.
137. Griffiths, J. E., Sunder, W. A., Wayda, A. L. *et al.* (1979) *J. Fluor. Chem.*, **14**, 299.
138. Christe, K. O. (1977) *Inorg. Chem.*, **16**, 937.
139. Christe, K. O., Wilson, W. W. and Schack, C. J. (1978) *J. Fluor. Chem.*, **11**, 71.
140. Bougon, R., Huy, T. B., Cadet, A. *et al.* (1974) *Inorg. Chem.*, **13**, 690.
141. Gillespie, R. J., Martin, D., Schrobiglen, G. J. and Slim, D. R. (1977) *J. Chem. Soc., Dalton Trans.*, 2234.
142. Gillespie, R. J., Boldrini, P., Ireland, P. R. and Schrobilgen, G. J. (1974) *Inorg. Chem.*, **13**, 1690.
143. Gillespie, R. J., Martin, D. and Schrobilgen, G. J. (1980) *J. Chem. Soc., Dalton Trans.*, 1898.
144. Bellard, S., Rivera, A. V. and Sheldrick, G. M. (1978) *Acta Crystallogr. Sect. B.*, **34**, 1034.
145. Torniporth-Oetting, I. C., Klapotke, T. M., Behrens, U. and White, P. S. (1992) *J. Chem. Soc., Dalton Trans.*, 2055.
146. Chen, S.-J., Behrens, U., Olbrich, F. and Mews, R. (1993) *Z. Anorg. Allg. Chem.*, **619**, 1725.
147. Apblett, A., Chivers, T. and Richardson, J. F. (1986) *Can. J. Chem.*, **64**, 849.

148. Tornieporth-Oetting, I. C., Klapotke, T. M., Cameron, T. S. *et al.* (1992) *J. Chem. Soc., Dalton Trans.*, 537.
149. Binas, H. (1967) *Anorg. Allg. Chem.*, **352**, 271.
150. Jochims, J. C., Abu-El-Halawa, R., Zsolinai, L. and Huttner, G. (1984) *C. Ber.*, **117**, 1161.
151. Hansen, H.-D., Shwarz, W., Rajca, G. and Weidlein, J. (1986) *Z. Naturforsch., Teil B*, **41**, 1223.
152. Webster, M. and Keats, S. (1971) *J. Chem. Soc., A.*, 836.
153. Karsch, H. H., Witt, E., Schneider, A. *et al.* (1995) *Angew. Chem. Int. Ed. Engl.*, **34**, 557.
154. Hulme, R. and Scruton, J. S., (1968) *J. Chem. Soc., A*, 2448.
155. Lipka, A. and Wunderlich, H. (1980) *Z. Naturforsch., Teil B*, **35**, 1548.
156. Battaglia, L. P., Corradi, A. B., Pelsi, G. *et al.* (1991) *J. Chem. Soc., Dalton Trans.*, 3153.
157. Holmes, R. R. and Bertant, E. F. (1958) *J. Am. Chem. Soc.*, **80**, 2980.
158. Summers, J. C. and Sisler, H. H. (1970) *Inorg. Chem.*, **9**, 862.
159. Clegg, W., Elsegood, M. R. J., Graham, V. *et al.* (1993) *J. Chem. Soc., Dalton Trans.*, 997.
160. Clegg, W., Elsegood, M. R. J., Norman, N. C. and Pickett, N. L. (1994) *J. Chem. Soc., Dlaton Trans.*, 1753.
161. Clegg, W., Elsegood, M. R. J., Graham, V. *et al.* (1994) *J. Chem. Soc., Dalton Trans.*, 1743.
162. Willey, G. R., Daly, L. T. and Drew, M. G. B. (1996) *J. Chem. Soc., Dalton Trans.*, 1063.
163. Hockless, D. C. R. and Norman, N. C. (1991) *Chem. Ber.*, **124**, 2457.
164. Siewert, B. and Muller, U. (1992) *Z. Naturforsch., Teil B.*, **47**, 680.
165. Borgias, B. A., Hardin, G. G. and Raymond, K. N. (1986) *Inorg. Chem.*, **25**, 1057.
166. Lundquist, I. and Branden, C. I. (1959) *Acta Crystallogr.*, **12**, 642.
167. Branden, C. I. and Lundquist, I. (1961) *Acta Chem. Scand.*, **15**, 167.
168. Knodler, F., Schwartz, W. and Schmidt, A. (1987) *Z. Naturforsch., Teil B.*, **42**, 1282.
169. Shihada, A. F. and Weller, F. (1981) *Z. Anorg. Allg. Chem.*, **472**, 102.
170. Gerrard, C., Lucken, E. A. C. and Bernardinelli, G. (1986) *J. Chem. Soc., Perkin Trans.*, **2**, 879.
171. Korte, L., Mootz, D., Scherf, M. and Wiebocke, M. (1988) *Acta Crystallogr., Sect. C.*, **44**, 1128.
172. Chevrier, B., Le Carpentier, J. M. and Weiss, R. (1972) *Acta Crystallogr., Sect. B*, **28**, 2667.
173. Yammamoto, J., Ito, S., Tsuldoi, T. *et al.* (1985) *Bull. Chem. Soc., Jpn.*, **58**, 470.
174. Koller, F. J., Schwartz, W. and Schmidt, A. (1979) *Z. Naturforsch., Teil B*, **34**, 563.
175. van Nuffle, P., Lenstra, A. T. H. and Teise, H. J. (1982) *Acta Crystallogra., Sect. B*, **38**, 3089.
176. Alcock, N. W., Ravindran, M. and Willey, G. R. (1993) *Acta Crystallogra., Sect. B*, **49**, 507.
177. Borgsen, B., Weller, F. and Dehnike, K. (1990) *Chem. Zeit.*, **114**, 111.
178. Beagley, B., Endregard, M. and Nicholson, D. G. (1991) *Acta Chem. Scand.*, **45**, 349.
179. Colapetro, M. Domenicano, A., Scaramuzza, L. and Vaciago, A. (1968) *J. Chem. Soc., Chem. Commun.*, 302.
180. Raston, C. L. and White, A. H. (1975) *J. Chem. Soc., Dalton Trans.*, 2425.
181. Hoskins, B. F., Tiekink, E. R. T. and Winter, G. (1985) *Inorg. Chim. Acta.*, **99**, 177.
182. Kisenji, J. M., Willey, G. R., Drew, M. G. B. and Wandiga, S. O. (1985) *J. Chem. Soc., Dalton Trans.*, 69.
183. Drager, M. (1985) *Z. Anorg. Allg. Chem.*, **411**, 79.
184. Drager, M. (1984) *Chem. Ber.*, **107**, 2601.
185. Poore, M. C. and Russell, D. R. (1971) *J. Chem. Soc., Chem. Commun.*, 18.
186. Golic, L. and Miliceu, S. (1978) *Acta Crystallogra., Sect C*, **34**, 3379.
187. Lundgvist, I. (1963) *Inorganic Adduct Molecules of Oxo-Compounds*, Springer, Berlin.
188. Raston, C. L. and White, A. H. (1976) *J. Chem. Soc., Dalton Trans.*, 791.
189. Snow, M. R. and Tienkink, E. R. T. (1987) *Aust. J. Chem.*, **40**, 743.
190. McKie, G., Raston, C. L., Rowbottom, G. L. and White, A. H. (1981) *J. Chem. Soc., Dalton Trans.*, 1360.

191. Kello, E., Kettman, V. and Garaj, J. (1985) *Acta Crystallogr., Sect. C*, **41**, 520.
192. Hough, E. and Nicholson, D. G. (1981) *J. Chem. Soc., Dalton Trans.*, 2083.
193. Rubin, B., Heldrich, F. J., Dean, W. K. *et al.* (1981) *Inorg. Chem.*, **20**, 4434.
194. Drew, M. G. B., Kisenji, J. M., Willey, G. R. and Wandiga, S. O. (1984) *J. Chem. Soc., Dalton Trans.*, 1717.
195. Pohl, S., Saak, W., Lotz, R. and Haase, D. (1990) *Z. Naturforsch., Teil B.*, **45**, 1355.
196. Massiani, M.-C., Papiernik, R., Hubert-Pfalzgraf L.G. and Daran, J.-C. (1990) *J. Chem. Soc., Chem. Commun.*, 301.
197. Stalhandske, C.-I. (1969) *Acta Chem. Scand.*, **23**, 1525.
198. Troyanov, S. I. and Pisarevsky, A. P. (1991) *Koord. Khim.*, **17**, 909.
199. Troyanov, S. I. and Pisarevsky, A. P. (1993) *J. Chem. Soc., Chem. Commun.*, 335.
200. Bharadwaj, P. K., Lee, A. M., Mandal, S. *et al.* (1994) *Aust. J. Chem.*, **47**, 1799.
201. Lazarini, F. and Milicev, S. (1976) *Acta Crystallogra. Sect. B.*, **31**, 2863.
202. Lazarini, F., Golic, L. and Pelizzi, G. (1975) *Acta. Crystallogra., Sect A.*, 5140.
203. Clegg, W., Farrugia, L.J., McCauley, A. *et al.* (1993) *J. Chem. Soc., Dalton Trans.*, 2579.
204. Willey, G. R., Rudd, M. D., Samuel, C. J. and Drew, M. G. B. (1995) *J. Chem. Soc., Dalton Trans.*, 59.
205. Battaglia, L. P., Corradi, A. B., Nardelli M. and Tani, V. (1978) *J. Chem. Soc., Dalton Trans.*, 583.
206. Raston, C. L., Rowbottom, G. L. and White, A. H. (1981) *J. Chem. Soc., Dalton Trans.*, 1352, 1366, 1369, 1379, 1383.
207. Drew, M. G. B., Kisenyi, J. M., Willey, G. R. and Wandiga, S. O. (1984) *J. Chem. Soc., Dalton Trans.*, 1707.
208. Drew, M. G. B., Kisenyi, J. M. and Willey, G. R. (1984) *J. Chem. Soc., Dalton Trans.*, 1723.
209. Battaglia, L. P. and Corradi, A. B. (1986) *J. Chem. Soc., Dalton Trans.*, 1513.
210. Bharadwaj, P. K., Lee, A. M., Skelton, B. W. *et al.* (1994) *Aust. J. Chem.*, **47**, 405.
211. Battaglia, L. P., Corradi, A. B., Pelizzi, G. and Vidoni, E. (1975) *Cryst. Struct. Comm.*, **4**, 399.
212. Battaglia, L. P., Corradi, A. B., Pelizzi, G. and Vidoni, E. (1977) *J. Chem. Soc., Dalton Trans.*, 1141.
213. Battaglia, L. P. and Corradi, A. B. (1981) *J. Chem. Soc., Dalton Trans.*, 23.
214. Williams, D. J., Quicksall, C. O. and Barkigia, K. M. (1982) *Inorg. Chem.*, **21**, 2097.
215. Praeckel, V. U., Huber, F. and Preut, H. (1982) *Z. Anorg. Allg. Chem.*, **494**, 67.
216. Bharadwaj, P. K. and Musker, W. K. (1987) *Inorg. Chem.*, **26**, 1453.
217. Farrugia, L. J., Lawlor, F. J. and Norman, N. C., (1995) *Polyhedron*, **14**, 311.
218. Farrugia, L. J., Lawlor, F. J. and Norman, N. C. (1995) *J. Chem. Soc., Dalton Trans.*, 1163.
219. Clegg, W., Norman, N. C. and Pickett, N. L. (1993) *Polyhedron*, **12**, 1251.
220. Bricklebank, N., Godfrey, S. M., Lane, H. P. *et al.* (1994) *J. Chem. Soc., Dalton Trans.*, 1759.
221. Willey, G. R., Barras, J. R., Rudd, M. D. and Drew, M. G. B. (1994) *J. Chem. Soc., Dalton Trans.*, 3025..
222. Gaskel, T. F. (1971) *Chem. Ind.*, 1149.
223. Vanysek, P. (1995) in *Handbook of Chemistry and Physics* (ed. D. R. Linde), CRC Press, New York, pp. 8–12 *et seq.*
224. Horner, L. and Winkler, H. (1964) *Tetrah. Sett.*, 3271.
225. Daniels, M. and Weiss, J. (1958) *J. Chem. Soc.*, 2467.
226. Rastogi, R. P., Das, I. and Singh, A. R. (1984) *J. Phys. Chem.*, **88**, 5132.
227. Dorfman, Y. A., Aleshkova, M. M., Doroshkevich, D. M. and Zilberberg, I. L. (1988) *Russ. J. Inorg. Chem.*, **33**, 782.
228. Dorfman, Y. A., Polimbetova, G. S., Mansurov, B. A. *et al.* (1988) *Koord. Khim.*, **14**, 1219.
229. Dorfman, Y. A., Polimbetova, G. S., Mansurov, B. A. and Doroshkevich, D. M. (1989) *Z. Obshch. Khim.*, **59**, 2007.
230. Graziano, J. H., Cuccia, D. and Friedheim, E. (1978) *J. Pharmacol. Exp. Ther.*, **207**, 1051.
231. Aposhian, H. V., Carter, D. E., Hoover, T. D. *et al.* (1984) *Fund. Appl. Toxicol.*, **4**, 558.

232. Holmes, R. R., Poutasse, C. A., Day, R. O. and Holmes, J. M. (1985) *Organometallics*, **4**, 714.
233. Juniere, P. and Sigwalt, M. (1964) *Aluminium*, Crosby Lockwood, London.
234. Copur, M., Colak, S. and Yapici, S. (1995) *Ind. Eng. Chem. Res.*, **34**, 3995.
235. Fedorov, V. A., Kalosh, T. N. and Shmydko, L. I. (1974) *Russ J. Inorg. Chem.*, **19**, 991.
236. Greenwood, N. N. and Earnshaw, E. (1986) *Chemistry of the Elements*, Pergamon, Oxford.
237. Sundvall, B. (1979) *Acta Chem. Scand.*, **33A**, 219.
238. Sundvall, B. (1983) *Inorg. Chem.*, **22**, 1906.
239. Olin, A. (1975) *Acta Chem. Scan.*, **29**, 907.
240. Laudise, R. A. (1987) *Chem. Eng. News*, **65**, 39.
241. Laudise, R. A., Cava, R. J. and Caporaso, A. J. (1986) *J. Cryst. Growth*, **74**, 275.
242. Bierlein, J. D. and Gier, T. E. (1976) *US Patent* 3 949 323.
243. Phillips, M. L. F. and Stucky, G. D. (1995) *Inorg. Synth.*, **30**, 143.
244. Seidell, A. (1958) *Solubilities of Inorganic and Metal Organic Compounds*, vol I, van Nostrand Reinhold, New York.
245. Seidell, A. (1965) *Solubilities of Inorganic and Metal Organic Compounds*, vol II, van Nostrand Reinhold, New York.
246. Robins, R. G. (1981) *Metall. Trans. B.*, **12B**, 103.
247. Holme, T. R., Anderson, M. A., Iverson, D. G. and Stanforth, R. S. (1979) *American Chemical Society Symposium Series, Number 93: Chemical Modelling in Aqueous Systems* (ed. E. A. Jenne), American Chemical Society, Washington, DC, paper 31.
248. Gulens, J. and Champ, D. R. (1979) in *American Chemical Society Symposium Series, Number 93: Chemical Modelling in Aqueous Systems* (ed. E. A. Jenne), American Chemical Society, Washington, DC, paper 4.
249. Sharma, S. A., Sharma, T. A. and Sethi, B. M. (1983) *J. Liq. Chromatog.*, **6**, 1253.
250. Lagaly, G. and Beneke, K. (1987) *Inorg. Chem.*, **26**, 2537.
251. Qureshi, S. Z., Khan, M. A. and Rahman, N. (1995) *Bull. Chem. Soc. Jpn.*, **68**, 1613.
252. Burgess, J., Blandammer, M. J. and Peacock, R. D. (1974) *J. Chem. Soc., Dalton Trans.*, 1084.
253. Alfassi, Z. B. and Wai, C. M. (eds) (1992) *Preconcentration Techniques for Trace Analysis*, CRC Press, Boca Raton, FL.
254. Apte, S. C., Howard, A. G. and Campbell, A. T. (1989) in *Arsenic and Antimony in Environmental Analysis using Chromatography Interfaced with Atomic Spectroscopy* (eds R. M. Harrison and S. Rapsomanikis), Ellis Horwood, Chichester, Sussex, ch. 9.
255. Hemmings, M. J. and Jones, E. A. (1991) *Talanta*, **38**, 151.

5 Organoarsenic and organoantimony compounds: synthesis, properties and uses in organic transformations

W.I. CROSS, S.M. GODFREY, C.A. McAULIFFE,
A.G. MACKIE and R.B. PRITCHARD

5.1 Introduction: synthesis of organoarsenic and organoantimony compounds

This chapter is dominated by two main sections, one on the tervalent species (Section 5.2) and one on the pentavalent species (Section 5.3). The tervalent species are used mainly as ligands in transition metal coordination chemistry, and as arsenic or antimony precursors in metallo-organic chemical vapour deposition (MOCVD) processes. The pentavalent species find most use in organic syntheses, for example the arsenic ylides. These themes will be dominant in their respective subsections.

5.2 The synthesis of tervalent organoarsenic and organoantimony compounds

The various methods available for the formation of C−As, and C−Sb bonds have been described in a number of reviews [1–5], and comprehensive lists of the known arsines [3, 6] and stibines [3, 6] have been published. Despite the very rapid growth of the coordination chemistry of these ligands the vast majority of work has still been concerned with relatively few ligands. The aim of the present chapter is neither to produce a comprehensive coverage of potential arsenic and antimony donor ligands nor to discuss exhaustively all the possible synthetic routes, either of which would fill this volume alone. Rather, this chapter will review the syntheses of the more important ligands and attempt to show how these can be extended to produce other ligands, as well as pointing out some of the limitations of certain of these syntheses. Where a number of routes is available to a particular compound, the one(s) to which reference(s) is (are) made has (have) been selected on the basis of their ease of synthesis from readily available starting materials and/or because high yields are obtainable. This selection of routes has involved considerable subjective judgement on our part and apologies are made here to workers whose efforts are not acknowledged.

In contrast to the monodentate arsines and stibines, the majority of which were first prepared by organogroup 15 chemists before the advent of modern coordination chemistry, the greatest impetus for the synthesis of bidentate and multidentate chelating ligands has been their novel and complex coordination chemistry. The synthesis of multidentates approaches something of an art form and almost all the work is based on the schools of Nyholm, Venanzi, Meek, Sacconi, King, and their co-workers.

Throughout this chapter emphasis is placed upon the chemical principles rather than on the experimental details; however, a few words of caution to the worker new to this field may not be out of place. The physiological properties of the intermediates and ligands are for the most part unknown, but it is to be expected that all are toxic to varying degrees and they should be treated as such. The lower alkyl arsines and alkyl stibines are malodorous, air-sensitive (in a few cases spontaneously inflammable) and highly toxic; chlorodiphenylarsine is a powerful sternutator; and 2-chloroethylsulphide derivatives and chloroalkylamines are vesicants. Thus, as a general rule the preparations of arsenic and antimony ligands must be conducted in adequate fume hoods, usually under a dry dinitrogen atmosphere and with adequate fire precautions. The more toxic and malodorous biproducts should be treated chemically before release into the environment. Suitable reagents include bromine water, aqueous permanganate, aqueous calcium hypochlorite and dilute nitric acid – of which the last two are the most commonly used.

Although the more widely used monodenate and bidentate, and even a few multidentate, ligands are commercially available, their costs are high, and most chemists will generally be compelled to synthesize their own ligands.

5.2.1 Precursors

There are a number of compounds which are used in the synthesis of a variety of ligands, and it is convenient to discuss the synthesis of these compounds first. The most important precursors are those containing halogen bonds – group 15, R_nEX_{3-n}, and alkali metal bonds – group 15, MER_2. [Note: throughout this chapter, E = As, Sb; X = Cl, Br, I; R = organic group; and M = Li, Na, K, unless otherwise stated.)

Halogroup 15 compounds, R_nEX_{3-n}. The trichlorides ECl_3, and a few others, for example $PhAsCl_2$, are commercially available and review articles on the preparation of haloarsines and halostibines are available [2, 3, 5, 7].

Chloroarsines are considerably easier to obtain than the phosphorus analogues, since arsenic(V) (as arsonic or arsinic acids) is readily reduced to arsenic(III) (as R_nAsX_{3-n}) by SO_2 in concentrated aqueous HX [2, 3]. Thus, $MeAsI_2$ is formed by the reaction sequence [8]

$$\mathrm{MeI} \xrightarrow{As_2O_3/NaOH} \mathrm{MeAsO(OH)_2} \xrightarrow{SO_2/HCl/KI} \mathrm{MeAsI_2} \quad [5.1]$$

The use of the alkylarsenites (from $RAsX_2$ + NaOH) in place of As_2O_3/NaOH leads to RR′AsX compounds [8]; for example cacodylic acid, $Me_2As(O)OH$, and its sodium salt are commercially available and are more usually used as the starting material to prepare Me_2AsI [9]. Similarly, $PhAsCl_2$/NaOH/MeI yields PhMeAsI on sulphur dioxide reduction. The much less reactive aryl halides generally fail to undergo this reaction. Dichlorophenylarsine, $PhAsCl_2$, is readily prepared from aniline [10].

$$\mathrm{PhNH_2} \xrightarrow{NaNO_2/HCl} \mathrm{PhN_2^+Cl^-} \xrightarrow{As_2O_3/NaOH} \mathrm{PhAs(O)(OH)_2} \xrightarrow{SO_2/HCl} \mathrm{PhAsCl_2} \quad [5.2]$$

2-Bromophenyldichloroarsine is obtained similarly, except that the diazotization is conducted in glacial acetic acid solution and $AsCl_3$ is the source of the arsenic [11, 12]:

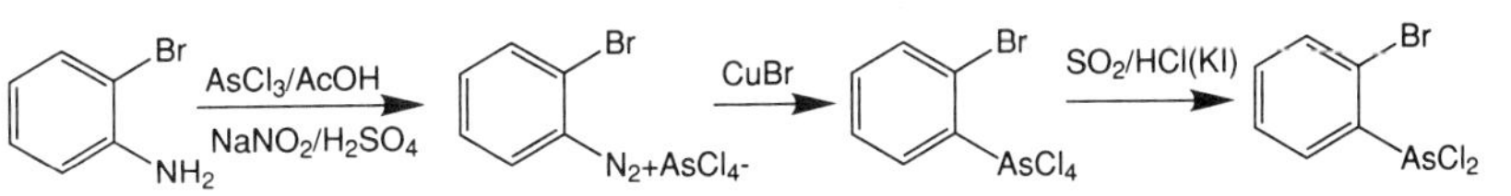

[5.3]

Diphenylchloroarsine is best made by comproportionation of $PhAsCl_2$ and PhAsO, although the product is contaminated with $PhAsCl_2$ and Ph_3As [10].

The ω-chloroalkyldichloroarsines, $Cl(CH_2)_nAsCl_2$ ($n = 2$, 3) are prepared [13] from $Cl(CH_2)_nOH$:

$$\mathrm{Cl(CH_2)_nOH} \xrightarrow{As_2O_3/NaOH} \mathrm{(HO)_2(O)As(CH_2)_nOH} \xrightarrow{SO_2/HCl} \mathrm{Cl_2As(CH_2)_nCl} \quad [5.4]$$

Other chloroarsines of considerable synthetic value are $Cl_2As(CH_2)_2AsCl_2$ (from $Ph_2As(CH_2)_2AsPh_2$ + $AsCl_3$ under pressure) [14]; $RAs(Cl)(CH_2)_n As(Cl)R$ can be obtained as follows [14–17]:

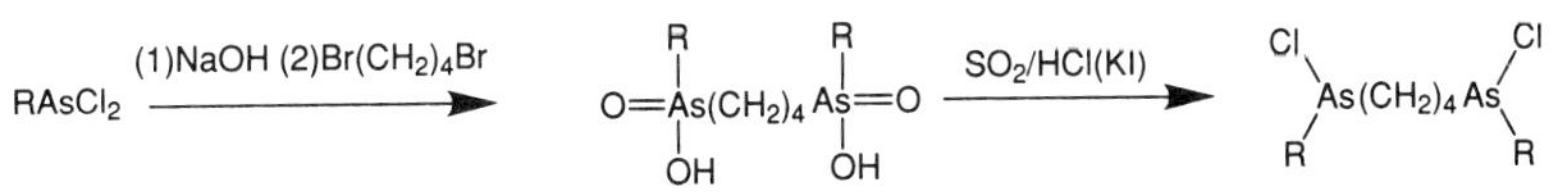

[5.5]

The aryl bis(dichloroarsine) ,1,2-$C_6H_4(AsCl_2)_2$ is prepared as follows [15]:

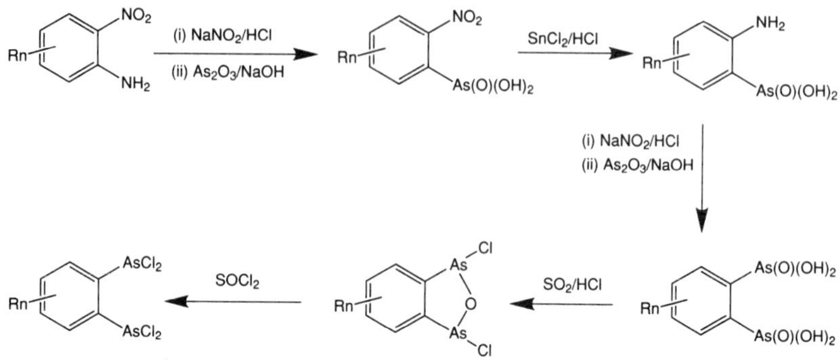

[5.6]

Chloroarsines are much more stable to water than are chlorophosphines, as can be judged by their preparation from concentrated aqueous HX solutions. They are highly toxic and corrosive but are generally air stable; they hydrolyse in neutral or dilute acid solution.

Antimony differs from arsenic in that the alkylstibonic or dialkylstibinic acids are poorly characterized and not useful starting materials for alkylhalostibines. In contrast, arylstibonic acids are useful synthetically, the chief disadvantages being the low yields (often < 20%) and their flocculent nature which makes manipulation difficult.

The compounds Me_2SbX (X = Cl, Br) are formed by halogenation of trimethylstibine followed by thermolysis of the resulting Me_3SbX_2 under reduced pressure (to Me_2SbX + MeX) [18]:

$$Me_2SbCl \xrightarrow{Cl_2} Me_2SbCl_3 \xrightarrow{\Delta} MeSbCl_2 + MeCl \qquad [5.7]$$

Similarly, R_2SbX (R = Me, Et, Pr^i; X = Br, I) were prepared in useful yields (*c.* 70%–75%) by Hartmann and Kühl [19, 20], and $Pr^i{}_2SbX$ and Pr^iSbX_2 in similar yields by Breunig and Kanig [21]. The reaction of $Me_2SbSbMe_2$ with SO_2Cl_2 [22] yields Me_2SbCl. The most attractive preparation for $MeSbCl_2$ seems to be from $PbMe_4$ and $SbCl_3$ (unpublished work, quoted in [4], page 390) since the original route gives only poor yields [18]. Yields of less than 20% are also obtained for the preparation of $PhSbCl_2$ from aniline [23]:

$$PhNH_2 \xrightarrow[(2)NaOH]{(1)NaNO_2/HCl/Sb_2O_3} PhSb(O)(OH)_2 \xrightarrow{SO_2/HCl} PhSbCl_2 \qquad [5.8]$$

which is more conveniently obtained by comproportionation of Ph_3Sb + $2SbCl_3$ in Et_2O [24, unpublished observations of C. A. McAutliffe, I. E. Niven and R. V. Parish]. 2-Bromophenyldichlorostibine, 2-$C_6H_4(Br)SbCl_2$, is made by the diazonium route [23, 24] from 2-bromoaniline; diphenylchlorostibine is readily available from $SbPh_3$ [25]:

$$Ph_3Sb \xrightarrow{HCl/MeOH} Ph_2SbCl + C_6H_6 \xrightarrow{NaOH} (Ph_2Sb)_2O \xrightarrow{HCl/EtOH} Ph_2SbCl \qquad [5.9]$$

which is much superior to the thermal decomposition of Ph_3SbCl_2.

Alkali metal derivatives. The formation and synthetic uses of alkali metal [26] arsenides and stibides [27] have been reviewed elsewhere. Alkali metal diphenylarsenides are readily obtained by cleavage of a phenyl group from $AsPh_3$ by lithium in tetrahydrofuran (thf), sodium in liquid ammonia, or potassium in dioxan:

$$2M + AsPh_3 \rightarrow MAsPh_2 + MPh \qquad [5.10]$$

The metal phenyl produced can be destroyed by addition of a stoichiometric amount of Bu^tCl or NH_4X [28–32]. Generally, the $MEPh_2$ are prepared *in situ* and used without isolation, although the solid compounds (usually solvated) can be obtained if required. Sodium diphenylstibide is best prepared in liquid ammonia [29].

The commercial availability of EPh_3 makes the route shown in reaction [5.10] the most widely used, but $MEPh_2$ are also obtainable from Ph_2ECl (E = As, Sb) and alkali metal or from Ph_2MH (E = As) and alkyllithium [33–35]. The thermal instability of Ph_2SbH makes it a less attractive starting point than Ph_3Sb, although it has been used with some success [25]. The elements themselves may be used to generate the synthons E^{3-} by reaction with alkali metals, usually sodium, in liquid ammonia. Rossi and co-workers [36–40] have developed a one-pot synthesis for triaryl arsines and stibines, in which the reactions are irradiated with light with a maximum emission at 350 nm [40]:

$$\begin{array}{ccccc} E & \xrightarrow{Na/NH_{3(l)}/h\nu} & E^{3-} & \xrightarrow{ArX/NH_{3(l)}/h\nu} & EAr_3 \\ & & & & \downarrow{\scriptstyle Na/NH_{3(l)}/h\nu} \\ & & Ar_2EAr^1 & \xleftarrow{Ar^1X/NH_{3(l)}} & Ar_2E^- \end{array} \qquad [5.11]$$

Dimethylarsenide, $[Me_2As]^-$, can be prepared by cleavage of $Me_2AsAsMe_2$ but is most conveniently obtained either from sodium and Me_2AsI or sodium and Me_2AsH in thf [9, 41], when it is generally used *in situ*. It can be isolated as the lithium salt by reaction of Me_2AsH and Bu^nLi in petroleum ether [42]; other dialkylarsenides are obtained in a similar manner [42]. Sodium dimethylstibide is best obtained from Me_2SbBr and Na/liquid ammonia [43], although direct cleavage of a methyl group from Me_3Sb by the same reagent is also possible [22]. Other dialkylstibides have been obtained from R_2SbH and RLi at low temperatures [44, 45]. Electrochemically generated $[Ph_2Sb]^-$ has been shown to react with $[Bu^n_4N]^+$ to afford Ph_2SbBu^n [46].

Bidentate phenyl-substituted ligands, $Ph_2E(CH_2)_nEPh_2$, have also been cleaved by Na or Li in liquid ammonia or thf to yield $MPhE(CH_2)_nEPhM$ (E = As [31], Sb [47]), and the products successfully used synthetically. However, it appears that Ph–E cleavage is accompanied by $E-CH_2$ fission in some cases and a mixture of products may be obtained.

The MAsRH and M_2AsR compounds are formed from $AsRH_2$ and R′M or M/liquid ammonia, although the products have been little used synthetically (KAsPhH is described eslewhere [48, 49]).

The alkali metal derivatives are usually deeply coloured, soluble in Et_2O, thf, dioxan and liquid ammonia (a few ammonolyse). They are extremely readily hydrolysed and oxidized and must be handled under dry dinitrogen. All are stong nucleophiles and react with many functional groups [27, 50]; they also slowly ring-open thf under reflux.

5.2.2 *Primary and secondary arsines and stibines*

Dimethylarsine, Me_2AsH, is produced by Zn/HCl reduction of the commercially available cacodylic acid [9, 41]; it is a highly toxic, malodorous, spontaneously inflammable liquid [boiling point (b.p.) 35°C–37°C]. Diphenylarsine, Ph_2AsH, is available from Ph_3As/Na/liquid ammonia, similarly to the phosphine analogue [51]. The main contemporary interest in these compounds is in the field of the MOCVD synthesis of electronic materials such as GaAs.

Secondary stibines are unstable and of very limited synthetic value although one example which has been structurally characterized is $SbH(2,4,6\text{-}Me_3C_6H_2)_2$.

5.2.3 *Tertiary arsines and stibines*

Tertiary arsines are prepared in a similar manner to their phosphine analogues – from Grignard or lithium reagents and R_nAsCl_{3-n} ($n = 0, 1, 2$) or from RX and $MAsR_2$. The yields of aliphatic arsines from the Grignard route are often fairly poor, but for convenience it remains the most widely used: Me_3As (b.p. 52°C) [52], Et_3As (b.p. 48°C at 24 mm) [53], and $Bu^n{}_3As$ (113°C at 10 mm) [54, 55]. Triphenylarsine [melting point (m.p.) 60°C] is usually purchased, but if necessary it is readily obtained by the Grignard route [56]. Mixed arsines, $Ph_{3-n}R_nAs$ (R = alkyl group, $n = 1, 2$) are usually prepared from RMgX and $Ph_{3-n}AsCl_n$, but metal arsenide + RX is also a viable route, for example to give Ph_2MeAs (b.p. 112°C–115°C at 0.35 mm) [57], $PhMe_2As$ (b.p. 88°C–90°C at 18 mm) [57, 58].

Trialkylarsines are highly toxic liquids having a repulsive odour and are readily oxidized in air (Me_3As is spontaneously inflammable); phenylalkylarsines are also air sensitive, but Ph_3As is air-stable. A selection of arsines is given in Table 5.1.

Table 5.1 Boiling and melting points of (a) symmetrical tertiary arsines (AsR_3); (b) asymmetrical tertiary arsines (R_2AsR'); (c) chiral tertiary arsines ($AsRR'R''$).

Molecular formula	Compound	Melting point (°C)	Boiling point (°C)[a]	Reference
(a)				
C_3AsF_9	$As(CF_3)_3$	—	33.3	[i]
C_3H_9As	$AsMe_3$	—	53–55	[ii]
C_6F_9As	$As(CF{=}CF_2)_3$	—	57.5 (83)	[iii]
$C_6F_{15}As$	$As(CF_2CF_3)_3$	—	92–93	[iv]
C_6H_9As	$As(CH{=}CH_2)_3$	-125.6	129.8	[v]
$C_6H_{15}As$	$AsEt_3$	—	25–26 (10)	[vi]
			140 (76)	[vii]
C_9H_9As	$As(C{\equiv}CMe)_3$	131		[viii]
$C_9H_{15}As$	$As(CH_2CH{=}CH_2)_3$	—	111 (50)	[vii]
$C_9H_{15}As$	$As(\textit{trans}\text{-}CH{=}CHMe)_3$	—	71–72 (6)	[ix]
$C_9H_{15}As$	$As(\textit{cis}\text{-}CH{=}CHMe)_3$	—	87–88 (18)	[ix]
$C_9H_{15}As$	$As(C(Me){=}CH_2)_3$	—	66–67 (15)	[ix]
$C_9H_{21}As$	$AsPr_3^n$	—	72–73 (8–9)	[vi]
$C_9H_{21}As$	$AsPr_3^i$	—	96 (38)	[vii]
$C_{12}H_{21}As$	$As(CH_2C(Me){=}CH_2)_3$	—	114 (15)	[x]
$C_{12}H_{27}As$	$AsBu_3^n$	—	97–99 (1)	[vi]
$C_{12}H_{27}As$	$AsBu_3^s$	—	97 (8)	[vii]
$C_{12}H_{27}As$	$AsBu_3^i$	—	119 (31)	[xi]
$C_{12}H_{27}As$	$AsBu^t_3$	—	50–52 (0.2)	[vii]
$C_{15}H_{15}As$	$As(cyclo\text{-}C_5H_5)_3$	—	—	[xii]
$C_{15}H_{27}As$	$As(cyclo\text{-}C_5H_9)_3$	>20	—	[xiii]
$C_{15}H_{33}As$	$AsAm_3^n$	—	146–149 (10)	[xi]
$C_{15}H_{33}As$	$As(CH_2Bu^t)_3$	68–71	—	[xiv]
$C_{18}H_{12}AsBr_3$	$As(2\text{-}BrC_6H_4)_3$	177	—	[xv]
$C_{18}H_{12}AsBr_3$	$As(3\text{-}BrC_6H_4)_3$	65–66	—	[xv]
$C_{18}H_{12}AsBr_3$	$As(4\text{-}BrC_6H_4)_3$	139	—	[xiv]
$C_{18}H_{12}AsCl_3$	$As(2\text{-}ClC_6H_4)_3$	156–157	—	[xvi]
$C_{18}H_{12}AsCl_3$	$As(3\text{-}ClC_6H_4)_3$	72.5–74.5	—	[xv]
$C_{18}H_{12}AsCl_3$	$As(4\text{-}ClC_6H_4)_3$	104–107	—	[xv]
$C_{18}H_{12}AsF_3$	$As(3\text{-}FC_6H_4)_3$	41	—	[xvii]
$C_{18}H_{12}AsF_3$	$As(4\text{-}FC_6H_4)_3$	74	—	[xvii]
$C_{18}H_{15}As$	$AsPh_3$	60.8	233 (14)	[xviii]
$C_{18}H_{27}As$	$As(C{\equiv}CBu^t)_3$	62–63	—	[viii]
$C_{18}H_{33}As$	$AsCy_3$	49	—	[vi]
$C_{18}H_{39}As$	$AsHx_3$	—	125–128 (0.05–0.01)	[vi]
$C_{12}H_{21}As$	$AsBz_3$	105–108	—	[ix]
$C_{21}H_{45}As$	$As(n\text{-}C_7H_{15})_3$	—	197–199 (9)	[ix]
$C_{24}H_{51}As$	$As(n\text{-}C_8H_{17})_3$	—	170–173 (0.05)	[vi]
$C_{27}H_{33}As$	$AsMes_3$	170	—	[vii]
$C_{36}H_{75}As$	$As(n\text{-}C_{12}H_{25})_3$	—	200 (0.009)	[xix]
$C_{42}H_{87}As$	$As(n\text{-}C_{14}H_{29})_3$	liquid	decomposes	[xix]
(b)				
$C_3H_3AsF_6$	$(CF_3)_2AsMe$	—	52	[xx]
$C_3H_6AsF_3$	Me_2AsCF_3	—	58	[xx]
C_4H_9As	$Me_2AsCH{=}CH_2$	—	79–80	[xxi]
$C_4H_{11}As$	Me_2AsEt	—	86	[xi]
$C_5H_6AsF_5$	$Me_2AsCF{=}CFCF_3$	—	108–109	[xxii]
C_5H_9As	$(CH_2{=}CH)_2AsMe$	—	102	[xxi]
$C_5H_{10}AsN$	Et_2AsCN	-50	80–81 (13)	[xxiii]
$C_5H_{11}As$	$Me_2AsCH_2CH{=}CH_2$	—	108–110	[xxiv]
$C_5H_{11}As$	$Me_2AsCH_2CH{=}CH_2$	—	108–110	[xxiv]
$C_5H_{13}As$	Me_2AsPr^n	—	27 (17)	[xi]
$C_5H_{13}As$	Et_2AsMe	—	107 (755)	[xxv]
$C_6H_{11}As$	$Et_2AsC{\equiv}CH$	—	133–133.5	[viii]
$C_6H_{11}As$	$(CH_2{=}CH)_2AsEt$	—	54 (40)	[xxii]

Table 5.1 *Continued*

Molecular formula	Compound	Melting point (°C)	Boiling point (°C)[a]	Reference
$C_6H_{13}As$	$Et_2AsCH{=}CH_2$	—	58 (39)	[xxii]
$C_6H_{15}As$	Me_2AsBu^n	—	51 (35)	[xi, xxv]
$C_7H_6AsF_7$	$(CH_2{=}CH)_2As(n\text{-}C_3F_7)$	—	44 (40)	[xxii]
$C_7H_{11}As$	$Me_2As(\eta'\text{-}C_5H_5)$	—	61 (18)	[xxvi]
$C_7H_{11}AsN_2$	$(NCCH_2CH_2)_2AsMe$	—	148–149	[xxvii, xxviii]
$C_7H_{15}As$	$Me_2As(CH_2)_3CH{=}CH_2$	—	148 (474)	[ix]
$C_7H_{17}As$	Me_2AsAm^n	—	50 (10)	[x, xi]
$C_7H_{17}As$	Pr_2^nAsMe	—	66 (34)	[xi]
$C_8H_{10}AsBr$	$Me_2As(2\text{-}Br{-}C_6H_4)$	—	85–86 (0.8)	[xxix]
$C_8H_{10}AsBr$	$Me_2As(4\text{-}Br{-}C_6H_4)$	—	134–136 (9)	[xi]
$C_8H_{11}As$	Me_2AsPh	—	30–32 (0.5)	[xxx]
$C_8H_{15}As$	$Pr_2^nAsC{\equiv}CH$	—	66.5–67.5 (17)	[viii]
$C_8H_{15}As$	$(CH_2{=}CH)_2AsBu^n$	—	64 (18)	[xxi]
$C_8H_{17}As$	Me_2AsCy	—	65 (9)	[xi]
$C_8H_{19}As$	Pr_2^nAsEt	—	60–64 (14)	[ix]
$C_8H_{19}As$	Et_2AsBu^n	—	83–84 (29)	[xi, xxv]
$C_9H_{10}AsF_3$	$Me_2As(4\text{-}CF_3C_6H_4)$	—	202–204	[ix]
$C_9H_{13}As$	$Me_2As(2\text{-}MeC_6H_4)$	—	93 (10)	[xi]
$C_9H_{13}As$	$Me_2As(3\text{-}MeC_6H_4)$	—	102–107 (19)	[xi, xxxi]
$C_9H_{13}As$	Me_2AsBz	—	110 (8)	[xxxii]
$C_9H_{21}As$	Bu_2^nAsMe	—	77 (10)	[xi]
$C_9H_{21}As$	Bu_2^iAsMe	—	81 (33)	[xi]
$C_{10}H_{11}As$	$(CH_2{=}CH)_2AsPh$	—	118–122 (20–25)	[ix]
$C_{10}H_{15}As$	Et_2AsPh	—	111–115 (14)	[ix]
$C_{10}H_{15}As$	$Me_2AsCH_2CH_2Ph$	—	119 (20)	[xi]
$C_{10}H_{23}As$	Bu_2^nAsEt	—	120 (34)	[xi]
$C_{10}H_{23}As$	Bu_iAsEt	—	86 (12)	[ix]
$C_{10}H_{23}As$	$Pr_2^nAsBu^n$	—	112 (23)	[xi, xxv]
$C_{11}H_{25}As$	Am_2^nAsMe	—	121 (21)	[xi]
$C_{12}H_{13}AsN_2$	$(NCCH_2CH_2)_2AsPh$	59–60	200–210 (0.1)	[xxvii, xxviii]
$C_{12}H_{17}AsCl_2$	$(ClCH_2CH_2CH_2)_2AsPh$	—	165 (0.02)	[xxxiii]
$C_{12}H_{19}As$	Pr_2^nAsPh	—	141 (19)	[x, xi]
$C_{12}H_{19}As$	$Me_2As(CH_2)_4Ph$	—	150 (14)	[xxxii]
$C_{12}H_{27}As$	Am_2^nAsEt	—	152 (39)	[xi]
$C_{13}H_{13}As$	Ph_2AsMe	—	95–100 (0.5)	[xi, xxx]
$C_{13}H_{25}As$	Cy_2AsMe	—	127 (4)	[xi]
$C_{13}H_{29}As$	Hx_2AsMe	—	134 (10)	[xi]
$C_{14}H_{15}As$	Ph_2AsEt	—	162–165 (10)	[xxxiv, xxxv]
$C_{14}H_{27}As$	Cy_2AsEt	—	183 (23)	[xi]
$C_{14}H_{31}As$	$Me_2As(CH_2)_{11}Me$	—	149–150 (10)	[xxxvi]
$C_{15}H_{14}AsN$	$Ph_2AsCH_2CH_2CN$	37–39	—	[xxvii, xxviii]
$C_{15}H_{17}As$	Bz_2AsMe	*c.* 36	169 (5)	[xi]
$C_{15}H_{17}As$	Ph_2AsPr^n	—	177 (10)	[xi]
$C_{16}H_{19}As$	Ph_2AsBu^n	—	197 (17)	[xi]
$C_{16}H_{19}As$	Ph_2AsBu^i	—	87 (0.05)	[xxxvii]
$C_{16}H_{27}As$	Am_2^nAsPh	—	154.5–156.5 (3)	[xxxviii, xxxix]
$C_{17}H_{21}As$	$(PhCH_2CH_2)_2AsMe$	—	190 (4)	[xi]
$C_{17}H_{21}As$	Ph_2AsAm^n	—	194 (10)	[xi]
$C_{17}H_{21}As$	$Ph_2As(2\text{-}n\text{-}C_5H_{11})$	—	195 (10)	[xi]
$C_{18}H_{21}As$	Ph_2AsCy	—	200–203 (4)	[ix]
$C_{18}H_{23}As$	Ph_2AsHx	—	102 (0.6)	[xxxvii]
$C_{18}H_{27}As$	Cy_2AsPh	—	220 (14)	[xxxii]
$C_{18}H_{31}As$	Hx_2AsPh	—	156–157 (3.2)	[xxxviii, xxix]
$C_{18}H_{39}As$	$Me_2As(CH_2)_{15}Me$	—	200–202 (11)	[xxxvi]
$C_{20}H_{17}As$	$Ph_2As(4\text{-}CH_2{=}CHC_6H_4)$	50–51	—	[ix]
$C_{20}H_{35}As$	$(n\text{-}C_7H_{15})_2AsPh$	—	134 (0.15)	[xxxviii, xxxix]
$C_{21}H_{16}AsN$	$Ph_2As(2\text{-}C_9H_6N)$	—	101–102	[xl]
$C_{22}H_{39}As$	$(n\text{-}C_8H_{17})_2AsPh$	—	148–150 (0.005)	[xxxviii, xxxix]

Table 5.1 *Continued*

Molecular formula	Compound	Melting point (°C)	Boiling point (°C)[a]	Reference
$C_{24}H_{43}As$	$(n\text{-}C_9H_{19})_2AsPh$	—	187–190 (0.008)	[xl]
$C_{26}H_{47}As$	$(n\text{-}C_{10}H_{21})_2AsPh$	—	198–200 (0.005)	[xxxviii, xxxix]
(c)				
$C_9H_{13}As$	AsMeEtPh	—	85–86 (8–10)	[xli]
$C_9H_{13}AsS$	$AsMePh(CH_2CH_2SH)$	—	84–85 (0.05)	[xlii]
$C_{10}H_{15}As$	$AsMeEt(4\text{-}MeC_6H_4)$	—	109–110 (10)	[xliii]
$C_{10}H_{15}As$	$AsMePr^nPh$	—	105–106 (12)	[xi]
$C_{13}H_{12}AsBr$	$AsMePh(4\text{-}BrC_6H_4)$	—	170–200 (15)	[xi]
$C_{13}H_{19}As$	AsMePhCy	—	151–153 (12)	[xliv]
$C_{13}H_{21}As$	AsEtAmPh	—	133–134 (8–10)	[xli]
$C_{14}H_{15}As$	AsMePhBz	—	174–177 (17)	[xi]
$C_{15}H_{17}As$	$AsMePh(CH_2CH_2Ph)$	—	187 (12)	[xliv]
$C_{15}H_{17}As$	$AsEtPh(4\text{-}MeC_6H_4)$	—	135 (2)	[xxxviii, xxxix]
$C_{17}H_{29}As$	$AsEtPh(n\text{-}C_9H_{19})$	—	181–182 (8–10)	[xli]
$C_{18}H_{13}AsBrCl$	$AsPh(4\text{-}ClC_6H_4)(4\text{-}BrC_6H_4)$	64–65	248–252 (0.5)	[xlv]

[a]Pressures (torr) are given in parentheses.

References

i Emeleus, H. T., Bennett, F. W., Brandt, G. R. A. and Haszeldine, R. N. (1950) *Nature*, **166**, 225.
ii Irgolic, K. J., Kos, C. and Liu, J. (1988) *Syn.*, **4**, 582.
iii Stone, F. G. A., Kaesz, H. D. and Stafford, S. L. (1959) *J. Am. Chem. Soc.*, **81**, 6336.
iv Tittle, B. (1972) *J. Fluorine Chem.*, **2**, 449.
v Seyferth, D., Maier, L., Stone, F. G. A. and Rochon, E. G. (1957) *J. Am. Chem. Soc.*, **79**, 5884.
vi Appel, R. and Rebhan, D. (1969) *C. Ber.*, **102**, 3955.
vii Krannich, L. K., Srivastava, D. K. and Watkins, C. L. (1990) *Inorg. Chem.*, **29**, 3502.
viii Azerbaev, I. N., Abramova, Z. A. and Bosyakov, Y. G., (1974) *Russ. Chem. Rev.*, **43**, 657.
ix Samaan, S. (1978) *Methoden Der Organischen Chemie (Houben-Weyle)*, vol. **XIII**/8, Georg Thieme, Stuttgart.
x Jones, W. J., Davies, W. C., Bowden, S. T. *et al.* (1947) *J. Chem. Soc.*, 1446.
xi Jones, W. J., Dyke, W. J. C., Davies, G. *et al.* (1932) *J. Chem. Soc.*, 2284.
xii Fischer, E. O., Deubzer, B., Elian, M. and Fritz, H. P. (1970) *C. Ber.*, **103**, 799.
xiii Jardine, F. H. (1990) *Inorg. Chem.*, **29**, 3042.
xiv Pazik, J. C. and George, C. (1989) *Organometallics*, **8**, 482.
xv Brill, T. B. and Lung, G. G. (1972) *Inorg. Chem.*, **11**, 225.
xvi Massey, A. G. and Al-Jabar, N .A. A. (1985) *J. Organomet. Chem.*, **288**, 145.
xvii van Der Kelen, G.P., De Ketellaere, R.F. and Delbeke, F.T. (1971) *J. Organomet. Chem.*, **28**, 217.
xviii Jones, W. J., Forward, M. V. and Bowden, S. T. (1949) *J. Chem. Soc.*, s121.
xix Meals, R. N. (1944) *J. Organomet. Chem.*, **9**, 211.
xx Banks, R. E. (1979) *Fluorocarbons and their Derivatives*, 2nd edn, MacDonald, London.
xxi Weiner, M. A. and Pasternack, G. (1967) *J. Organomet. Chem.*, **32**, 3707.
xxii Cullen, W. C., Dhazilwal, P. S. and Styan, G. E. (1966) *J. Organomet Chem.*,**6**, 364.
xxiii Hamilton, C.S. (1947) *J. Am. Chem. Soc.*, **69**, 927.
xxiv Chellenger, F. (1935) *J. Chem. Soc.*, 396.
xxv Zingaro, R. A. and Meriganian, A. (1966) *Inorg. Chem.*, **5**, 187.
xxvi Jutzi, P. and Kuhn, M. (1974) *C. Ber.*, **107**, 1228.
xxvii Mann, F. G. and Cookson, R. C. (1947) *J. Chem. Soc.*, 618.
xxviii Mann, F. G. and Cookson, R. C. (1949) *J. Chem. Soc.*, 67.
xxix Jones, E. R. H. and Mann, F. G. (1955) *J. Chem. Soc*, 4472.
xxx Wild, S. B. and Henrick, K. (1985) *J. Chem. Soc., Dalton Trans.*, 1506.
xxxi Phillips, J. R. and Vis, J. H. (1967) *Can. J. Chem.*, **45**, 675.
xxxii Roberts, E., Turner, E. E. and Bury, F. W. (1926) *J. Chem. Soc.*, 1445.
xxxiii Kauffmann, T. and Ennen, J. (1985) *C. Ber.*, **118**, 2692.
xxxiv Titon, A. I. and Levin, B. B. (1953) *Sbornik Statei Obshchei Khim.*, **2**, 1478.
xxxv Titon, A. I. and Levin, B. B. (1955) *Chem. Abs.*, **49**, 4504.
xxxvi Jerchel, D. (1943) *C. Ber.*, **76**, 600.
xxxvii Kauffmann, T., Ahlers, H., Tilhard, H. J. and Woltermann, A. (1977) *Angew. Chem. Int. Ed. Engl.*, **16**, 710.
xxxviii Kamai, G. and Usacheva, G. M. (1966) *J. Gen. Chem. USSR*, **36**, 1991.
xxxix Kamai, G. and Usacheva, G. M. (1966) *J. Gen. Chem. USSR*, **38**, 364 [in English]
xl Rossi, R. A., Bornancini, E. R. and Alonso, R. A. (1984) *J. Organomet. Chem.*, **270**, 177.
xli Mann, F.G. and Cookson, R. C. (1949) *J. Chem. Soc.*, 288.
xlii Wild, S. B., Leung, P. H., McLaughlin, G. M. and Martin, J. W. L. (1986) *Inorg. Chem.*, **25**, 3392.
xliii Kamai, G. and Gatilov, Y. F. (1961) *J. Gen. Chem. USSR,*, **31**, 2685 [in English].
xliv Roberts, E., Turner, E. E. and Bury, F. W. (1926) *J. Chem. Soc.*, 1445.
xlv Mann, F. G. and Watson, J. (1947) *J. Chem. Soc.*, 505.

Tertiary stibines have usually been prepared by the Grignard route [2, 3] although the lower trialkyls are difficult to separate from the solvent (Et_2O) and are often purified by halogenation to R_3SbX_2 followed by subsequent reduction with zinc or $LiAlH_4$. The higher alkyls [59] and the triarylstibines can be obtained directly by the Grignard route. Alkali metal stibides have been less used, but this is more a result of their unavailability, until relatively recently, than to problems inherent in the reactions [47]. The more common stibines are Me_3Sb (b.p. 80°C–81°C) [18] (forms an azeotrope with Et_2O) (b.p. 37°C), Et_3Sb (b.p. 158°C) [18], Ph_3Sb (m.p. 50°C) [60]. Mixed alkylphenylstibines are easily prepared from Ph_nSbCl_{3-n} ($n = 1, 2$) and RMgX [47, 60]. Gedridge [62] has described a method of preparing mixed stibines R_2SbR' in a one-pot synthesis from $SbCl_3$:

$$SbCl_3 \xrightarrow[Et_2O/-50°C]{Bu^tMgCl} \xrightarrow[Et_2O/0°C]{2MeMgBr} \underset{(64\%)}{Bu^tSbMe_2} + \underset{(6\%)}{Bu^t_2SbMe} \qquad [5.12]$$

Trimethylstibine is spontaneously inflammable and the other trialkylstibines are readily oxidized in air; triphenylstibine is air-stable. A selection of stibines are given in Table 5.2.

The structure of acyclic tertiary arsines. The lower alkyl tertiary arsines are liquids at ambient temperature as are the olefinic arsines, but the acetylenic arsines are solids. Their structures have been shown to be pyramidal at arsenic by electron diffraction for $AsMe_3$ [63] and $As(CF_3)_3$ [64] and spectroscopically for $As(CH{=}CH_2)_3$ [65] and $As(C{\equiv}CH)_3$ [66].

The substitution of a methyl group by a cyano group results in a solid material, Me_2AsCN, which is unusual in being the only solid with the formula Me_2AsX (X = halogen, CF_3Ph, *inter alia*). The crystal structure revealed an As $\cdots$ N intermolecular distance of 3.18 Å which is less than the van der Waals radii sum for As and N (3.50 Å) [67]. As the methyl groups are progressively replaced by cyano groups, the intermolecular separation reduces to 2.74 Å through 2.94 Å (Scheme 5.1, Table 5.3).

The conformations of the lower alkyl tertiary arsines have been exam-

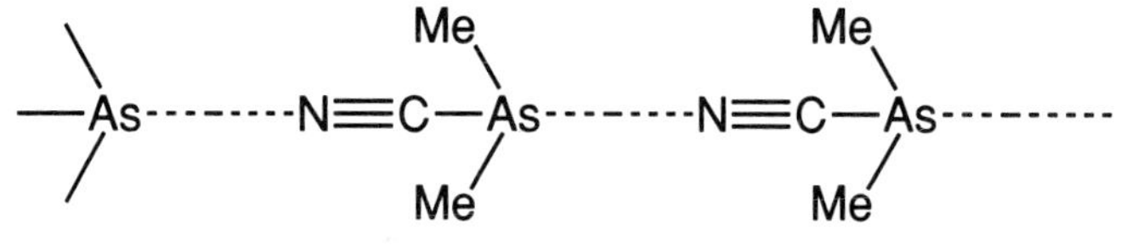

Scheme 5.1 The structure of Me_2AsCN. Redrawn from Trotter, J. and Camerman, N. (1963) *Can. J. Chem.*, **41**, 460.

Table 5.2 Boiling and melting points of a selection of (a) symmetrical stibines (SbR_3); (b) asymmetrical stibines (R_2SbR'); (c) chiral stibines (RR′R″Sb).

Molecular formula	Compound	Melting point (°C)	Boiling point (°C)[a]	Reference
(a)				
C_3F_9Sb	$Sb(CF_3)_3$	—	72	[i]
C_3H_9Sb	$SbMe_3$	—	79	[ii]
$C_6H_{15}Sb$	$SbEt_3$	—	160	[ii]
$C_9H_{21}Sb$	$SbPr_3{}^n$	—	100 (25)	[ii]
$C_9H_{21}Sb$	$SbPr^i{}_3$	—	55–58 (4)	[iii]
$C_{12}H_{27}Sb$	$SbBu_3{}^n$	—	131 (12)	[ii]
$C_{18}F_{15}Sb$	$Sb(C_6F_5)_3$	74.5	—	[iv]
$C_{18}H_{12}Cl_3Sb$	$Sb(4\text{-}ClC_6H_4)_3$	110–111	—	[iv]
$C_{18}H_{12}F_3Sb$	$Sb(4\text{-}FC_6H_4)_3$	83–84.5	—	[iv]
$C_{18}H_{15}Sb$	$SbPh_3$	50	—	[iv]
$C_{21}H_{21}Sb$	$Sb(2\text{-}MeC_6H_4)_3$	105–106	—	[iv]
$C_{21}H_{21}Sb$	$Sb(3\text{-}MeC_6H_4)_3$	67.5–68	—	[iv]
$C_{21}H_{21}Sb$	$Sb(4\text{-}MeC_6H_4)_3$	126	—	[iv]
$C_{27}H_{33}Sb$	$SbMes_3$	—	—	[v]
(b)				
C_4H_7Sb	$Me_2SbC{\equiv}CH$	—	86–88 (388	[vi]
$C_4H_{11}OSb$	$Me_2SbCH_2CH_2OH$	—	97–98 (6)	[vii]
$C_5H_{13}OSb$	$Me_2SbCH_2CH(Me)OH$	—	39–41 (0.05)	[vii]
$C_6H_{11}Sb$	$Et_2SbC{\equiv}CH$	—	87 (79)	[vi]
$C_6H_{15}Sb$	Me_2SbBu^t	12–13	44 (11)	[viii]
$C_8H_{11}Sb$	Me_2SbPh	—	74–75 (3.5)	[ix]
$C_8H_{19}Sb$	Et_2SbBu^n	—	73–76 (9)	[vii]
$C_8H_{19}OSb$	$Pr^n{}_2SbCH_2CH_2OH$	—	73–76 (0.2)	[vii]
$C_{10}H_{23}OSb$	$Bu^n{}_2SbCH_2CH_2OH$	—	100–104 (0.4)	[vii]
$C_{13}H_{13}Sb$	Ph_2SbMe	34	115–120 (1)	[ix]
$C_{14}H_{15}Sb$	Ph_2SbEt	—	123–124 (0.5)	[x]
$C_{15}H_{17}Sb$	Ph_2SbPr^i	—	128–128 (0.5)	[x]
$C_{16}H_{19}Sb$	Ph_2SbBu^n	—	—	[xi]
$C_{17}H_{14}NSb$	$Ph_2Sb(2\text{-}Py)$	51–52	—	[xii]
$C_{17}H_{14}NSb$	$Ph_2Sb(4\text{-}Py)$	63–65	—	[xii]
$C_{19}H_{17}Sb$	Ph_2SbBz	—	179–180 (0.5)	[x]
(c)				
$C_9H_{13}Sb$	PhEtMeSb	—	74–75 (2.5)	[ix]
$C_9H_{21}Sb$	Bu^nPr^nEtSb	—	83–87 (8)	[vii]
$C_{10}H_{15}Sb$	$PhPr^iMeSb$	—	84–85 (4)	[ix]
$C_{14}H_{15}Sb$	BzPhMeSb	—	135–136 (2)	[ix, x]
$C_{16}H_{14}Sb$	Q*PhMeSb	—	156–157 (0.2)	[xii]

[a]Pressures (torr) are given in parentheses.

References

i Mann, F. G. and Watson, J. (1947) *J. Chem. Soc.*, 505.
ii Dyke, W. J. C., Davies, W. C. and Jones, W. J. (1930) *J. Chem. Soc.*, 463.
iii Gedridge, R. W., Stringfellow, G. B., Chen, C. H. and Fang, Z. M. (1991) *J. Appl. Phys.*, **69**, 7605.
iv Brill, T. B. and Long, G. G. (1972) *Inorg. Chem.*, **11**, 225.
v Benmalek, M., Charmette, H., Martclet, C. *et al.* (1974) *J. Organomet. Chem.*, **67**, 53.
vi Hartmann, H. and Kühl, G. (1961) *Z. Anorg. Allg. Chem.*, **312**, 186.
vii Noltes, J. G., Meinema, H. A. and Martins, H. F. (1973) *J. Organomet. Chem.*, **51**, 223.
viii Gedridge R. W. Jr (1992) *Organometallics*, **11**, 969.
xix Sato, S., Matsumura, Y. and Okawara, R. (1972) *J. Organomet. Chem.*, **43**, 333.
x Sato, S., Matsumura, Y. and Okawara, R. (1982) *Inorg. Nucl. Chem. Lett.*, **8**, 837.
xi Yusopov, F. Y., Manulkin, Z. M. and Tuutorenko, A. N. (1953) *Dokl. Akad. Nank. SSSR.*, **88**, 687.
xii Tolmachev, A. A., Kozlor, E. S. and Pinchuk, A. M. (1984) *Z. Obshch. Khim.*, **54**, 971.

Table 5.3 The cyano arsines

Compound	Intermolecular (As . . . N) distance (Å)	Reference
Me_2AsCN	3.18	[i]
$MeAs(CN)_2$	2.94	[ii]
$As(CN)_3$	2.74	[ii, iii]

References

i Trotter, J. and Camerman, N. (1963) *Can. J. Chem.*, **41**, 460.
ii Britton, D. and Schlemper, E. O. (1966) *Acta Cryst.*, **20**, 777.
iii Britton, D. and Emerson, K. (1963) *Acta Cryst.*, **16**, 113.

ined by vibrational spectroscopy [68, 69]. The members of the sequence Me_3As, Me_2AsEt, $MeAsEt_2$ and $AsEt_3$ are all essentially pyramidal with respect to the arsenic centre, but the ethyl group increases the conformational complexity. In going from Me_3As to Me_2AsEt a methyl hydrogen is replaced by a CH_3 group, and this substitution can be *gauche* (*g*) or *trans* (*t*) to the arsenic lone pair giving rise to two vibrationally distinguishable conformations. Goggin and Crocker [68] showed that in the liquid phase both isomers occurred but that on cooling the concentration of *trans* isomer falls away untimately to zero in the crystalline material.

In the same way it can be seen that $MeAsEt_2$ has six and $AsEt_3$ seven differentiable conformations. However, Goggin and Crocker [68] exclude, on the basis of molecular modelling, configurations having *gauche* methyl groups inclined towards each other and those with more than one *trans* methyl group. In $MeAsEt_2$, IR evidence was adduced for four confomers in the liquid state. Interestingly, if the liquid is cooled, a glass containing two conformers is formed, whereas condensation from the vapour phase results in a solid containing only one species.

There is an increase in the C–As–C angle in aromatic tertiary arsines, $AsAr_3$, when the rings are substituted; thus, in tris(2,4,6-trimethyl)arsine the mean C–As–C angle is 107.6° (with a As–C bond length of 1.976 Å) with a loss of C_{3v} symmetry [69].

The electronic effects of ring substitution were investigated by studying the effect of *para* substituents which would not affect the bulk properties of the ring. For the three arsines $(4\text{-}RC_6H_4)_3As$ (R = Cl, MeO, Me), the C–As–C angles (99.5°, 98.3°, 99.3°, respectively) were found to be almost constant showing no inductive effect on the geometry [70]. A correlation between electric dipole moments and the geometry of arsines has been claimed by Claeys, van der Kelen and de Ketelaere and a method for estimating unknown C–As–C angles has been proposed [71].

Chiral arsines. The inversion barriers for tertiary arsines are in general higher than those of the corresponding phosphines; indeed chiral arsines will often survive refluxing in toluene [other data for chiral tertiary arsines

Table 5.4 Arsenic pyramidal inversion barriers of tertiary arsines.

Compound	Inversion barrier ($kJ\ mol^{-1}$)	
	Calculated	Experimental
$AsMe_3$	211	—
$PhAsMe_2$	183	177.5
$AcAsMe_2$	157	—
MeEtAsPh	—	177

Source: Mislow, K., Andose, J. D. and Rauk, A. (1972) *J. Am. Chem. Soc.*, **96**, 6904.

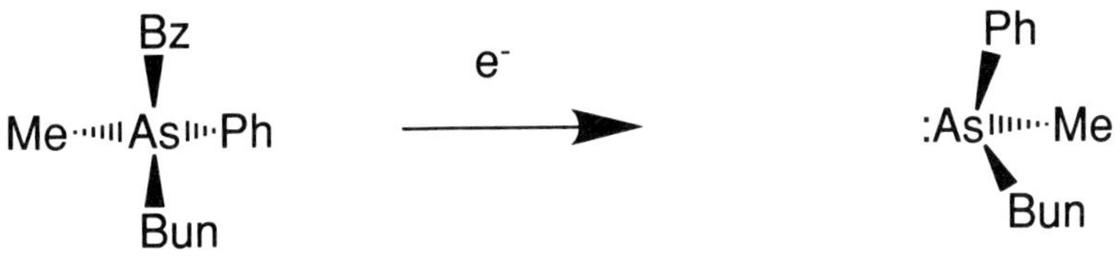

Scheme 5.2 The electrochemical reduction of chiral arsonium salts.

are given in Table 5.1(c)]. The inversion barrier for MeEtPhAs is 177 kJ mol^{-1}, with an estimated half life of *c.* 740 h at 200°C [72]. Historically, the first chiral tertiary arsines of the type AsR′R″R″ were prepared by Horner and Fuchs [73] by the electrochemical reduction of chiral arsonium salts (Scheme 5.2).

Suitable functionalized tertiary arsines can be resolved by conventional organic chemistry techniques, that is, by the formation of acid–base pairs with an appropriate optically active moiety; for example (±) 4-(ethylphenylarsinyl)benzoic acid was resolved via the quinine salt [74] and (±) ethyl-4-methylphenyl-4-aminophenylarsine was resolved via the tartarate salt [75].

Wild and co-workers have used chiral palladium [76] and iron [77] complexes to resolve chiral tertiary arsines. (±) (Methylphenyl-8-quinolyl)arsine forms a pair of internally diastereoisomeric palladium(II) complexes with $(\pm)_{589}$-di-μ-chloro-bis[(R)-dimethyl(1-ethyl-1-naphthyl)-aminato–C^2,N [dipalladium(II)], which is resolved by fractional crystallization. The chiral arsine is released by ligand exchange in two stages by treatment with diaminoethane followed by cyanide anion.

Bis(tertiary arsines) with chiral arsenic centres are more complex. Wild and co-workers separated meso- and racemic-1,2-bis(methylphenyl-arsino) benzene (Scheme 5.3) [78].

Aromatic tertiary arsines are all chiral inasmuch as the conformation

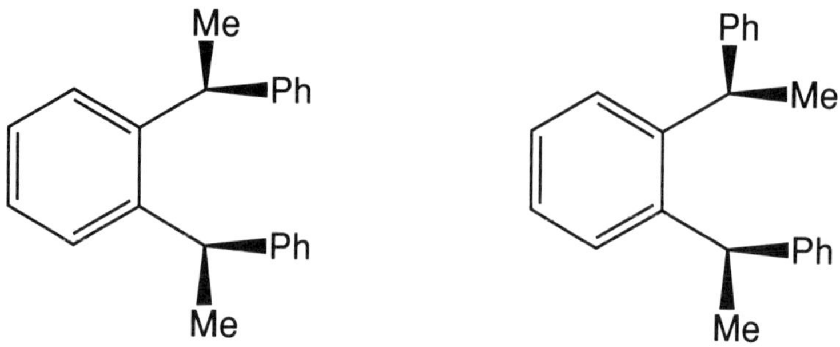

Scheme 5.3 The optical isomers of 1,2-bis(methylphenylarsino)benzene. Redrawn from Wild, S. B. and Henrick, K. (1975) *J. Chem. Soc., Dalton Trans.*, 1506.

of the rings about the central arsenic atom describes a left-handed or right-handed propeller, albeit that free rotation in the liquid state or in solution results in loss of the solid state structure. The conformations of the rings with respect to arsenic are stabilized by bulky substituents in the 2- and 6- positions. The case of $AsArAr^1Ar^2$, where Ar, Ar^1 and Ar^2 each possess *inter alia* bulky 2,6-substituents, provides a rich stereochemistry. In solution rotation is possible, reversing the chiral propeller. The mechanism for this inversion was shown for tris[2,6-bis-(difluoromethyl)phenyl] arsine to be via the colloquially termed 'two-ring flip' [79].

5.2.4 Organobis(arsines) and bis(stibines), $R(ER'_2)_2$

Diarsines. Alkanediarsines are usually obtained from R_2AsM and the appropriate organohalogen compound. Elimination occurs instead of substitution in a number of cases, including $LiAsPh_2 + BrCH_2CH_2Br$, although $ClCH_2CH_2Cl$ reacts normally to afford $Ph_2AsCH_2CH_2AsPh_2$ (m.p. 100°C) [31]. Bis(diphenylarsino)methane can be prepared from $LiAsPh_2$ and CH_2Cl_2, although it is somewhat difficult to crystallize (m.p. 96°C) [80]. Other ligands prepared in this way include $Ph_2As(CH_3)_3AsPh_2$ [31], *p*-$C_6H_4(AsPh_2)_2$ [34] and *cis*-$Ph_2AsCH{=}CHAsPh_2$ (m.p. 112°C) [81], the last ligand resulting in a stereospecific preparation from *cis*-$ClCH{=}CHCl$. Contrary to the original report [80], the *trans*-$Ph_2AsCH{=}CHAsPh_2$ is obtained in high yield from *trans*-$ClCH{=}CHCl$ and $LiAsPh_2$ in thf [81]. Attempts to prepare *o*-phenylenebis(diphenylarsine) from $LiAsPh_2$/thf or $NaAsPh_2$/liquid ammonia and *o*-$C_6H_4Cl_2$ failed (Chow, K. K., Levason, W. and McAuliffe, C.A. unpublished results), but good yields of the ligand are obtainable from *o*-$C_6H_4Br(AsPh_2)$ + $LiAsPh_2$ in thf [82].

Bis(dimethylarsino)- ligands are usually synthesized from $NaAsMe_2$/thf and the appropriate dichloro-organic compound. The reaction works smoothly with $Cl(CH_2)_nCl$ ($n > 3$) [42, 83], but for $n = 2$ the products are

$Me_2AsAsMe_2$ and C_2H_4 [81], although a low yield of $Me_2AsCH_2CH_2AsMe_2$ is claimed from $NaAsMe_2$ and $ClCH_2CH_2Cl$ in liquid ammonia at –78°C [84]. Unlike the reactions of the $[Ph_2As]^-$ ion, $NaAsMe_2$ and *cis*-ClCH=CHCl produces both isomers of $Me_2AsCH{=}CHAsMe_2$ [58, 85, 86] in 35% yield, and an isomer ratio *cis*:*trans* of 1:9:

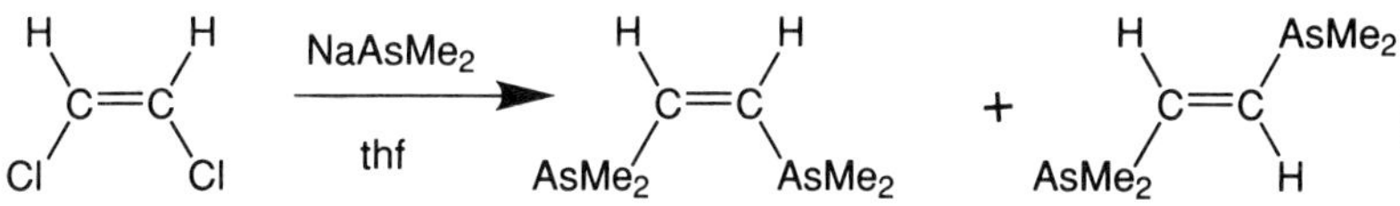

[5.13]

Different alkali metals and a variety of reaction conditions produce essentially the same distribution. *Trans*-ClCH=CHCl and $NaAsMe_2$ afford $Me_2AsAsMe_2$ and some $Me_2AsCH{=}CHAsMe_2$ (isomer uncertain) [58]. *Cis*-1,2-bis(dimethylarsino)ethylene (edas) may be separated from the mixture as $[Fe(edas)_2Cl_2][FeCl_4]$, which can subsequently yield the pure *cis* ligand on reaction with PPh_3 [85, 86]. It can also be prepared by hydroboration [86] of $Me_2AsC{\equiv}CAsMe_2$, itself formed from $XMgC{\equiv}CMgX + Me_2AsI$.

The most widely used diarsine, *o*-phenylenebis(dimethylarsine), was originally formed in a six-stage synthesis from *o*-nitroaniline [15], but this has been superseded by the reaction of $NaAsMe_2$ with *o*-$C_6H_4Cl_2$ [9, 41, 86] which produces the ligand in 45% yield (b.p. 100°C/1 mm) [35]. Particularly interesting is the identification of the byproducts of this reaction, including Me_2AsH, Me_3As, Me_2PhAs and $MeAs(2\text{-}C_6H_4AsMe_2)_2$, obtain by Feltham and Metzger [86]. The tetrafluoro analogue, $1,2\text{-}C_6F_4(AsMe_2)_2$ has been prepared from $1,2\text{-}C_6F_4Br_2$ by successive treatments with Mg and Me_2AsI [87]:

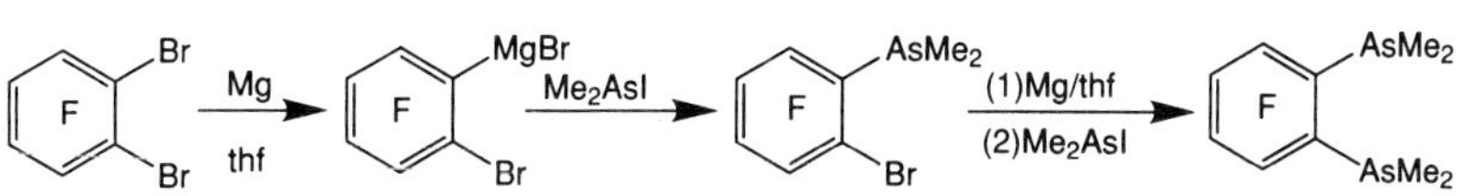

[5.14]

This synthesis may also afford diarsines of the type $1,2\text{-}C_6F_4(AsMe_2)AsR_2$.

Fluoroalicyclic diarsines (Scheme 5.4) are prepared from Me_2AsH and the appropriate perfluorocycloalkene by addition–elimination [88–90]. Fluoroaliphatic diarsines are often prepared by addition of tetramethyldiarsine to fluoro-olefins or acetylenes [91, 92]:

$$F_2C{=}CFX + Me_2AsAsMe_2 \rightarrow Me_2AsCF_2CFXAsMe_2 \quad [5.15]$$

$$FHC{=}CHF + Me_2AsAsMe_2 \rightarrow Me_2AsCFHCFHAsMe_2 \quad [5.16]$$

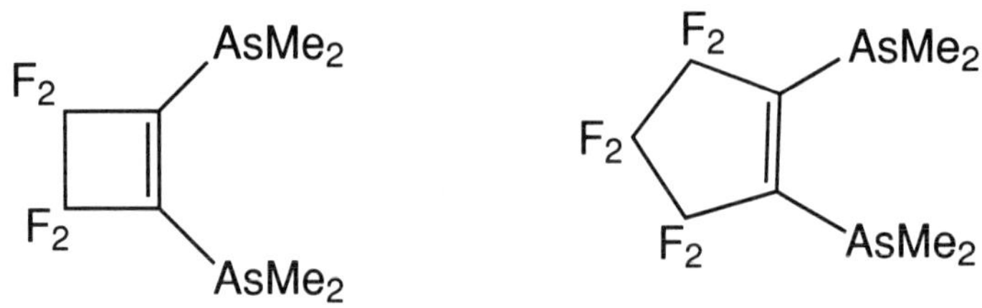

Scheme 5.4 Two examples of fluoroalicyclic diarsines.

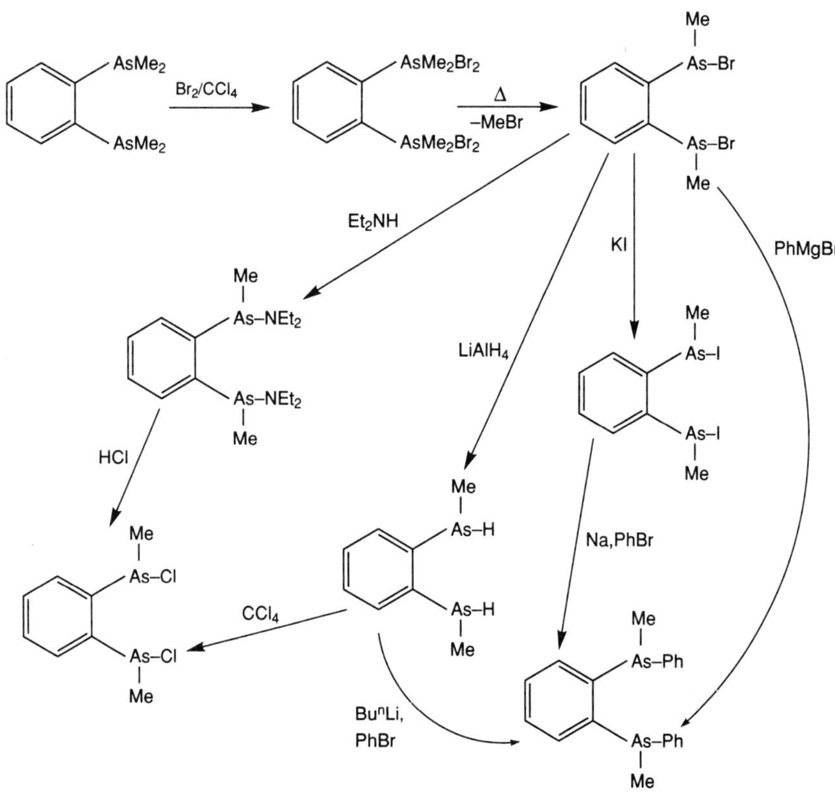

Scheme 5.5 Some useful synthetic reactions leading to 1,2-phenylenediarsines. Redrawn from Henrick, K. and Wild, S. B. (1975) *J. Chem. Soc., Dalton Trans.*, 1506.

Increasing interest is being shown in bis(phenylalkylarsino)-ligands. 1,2-Bis(phenylbutylarsino)ethane was prepared in 1939 from PhAs(Cl) $CH_2CH_2As(Cl)Ph$ and Bu^nMgBr [15], and more recently $PhMeAsCH_2CH_2AsMePh$ has been obtained [93] and separated into *meso* and *rac* isomers by chromatography of its $PdLCl_2$ complexes, which were subse-

Table 5.5 Bis(diorganylarsinyl) compounds.

Molecular formula	Arsine	Melting point (°C)	Boiling point (°C)[a]	Reference
$C_5H_{14}As_2$	$CH_2(AsMe_2)_2$		64 (15)	[i]
$C_6H_{12}As_2$	$C_2(AsMe_2)_2$		84.5	[ii]
$C_6H_{14}As_2$	*cis*-$Me_2AsCH{=}CHAsMe_2$		90–93 (24)	[ii]
$C_6H_{14}As_2$	*trans*-$Me_2AsCH{=}CHAsMe_2$		66–70 (6)	[iv]
$C_7H_{18}As_2$	$(CH_2)_3(AsMe_2)_2$		96 (15)	[v, vi]
$C_8H_{12}As_2F_6$	$Me_2AsC(CF_3){=}C(CF_3)AsMe_2$			[vii]
$C_8H_{20}As_2$	$(CH_2)_4(AsMe_2)_2$		110–112 (15)	[v, vi]
$C_9H_{22}As_2$	$CH_2(AsEt_2)_2$		85 (5)	[i]
$C_9H_{22}As_2$	$(CH_2)_5(AsMe_2)_2$		126 (15)	
$C_{10}H_{16}As_2$	1,2-$C_6H_4(AsMe_2)_2$		100–101 (1.0)	[viii, ix]
$C_{10}H_{20}As_2$	$C_2(AsEt_2)_2$		139 (8)	[ii]
$C_{13}H_{30}As_2$	$CH_2(AsPr_2^n)_2$		117 (4)	[i]
$C_{13}H_{30}As_2$	$CH_2(AsPr_2^i)_2$		115–117 (0.03)	[i]
$C_{14}H_{28}As_2$	$C_2(AsPr_2^n)_2$		118.5–119.5 (2)	[ii]
$C_{17}H_{38}As_2$	$CH_2(AsBu_2^n)_2$		159 (4)	[i]
$C_{18}H_{36}As_2$	$C_2(AsBu_2^i)_2$		134–136 (2)	[ii]
$C_{18}H_{36}As_2$	$C_2(AsBu_2^n)_2$		159–160 (1.5)	[ii]
$C_{20}H_{20}As_2$	1,2-$(C_6H_4)(AsPhMe)_2$	88–88.5[b]	180–185 (0.05)[c] 180–186 (0.05)[d]	[x]
$C_{25}H_{18}As_2Cl_4$	$CH_2\{As(4\text{-}ClC_6H_4)_2\}_2$	89		[xi]
$C_{25}H_{18}As_2F_4$	$CH_2\{As(4\text{-}FC_6H_4)_2\}_2$	52		[xi]
$C_{25}H_{22}As_2$	$CH_2(AsPh_2)_2$	97		[xi]
$C_{26}H_{22}As_2$	$Ph_2AsCH{=}CHAsPh_2$	112–113[e] 103–104[f]		[iii] [iii]
$C_{26}H_{24}As_2$	$Ph_2AsCH_2CH_2AsPh_2$	100		[v, vi, xii]
$C_{27}H_{26}As_2$	$(CH_2)_3(AsPh_2)_2$	70–71		[xii]
$C_{28}H_{28}As_2$	$(CH_2)_4(AsPh_2)_2$	122–123		[xii, xiii]
$C_{28}H_{52}As_2$	$(CH_2)_4(AsCy_2)_2$	66		[xiii]
$C_{29}H_{30}As_2$	$(CH_2)_5(AsPh_2)_2$	viscous liquid	278–281 (1)	[xii]
$C_{30}H_{32}As_2$	$(CH_2)_6(AsPh_2)_2$	117–118		[xii]
$C_{33}H_{42}As_2N_2$	$CH_2\{As(4\text{-}Me_2NC_6H_4)_2\}_2$	165		[xi]
$C_{44}H_{28}As_2$	$(C{\equiv}C)_2\{As(1\text{-}C_{10}H_7)_2\}_2$			[xiv]

[a]Pressures (torr) are given in parentheses. [b]Melting point of racemic isomers. [c]Mixed diasterioisomers. [d]*Meso* isomer. [e]*Cis* isomer. [f]*Trans* isomer

References

i Gavrilov, V. I., Garieva, F. R., Galyametdinov, Y. G. and Gimaletdinova, N. M. (1981) *J. Gen. USSR*, **51(7)**, 1591.
ii Azerbaev, I. N., Abramova, Z. A. and Bosyakov, Y. G. (1974) *Russ. Chem. Rev.*, **43**, 657.
iii Alyea, E. C. (1971) in *Transition Metal Complexes of Phosphorus, Arsenic and Antimony Ligands* (ed. C.A. McAuliffe), Macmillan, London.
iv Wild, J. D., Bennett, M. A. and Erskine, G. J. (1968) *Inorg. Chim. Acta.*, **2**, 379.
v Sommer, K. (1970) *Z. Anorg. Allg. Chem.*, **376**, 150.
vi Sommer, K. (1970) *Z. Anorg. Allg. Chem.*, **377**, 278.
vii Cullen, W. R., Crow, J. P., Herring, F. G. *et al.* (1971) *Inorg. Chem.*, **10**, 1616.
viii Feltham, R. D. and Silverthorn, W. (1967) *Inorg. Syn.*, **10**, 159.
ix Nyholm, R. S., Feltham, R. D. and Kasenally, A. (1967) *J. Organomet. Chem.*, **7**, 285.
x Wild, S. B. and Henrick, K. (1985) *J. Chem. Soc., Dalton Trans.*, 1506.
xi Chernokalskii, B. D. and Garieva, F. R. (1980) *J. Ge. Chem. USSR*, **50(6)**, 1329.
xii Tzschach, A. and Lange, W. (1962) *C. Ber.*, **95**, 1360.
xiii Issleib, K. and Tzschach, A. (1961) *Angew. Chem.*, **73**, 26.
xiv Hartmann, H. (1968) *Liebigs Ann. Chem.*, **714**, 1.

quently decomposed with aqueous KCN to yield the pure diarsine isomers [93, 94]. 1,2-Phenylenebis(methylphenylarsine) has been prepared [95] from NaAsMePh and 1,2-$C_6H_4Cl_2$ and separated into diastereoisomers. The steric course of some reactions leading to 1,2-phenylenediarsines has also been studied and illustrates several useful synthetic reactions (Scheme 5.5) [95].

Asymmetrical diarsines, $R_2As{-}AsR_2'$ have been little studied as ligands, although they can be prepared from $NaAsR_2$ and $Cl{-}AsR_2'$ analogously to hybrid ligands. One particularly interesting case is the ligand *cis*-$Ph_2AsCH{=}CHAsMe_2$ which can be obtained stereospecifically in good yield (*c.* 40%) from *cis*-$ClCH{=}CHAsPh_2$ and $NaAsMe_2$ [81] (unlike $Me_2AsCH{=}CHAsMe_2$ which is obtained as a *cis*/*trans* mixture from *cis*-$ClCH{=}CHCl$ and $NaAsMe_2$). A selection of diarsines is given in Table 5.5.

Distibines. Ditertiary stibinoalkanes are obtained by the reaction of the appropriate dihaloalkane with lithium dialkyl- or diphenylstibide in thf at low temperatures:

$$X(CH_2)_nX + 2LiSbR_2 \xrightarrow{\text{thf/}-95^\circ\text{C}} R_2Sb(CH_2)_nSbR_2 + 2LiX \qquad [5.17]$$

R = Cy, $n = 3, \ldots, 6$; [44]; R = Bu^t, $n = 4$ [96]; R = Et, $n = 4, \ldots, 6$ [45]; R = Ph, $n = 4, \ldots, 6$ [50]. Alternatively, sodium stibides in liquid ammonia have been used:

$$X(CH_2)_nX + 2NaSbR_2 \xrightarrow{NH_{3(l)}/-33^\circ\text{C}} R_2Sb(CH_2)_nSbR_2 + 2NaX \qquad [5.18]$$

R = Me, $n = 1, 3$ [43, 97]; R = Ph, $n = 1, 3$ [47, 98]; R = Et, Pr^n, $n = 1$ [22]. 1,2-Dichloroethane in these reactions yields ethylene and R_2SbSbR_2 [22, 25, 29, 44, 45]; $R_2SbCH_2CH_2SbR_2$ are unavailable by this route. Asymmetrical distibines, $PhMeSb(CH_2)_nSbMePh$, have been prepared from PhMeSbNa, or by the reaction [47]

$$Ph_2Sb(CH_2)_3SbPh_2 \xrightarrow[\text{(2) MeCl}]{\text{(1) Na/NH}_{3(l)}} PhMeSb(CH_2)_3SbMePh \qquad [5.19]$$

The alkyl substituted compounds are very air-sensitive – some are spontaneously inflammable – whereas the phenyl compounds appear to be air-stable in the solid state. *Cis*- and *trans*-1,2-dichloroethylene and sodium diphenyl stibide afford $Ph_2SbSbPh_2$ rather than the desired $Ph_2SbCH{=}CHSbPh_2$ [81, 99], although the addition of Ph_2SbH to $Ph_2SbCH{=}CH_2$ has been more successful [98], and the product has the expected *trans* configuration. Sodium acetylide and Me_2SbBr yield the expected $Me_2SbC{\equiv}CSbMe_2$ [20]. *o*-Phenylenedi(tertiarystibines) are obtained only with considerable difficulty and in poor yield according to the following reaction [100]:

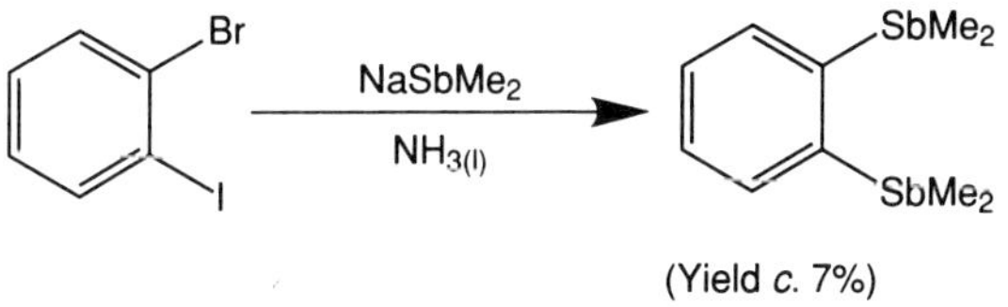

[5.20]

1,2-Phenylenebis(diphenylstibine) cannot be obtained similarly, but is produced as follows [35]:

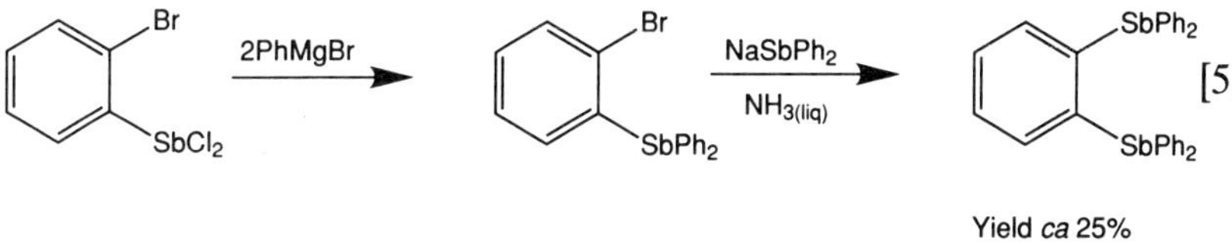

[5.21]

1,2-Phenylenebis(dimethylstibine) is also obtained in higher yield by this route (Levason, W., McAuliffe, C. A. and Murray, S. G., unpublished results). A selection of distibines is given in Table 5.6.

Table 5.6 Bis(diorganylstibinyl) compounds.

Molecular formula	Stibine	Melting point (°C)	Boiling point (°C)[a]	Reference
$C_6H_{12}Sb_2$	$C_2(SbMe_2)_2$	—	116 (17)	[i, ii]
$C_{10}H_{20}Sb_2$	$C_2(SbEt_2)_2$	—	122 (1.5)	[i, ii]
$C_{12}H_{28}Sb_2$	$(CH_2)_4(SbEt_2)_2$	—	—	[iii]
$C_{13}H_{30}Sb_2$	$(CH_2)_5(SbEt_2)_2$	—	—	[iv]
$C_{14}H_{28}Sb_2$	$C_2(SbPr^i_2)_2$	—	120 (1)	[i, ii]
$C_{18}H_{36}Sb_2$	$C_2(SbBu^t_2)_2$	85.5	—	[i, ii]
$C_{20}H_{44}Sb_2$	$(CH_2)_4(SbBu^t_2)_2$	—	—	[iv]
$C_{27}H_{50}Sb_2$	$(CH_2)_3(SbCy_2)_2$	—	—	[v]
$C_{28}H_{28}Sb_2$	$(CH_2)_4(SbPh_2)_2$	—	—	[vi]
$C_{29}H_{30}Sb_2$	$(CH_2)_5(SbPh_2)_2$	—	—	[vi]
$C_{30}H_{24}Sb_2$	$1,2\text{-}C_6H_4(SbPh_2)_2$	—	—	[vii]
$C_{30}H_{32}Sb_2$	$(CH_2)_6(SbPh_2)_2$	—	—	[vi]
$C_{30}H_{56}Sb_2$	$(CH_2)_6(SbCy_2)_2$	—	—	[v]

[a]Pressures (torr) are given in parentheses.

References

i Hartmann, H. and Kühl, G. (1956) *Z. Anorg. Allgem. Chem.*, **68**, 619.
ii Hartmann, H. and Kühl, G. (1969) *Angew. Chem.*, **312**, 186.
iii Issleib, K. and Hamann, B. (1964) *Z. Anorg. Allgem. Chem.*, **332**, 179.
iv Issleib, K., Hamann, B. and Schmidt, L. (1965) Z. Anorg. Allgem. Chem., **339**, 298.
v Issleib, K. and Hamann, B. (1965) *Z. Anorg. Allgem. Chem.*, **339**, 289.
vi Issleib, K. and Krech, F. (1961) *Chem. Ber.*, **94**, 2656.
vii Levason, W., McAuliffe, C. A. and Murray, S. G. (1975) *J. Organomet. Chem.*, **88**, 171.

Mixed digroup 15-donor bidentates. In this section we consider some routes to ligands of the type $R_2E \cdots E'R'_n$ ($E \neq E'$), where E is a group 15 element (As, Sb) and $E'R'_n = NR_2$, PR_2, AsR_2, SbR_2 or, SR, SeR, $P(O)R_2$, etc.; the preparation of this type of ligand usually involves the synthesis of an intermediate. The order in which E and E′ are added to the ligand backbone is often crucial. As an example of this *cis*-l-diphenylarsino-2-diphenylphosphino-ethylene, *cis*-$Ph_2AsCH{=}CHPPh_2$, is prepared as follows [81]:

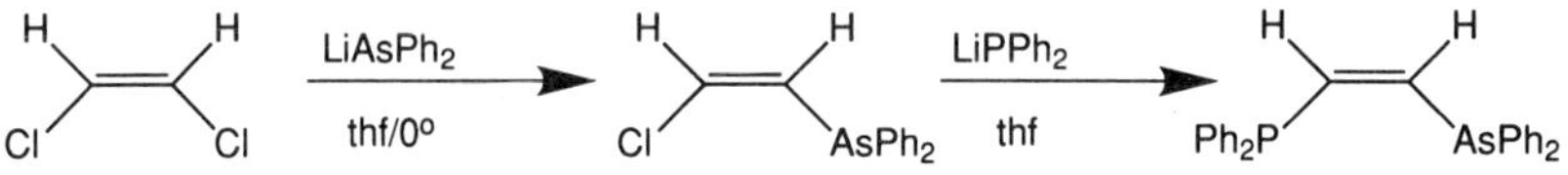

[5.13]

although attempts to prepare *cis*-$Ph_2PCH{=}CHCl$ resulted only in *cis*-$Ph_2PCH{=}CHPPh_2$ [81].

A number of intermediates are obtainable by reaction of $LiEPh_2$ (E = P, As) with dihaloalkanes under conditions which favour monosubstitution, that is excess dihalo compound diluted with thf at low temperatures (< 0°C) with the $MEPh_2$ added dropwise. In this way $Ph_2PCH_2CH_2Cl$ [81, 101, 102], $Ph_2PCH_2CH_2CH_2Cl$ [101], $Ph_2AsCH_2CH_2Cl$ [12, 81], *cis*-$Ph_2AsCH{=}CHCl$ [12, 81], *trans*-$Ph_2AsCH{=}CHCl$ [81], $R_2As(CH_2)_nCl$ ($n = 3$, 4) [42] and the unstable Ph_2PCH_2Cl [103] are obtainable. The products are generally isolated by hydrolysis and subsequent removal of the volatile impurities *in vacuo*, although distillation usually promotes extensive self-quaternization and should be avoided. This method fails to yield $Ph_2SbCH_2CH_2Cl$, or *cis*- or *trans*-$Ph_2PCH{=}CHCl$ [81]. The synthon $Me_2AsCH_2CH_2CH_2Cl$ is readily prepared from $ClCH_2CH_2CH_2AsCl_2$ and MeMgI [104]. Once obtained the ω-chloroalkyl or 2-chlorovinyl compound can be reacted with MER_2 in thf to yield the required hybrid ligand:

$$ClCH_2CH_2AsPh_2 \xrightarrow{LiPPh_2} Ph_2PCH_2CH_2AsPh_2 \text{ [12, 81]} \qquad [5.23]$$

$$cis\text{- or } trans\text{-}ClCH{=}CHAsPh_2 \xrightarrow{LiPPh_2} cis\text{- or } trans\text{-}Ph_2PCH{=}CHAsPh_2 \text{ [12, 81, 105]} \qquad [5.24]$$

$$Cl(CH_2)_3AsMe_2 \xrightarrow{LiPMe_2} Me_2P(CH_2)_3AsPh_2 \text{ [106]} \qquad [5.25]$$

Alternatively, from $Ph_2PCH_2CH_2CH_2Cl$ or $Me_2AsCH_2CH_2CH_2Cl$, the Grignard may be prepared and subsequently reacted with R_2ECl [103], but $Ph_2PCH_2CH_2Cl$ [99] and $Cl_2AsCH_2CH_2Cl$ [104] undergo β-elimination when reacted with Mg or MeMgI, respectively.

King and Kapoor [107] have prepared $Ph_2PCH_2CH_2AsPh_2$ (also [108]) and *trans*-$Ph_2PCH{=}CHAsPh_2$ by base-catalysed addition of Ph_2AsH to

$Ph_2PCH{=}CH_2$ or $Ph_2PC{\equiv}CH$, respectively. Acetylenic phosphine-arsines are readily prepared according to the following reaction [109]:

$$Ph_2PCl + XMgC{\equiv}CH \longrightarrow Ph_2PC{\equiv}CH \xrightarrow{Bu^nLi} Ph_2PC{\equiv}CLi \xrightarrow{Ph_2AsCl} Ph_2PC{\equiv}CAsPh_2 \qquad [5.26]$$

1,2-Phenylene ligands are most easily prepared from the 1,2-$C_6H_4Br(ER_2)$ derivative as shown below for $ER_2 = PR_2$ [110], SbR_2 [111] and AsR_2 [24] in reactions [5.27]–[5.29], respectively:

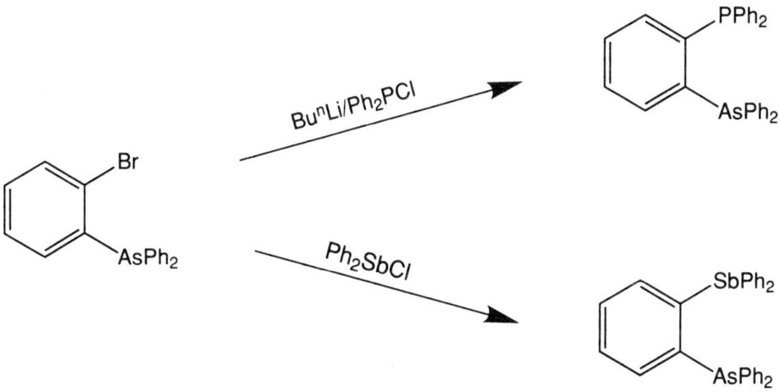

[5.27]

[5.28]

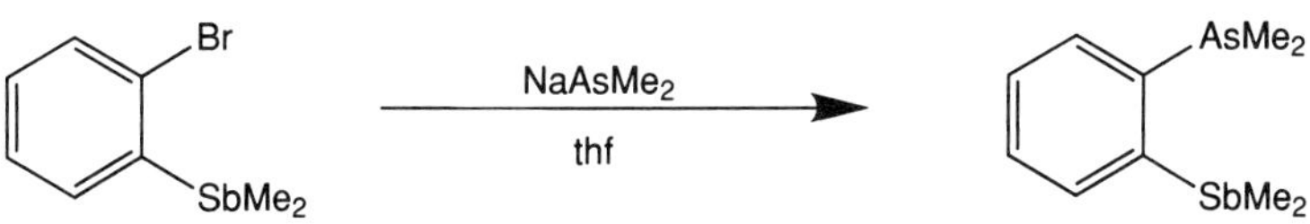

[5.29]

The compounds 1,2-$C_6H_4(PPh_2)(SbPh_2)$ [111], 1,2-$C_6H_4(PMe_2)(AsMe_2)$ [112] and 1,2-$C_6H_4(AsMe_2)(PEt_2)$ [11] have been prepared similarly.

Hybrid ligands containing a nitrogen donor are usually prepared from commercially available ω-chloroalkylamines, whilst aziridine is a useful reagent for introducing the $-CH_2CH_2NH_2$ group. A range of ligands containing a heterocyclic N donor atom have been prepared; the ligand tris(2-pyridyl)arsine (Scheme 5.6) [113, 114] is well known.

Sulphur-containing [either thiol (−SH) or thioether (−SR)] hybrids are obtained fairly easily [115, 116] (see p. 228 (top)).

For example 1,2-$C_6H_4(SMe)(AsPh_2)$ is obtained from 1,2-C_6H_4Br (SMe) [117] and the free thiol [118]. A point to note however, is that reaction [5.32] fails with 1,2-$C_6H_4Br(PPh_2)$ and with 1,2-$C_6H_4Br(AsMe_2)$

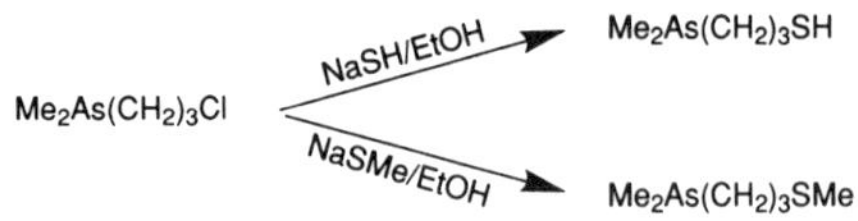

[5.30]

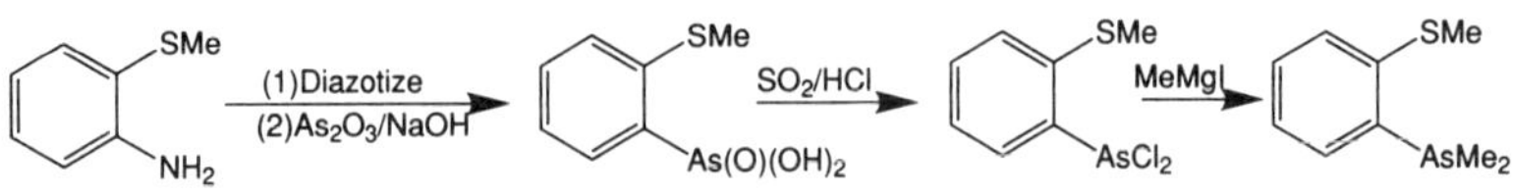

[5.31]

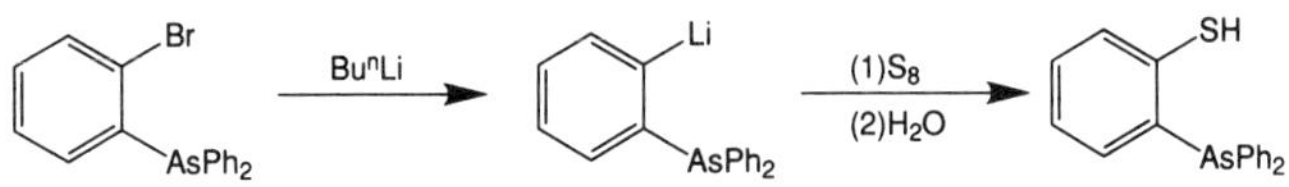

[5.32]

because the sulphur adds to the group 15 element (McAuliffe, C. A. and Murray, S. G., unpublished observations).

Phosphine sulphide donors are also readily incorporated [110]:

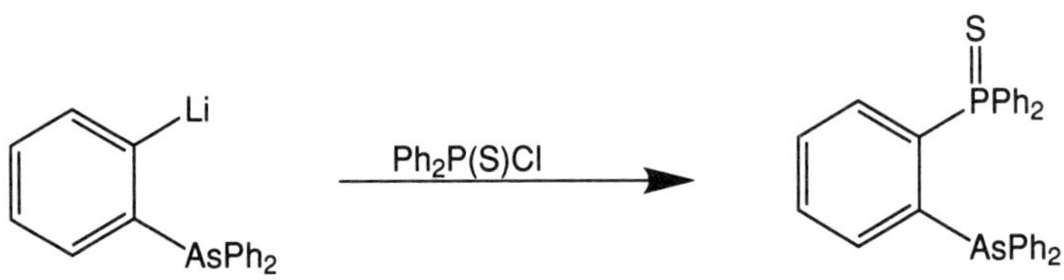

[5.33]

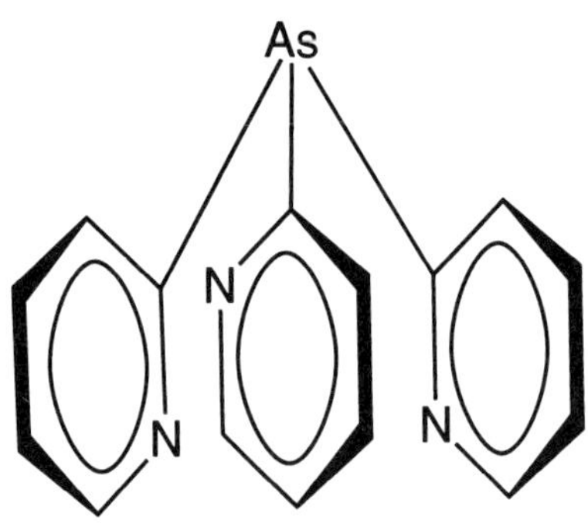

Scheme 5.6 The ligand tris(2-pyridyl)arsine.

5.2.5 Multidentate arsine and stibine ligands

The synthesis of multidentate ligands generally utilizes the methods and intermediates used to prepare asymmetrical and mixed-donor bidentates, although considerable difficulties are often experienced in attempts to isolate a particular product in reasonable yield and purity. The order in which the various donors and linkages are introduced into the multidentate ligand is often crucial. The reactivity of the donor grouping introduced first or the weakness of the carbon–donor bond (e.g. where the donor is Sb or Se) considerably restrict the types of reaction which can be performed in order to introduce further constituents into the multidentate compound.

Multidentate arsines. In contrast to the variety of methods used to prepare polytertiary phosphines, the arsenic analogues are almost invariably obtained from $R_2As{-}Cl$ intermediates.

The simplest triarsine, bis(3-dimethylarsinopropyl)methylarsine [24], is prepared from $Me_2AsCH_2CH_2MgCl$ and $MeAsCl_2$ [104], and $PhAs(CH_2CH_2CH_2AsMe_2)_2$ [16] is obtained by substituting $PhAsCl_2$ for $MeAsCl_2$ [119]. Similarly, 1,2-lithiophenyldiphenylarsine and $PhAsI_2$ produce $PhAs(1,2\text{-}C_6H_4AsPh_2)_2$ [120], and $1,2\text{-}C_6H_4(AsMe_2)Li$ and $MeAsI_2$ produce $MeAs(1,2\text{-}C_6H_4AsMe_2)_2$ [121]. The triarsine $MeC(CH_2AsMe_2)_3$ is formed from $MeC(CH_2Br)_3$ and $NaAsMe_2$ in thf [121] and mixed tridentate phosphine-arsines have also been isolated [122].

Tripod tetradentates are prepared by essentially similar routes by using $AsCl_3$ in place of $RAsCl_2$ [120, 123, 124]:

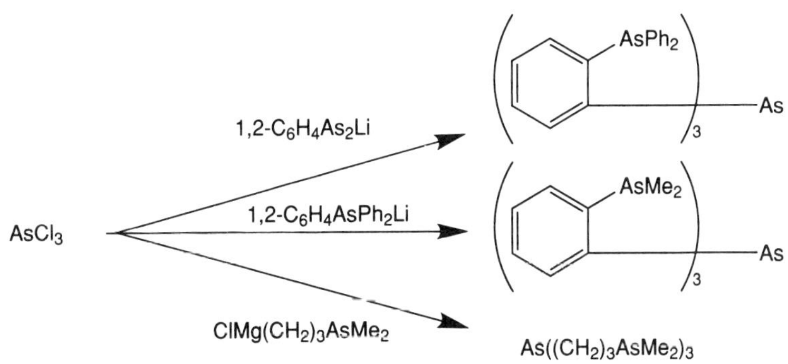

[5.34]

[5.35]

[5.36]

Facultative tetradentate arsines have been prepared by similar routes for $n = 2$ [16] and $n = 3$ [123] (see reaction [5.37].

o-Phenylene linkages are introduced via $1,2\text{-}C_6H_4(Br)AsMe_2$ and $1,2\text{-}C_6H_4(AsMeBr)_2$ to form the compounds shown in Scheme 5.7(a) and 5.7(b), respectively [124].

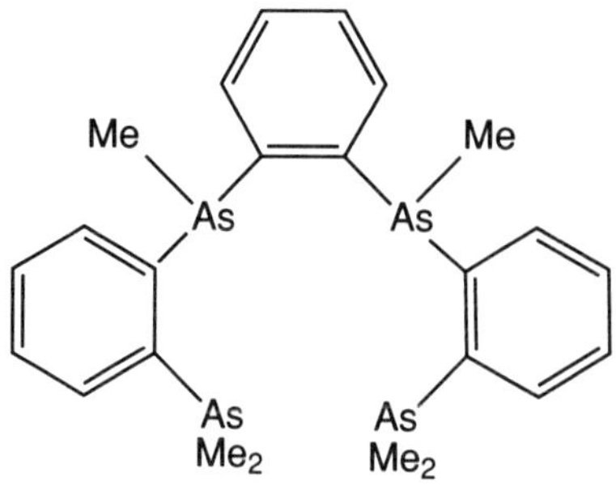

(a)

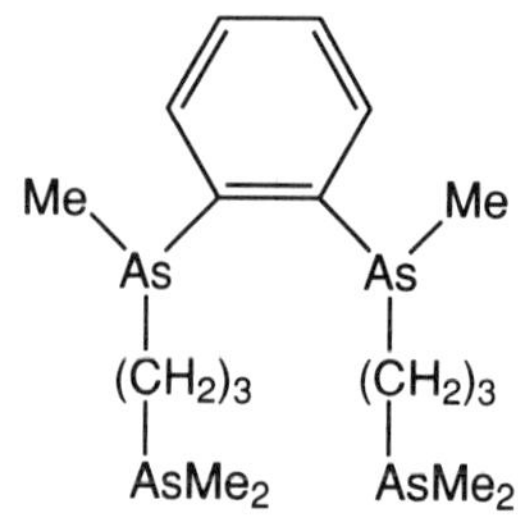

(b)

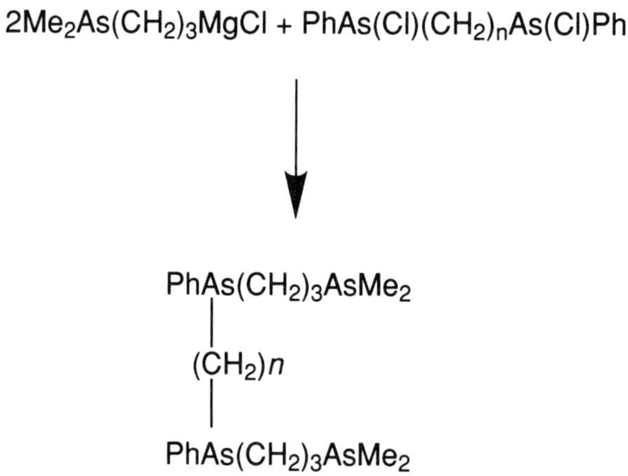

[5.37]

Scheme 5.7 Examples of tetradentate arsine ligands.

The compound $C(CH_2AsPh_2)_4$ is formed from $C(CH_2Br)_4$ and $NaAsPh_2$ [32]. The facultative tetraarsines can in many cases exist in isomeric forms and the separation of these isomers via their metal complexes has been described by Bosnich and co-workers [16, 124]. The only hexadentate arsine (Scheme 5.8) is prepared from o-$C_6H_4(AsCl_2)_2$ and $4Me_2AsCH_2CH_2CH_2MgCl$ [125].

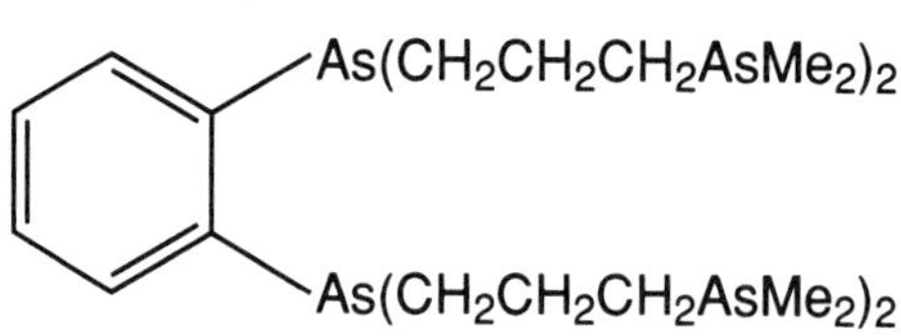

Scheme 5.8 The hexadentate arsine ligand.

Multidentate stibines. Multidentate stibines are difficult to prepare and an attempt to produce $Sb(2\text{-}C_6H_4SbPh_2)_3$ from $1,2\text{-}C_6H_4(SbPh_2)Br$, Bu^nLi and $SbCl_3$ failed [126]. The great obstacle to overcome in the synthesis of multidentate antimony ligands is the weakness of the carbon–antimony bond; indeed it is possible that in the above reaction of Bu^nLi with $o\text{-}C_6H_4(SbPh_2)Br$ the nucleophile attacks the C−Sb bond as well as, or instead of, the C−Br bond [35].

Ellermann and Veit [127] prepared the tridentate stibine 1,1,1-tris((diphenylstibino)methyl)ethane $MeC(CH_2SbPh_2)_3$ from $MeC(CH_2Cl)_3$ and $[Ph_2Sb]^-$, and Cheffey *et al.* have also utilized this ligand [128]

Multidentate mixed donors. The preparation of multidentate ligands containing different group 15 atoms is usually accomplished by trivial extensions of the methods used to prepare multidentate phosphines and arsines. The base catalysed ($KOBu^t$) addition of As−H bonds (Ph_2AsH) to vinylphosphines has yielded $PhP(CH_2CH_2AsPh_2)_2$, [107] and $PhPCl_2$ reacts with $o\text{-}C_6H_4(AsPh_2)Li$ and $ClMgCH_2CH_2CH_2AsMe_2$ to form $PhP(o\text{-}C_6H_4AsPh_2)_2$ [120] and $PhP(CH_2CH_2CH_2AsMe_2)_2$ [129], respectively. In the same way, the apical donor atom (usually P or As) of tripodal tetradentates has been replaced by antimony as in the compounds $Sb(2\text{-}C_6H_4PPh_2)_3$ [123, 126], $Sb(2\text{-}C_6H_4AsPh_2)_3$ [126, 130], $Sb(2\text{-}C_6H_4AsMe_2)_3$ [131], and $Sb(CH_2CH_2CH_2AsMe_2)_3$ [132].

The yields of the tripodal 1,2-phenylene ligands, $E(1,2\text{-}C_6H_4E'Ph_2)_3$, increase as E varies with Sb > As > P, as do those of $E''(1,2\text{-}C_6H_4E'Me_2)_3$ as E″ varies, with As > Sb > Bi, which can be attributed to steric hindrance and to increasing weakness of the C−E″ bond, respectively. The extreme case is tris(3-dimethylarsino-propyl)bismuthine, where low yields are obtained from the filtered Grignard reagent of 3-chloropropyldimethylarsine added to $BiCl_3$ at −30°C (Levason, W., McAuliffe, C. A. and Murray, S. G., unpublished work). Inverse addition or conduction of the reaction at room temperature results in essentially zero yields of the ligand.

Dawson and Venanzi [133] have devised a route to tripodal ligands of C_s symmetry in which the blocking effect of $>PNMe_2$ groups is used:

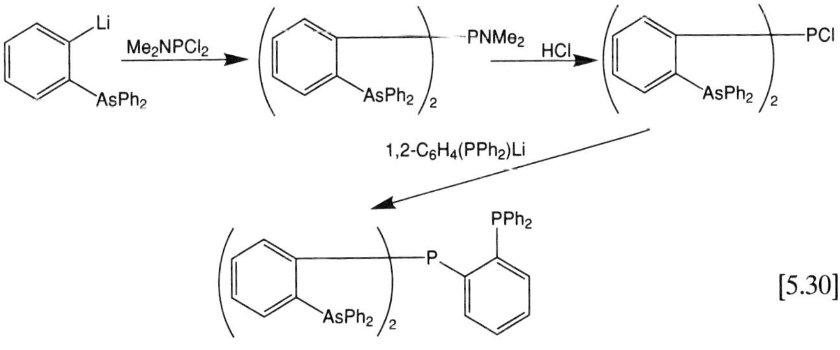

[5.30]

Nitrogen-containing multidentate ligands. The preferred route, if the starting material is available, is from the chloroalkylamine and MER_2, as shown in reactions [5.39] and [5.40] for $ClCH_2CH_2NEt_2$ [134] and $N(CH_2CH_2Cl)_3$ [135, 136], respectively:

$$2ClCH_2CH_2NEt_2 + Na_2AsPh \rightarrow Et_2NCH_2CH_2As(Ph)CH_2CH_2NEt_2 \qquad [5.39]$$

$$N(CH_2CH_2Cl)_3 + 3KAsPh_2 \rightarrow N(CH_2CH_2AsPh_2)_3 \qquad [5.40]$$

The method can also be used to yield facultative tetradentates [137]:

$$ClCH_2CH_2N(Me)CH_2CH_2N(Me)CH_2CH_2Cl \xrightarrow{2KAsPh_2/dioxan} Ph_2AsCH_2CH_2N(Me)CH_2CH_2N(Me)CH_2CH_2AsPh_2 \qquad [5.41]$$

and when extended to asymmetrical chloroalkylamines it yields yet further ligand varieties [137, 138]:

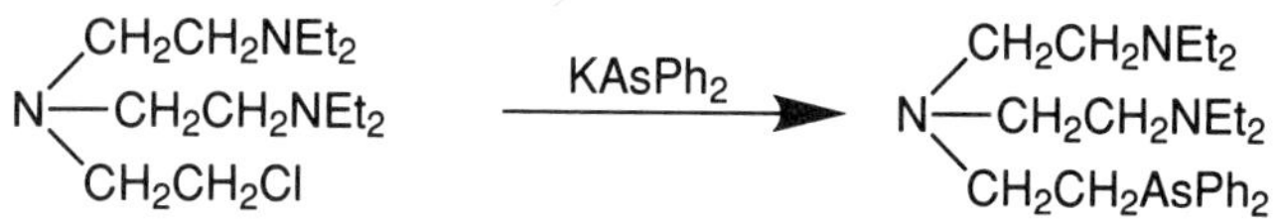

[5.42]

Sulphur- and selenium-containing multidentate ligands. Although facultative tetrathioether [139, 140] and tetraselenoether [141] ligands have been prepared, their preparation was preceded by that of mixed group 15 sulphur and selenium ligands. The simplest As−S−As tridentate, $Ph_2AsCH_2CH_2SCH_2CH_2AsPh_2$, is prepared from $KAsPh_2$ and mustard gas, $S(CH_2CH_2Cl)_2$ [135]. Facultative diarsine–dithioether ligands are quite easily prepared, as shown in reaction [5.43] for an acyclic species [105] (Levason, W., McAuliffe, C. A. and Murray, S. G., unpublished results) and in reaction [5.44] for an aryl species [118]:

$$Me_2As(CH_2)_3Cl \xrightarrow{NaS(CH_2)_3SNa} Me_2As(CH_2)_3S(CH_2)_3S(CH_2)_3AsMe_2 \qquad [5.43]$$

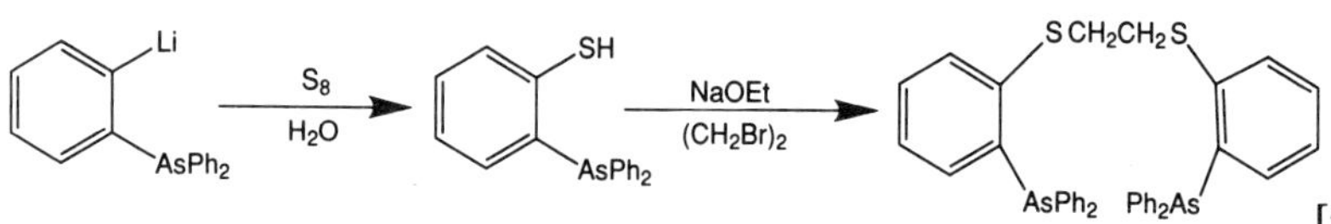

[5.44]

Dithioether–diarsine tetradentates are synthesized by combining 2-bromo-thioanisole, Bu^nLi and the required bis(chloroarsino)alkane [17, 142].

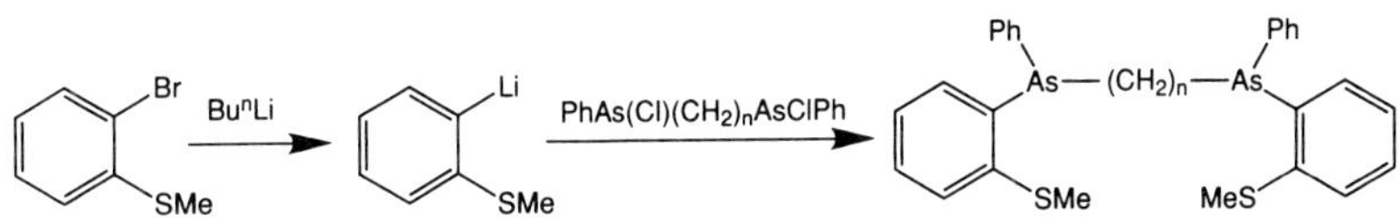

[5.45]

Oxygen-containing multidentate ligands. Arylalkoxy donors are usually introduced via the Grignard route [143, 144]:

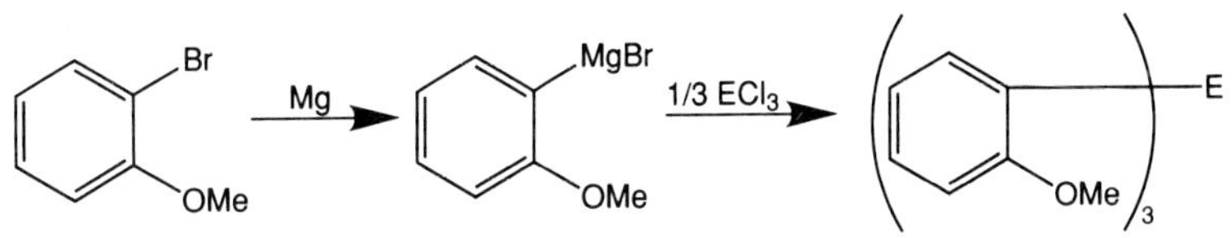

[5.46]

and can be converted to the corresponding phenol by HI cleavage [143]. Metal arsenides dealkylate anisole (PhOMe) and thus cannot be used to prepare arsine-substituted arylethers [145]. Alkyl derivatives are obtained from the corresponding chloroalkylethers according to reactions [5.47] and [5.48] [118, 135, 137]:

$$ClCH_2(CH_2OCH_2)_nCH_2Cl \xrightarrow{KAsPh_2} Ph_2AsCH_2(CH_2OCH_2)_nCH_2AsPh_2$$
$n = 1$ [135], $n = 2$ [137] [5.47]

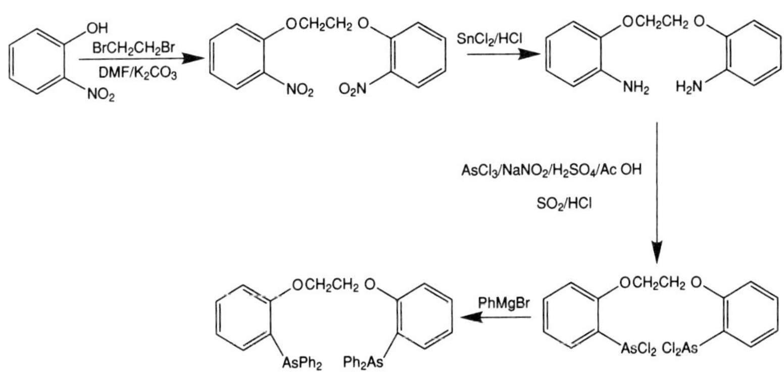

[5.48]

Group 15–olefin ligands. Olefinic groups are almost always introduced via the Grignard reagent of the corresponding haloalkene, which is then reacted with R_nPX_{3-n}. Examples are given for acyclic [146] and aryl species [147, 148] in reactions [5.49]–[5.53] (see p. 234 (top)).

Ligands with more than one group 15 donor atom require more intricate syntheses, and the success of these preparations is dependent upon the different reactivities of C–Cl and As–Cl bonds towards Grignard

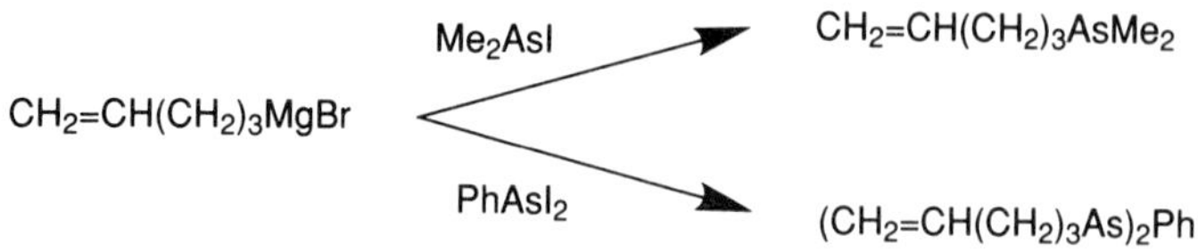

[5.49]

[5.50]

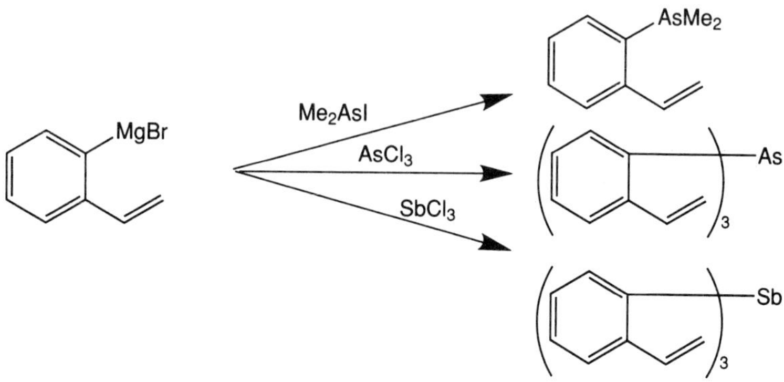

[5.51]

[5.52]

[5.53]

reagents as shown in reactions [5.54]–[5.56] (McAuliffe, C. A. and Watson, D. G., unpublished results) [149, 150]:

$$3CH_2{=}CHCH_2CH_2MgCl + AsCl_3 \rightarrow As(CH_2CH_2CH{=}CH_2)_3 \qquad [5.54]$$

$$\begin{array}{l} 2CH_2{=}CHCH_2CH_2MgCl + ClCH_2CH_2CH_2AsCl_2 \rightarrow \\ \quad ClCH_2CH_2CH_2As(CH_2CH_2CH{=}CH_2)_2 \\ \qquad \downarrow \text{Me}_2\text{AsNa} \\ \quad Me_2AsCH_2CH_2CH_2As(CH_2CH_2CH{=}CH_2)_2 \end{array} \qquad [5.55]$$

$$\begin{array}{l} 2ClMgCH_2CH_2CH_2AsMe_2 + ClCH_2CH_2AsCl_2 \rightarrow \\ \quad ClCH_2CH_2As(CH_2CH_2CH_2AsMe_2)_2 \\ \qquad \downarrow \text{LiCH=CH}_2 \\ \quad CH_2{=}CHCH_2CH_2As(CH_2CH_2CH_2AsMe_2)_2 \end{array} \qquad [5.56]$$

Completely hybrid ligands. The hybrid ligands, containing three or more different donor atoms or groups for example 2-di(R)arsinoanilines can be condensed with aldehydes [151]:

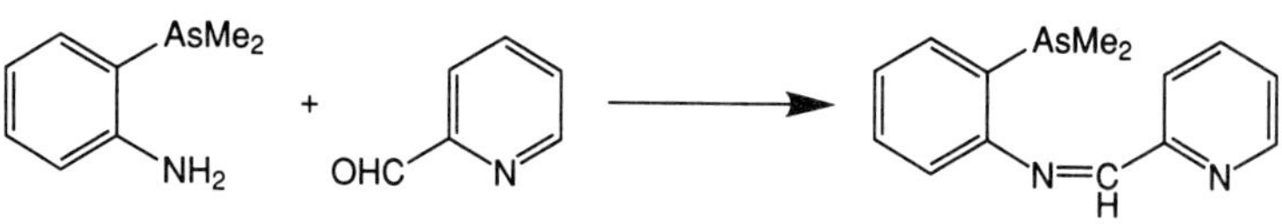

[5.46]

Table 5.7 IUPAC-recommended nomenclature for cyclic organoarsenic compounds.

Ring size	Unsaturated ring[a]	Saturated ring[a]
3	Arsirene	Arsirane
4	Arsete	Arsetane
5	Arsole	Arsolane
6	Arsinin(e)	Arsinan(e)
7	Arsepin(e)	Asepan(e)
8	Arsocin(e)	Arsocan(e)
9	Arsonin(e)	Arsonan(e)
10	Arsecin(e)	Arsecan(e)

[a]For ring size 6 and over the terminal 'e' is optional.

5.2.6 *Cyclic organoarsenic and organoantimony compounds*

Cyclic organoarsenic(III) compounds have been prepared with ring sizes from three upward and with varying numbers of arsenic ring members. Only those compounds containing the C–As–C moiety are considered here. The organoarsenic macrocycles are described later. The nomenclature used for the organoarsenic rings is the extended Hantzsh–Widman system approved by IUPAC in 1990 [152] and presented in Table 5.7.

Three-membered organoarsenic and organoantimony rings. For three-membered organoarsenic rings the first isolated arsirane was synthesized by Appel, Gaitzsch and Knoch [153]:

$$PhAsCl_2 + LiC(Cl)(SiMe_3)_2 \xrightarrow{thf/-110°C} PhAs[C(SiMe_3)_2]_2 \qquad [5.58]$$

The thermally unstable three-coordinate arsenic(V) intermediate spontaneously rearranges to the arsirane at room temperature:

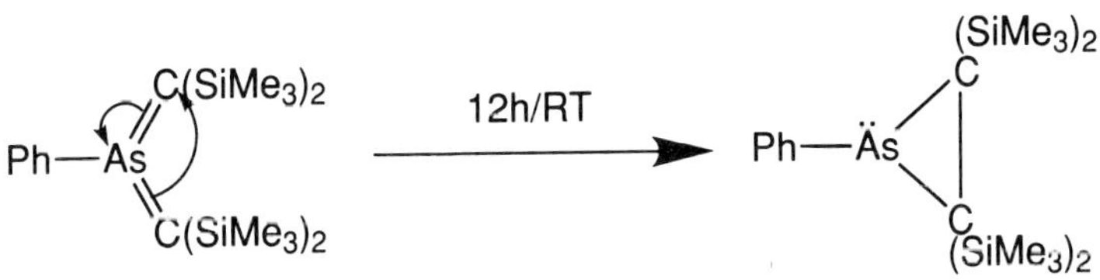

[5.59]

The colourless crystalline arsirane is stable at room temperature in contrast to the phosphorus analogues. The structure was determined by X-ray crystallography and showed the angle of pitch of the 1-phenyl group to the three-membered plane to be 113°, underlining the pyramidal geometry of arsenic.

The fused arsole-arsirane ring system has been prepared by Dianova *et al.* [154] by a carbene insertion reaction:

[5.60]

A similarly fused ring system was synthesized by Hamelin and Himidi-Kabbab [155], but in which all the ligands to arsenic were carbon (R^1 = CO_2Et, R = $CHPh_2$):

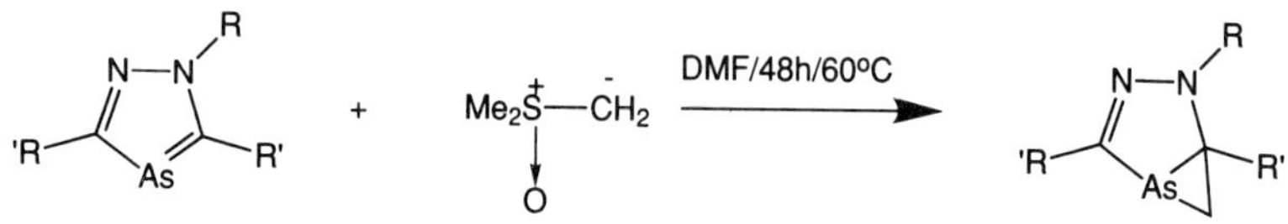

[5.61]

Four-membered organoarsenic and organoantimony rings. Four-membered organoarsenic rings, the arsetanes, were first synthesized by Wild and Mickiewicz [156] by an intramolecular nucleophilic attack on carbon by arsenic:

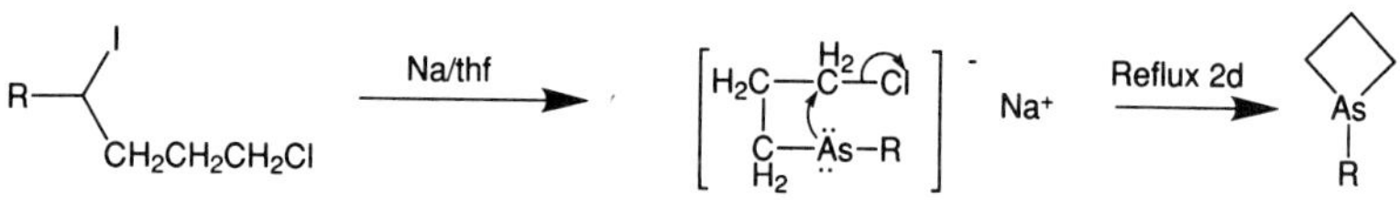

[5.62]

Five-membered organoarsenic and organoantimony rings. Five-membered organoarsenic rings, the arsolanes, are relatively strain-free rings and are readily formed by reacting difunctional organo precursors with a difunctional arsenic reactant. A number of representative examples are given in reactions [5.63]–[5.65] [157–159]:

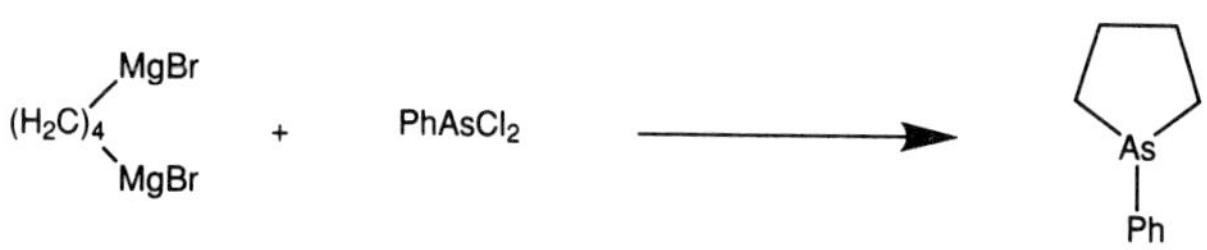

[5.63]

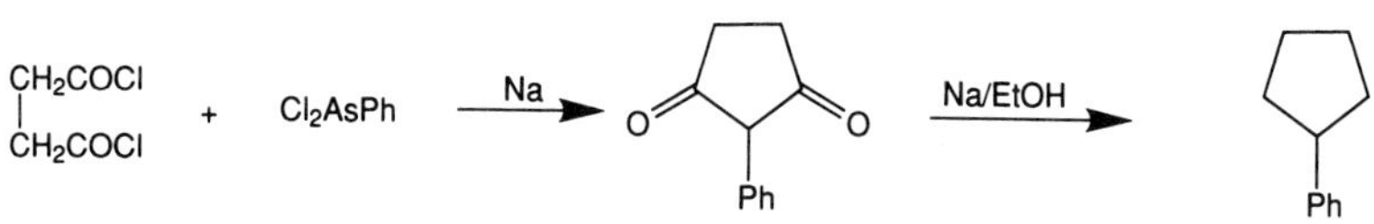

[5.64]

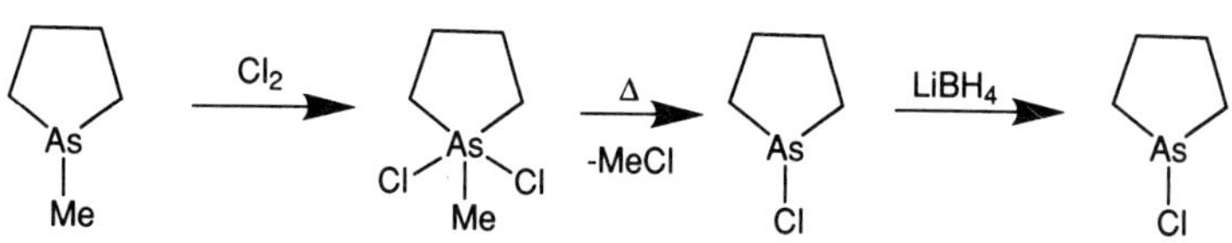

[5.65]

Fused-ring derivatives of 2-arsolenes and 3-arsolenes have also been prepared, as shown in Scheme 5.9(a) [160] and 5.9(b) [161], respectively.

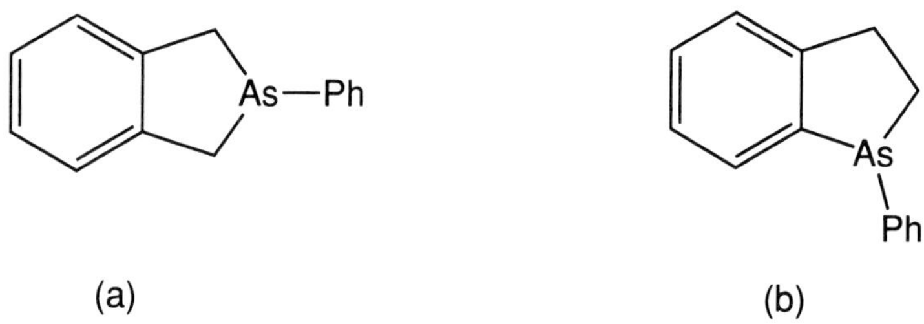

Scheme 5.9 Fused-ring derivatives of (a) a 2-arsolene; (b) a 3-arsolene.

A selection of reactions of the arsolanes is presented in Scheme 5.10.

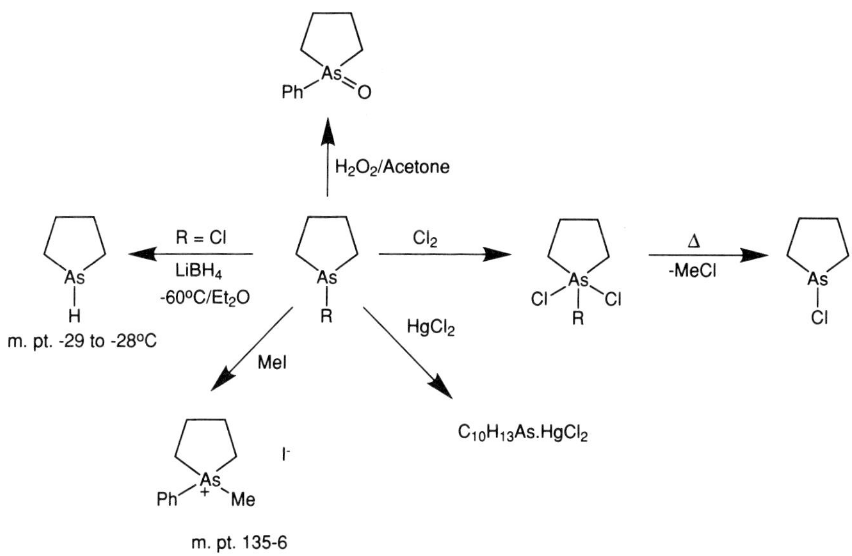

Scheme 5.10 Some reactions of the arsolanes.

The stibolanes are in general similar to the arsolanes. The first stibolane (1-phenyl derivative) prepared was reported in 1916 by Grütter and Krause [162]:

 [5.66]

Table 5.8 The 3-arsolenes:

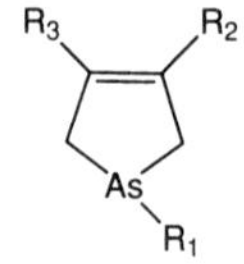

Empirical formula	Substituents R^1	R^2	R^3	Boiling point (°C)[a]	MeI derivative melting point (°C)	Reference
$C_6H_{11}As$	Me	Me	H	—	159	[i]
$C_7H_{13}As$	Me	Me	Me	—	162	[i]
$C_9H_{17}As$	Bu^t	Me	H	—	142	[i]
$C_{10}H_{11}As$	Ph	H	H	59–60 (0.6)	—	[ii]
$C_{10}H_{19}As$	Bu^t	Me	Me	—	181	[i]
$C_{11}H_{13}As$	PL	Me	H	61–62 (0.3)	—	[ii]
$C_{12}H_{15}As$	Ph	Me	Me	72–73 (0.3)	174	[i, ii]

[a]Pressures (torr) are given in parentheses.

References

i Levason, W., McAuliffe, C. A. and Murray, S. G. unpublished results.
ii Chow, K. K., Levason, W. and McAuliffe, C. A. (1976) *J. Chem. Soc., Dalton Trans.*, 1429.

It is a malodorous, colourless oily liquid (b.p. 156°C–158°C at 20–22 torr) readily oxidized by atmospheric oxygen.

Unsaturated arsolanes fall into two classes, the singly unsaturated arsolenes (of types I and II) and the doubly unsaturated arsoles. Arsolenes of type I (3-arsolenes) have been prepared by Mathey and Thiollet [163] and Schenk and Voss [164] (Table 5.8):

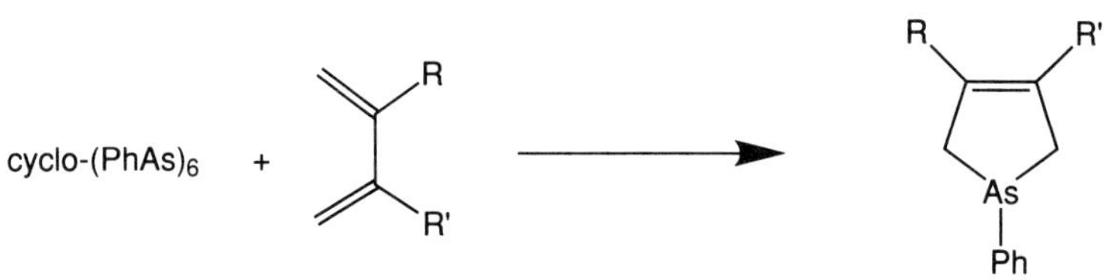

[5.67]

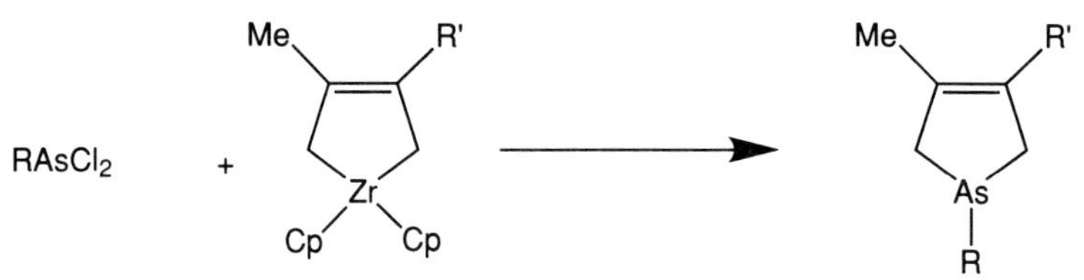

[5.68]

3-Arsolenes are isomerized by strong heating [163], the composition of the product indicating the 2-isomer (Type II) is the more themodynamically stable. Märkl and Hauptmann [165] prepared a series of arsolenes of type II in modes and yields as shown in reactions [5.69]–[5.71]:

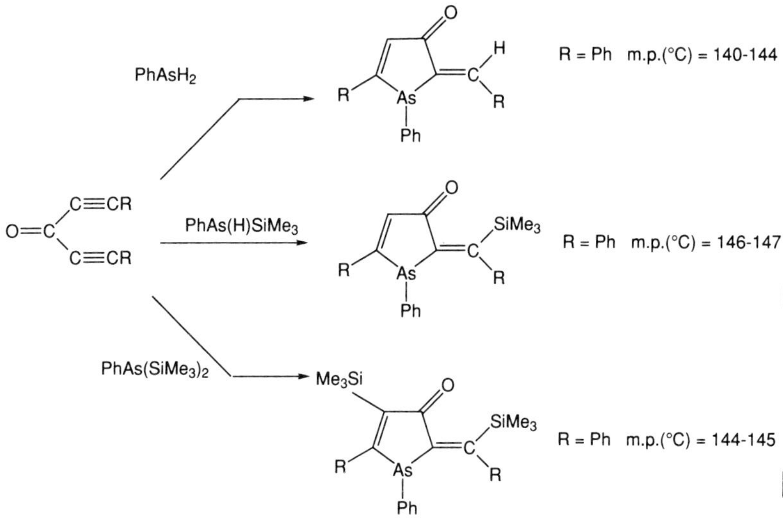

[5.69]

[5.70]

[5.71]

These keto-arsolenes can be converted to condensed-ring systems [165] as shown in reaction [5.72]:

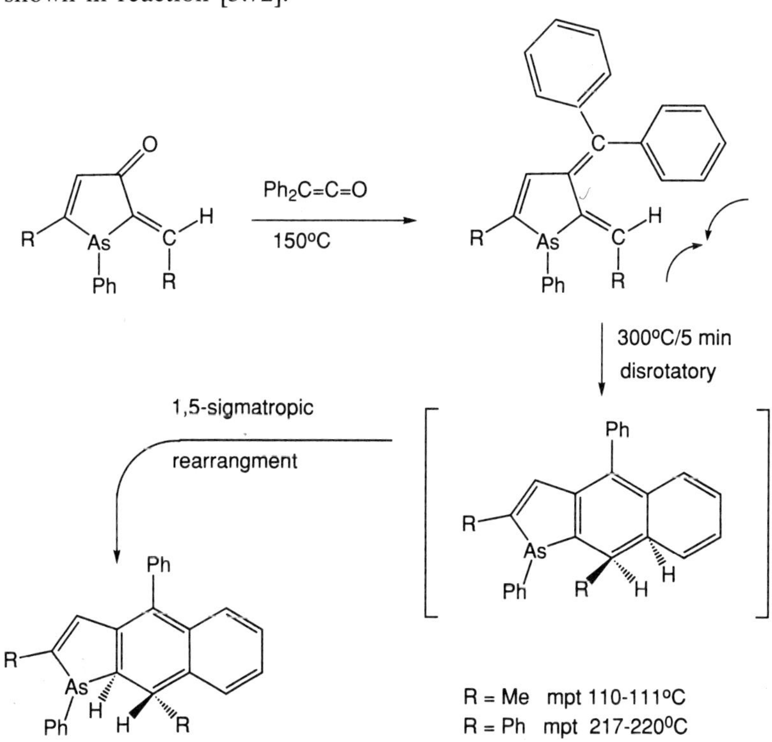

[5.71]

although the thermal cyclization pathway is blocked for R = Bu^t.

The arsoles are the unsaturated derivatives of the arsolanes and are much more interesting having a richer chemistry than the arsolanes. The synthesis of arsoles follows broadly similar principles to that of the arsolanes [166, 167], examples of which are given in reactions [5.73] and [5.74]:

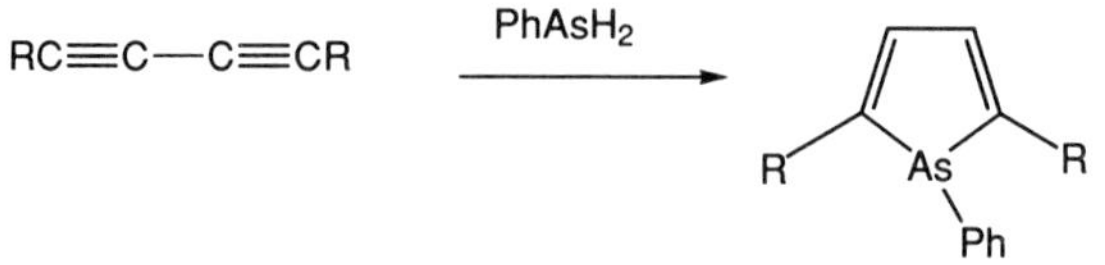

[5.73]

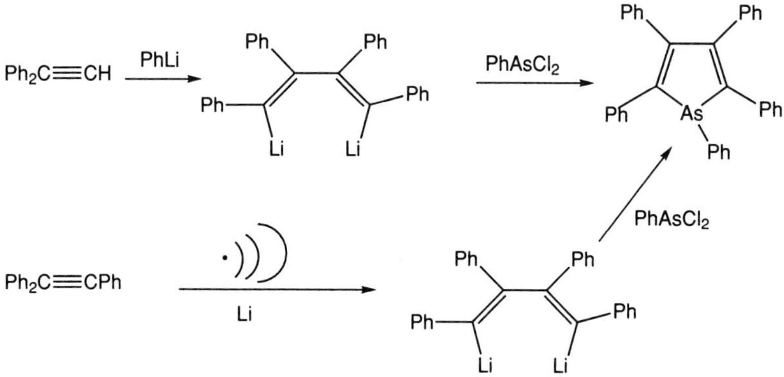

[5.74]

An improved synthesis of the lithio derivatives using ultrasound has been developed [168] which could usefully improve this route to arsoles.

Fagan and Nugent [169] have disclosed an interesting general synthesis for p-block five-membered rings involving zirconocene precursors which, in the case of arsenic, affords a 76% yield of the desired arisole:

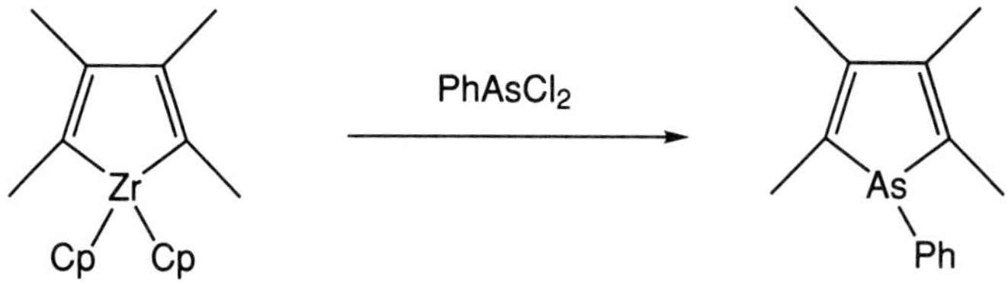

[5.75]

Some reactions of arsoles are given in Scheme 5.11 [170, 171] and some physical data for selected arsoles are presented in Table 5.9.

The fused-ring benzarsoles [Scheme 5.12(a)] and dibenzarsoles [Scheme 5.12(b)] (referred to in the classical literature as arsindoles and 9-arsa-fluorines, respectively) are prepared by use of similar techniques to the

Table 5.9 Selection of arsoles (the structure, showing R[1]-R[5], is given in Scheme 5.11)

Empirical formula	Substituent R^1	R^2	R^3	R^4	R^5	Melting point (°C)	Boiling point (°C)[a]	References
$C_{10}H_9As$	Ph	H	H	H	H	—	95 (0.005)	[i]
$C_{12}H_{13}As$	Ph	Me	H	H	Me	—	81 (0.5)	[ii]
$C_{14}H_{17}As$	Ph	Me	Me	Me	Me	—	—	[iii]
$C_{15}H_{19}As$	But	Ph	H	H	Ph	—	—	[iv]
$C_{17}H_{15}As$	Me	Ph	H	H	Ph	101–102	—	[ii, v]
$C_{18}H_{17}As$	Et	Ph	H	H	Ph	—	—	[ii]
$C_{22}H_{17}As$	Ph	Ph	H	H	Ph	160–161	—	[ii]
$C_{23}H_{19}As$	Bz	Ph	H	H	Ph	90–91	—	[v]
$C_{23}H_{19}As$	Ph	Ph	Me	H	Ph	122–124	—	[vi]
$C_{28}H_{21}As$	Ph	Ph	Ph	H	Ph	151–152.5	—	[vi]
$C_{29}H_{23}As$	Me	Ph	Ph	Ph	Ph	—	—	[vii]
$C_{30}H_{21}As$	Ph	1-$C_{10}H_7$	H	H	1-$C_{10}H_7$	—	—	[ii]
$C_{34}H_{25}As$	Ph	Ph	Ph	Ph	Ph	215–216	—	[viii]

[a]Pressures (torr) are given in parentheses.

References

i Ashe, A. J., Mahmond, S., Elschenboich, C. and Wünsch, M. (1987) *Angew. Chem. Int. Ed. Engl.*, **26**, 229.
ii Märkl, G. and Hauptmann, H. (1968) *Tetrahedon Lett.*, 3257.
iii Fagan, P. F. and Nugent, W. A. (1988) *J. Am. Chem. Soc.*, **110**, 2310.
iv Märkl, G. (1974) *Tetrahedon Lett.*, 203.
v Hauptmann, H. and Märkl, G. (1982) *Angew. Chem. Int. Ed. Engl.*, **11**, 439.
vi Braye, E. H., Caplier, I. and Saussez, R. (1971) *Tetrahed.*, **27**, 5523.
vii Leavitt, F. C., Manuel, T. A. and Johnson, F. (1959) *J. Am. Chem. Soc.*, **81**, 3163.

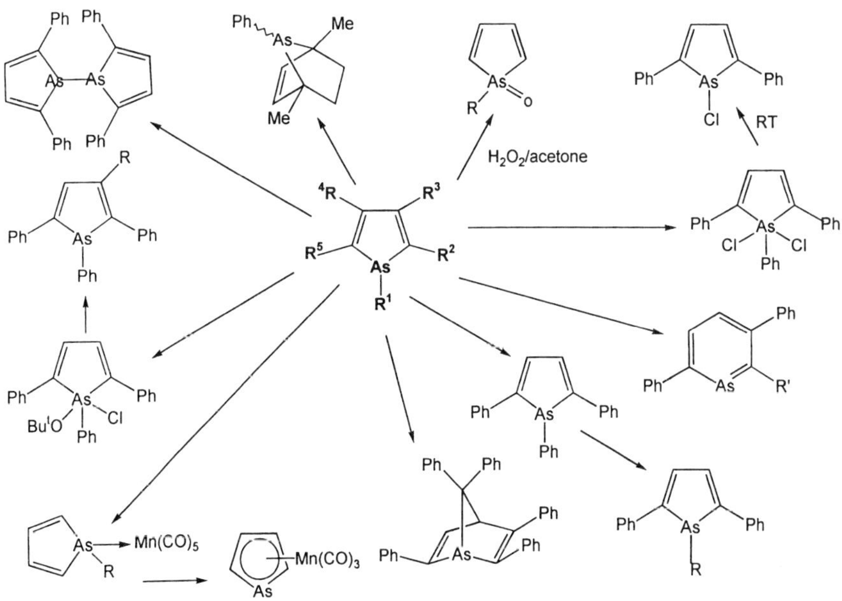

Scheme 5.11 Some reactions of the arsoles.

Scheme 5.12 (a) Fused-ring benzarsole and (b) fused-ring dibenzarsole.

arsolanes and arsoles but, in addition, intramolecular aromatic substitution routes are available. The arsolenes prepared by Märkl and Hauptmann [165] can be dehydrogenated by treatment with base:

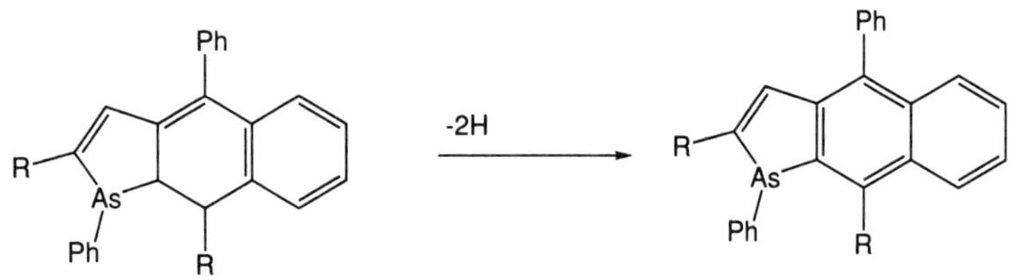

[5.76]

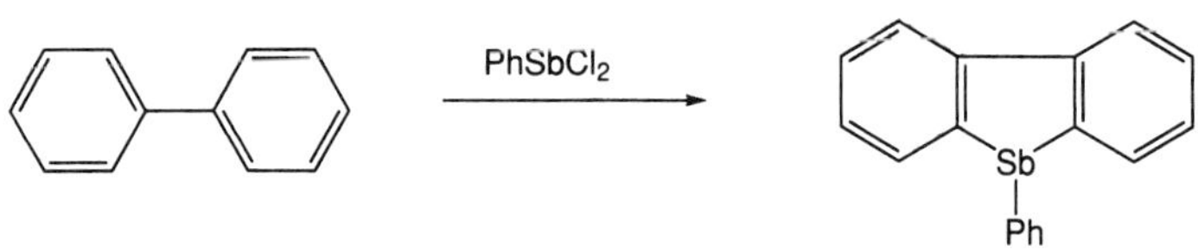

[5.77]

In the case of the tetraphenyl derivative, a direct synthesis is possible [165]. The substituent on arsenic may be replaced via a potassium salt by another group, for example, Me.

The syntheses and properties of the benzstiboles follow the same broad principles of the benzarsoles [172] as shown in Scheme 5.13 [173].

Buchwald, Fisher and Foxman [174] have extended the zirconocene-based syntheses of heterocycles to the stibaindoles (Scheme 5.14). The 1,3-benzarsoles substituted at arsenic are prepared by condensing a secondary arsine with an imine [175]:

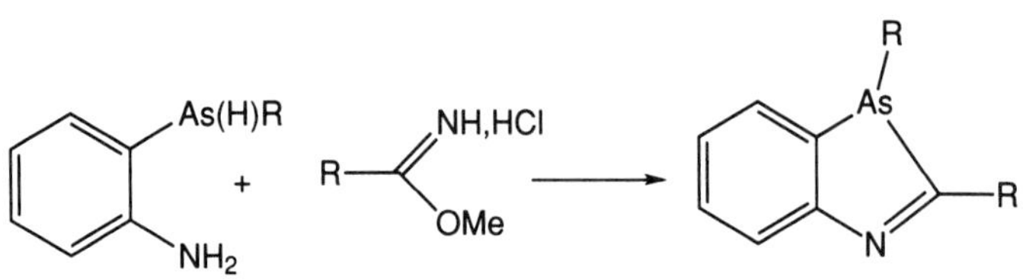

[5.78]

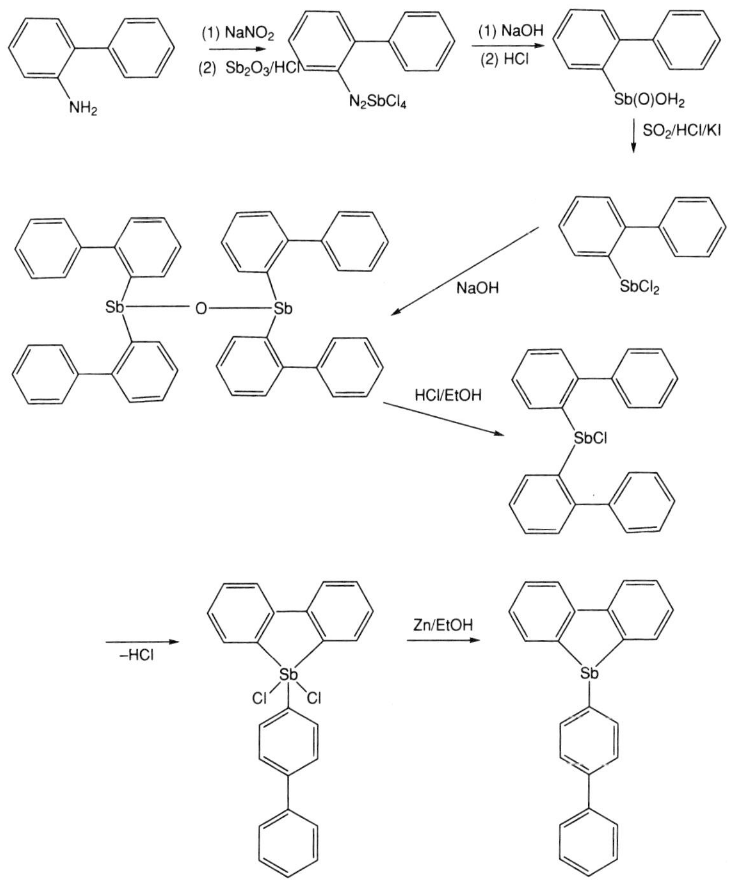

Scheme 5.13 Syntheses of the benzstiboles.

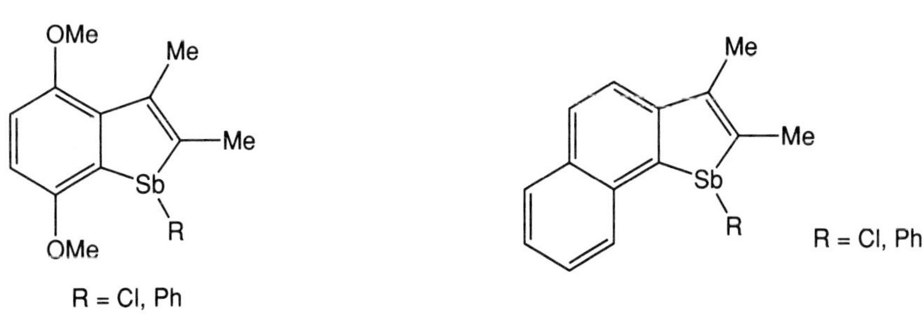

Scheme 5.14 Two examples of stibaindoles.

A few examples of benzarsoles are given in Table 5.10. Photoelectron spectroscopy has shown [176] that the arsenic lone pair has minimal interaction with the π-electron systems in these molecules. Many other azarsoles have been reported, one example of which is shown in Scheme 5.15 [177].

The zirconocene-based synthesis of stibaindoles is as follows:

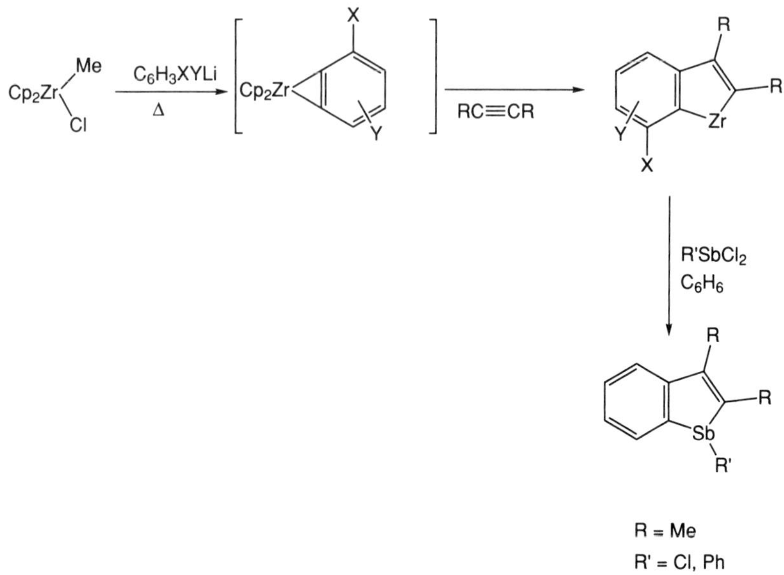

[5.79]

Some routes to benzarsoles are as follows:

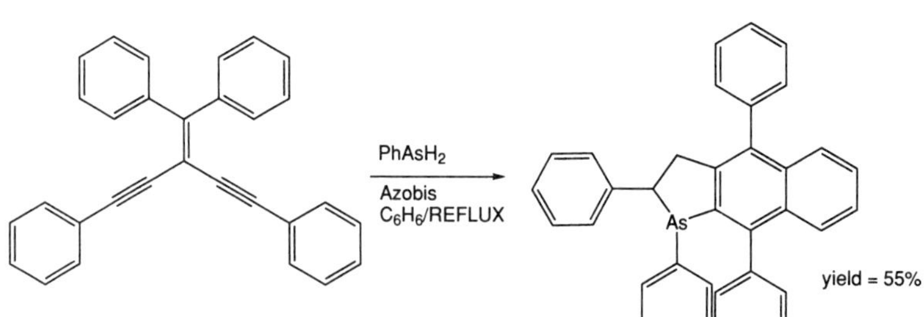

[5.80]

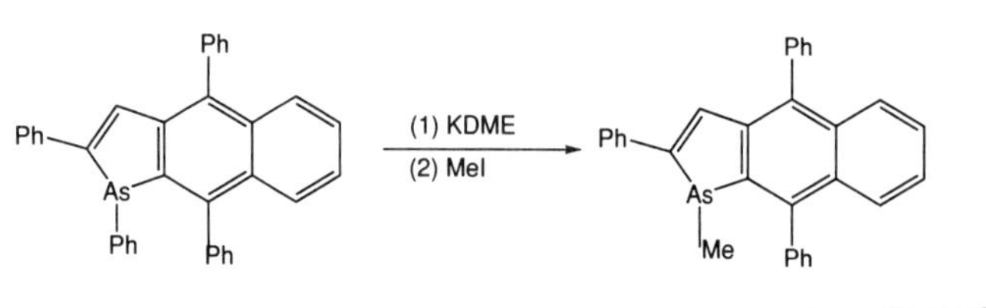

[5.81]

Table 5.10 Arsinic-substituted 1,3-benzarsoles

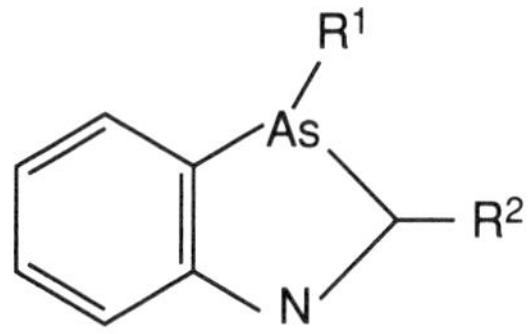

Empirical formula	Substituents R[1]	R[2]	Boiling point (°C)[a]	Reference
$C_{11}H_{14}AsN$	H	Bu^n	80–82 (0.01)	[i]
$C_{12}H_{16}AsN$	Me	Bu^n	88–89 (0.01)	[i]
$C_{12}H_{16}AsN$	Bu^t	Me	90–93 (0.1)	[i]
$C_{14}H_{12}AsN$	Ph	Me	125–130 (0.1)	[ii]
$C_{15}H_{14}AsN$	Ph	Et	145–150 (0.1)	[ii]
$C_{17}H_{18}AsN$	Ph	Bu^s	120–130 (0.02)	[ii]

[a]Pressures (torr) are given in parentheses.

References

i Heinicke, J., Nyulaszi, L., Gsonka, G. *et al.* (1989), *J. Organomet. Chem.*, **373**, 49.

ii Tzschach A. and Heinicke, J. (1978) *J. Organomet. Chem.*, **154**, 1.

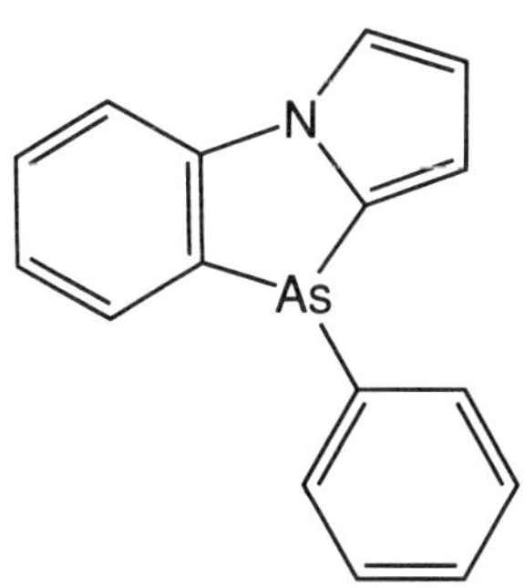

Scheme 5.15 An example of an azarsole.

Six-membered organoarsenic and organoantimony rings. Arsinanes are synthesized by methods analogous to the arsolanes with suitably lengthened alkyl chains, a few examples of which are listed in Table 5.11.

The barrier to ring reversal in 1-methyl-1-arsinane

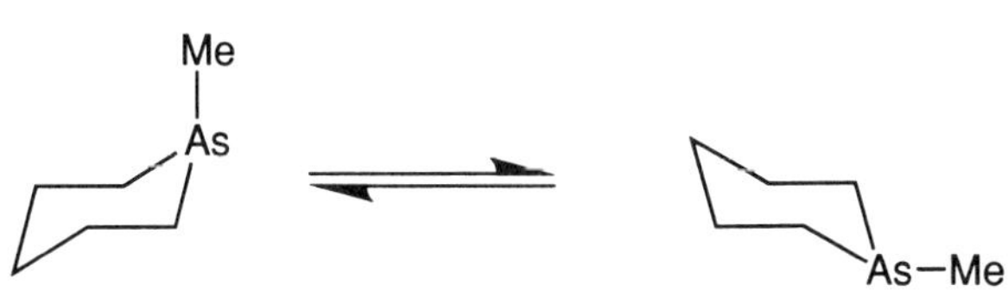

is, from ¹H NMR data, 7.8 kcal mol^{-1}, with a coalescence temperature of −128°C.

The triply bridged triarylarsine shown in [5.82] reveals only slight deviations from C_{3v} symmetry [178], although the C–As–C angles (average 93.0°) are notably less than in $AsPh_3$ (100.1°), leading to a more pronounced AsC_3 pyramid. It may by synthesized as follows:

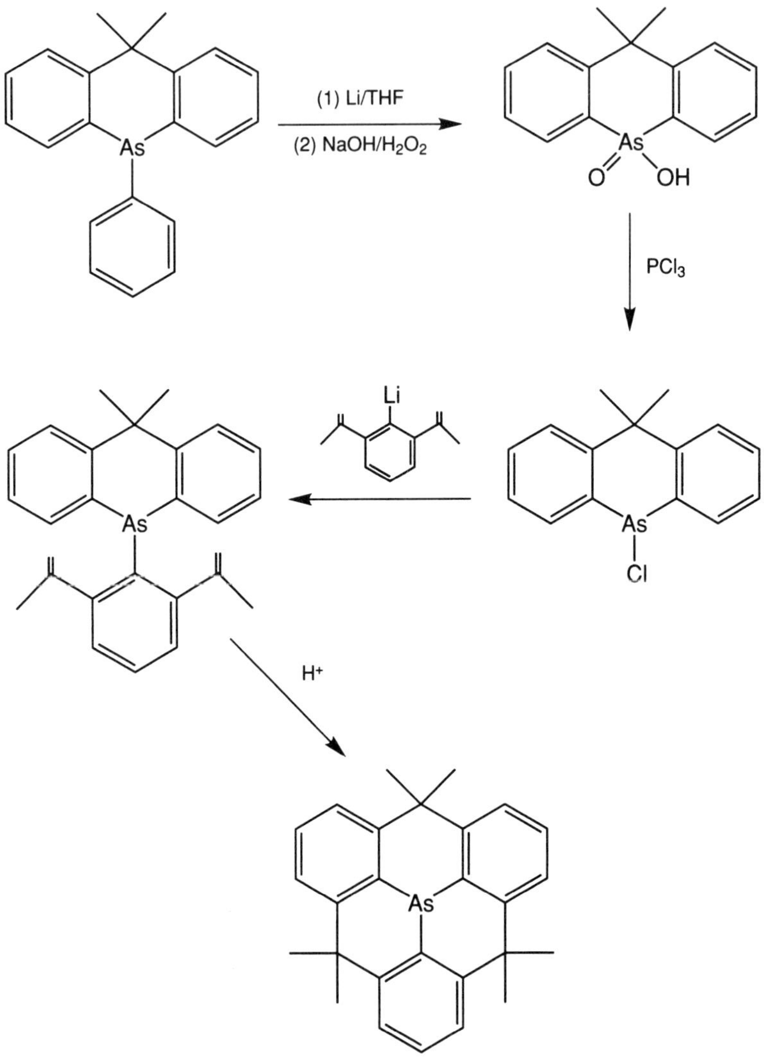

[5.82]

The arsatriptycene (Scheme 5.16) cage compounds are structurally related [179] and are synthethized as follows:

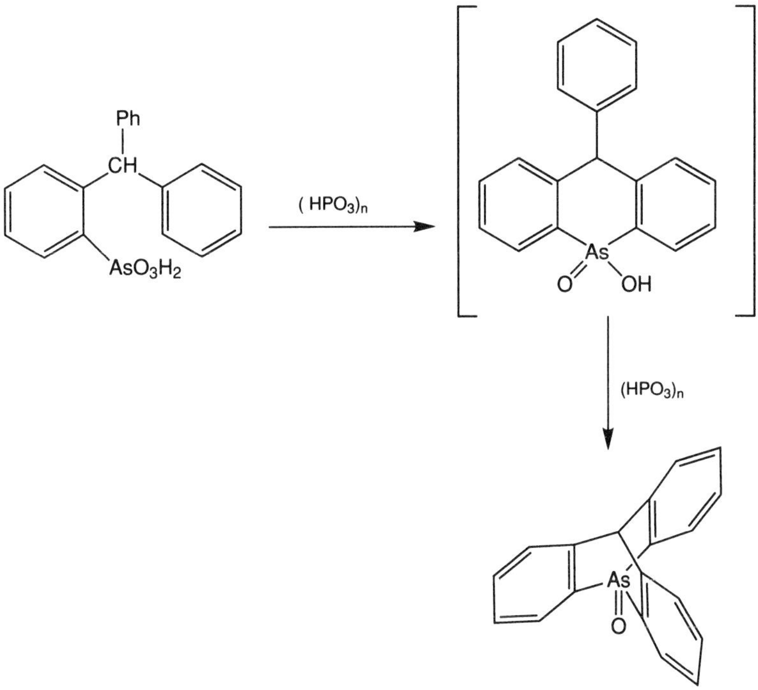

[5.83]

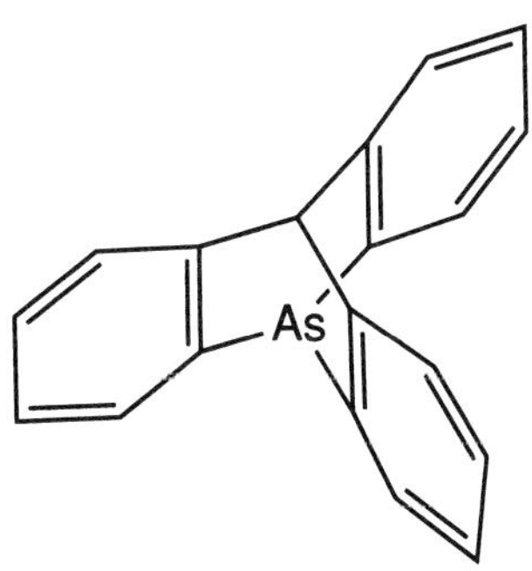

Scheme 5.16 Arsatriptycene.

An X-ray study [180] showed the arsenic was pyramidal [C–As–C angle 90°, As–C length 1.94 Å (average)]. The structures are disordered with a fraction of the molecules with the bridgehead carbon and arsenic atoms interchanged. Improved data were obtained from the crystal structure of 9,10-dihydro-1-phenyl-9,10-benzeno-9-arsanthiacene [181].

With regard to six-membered rings with two or more arsenic atoms, only those ring systems with interjacent carbons are considered here. Within these limits, there are only three basic structures, the 1,3-diarsinanes, the 1,4-diarsinanes and the 1,3,5-triarsinanes, shown in Scheme 5.17(a), (b) and (c), respectively. All these compounds are well known and their syntheses present no special problems other than that of complex chirality if a specific isomer is required. Bis-arsine cage structures based on 1,4-diarsinanes (diarsatryptacenes) are also well known {Schemes 5.17(d) [182] and 5.17(e) [183]}.

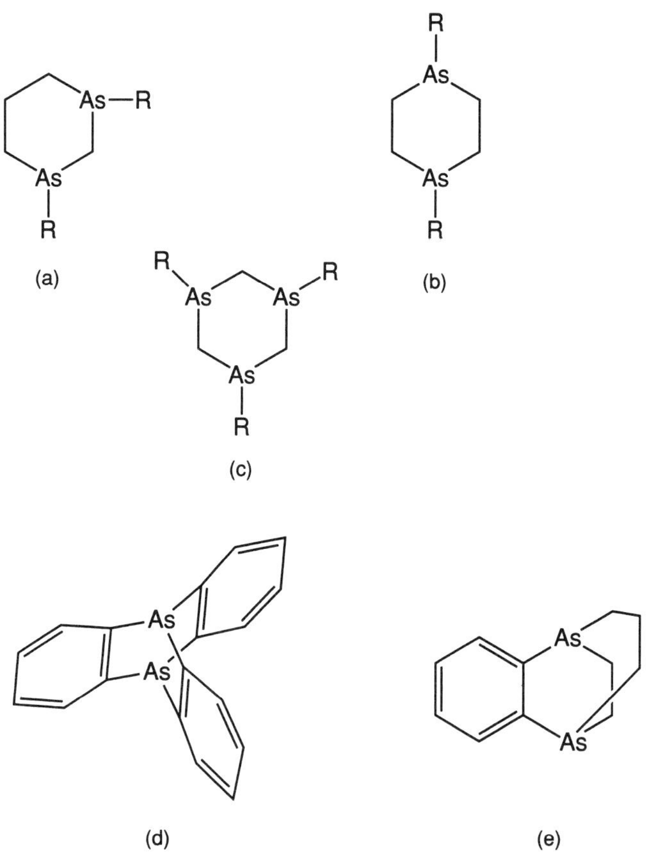

Scheme 5.17 General sructural formulae of (a) 1,3-diarsinanes; (b) 1,4-diarsinanes; (c) 1,3,5-triarsinanes; and (d), (e) bis-arsine cage structures based on the 1,4-diarsinanes.

The perfluoro-analogue of a diarsatryptacene has been prepared [184] as has 2,3,5,6,7,8-hexakis(trifluoromethyl)-1,4-diarsabicyclo[2,2,2]octa-2,5,7-triene [185] (m.p. 139°C–140°C):

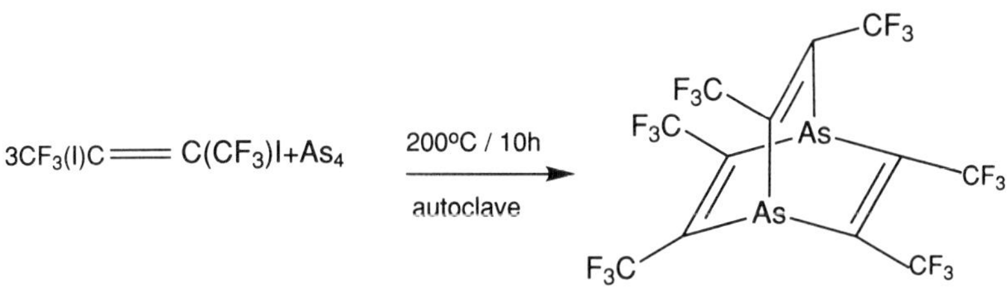

[5.84]

Larger organoarsenic rings. The compounds considered here are those with seven- , eight- and nine-membered rings. The rings with ten or more members are termed macrocyclic and are discussed later although the distinction is arbitrary. One synthesis of each type is presented. All have the spectroscopic properties expected for the pyramidal AsC_3 moiety.

The seven-membered organoarsenic rings, the arsepins (unsaturated) and the arsepans (saturated) are known through their derivatives. The unsubstituted ring at carbon is formed by condensing a primary arsine with 1,5-hexadine [186]:

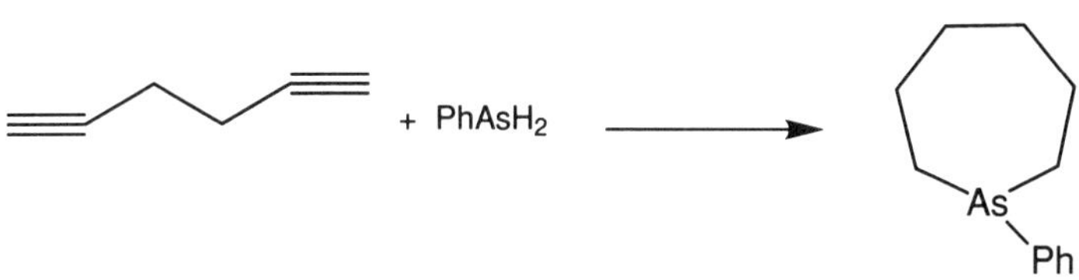

[5.85]

Fused-ring systems featuring the arsepin/arsepan ring have been prepared by Mann and co-workers [187–189]:

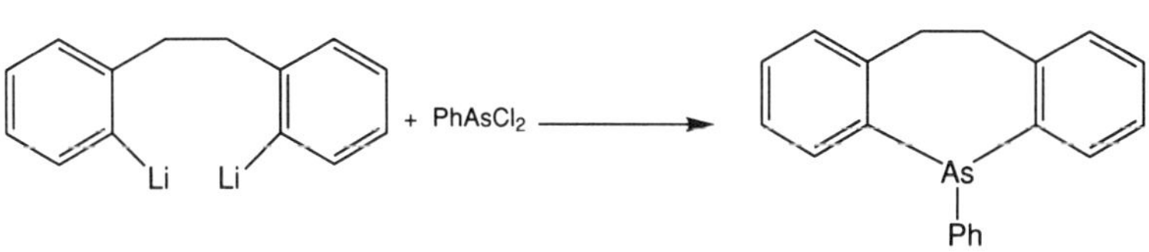

[5.86]

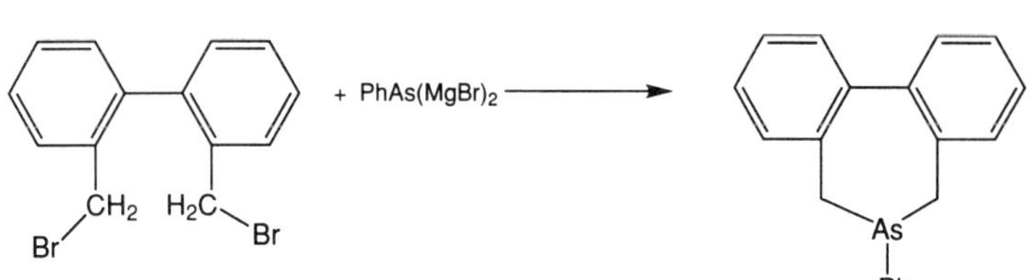

[5.87]

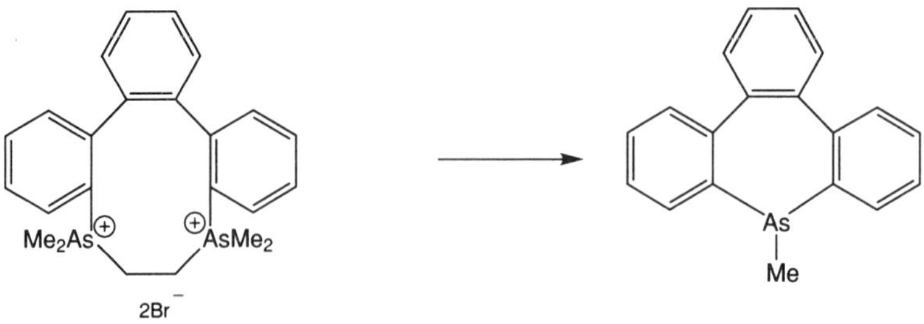

[5.88]

The nine-membered organoarsenic rings, the arsonin (unsaturated) and arsonan (saturated) systems are known only through their derivatives [190], one example of which is illustrated in reaction [5.89]:

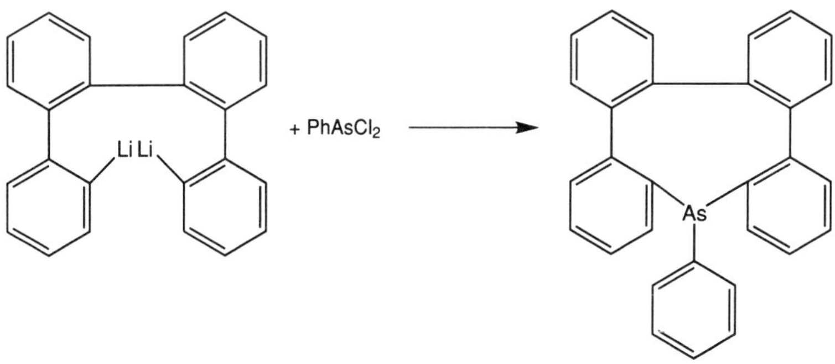

[5.89]

Arsenic(III) macrocycles. Arsenic–carbon macrocycles are arbitrarly defined here as those rings with more than ten members, the interest in this class of compounds being mainly as ligands for transition metals. In this section the synthesis of the macrocycles is covered first, followed by their structural and conformational aspects. The short form of macrocyclic nomenclature is used, that is [14]ane-As_4 describes a fourteen-membered ring, four of which are arsenic atoms. First, the arsenic–carbon macrocycles will be described, followed by the macrocycles containing other heteroatoms.

Arsenic(III)–carbon macrocycles. In broad terms there are two synthetic approaches employed in the synthesis of macrocycles. First is the simple condensation of two bi-functional percursor molecules to give the target compound, which is an extension of the synthesis of small to medium-size ring compounds. The other main strategy employs a metal, usually a transition metal, to ligate the precursors followed by an interligand cyclization reaction resulting in a metal-macrocycle complex; this type of route is termed a template synthesis. Kyba and Chou [191] used the first strategy to prepare [11]ane-As_3 by employing a high-dilution technique to minimize the competing linear oligomerization reaction. Subsequently, they extended this method to [14]ane-As_4 [192]. Kauffmann and Ennen [193] used similar techniques to synthesize [12]ane-As_3,

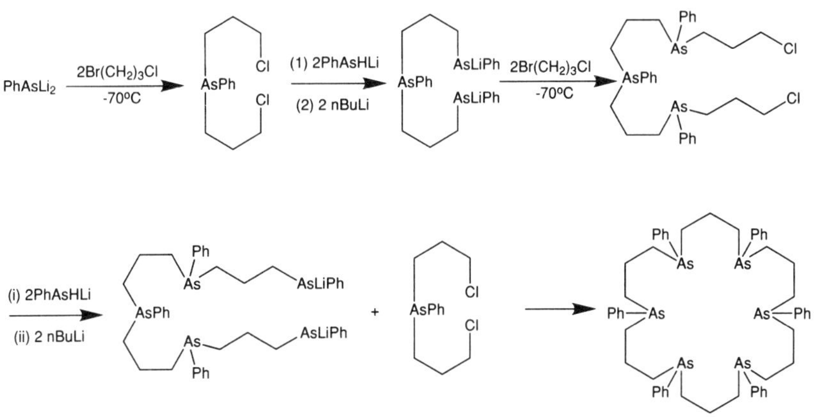

[5.90]

[14]ane-As_4 and [24]ane-As_6:

Arsenic(III)–element–carbon macrocycles. Arsenic(III)–carbon macrocycles containing heteroatom ring members, usually a pnictide or chalcogen, are prepared by similar techniques to the parent [*x*]ane-As_y macrocycles. Kyba and Chou [192] prepared, *inter alia*, [14]ane-As_2P_2 and [14]ane-As_2E_2 (E = O, S). Kauffmann and Ennen [193] extended their method to prepare [28]ane-As_2S_6 and [28]ane-$As_2O_2S_4$, albeit as low-yield by-products. They also prepared a macrocycle [14]ane-AsS_3 with a functionalized side chain with potential for attaching to a polymer or other solid support.

The preparation of [14]ane-*trans*-As_2S_2 [194] is an example of a template synthesis, the complexed ligand, 2-$MeOCH_2C_6H_4As(Mc)CH_2CH_2SH$, being dimerized on Pd^{2+} to give the macrocycle:

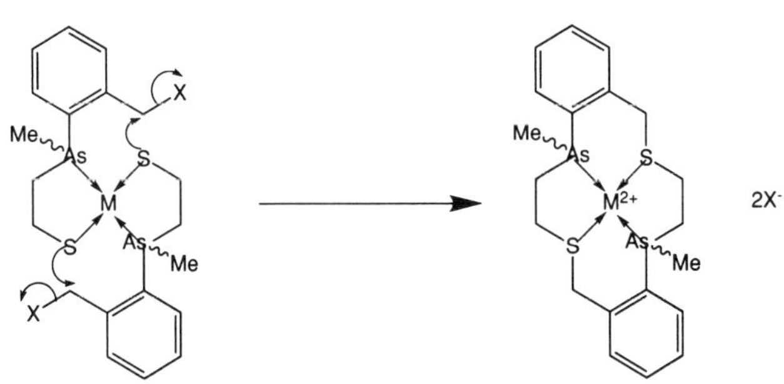

[5.91]

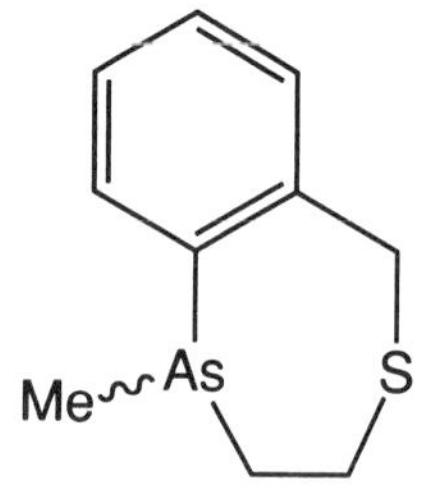

Scheme 5.18 A thiaarsepan.

There is a degree of steric control inasmuch as the *cis*-As_2S_2 macrocycle is excluded mechanistically, but the alternative reaction pathway results in an intramolecular cyclization giving a thiaarsepan (Scheme 5.18).

The conformations of arsenic macrocycles. The arsenic atoms in the macrocycles have an inversion barrier of the order of 180 kJ mol^{-1} and are thus in principle isolable, although of course in practice this is extremely difficult; the exocyclic arsenic bonds in, say, a tetraarsamacrocycle can adopt the four possible conformations shown in Scheme 5.19 [195].

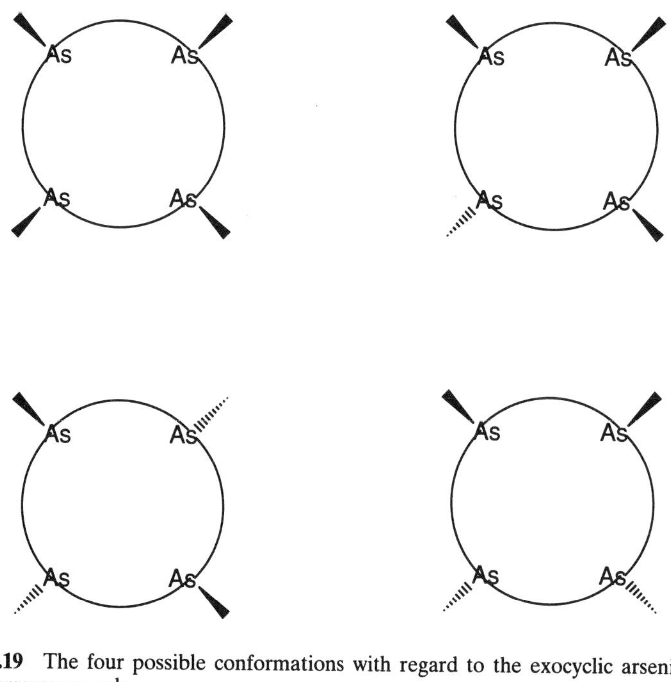

Scheme 5.19 The four possible conformations with regard to the exocyclic arsenic bonds in a tetraarsamacrocycle.

As with much group 15 chemistry, the phosphorus analogues have been better researched, with isomers isolated and characterized [1]. Where the arsenic centres are chiral, additional isomerization occurs. Wild, Kerr and Leung [194] in their template synthesis found that the racemic and *trans* [14] -As_2S_2 (R*R*) were favoured in the case of the structure shown in Scheme 5.20, although the mechanism underlying this bias has yet to be elucidated.

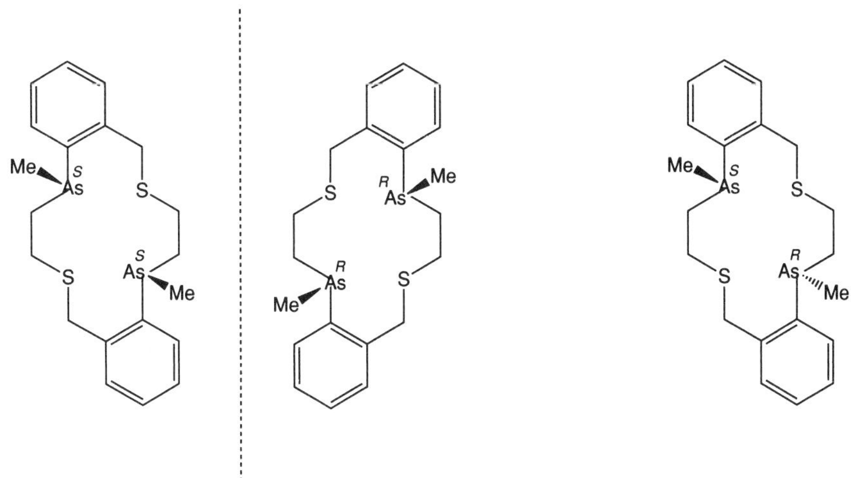

(R,R*)*-I *(R*,S*)*-I

Scheme 5.20 The racemic mixture and *trans* isomers preferred by the As_2S_2 arsenic macrocycle illustrated. Source: Wild, S. B., Kerr, P. G. and Leung, P. H. (1987) *J. Am. Chem. Soc.*, **109**, 4321.

5.3 Pentavalent organoarsenic

In this section the pentavalent organoarsenic compounds are discussed in order of increasing coordination to carbon.

5.3.1 Low-coordinate pentavalent organoarsenic compounds

Pentavalent organoarsenic compounds have a coordination sphere AsC_n, omitting the other substituents on carbon. Low-coordinate compounds are defined as those where $n < 4$, which therefore entails arsenic–carbon multiple bonding.

Arsanthracene does not exist in the free state [196, 197] and attempts to prepare it result in a dimer; however, the bisarsonium salt has been prepared and the crystal structure established (Scheme 5.21) [89].

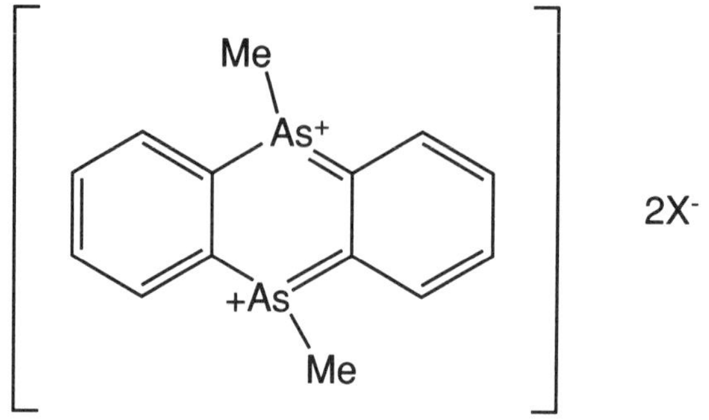

Scheme 5.21 Bisarsonium salt: when X = Br, one obtains cream crystals, melting point = 256°C (decomposes); when X = dipicrate, one obtains yellow crystals, melting point = 244°C (decomposes).

5.3.2 Four-coordinate pentavalent organoarsenic compounds

Four-coordinate compounds are the commonest and most important class of the pentavalent organoarsenic species and comprise both molecular (the arsonium ylides) and ionic (arsonium salts) species.

Arsenic and antimony ylides. The use of arsonium ylides in the organic synthesis is of growing significance [198]. The practice of writing ylide structures as $R_3As{=}CR_2$ will be used, although this may not correspond accurately to the arsenic–carbon bonding (discussed later).

Arsonium ylides are most commonly prepared by the deprotonation of an arsonium salt by means of a strong base:

$$R_3As + R'CH_2X \rightarrow [R_3AsCH_2R']^+X^- \xrightarrow[-H^+]{\text{base}} R_3As{=}CHR' \qquad [5.92]$$

and Schmidbaur and Nusstein [199] have shown that the central carbon atom can service two arsenic π-bonds:

$$MePh_2As^+CH_2As^+Ph_2Me \xrightarrow{NaNH_2} MePh_2As{=}C{=}AsPh_2Me \qquad [5.93]$$

The use of diazo derivatives enables tertiary arsines to be directly converted to ylides, although a copper catalyst is often necessary [203]:

$$Ph_3As + N_2{=}C(CO_2Et)_2 \xrightarrow{\text{Cu/150°C}} Ph_3As{=}C(CO_2Et)_2 \qquad [5.94]$$

This route has been much improved by the use of bis(hexafluoroacetylacetonato) copper(II) which enables much milder conditions to be used [200]:

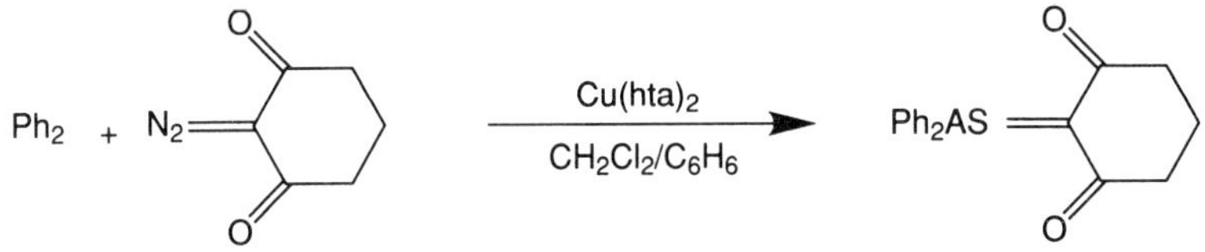

[5.95]

Other arsenic(V) precursors have also been used in the preparation of ylides, for example R_3AsX_2 [201] and R_3AsO [202, 203], the former case being shown in reaction [5.96] in which R′ = CN, SO_2Ph, CO_2Me, NO_2 ; R″ = CN, SO_2Ph, Ph:

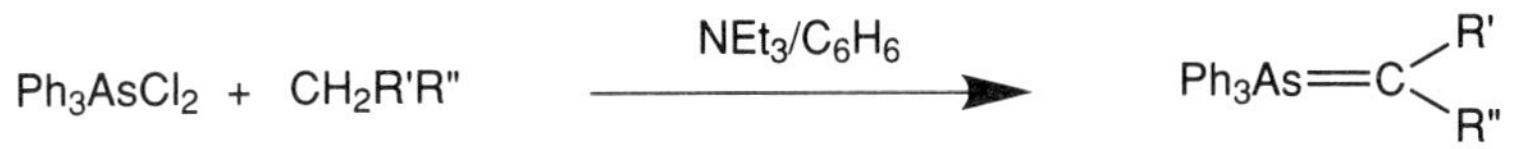

[5.96]

Some examples of arsonium ylides are given in Table 5.12.

The main use of arsonium ylides is in reactions to form olefins, their use being to complement rather than replace the phosphorus ylides. When the alkyl precursor has electron-withdrawing groups the phosphorus ylide

Table 5.11 A selection of arsinanes:

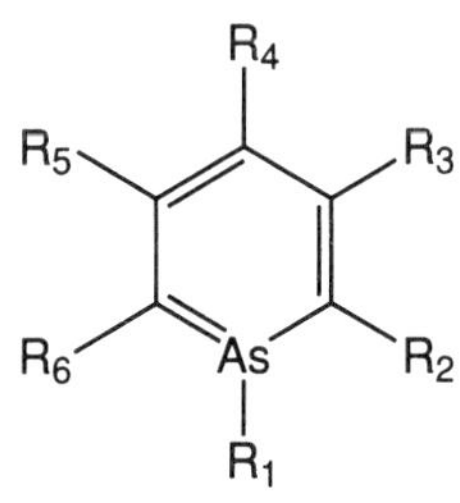

Substituents						Boiling	Melting	References
R^1	R^2	R^3	R^4	R^5	R^6	point (°C)[a]	point (°C)	
Me	H	H	H	H	H	153–155	—	[i]
Et	H	H	H	H	H	62–64 (12.5)	—	[ii]
Ph	H	H	H	H	H	153–154 (20)	—	[iii]
$2MeC_6H_4$	H	H	H	H	H	162–163 (20)	—	[iii]
CH–CHCl	H	H	H	H	H	89–91 (5)	—	[iv]
Me	Me	H	H	H	H	169	—	[v, vi]
Ph	H	H	O	H	H	114–115 (0.3)	9–10	[vii]

[a]Pressures (torr) are given in parentheses.

References

i Lambert, J. B. and Sun, H. N. (1976) *J. Organomet. Chem.*, **117**, 17.
ii Steinkopf, W. (1922) *C. Ber.*, **55**, 2579.
iii Gruttner, G. (1915) *C. Ber.*, **48**, 1473.
iv Gorski, E. (1934) *C. Ber.*, **67**, 730.
v Zappi, E. V. and Simonin, L. M. (1942) *Ciencia (Mex.)*, **3**, 160.
vi Zappi, E. V. and Simonin, L. M. (1942) *Chem. Abs.*, **37**, 2009.
vii Gallagher, M.H. and Mann, F.G. (1962) *J. Chem. Soc.*, 5110.

Table 5.12 Arsonium ylides: $R_3As{=}CR_2'$

Empirical formula	Compound	Melting point (°C)	Reference
$C_4H_{11}As$	$Me_3As{=}CH_2$	33–35	[i]
$C_{19}H_{17}As$	$Ph_3As{=}CH_2$	74	[ii]
$C_{20}H_{15}AsO$	$Ph_3As{=}C{=}C{=}O$	141–143	[iii]
$C_{21}H_{19}AsO_2$	$Ph_3As{=}CHCOOMe$	181–183	[iv]
$C_{22}H_{21}AsO_2$	$Ph_3As{=}CHCO_2Et$	162	[iv]
$C_{24}H_{23}AsO_2$	$Ph_3As{=}CHCH{=}CHCO_2Et$	130	[v]
$C_{24}H_{23}AsO_2$	$Ph_3As{=}CHC(Me){=}CHCO_2Me$	108–111	[vi]
$C_{26}H_{21}AsO$	$Ph_3As{=}CHCOPh$	154–156	[vii]
$C_{26}H_{24}As_2$	$MePh_2As{=}CHAsPh_2$	—	[viii]
$C_{27}H_{26}As_2$	$MePh_2As{=}C{=}AsPh_2Me$	—	[viii]
$C_{38}H_{33}As_3$	$MePh_2As{=}C(AsPh_2)_2$	151	[ix]
$C_{41}H_{31}As$		198–202	[x]
$C_{47}H_{35}As$		228–230	[xi]
		208–212	[x]

References

i Schmidaur, H., Richter, W., Wolf, W. and Köhler, F. H. (1975) *C. Ber.*, **108**, 2649.
ii Kendarker, P. S. and Tewari, R. S. (1975) *J. Organomet. Chem.*, **102**, 141.
iii Bestmann, H. J. and Bansal, R. K. (1981) *Tetrahedron Lett.*, 3839.
iv Depoorter, H. (1968) *Bull. Soc. Chim. Belg.*, **77**, 521.
v Huang, Y. Z., Shen, Y., Zhang, J. and Zhang, S. (1985) *Synthesis*, 57
vi Shi, L., Xiao, W., Wen, X. and Huang, Y. (1987) *Synthesis*, 370.
vii Nesmeyanov, A. N., Mikulshina, V. V. and Reutov, O. A. (1968) *J. Organomet. Chem.*, **13**, 263.
viii Schmidbaur, H. and Nusstein, P. (1985) *Organometallics*, **4**, 344.
ix Schmidbaur, H. (1987) *C. Ber.*, **120**, 1281.
x Lloyd, D., Glidewell, C. and Metcalfe, S. (1988) *Synthesis*, 319.
xi Freeman, B. H., Lloyd, D. and Singer, M. I. C. (1972) *Tetrahed.*, **28**, 343.

is relatively unreactive (e.g. with ketones) whereas the corresponding arsenic ylide is more active.

An interesting example of the superior stereospecificity of arsenic compared with phosphorus ylides has been provided by Shen and Liao [204]. The phosphorus ylide gives low yields and mixed isomers of cyclopropanes whereas the arsenic analogues give high yields and are often 100% stereospecific for isomer A in the reaction between $Ph_3As{=}CHC{\equiv}SiMe_3$ and $ArCH{=}CHCOPh$:

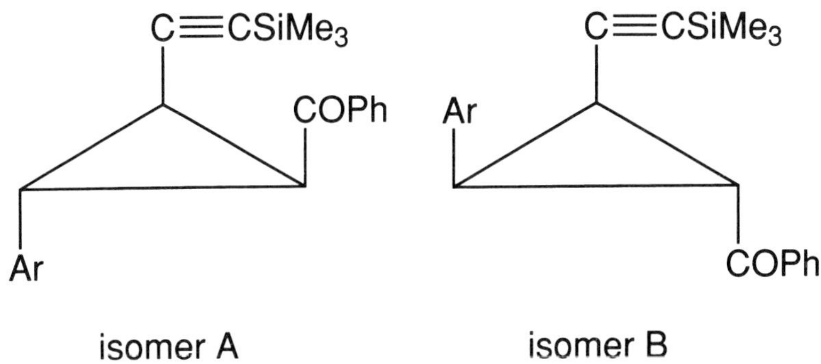

Structurally, the arsonium ylides consist of a central arsenic atom tetrahedrally surrounded by four carbon atoms in which the tetrahedral disposition of the carbon ligands is always distorted to some extent. The ylides have a resonance structure within the following two extremes:

$$R_3As{=}CR_2 \longleftrightarrow R_3\overset{+}{A}s{-}\overset{-}{C}R_2$$

although the resonance structures can be more complex (more contributing forms) in the case of triphenylarsonium 3,4-dibenzoylcyclopentadienylide (Scheme 5.22) [205]; calculations based on the cyclopentadiene ring C–C bond lengths are compatible with the percentage contributions shown.

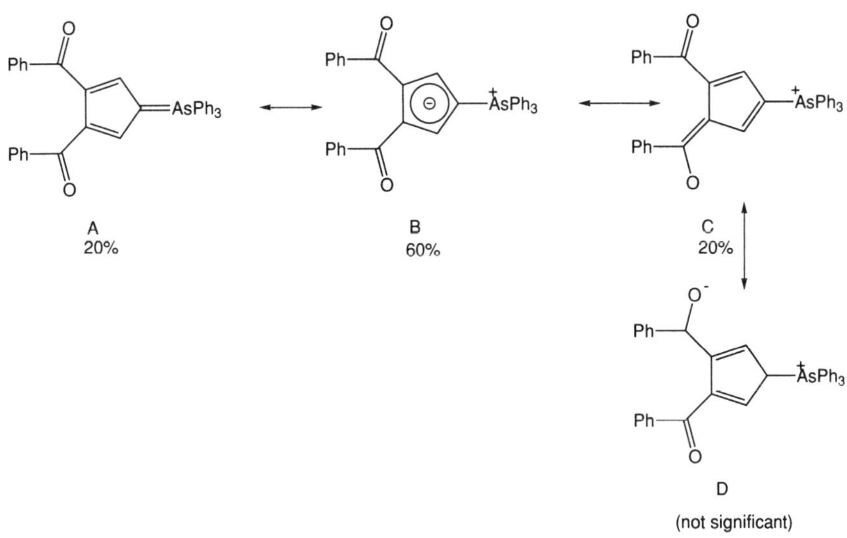

Scheme 5.22 Resonance forms of triphenylarsonium 3,4-dibenzoylcyclopentadienylide.

The arsonium ylide shown in the centre of Scheme 5.23 is of interest structurally since it was shown by Ferguson and Rendle [206] to have an As . . . O interaction (2.77 Å) resulting in a distorted trigonal bipyramidal geometry about arsenic; the main canonical forms are shown in Scheme 5.23.

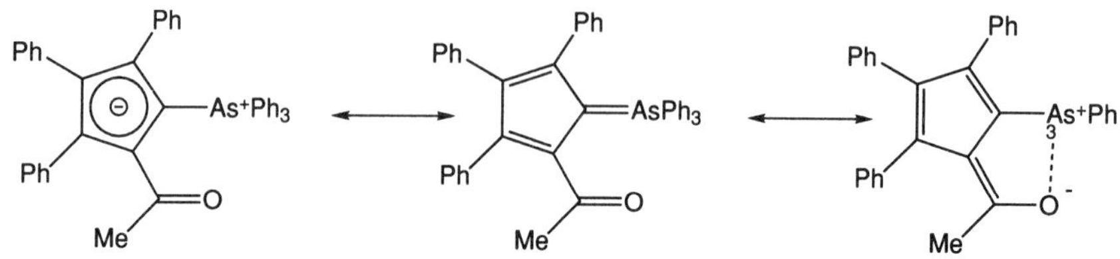

Scheme 5.23 Canonical forms of an arsonium ylide (centre).

The ylide is protonated at position 1 of the cyclopentadiene ring resulting in the loss of the ylide structure. The As . . . O distance increases to 3.10 Å, which is still less than the van der Waals contact distance of 3.4 Å [207].

Shen, Fan and Qui [208] calculated the bond order of the As=C unit as 1.6 in $Ph_3As{=}CHCOCF_3$, from the crystal structure. Huang *et al.* [209] cite crystal structure data to support an ylide structure as the major contributor to the resonance structure of $Ph_3AsCHCOPh$ (Scheme 5.24).

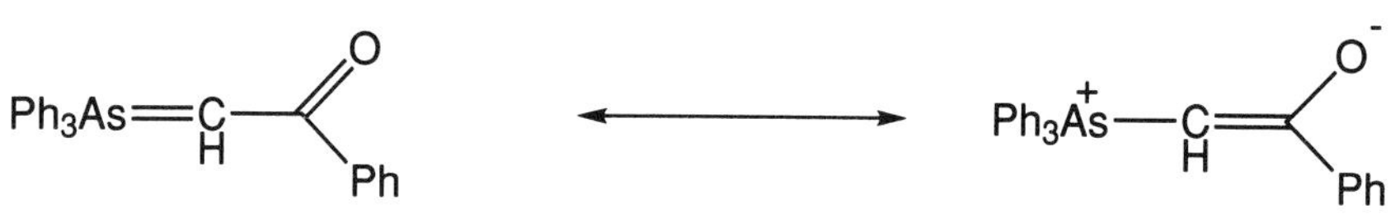

Scheme 5.24 The resonance structure of $Ph_3AsCHCOPh$.

Tetraorganoarsonium compounds, $[R_4As]^+X^-$. The tetraorganoarsonium species are normally classed as members of the series $[AsH_{4-n}R_n]^+$ ($n = 0, \ldots, 4$), although only the members of the $n = 4$ series will be discussed in this subsection. As with so much arsenic chemistry, the classical work on arsonium compounds is in need of revision and re-presentation. Arsonium salts are usually water soluble and the solids are very hygroscopic. Some of the classical data unwittingly refer to the hydrates of arsonium salts [210].

The quaternerization of tertiary arsines with methyl iodide to afford $[R_3AsMe]^+I^-$ salts as a means of characterizing arsines was and is widely used. Arsines, being basic compounds with an uncoordinated lone pair, are nucleophilic reagents, and they react with alkyl halides by attacking the carbon of the carbon–halogen bond. The reaction (S_N2) is facile with R_3As (R = alkyl), but as the alkyl groups are replaced by aryl groups (Ar) in the series $R_nAr_{3-n}As$, the arsines become much less reactive, Ph_3As being a hundredfold less reactive to alkyl halides than is Ph_3P. In a given aromatic arsine, $RC_6H_4AsPh_2$, the rate of reaction with alkyl halides is increased if R is electron-releasing and is decreased if it is electron-withdrawing. For a given halide the rates follow the order MeX > EtX >> ArX, vinylX, and, for a given alkyl group, for example methyl, the rates follow the order MeI > MeBr. Perfluoromethyl groups also slow the quaternization reaction. Thus Me_2AsCF_3 reacts with MeI under forcing conditions:

$$Me_2AsCF_3 + MeI \xrightarrow{85°C/14\ h} [Me_3AsCF_3]I \qquad [5.97]$$

but Ph_2AsCF_3, $MeAs(CF_3)_2$ and $As(CF_3)_3$ do not react with MeI [211].

The alternative approach to perfluoroarsonium compounds, using organoperfluorohalides and arsines, gave only decomposition products [212]:

$$AsPh_3 + CF_3I \xrightarrow{\text{Sealed tube}} CHF_3 + C_6H_6 + PhCF_3 + AsI_3 \qquad [5.98]$$

Trivinylarsine reacts with alkyl halides:

$$(CH_2{=}CH)_3As+RX \longrightarrow [(CH_2{=}CH)_3AsR]X \qquad [5.99]$$

but not with the vinyl halides, such as $CH_2{=}CHBr$ [213].

The rate of reaction of Et_2EPh (E = N, P, As) with MeI is in the ratio N:P:As = 1:520:70.

An exception to the reaction of MeI with tertiary arsine centres before tertiary amines was observed by Gavrilov *et al.* [214]. An interesting compound reversing the usual AsC_4 coordination to CAs_4 was prepared by Nyholm, Collinge and Tobe [215].

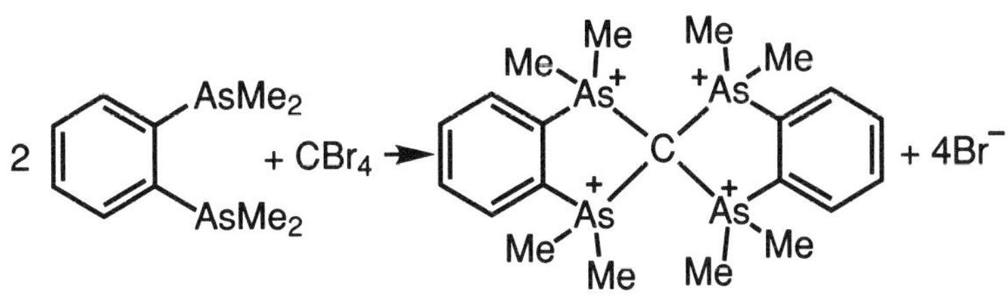

[5.100]

Tetraorganoarsonium salts have also been prepared with bulky organic ligands, for example Me_3SiCH_2 [216]. Aromatic arsonium salts are best prepared by the reaction between arsine oxides and Grignard reagents, addition of HX′ in the work-up giving the desired salt. The procedure given by Wolf and Shriner in *Organic Synthesis* [217] is claimed to yield $[Ph_4As]^+[ClHCl]^-$ with aqueous hydrochloric acid, although Harmon and Lake [218] demonstrated that in aqueous environments the isolated product is a dihydrate. If the anhydrous hydrogen dihalogen salt is required then anhydrous liquid HCl should be used [219].

$[Ph_4As]^+X^-$ salts are conveniently obtained by Friedel–Crafts reactions, and when potassium iodide is added to the reaction mixture the iodide is obtained. The C–X bond attacked may be part of the arsine, resulting in intramolecular quaternarization and in attempting to isolate $Cy_2AsCH_2CH_2Cl$ (Cy = cyclohexyl) {structure (i) in reaction [5.101]}, structure (iii) was obtained via structure (ii) [220]:

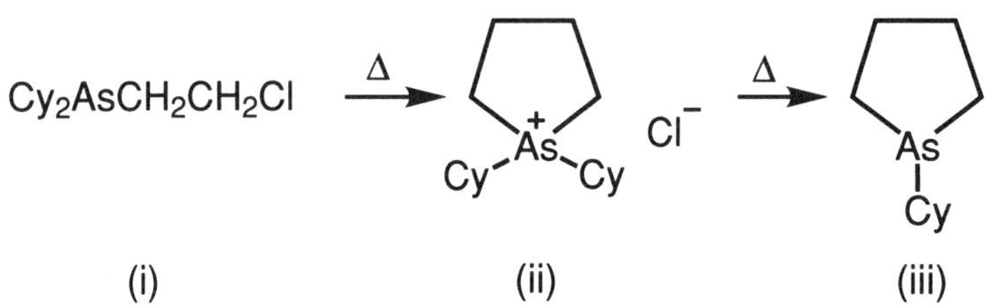

[5.101]

Mann, Hollimann and Thornton [221] used an intramolecular reaction to prepare the spiroarsonium bromide:

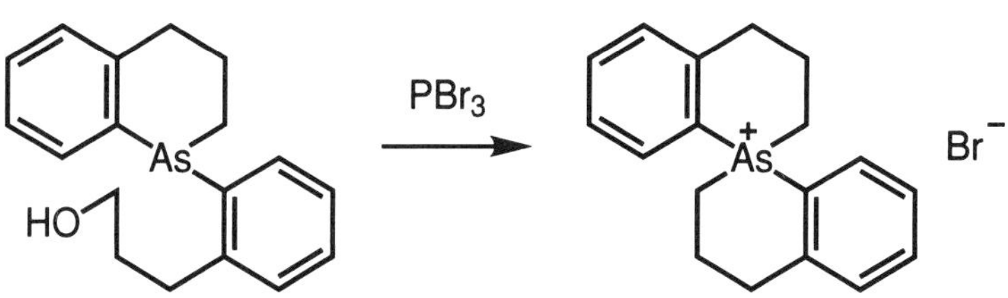

[5.102]

Intramolecular condensation reactions were also used by Wittig and Helliwinkel [222] to prepare spiroarsonium salts.

The anionic moiety of an arsonium salt may be bridged to the arsenic centre by a carbon chain forming a zwitterion, for example [223, 224]:

$$Me_3\overset{+}{As}CH_2\overset{-}{C}OO \quad \text{and} \quad Me_3\overset{+}{As}CH_2CH(OH)\overset{-}{C}OO$$

which occur as metabolites in some biological systems.

Quaternarization is in general reversible, but the leaving group is not necessarily the same as the entering group since factors such as the stability of the leaving group are decisive. The formation and thermal decomposition of arsonium salts is a useful synthetic route particularly in the formation of organoarsenic heterocycles [220]:

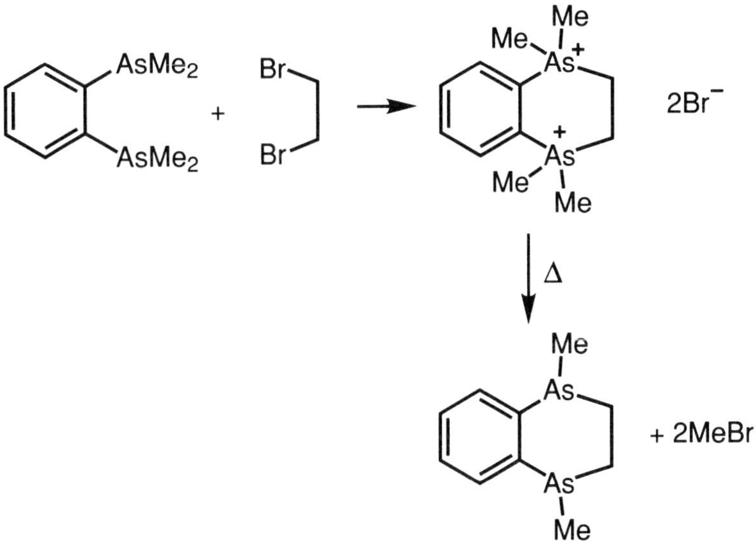

[5.103]

Physical data for selected arsonium salts are given in Table 5.13.

Substitution in arsonium ions leads to loss of symmetry, which in the case of R′ replacing R regains T_d symmetry on complete substitution:

$$\underset{T_d}{^{+}AsR_4} \rightarrow \underset{C_{3v}}{^{+}AsR_3R'} \rightarrow \underset{C_{2v}}{^{+}AsR_2R'_2} \rightarrow \underset{C_{3v}}{^{+}AsRR'_3} \rightarrow \underset{T_d}{^{+}AsR'_4}$$

If the substitution is by different groups, leading to [AsRR′R″R‴], then the symmetry changes with respect to arsenic are:

$$\underset{T_d}{AsR_4} \rightarrow \underset{C_{3v}}{AsR_3R'} \rightarrow \underset{C_s}{AsR_2R'R''} \rightarrow \underset{C_1}{AsRR'R''R'''}$$

The carbon ligands in the arsonium moiety are tetrahedrally disposed about arsenic with almost perfect T_d symmetry when the organo groups are identical, $[AsR_4]^+$, R = Me, Ph [225], and this symmetry is held in solution which in the case of $[Ph_4As]^+Cl^-$ in methanol has been shown spectroscopically [226].

The salts of large cations decrease in solubility as the size of the anion increases, and this together with their stability has led to the widespread use of arsonium cations as counter ions in the chemistry of complex anions. Müller [225] has systematically reviewed the structural implications of $[EPh_4]^+$ salts (E = P, As, Sb, Bi), and the change from $[PR_4]^+$ to $[AsR_4]^+$ can produce surprising changes in crystal structure. For example, the study by Anderson and Jagner [227] of the dihalocurpriate(I) anions using $[AsPh_4]^+$ and $[PPh_4]^+$ shows significant differences.

Table 5.13 Selected data on tetraorganoarsonium halides ($[R_4As]X$, X = halide): (a) $[R_4As]Cl$; (b) $[R_4As]Br$; (c) $[R_4As]I$.

Empirical formula	Compound	Melting point (°C)[a]	Reference
(a)			
$C_{19}H_{18}AsCl$	$[Ph_3AsMe]Cl$	118–120	[i]
$C_{19}H_{36}AsCl$	$[Cy_3AsMe]Cl$	93–94	[i]
$C_{24}H_{20}AsCl(2H_2O)$	$[Ph_4As]Cl{\cdot}2H_2O$	258–260[a]	[ii]
$C_{25}H_{22}AsCl$	$[Ph_3AsBz]Cl$	180–181	[iii, iv]
(b)			
$C_4H_{12}AsBr$	$[Me_4As]Br$	—	[v]
$C_8H_{20}AsBr$	$[Et_4As]Br$	308	[v, vi]
$C_{12}H_{28}AsBr$	$[Pr^n_4As]Br$	—	[vii]
$C_{17}H_{22}AsBr$	$[BzMeAsPr^nPh]Br$	—	[viii]
$C_{19}H_{18}AsBr$	$[Ph_3AsMe]Br$	193[b]	[i]
$C_{20}H_{20}AsBr$	$[Ph_3AsEt]Br$	—	[ix]
$C_{21}H_{20}AsBr$	$[Ph_3AsCH_2CH{=}CH_2]Br$	180–181	[iv, ix]
$C_{24}H_{20}AsBr$	$[Ph_4As]Br$	—	—
$C_{25}H_{22}AsBr$	$[Ph_3AsBz]Br$	168–169	[iii, iv]
(c)			
$C_4H_{12}AsI$	$[Me_4As]I$	333–336	[x]
$C_8H_{20}AsI$	$[Et_4As]I$	307	[vi, xi]
$C_9H_{14}AsI$	$[Me_3AsPh]I$	248–249	[i]
$C_{12}H_{28}AsI$	$[Pr^n_4As]I$	223	[xii]
$C_{14}H_{16}AsI$	$[Ph_2AsMe_2]I$	189–190; 209–211	[i]
$C_{19}H_{18}AsI$	$[Ph_3AsMe]I$	176	[xiii]
$C_{19}H_{42}AsI$	$[n\text{-}C_{16}H_{33}AsMe_3]I$	132	[xiv]
$C_{20}H_{20}AsI$	$[Ph_3AsEt]I$	157–158	[xi]
$C_{21}H_{20}AsI$	$[Ph_3AsCH_2{=}CH_2]I$	163–164	[iv]
$C_{24}H_{20}AsI$	$[Ph_4As]I$		[ii]
$C_{25}H_{22}AsI$	$[Ph_3AsBz]I$	168–170	[iii, iv]

[a]–H_2O at 100°C.
[b]Dissociates.

References

- i Gibson, N. A., Hosking, J. W. and Weatherburn, D. C. (1967) *Aust. J. Chem.*, **20**, 2771.
- ii Wolf, C. N. and Shriner, R. L. (1939) *Org. Syn. Coll.*, **4**, 910.
- iii Schiemenz, G. P. (1973) *J. Organomet. Chem.*, **52**, 349.
- iv Blick, F. F., Willard, H. H. and Taras, J. T. (1939) *J. Am. Chem. Soc.*, **61**, 88.
- v Collins, E., Sutor, D. J. and Mann, F. G. (1963) *J. Chem. Soc.*, 4051.
- vi Baig, M. M. and Cullen, W. R. (1961) *Can. J. Chem.*, **39**, 420.
- vii Balimann, G. and Pregosin, P. S. (1975) *Helv. Chim. Acta.*, **58**, 1913.
- viii Horner, L. and Fuchs, H. (1963) *Tetrahed. Lett.*, 1573.
- ix Horner, L. and Ernst, M. (1970) *C. Ber.*, **103**, 318.
- x von Doering, W. and Hoffmann, A. K. (1955) *J. Am. Chem. Soc.*, **77**, 521.
- xi Zykova, T. V., Kamai, G. K., Chernokalskii, B. D. *et al.* (1971) *J. Gen. Chem. USSR.*, **41**, 1049, 1513.
- xii Abalonin, B. E., Kamai, G. K., Chernokalskii, B. D. *et al.* (1970) *J. Gen. Chem. USSR.*, **40**, 1796.
- xiii Deacon, G. B. and Jones, R. A. (1963) *Aust. J. Chem.*, **16**, 499.
- xiv Jerchel, D. (1943) *C. Ber.*, **76**, 600.

Table 5.14 Selected arsoranes (AsR_5).

Empirical formula	Compound	Melting point (°C)	Reference
$C_5H_{15}As$	Me_5As	−6 to −7	[i]
$C_{30}H_{20}AsCl_5$	$(4\text{-}ClC_6H_4)_5As$	149	[ii]
$C_{30}H_{25}As$	Ph_5As	139.5	[iii]
$C_{35}H_{35}As$	$(4\text{-}MeC_6H_4)_5As$	139–140	[ii]

References

i Eberwein, B., Ott, R. and Weidlein, J. (1977) *Z. Anorg. Allg. Chem.*, **431**, 95.
ii Hellwinkel, D. and Kilthau, G. (1968) *C. Ber.*, **101**, 121.
iii Wittig, G. and Torsell, E. K. (1953), *Acta Chem. Scand.*, **7**, 1293.

Molecular dimensions for $[(Me_4)As]Br$ are d(As−C) 1.86 Å (average) and C−As−C 107°–112° (109.5° average) [228], and for Ph_4AsI d(As−C) 1.95 Å and C−As−C 109° [229].

5.3.3 *Five-coordinate pentavalent organoarsenic compounds*

Pentavalent organoarsenic compounds were first prepared by Wittig and Torsell in the 1950s [230]. All pentaalkyl arsoranes except $AsMe_5$ are unstable with respect to β-elimination. Siginficantly, the attempted preparation of Ph_4AsMe resulted in the ylide $Ph_3As{=}CH_2$ [221]. A few examples of the arsoranes, AsR_5, are given in Table 5.14.

The arsoranes are prepared from arsenic(V) compounds such as R_3AsO and R_3AsX_2 (X = halogen) and an organometallic synthon, for example organolithium compounds [232]:

$$Me_3AsCl_2 + 2MeLi \xrightarrow{Me_2O/-60°C} Me_5As + LiCl \qquad [5.104]$$

Improved yields are obtained with use of arsenic tosylimines, which also give access to cyclic arsoranes [190].

A trigonal bipyramidal structure has been proposed for Me_5As based on spectroscopic evidence [232], but the methyl groups are magnetically equivalent on the ^{1}H NMR time scale down to −95°C. The equatorial and axial methyl groups are thought to be rapidly exchanging via a Berry turnstyle process. The structure of Ph_5As proved difficult to resolve by X-ray crystallographically; however, the hexane solvate $Ph_5As{\cdot}0.5C_6H_{12}$ was shown to have an undistorted trigonal bipyramidal geometry (D_{3h}) [233]. Vibrational solution spectra indicate retention of the structure in solution [224] but that interconversion of axial and equatorial sites is rapid (^{13}C NMR) down to 173 K [235].

Spiroarsoranes. Spiroarsoranes have a core structure of spiro five-membered rings as shown in Scheme 5.25. Spiroarsoranes are conveniently prepared from bis(biphenyl)chloroarsoranes via spiroarsonium salts:

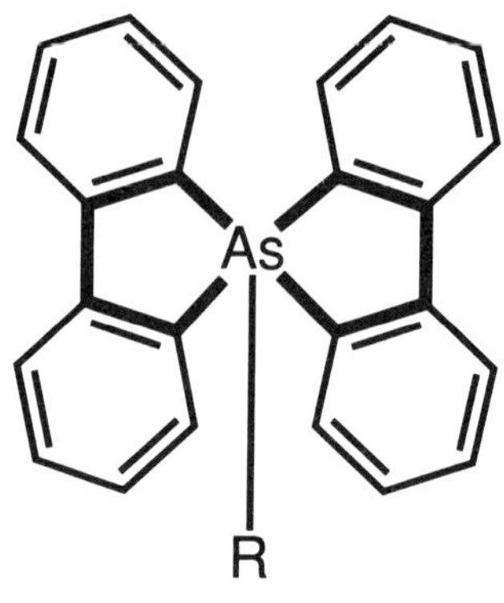

Scheme 5.25 General structural formula of a spiroarsorane.

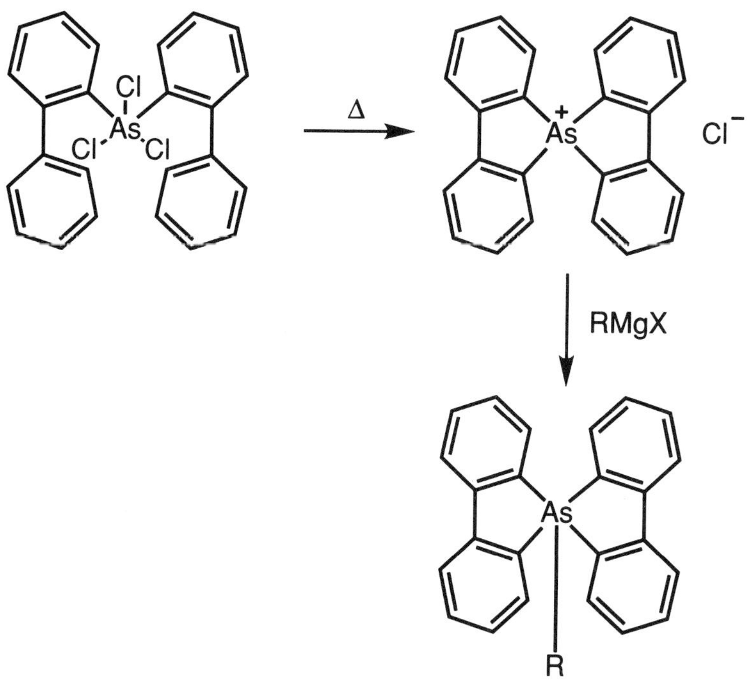

[5.105]

A selection of reactions is given in Scheme 5.26.

The parent compound where R = H has proved to be elusive despite the stability of the spirocyclic structure. Reduction of bis-(2′2′biphenylene)arsonium iodide affords mainly a dimer and 2-biphenyl-2,2′-biphenylenearsine [236]. The spiroarsorane trigonal bipyramidal structure is less rigid than the phosphorane analogues as shown by the somewhat

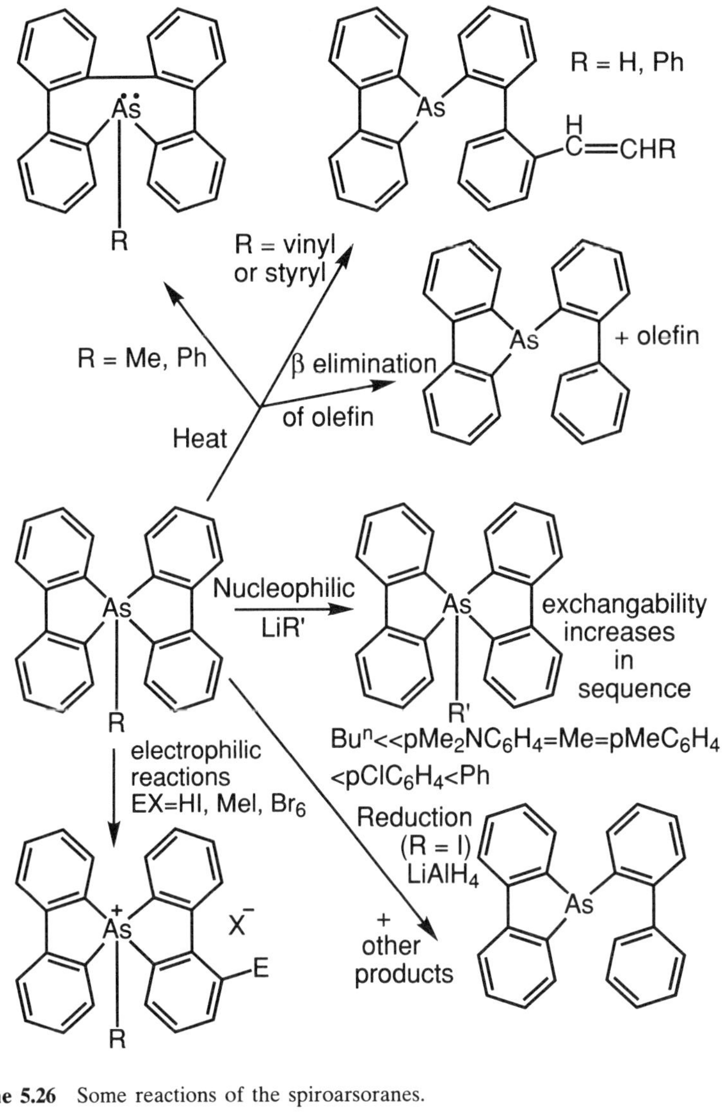

Scheme 5.26 Some reactions of the spiroarsoranes.

lower pseudo-rotation barriers which have been ascribed to the longer As–C bond.

5.3.4 Five-coordinate pentavalent organoantimony compounds

The antimony analogues of the previously discussed arsenic compounds are synthesized by broadly similar routes [237]:

$$Me_3SbCl_2 + 2MeLi \xrightarrow{Et_2O} Me_5Sb + 2LiCl \qquad [5.106]$$

D_{3h} symmetry is indicated by the IR and Raman spectra of the neat liquid [238], and the ^{1}H NMR spectra [239] in carbon disulphide solution shows only one narrow line down to 173 K, indicating the equivalence of the methyl groups on the NMR time-scale. This is interpreted as a rapid exchange between axial and equatorial methyl groups via the Berry turnstyle mechanism [240].

The D_{3h} symmetry was confirmed in the gas phase by electron diffraction [241] [Fig. 5.1(a)] and *ab initio* molecular orbital calculations indicate the C_{4v} model is only 12.1 kJ mol^{-1} higher in energy than the D_{3h} model. In contrast, pentaphenylantimony was shown [242] to have C_{4v} symmetry by X-ray crystallography with apical Sb–C bond lengths 2.12 Å and basal Sb–C bond lengths 2.22 Å [Fig. 5.1(b)]. Evidence has been presented indicating that the square-based pyramidal structure of Ph_5Sb is retained in solution [243]. However, a ^{13}C NMR study showed that the α-carbons are magnetcially equivalent down to 173 K, implying a non-rigid structure.

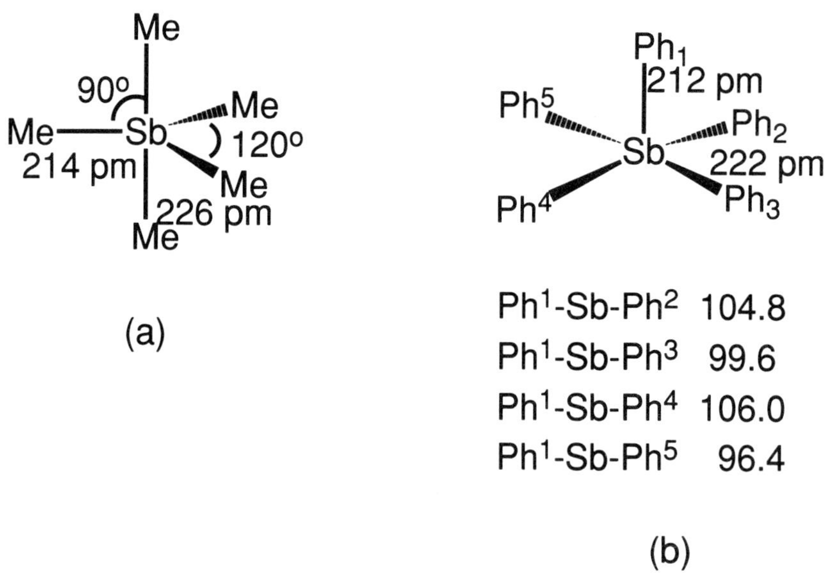

Figure 5.1 (a) The structure of pentamethylantimony, deduced from gas phase electron diffraction data. Source: Pulham, C., Haaland, A., Hammel, A. *et al.* (1992) *Angew. Chem. Int. Ed. Engl.*, **31**, 1464. (b) The structure of pentaphenylantimony.

5.3.5 *Six-coordinate pentavalent organoarsenic compounds*

The hexaarylarsenate anion is formed on treating spiroarsonium salts with 2,2′-dilithio-biphenyls in ether [169], although the anion is moisture sensitive and is decomposed by alcohols. The octahedral structure of this chiral anion was demonstrated by an optical resolution via its stable methylbucinium salt. The optical activity is lost when the anion is hydrolysed to the spiroarsorane, showing the non-rigid fluxional nature of this molecule.

5.4 Some uses of organoarsenic and antimony compounds in organic synthesis

The following brief description of the uses of arsenic and antimony compounds in organic synthesis is divided into two sections based on the tervalent and pentavalent species.

5.4.1 *Arsenic ylides and related compounds*

The molecule-building step of converting a ketone to an olefin ($RR'C{=}O \rightarrow RR'C{=}CR''R'''$) is widely used in organic synthesis, the usual 'method of choice' to effect this transformation being the Wittig synthesis which was originally developed with phosphoranes of the type $R_3P{=}CR''R'''$:

$$RR'C{=}O + Ar_3P{=}CR''R''' \rightarrow RR'C{=}CR''R''' + Ar_3PO \qquad [5.107]$$

The range of reagents was extended to include arsenic and antimony when it was found that they had in many cases advantages over their phosphine analogues; for example, arsonium ylides bearing an electron-withdrawing substituent on the alkylidene moiety have a greater activity than do their phosphorane analogues [198, 244–247].

The arsonium ylide $Ph_3As{=}CHCH_2CH(OPr^i)_2$ is available for the addition of the CH=CHCHO moiety and this useful synthon is itself prepared from an arsonium ylide [247]:

$$\begin{array}{c} Ph_3As{=}CH_2 + ICH_2CH(OPr^i)_2 \rightarrow [Ph_3As^+CH_2CH_2CH(OPr^i)_2]I^- \\ \downarrow \\ Ph_3As{=}CHCH_2CH(OPr^i)_2 \end{array} \qquad [5.108]$$

An application in terpene synthesis is shown in reaction [5.109] [247]:

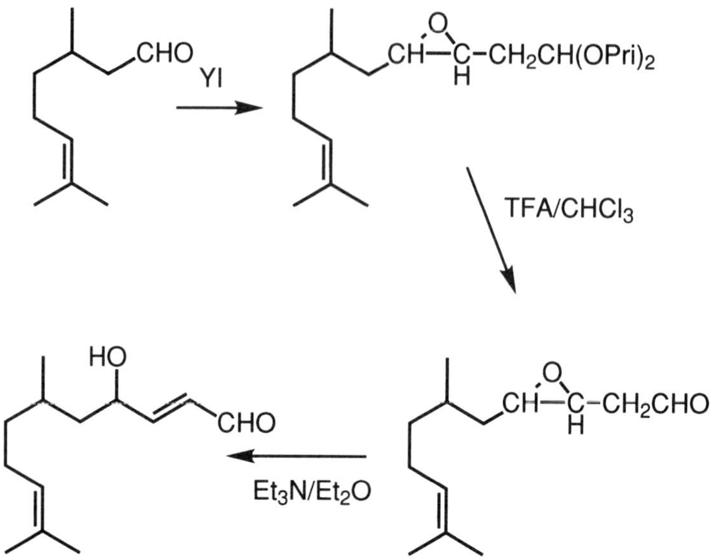

[5.109]

A metabolite of arachidonic acid (hepoxilin A_3) has been synthesized [248], further illustrating the use of an arsonium ylide in organic synthesis:

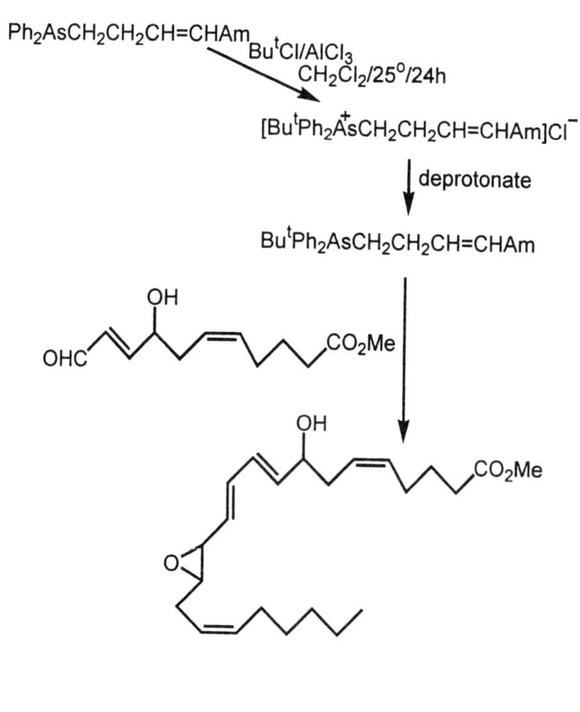

[5.110]

Triphenylarsinimines react under mild conditions with carbonyl compounds to afford the corresponding Schiff's bases in high yields [249–251]:

$$Ph_3As{=}NR + RCHO \longrightarrow RC(H){=}NR \qquad [5.111]$$

the arsinimines being prepared by the reaction of Ph_3AsO with organoisocyanates:

$$Ph_3AsO + RNCO \longrightarrow Ph_3As{=}NR + CO_2 \qquad [5.112]$$

Froyen [252] has shown that these reactions can be combined, that is Ph_3AsO need only be added in catalytic amounts to afford Schiff bases from organoisocyanates and carbonyl compounds:

$$RCHO + R'NCO \longrightarrow RC(H) = NR' + CO_2 \qquad [5.113]$$

In the case of quinones, the classical synthesis of ketimines from the appropriate quinone and RNH_2 fails [253–255]. However, using RNCO, the ketone precursor and Ph_3AsO in catalytic amounts monoketimines or diketimines are produced. In the case of 9,10-phenanthrenequinone when a vinylisocyanate is used an intramolecular rearrangement occurs resulting in 2H-phenanthro(9,10)-1,4-oxazines.

In principle, diisocyanates and diketones should afford high polymers in the presence of Ph_3AsO, and Froyen [252] has obtained an uncharacterized polymer from terephthaldehyde and phenylene-1,3-diisocyanate.

Froyen and Juvvik [253] have described a convenient spot test for ylides (N, P, As, S) based on their reaction with 2,4-dinitrobenzenediazonium tetrafluoroborate (2,4-dnbt) to form intensely coloured hydrazones. This test is useful in synthesis for monitoring the consumption or formation of ylides and in analytical techniques such as column chromatography and thin-layer chromatography (TLC). The colours range from red-orange through bright-red to various dark-red shades, depending on the nature of the CRR′ group. These colours are readily detected with ylide concentrations down to 0.3–0.5 $\mu g\ \mu l^{-1}$.

5.4.2 *Anchored-arsenic and arsenic-containing polymers*

There are often considerable advantages in 'anchoring' an active species in a homogeneous catalytic system to a polymer [254], *inter alia* increased selectivity, retention of expensive metals and reduced toxicity. Typically, a halogenated styrene/divinylbenzene copolymer is functionalized by one of two processes, and these arsenical polymers may be used to form an anchored complex with the desired transition metal, for example iridium [255].

The main difficulty in forming arsenic(III) organopolymers directly is that the arsenic monomers act as chain transfer agents reducing the molecular weight of the polymer. Hay [256] synthesized a low molecular weight polymer which forms brittle films, but the molecular weight was increased by the addition of a Cu(I)Cl catalyst in the presence of dioxygen.

Anchored-arsenic chemistry is not as well developed as that of phosphorus, as described by, for example, Akelah and Moet [257]. However, in general, most systems cited for phosphorus could be developed for the arsenic analogue.

5.4.3 Antimony–Wittig and allied reactions

Stibine oxides, R_3SbO, have been shown to catalyse the conversion of organoisocyanates to carbodiimides or to oligomers. The product obtained depends on the stibine substituent R groups. For R = alkyl the product is a trimer (RNCO $\rightarrow$ $[RNCO]_3$) [258]. When the substituents are aryl, for example Ph_3SbO, the reaction is much more complex, being mediated by the inferred ylide-like intermediate $Ph_3Sb{=}NPh$ [259]:

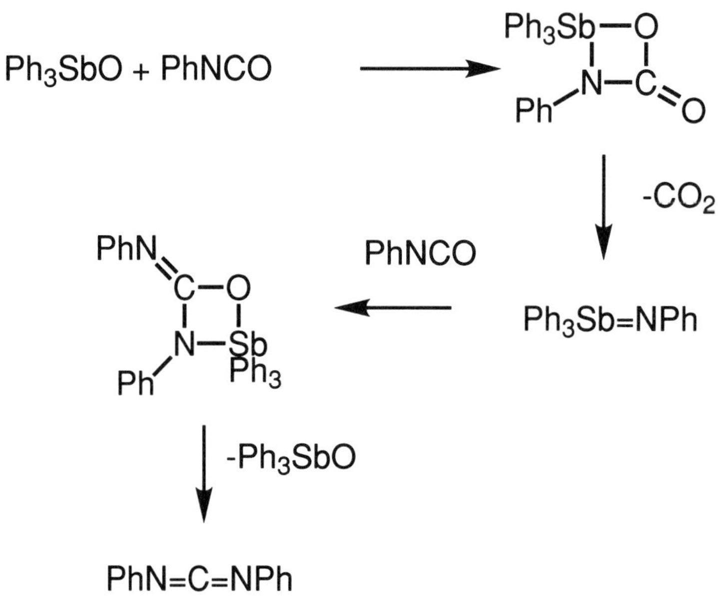

[5.114]

Wittig and Laib's first attempts to prepare the antimony analogues of the phosphoranes failed [260], and although Wittig and Henry [261] later synthesized $Ph_3Sb{=}CH_2$ it did not yield the target molecule but gave rearranged starting material.

Attempts to prepare stiboranes with electron-withdrawing substituents also failed; for example $PhCOCH_2Br$ with R_3Sb (R = alkyl or aryl group) gave only debrominated products [262].

Huang, Shen and Chen [263] harnessed the instability of the stiboranes by using a stibine *in situ* through consuming any intermediate stiborane as it was formed to produce the target olefin:

$$R^1R^2C{=}O \xrightarrow{BrCH_2CO_2R^3/Bu_3{}^nSb} R'R^2C{=}CHCO_2R^3 \qquad [5.115]$$

where R' = alkyl, Ar, Vy; R^2 = alkyl, H; R^3 = Me, Et. This is a one-pot reaction which may be carried out with or without solvents such as MeCN, thf, hexane and toluene.

Singer and Lloyd [264] were the first to characterize a stibonium ylide in thc solid state and use it to form an olefin in high yield.

5.4.4 The uses of dihalogen arsoranes and stiboranes

The thermolysis of dihalogen arsoranes and stiboranes affords $R_2E(III)X$ and usually RX, although these products may themselves further decay:

$$R_3EX_2 \xrightarrow{\Delta} R_2EX + RX \qquad [5.116]$$

This reaction is used as a route to R_2EX compounds (Section 5.2.2) and less commonly to prepare RX.

The thermal stability of the Ph_2Sb moiety has been used in the synthesis of RCH_2X (R = alkyl; X = Br, I), that is, $RX \rightarrow RCH_2X$. The methodology, which has mainly been developed by Kauffmann and co-workers [265–267], is as follows:

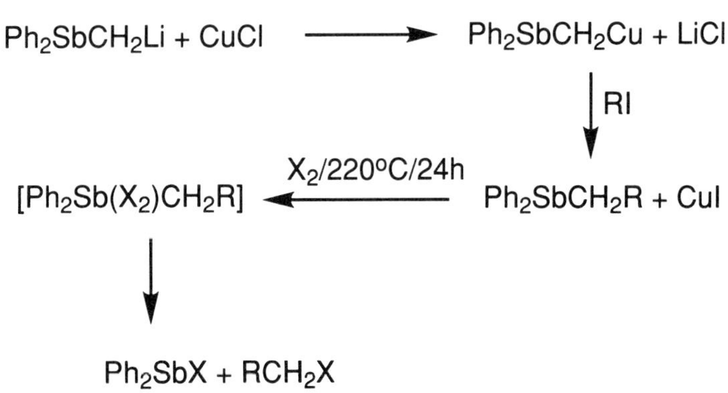

[5.117]

One-pot syntheses are always of interest, so the new synthetic strategy, developed by Godfrey and co-workers, is included here, although strictly speaking it is an organometallic process. The normal decomposition of R_3EX_2 compounds affords R_2EX and RX, as discussed above. In their studies on the preparation and properties of R_3EX_2 (R = aryl, alkyl group;

E = group 15 element P, As, Sb, Bi; X = Cl, Br, I) compounds it was found that the addition of metal powder induced a new decomposition pathway. The group 15 element was reduced and the metal oxidized, followed usually by the formation of a complex, often of novel structure. In a wide-ranging series of papers [268–281] the reaction was shown to be general for transition metals, lanthanides and p-block metals:

$$aR_3EX_2 + bM \longrightarrow (R_3E)_xM_yX_z \quad [5.118]$$

However, the stoichiometry is often complex and it is not yet easy to predict the products except in a few cases. Three examples are given below for the formation of arsenic and antimony donor complexes:

$$Ph_3SbI_2 + Co \rightarrow CoI_3(SbPh_3)_2 + \tfrac{1}{2}I_2 \quad [5.119]$$

$$Me_3AsI_2 + Ni \rightarrow NiI_3(AsMe_3)_2 + \tfrac{1}{2}I_2 \quad [5.120]$$

$$3Me_3AsI_2 + 2Au \rightarrow 2AuI_3(AsMe_3) + AsMe_3 \quad [5.121]$$

5.5 Multiple bonds in organoarsenic compounds

This section is divided into three parts. First, arsenic–carbon multiple bonds, that is, compounds of type As≡CR are described; second, compounds of type RAs=AsR are described; and, finally, compounds of type $RAs{=}CR_2$ are described.

5.5.1 One-coordinate organoarsenic compounds

The first stable one-coordinate organoarsenic compound, 2-(2,4,6-$Bu^t_3C_6H_2$)-1-arsaethyne, was prepared by Märkl and Sejpka [282] as follows:

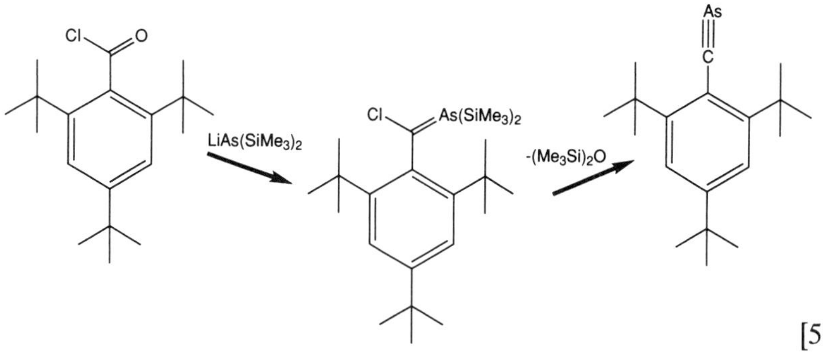

[5.122]

5.5.2 Compounds with arsenic–arsenic double bonds

In the early part of this century it was common practice to assign structures to compounds of formula $(RAs)_n$ (where n was not known) of RAs=AsR, for example Erlich's (1907) 'Salvarsan 606' [Scheme 5.27(a)], an early chemotherapeutic agent for syphilis [3, 283]. In the 1960s, with the elucidation of the structures of cycloarsines, for example $(PhAs)_6$ [284, 285], the earlier claims to have synthesized arsenic–arsenic double bonds were viewed with scepticism. The consensus view was expressed as the so-called 'classical double bond rule' – the assertion that elements with principal quantum number greater than two are unable to form stable pπ–pπ bonds with themselves or with other elements [286]. However, although free-standing RAs=AsR molecules could not be synthesized several workers used coordination to transiton metals to stabilize these systems, for example Elms, Leverett and West [287] isolated $[Fe(CO)_4\{As(C_6F_5)\}_2]$ with As–As = 2.388 Å (cf. As–As single bond length 2.46 Å) [Scheme 5.27(b)] and thus reasonably supposed this molecule to contain an As–As bond order greater than one.

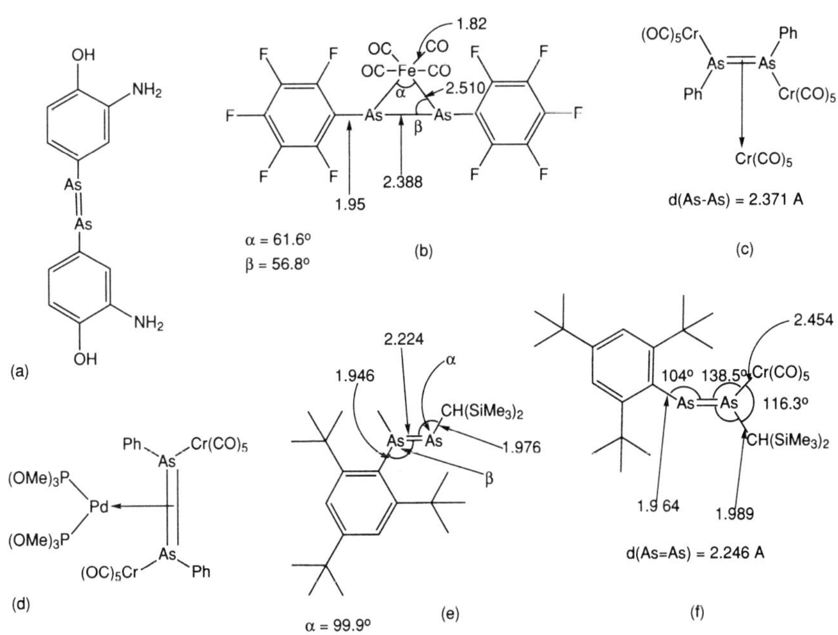

Scheme 5.27 Selected organoarsenic compounds with arsenic–arsenic double bonds. Individual structures (a)–(f) are referred to in text, Section 5.5.2.

Huttner *et al.* [288] also developed a series of complexed arsenic double bond compounds with As−As length of *c.* 2.37 Å [Scheme 5.27(c) and 5.27(d)].

The situation was transformed in the early 1980s by Yoshifuji *et al.* [289] with phosphorus–phosphorus multiple bonding and later by Cowley's group [290, 291]. They discovered that, by employing serically demanding ligands for pnictides, they were able to stabilize the two-coordinate state, that is, to form a pnictide–pnicide double bond with itself or with another group 15 element. The arsenic–arsenic double bonds are kinetically stabilized since the mechanism to oligomerization is blocked. The compounds are prepared by condensing a sterically hindered primary arsine with a sterically hindered dichloroarsine in the presence of a base:

$$(Me_3Si)_2CHAsCl_2 + H_2AsR \xrightarrow{\text{DBU}} (Me_3Si)_2CHAs{=}AsR \qquad [5.123]$$

(DBU = 1,5-diazabicyclo[5,4,0]undec-5-ene; $R = 2,4,6\text{-Bu}^t{}_3C_6H_2$). [The product is also illustrated in Scheme 5.27(e)].

The arsenic–arsenic bonding is not thought to be olefinic (i.e. sp^2), but more likely to be lacking any hybridization owing to the emergence of the inert s-pair effect on going down group 15.

Couret *et al.* [292, 293] have developed an alternative synthesis:

$$2(Me_3Si)_3CAsCl_2 + Bu^tLi \longrightarrow (Me_3Si)_3CAs{=}AsC(SiMe_3)_2 \qquad [5.124]$$

but these compounds have not been as well characterized as those of Cowley *et al.* [290, 291].

The formation of an As=As bond in the coordination sphere of transition metal results in η^2-coordination, and this side-on coordination results in As−As bond lengthening. It may be anticipated that a polymer backbone could be one of the arsenic's bulky 'substituents', enabling an As=As system to be built into a polymer. Huttner and Jibril [294] have demonstrated mixed-metal coordination to RAs=AsR ligands, and thus sophisti-cated catalyses with two active centres may be developed.

There are several useful reviews of arsenic–arsenic double bonds and their transition metal complexes [e.g. the structure in Scheme 5.27(f)] [295–297].

5.5.3 *Compounds with arsenic–carbon double bonds*

The arsenic–carbon (4p, 2p)π bond has attracted much attention in recent years for its utility, for example in ylide synthesis, and for its theoretical interest.

The analogous arsenic(III) compounds require bulky substituents on carbon to stabilize them. Several cationic species with a C−As=C moiety in the proposed structures were isolated by Märkl and Sejpka [282]

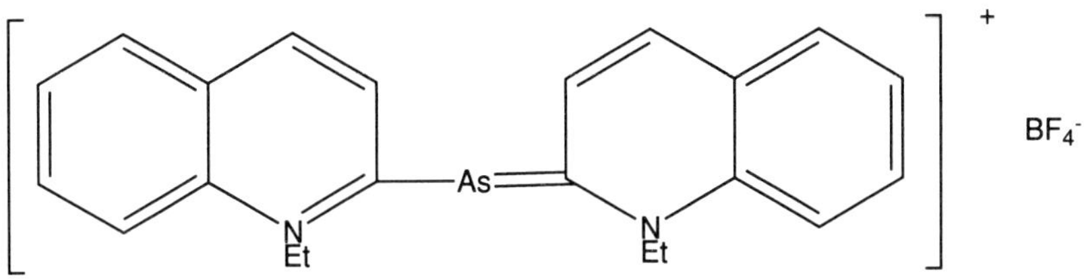

Scheme 5.28 Cationic species with a C–As=C moiety as isolated by Märkl and Sejpka. Source: Märkl, G. and Sejpka, J., *Angew. Chem. Int. Ed. Engl.*, **25**, 262, 1986.

(Scheme 5.28), but the compound Ph–As=C(CMe_3)$OSiMe_3$ (a reddish liquid, b.p. 82°C–86°C at 0.5 mm pressure) has the distinction of being the first neutral molecular compound isolated:

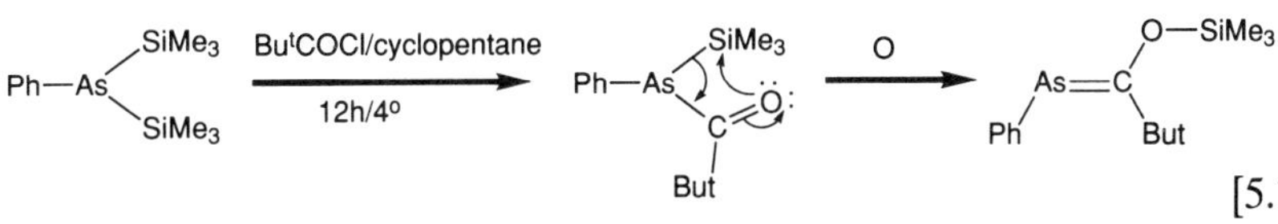

[5.125]

Either the arsenic or the carbon atom of the As=C group may have the stabilizing bulky substituent (Scheme 5.29). Selected data of compounds with arsenic–arsenic double bonds are given in Table 5.15.

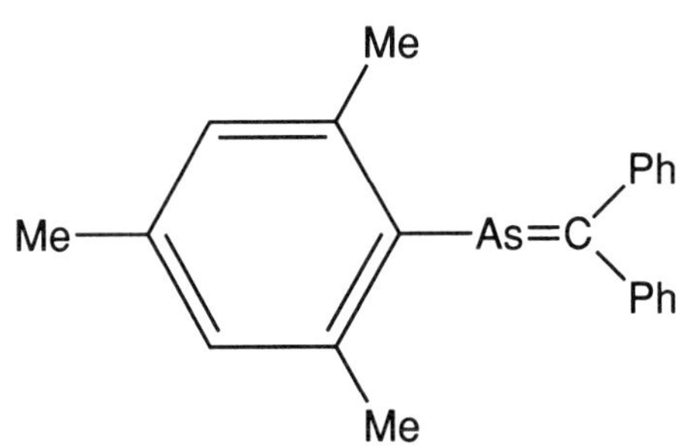

Scheme 5.29 An example of a stabilizing bulky substituent.

Table 5.15 Arsenic–arsenic double bonds and selected transition metal complexes.

Compound	As=As bond length (Å)	As=As−R bond angle (°) [a]	Reference
$R'As{=}AsCH(SiMe_3)_2$ [b]	2.224	93.6 (CH) 99.9 (Ar)	[i]
$(Me_3Si)_3CAs{=}AsC(SiMe_3)_3$	2.244	106.3[c]	[ii]
$R'As{=}AsCH(SiMe_3)_2]Cr(CO)_5$ [b]	2.246	—	[iii]
$\{(PhAs{=}AsPh)[Cr(CO_5]_3\}$	2.371	—	[iv]
$[(C_6F_5As{=}AsC_6F_5)Fe(CO)_4]$	2.388	56.8[d]	[v]

[a]R is shown in parentheses. Ar = aryl group.
[b]$R' = 2,4,6\text{-}Bu^t_3C_6H_2$.
[c]Average.
[d]As−Fe−As.

References

i Yoshifuji, M. Shima, I., Inamoto, N. *et al.* (1981) *J. Am. Chem. Soc.*, **103**, 457.
ii Cowley, A. H. and Norman, N. C. (1986) *Prog. Inorg. Chem.*, **34**, 1.
iii Norman, N. C., Cowley, A. J., Lasch, J. G. and Pakulski, M. (1983) *Angew. Chem. Int. Ed. Engl.*, **22**, 978.
iv Huttner, G., Schmid, H.-G., Frank, A. and Orama, O. (1976) *Angew. Chem. Int. Ed. Engl.*, **15**, 234.
v Elms, P. S., Leverett, P. and West, B. O. (1971) *J. Chem. Soc., Chem. Commun.*, 747.

References

1. McAuliffe, C. A. (1987) in *Comprehensive Coordination Chemistry* (ed. G. Wilkinson), Pergamon Press, Oxford, Ch. 14.
2. Doak, G. O. and Freedman, L. D. (1970) *Organometallic Compounds of Arsenic, Antimony, and Bismuth*, John Wiley, New York.
3. Dub, M. (1968) *Organometallic Compounds*, 2nd edn, vol. III, Springer, New York.
4. Levason, W. and McAuliffe, C. A. (1979) *Phosphine, Arsine and Stibine Complexes of the Transition Elements*, Elsevier, Amsterdam.
5. Patai, S. and Rapport, Z. (eds) (1994) *The Chemistry of Organic Arsenic, Antimony and Bismuth Compounds*, John Wiley, Chichester, Sussex.
6. Buckingham, J. (ed.) (1984) *Dictionary of Organometallic Compounds*, Chapman & Hall, London.
7. King, R. B. and Eisch, J. J. (eds) (1986) *Organometallic Synthesis*, vol. 3, Elsevier, Oxford.
8. Millar, I. T., Heaney, H., Heinekey, D. M. and Ferrnelius W. C. (1960) *Inorg. Synth.*, **6**, 113.
9. Feltham, R. D., Kasenally, A. and Nyholm, R. S. (1967) *J. Organomet. Chem.*, **7**, 285.
10. Barker, R. L., Booth, E., Jones, W. E. *et al.* (1949) *Chem. Ind.*, **68**, 285, 289, 295.
11. Jones, E. R. H. and Mann, F. G. (1955) *J. Chem. Soc.*, 4472.
12. Levason, W., McAuliffe, C. A., Plaza, A. I. and Grim, S. O. (1976) *Inorg. Synth.*, **16**, 184.
13. Gough, G. A. and King, H. (1928) *J. Chem. Soc.*, 2426.
14. Sommer, K. (1970) *Z. Anorg. Allgem. Chem.*, **376**, 150.
15. Chatt, J. and Mann, F. G. (1939) *J. Chem. Soc.*, 610.
16. Bosnich, B., Jackson, W. G. and Wild, S. B. (1973) *J. Amer. Chem. Soc.*, **95**, 8269.
17. Dutta, R. L., Meek, D. W. and Busch, D. H. (1970) *Inorg. Chem.*, **9**, 1215.
18. Morgan, G. T. and Davies, G. T. (1926) *Proc. Roy. Soc. A.*, **110**, 523.
19. Hartmann, H. and Kühl, G. (1956) *Angew. Chem.*, **68**, 619.
20. Hartmann, H. and Kühl, G. (1969) *Z. Anorg. Allgem. Chem.*, **312**, 186.
21. Breunig, H. J. and Kanig, W. (1982) *Phos. Sulph.*, **12**, 149.

22. Meinema, H. A., Martcns, H. F. and Noltes, J. G. (1973) *J. Organomet. Chem.*, **51**, 223.
23. Schmidt, H. (1920) *Liebig's Ann. Chem.*, **421**, 174.
24. Cook, B. R., McAuliffe C. A. and Meek, D. W. (1972) *Inorg. Chem.*, **10**, 2676.
25. Isslieb, K. and Hamann, B. (1966) *Z. Anorg. Allgem. Chem.*, **343**, 196.
26. Maier, L. (1972) *Organic Phosphorus Compounds,*, vol. 1, John Wiley, New York, p. 289.
27. Doak, G. O. and Freedman, L. D. (1974) *Synthesis,*, 328.
28. Levason, W., McAuliffe, C. A., Barth, R. C. and Grim, S. C. (1976) *Inorg. Synth.*, **16**, 188.
29. Hewertson, W. and Watson, H. R. (1962) *J. Chem. Soc.*, 1490.
30. Aguair A. M., and Archibald, T. G. (1967) *J. Org. Chem.*, **32**, 2627.
31. Tzschach, A. and Lange, W. (1962) *Chem. Ber.*, **95**, 1360.
32. Ellerman, J. and Dorn, K. (1967) *Chem. Ber.*, **100**, 1230.
33. Kuchen, W. and Buckwald, H. (1959) *Chem. Ber.*, **92**, 227.
34. Zorn, H., Schindbauer, H. and Hammer, D. (1967) *Monatsch.*, **98**, 731.
35. Levason, W., McAuliffe, C. A. and Murray, S. G. (1975) *J. Organomet. Chem.*, **88**, 171.
36. Rossi, R. A. and de Rossi, R. H. (1983) Aromatic Substitution by the $S_{RN}1$ mechanism, *ACS Monograph*, **178**.
37. Rossi, R. A., Alonso, R. A. and Palacios, S. M. (1981) *J. Org. Chem.*, **46**, 2498.
38. Rossi, R. A. Alonso, R. A. (1982) *J .Org. Chem.*, **47**, 77.
39. Rossi, R. A. (1982) *Acc. Chem. Res.*, **15**, 164.
40. Rossi, R. A., Alonso, R. A. and Bornancini, E. R. (1984) *J. Organometal. Chem.*, **270**, 177.
41. Feltham, R. D. and Silverthorn, W. (1967) *Inorg. Synth.*, **10**, 159.
42. Tzschach, A. and Nindel, H. (1970) *Z. Chem.*, **10**, 118.
43. Matsumura, Y. and Okawara, R. (1969) *Inorg. Nucl. Chem. Letts.*, **5**, 449.
44. Isslieb, K. and Hamann, B. (1964) *Z. Anorg. Allgem. Chem.*, **332**, 179.
45. Isslieb, K. and Hamann, B. (1965) *Z. Anorg. Allgem. Chem.*, **339**, 289.
46. Breunig, H. J., Ates, M., Mourad, M. and Mugnier, Y. (1990) *J. Organomet. Chem.*, **398**, 85.
47. Sato, S., Matsumura, Y. and Okawara, R. (1972) *J. Organomet. Chem.*, **43**, 333.
48. Isslieb, K. and Kummel, R. (1965) *J. Organomet. Chem.*, **3**, 84.
49. Tzschach, A. and Pacnolke, G. (1964) *Chem. Ber.*, **97**, 419.
50. Isslieb, K. and Krech, F. (1961) *Chem. Ber.*, **94**, 2656.
51. Mann, F. G. and Pragnell, M. J. (1965) *J. Chem. Soc.*, 4120.
52. Claeys, E. G. (1966) *J. Organomet. Chem.*, **5**, 446.
53. Durand, M. and Laurent, J. P. (1965) *C. R. Acad. Sci., C,*, **261**, 3793.
54. Dyke, W. J. C. and Jones, W. J. (1930) *J. Chem. Soc.*, 2426.
55. Seifter, J. (1939) *J. Amer. Chem. Soc.*, **61**, 530.
56. Schriner, R. L. and Wolf, C. N. (1950) *Org. Synth.*, **30**, 95.
57. Burrows, G. J. and Turner, B. B. (1920) *J. Chem. Soc.*, 1337.
58. Phillips, J. R. and Vis, J. H. (1967) *Can. J. Chem.*, **45**, 675.
59. Dyke, W. J. C., Davies, W. C. and Jones, W. J. (1930) *J. Chem. Soc.*, 463.
60. Hiers, G. S. (1944) *Org. Synth.,* Coll. Vol. 1, 550.
61. McAuliffe, C. A., Niven, I. E. and Parish, R. V. (1975) *Inorg. Chim. Acta*, **15**, 67.
62. Gedridge, R. W. Jr (1992) *Organometallics,* **11**, 969.
63. Beagley, B. and. Medwid, A. R (1977) *J. Mol. Struct.*, **38**, 229.
64. Bowen, H. J. M. (1954) *Trans. Farad. Soc.*, **50**, 463.
65. Sille, K., Weidlein, J. and Haaland, A. (1982) *Spectrochim. Acta*, **38A**, 475.
66. Miller, F. A. and Lemmon, D. H. (1967) *Spectrochim. Acta*, **23A**, 1099.
67. Trotter, J. and Camerman, N. (1963) *Can. J. Chem.*, **41**, 460.
68. Goggin, P. L. and Crocker, C. (1978) *J. Chem. Soc., Dalton Trans.*, 388.
69. Sobolev, A. N., Romm, I. P., Chernikova, N. Y. *et al.* (1981) *J. Organomet. Chem.*, **219**, 35.
70. Sovolev, A. N. and Belsky, V. K. (1981) *J. Organomet. Chem.*, **214**, 41.
71. Claeys, E. G., van der Kelen, G. P. and de Ketelaere, R. F. (1977) *J. Mol. Struct.*, **40**, 89.

72. Mislow, K. and Senkler G. H. Jr. (1972) *J. Am. Chem. Soc.*, **94**, 291.
73. Horner, L. and Fuchs, H. (1962) *Tet. Lett.*, 203.
74. Gatilov, Y. F., Kamai, G. K. and Ionov, L. B. (1968) *J. Gen. Chem., USSR*, **38**, 368.
75. Gatilov, Y. F., Yambushev, F. D. and Tenisheva, N. K. (1973) *J. Gen. Chem., USSR*, **43**, 2659.
76. Wild, S. B., Allen, D. G., McLaughlin, G. M. *et al.* (1982) *Inorg. Chem.*, **21**, 1007.
77. Wild, S. B. and Salem, G. (1989) *J. Organomet. Chem.*, **370**, 33.
78. Wild, S. B. and Henrick, K. (1975) *J. Chem. Soc. Dalton Trans.*, 1506.
79. Binsch, G., Wille, E. E., Stephenson, D. S. and Capriel, P. (1982) *J. Am. Chem. Soc.*, **104**, 405.
80. Aguiar, A. M., Mague, J. T., Aguiar, H. J. *et al.* (1968) *J. Org. Chem.*, **33**, 1681.
81. Chow, K. K., Levason, W. and McAuliffe, C. A. (1976) *J. Chem. Soc., Dalton Trans.*, 1429.
82. Levason, W. and McAuliffe, C. A. (1974) *J. Chem. Soc., Dalton Trans*, 2238.
83. Thornhill, D. J. and Manning, A. R. (1973) *J. Chem. Soc., Dalton Trans.*, 2086.
84. Somer, K. (1970) *Z. Anorg. Allgem. Chem.*, **377**, 278.
85. Feltham, R. D., Metzger, H. G. and Silverthron, W. (1968) *Inorg. Chem.*, **7**, 2003.
86. Feltham, R. D. and Metzger, H. G. (1971) *J. Organomet. Chem.*, **33**, 347.
87. Duffy, N. V., Layton, A. J., Nyholm, R. S. *et al.* (1966) *Nature*, **212**, 177.
88. Cullen, W. R. (1972) *Adv. Inorg. Chem. Radiochem.*, **15**, 323.
89. Cullen, W. R., Dhaliwal, P. S. and Styas, G. E. (1966) *J. Organomet. Chem.*, **6**, 432.
90. Crow, J. R. and Cullen, W. R. (1967) *Inorg. Chem.*, **6**, 2165.
91. Cullen, W. R. and Hota, N. K. (1964) *Can. J. Chem.*, **42**, 1123.
92. Cullen, W. R., Hall, L. D. and Ward, J. E. H. (1974) *J. Amer. Chem. Soc.*, **96**, 3422.
93. Bosnich, B. and Wild, S. B. (1970) *J. Amer. Chem. Soc.*, **92**, 459.
94. Cheney, A. J. and Shaw, B. L. (1971) *J. Chem. Soc. (A)*, 3545.
95. Henrick, K. and Wild, S. N. (1975) *J. Chem. Soc., Dalton Trans.*, 1506.
96. Isslieb, K., Hamann, B. and Schmidt, L. (1965) *Z. Anorg. Allgem. Chem.*, **339**, 298.
97. Dickinson, R. J., Levason, W., McAuliffe, C. A, and Parish, R. V. (1975) *J. Chem. Soc. Chem. Commun.*, 272.
98. Nesmayanov, A. N., Vorisov, A. E. and Novikova, N. V. (1967) *Dokl. Adad. Nauk. S.S.S.R.*, **172**, 1329.
99. Isslieb, K. (1964) *Pure Appl. Chem.*, **9**, 205.
100. Schewchuk, B. and Wild, S. B. (1974) *J. Organomet. Chem.*, **71**, C1.
101. Grim, S. O. and Borth, R. C. (1975) *J. Organomet. Chem.*, **94**, 327.
102. Lloyd, J. C. and Meek, D. W. (1972) *Inorg. Chim. Acta.*, **6**, 607.
103. Grim, S. O., Del Gaudio, J., Molena R. P. *et al.* (1974) *J. Amer. Chem. Soc.*, **96**, 3416.
104. Barchaly, G. A., Nyholm, R. S. and Parish, R. V. (1961) *J. Chem. Soc.*, 4433.
105. Murray, S. G. (1975) PhD thesis, Department of Chemistry, University of Manchester, Manchester.
106. Kordosky, G., Cook, B. R., Lloyd, J. C. and Meek, D. W. (1973) *Inorg. Syn.*, **14**, 14.
107. King, R. B. and Kapoor, P. N. (1971) *J. Amer. Chem. Soc.*, **93**, 4158.
108. Chow, K. K., Halfpenny, M. T. and McAuliffe, C. A. (1973) *J. Chem. Soc., Dalton Trans.*, 147.
109. Carty, A. J., Hota, N. K., Ng, T.W. *et al.* (1971) *Can. J. Chem.*, **49**, 2706.
110. Nicpon, P. E. and Meck, D. W. (1967) *Inorg. Chem.*, **6**, 145.
111. Levason, W. and McAuliffe, C. A. (1974) *Inorg. Chim. Acta*, **11**, 33.
112. Carlton, T. R. and Cook, C. D. (1971) *Inorg. Chem.*, **10**, 2628.
113. Plazek, E. and Tyka, R. (1957) *Zesz. Nauk. Polytech. Wroclaw. Chem.*, **4**, 79.
114. Gregorzik, R., Wirbsor, J. and Vahrenkamp, H. (1992) *Chem. Ber.*, **125**, 1575.
115. Livingstone, S. E. (1956) *J. Chem. Soc.*, 437.
116. Livingstone, S. E. and Chiswell, B. (1961) *J. Inorg. Nucl. Chem.*, **23**, 37.
117. Lockyer, T. N. (1974) *Aust. J. Chem.*, **27**, 259.
118. Cannon, R. D., Chiswell, B. and Venanzi, L. M. (1967) *J. Chem. Soc. (A)*, 1277.
119. Levason, W., McAuliffe, C. A. and Watson, D. G. (1975) *J. Coord. Chem.*, **4**, 173.
120. Howell, T. E. W., Pratt, S. A. J. and Venanzi, L. M. (1961) *J. Chem. Soc.*, 3167.
121. Cunningham, R. G. Nyholm, R. S. and Tobe, M. L. (1964) *J. Chem. Soc.*, 580.
122. Hill, W. E., Dalton, J. and McAuliffe, C. A. (1973) *J. Chem. Soc., Dalton Trans.,* 143.
123. Chow, S. T. and McAuliffe, C. A. (1974) *J. Organomet. Chem.*, **77**, 401.

124. Bosnich, B., Lo, S. T. D. and Sullivan, A. E. (1975) *Inorg. Chem.*, **14**, 2305.
125. Barclay, G. A., Harris, C. M. and Kingston, J. V. (1968) *J. Chem. Soc., Chem. Commun.*, 965.
126. McAuliffe, C. A. (1967) DPhil thesis, Inorganic Chemistry Laboratory, Oxford University.
127. Ellermann, J., and Veit, A. (1985) *J. Organomet. Chem.*, **290**, 307.
128. Cheffey, A. F., Evans, J., Levason, W. and Webster, M. (1996) *Organometallics*, **15**, 1280.
129. Workman, M. O., McAuliffe, C. A. and Meek, D. W. (1972) *J. Coord. Chem.*, **2**, 137.
130. Higginson, B. R., McAuliffe, C. A. and Venanzi, L. M. (1971) *Inorg. Chim. Acta*, **5**, 37.
131. Baracco, L. and McAuliffe, C. A. (1972) *J. Chem. Soc., Dalton Trans.*, 948.
132. McAuliffe, C. A. and Meek, D. W. (1971) *Inorg. Chim. Acta*, **5**, 270.
133. Dawson, J. W. and Venanzi, L. M. (1971) *J. Chem. Soc.(A)*, 2897.
134. Morris, T. L. and Taylor, R. C. (1973) *J. Chem. Soc., Dalton Trans.*, 175.
135. Sacconi, L., Bertini, I. and Mani, F. (1968) *Inorg. Chem.*, **7**, 1417.
136. Morassi, R., Sacconi, L., McAuliffe, C.A. and Levason, W. (1976) *Inorg. Synth.*, **16**, 174.
137. Sacconi, L. and Gatteschi, D. (1972) *J. Coord. Chem.*, ,**2**, 107
138. Sacconi, L. and Morassi, R. (1969) *J. Chem. Soc., (A)*, 2904.
139. Levason, W., McAuliffe, C. A. and Murray, S. G. (1975) *J. Chem. Soc., Dalton Trans.*, 1556.
140. Levason, W. McAuliffe, C. A. and Murray, S. G. (1976) *Inorg. Chim. Acta*, **17**, 247.
141. Levason, W., McAuliffe, C. A. and Murray, S. G. (1976) *J. Chem. Soc., Dalton Trans.*, 269.
142. Dutta, R. L., Meek, D. W. and Busch, D. H. (1970) *Inorg. Chem.*, **9**, 2098.
143. Neunhoeffer, O. and Lamza, L. (1961) *Chem. Ber.*, **94**, 2514.
144. Mann, F. G. and Chaplin, B. J. (1937) *J. Chem. Soc.*, 527.
145. Mallion, K. B. and Mann, F. G. (1964) *J. Chem. Soc.*, 6121
146. Bennett, M. A. Konwenhoven, H. W., Lewis, J. and Nyholm, R. S. (1964) *J. Chem. Soc.*, 4570.
147. Bennett, M. A., Chatt, J., Erskine, G. J. *et al.* (1967) *J. Chem. Soc. (A)*, 501.
148. Hall, D. T. and Nyholm, R. S. (1971) *J. Chem. Soc. (A)*, 1491.
149. McAuliffe, C. A. and Watson, D. G. (1974) *J. Chem. Soc., Dalton Trans.*, 153.
150. McAuliffe, C. A. and Watson, D. G. (1974) *J. Organomet. Chem.*, **78**, C51.
151. Lee, K. W. (1971) PhD thesis, Department of Chemistry, University of Queensland.
152. Leigh, G. J. (ed.) (1990) *Nomenclature of Inorganic Chemistry: Recommendations 1990*, IUPAC, Blackwell Scientific Publications, Oxford.
153. Appel, R., Gaitzsch, T. and Knoch, F. (1985) *Angew. Chem. Int. Ed. Engl.*, **24**, 419.
154. Dianova, E. N., Litvinov, I. A., Struchkov, Y. T. *et al.* (1983) *Dokl. Akad. Nauk., SSSR*, **268**, 885.
155. Hamelin, J. and Himidi-Kabbab, S. (1990) *Tetrahed. Lett.*, **31**, 7607.
156. Wild, S. B. and Mickiewicz, M. (1977) *J. Chem. Soc., Dalton Trans.*, ,704.
157. Monagle, J. J. (1962) *J. Org. Chem.*, **27**, 385.
158. Dasogupta, H. N. (1938) *J. Ind. Chem. Soc.*, **15** ,498.
159. Wiberg, E. and Modritzer, K. (1957) *A. Naturforsch.*, **12B**, 135.
160. Lyon, D. R. and Mann, F. G. (1945) *J. Chem. Soc.*, 30.
161. Jones, E. R. H. and Mann, F. G. (1958) *J. Chem. Soc.*, 1719.
162. Grütter, G. and Krause, E. (1916) *Chem. Ber.*, **49**, 437.
163. Mathey, F. and Thiollet, G. (1979) *Tetrahed. Lett.*, 3157.
164. Schenk, W. A. and Voss, E. (1990) *J. Organomet. Chem.*, **396**, C8.
165. Märkl, G. and Hauptmann, H. (1976) *Tetrahed.*, **32**, 2131.
166. Märkl, G. and Hauptmann, H. (1968) *Tetrahedron Lett.*, 3257.
167. Leavitt, F. C., Manuel, T. A., Johnson, F. *et al.* (1960) *J. Am. Chem. Soc.*, **82**, 5099.
168. Boudjouk, P., Sooriyakumaran, P. and Han, B. H. (1986) *J. Org. Chem.*, **51**, 2818.
169. Fagan, P. F. and Nugent, W. A. (1988) *J. Am. Chem. Soc.*, **110**, 2310.
170. Mathey, F. (1980) *Topics Phos. Chem.*, **10**, 1.
171. Mathey, F. and Sennyey, G. (1981) *Tetrahed. Lett.*, 4713.
172. Heinekey, D. M. and Miller, I. T. (1959) *J. Chem. Soc.*, 3101.

173. Morgan, G. T. Davies, G .R. (1930) *Proc. Roy. Soc. London, A*, **127**, 1.
174. Buchwald, S. L., Fisher, R. A. and Foxman, B. M. (1990) *Angew. Chem. Int. Ed. Engl.*, **29**, 771.
175. Tzschach A. and Heinicke, J. (1978) *J. Organomet. Chem.*, **154**, 1.
176. Heinicke, J., Nyulaszi, L., Gsonka, G. *et al.* (1989) *J. Organomet. Chem.*, **373**, 49.
177. Cheeseman, G. W. H. and Greenberg, S. G. (1979) *J. Organomet. Chem.*, **166**, 139.
178. Helliwinkel, D., Wiel, A., Sattler, G. and Nuber, B. (1990) *Angew. Chem. Int. Ed. Eng.*, **29**, 689.
179. Vermeer, H. Kevenaar, P. C. J. and Bickelhaupt, F. (1972) *Justus Liebigs Ann. Chem.*, **763**, 155.
180. Stam, C. H. and Freijee, F. J. M. (1980) *Acta. Cryst.*, **B36**, 1247.
181. Stam, C. H. and Van Rooyen-Reiss, C. (1980) *Acta Cryst.*, **B36**, 1252.
182. Massey, A. G., Humphries, R. E., Al-Jabar, N. A. A. *et al.* (1987) *J. Organomet. Chem.*, **319**, 59.
183. Mann, F.G. *et al.* (1952) *J. Chem. Soc.*, 4142.
184. Massey, A.G., Hughes, G. and Woodward, C. M. (1976) *J. Organomet. Chem.*, **112**, 9.
185. Krespan, C.G. (1961) *J. Am. Chem. Soc.*, **83**, 3432.
186. Märkl, G. and Dannhardt, G. (1973) *J. Chem. Soc.*, 1130.
187. Mann, F. G., Miller, I. T. and Smith, B. B. (1953) *J. Chem. Soc.*, 1130.
188. Mann, F. G., Beeby, M. H. and Turner, E. E. (1950) *J. Chem. Soc.*, 1923.
189. Mann, F. G., Miller, I. T. and Baker, F. C. (1965) *J. Chem. Soc.*, 6342.
190. Wittig, G. and Hellwindel, D. (1964) *C. Ber.*, **106** ,3645.
191. Kyba, E. P. and Chou, S. S. P. (1980) *Chem. Commun.*, 449.
192. Kyba, E. P. and Chou, S. S. P. (1981) *J. Org. Chem.*, **46**, 860.
193. Kauffmann, T. and Ennen, J. (1985) *C. Ber.*, **118**, 2692, 2714.
194. Wild, S. B., Kerr, P. G. and Leung, P. H. (1987) *J. Am. Chem. Soc.*, **109**, 4321.
195. Ciampolini, M. (1986) *Pure Appl. Chem.*, **58**, 1429.
196. Vermeer, H. and Bickelhaupt, F. (1970) *Tetrahedron Lett.*, 1007, 3255.
197. Van Der Meer, H. and Jongsma, C. (1970) *Tetrahedron Lett.*, 1323.
198. Huang, Y. Z. and Shen, Y. C. (1982) *Adv. Organomet. Chem.,* **20**, 115.
199. Schmidbaur, H. and Nusstein, P. (1985) *Organometallics*, **4**, 344.
200. Lloyd, D., Glidewell, C. and Metcalfe, S. (1988) *Synthesis*, 319.
201. Horner, L. and Oediger, H. (1958) *C. Ber.*, **91**, 437.
202. Lloyd, D. and Singer, M.I.C. (1971) *J. Chem. Soc. (C)*, 2941.
203. Dale, A. J. and Froyen, P. (1971) *Acta. Chem. Scand.*, **25**, 1452.
204. Shen, Y. C. and Liao, Q. (1988) *Synthesis*, 321.
205. Ferguson, G. and Rendle, D. F. (1986) *J. Chem. Soc., Dalton Trans.*, 171.
206. Ferguson, G. and Rendle, D. F. (1975) *J. Chem. Soc., Dalton Trans.*, 1284.
207. Ferguson, G., March, F. C. and Lloyd, D. (1985) *J. Chem. Soc., Dalton Trans.*, 1377./
208. Shen, Y. C., Fan, Z. and Qui, W. (1987) *J. Organomet. Chem.*, **320**, 21.
209. Huang, Q., Shao, M., Jin, X. and Tang, Y. (1982) *Tetrahedron Lett.*, 5343.
210. Armstrong, R., Gibson, N. A., Hosking, J. W. and Weatherburn, D. C. (1967) *Aust. J. Chem.*, **20**, 2771.
211. West, B. O. and Haszeldine, R. N. (1956) *J. Chem. Soc.*, 3631.
212. Cullen, W. R. (1960) *Can. J. Chem.*, **38**, 445.
213. Stone, F. G. A., Maier, L., Seyferth, D. and Rochow, E. G. (1957) *J. Am. Chem. Soc.*, **79**, 5884.
214. Gavrilov, V. I., Khlebnikov, V. N., Galyametdinov, Y. G. *et al.* (1977) *J. Gen. Chem. USSR.*, **47**, 373.
215. Nyholm, R. S., Collinge, R. N. and Tobe, M. L. (1964) *Nature*, **201**, 1322.
216. Seyferth, D. (1958) *J. Am. Chem. Soc.*, **80**, 1336.
217. Wolf, C. N. and Shriner, R. L. (1950) *Org. Syn.*, **30**, 95.
218. Harmon, K. M. and Lake, R. R. (1968) *Inorg. Chem.*, **7**, 1921.
219 Waddington, T. C. and White, J. A. (1963) *J. Chem. Soc.*, 2701.
220. Tzschach, A. and Fescher, W. (1967) *Z. Chem.*, **7**, 196.
221. Mann, F. G., Hollimann, F. G. and Thornton, D. A. (1960) *J. Chem. Soc.*, 9.
222. Wittig, G. and Helliwinkel, D. (1962) *Angew. Chem. Int. Ed. Engl.*, **1**, 598.

223. Edmons, J. S., Francesconi, K. A., Cannon, J. R. *et al.* (1977) *Tetrahedron Lett.*, 1543.
224. Simmons, R. E. (1982) *Phos. Sulf.*, **13**, 13.
225. Müller, U. (1980) *Acta. Cryst.*, **B36**, 1075.
226. Orenber, J. B., Morris, M. D. Long, T. V. (1971) *Inorg. Chem.*, **10**, 933.
227. Anderson, S. and Jagner, S. (1985) *Acta. Chem. Scand.*, **39A**, 297.
228. Mann, F. G., Collins, E. and Sutor, D. J. (1963) *J. Chem. Soc.*, 4051.
229. Mooney, R. C. L. (1940) *J. Am. Chem. Soc.*, **62**, 2955.
230. Wittig, G. and Torsell, E. K. (1953) *Acta. Chem. Scand.*, **7**, 1293.
231. Seyferth, D. and Cohen, H. M. (1961) *J. Inorg. Nucl. Chem.*, **20**, 73.
232. Schmidbaur, H. and Mitschke, K. H. (1973) *C. Ber.*, **106**, 3645..
233. Brock, C. P. and Webster, D. F. (1976) *Acta. Cryst.*, **32B**, 2089.
234. Beattie, I. R., Livingstone, K. M., Ozin, K. M. S. and Sabine, R. (1972) *J. Chem. Soc., Dalton Trans.*, 784.
235. Roesky, H. W. and Bohra, R. (1984) *Adv. Inorg. Chem. Radiochem.*, **28**, 203.
236. Helliwinkel, D. and Kilthau, G. (1966) *Angew. Chem. Int. Ed. Eng.*, **5**, 969.
237. Wittig, G. and Torscll, K. (1953) *Acta. Chem. Scand.*, **7**, 1293.
238. Downs, A. J., Schmutzler, R. and Steer, I. A. (1966) *Chem. Commun.*, 221.
239. Mahler, W., Muetterties, E. H., Packer, K. J. and Schmutzler, R. (1964) *Inorg. Chem.*, **3**, 1298.
240. Berry, R. S. (1969) *J. Chem. Phys.*, **32**, 933.
241. Pulham, C., Haaland, A., Hammel, A. *et al.* (1992) *Angew. Chem. Int. Ed. Engl.*, **31**, 1464.
242. Cotton, F. A., Beauchamp, A. L. and Bennett, M. J. (1968) *J. Am. Chem. Soc.*, **90**, 6675.
243. Beattie, I. R., Livingstone, K. M. S., Ozin, G. A. and Sabine, R. (1972) *J. Chem. Soc., Dalton Trans.*, 784.
244. Huang, Y. Z., Xu, Y. Y. and Li, Z. (1982) *Org. Prep. Proc. Int.*, **14**, 373.
245. Huang, Y. Z., Shen, Y. C. and Zheng, J. H. (1985) *Synthesis*, 57.
246. Yang, J. H., Shi, L. L. and Huang, Y. Z. (1985) *Tetrahedron Lett.*, **26**, 6447.
247. Mioskowski, C., Chabert, P. and Ousset, J. B. (1989) *Tetrahedron Lett.*, **30**, 179.
248. Chabert, P., Falck, J. R. and Mioskowski, C. (1989) *Tetrahedron Lett.*, **30**, 2545.
249. McCoy, G. and Day, A. R. (1943) *J. Am. Chem. Soc.*, **65**, 1956.
250. Froyen, P. (1971) *Acta. Chem. Scand.*, **25**, 983.
251. Froyen, P. (1973) *Acta Chem. Scand.*, **27**, 141.
252. Froyen, P. (1993) *Phys. Sulf. Silicon.*, **81**, 37.
253. Froyen, P. and Juvvik, P. (1992) *Phos. Sulf. Silicon.*, **69**, 83.
254. Hartley, F. R. (1985) *Supported Metal Catalysts*, D. Reidel, Dordrecht.
255. Parish R. V. and Vania, M. L. (1984) *J. Organomet. Chem.*, **263**, 139.
256. Hay, A. S. (1969) *J. Polym. Sci.*, **A7**, 1625.
257. Akelah, A. and Moet, A. (1990) *Funtionalised Polymers and their Applications*, Chapman & Hall, London.
258. Herbstman, S. (1965) *J. Org. Chem.*, **30**. 1259.
259. Monagle, J. J. (1962) *J. Org. Chem.*, **27**, 3851.
260. Wittig, G. and Laib, H. (1953) *Justus Liebigs Ann. Chem.*, **580**, 57.
261. Wittig, G. and Henry, M. C. (1960) *J. Am. Chem. Soc.*, **82**, 563.
262. Shimizu, A., Ohkata, K., Ohnari, H. and Akiba, K. Y. (1985) *Tetrahedron Lett.*, **26**, 3211.
263. Huang, Y., Shen, Y. and Chen, C. (1986) *Tetrahedron Lett.*, **27**, 2903.
264. Singer, M. I. C. and Lloyd, D. (1967) *Chem. Ind. (London)*, 787.
265. Kauffmann, R., Hamsen, A., Kriegesmann, R. and Vahrenhorst, A. (1978) *Tetrahedron Lett.*, **19**, 4395.
266. Kauffmann, T. (1982) *Angew. Chem. Int. Ed. Eng.*, **21**, 410.
267. Kauffmann, R., Joussen, R., Klas, N. and Vahrenhorst, A. (1983) *C. Ber.*, **116**, 473.
268. Godfrey, S. M., Kelly, D. G., Mackie, A. G. *et al.* (1991) *J. Chem. Soc., Chem. Commun.*, 1991 1447.
269. Godfrey, S. M., McAuliffe, C. A. and Pritchard, R. G. (1993) *J. Chem. Soc., Dalton Trans.*, 1993 2875.
270. Godfrey, S. M., Mackie, A. G., McAuliffe, C. A. and Pritchard, R. G. (1992) *J. Chem. Soc., Chem Commun.*, 483.

271. Godfrey, S. M., McAuliffe, C. A. and Pritchard, R. G. (1993) *J. Chem. Soc., Dalton Trans.*, **371**, 2229.
272. Bricklebank, N., Godfrey, S. M., McAuliffe, C. A. and Pritchard, R. G. (1994) *J. Chem. Soc., Chem. Commun.*, 695.
273. Bricklebank, N., Godfrey, S. M., McAuliffe, C.A. and Molloy, K.C. (1995) *J. Chem. Soc., Dalton Trans.*, 1593.
274. Bricklebank, N., Godfrey, S. M., McAuliffe, C. A. *et al.* (1992) *J. Chem. Soc., Chem. Commun.*, 944.
275. Godfrey, S. M., McAuliffe, C. A., Pritchard, R. G. and Sheffield, J. M. (1996) *J. Chem. Soc., Dalton Trans.*, 3309.
276. Beagley, B., Godfrey, S. M., Kelly, K. J. *et al.* (1996) *J. Chem. Soc., Chem. Commun.*, 2179.
277. McAuliffe, C. A., Godfrey, S. M., Mackie, A. G. and Pritchard, R. G. (1992) *Angew. Chem. Int. Ed. Engl.*, **31**, 919.
278. Godfrey, S. M., Lane, H. P., McAuliffe, C. A. and Pritchard, R. G. (1993) *J. Chem. Soc., Dalton Trans.*, 1599.
279. Godfrey, S. M., McAuliffe, C. A. and Pritchard, R. G. (1994) *J. Chem. Soc., Chem. Commun.*, 45.
280. Bricklebank, N., Godfrey, S. M., McAuliffe, C. A. and Pritchard, R. G. (1996) *J. Chem. Soc., Dalton Trans.*, 157.
281. Godfrey, S. M., Ho, N., McAuliffe, C. A. and Pritchard, R. G. (1989) *Angew. Chem. Int. Ed. Engl.*, **35**, 2343.
282. Märkl. G. and Sejpka, J. (1986) *Angew. Chem. Int. Ed. Engl.*, **25**, 262.
283. Christiansen, W. G. (1920) *J. Amer. Chem. Soc.*, **42**, 2402.
284. Elms, P. S., Middleton, S. and West, B.O. (1970) *Aust. J. Chem.*, **23**, 1559.
285. Rheingold, A. L. and Sullivan, P.J. (1983) *Organometallics*, **2**, 327.
286. Dasent, W. E. (1965) *Non-existent Compounds,* Marcel Dekker, London.
287. Elms, P. S., Leverett, P. and West, B.O. (1971) *J. Chem. Soc., Chem. Commun.*, 747.
288. Huttner, G., Schmid, H.-G., Frank, A. and Orama, O. (1976) *Angew. Chem. Int. Ed. Engl.*, **15**, 234.
289. Yoshifuji, M., Shima, I., Inamoto, N. *et al.* (1981) *J. Am. Chem. Soc.*, **103**, 457.
290. Cowley, A. H., Kilduff, J. E., Newman, T. H. and Pakulski, M. (1982) *J. Am. Chem. Soc.*, ,**104**, 5820.
291. Cowley, A. H., Kilduff, J. E., Lasch, J. G. *et al.* (1984) *Inorg. Chem.*, **23**, 252.
292. Couret, C., Escudie, J., Madaule, Y. *et al.* (1983) *Tetrahedron Lett.*, 2769.
293. Couret, C., Escudie, J., Madaule, Y. *et al.* (1983) *Tetrahedron Lett.*, 3625.
294. Huttner, G. and Jibril, I. (1984) *Angew. Chem. Int. Ed.*, **23**, 740.
295. Huttner, G. (1986) *Pure Appl. Chem.*, **58**, 585.
296. Cowley, A.H. and Norman, N.C. (1986) *Prog. Inorg. Chem.*, **34**, 1.
297. Nagase, S., Sasuki, S. and Kurakake, T. (1990) *J. Chem. Soc., Chem. Commun.*, 1724.

6 Organobismuth compounds: synthesis, properties and uses in organic transformations

H. SUZUKI and Y. MATANO

6.1 Introduction

The first synthesis of organobismuth compounds dates back to the early middle of the eighteenth century [1], although the chemistry of organobismuth compounds remained outside the attention of organic chemists until the 1980s, when Barton and co-workers started their pioneering work on the use of bismuth compounds as reagents for organic transformations. The slow development of organobismuth chemistry may be attributed to limited access to many compounds as well as to the thermal or chemical instability of organobismuth compounds. Since the 1970s, however, organobismuth compounds have come to be studied intensively and the uniqueness of bismuth in organic chemistry has become widely recognized. In particular, they have been receiving increasing attention for medicinal and electronic applications. The objective of the present review is to provide an overview of the synthesis, properties and transformations of organobismuth compounds, particular emphasis being placed on the recent progress relevant to the subject. The review is organized on the basis of the type of compound, and, for each type, brief descriptions of properties, typical reaction modes and some synthetic applications are included. Illustrative examples of reactions are mostly taken from recent publications and almost all pertinent references which have appeared since the 1960s are included in the present review.

Aspects of organobismuth chemistry have been reviewed extensively by Gilman [2] and Freedman and Doak [3]. Besides these two systematic review articles, a number of surveys and monograph chapters are devoted to general and more or less specialized topics in this and related areas by Freedman and Doak [4, 5], Dub [6], Hellwinkel [7], Kitchin [8], Barton and co-workers [9–11], Finet [12], Norman and co-workers [13, 14], Wada and Ohki [15], Seppelt [16, 17], Ashe [18], Arduengo and Dixon [19], Akiba and Yamamoto [20], Wardell [21], Huang and Zhou [22] and Suzuki, Ikegami and Matano [23]. Progress in organobismuth chemistry has been reviewed annually in the *Journal of Organometallic Chemistry* (e.g. refs [24–26]), *Organometallic Chemistry* (e.g. refs [27, 28]) and *Annual Reports Section A, The Royal Society of Chemistry* (e.g. refs [29–31]).

Gmelin Handbuch der Anorganischen Chemie, volume 47 [32], provides a comprehensive compilation of the preparation and properties of known organobismuth compounds through 1977, and *Houben-Weyl Methoden der Organischen Chemie,* volume XIII, part 8 [33], mainly concerned with the preparation and reactions of organobismuth compounds, does so through 1978. For references published before 1977 or 1978, therefore, these two large compilations are best consulted.

6.2 Organobismuth(III) compounds

6.2.1 *Triorganylbismuth(III) compounds*

Preparation

From organomagnesium halides (Grignard reagents). The most important means by which tertiary bismuthines, R_3Bi, can be prepared is by the reaction of bismuth trihalides such as $BiCl_3$ and $BiBr_3$, with organomagnesium halides. A wide variety of symmetrical tertiary bismuthines containing such groups as alkyl, alkenyl and aryl have been prepared via this route [34]:

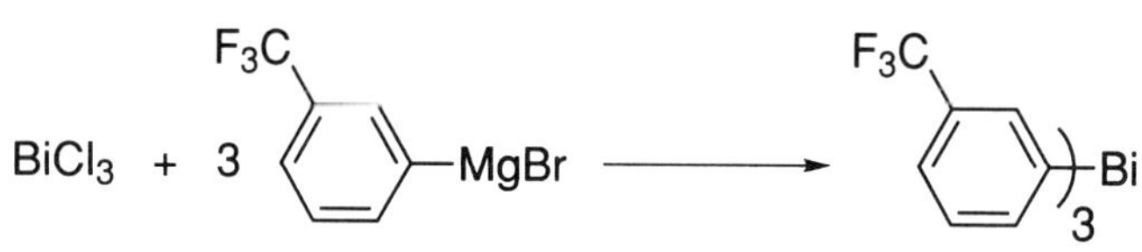

[6.1]

Asymmetrical bismuthines with two or three different organyl groups ($R_2R'Bi$ and $RR'R''Bi$) are best prepared by the reaction between a bismuth compound R_2BiX or $RR'BiX$ (X = halide) and a Grignard reagent $R''MgBr$ [35, 36]:

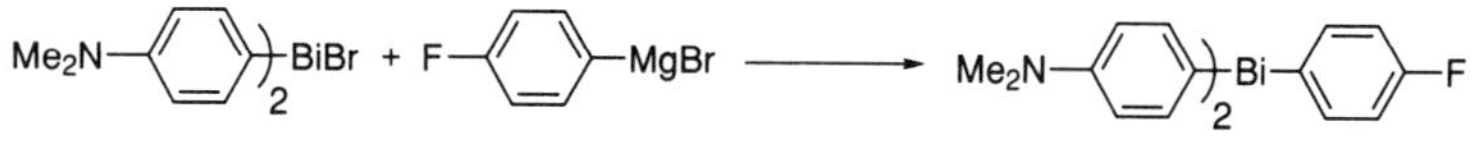

[6.2]

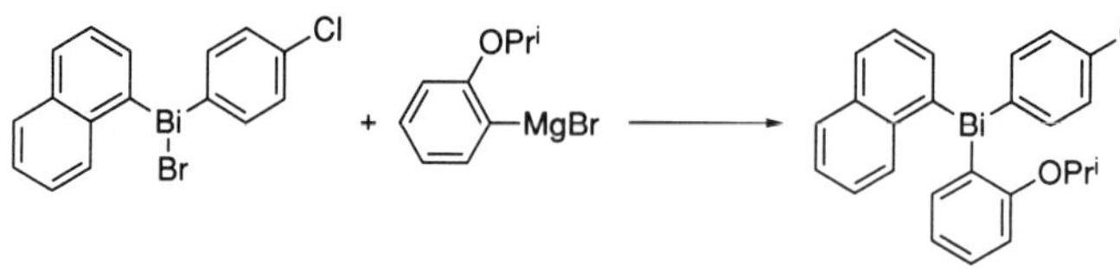

[6.3]

A diarylbismuth triflate-HMPA complex (HMPA = haxamethylphosphoric triamide) is also convenient for this purpose owing to its ease of preparation and handling [37].

From organolithium reagents. The reaction between bismuth trihalides and organolithium compounds is an alternative convenient method for the synthesis of symmetrical tertiary bismuthines [38]:

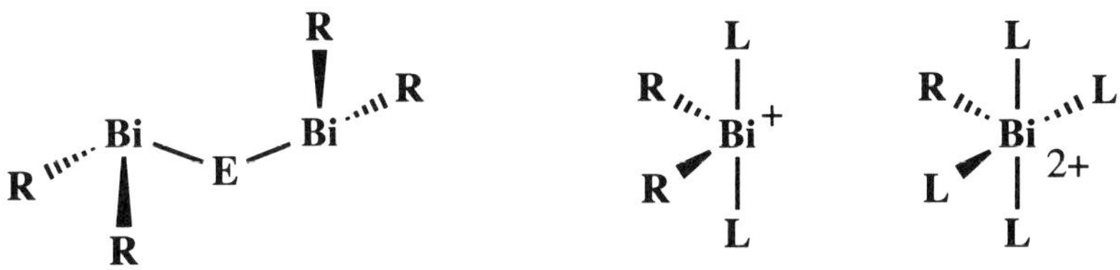

(a) (b) (c) [6.4]

Asymmetrical bismuthines with two different organyl groups, $R_2R'Bi$, can be obtained by the reaction between a bismuth compound R_2BiX and an organolithium R′Li [39–41]:

$$Ph_2BiCl + MeLi \longrightarrow Ph_2BiMe \qquad [6.5]$$

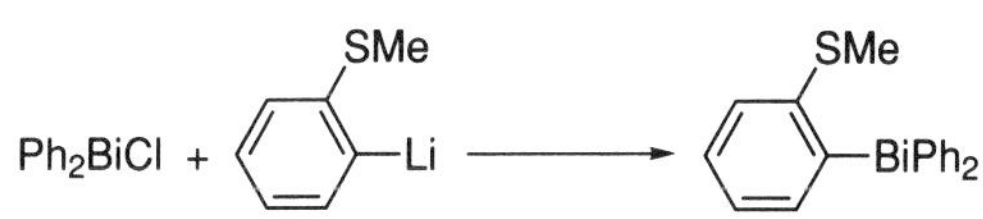

[6.6]

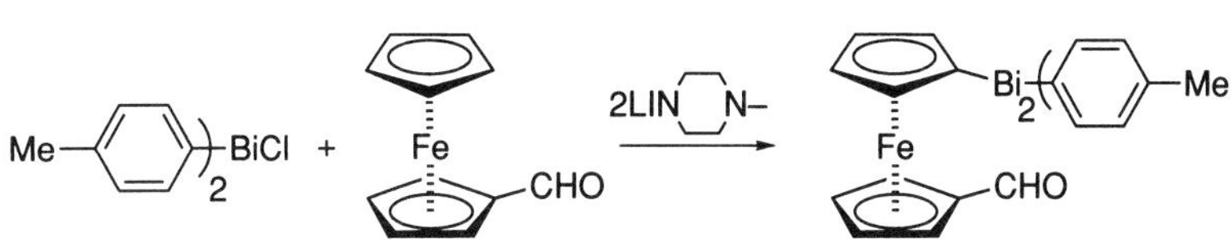

[6.7]

Tris-*ortho*-lithiation of tris[(2-diethylaminosulphonyl)phenyl]bismuthine with tert-butyllithium and subsequent treatment with three equivalents of bis[(2-diethylaminosulphonyl)phenyl]bismuth iodide at low temperature yields a branched tetrameric bismuthine, which, on a similar treatment, leads to a dendrimer-type bismuthine (Suzuki, H., Kurata, H. and Matano, Y., *J. Chem. Soc., Chem. Commun.*, in press):

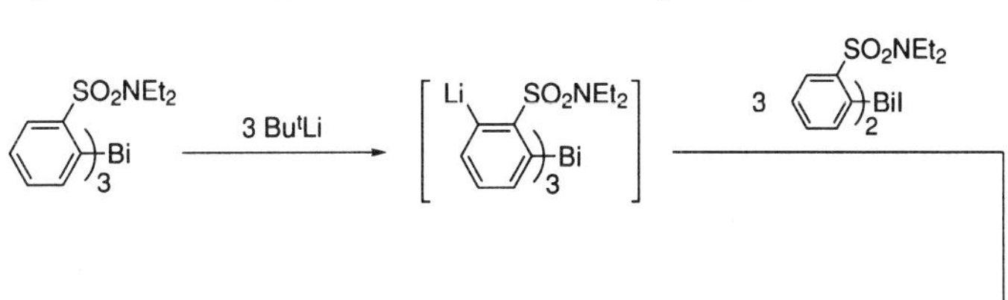

continued overleaf

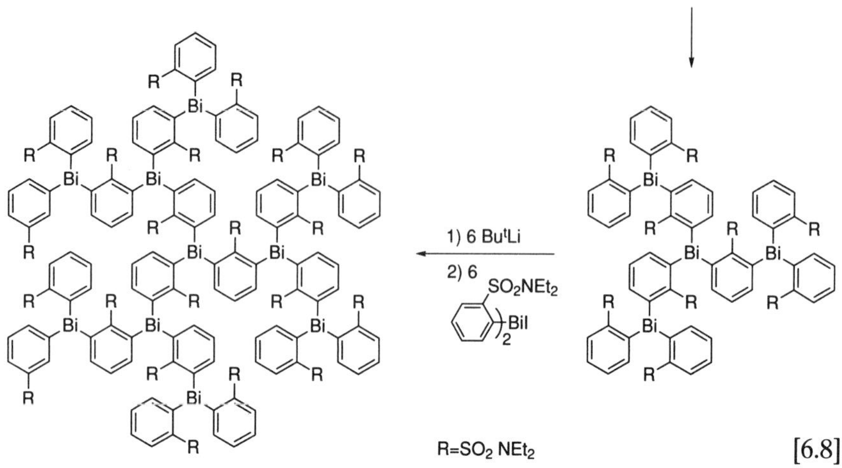

[6.8]

From other organometallic reagents. For the preparation of triorganylbismuthines, organometallic reagents other than organomagnesium and organolithium have also been used, which include organosodium, -zirconium, -cadmium, -silver, -zinc, -aluminum, -mercury and -silicon compounds [42–45]:

$$Ph_2BiBr + NaC{\equiv}CMe \longrightarrow Ph_2BiC{\equiv}CMe \qquad [6.9]$$

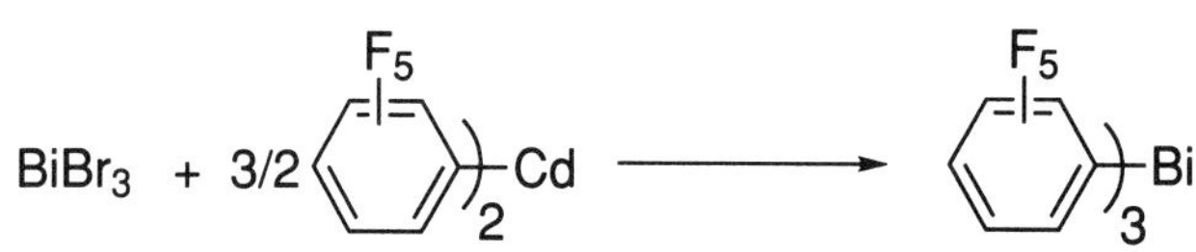

[6.10]

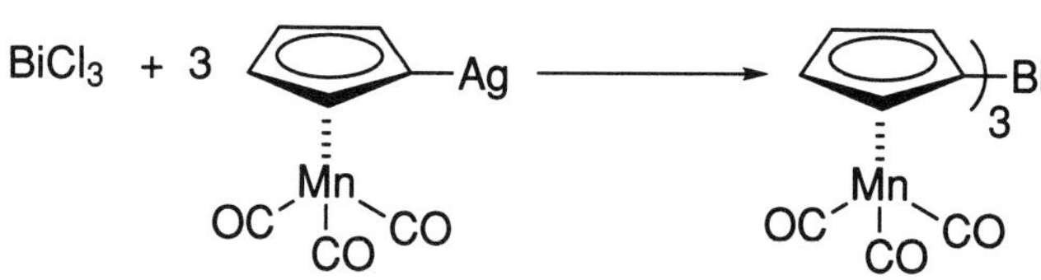

[6.11]

From onium compounds. Complete replacement of the halogen atoms in bismuth trihalides by aryl radicals generated from the decomposition of a diazonium compound in the presence of a metal catalyst leads to triarylbismuthines, and the decomposition of iodonium salts in the presence of bismuth trichloride and metallic bismuth also produces triarylbismuthines [46, 47]:

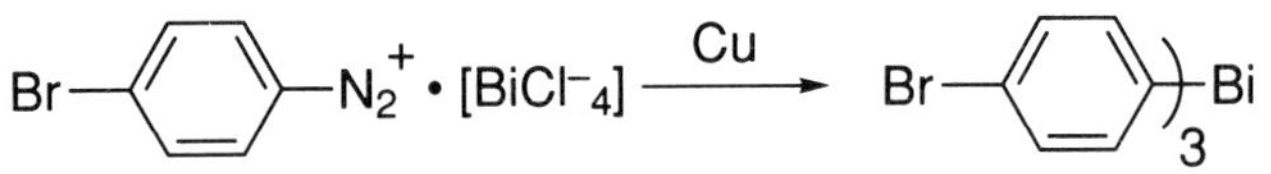 [6.12]

Radical-induced aryl ligand exchange with triarylbismuthines leads to the formation of mixed triarylbismuthines [48]:

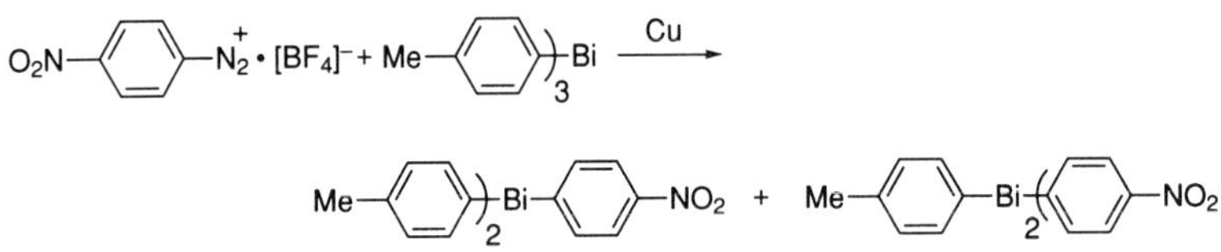 [6.13]

From bismuth metal. The action of a sodium–bismuth or potassium–bismuth alloy on alkyl and aryl halides results in the formation of the corresponding trialkyl- and triarylbismuthines [49–51]:

 [6.14]

and organyl radicals generated in the gas phase react with a bismuth mirror to form organylbismuthines [52].

From bismuthides. Alkali metal diaryl- and dialkylbismuthides, R_2BiM (M = Li, Na, K), generated *in situ* from the corresponding diorganylhalobismuthines and alkali metal in liquid NH_3 or tetrahydrofuran (thf) at low temperatures readily react with alkyl and aryl halides to give asymmetrical triorganylbismuthines [52–56]:

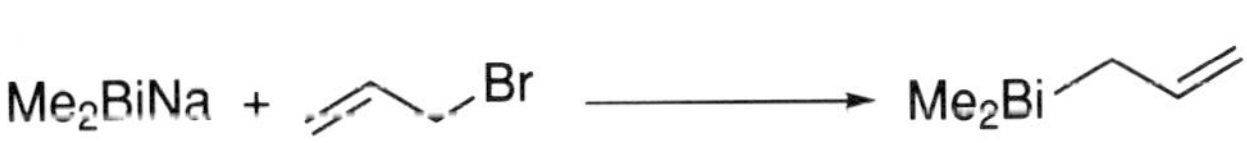 [6.15]

$$Me_2BiNa + Me_3MCH_2Cl \longrightarrow Me_3BiCH_2MMe_3 \quad M = Si, Ge \qquad [6.16]$$

$$2\,Ph_2BiNa + ClCH_2Cl \longrightarrow Ph_2BiCH_2BiPh_2 \qquad [6.17]$$

Triarylbismuthines bearing a substituent group incompatible with Grignard or organolithium reagents can be obtained by the reaction between sodium diarylbismuthide and acyl-, ethoxycarbonyl- or nitrophenyldiazonium tetrafluoroborate (Rahman, M. M., Matano, Y. and Suzuki, H. (1997) Organometallics, **16**, 3565).

From bismuth amides. Tris(cyclopentadienyl)bismuthine is prepared by treating a tris(dialkylamino)bismuthine with an excess of cyclopentadiene at room temperature [57].

$$Bi(NMe_2)_3 + 3\ C_5H_6 \longrightarrow Bi(C_5H_5)_3 + 3\ HNMe_2 \qquad [6.18]$$

A similar Bi–N bond cleavage reaction is employed for the synthesis of dimethyl(cyclopentadienyl)bismuthine [58]:

$$Me_2BiN(Me)SiMe_3 + C_5H_6 \longrightarrow Me_2BiC_5H_5 + HN(Me)SiMe_3 \qquad [6.19]$$

Miscellaneous. Electrolysis of organoaluminum or organoboron compounds with use of a sacrificial bismuth anode gives the corresponding tertiary bismuthines [59], and a similar electrolysis of vinyl halides, alkyl nitriles, carboxylic acids or acrylonitrile at a sacrificial bismuth cathode yields functionalized trialkylbismuthines [60]:

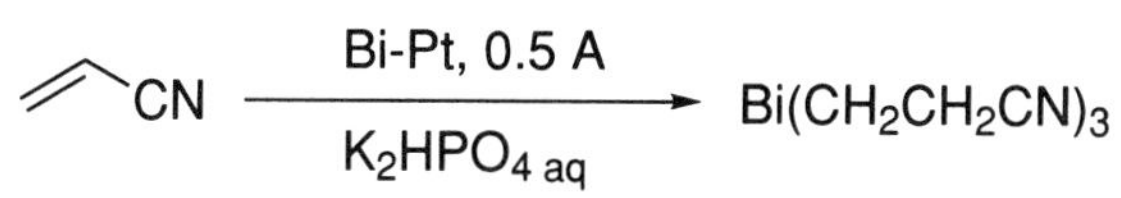

[6.20]

Thermolysis of bismuth tris(arenesulphinate) leads to the formation of triarylbismuthines [61, 62]:

$$(PhSO_2)_3Bi \xrightarrow{\Delta} Ph_3Bi + 3\ SO_2 \qquad [6.21]$$

and a thermal redistribution reaction between a trialkylbismuthine and a perfluoroalkyl iodide gives mixed alkyl(perfluoroalkyl)bismuthines [63]:

$$Me_3Bi + CF_3I \xrightarrow{\Delta} MeBi(CF_3)_2 + Me_2BiCF_3 \qquad [6.22]$$

Triarylbismuth dichlorides are reduced to the parent triarylbismuthines by treatment with hydrazine hydrate or sodium hydrosulphite [64, 65]:

$$Ph_3BiCl_2 \xrightarrow{H_2NNH_2} Ph_3Bi \qquad [6.23]$$

An aryl ligand with a coordinating substituent, for example, the 2-tert-butylsulphonylphenyl group, when presented in asymmetrical triarylbismuthines, can be preferentially displaced by another aryl group through transmetallation with an aryllithium reagent [66]:

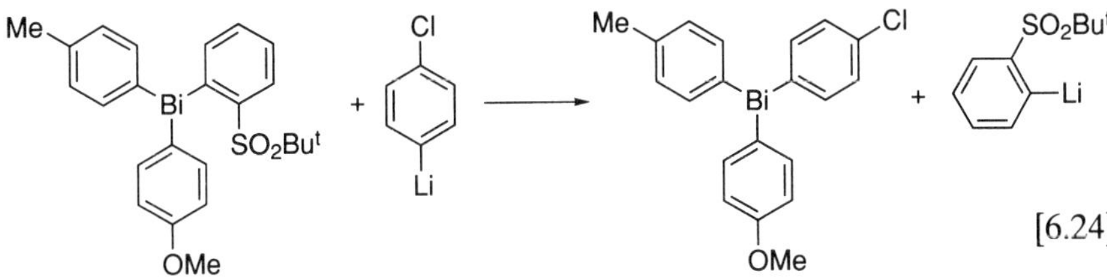

[6.24]

The resulting asymmetrical tertiary bismuthines have been resolved by chiral high-pressure liquid chromatography (HPLC).

Diazomethane inserts its methylene moiety into the Bi–Bi bond of tetraphenyldibismuthine to form bis(diphenylbismuthyl)methane (cf. Section 6.2.5).

Properties. Trialkylbismuthines are mostly liquids at room temperature and can be purified by distillation under atmospheric or reduced pressure. Lower members of the trialkylbismuthines are spontaneously inflammable in air, suggesting the ease of oxidative cleavage of the Bi–C bond by molecular oxygen. However, although the Bi–C_{alkyl} bond in alkyldiphenylbismuthines is readily cleaved by air or light, coordinative or bulky aryl groups can stabilize this labile Bi–C_{alkyl} bond enough such that the compounds may be handled in air without decomposition. Mixed tertiary alkylarylbismuthines tend to undergo dismutation on storage to produce a mixture of trialkyl-, alkyldiaryl-, diarylalkyl- and triarylbismuthines. In contrast to the alkyl counterparts, triarylbismuthines are generally stable solids and can be purified by recrystallization from organic solvents. They are soluble in benzene, dichloromethane, chloroform, thf, diethyl ether, DMSO and EtOAc, slightly soluble in hexane, and most of them nearly insoluble in MeOH. Highly hindered tertiary bismuthines become unstable in solution and in the solid state when exposed to air, probably owing to the steric crowding around the bismuth atom [67].

Electron diffraction analysis of gaseous trimethylbismuthine [68] as well as X-ray diffraction analysis of some triorganylbismuthines have established that the geometry of the bismuth atom in these compounds is essentially trigonal pyramidal [Scheme 6.1]. Triphenylbismuthine has a mean Bi–C bond distance of 2.24 Å and a mean C–Bi–C bond angle of 94°, indicating a very small contribution of 6s6p orbital hybridization at the bismuth atom [69]. The mean Bi–C bond lengths are considerably lengthened and the mean C–Bi–C bond angles are widened in crowded triarylbismuthines such as tris[bis(trimethylsilyl)methyl]bismuthine (2.328 Å and 102.9°) [70], tris(2,4,6-trimethylphenyl)bismuthine (2.29 Å, 2.31 Å and 102°, 103°) [71, 72], tris[2,4,6-tris(trifluoromethyl)phenyl]bismuthine (2.367 Å and 106.0°) [67], and tris(2,6-dimethoxyphenyl)bismuthine (2.263 Å and 99.4°) [72], as compared with those of triphenylbismuthine. The crystal structure of the deep-red tris(cyclopentadienyl)bismuthine has been determined at –70°C, and the bismuth atom in this molecule is found to possess a distorted trigonal pyramidal geometry being directly bonded to three carbon atoms of the cyclopentadienyl ligands in a monohapto fashion; the mean Bi–C bond length and C–Bi–C bond angle are 2.38 Å and 94.7°, respectively [57]. The bismuth centre is also weakly coordinated to the cyclopentadienyl ring of a neighbouring molecule with a Bi–Cp (centroid) distancc of 3.17 Å.

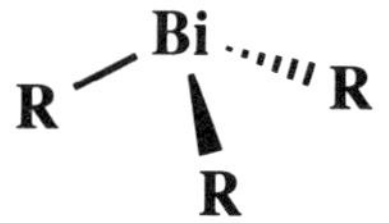

Scheme 6.1 A triorganylbismuthine.

Reactions. Almost all reactions of trialkylbismuthines begin with the cleavage of the carbon–bismuth bond. They are rapidly oxidized in air, affording R_2BiOR, RBiO, and polymeric bismuth oxides containing alkyl groups. Trialkylbismuthines react with one equivalent of halogens at room temperature to form the corresponding dialkylhalobismuthines and halogenoalkanes. In contrast to the alkyl derivatives, however, most triarylbismuthines are stable toward molecular oxygen but add halogens onto the bismuth atom to yield the corresponding triarylbismuth dihalides. Such oxidative halogenation can be easily accomplished by using a variety of halogenating agents, which include sulphur dichloride, sulphur monochloride, sulphuryl chloride, thionyl chloride, iodine trichloride, phenyliodine dichloride, and others (cf. Section 6.3.1). However, the reaction between triarylbismuthines and iodine usually leads to diaryliodobismuthines and an iodoarene via the reductive coupling of initially formed triarylbismuth diiodide. Compounds of the general formula Ar_3BiX_2, where X is an anionic group other than halogen, have also been prepared by the oxidative addition of electrophiles to triarylbismuthines (cf. Section 6.3.1). The addition of carbene, nitrene or oxygen molecules to triphenylbismuthine produces the corresponding ylide, imide and oxide (cf. Sections 6.3.5, 6.3.6 and 6.3.7, respectively).

The unshared pair of electrons in tertiary bismuthines possesses considerable s character owing to the small extent of hybridization between the 6s and 6p orbitals. Thus the number of known coordination compounds in which bismuthines are used as donor ligands for transition metals is limited as compared with the lighter pnictogen counterparts (chapter 8).

The Bi–C bond of bismuthines is readily cleaved by acid reagents such as hydrochloric acid, sulphuric acid, nitric acid, hydrazoic acid, hydrogen sulphide, carboxylic acids, acyl chlorides, sulphonic acids, thiophenols, selenols, and perfluoroalkyl iodides. A wide variety of metals and metal salts can also cleave the Bi–C bond of bismuthines, and these reactions are employed in the synthesis of compounds such as R_2BiX and $RBiX_2$, where X denotes halogen atoms or other anionic groups (cf. Sections 6.2.2, 6.2.3 and 6.2.4). Carbon acids such as 1,3-diketones also cleave the Bi–C bonds of triphenylbismuthine to yield the corresponding bismuth tris-(enolate) complexes, which find use as materials for gas phase deposition of Bi-containing superconducting films [73].

Ultraviolet irradiation of triphenylbismuthine in organic solvents causes the homolytic cleavage of the Bi–C bond; benzene, biphenyl and phenylation products of the solvent as well as metallic bismuth are obtained [74]. The Bi–C bond of alkyldiphenylbismuthines is cleaved even by ambient light [39].

Action of sodium or potassium metal on triorganylbismuthines in liquid ammonia generates the corresponding diorganylbismuthides, R_2BiM (M = Na, K), as highly reactive species (cf. Section 6.2.6). Although titanium tetrachloride alone induces the polymerization of olefins, a combination of it with triphenylbismuthine is found to accelerate the Diels–Alder reaction between alkyl acrylate and cyclopentadiene without the accompanying polymerization of both olefins [75].

Triarylbismuthines have been used as oxidizing and arylating agents [9–12]. Combined use of a catalytic amount of triphenylbismuthine and a stoichiometric amount of NBS (N-bromosuccinimide) converts 1,2-diols to carbonyl compounds via oxidative cleavage of the central C–C bond, where the Bi(V) species, generated *in situ*, is assumed to act as the oxidant [76, 77]:

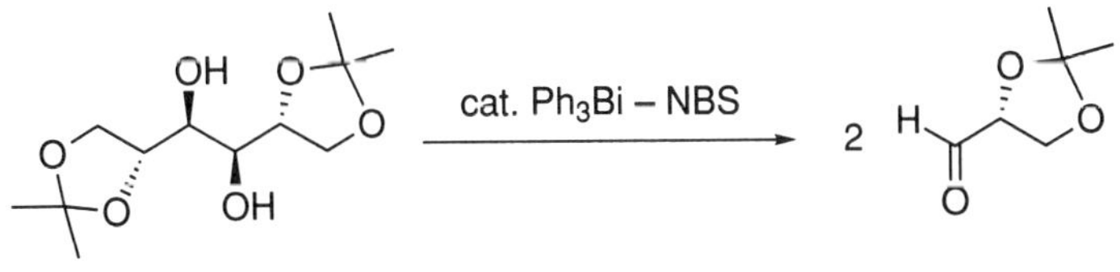

[6.25]

In the presence of a copper catalyst, triarylbismuthines can convert alcohols and amines directly to the corresponding aryl ethers and aryl amines [9–12, 78]:

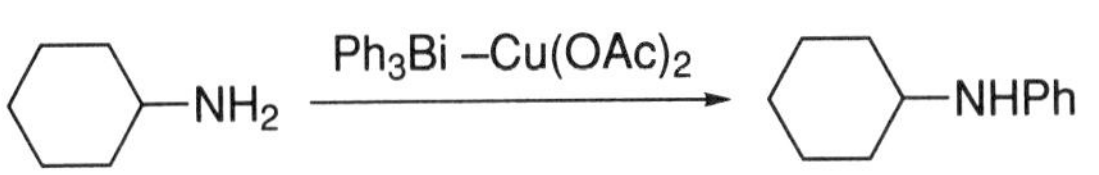

[6.26]

The Pd-catalyzed Heck-type arylation of olefins with triarylbismuthines has been reported, and the transmetallation between a Pd catalyst and triarylbismuthines leads to the formation of biaryls [79–81]:

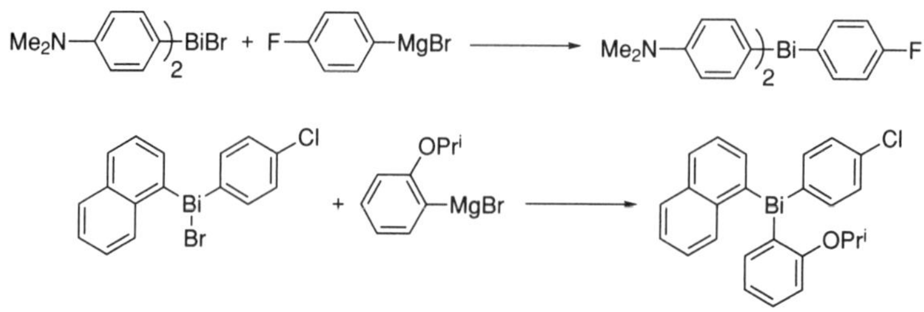

[6.27]

A similar transmetallation was used for the synthesis of ketones from acyl chlorides and triphenylbismuthine [81], and the Rh-catalysed or Pd-catalysed carbonylation of triarylbismuthines with carbon monoxide gives diaryl ketones, methyl benzoates and biaryls, where the distribution of the products is strongly dependent on the catalysts employed [82, 83]:

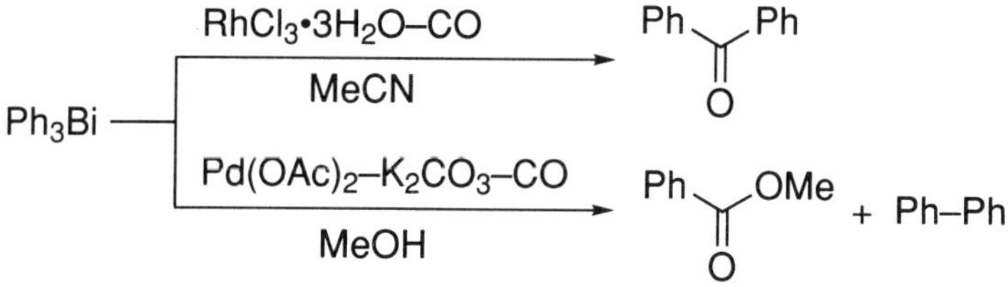

[6.28]

Tris(2,6-dimethoxyphenyl)bismuthine promotes dehydration between α-unbranched carboxylic acids and alcohols or amines under neutral conditions to afford the corresponding esters or amides, although triphenylbismuthine lacks such an ability [84, 85]. This method has been applied successfully to the template synthesis of macrocyclic diesters [86]:

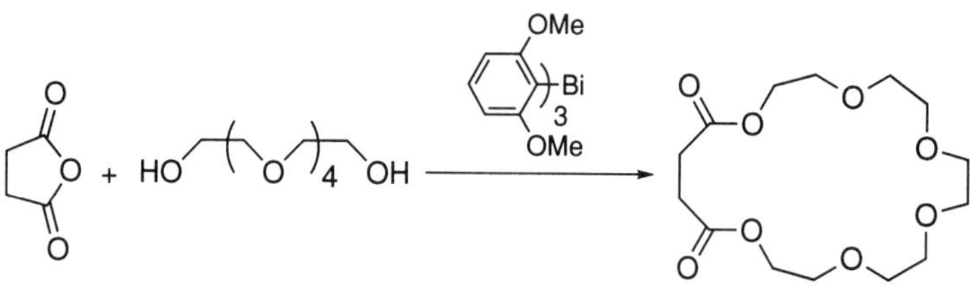

[6.29]

Alkylbismuthines have so far found little use in organic synthesis, but treatment of allyl bromide with carbonyl compounds in the presence of tributylbismuthine has been reported to yield homoallyl alcohols and allyl homoallyl ethers [87]:

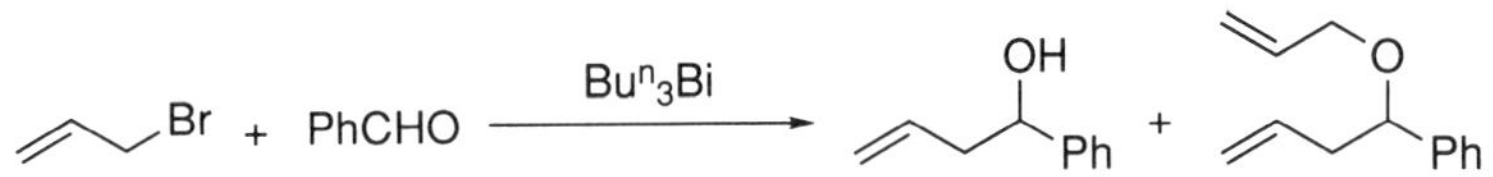 [6.30]

In the presence of an equimolar amount of copper(II) acetate, primary and secondary amines are easily *N*-alkylated by alkyldiphenyl- and trialkylbismuthines [88]:

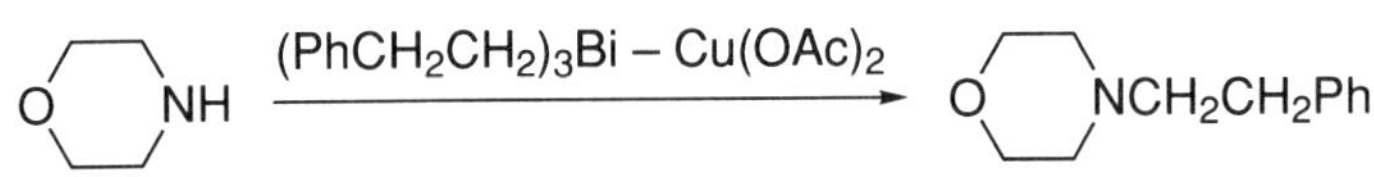 [6.31]

On treatment with benzoyl chloride, a sterically crowded alkynyldiarylbismuthine releases its alkynyl moiety to yield the corresponding diarylbismuth chloride and an alkynyl ketone [89]:

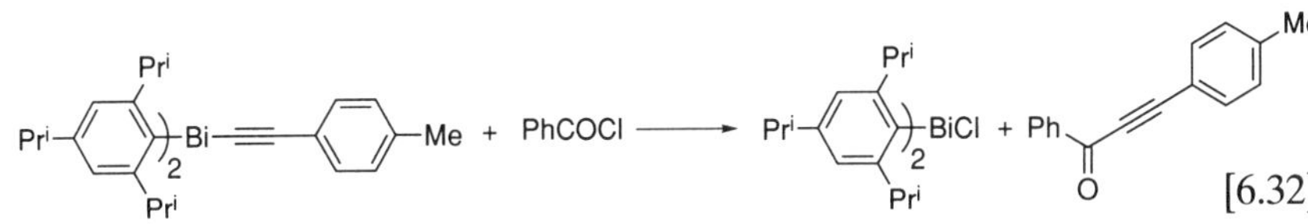 [6.32]

6.2.2 Organobismuth(III) compounds with bismuth–halogen bonds

Preparation

Redistribution reactions. A number of mono- and dihalobismuthines (R_2BiX and $RBiX_2$) have been prepared by the redistribution reaction between tertiary bismuthines (R_3Bi) and bismuth trihalides (BiX_3). The monohalobismuthine is obtained when 1 equivalent of BiX_3 and 2 equivalents of R_3Bi are allowed to react in an inert organic solvent, but if the molar ratio of reactants is reversed, the dihalobismuthine is obtained as the major product [90]:

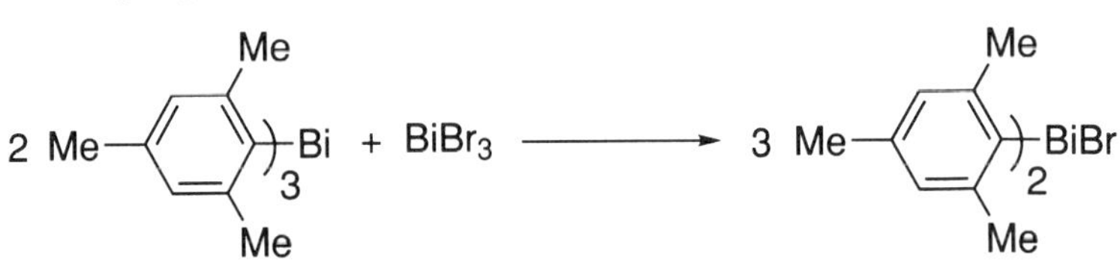 [6.33]

Metathetical reactions. Mono- and diiodobismuthines are easily prepared by treating the corresponding chloro-, bromo-, dichloro-, or dibromobismuthines with an alkali metal iodide [91, 92]:

$$Ph_2BiCl + NaI \longrightarrow Ph_2BiI \qquad [6.34]$$

From organometallic reagents. The reaction between an appropriate molar ratio of an organometallic compound and a bismuth trihalide can be used for the preparation of mono- and dihalobismuthines. Sterically crowded aryllithium and aryl Grignard reagents allow the introduction of only two aryl moieties onto the bismuth centre [67, 89, 93]:

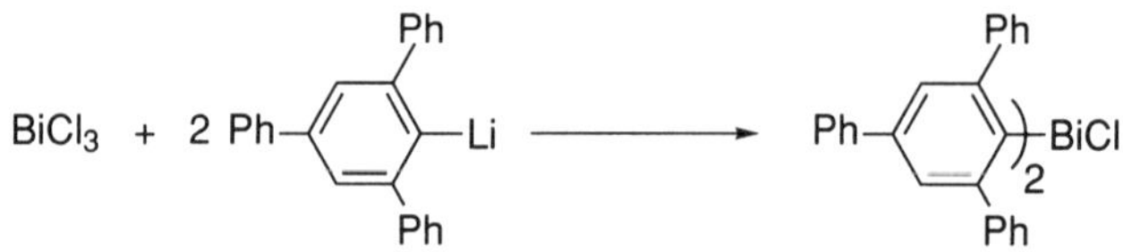

[6.35]

Bismuth trichloride readily induces the cleavage of the C–C bond of siloxycyclopropanes and the Si–C bond of trimethylsilylcyclopentadiene to give the corresponding functionalized alkylbismuth dichlorides [94, 95]:

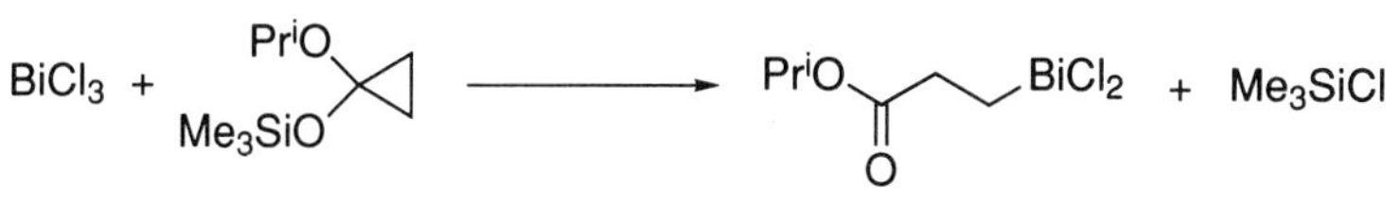

[6.36]

Asymmetrical diarylhalobismuthines have been synthesized by the reaction of aryldihalobismuthines with 1 equivalent of a lithium reagent [96]. Alkyl- and aryllithium reagents having an alkylamino group at a suitable position in the molecule couple with 1 equivalent of bismuth trichloride to afford stable monoorganylbismuth dichlorides, in which the amino group is intramolecularly coordinated to the Lewis acidic bismuth centre [97, 98]:

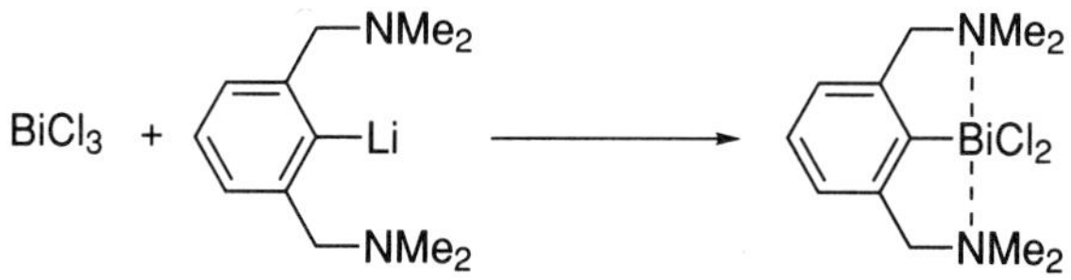

[6.37]

Halogenolysis of tertiary bismuthines. When treated with 1 equivalent of halogen at ambient temperature trialkylbismuthines are converted to the corresponding dialkylhalobismuthines, and a similar reaction between triarylbismuthines and iodine affords diaryliodobismuthines. When different organyl ligands are attached to the bismuth atom in asymmetrical triarylbismuthines, the most electron-donating ligand is preferentially removed [99]:

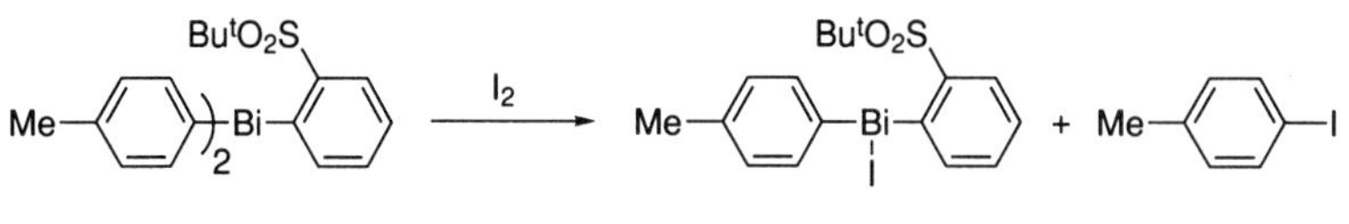

[6.38]

Acidolysis of tertiary bismuthines. Lewis acids such as BF_3, $AlCl_3$, $TlCl_3$, $SiCl_4$, $SnCl_4$, PCl_3, $AsCl_3$, $SbCl_3$, $TiCl_4$, $FeCl_3$, $HgCl_2$, $CuCl_2$ and $ZnCl_2$ as well as protonic acids such as HCl and HBr all cleave the Bi–C bond of tertiary bismuthines to afford the corresponding mono- or dihalobismuthines. When different organyl ligands attached to a bismuth atom are present in asymmetrical triarylbismuthines, the most electron-donating ligand is preferentially removed [36, 41, 66, 99–102]:

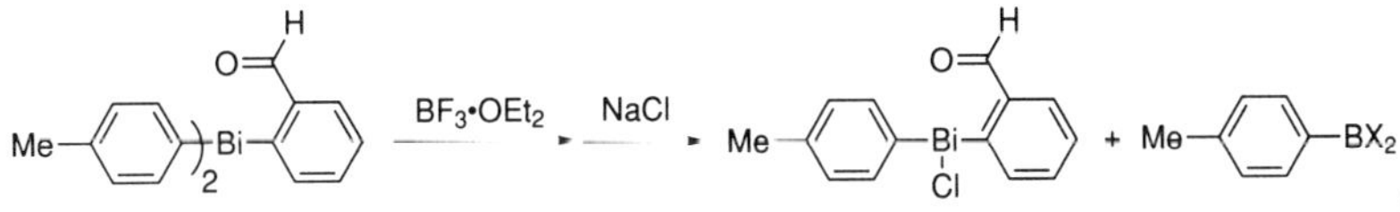

[6.39]

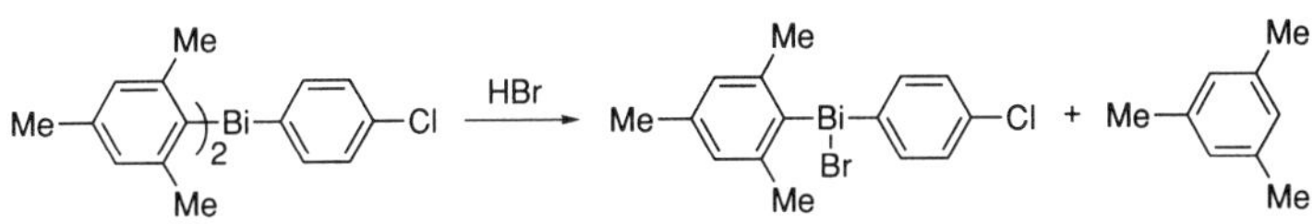

[6.40]

From bismuth metal. Activated bismuth metal reacts with alkyl halides or trifluoromethyl iodide in the presence or absence of copper at high temperatures to form a mixture of halobismuthines [63, 103, 104]:

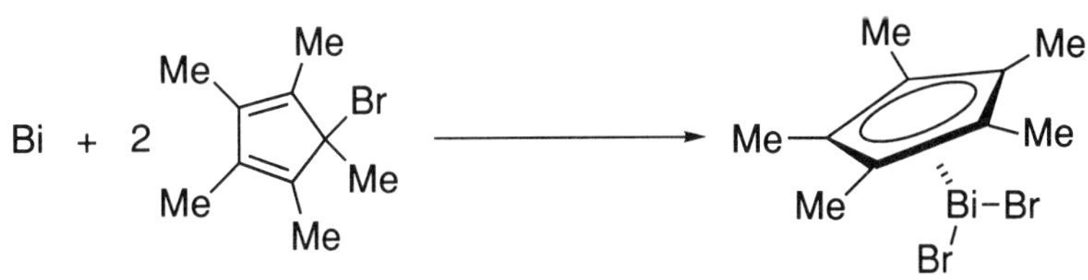

[6.41]

When allylic halides are used bismuth is inserted into the carbon–halogen bond to generate reactive allylic bismuth(III) halides *in situ* [105–108]. Electrolysis of alkyl halides at a sacrificial bismuth cathode forms halobismuthines [60].

Properties. Lower members of the dialkylhalobismuthines are extremely reactive and spontaneously inflammable in air. Crystalline mono- and dihalobismuthines are poorly soluble in non-polar organic solvents owing to the intermolecular polymeric association between the bismuth and halogen atoms in the solid state. Halobismuthines other than iodides are generally moisture-sensitive and gradually hydrolysed in open air. However, complexation with an appropriate donor molecule makes halobismuthines soluble in common organic solvents, and their Bi–X bond becomes less sensitive toward moisture. The presence of a coordinating functional group or sterically demanding group in the organyl ligand also stabilizes the labile Bi–Cl and Bi–Br bonds considerably. In solution, some diorganylhalobismuthines are known to undergo slow disproportionation to yield an equilibrium mixture of corresponding tri-, di- and monoorganylhalobismuthines.

X-ray diffraction analyses have been performed for a variety of halobismuthines. The coordination geometry of the bismuth atom is strongly dependent on the ligand attached. The bismuth atom in bromodimesitylbismuthine has a trigonal pyramidal geometry with the bond angles around the bismuth atom of 89.4°–106.4° [90]. The mean Bi–C bond length is 2.269 Å and the intramolecular Bi–Br bond length is 2.696 Å, and a significantly longer intermolecular Bi···Br distance is 3.795 Å. Chlorobis[2,4,6-tris(trifluoromethyl)phenyl]bismuthine also possesses a trigonal pyramidal bismuth centre with bond angles around the bismuth atom of 87.8°–106.9°, a Bi–Cl bond length of 2.463 Å and relatively longer Bi–C bond lengths of 2.338–2.356 Å [67]. When a coordinative group such as a carbonyl, a sulphonyl or an amino moiety is placed at a spatially suitable position in the molecule, the bismuth centre adopts a disphenoidal or pseudotrigonal bipyramidal geometry [97–102], wherein the coordinating oxygen or nitrogen atom and the halogen atom occupy the axial positions, whereas the two *ipso* carbons of the aryl groups and a pair of unshared electrons are located at the equatorial positions [Scheme 6.2(a)]. For example, in chloro[2-(*N*,*N*-dimethylaminomethyl) phenyl](4-methylphenyl)bismuthine, the Bi–N distance (2.525 Å) is much shorter than the sum of their van der Waals radii [101]. The relatively long Bi–Cl bond length (2.700 Å) as well as the wide N–Bi–Cl bond angle (162.9°) suggests the operation of a transannular interaction between the bismuth and nitrogen atoms, which makes this molecule adopt a pseudo-trigonal bipyramidal structure with the chlorine and nitrogen atoms at both axial sites [101]. The thf complexes of the formula [$PhBiX_2$(thf)], where X denotes chlorine, bromine or iodine, form halogen-bridged polymeric chains in which each bismuth centre is located in a square-based pyramidal coordination environment [Scheme 6.2(b)] [109–111]. In these molecules, the phenyl group occupies the apical site, and three halogen atoms and the oxygen atom of thf occupy the four basal positions. Some donor adducts of the type $PhBiX_2 \cdot L$ have been shown to have a dimeric structure with a similar square-based pyramidal geometry around each bismuth centre [112]. In both cases, L is *trans* to X as discussed briefly in ref. [112] and more fully in chapter 1.

In general, therefore, R_2BiX compounds usually coordinate one two-electron donor (L) to give four-coordinate disphenoidal species [R_2BiX(L)] as shown in Scheme 6.2(a) whereas $RBiX_2$ compounds coordinate two two-electron donors affording five-coordinate, square-based pyramidal species [$RBiX_2(L)_2$] as shown in Scheme 6.2(b).

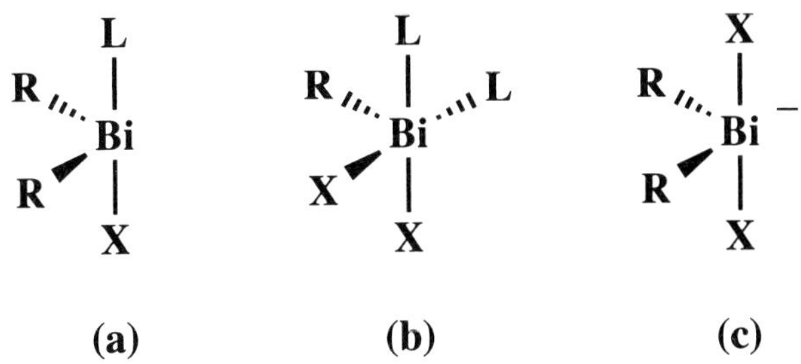

Scheme 6.2 Some organylhalobismuthines. R = organyl group; X = halide; L = two-electron donor group. Structures (a)–(c) are described separately in text.

However, four-coordinate, ten-electron bismuthate complexes of the formula $[R_2BiX_2][Ph_4P]$ have a pseudo-trigonal bipyramidal geometry at the bismuth centre with two halogen atoms at the axial sites and two organyl ligands and a pair of unshared electrons in the equatorial sites [Scheme 6.2(c)] [109, 110]. The X-ray analysis of a red-coloured thermolabile, light-sensitive compound, $C_5H_5BiCl_2$, has revealed that the cyclopentadienide ring is π-bonded to the bismuth atom, in which pseudo-rotational disorder arising from packing requirements results in a mixture of η^2-bonded states (Bi–C bond lengths 2.47 Å and 2.48 Å) and η^3-bonded states (Bi–C bond lengths 2.38 Å, 2.61 Å and 2.64 Å).

Reactions. Diorganylhalobismuthines and monoorganyldihalobismuthines undergo facile metathetical reactions with a variety of metal salts. This reaction is useful for obtaining iodo- and diiodobismuthines as well as the compounds of the general formula R_2BiY and $RBiY_2$, where Y denotes an anionic group other than halogen (cf. Sections 6.2.3 and 6.2.4). When treated with organolithium or Grignard reagents, halobismuthines undergo Bi–C coupling to give the corresponding asymmetrical triorganylbismuthines (cf. Section 6.2.1). The metathetical reaction of mono- or dihalobismuthines with transition metal anions gives organobismuthines having the bismuth–transition metal bond(s) (cf. Section 6.2.6) [113–115]:

$$PhBiBr_2 + [Co(Co)_3(PPh_3)]K \longrightarrow [PhBi\{Co(CO)_3PPh_3\}_2] \qquad [6.42]$$

In liquid ammonia or thf, diorganylhalobismuthines are reduced by 2 equivalents of alkali or alkaline earth metals such as Li, Na, K, Ca and Ba to give a highly reactive compound of the formula R_2BiM or $(R_2Bi)_2M$ as deep-red solutions (cf. Section 6.2.6). When one equivalent of sodium or cobaltocene is used for the reduction, tetraorganyldibismuthines are obtained as orange-red crystals (cf. Section 6.2.5). The reduction of methyldichlorobismuthine and dimethylchlorobismuthine with $LiAlH_4$

at low temperatures gives methylbismuthine and dimethylbismuthine, respectively, but both of these compounds rapidly decompose at room temperature [116]:

$$Me_2BiBr + LiAlH_4 \xrightarrow[-60\ ^\circ C]{} [Me_2BiH] \qquad [6.43]$$

Diphenylhalobismuthines and phenyldihalobismuthines react with tetraorganylarsonium salts to give compounds of the formula $[R_4As][Ph_2BiX_2]$ and $[R_4As][PhBiX_3]$, where the bismuth moiety exists as the anionic species [117]:

$$Ph_2BiCl + [Ph_4AsCl] \longrightarrow [Ph_4As][Ph_2BiCl_2] \qquad [6.44]$$

Similar types of ammonium salts have also been reported [118]. Cationic species $[Ph_2Bi(L)_2]^+$, where L denotes triphenylphosphine oxide, triphenylarsine oxide or hexamethylphosphoric triamide, have been prepared by halide abstraction with use of $AgBF_4$ of the corresponding halobismuthines and subsequent complexation with donor molecules [119]:

$$Ph_2BiBr + Ag[BF_4] \xrightarrow{2\ HMPA} [Ph_2Bi(HMPA)_2][BF_4] \qquad [6.45]$$

Similarly, pyridine, 2,2′-bipyridine and 1,10-phenanthroline adducts of Ph_2BiX and $PhBiX_2$ are obtained by the reaction of the corresponding halobismuthines with donor molecules [120, 121].

Although a definite reactive species remains to be identified, the assumed mono- or diallylbismuth(III) halides generated *in situ* from allylic halides and activated bismuth metal readily undergo the Barbier-type or Grignard-type reaction with aldehydes to give homoallylic alcohols in good yield [105–108]:

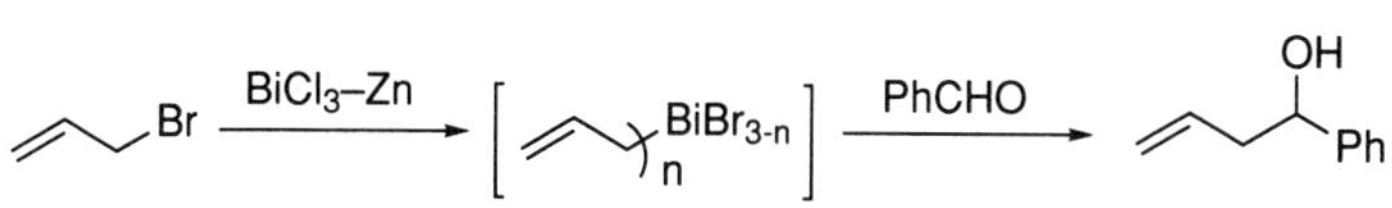

[6.46]

Activated bismuth metal is prepared by reducing a bismuth trihalide with aluminum, zinc, or iron or at an electrode [122].

6.2.3 *Organobismuth(III) compounds with bismuth–chalcogen bonds*

Preparations

Metathetical reactions. Mono- and dihalobismuthines undergo metathetical reactions with alkoxides, aryloxides, siloxides, carboxylates, sulphonates, sulphides, thiolates, thiocarbamates or xanthates to afford the corresponding bismuthines bearing one or two Bi–O/Bi–S bond(s) [123]:

$$Ph_2BiBr + NaOPr^i \longrightarrow Ph_2BiOPr^i \qquad [6.47]$$

When treated with sodium hydroxide, μ-oxobis(diarylbismuth) or polymeric organobismuth oxides are obtained instead of the expected hydroxybismuthines [124]:

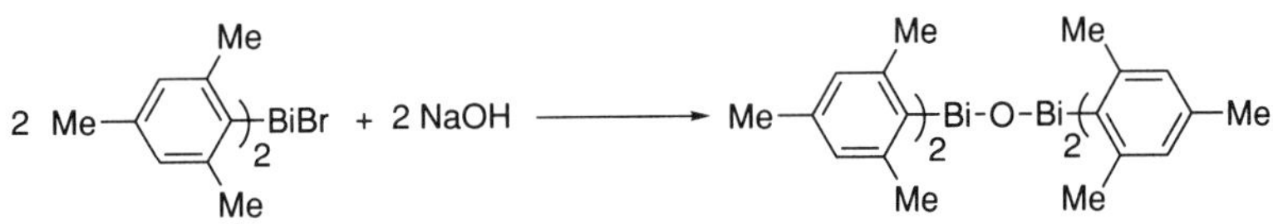

[6.48]

Alkoxybismuthines are converted to thiobismuthines simply by treating with thiols [125]:

$$MeBi(OPr^i)_2 + 2\,Pr^iSH \longrightarrow MeBi(SPr^i)_2 + 2\,Pr^iOH \qquad [6.49]$$

Redistribution reactions. Tetraorganyldibismuthines readily undergo redistribution reactions with diaryl diselenides or ditellurides, affording the corresponding arylseleno- and aryltellurodiorganylbismuthines (cf. Section 6.2.5) [126–129]:

$$Ph_2BiBiPh_2 + PhSeSePh \longrightarrow 2\,Ph_2BiSePh \qquad [6.50]$$

Elemental chalcogens also cleave the Bi–Bi bond of dibismuthines to give the corresponding bis(diorganylbismuth) oxide, sulphide, selenide or telluride (cf. Section 6.2.5) [124, 127–132]:

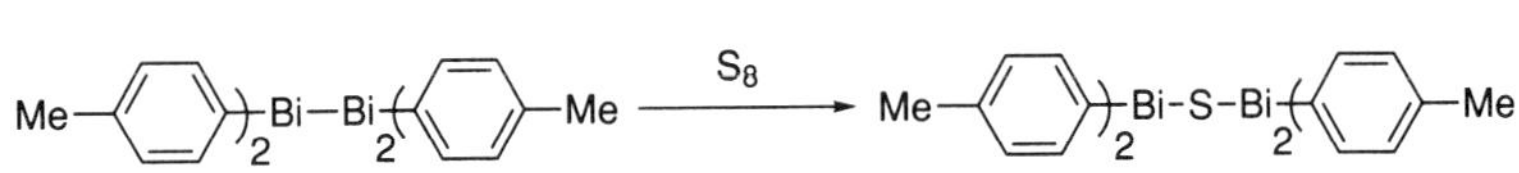

[6.51]

Bismuth tris(phenylsulphide) reacts with free carbon radicals generated from thiohydroxamate esters to give organylbismuth disulphides as the reactive species [133]:

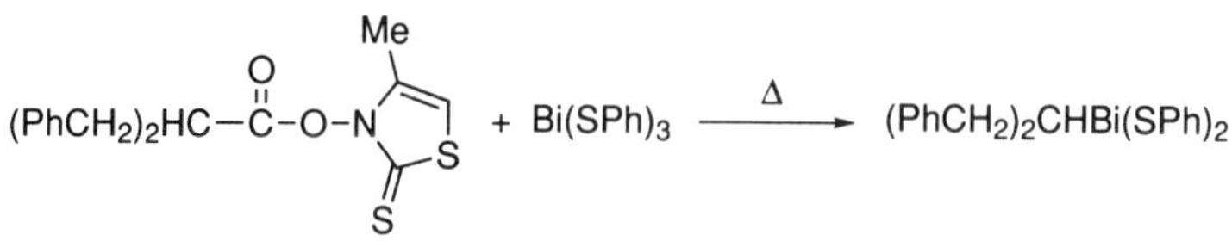

[6.52]

Bi-C bond cleavage. Carboxylic acids, sulphinic acids and sulphonic acids readily cleave the Bi−C bond of triorganylbismuthines to give the corresponding mono- or diorganylbismuth carboxylates, sulphinates and sulphonates [134]. This approach is convenient for the synthesis of asymmetrical diarylbismuth compounds of the formula ArAr′BiY [37]. Hydrogen thiocyanate and selenocyanate also cleave the Bi−C bond of bismuthines to yield the corresponding diorganylbismuth thiocyanate and selenocyanate [135, 136]. On prolonged refluxing in toluene, triethylbismuthine reacts slowly with phenol to yield a chiral chain polymer of the formula $[Et_2Bi(OAr)]_\infty$, in which bridging aryloxides join Et_2Bi groups containing a stereoactive lone pair of electrons [137]:

$$Et_3Bi + C_6F_5OH \longrightarrow [Et_2Bi(OC_6F_5)] + C_2H_6 \qquad [6.53]$$

Properties and reactions. The stability of the organobismuth compounds R_2BiY and $RBiY_2$ bearing one or two bismuth–chalcogen bond(s) is greatly dependent on the nature of the anionic group Y attached. Thus alkoxybismuthines are relatively unstable and sensitive to water and alcohols, but they are soluble in benzene and dichloromethane and slowly decompose in solution to give bismuth oxides and triorganylbismuthines. A similar disproportionation reaction is also observed with bis(diorganylbismuth) oxides ($R_2BiOBiR_2$) in solution. Thiobismuthines are usually yellow oils or solids and are insensitive to moisture, being more stable than alkoxybismuthines in solutions of organic solvents such as acetone, acetonitrile, DMSO and DMF (*N*,*N*-dimethylformamide). Bismuthines with an electron-withdrawing group such as a carboxylate, sulphonate, thiocarbamate and xanthate are relatively sensitive to moisture and undergo gradual redistribution reactions in solution to form an equilibrium mixture of all possible Bi(III) species [134]. As is the case with halobismuthines, however, donor molecules such as DMSO and HMPA or intramolecular coordinating groups stabilize this type of bismuthine through coordination, making them insensitive toward moisture.

A number of X-ray diffraction analyses have been carried out to determine the structures of this class of compound. The coordination geometry

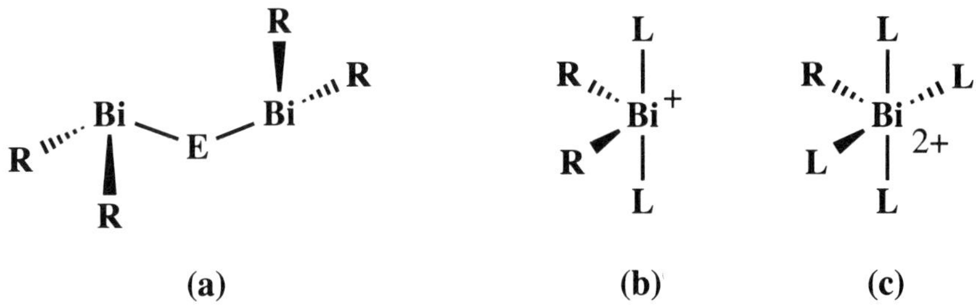

Scheme 6.3 Some organobismuthines with bismuth–chalocogen bonds. R = organyl group; L = two electron donor group; E = O, S, Se. Structures (a)–(c) are described separately in text.

about the bismuth atom in $Ph_2BiSePh$ is trigonal pyramidal with the bond angles around the bismuth of 87.4(3)°–97.4(3)° and a Bi–Se bond length of 2.74 Å [129]. The bismuth atom in the bis(dimesitylbismuth) chalcogenides ($Mes_2BiEBiMes_2$) also has a pyramidal geometry, with mean bond angles around the bismuth of 95.9° (E=O), 98.1° (E=S) and 98.6° (E=Se), respectively, wherein the carbon atoms adjacent to the bismuth are in a staggered conformation *trans* to each other over the bridging chalcogen atom [Scheme 6.3(a)] [124]. The mean Bi–E bond length and the Bi–E–Bi bond angle of these compounds ($Mes_2BiEBiMes_2$) are 2.106 Å and 117.1(8)° (E=O), 2.533 Å and 98.7(3)° (E=S), and 2.651 Å and 91.2(1)° (E=Se). The coordination geometry around the bismuth centre in bis(diethyldithiocarbamato)methylbismuthine is approximately pentagonal pyramidal with a methyl carbon at the apical site and a free pair of electrons presumably directed opposite the apical carbon atom [138]. The structures of related organobismuth(III) derivatives with dithiocarbamato and xanthato ligands have also been determined [139–142]. The cationic bismuth atom in $[Ph_2Bi(HMPA)_2][BF_4]$ has a four-coordinate, disphenoidal or pseudo-trigonal bipyramidal geometry with two phenyl groups in the equatorial sites and two oxygen atoms of the HMPA ligands in the axial sites (mean Bi–O bond length and O–Bi–O bond angle are 2.364 Å and 167.5°, respectively), and no close interionic contacts are observed between the bismuth atom and tetrafluoroborate anion [Scheme 6.3(b)] [119]. Similar types of cationic bismuth complexes $[Ar_2BiL_2][Y]$ (L = HMPA, triphenylphosphine oxide or pyridine; Y = BF_4 or PF_6) have also been characterized structurally, wherein each bismuth centre is found to adopt an essentially four-coordinate disphenoidal structure [143]. The coordination geometry at the dicationic bismuth centre in $[PhBi(HMPA)_4][PF_6]_2$ is square-based pyramidal with the *ipso* ring carbon atom in the apical position and the four oxygen atoms in basal positions, the bismuth atom lying 0.135 Å below the mean O_4 plane with respect to the phenyl group [Scheme 6.3(c)] [143]. The bismuth atom in

$[Me_2NH(CH_2)_3NHMe_2][PhBi(OCOCF_3)_4]$, where the bismuth may formally constitute a dianion, has a square pyramidal geometry with the bismuth atom 0.27 Å above the plane formed by four oxygen atoms bonded to the bismuth [144]. The coordination geometry at the bismuth atom in $[Et_2Bi(OC_6F_5)]_\infty$ is best described as a pseudotrigonal bipyramid in which the bridging aryloxide ligands occupy the axial positions (the O–Bi–O and Bi–O–Bi bond angles are 179.54° and 113.57°, respectively) [137].

Diarylbismuth triflate–HMPA complexes readily react with an aryl Grignard reagent to yield asymmetrical triarylbismuthines and this methodology has been employed successfully for the preparation of bismuthines bearing three different aryl groups (cf. Section 6.2.1) [37]:

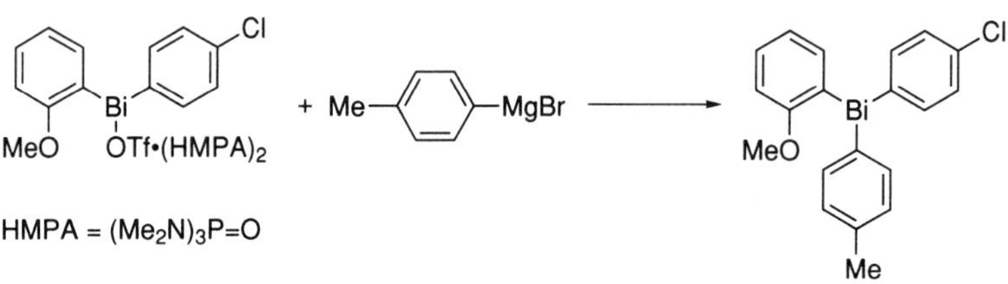

[6.54]

It is highly likely that a mono- or diarylbismuth carboxylate is the real active species in the bismuthine-promoted condensation of carboxylic acids with alcohols or amines (cf. reaction [6.29]) [85, 86].

6.2.4 *Organobismuth(III) compounds with bismuth–nitrogen bonds*

Mono- and dihalobismuthines undergo metathetical reactions with lithium or sodium amide or azide to afford the corresponding bismuthines bearing the Bi–N bond [58, 145–147]:

$$Me_2BiBr + LiN(SiMe_3)Me \longrightarrow Me_2BiN(SiMe_3)Me \qquad [6.55]$$

whereas hydrazoic acid selectively cleaves the Bi–C or Bi–N bond of trimethylbismuthine or bismuth amides [148]:

$$Me_2BiN(SiMe_3)Me + HN_3 \longrightarrow Me_2BiN_3 + HN(SiMe_3)Me \qquad [6.56]$$

Azidodimethylbismuthine is insoluble in common organic solvents such as thf, EtOH and pyridine, and at around 150°C, it turns gray with decomposition. X-ray diffraction analysis of this compound revealed that the bismuth atoms are linked together via the α-N atoms of the azido groups to form the polymeric zig-zag chain [149]. The bismuth centre adopts a distorted pseudotrigonal bipyramidal geometry with two methyl carbons in equatorial sites and two nitrogen atoms in axial positions, wherein the Bi–N bond lengths are 2.49–2.50 Å and the N–Bi–N bond angle is 169°.

Diazomethane reacts with an aminobismuthine to form the corresponding diazo compound [145, 146]:

$$2\ Me_2BiN(Me)SiMe_3 \xrightarrow{CH_2N_2} (Me_2Bi)_2CN_2 + 2\ HN(SiMe_3)Me \quad [6.57]$$

6.2.5 *Organobismuth(III) compounds with bismuth–bismuth bonds*

Two methods have been reported for the synthesis of tetraorganyldibismuthines. Thus, reduction of diorganylbismuth halides with an alkali metal or with cobaltocene, Cp_2Co, in liquid ammonia or thf gives the corresponding tetraorganyldibismuthines; and treatment of sodium dialkylbismuthides with 1,2-dibromoethane also yields tetraalkyldibismuthines [124, 126, 127, 129, 130, 132, 150–159]:

$$2\ Ph_2BiI + 2\ CoCp_2 \xrightarrow[THF]{} Ph_2BiBiPh_2 + 2\ Cp_2CoI \quad [6.58]$$

$$2\ Me_2BiBr + 2\ Na \xrightarrow[liq\ NH_3]{} Me_2BiBiMe_2 \quad [6.59]$$

Asymmetrical dibismuthines are currently unknown.

Recently, the first dibismuthene was reported from the reaction of a cyclic trimer of arylbismuth selenide with tris(dimethylamino)phosphine (Tokitoh, N., Arai, Y., Okazaki, R. and Nagase, S. (1997) *Science*, 277, 78):

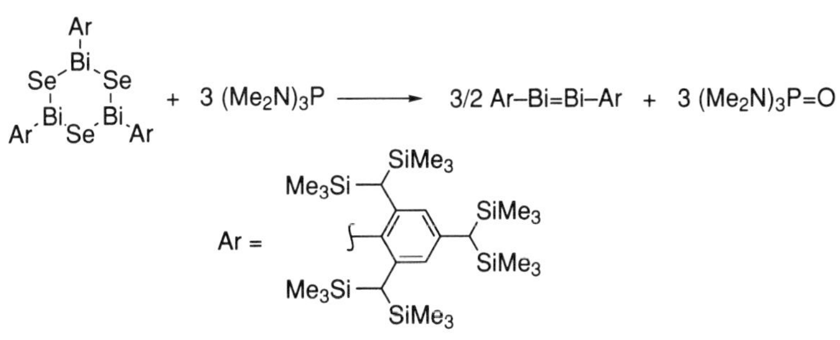

[6.60]

Dibismuthines are very labile and readily react with radical, electrophilic or nucleophilic reagents through cleavage of the Bi–Bi bond; ultraviolet irradiation also induces Bi–Bi bond cleavage. Since dibismuthines are sensitive to oxygen they need to be handled under an inert atmosphere, but even in an inert atmosphere tetramethyldibismuthine decomposes rapidly at room temperature to trimethylbismuthine and bismuth metal. Other known tetraalkyldibismuthines also undergo facile disproportionation. Tetraphenyldibismuthine is relatively stable and decomposes only above 100°C. Dibismuthines are orange to red in colour both in the solid state and in solution, and thermochromism is observed with some compounds, for which structural, spectroscopic and theoretical studies have been made in considerable detail [18]. Several dibismuthines have been structurally characterized by X-ray diffraction analysis. Tetraphenyldibismuthine has a staggered *trans* conformation with a Bi–Bi distance of 2.990(2) Å [152], the geometry of each bismuth atom being trigonal pyramidal with the mean bond angle around bismuth 93.3° and the mean Bi–C bond length 2.27 Å [Scheme 6.4]. The intramolecular Bi–Bi bond distances in tetrakis(trimethylsilyl)dibismuthine [160], tetramethyldibismuthine [161], and 2,2′,5,5′-tetramethyldibismole [155, 162] arc 3.04 Å, 3.12 Å and 2.99 Å, respectively, and the corresponding intermolecular Bi ··· Bi distances are 3.80 Å, 3.58 Å and 3.66 Å, respectively.

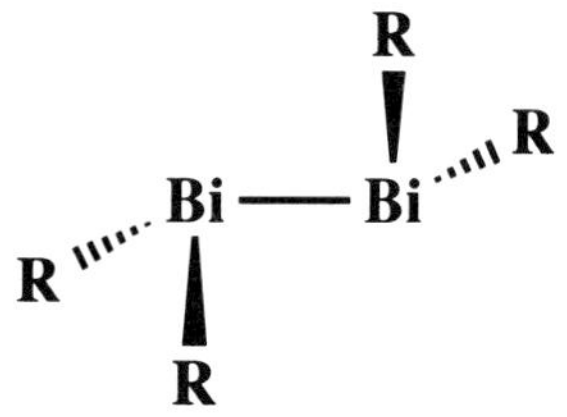

Scheme 6.4 Tetraorganyldibismuthine.

Tetramethyldibismuthine is cleaved with hydrogen chloride to give hydrogen gas, bismuth metal, trimethylbismuthine and chlorodimethylbismuthine [150], the first three products probably deriving from initially formed dimethylbismuthine [116]. Reaction of tetramethyldibismuthine with butyllithium leads to all possible mixed alkylbismuthines [150], and a similar reaction with phenyllithium gives dimethylphenylbismuthine [130]:

$$Me_2BiBiMe_2 + PhLi \longrightarrow Me_2BiPh + Me_2BiLi \qquad [6.61]$$

The Bi–Bi bond is easily cleaved by the action of molecular chalcogens or halogens. Thus the reaction of dibismuthines with elemental chalcogens affords the corresponding bis(diorganylbismuth) oxides, sulphides, selenides or tellurides (cf. Section 6.2.3) [124, 127, 132, 152]. Diorganyldichalcogenides also cleave the Bi–Bi bond to form chalcogenobismuthines quantitatively (cf. Section 6.2.3) [126–129, 132]:

$$Ph_2BiBiPh_2 + PhTeTePh \longrightarrow 2\ Ph_2BiTePh \qquad [6.62]$$

However, (dimethylarsino)- and (dimethylstibino)dimethylbismuthine are found to exist in equilibrium with an original tetramethyldipnictogen and tetramethyldibismuthine [129, 131, 163, 164]. Elemental iodine reacts with tetraphenyldibismuthine to yield diphenylbismuth iodide as the major product [132, 152].

Diazomethane and *p*-benzoquinone can also insert themselves into the Bi–Bi bond of dibismuthines [124, 127, 129, 132, 152]:

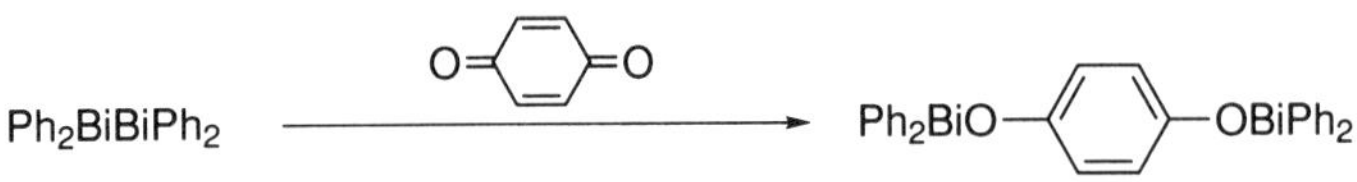

[6.63]

$$Ph_2BiBiPh_2 \xrightarrow{CH_2N_2} Ph_2BiCH_2BiPh_2 \qquad [6.64]$$

Dibismuthines also react with transition metal complexes to give compounds having bismuth–transition metal bonds [165]:

$$Ph_2BiBiPh_2 + [Cr_2Cp_2(CO)_6] \longrightarrow 2\,[Ph_2Bi\{CrCp(CO)_3\}] \qquad [6.65]$$

The thermally stable dibismuthene bearing two bulky aryl moieties has been obtained as a reddish-purple crystalline solid, insoluble in common organic solvents and with a structure characterized by an X-ray diffraction analysis. The bond length between two bismuth atoms is 2.823 Å and the C–Bi–Bi bond angle is 99.6°. Both values are a little bigger than those calculated for HBi=BiH (2.719 Å and 91.8°) [166]. The dibismuthene easily absorbs one oxygen molecule in the solid state to form a cyclic μ-oxobismuth(III) compound without affecting the lattice structure (Tokitoh, N., Arai, Y., Okazaki, R. and Nagase, S. (1997) *Science*, **277**, 78):

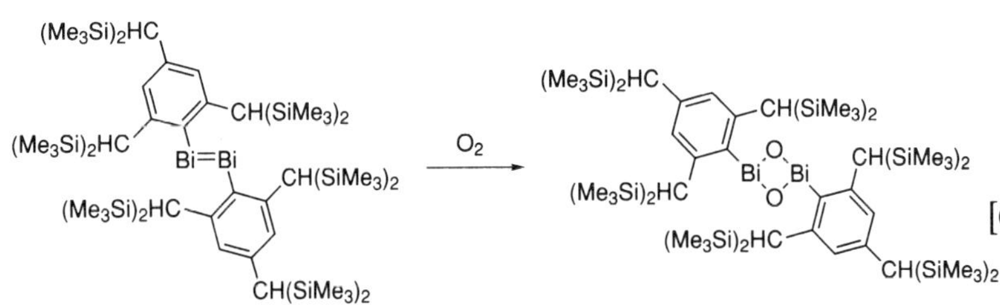

[6.66]

6.2.6 *Organobismuth(III) compounds with bismuth–metal/metalloid bonds*

Alkali or alkaline earth metals such as Li, Na, K, Ca and Ba reduce halodiorganylbismuthines in liquid NH_3 or in thf at low temperatures to generate the corresponding diorganylbismuthides, R_2BiM or $(R_2Bi)_2M$, as deep green to red solutions [151, 167–169]:

$$Ph_2BiCl + 2\,Na \xrightarrow[\text{liq } NH_3]{} Ph_2BiNa \qquad [6.67]$$

and direct conversion of triorganylbismuthines to diorganylbismuthides by the use of sodium or potassium metal has also been reported [150, 153–158, 170]:

$$Ph_3Bi + 2\,K \xrightarrow[\text{THF}]{} Ph_2BiK + PhK \qquad [6.68]$$

The diarylbismuthides thus obtained readily decompose to the corresponding tertiary bismuthines and inorganic bismuth under ambient conditions, but at low temperatures they are stable enough to react with aryl and alkyl halides to form the corresponding asymmetrical tertiary bismuthines (cf. Section 6.2.1) (Rahman, M. M., Matamo, Y. and Suzuki, H. (1997) *Organometallics*, **16**, 3565) [53–56, 167]:

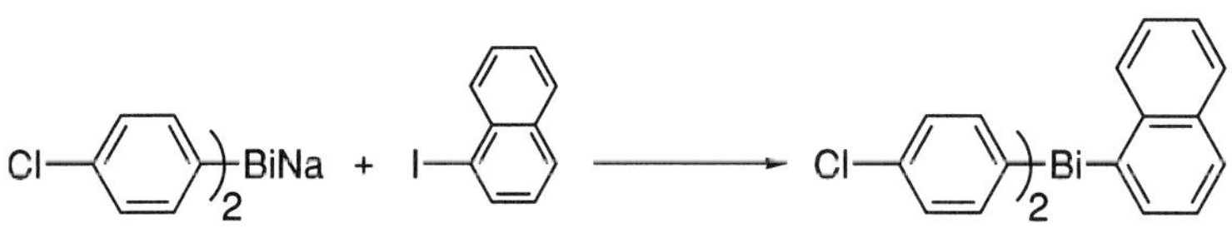

[6.69]

When treated with 1,2-dibromoethane or 1,2-dichloroethane, they undergo bimolecular coupling to yield dibismuthines (cf. Section 6.2.5) [150, 153–157]:

$$2\,Pr_2BiNa + BrCH_2CH_2Br \longrightarrow Pr_2BiBiPr_2 + CH_2{=}CH_2 \qquad [6.70]$$

Triethylbismuthine reacts with a triorganylgermane to give diethyl- and monoethylbismagermanes with one or two Bi−Ge bond(s) depending on the amount of the germane reagent used [171, 172]:

[6.71]

The Bi−Ge bond(s) of these compounds is (are) selectively cleaved by hydrogen chloride to give the corresponding mono- or dichlorobismuthine [171, 172]:

$$[(C_6F_5)_3Ge]_2BiEt + 2\,HCl \longrightarrow EtBiCl_2 + 2\,GeH(C_6F_5)_3 \qquad [6.72]$$

When heated with $CpMH(CO)_3$ (M = Mo or W) one methyl group is cleaved from trimethylbismuthine to yield compounds of the formula $Me_2BiM(CO)_3Cp$ [173]. This class of compounds bearing the bismuth–transition metal bond(s) have also been prepared by the metathetical reaction between diorganylhalobismuthines and salts of the type $Na[ML_n]$ (M = Mo, W, Mn, Fe, Re and Co; L denotes a ligand such as CO, Cp or PPh_3) [113–115, 165, 173, 174]:

$$Ph_2BiI + Na[Mn(CO)_5] \longrightarrow [Ph_2Bi\{Mn(CO)_5\}] \qquad [6.73]$$

$$[\{Cp(CO)_3Mo\}_2BiCl] + KBHEt_3 \longrightarrow [EtBi\{MoCp(CO)_3\}_2] \qquad [6.74]$$

Reaction of halobismuthines having a Bi–M bond with organoaluminium or organoboron reagents gives similar compounds [115, 175, 176]. Cleavage of the Bi–Bi bond with neutral transition metal complexes also forms the corresponding bismuth–transition metal bonds (cf. Section 6.2.5, also chapter 8) [163, 177].

6.3 Organobismuth(V) compounds

6.3.1 Triorganylbismuth(V) compounds

Preparation

Oxidative addition to tertiary bismuthines. The oxidation of triarylbismuthines to pentavalent triarylbismuth dihalides can be accomplished by means of a variety of oxidizing agents, which include halogens, sulphur dichloride, disulphur dichloride, thionyl chloride, sulphuryl chloride, phenyliodine dichloride, xenon difluoride, iodine pentafluoride and iodine trichloride [178–186]. Tris(1-propenyl)bismuthine and trimethylbismuthine can be halogenated at low temperatures by molecular bromine and sulphuryl chloride, respectively, to the corresponding trialkylbismuth dihalides [185, 186]:

$$Me_3Bi + SO_2Cl_2 \longrightarrow Me_3BiCl_2 + SO_2 \qquad [6.75]$$

Lead tetraacetate, benzoyl peroxide, benzoyl nitrate and ozonized oxygen convert triarylbismuthines to the corresponding triarylbismuth(V) dicarboxylates and dinitrates [187–190]:

$$Ph_3Bi + 2PhCOOH \xrightarrow{Bu^tOOH} Ph_3Bi(OCOPh)_2 \qquad [6.76]$$

Oxidative addition of *N*-bromosuccinimide to triphenylbismuthine gives Ph_3BiBr(succinimide) [77, 191].

Metathetical reactions of triarylbismuth dihalides. The metathetical reactions of triarylbismuth dihalides and derivatives have been widely used for the preparation of pentavalent bismuth compounds of the type Ar_3BiX_2, sodium, potassium, silver and lead salts usually being used for this purpose [192, 193]:

$$Ph_3BiCl_2 + 2\ AgOCOCH_2F \longrightarrow Ph_3Bi(OCOCH_2F)_2 \qquad [6.77]$$

Triphenylbismuth molybdate has been prepared by the metathesis of the corresponding dibromide with ammonium molybdate [194]:

$$Ph_3BiBr_2 + (Bu_4N)_2MoO_4 \longrightarrow Ph_3BiMoO_4 + 2\ Bu_4NBr \qquad [6.78]$$

Reactions with carboxylic acids. A number of triphenylbismuth dicarboxylates have been prepared from triarylbismuth dihalides and carboxylic acids in the presence of amines [195, 196]:

$$Ph_3BiBr_2 + 2\ EtSCH_2CO_2H \xrightarrow{2Et_3N} Ph_3Bi(OCOCH_2SEt)_2 + 2\ Et_3NHBr \qquad [6.79]$$

Reaction of triphenylbismuth carbonate with carboxylic acids or sulphonic acids also gives the corresponding dicarboxylates and disulphonates [197–199]:

$$Ph_3BiCO_3 + 2\ AcOH \longrightarrow Ph_3Bi(OAc)_2 \qquad [6.80]$$

Properties. Most triarylbismuth(V) compounds of the formula Ar_3BiX_2 are crystalline, non-hygroscopic solids, which are soluble in benzene, chloroform, dichloromethane, dioxane, thf or acetone, slightly soluble in alcohol or diethyl ether, and insoluble in hexane. The thermal stability of these compounds is highly dependent on the nature of the anionic groups attached such that the stability of triphenylbismuth dihalides decreases in the order fluoride>chloride>bromide>>iodide, the diiodide decomposing spontaneously to diphenyliodobismuthine and iodobenzene even at low temperature. Triphenylbismuth dicarboxylates, disulphonates and dinitrate are usually stable, whereas the diazide is relatively unstable and decomposes rapidly on heating to yield diphenylbismuth azide and phenyl azide [200, 201]. Triphenylbismuth dicyanide also undergoes a similar type of decomposition on heating. Trimethylbismuth dichloride is thermally unstable and decomposes rapidly at room temperature, although its crystal structure has been characterized by low-temperature X-ray diffraction analysis [186].

Triorganylbismuth dihalides have a non-ionic, five-coordinate structure in which the bonds around the bismuth atom occupy a trigonal bipyramidal arrangement. X-ray analyses of trimethylbismuth dichloride (mean Bi–C and Bi–Cl bond lengths 2.197 Å and 2.610 Å, respectively, and mean C–Bi–C and Cl–Bi–Cl bond angles 120.0° and 177.8°, respectively) and triphenylbismuth dichloride (mean Bi–C and Bi–Cl bond lengths 2.12 Å and 2.57 Å, respectively, and mean C–Bi–C and Cl–Bi–Cl bond angles 120° and 176°, respectively) have shown the location of three carbon atoms attached to the bismuth at the equatorial sites and of two chlorine atoms at the axial sites in these molecules (Scheme 6.5) [186, 202]. X-ray investigations of triphenylbismuth(V) dicarboxylates have revealed that the bismuth centre possesses a distorted trigonal (or pentagonal if both of the carboxylate oxygens are considered) bipyramidal structure weakly coordinated by the neighbouring carbonyl oxygen atoms [190, 203, 204]. Thus with $Ph_3Bi(OCOCF_3)_2$, the mean bonded and non-bonded Bi–O distances are 2.31 Å and 2.98 Å, respectively [203]. Intramolecular coordination is also observed in chloro(2-methyl-8-quinolinolato)triphenylbismuth(V), where the bismuth atom has a distorted octahedral geometry with the Bi–C, Bi–Cl, Bi–N and Bi–O bond lengths of 2.20 Å, 2.67 Å, 2.71 Å and 2.19 Å, respectively [205]. The oxygen atom is located at the opposite side of the chlorine atom (O–Bi–Cl 174.5°).

Determination of the molecular weights and electrical conductance as well as NMR and IR spectra of several Ph_3BiY_2-type compounds have clearly demonstrated that the anionic groups are covalently bound to the bismuth and that the molecules have a non-ionic, monomeric five-coordinate structure with a trigonal bipyramidal geometry [201, 206].

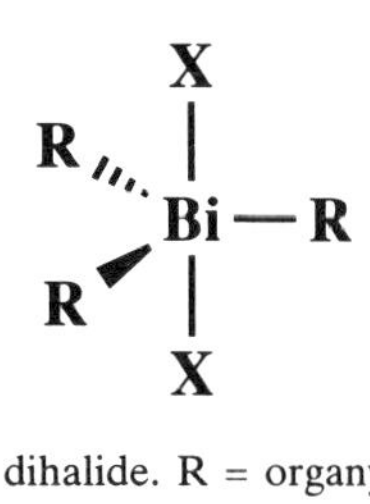

Scheme 6.5 A triorganyl bismuth dihalide. R = organyl group; X = halide.

Reactions. The reduction of triarylbismuth dihalides to the corresponding triarylbismuthines has been carried out by means of a variety of reducing agents, which include hydrazine hydrate, sodium hydrosulphite, $LiAlH_4$, $LiBH_4$ and sodium dialkyldithiocarbamates. In some cases this reduction has been used for the purification of tertiary bismuthines that are difficult to purify in the trivalent state [65]. A single-step, two-electron process has been suggested for the electrolytic reduction of triphenylbismuth dibromide [207].

Treatment of triarylbismuth(V) compounds with nitric acid causes nitration of the aromatic rings affording tris(nitroaryl)bismuth(V) compounds [189, 208]:

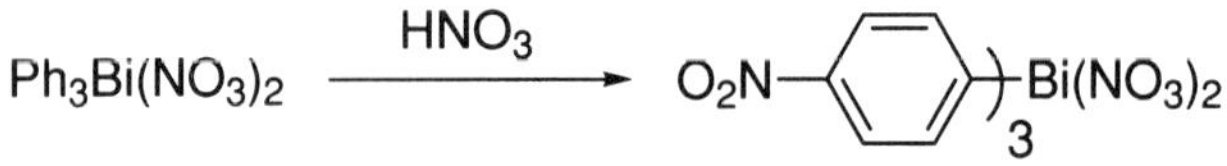

[6.81]

whilst oxidation of the methyl groups of tris(4-methylphenyl)bismuth dichlorides with potassium permanganate or chromic acid affords tris-(carboxyphenyl)bismuth dichlorides, which can subsequently be converted to the corresponding methyl esters [65]:

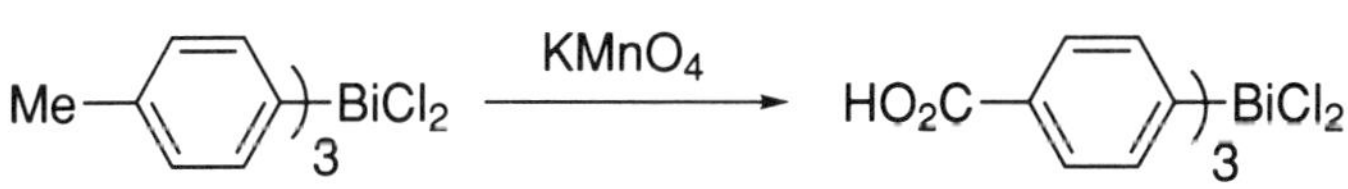

[6.82]

Reaction of triphenylbismuth dichloride with 2 equivalents of phenyllithium in thf at low temperatures affords pentaphenylbismuth (cf. Section 6.3.4), but attempts to prepare tetraarylbismuth halides by the reaction of triarylbismuth dihalides with arylmagnesium bromide have failed. In contrast, treatment of triarylbismuth difluorides with the appropriate organosilanes or organostannanes in the presence of a Lewis acid produces asymmetrical tetraorganylbismuthonium salts (cf. Section 6.3.3).

The reaction of triphenylbismuth carbonate with dimedone gives a stabilized bismuthonium ylide, and related ylides can also be prepared from triphenylbismuth dichloride and α,α′-dicarbonyl- or α,α′-disulphonylmethanes in the presence of a suitable base (cf. Section 6.3.5). A similar base-promoted reaction of triphenylbismuth dihalides with sulphonylamines has been conveniently used for the preparation of bismuthine imides (cf. Section 6.3.6).

The reaction of triphenylbismuth dichloride with mercuric oxide in an alkaline medium leads to triphenylbismuthine, and treatment of triphenylbismuth dicyanide with mercuric oxide gives triphenylbismuthine oxide (cf. Section 6.3.7) [201].

Barton and co-workers have made extensive studies of the oxidation and arylation of organic substrates by using a variety of tri-, tetra- and pentaarylbismuth(V) compounds, details of which have been summarized in several review articles [9–12]. Triphenylbismuth dihalides and related triphenylbismuth(V) compounds can be used for the oxidation of a wide range of primary, secondary, allylic and benzylic alcohols under basic conditions [9–12]. Triphenylbismuth carbonate has also been shown to oxidize alcohols efficiently under neutral conditions [209]:

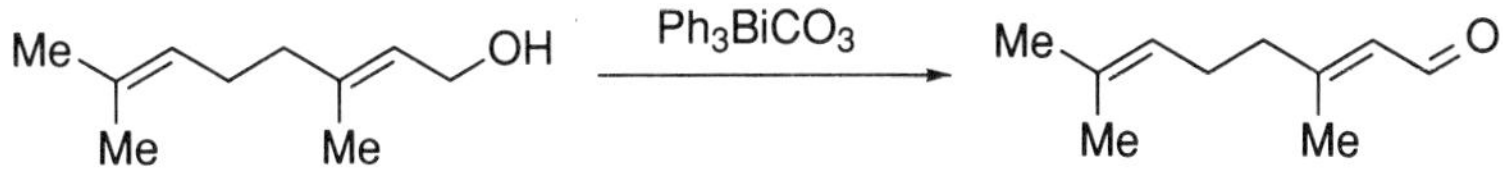

[6.83]

Triarylbismuth compounds of the type Ar_3BiX_2 can be utilized for the arylation of a variety of organic substrates, which include alcohols, enols, amines, phenols, indoles, thiols, hydrazine and nitroalkanes [9–12, 198, 210–212]:

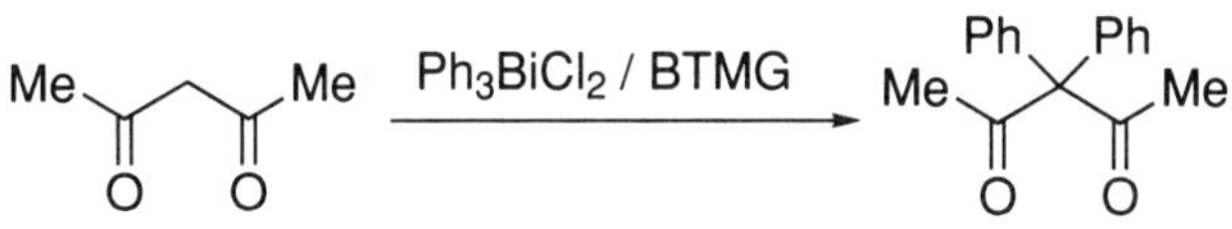

[6.84]

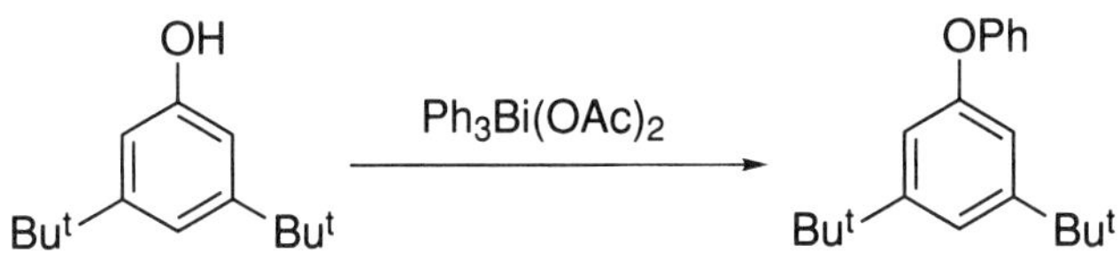

[6.85]

The arylation of alcohols, amines, phenols and indoles is accelerated in the presence of copper or its salts, and two different mechanisms have been proposed for the copper-promoted arylation [9–12, 78, 212–217]:

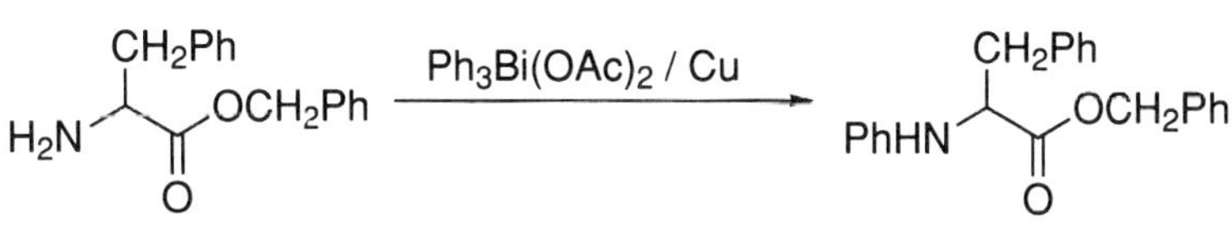

[6.86]

The Lewis acid-promoted reaction of triphenylbismuth diazide with phenyl isocyanate affords carbamoyl azide [218]:

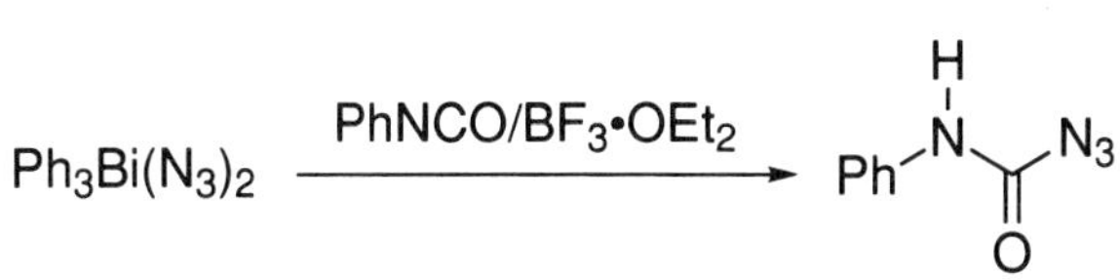

[6.87]

whereas irradiation in the presence of alkynes results in the transfer of one of the azido ligands to the triple bond, leading to *N*-unsubstituted 1,2,3-triazoles [219]:

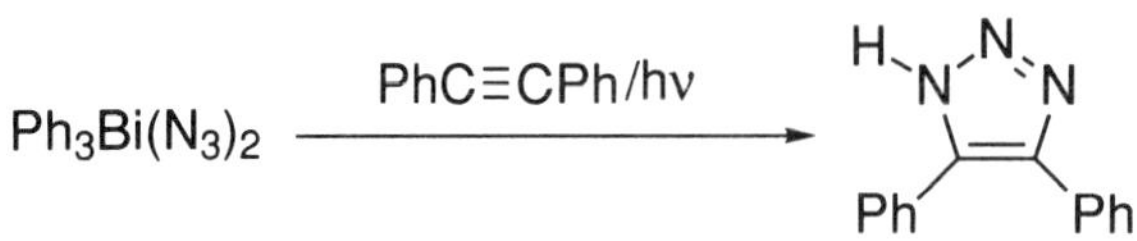

[6.88]

6.3.2 Oxybis[triorganobismuth(V)] compounds

Preparation

From tertiary bismuth dihalides. Metathetical reactions between triphenylbismuth dichloride and silver perchlorate or nitrate gives the corresponding oxybis(triphenylbismuth) compounds, and treatment of triphenylbismuth dihalides with sodium hydroxide in alcoholic solvents affords the corresponding oxybis(triphenylbismuth) dihalides [220]:

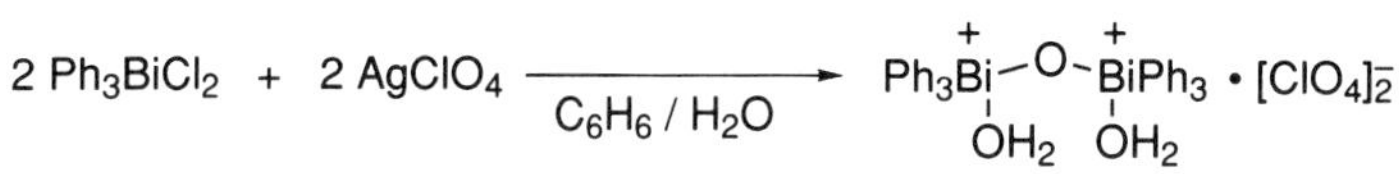

[6.89]

[6.90]

The reaction of triarylbismuth difluorides with hexamethyldisiloxane in the presence of trimethylsilyl triflate yields oxybis(triarylbismuth) ditriflates [221]:

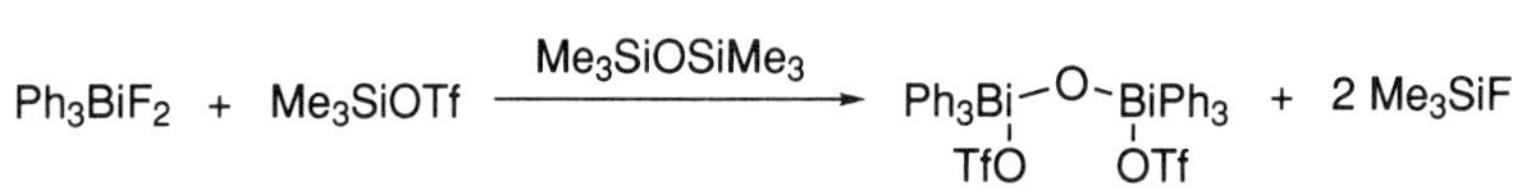

[6.91]

From oxybis(triarylbismuth) dihalides. Oxybis(triphenylbismuth) dichloride undergoes a metathetical reaction with silver salts, and a combination of silver(I) oxide and sulphonic acids can also be used for this purpose [222]:

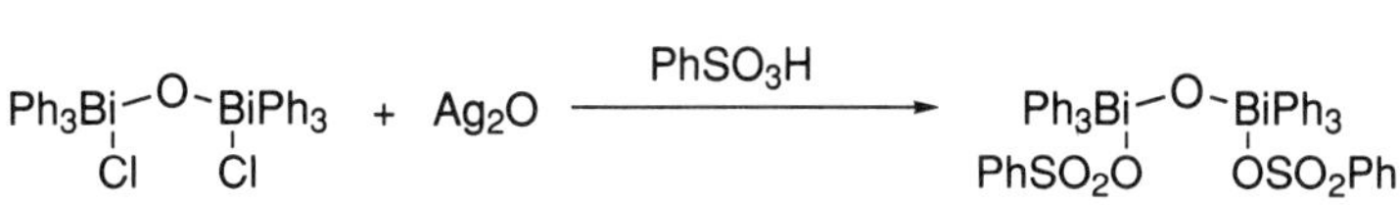

[6.92]

Properties and reactions. Oxybis(triphenylbismuth) compounds of the formula $(Ph_3BiX)_2O$ are stable and dissolve in dichloromethane, ethanol, acetonitrile and DMSO. Known oxybis(triphenylbismuth) compounds are mostly non-ionic in the solid state, the anionic group acting as a monodentate ligand. They have been shown by X-ray analysis to be pentacoordinate [221, 223, 224], where both bismuth atoms have a distorted trigonal bipyramidal geometry, with the *ipso* ring carbon atoms in equatorial sites and the bridging oxygen and the anionic or coordinative ligands in axial sites. Thus, for $[Ph_3Bi(OTf)]_2O$, the mean O−Bi−O and C−Bi−C angles are 177.9° and 119.0°, respectively, and each bismuth atom is located 0.20–0.24 Å above the plane formed by three *ipso* ring carbon atoms, being slightly bent toward the bridging oxygen atom (Scheme 6.6) [221]. The bridging Bi−O bond lengths, 2.038 Å and 2.091 Å, are consistent with the expected values of covalent single Bi−O bonds, whereas the Bi−O distances between the bismuth and triflate oxygen atoms, 2.58 Å and 2.53 Å, are too long to be covalent, indicating a contribution of some tetrahedral character of the bismuth centre to the distortion of whole molecular geometry. The Bi−O−Bi bond angle of 134.0° is considerably larger than the expected tetrahedral angle (109.8°), probably owing to steric repulsion between six phenyl groups.

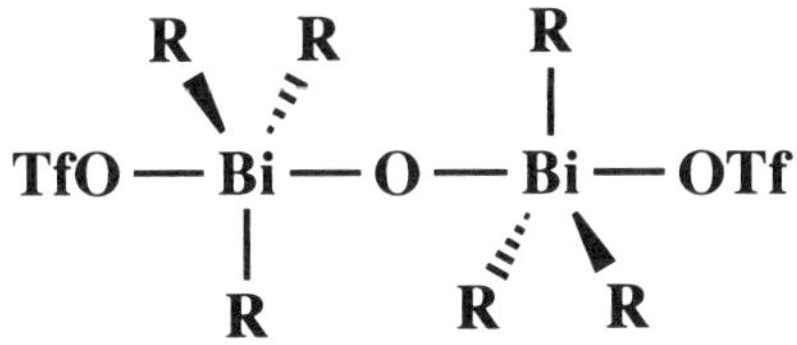

Scheme 6.6 An oxybis(triorganylbismuth) compound. R = organyl group.

Treatment of anhydrous oxybis(triphenylbismuth) diperchlorate with oxygen-donor bases results in the formation of pentacoordinate cationic complexes of the type $[(Ph_3BiL)_2O][ClO_4]_2$, where L is DMSO, pyridine *N*-oxide, triphenylphosphine oxide or triphenylarsine oxide [223]. These complexes can also be prepared directly from triphenylbismuth dichloride, the appropriate ligand and a silver salt [223]:

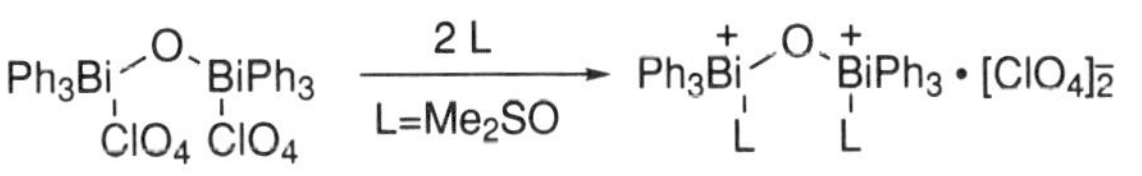

[6.93]

Oxybis(triarylbismuth) compounds can act as oxidizing and arylating reagents for organic substrates; oxybis(triphenylbismuth) dichloride has been used for the synthesis of some natural products [9–12, 225–227]:

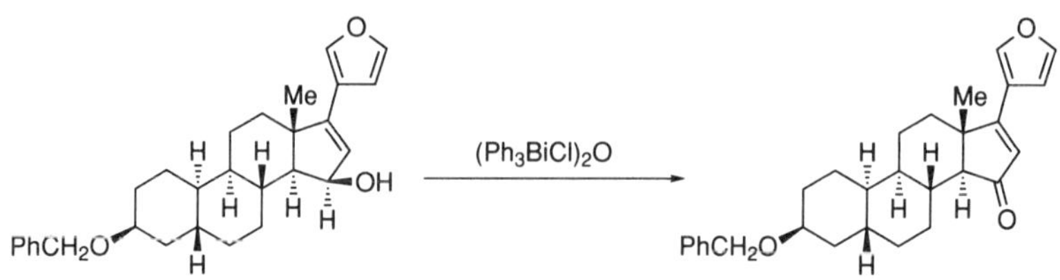
[6.94]

6.3.3 Tetraorganylbismuth(V) compounds

Preparation

From quinquenary bismuth compounds. The cleavage of one Bi–C bond of pentaarylbismuth with protic acids such as hydrogen chloride, fluoroboric acid, carboxylic acids, sulphonic acids and triphenylsilanol, as well as Lewis acids such as bromine, triphenylborane and phenylbismuth bis(trifluoroacetate), all result in the formation of quaternary bismuth(V) compounds, R_4BiX [198, 228–233]:

$$Ph_5Bi \xrightarrow{TfOH} [Ph_4Bi][OTf] + C_6H_6 \qquad [6.95]$$

$$Ph_5Bi \xrightarrow{Ph_3B} [Ph_4Bi][BPh_4] \qquad [6.96]$$

From triarylbismuth dihalides. When treated with silver perchlorate or tetrafluoroborate in acetylacetone or ethyl methyl ketone, triphenylbismuth dichloride is converted to tetraphenylbismuthonium salts [234]:

$$Ph_3BiCl_2 \xrightarrow[MeCOCH_2COMe]{AgClO_4} [Ph_4Bi][ClO_4] \qquad [6.97]$$

but when the same reaction is carried out in acetone, the acetonyltriphenylbismuthonium salt is the sole product obtained [235]:

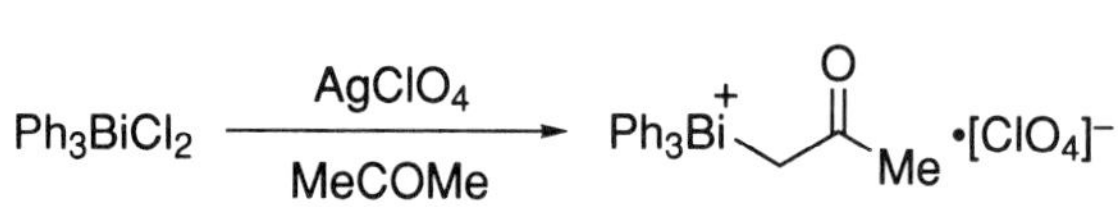
[6.98]

Treatment of triarylbismuth difluorides with organosilicon compounds such as silyl enol ethers, siloxycyclopropanes and allylsilanes in the presence of boron trifluoride or trimethylsilyl triflate gives 2-oxoalkyl- [221, 236], 3-oxoalkyl- [237], and allyltriarylbismuthonium [238] tetrafluoroborates and triflates, respectively, the sole by-product being volatile fluorotrimethylsilane [221]:

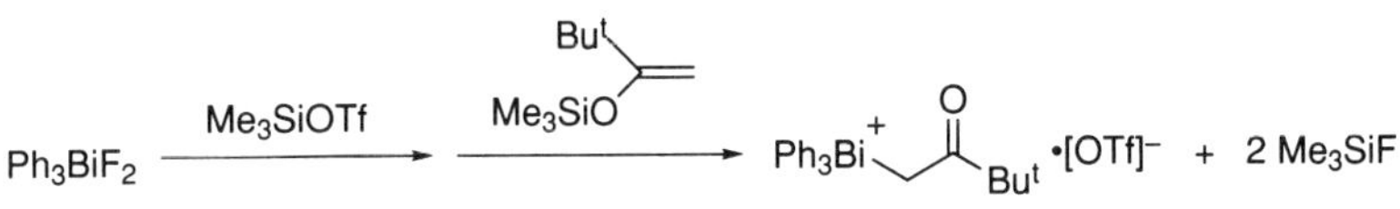

[6.99]

In this reaction, the Lewis acid plays an important dual role; initially it activates the difluoride and then it transforms itself to the soft counteranion to stabilize the resulting bismuthonium cation. Organotin compounds have been employed for the synthesis of asymmetrical alkenyl [239] and arylbismuthonium salts (Matano, Y., Miyamatsu, T. and Suzuki, H., unpublished results):

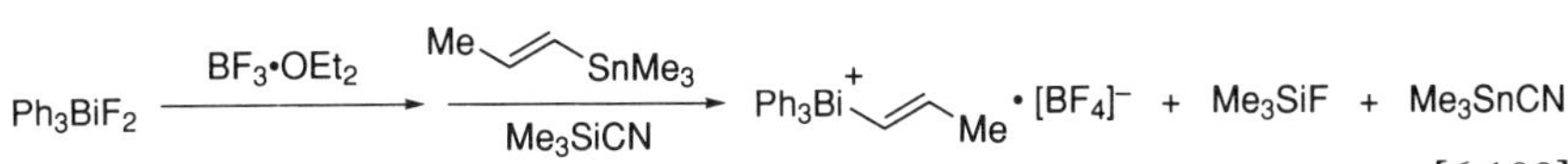

[6.100]

Metathesis. Treatment of tetraphenylbismuthonium chloride with sodium, potassium or silver perchlorate, nitrate, tetraphenylborate, cyanate, isocyanate and thiocyanate in an appropriate solvent such as water, acetone or methanol yields the corresponding tetraphenylbismuth(V) compounds [228, 229, 234, 240, 241]:

$$Ph_4BiCl \xrightarrow{NaNO_3} [Ph_4Bi][NO_3] \qquad [6.101]$$

Oxidative addition. In contrast to lighter pnictogen analogues, quaternary bismuth compounds are not generally accessible by direct interaction of triarylbismuthines with halogenoalkanes because of the low nucleophilicity of the bismuthines, whose unshared pair of electrons has a high s character. However, tetramethylbismuthonium triflate has been prepared by the reaction between trimethylbismuthine and methyl triflate in boiling acetonitrile [186]:

$$Me_3Bi + MeOTf \longrightarrow [Me_4Bi][OTf] \qquad [6.102]$$

Reaction of tris(2-methoxyphenyl)bismuthine with iodosylbenzene in dichloromethane results in the formation of tetrakis(2-methoxyphenyl)-bismuthonium chloride [242]:

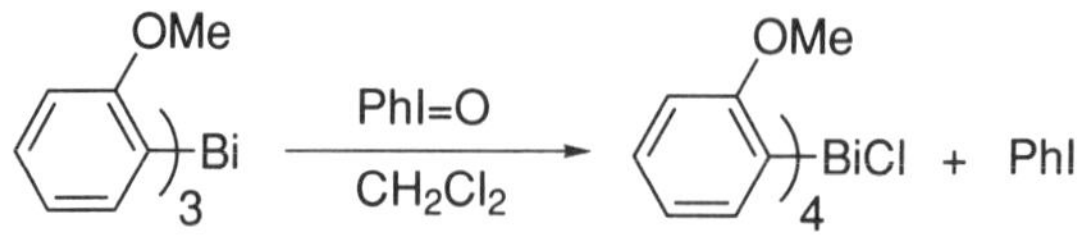

[6.103]

Properties. Synthesis, properties and reactions of bismuthonium compounds have been reviewed [243]. Quaternary bismuth halides (R_4BiX) are generally less stable than the corresponding lighter pnictogen analogues. Thus, tetraphenylbismuthonium chloride decomposes in the solid state rapidly at room temperature to yield triphenylbismuthine and chlorobenzene [228], although this chloride is stable in water for several days. However, tetrakis(2-methoxyphenyl)bismuthonium chloride, bromide and iodide are all highly stabilized through the intramolecular coordination of the neighbouring methoxy oxygen atoms with the bismuth centre [242]. In contrast to the halides, tetraphenylbismuthonium perchlorate, tetrafluoroborate, triflate, and hexafluorophosphate salts do not show any signs of decomposition under ambient conditions. Tetramethylbismuthonium triflate is also stable at room temperature [186]. Tetraphenylbismuth nitrate, carboxylates, sulphonates, cyanate and thiocyanate are moderately stable, wheareas the azide and selenocyanate are thermally unstable and decompose rapidly at room temperature. 2-Oxoalkyl-(triaryl)bismuthonium perchlorates, tetrafluoroborates and triflates are thermally stable mainly as a result of the low nucleophilicity of the counteranions [221, 236]. 3-Oxoalkyl(triaryl)bismuthonium triflates are stable in the solid state, but they slowly undergo reductive coupling in solution [237]. Alkenyl(triphenyl)bismuthonium salts are thermally stable [239], but allylic derivatives decompose rapidly at room temperature affording allylated arenes via ligand coupling [238].

Tetraorganylbismuth(V) compounds are soluble in chloroform, dichloromethane and acetonitrile, slightly soluble in benzene and thf, and almost insoluble in hexane and diethyl ether. Both protic and aprotic polar solvents such as MeOH, DMF and DMSO induce decomposition of some alkylbismuthonium salts through the nucleophilic attack of solvent molecule at the cationic bismuth centre.

The crystal structures of several quaternary bismuth compounds have been determined by X-ray diffraction analysis. Tetraorganylbismuth(V) compounds stabilized by soft counteranions such as perchlorate, tetrafluoroborate and triflate have a tetrahedral geometry around the bismuth centre, showing the onium nature of these compounds. Thus, tetramethylbismuthonium triflate has a tetrahedral geometry around the bismuth atom, with the mean Bi−C bond length of 2.22 Å and mean C−Bi−C bond angle of 110° [186]. The coordination geometry of the bismuth atom in tetraphenylbismuthonium perchlorate is also tetrahedral [244], but the bismuth centre of tetraphenylbismuth tosylate has a distorted trigonal bipyramidal structure with C_{eq}−Bi−C_{ax} bond angles in the range of 98°–103° and the C_{eq}−Bi−C_{eq} bond angles in the range of 113–123° [232]. The tosylate anion coordinates more strongly to the bismuth centre (the nearest Bi−O bond distance is 2.77 Å) in Ph_4BiOTs, which is therefore better described as a hypervalent compound [232]. The bismuth atoms in 2- and 3-oxoalkyl(triphenyl)bismuthonium tetrafluoroborates and triflates show a distorted tetrahedral geometry with the mean Bi−C bond length of 2.19–2.22 Å and mean C−Bi−C bond angle of 109.1°–109.2°, the carbonyl oxygen atom coordinating weakly to the bismuth centre (Bi···O distances range between 2.90 Å and 3.16 Å) [221, 237]. Alkenyltriphenylbismuthonium tetrafluoroborate also shows a distorted tetrahedral geometry with a mean C−Bi−C bond angle of 109.3° [239]. The observed relationship between the coordination geometry of the bismuth atom and the nucleophilicity of the counteranion described herein does not hold in the case of tetrakis(2-methoxyphenyl)bismuthonium bromide, which is found to consist of a distorted tetrahedral Ar_4Bi cation and a spatially separated bromide anion [242]. Intramolecular coordination between the methoxy oxygen and bismuth atoms is most likely to contribute to the thermal stabilization of this onium salt.

Several tetraphenylbismuth(V) compounds have been studied by IR and laser Raman spectroscopies as well as by molecular weight and conductance measurements [234]; the perchlorate, tetrafluoroborate and hexafluorophosphate salts behave as 1:1 electrolytes in acetonitrile or nitromethane. Although the nitrate and thiocyanate compounds are also 1:1 electrolytes in these solvents, they have a trigonal bipyramidal skeleton around the bismuth centre, with four phenyl groups occupying one axial and three equatorial positions and with the anionic group occupying the remaining axial position, but they are presumably dissociated in solution. The trichloroacetate and cyanate compounds are found to be non-ionic.

Reactions. An onium exchange reaction between tetraphenylbismuthonium tetrafluoroborate and triphenylphosphine occurs thermally and photochemically, yielding triphenylbismuthine and tetraphenylphosphonium tetrafluoroborate [245]:

$$[Ph_4Bi][BF_4] + Ph_3P \xrightarrow{h\nu} Ph_3Bi + [Ph_4P][BF_4] \qquad [6.104]$$

Tetraphenylbismuth(V) compounds have been used extensively as the oxidizing agents for alcohols under basic conditions [9–12, 233]:

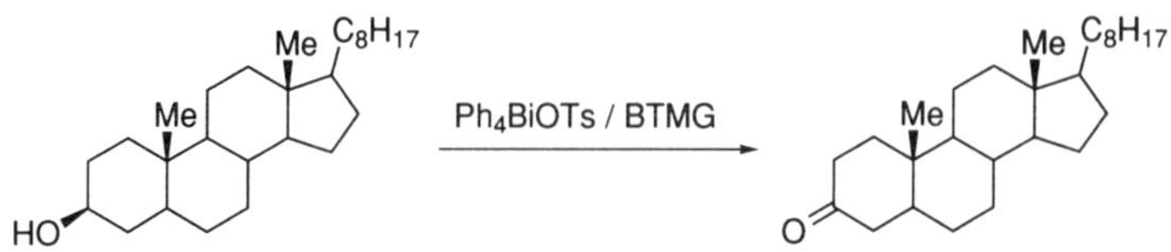
[6.105]

They have also been employed as the phenylating agents for alcohols, enols, amines, phenols, indoles, thiols, sulphinates, nitroalkanes and others [9–12]. The selectivity between *O*- and *C*-arylations is dependent on the reaction conditions; 2-naphthol is *O*-phenylated by tetraphenylbismuth trifluoroacetate under acidic conditions, whereas it is *C*-phenylated under basic conditions [198, 211]:

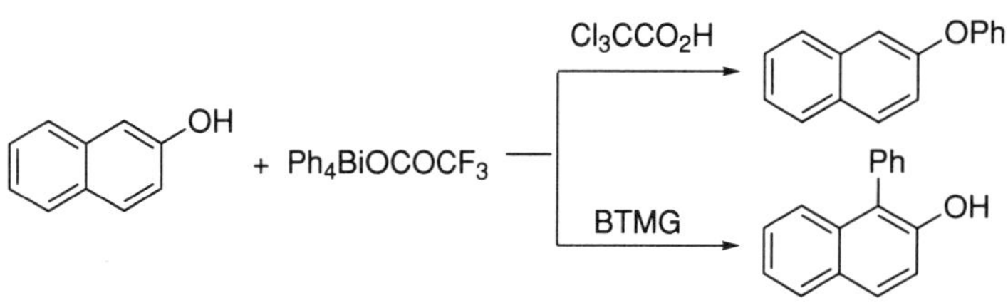
[6.106]

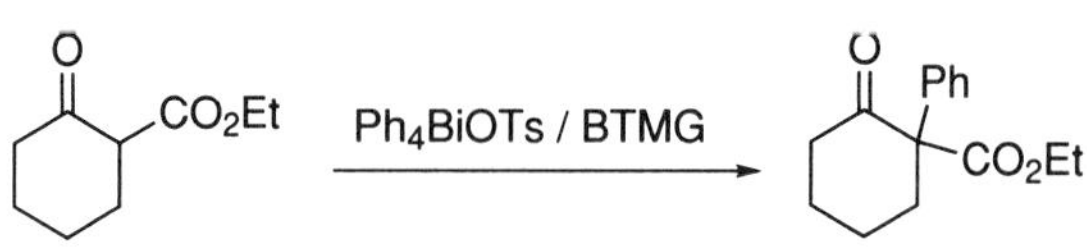
[6.107]

2- and 3-oxoalkyl(triphenyl)bismuthonium salts readily undergo onium exchange with triphenylphosphine, triphenylstibine and dimethyl sulphide to yield the corresponding phosphonium, stibonium and sulphonium salts [221, 236, 237]. They easily react with enolates, thiolates, phenoxides, sulphinates, piperidine and halides to give the corresponding α-substituted or β-substituted ketones with a good recovery of triphenylbismuthine [221, 237]:

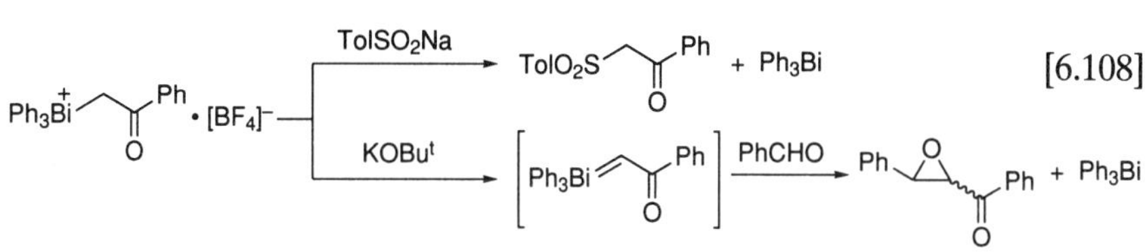
[6.108]

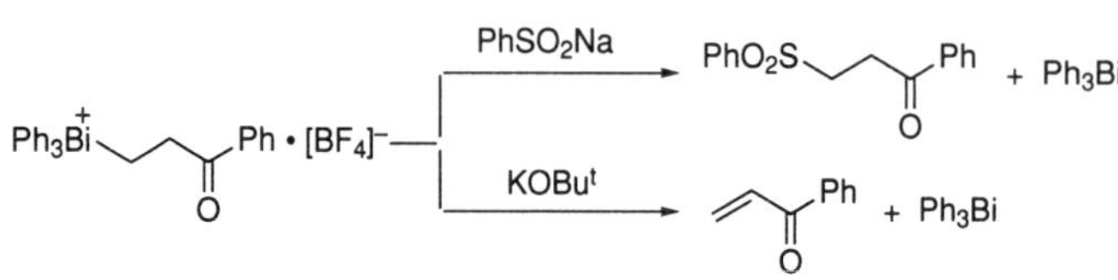
[6.109]

When treated with a base such as potassium tert-butoxide or LDA (lithium diisopropylamide), 2-oxoalkyl(triphenyl)bismuthonium salts are converted to the corresponding triphenylbismuthonium 2-oxoalkylides, whereas 3-oxoalkyl(triphenyl)bismuthonium salts undergo β-elimination to afford α,β-unsaturated ketones and triphenylbismuthine as shown in the lower pathways of reactions [6.108] and [6.109] [237, 246].

Alkenyltriphenylbismuthonium salts do not react with phosphines, sulphides and piperidine, but readily transfer one of their organyl ligands to sulphinates and thiolates to give the corresponding sulphones and sulphides as shown in the upper pathway of reaction [6.110] [239]:

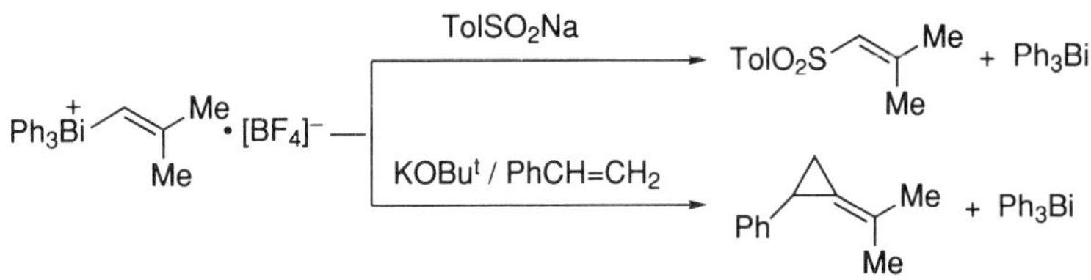

[6.110]

When treated with ethyl acrylate in the presence of a palladium catalyst they effect the Heck-type alkenylation and phenylation. Treatment of 2-methyl-1-propenyl(triphenyl)bismuthonium tetrafluoroborate with potassium tert-butoxide in the presence of an excess of styrenes at –78°C affords the corresponding alkylidene cyclopropanes in quantitative yield (lower pathway of reaction [6.110]) [239]. A Hammett study on this reaction indicates that a free alkylidene carbene is generated with simultaneous elimination of triphenylbismuthine.

Allyltriphenylbismuthonium compounds generated *in situ* from triphenylbismuth difluoride, $BF_3{\cdot}OEt_2$ and allylsilanes undergo the Friedel–Crafts-type allylation with electron-rich arenes to give allylated arenes, and they can also allylate phosphines, sulphides, thiols and enolates [238]:

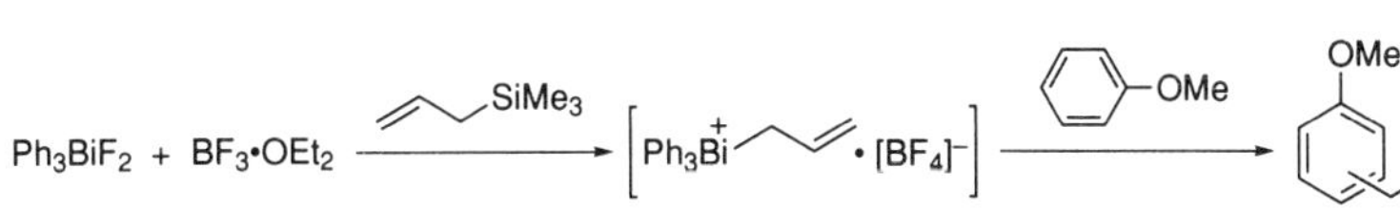

[6.111]

6.3.4 *Penta- and hexaorganylbismuth(V) compounds*

Coupling between triorganylbismuth dihalides and 2 equivalents of aryllithium or alkyllithium reagents constitutes the most convenient methodology for the preparation of the pentaarylbismuth and pentaalkylbismuth compounds [181, 186, 228, 229, 247–249]:

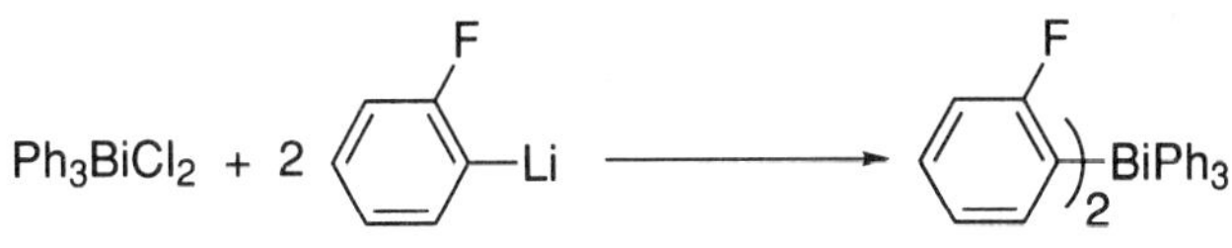

[6.112]

Pentaorganylbismuth compounds thus obtained can be transformed into lithium hexaorganylbismuthates by the additional action of a lithium reagent [186, 229, 250, 251]:

$$Ph_5Bi + PhLi \longrightarrow Li[BiPh_6] \qquad [6.113]$$

Triphenylbismuthine imide reacts with 2 equivalents of phenyllithium to give pentaphenylbismuth [187]:

$$Ph_3Bi{=}NTs + 2\,PhLi \longrightarrow Ph_5Bi \qquad [6.114]$$

Pentaphenylbismuth is less stable compared with the corresponding phosphorus, arsenic and antimony congeners but it is stable under an inert atmosphere at room temperature and can be crystallized from aqueous ethereal solvent. When heated to around 100°C under nitrogen, pentaphenylbismuth decomposes to give benzene, biphenyl and triphenylbismuthine. Compounds in this class are found mostly to be dichroic. The crystal structures of several symmetrical and asymmetrical pentaarylbismuth com-pounds have been determined by X-ray diffraction analysis [16, 17]; $BiPh_5$ (violet-blue), $Bi(C_6F_5)_3(4\text{-}MeC_6H_4)_2$ (yellow), $Bi(C_6F_5)_3(4\text{-}FC_6H_4)_2$ (yellow), $BiPh_3(2,6\text{-}F_2C_6H_3)_2$ (red), $Bi(4\text{-}MeC_6H_4)_3(2,6\text{-}F_2C_6H_3)_2$ (red) and $Bi(C_6F_5)_3(2\text{-}FC_6H_4)_2$ (violet) are all square-based pyramidal [Scheme 6.7(a)], whereas $Bi(4\text{-}MeC_6H_4)_5$ (violet) and $Bi(4\text{-}MeC_6H_4)_3(2\text{-}FC_6H_4)_2$ (orange) are trigonal bipyramidal [Scheme 6.7(b)]. In the square-based pyramidal Ph_5Bi, the four basal Bi$-$C bonds are longer (2.32 Å) than the axial Bi$-$C bond (2.21 Å), whereas in the trigonal bipyramidal $Bi(4\text{-}MeC_6H_4)_3(2\text{-}FC_6H_4)_2$, two axial Bi$-$C bonds are significantly longer (2.39 Å) than the three equatorial Bi$-$C bonds (2.18 Å). Pentamethylbismuth is obtained as violet crystals at –110°C, an X-ray diffraction analysis of which revealed that the bismuth atom is trigonal bipyramidal surrounded by five carbons with the mean $Bi{-}C_{eq}$ and $Bi{-}C_{ax}$ bond

Scheme 6.7 Symmetrical and asymmetrical pentaarylbismuth compounds: (a) square-based pyramidal; (b) trigonal pyramidal.

lengths 2.29 Å and 2.28 Å, respectively, and the mean $C_{eq}-Bi-C_{eq}$ and $C_{ax}-Bi-C_{ax}$ bond angles 120° and 179.8°, respectively [186].

The relationship between the colour and the coordination geometry of pentaorganylbismuth compounds has been discussed by Seppelt [16, 17]. The square-based pyramidal geometry observed for a number of pentaarylbismuth compounds is unusual, but calculations indicate that it is close in energy to the trigonal bipyramidal geometry. Also, the colours result from relativistic stabilization of the Bi 6s orbital. The LUMO has much Bi 6s character and so is lowered in energy so that the HOMO–LUMO gap is decreased and electronic transitions are in the visible region. For antimony there is no relativistic stabilization of the Sb 5s orbital and the compounds are colourless [17].

X-ray diffraction analyses of yellow crystalline-ate complexes $Li[BiPh_6]$ and $Li[BiMe_6]$ revealed that each bismuth centre in these molecules has an octahedral geometry, with the mean Bi–C bond lengths being 2.34 Å and 2.33 Å, respectively [186, 251]. The pyridine-$BiPh_5$ complex has a somewhat distorted octahedral geometry owing to a weak interaction between the bismuth and nitrogen atoms (Bi ··· N, 3.202 Å) [251].

In dry pyridine, chloroform or carbon tetrachloride, pentaphenylbismuth rapidly decomposes to form triphenylbismuthine, benzene and tarry products. The benzyne intermediate generated during decomposition can be trapped with furan, tert-butanol or phenol [231, 252]:

[6.115]

Treatment of pentaarylbismuth compounds with 1 equivalent of bromine, hydrogen chloride, hydrogen tetrafluoroborate, triphenylsilanol or triphenylboron all result in the cleavage of one of the C–Bi bonds, and the corresponding quaternary bismuth(V) compounds are formed (cf. Section 6.3.3).

Pentaphenylbismuth has been employed as an oxidizing and phenylating agent in organic synthesis; isopropanol, benzyl alcohol and cholestanol are oxidized to the corresponding carbonyl compounds, and phenols, thiophenols and nitroalkanes are phenylated [9–12]. These reactions are found to proceed through a pentavalent organobismuth intermediate without involving free radical species [198, 253, 254]:

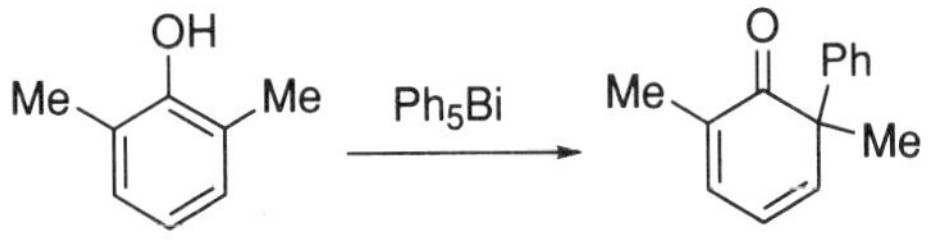

[6.116]

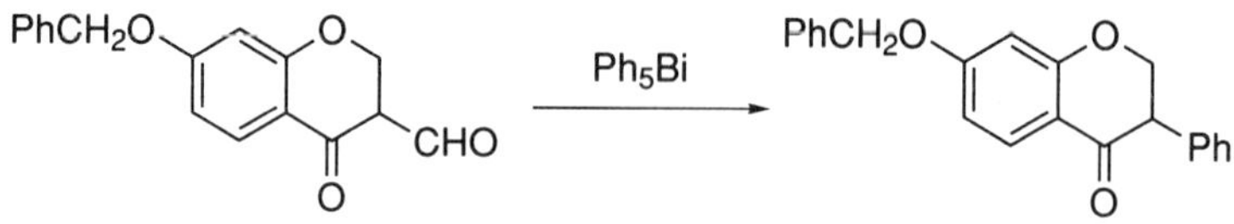

[6.117]

6.3.5 Bismuthonium ylides

Preparation

From tertiary bismuthines. In 1967 Lloyd and Singer reported the first synthesis of a bismuthonium ylide by heating triphenylbismuthine and diazotetraphenylcyclopentadiene together at 140°C:

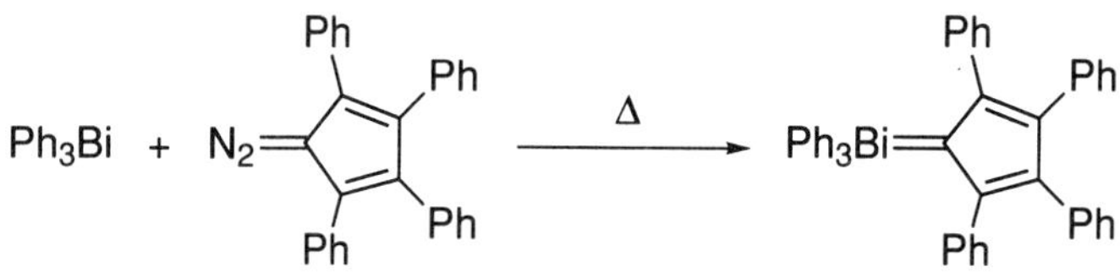

[6.118]

although the material obtained was not well characterized [255]. Bis(hexafluoroacetylacetonato) copper(II) is an effective catalyst for the preparation of bismuthonium ylides from diazo compounds [256, 257]:

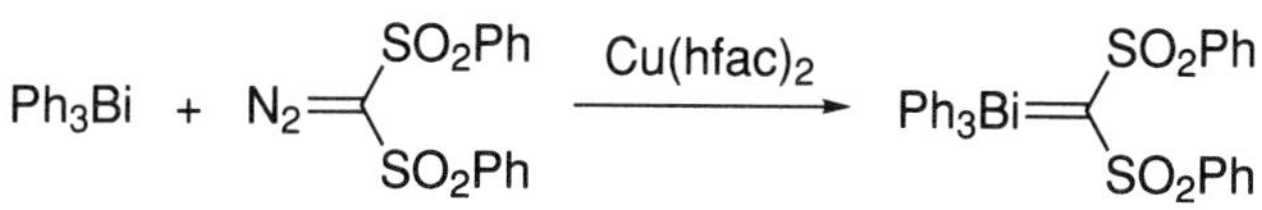

[6.119]

From triarylbismuth(V) compounds. The first stabilized bismuthonium ylide was synthesized from triphenylbismuth carbonate and dimedone [213, 258]:

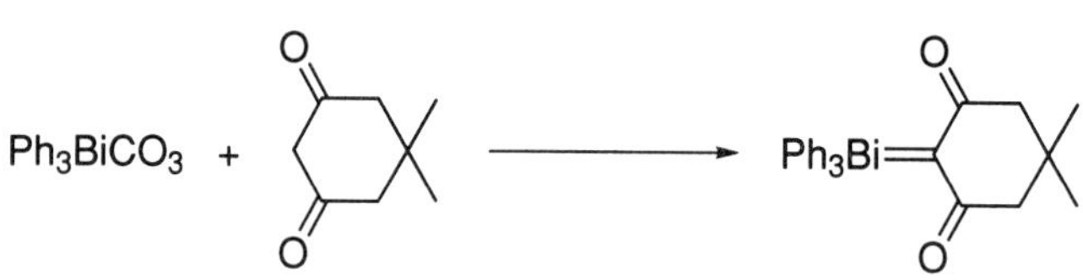

[6.120]

This type of bismuthonium ylides having an α,α'-dicarbonyl or α,α'-disulphonyl skeleton can also be prepared from triphenylbismuth dihalides and the corresponding activated methylene compounds in the presence of a suitable base [259, 260]:

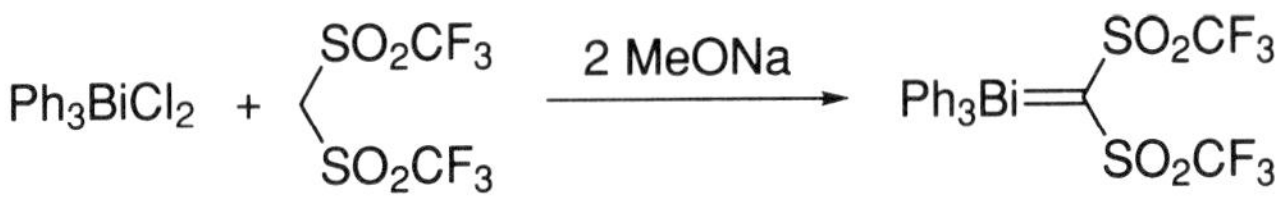

[6.121]

From bismuthonium salts. Proton abstraction from alkylbismuthonium salts by a base generates the corresponding ylides at low temperatures. The formation of triphenylbismuthonium 2-oxoalkylides has been observed by NMR, although they were not isolable under ambient conditions due to their thermal instability and moisture sensitivity [246]:

[6.122]

Properties and reactions. The first bismuthonium ylide reported in 1967 was claimed to be thermally stable, but the literature lacks detailed information on its properties [255]. Bismuthonium ylides containing a cyclic α,α′-dicarbonyl skeleton have been isolated as stable crystalline solids, however, and one of those ylides has been structurally characterized by spectroscopic and X-ray diffraction analyses, where the bismuth atom is found to have a distorted tetrahedral geometry and interacts with one of the carbonyl oxygen atoms (the Bi–O distances are 3.019 Å and 3.352 Å) [261]. The $Bi-C_{ylide}$ bond (2.156 Å) is only slightly shorter than the $Bi-C_{Ph}$ bond (2.21–2.22 Å), suggesting the lack of double-bond nature between the bismuth and ylidic carbon atoms. Bismuthonium ylides of an acyclic α,α′-dicarbonyl skeleton are not highly stabilized and decompose during chromatographic treatment over silica gel [259]. Bismuthonium ylides having an acyclic α,α′-disulphonyl skeleton are coloured from pale yellow to khaki to deep blue, depending on the structure of organyl moieties attached to the sulphonyl function [256, 257, 260]. They are thermally stable but readily decompose in open air. The moderately stabilized bismuthonium 2-oxoalkylides can be generated only at low temperatures and decompose rapidly at room temperature to form triphenylbismuthine along with small amounts of α-phenyl ketone, 1,2-diacylethene, 1,2,3-triacylcyclopropane, and others [246].

Stabilized bismuthonium ylides having a cyclic α,α′-dicarbonyl skeleton undergo a coupling reaction with aldehydes to give the tandem Wittig and Corey type addition products under gentle reflux in benzene or dichloromethane (upper pathway in reaction [6.123]) [262]. They also react with isothiocyanate [258], acetylene [263], phosphines and sulphides [264] in the presence of a copper catalyst to yield the corresponding heterocyclic compounds, phosphonium ylides, and sulphonium ylides, respectively (lower pathway of reaction [6.123]):

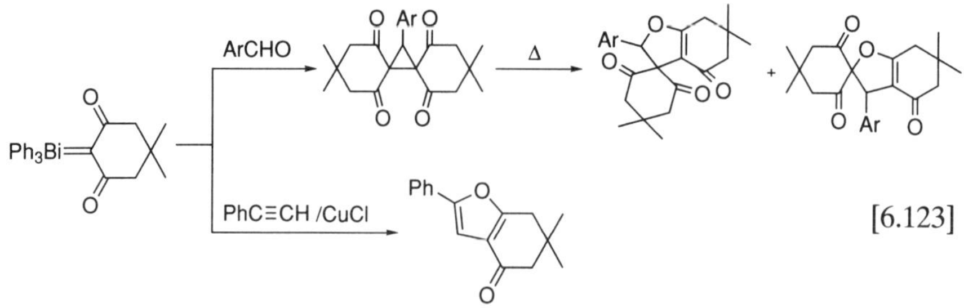

[6.123]

Bismuthonium ylides with an acyclic α,α′-dicarbonyl skeleton do not react with benzaldehyde, but readily couple with a sulphene generated *in situ* from methanesulphonyl chloride and triethylamine to afford oxathiol-*S*,*S*-dioxides [259]:

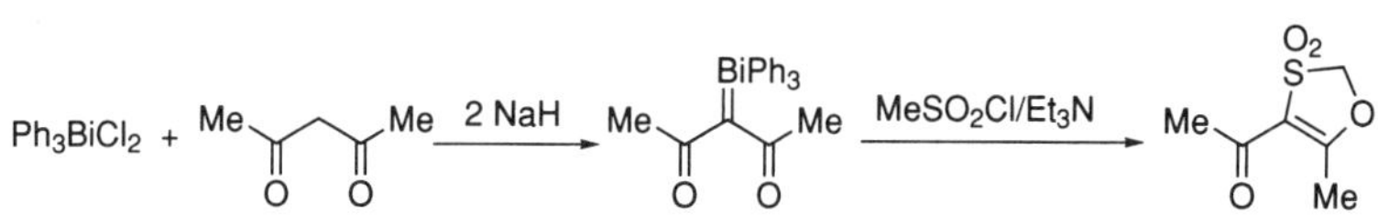

[6.124]

Moderately stabilized triphenylbismuthonium 2-oxoalkylides smoothly undergo coupling reactions with a variety of aldehydes at low temperatures to give α,β-epoxyketones with high *trans* selectivity, where the triphenylbismuthonio moiety acts as a good neutral leaving group (upper pathway of reaction [6.125]) [246]:

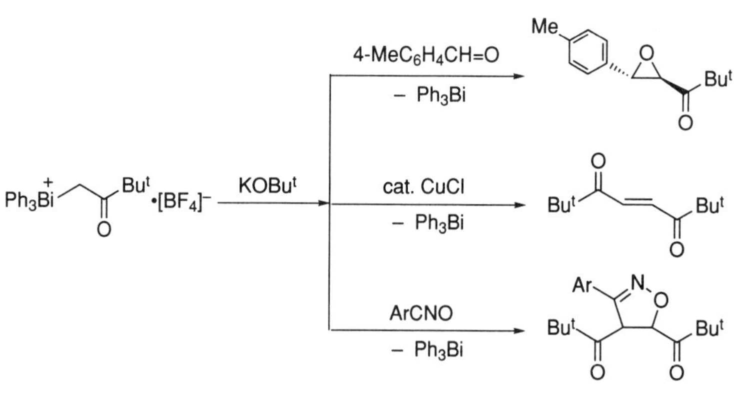

[6.125]

This mode of reaction leading to oxirane ring formation is in marked contrast to those of lighter pnictogen counterparts, which always result in olefin formation. Bismuthonium 2-oxoalkylides also react with dimedone, sulphonic acids and thiols to give the corresponding α-substituted ketones and triphenylbismuthine [246]. In the presence of a catalytic amount of copper(I) salt or nitrosobenzene, *trans*-1,2-diacylethene is formed via reductive coupling (middle pathway of reaction [6.125]), whereas in the presence of a nitrile oxide, 4,5-diacyl-3-isoxazolines are obtained together with triphenylbismuthine (lower pathway of reaction [6.125]) [246].

Bismuthonium 2-oxoalkylides react with activated *N*-sulphonylimines to give α,β-iminoketones and triphenylbismuthine. The *cis*/*trans* stereoselection in the aziridine ring formation can be controlled by proper choice of a base and an additive [265]:

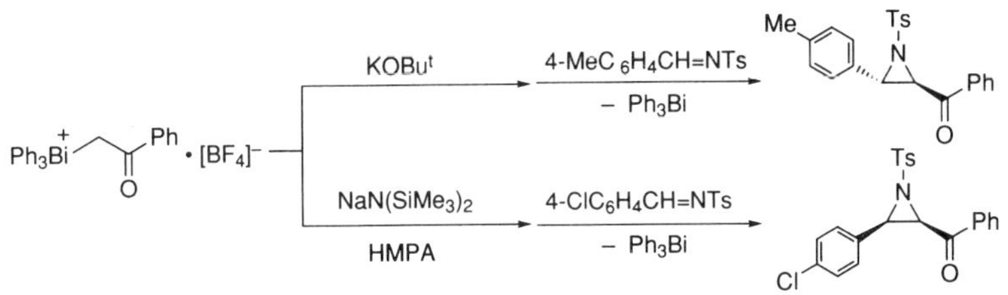

[6.126]

Although bismuthonium 2-oxoalkylides do not react with simple ketones and esters such as benzophenone and ethyl acetate they couple smoothly with α-diketones in two different reaction modes; with ethyl pyruvate, the epoxidation occurs to yield 2,3-bis-functionalized oxiranes (upper pathway of reaction [6.127]), whereas with *o*-quinones, the ring enlargement reaction proceeds to give vicinally substituted tropone derivatives, both in good yields (lower pathway of reaction [6.127]) [266]:

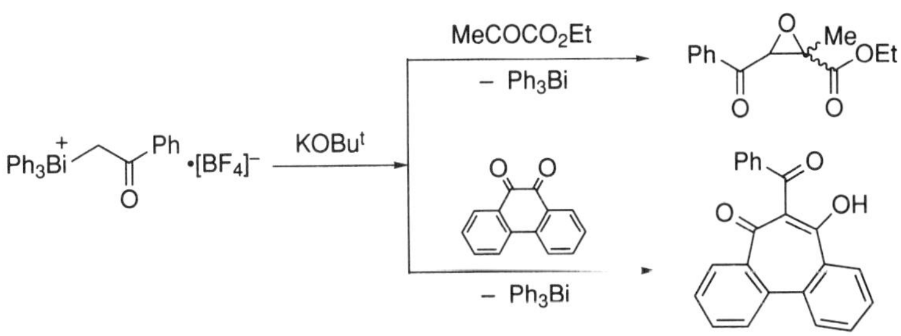

[6.127]

6.3.6 *Bismuthine imides*

A nitrene unit generated *in situ* from Chloramine-T® adds oxidatively to triarylbismuthines to give *N*-tosyltriarylbismuthine imides as gummy substances [187, 267]:

$$\mathrm{Ph_3Bi + TolSO_2NNaCl \longrightarrow Ph_3Bi{=}NSO_2Tol + NaCl} \qquad [6.128]$$

N-tosyliminoiodobenzene can also be used for this purpose [268]:

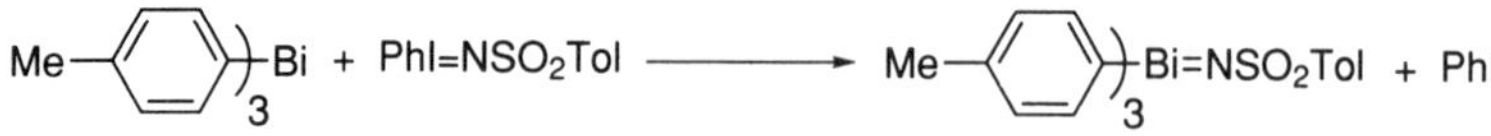

[6.129]

Crystalline bismuthine imides have recently been obtained from the defluorosilylation of triphenylbismuth difluoride with *N*,*N*-bis(trimethylsilyl)-sulphonamides and also from the base-promoted condensation of triarylbismuth dichloride with sulphonamides [269]:

$$Ph_3BiF_2 + (Me_3Si)_2NSO_2CF_3 \longrightarrow Ph_3Bi{=}NSO_2CF_3 + 2\,Me_3SiF \quad [6.130]$$

All bismuthine imides reported to date contain the *N*-sulphonyl group. They are thermally stable under dry nitrogen but are gradually decomposed by moisture on standing in open air. *N*-sulphonylated bismuthine imides are readily soluble in dichloromethane, chloroform, benzene and acetonitrile, but almost insoluble in hexane. No X-ray analysis has been reported to date for this class of compounds. *N*-Tosyltriarylbismuthine imides react with benzaldehyde (reaction [6.131]), benzoyl chloride and phenyl isocyanate to afford *N*-tosylimine, *N*-tosylamide and *N*-tosylurea derivatives, respectively [267]. They oxidize activated alcohols to carbonyl compounds, sometimes accompanied by the concurrent formation of an *N*-tosylaminodiarylbismuthine (reaction [6.132]) [268]:

$$PhCHO \xrightarrow{Ph_3Bi{=}NSO_2Tol} PhCH{=}NSO_2Tol \quad [6.131]$$

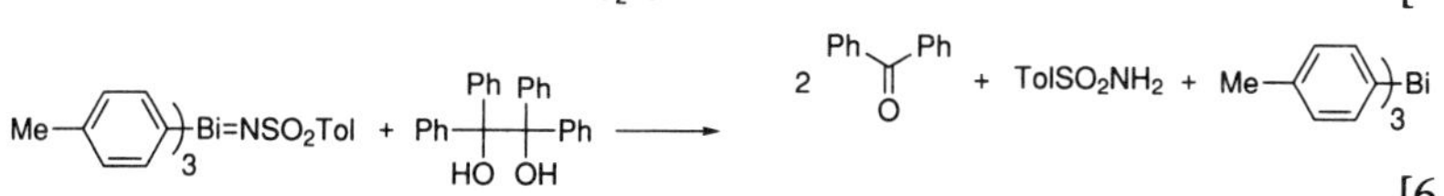

[6.132]

6.3.7 *Bismuthine oxides*

The metathesis reaction between triphenylbismuth dichloride and silver(I) oxide in benzene has been reported to give triphenylbismuthine oxide as a white polymeric powder [223]:

$$Ph_3BiCl_2 + 2\,AgO \longrightarrow (Ph_3Bi{=}O)_n \quad [6.133]$$

A combination of triphenylbismuth dicyanide and mercury(II) oxide can also be used for this purpose [201]. On the other hand, the sonochemical reaction between triarylbismuthine and iodosylbenzene in dichloromethane has been found to give the corresponding triarylbismuthine oxides as a yellow-coloured solution [270]:

 [6.134]

Photoirradiation of triphenylbismuthine under an oxygen-containing atmosphere in the presence of titanium(IV) oxide has been reported to yield the corresponding oxide; however, no characterization data are available [271].

Triphenylbismuthine oxide obtained from the metathetical reaction between Ph_3BiCl_2 and Ag_2O is a white solid, which is insensitive to moisture and is stable at room temperature [222]. It is moderately soluble in benzene, slightly soluble in acetonitrile and insoluble in chloroform and diethyl ether; in acetone and alcohol it is reduced to triphenylbismuthine. An IR spectrum has shown that this compound possesses a five-coordinate polymeric structure involving the Bi–O–Bi bonds. Soluble triarylbismuthine oxides prepared from the corresponding bismuthines and iodosylbenzene in CH_2Cl_2 are sensitive to carbon dioxide, giving the corresponding triarylbismuth carbonates [270]. They are transformed into triarylbismuth diacetate and ditosylate by treatment with benzoyl anhydride and *p*-toluenesulphonic acid, respectively [268]. The presumed bismuthine oxides can be obtained as a glass after removal of the solvent, but they are easily transformed into a polymeric substance when scratched by a spatula. Recently, however, some cyclic triarylbismuthines such as 10-(4-methylphenyl)thiaphenobismin have been found, when treated with iodosylbenzene in dichloromethane, to afford the corresponding oxides as a stable crystalline solid (cf. Section 6.4.2) (Suzuki, H. and Ikegami, T., unpublished results).

The soluble bismuthine oxides possess mild oxidizing abilities, converting alcohols to ketones or aldehydes, hydrazobenzene to azobenzene, and triphenylphosphine to triphenylphosphine oxide [270]:

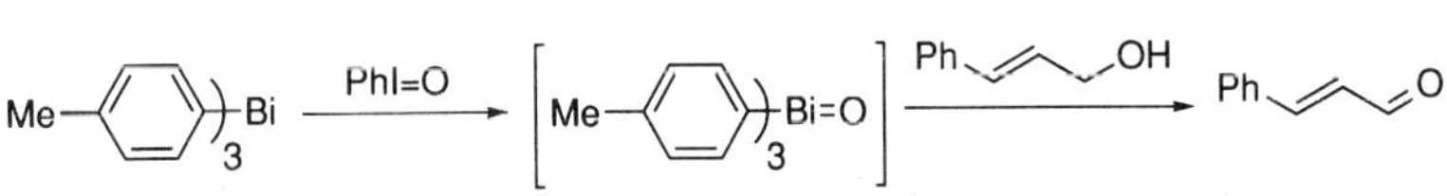 [6.135]

This type of oxidizing ability has not been observed for the corresponding lighter pnictogen oxides. Attempts to obtain the bismuthine oxides from the reaction of tris(2-methoxyphenyl)bismuthine or trimesitylbismuthine with iodosylbenzene in dichloromethane have resulted only in the formation of tetrakis(2-methoxyphenyl)bismuthonium chloride and trimesitylbismuth dichloride, respectively, as the major products [242]. Interestingly enough, bismuthine oxides generated from these bismuthines appear to have an ability to abstract the chlorine atom from chlorinated solvents.

6.3.8 *Bismuthinic acids and esters*

Only one report has been published for the synthesis of alkyl diarylbismuthinates and diarylbismuthinic acids, which are obtained by the reaction of triarylbismuthines with Chloramine-T® in a mixture of methanol and benzene [272]:

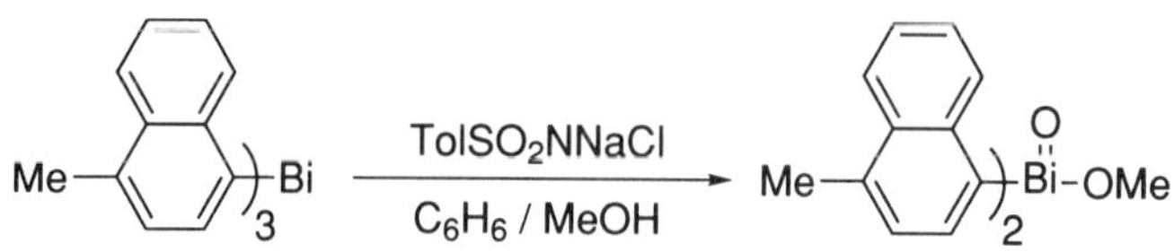

[6.136]

Ortho-substitution seems to be crucial for the stabilization of this class of compounds such that they can be handled without decomposition in air.

Methyl and ethyl diarylbismuthinates are mostly solids and are soluble in chloroform and benzene/MeOH. When a bulky group is absent at the *ortho* position of the aromatic ring, bismuthinates readily decompose to bismuthines.

The Bi−OMe bond of methyl bismuthinates is labile and the methoxy group is easily replaced by hydroxyl or other alkoxyl groups when treated with water or alcohols. Benzoin, tetraphenylethanediol, hydrazobenzene and 1,2-diphenyl-1,2-ethanedione dihydrazone are oxidized by methyl bis(1-naphthyl)bismuthinate to benzophenone, azobenzene and diphen ylacetylene, respectively [272]. Treatment of thioamides, ureas and thioureas with the same bismuthinate in a mixture of acetic acid and dichloromethane at room temperature leads to the formation of the corresponding *N*-acylamides in good yield [273]:

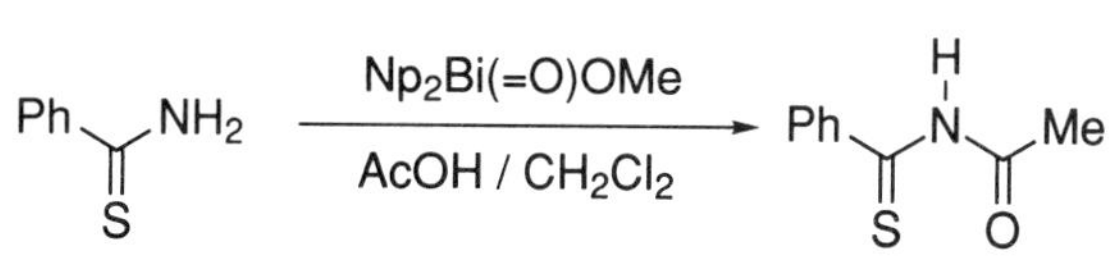

[6.137]

6.4 Bismuth-containing heterocycles

This section deals with ring compounds containing bismuth. Because of the thermal instability or air-sensitive nature of the bismuth–sp^3 carbon bond, most of the known bismuth heterocycles contain a bismuth-sp^2 carbon bond or bonds. Fusion with an aromatic ring remarkably enhances the stability of the bismuth-containing heterocycles.

6.4.1 *Bismacycles*

The term 'bismacycle' is used here to refer to cyclic compounds which contain bismuth and carbon atoms only. Most of the known bismacycles are five- and six-membered rings; three- and four-membered rings are currently unknown.

Five-membered rings. The bismuth analogue of pyrrole is commonly referred to as bismole. The parent 1H-bismole is not yet known, but derivatives having a substituent on the bismuth atom are comparatively stable coloured oils or crystalline solids. Bismole and its derivatives are synthesized by the reaction of a dihalobismuthine or a bismuth trihalide with an organolithium compound [154, 155, 274]:

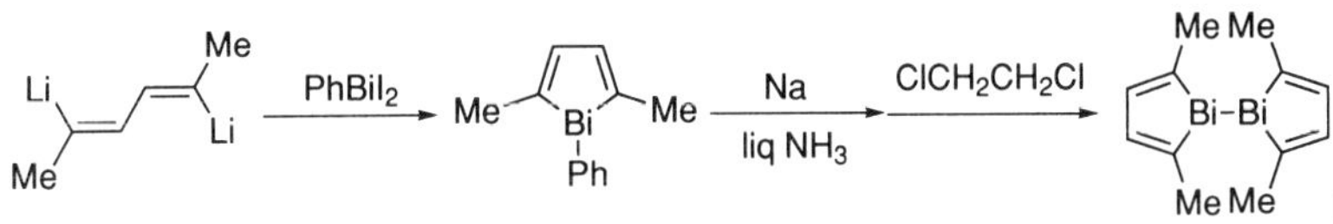

[6.138]

or by the transmetallation with appropriate metallacycles [151, 158, 275]:

[6.139]

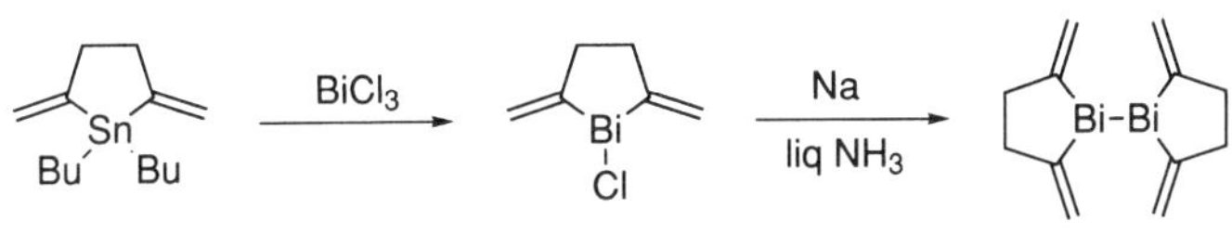

[6.140]

Some theoretical studies have been made for this type of ring compound [276–280].

By treating with sodium amide in hexane at low temperatures, 1-phenylbismole is cleaved to form 1-sodiobismole which, on further treatment with 1,2-dichloroethane, affords 1,1′-bibismole as a green crystalline solid [154, 155, 275]. When treated with sodium in liquid ammonia the 1-chlorobismole derivative undergoes reductive coupling to give the corresponding 1,1′-bibismole derivative. Cleavage of 1-phenylbismole with lithium in thf followed by successive treatment with ammonia and ferric chloride produces bismaferrocene as shiny black needles [155, 274, 281]:

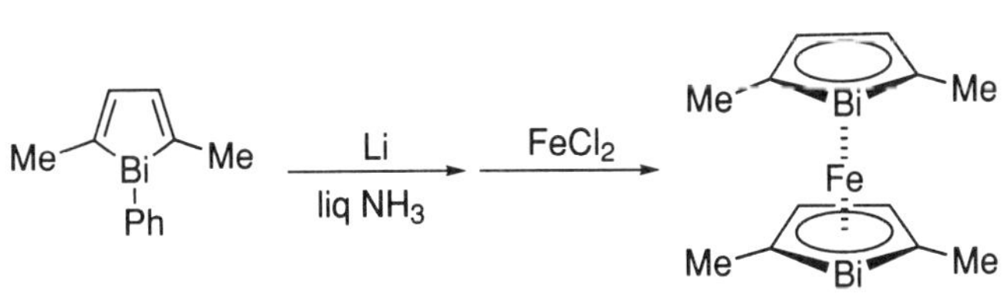

[6.141]

Treatment of 2,4-dimethyl-1-phenylbismole with lithium, liquid ammonia, and then $Mn(CO)_5Br$ followed by heating affords the corresponding Bi–Mn derivatives [282].

Reaction of 2,β-dilithiated styrene or 2,2′-dilithiated biphenyl with phenyldihalobismuthine or bismuth trichloride gives benzobismoles [283, 284] and dibenzobismoles [187, 285], respectively, although in some cases the use of organometallic compounds other than organolithiums gives better results [286]. Fusion of a benzene ring with the bismole unit results in a remarkable increase in the stability of bismuth-containing ring systems:

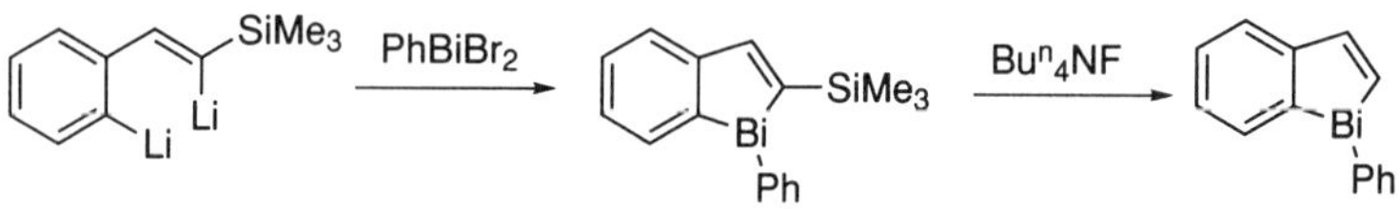

[6.142]

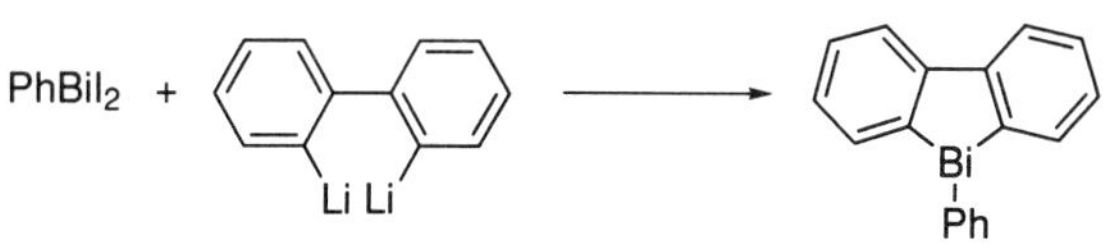

[6.143]

Treatment of 1-phenylbenzobismole with excess methyllithium leads to 1-methylbenzobismole in good yield.

2,2′-Dilithiobiphenyl reacts with triphenylbismuth dichloride to give 5,5,5-triphenyldibenzobismole, which reversibly combines with phenyllithium at low temperatures to form an ate complex [229]:

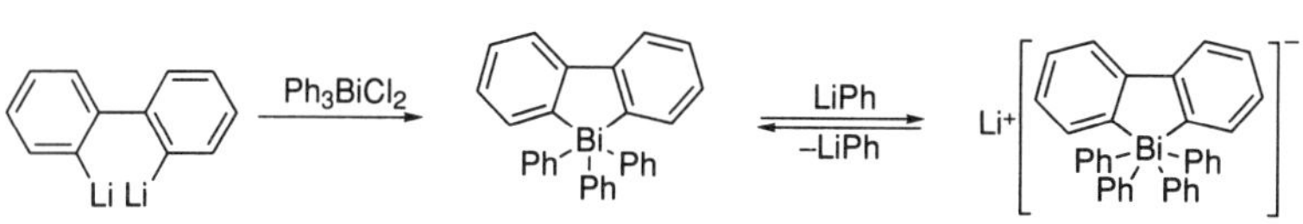

[6.144]

Six-membered rings. The parent heterocycle of this class of compound is known as bismabenzene or bismin, the chemistry of which has received some attention in connection with the aromaticity of heavy heteroatom benzene analogues. Bismabenzene is generated *in situ* as a coloured solution through the transmetallation reaction between stannabenzene and bismuth trichloride [287–289]. Although the weak aromatic nature of bismabenzene is confirmed by spectroscopic studies, the compound is very unstable and readily decomposes into polymeric substances under ambient conditions. At low temperatures, bismabenzene exists in a rapid thermal equilibrium between monomeric and dimeric forms, and the former can be trapped by an electron-deficient alkyne to give the corresponding Diels–Alder adduct [287–289]:

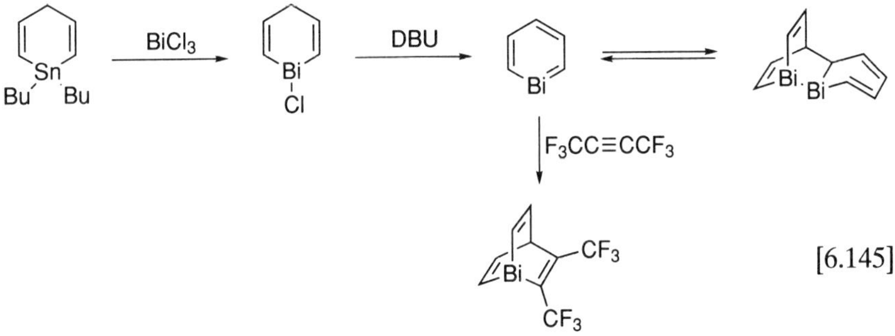

[6.145]

Fusion with a benzene ring stabilizes the bismabenzene framework considerably. Thus, the treatment of 2,2′-bis-metallated diphenylmethane with a bismuth trihalide or an aryldihalobismuthine affords the corresponding 5,10-dihydrodibenzo[*b*,*e*]bismabenzene as comparatively stable crystalline solids [290]:

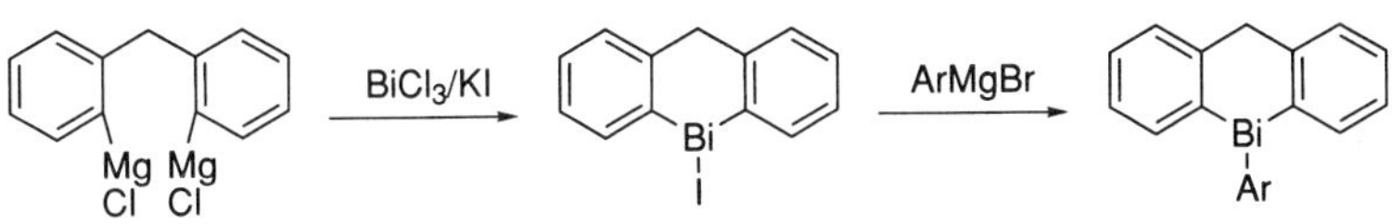

[6.146]

Derivatives of dihydrobismabenzenes and hexahydrobismabenzenes are similarly obtained by the reaction between an appropriate organometallic compound and a halobismuthine [54, 291]. Fully saturated compounds, which were originally called bisminanes, are all air-sensitive and readily decompose under ambient conditions [54]:

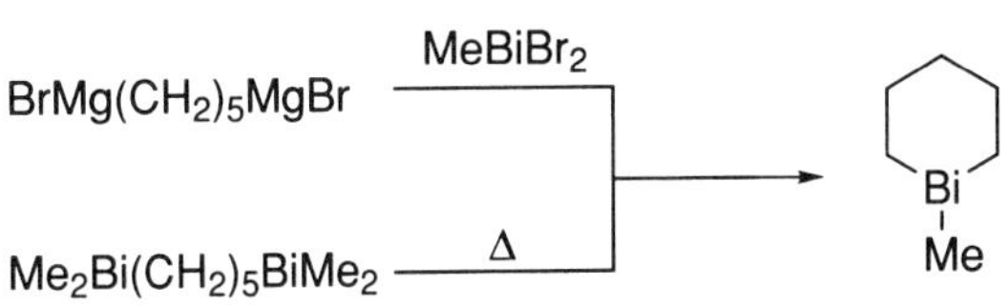

[6.147]

Synthesis of highly substituted bicyclic systems with the bismuth atoms at both bridge head positions has been reported, although they are not well characterized structually [292, 293]:

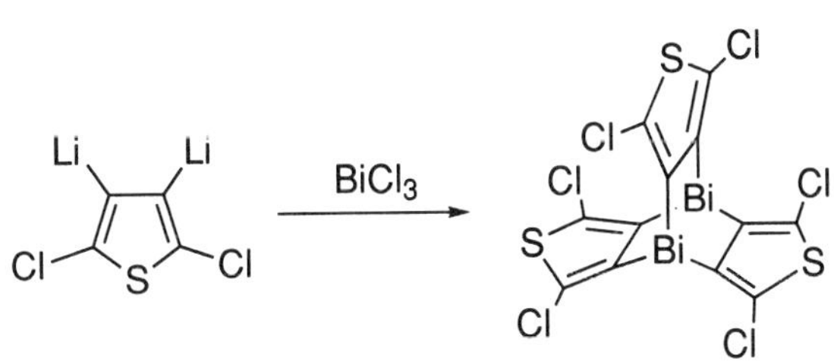

[6.148]

In situ trapping of an aryne intermediate with elemental bismuth is assumed to lead to this type of compound [294]:

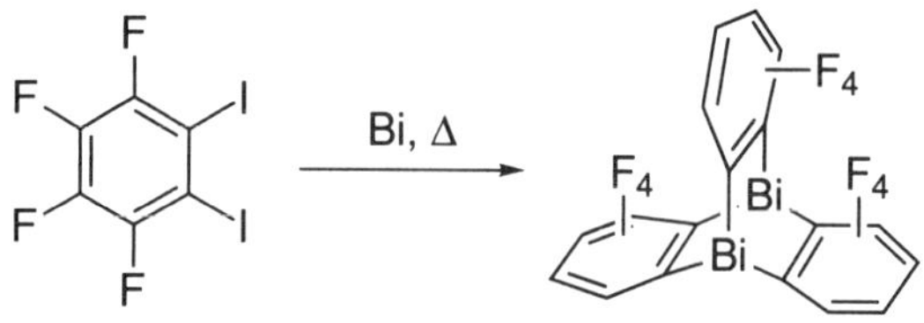

[6.149]

Sodium or potassium bismuthide is used for the preparation of two interesting cage compounds, $Bi_2(SiMe_2)_6$ and $Bi_2(SiMe_3)_5$ [295, 296].

Seven-membered rings. Dilithiation of (*Z*,*Z*)-1-bromo-4-(2-bromophenyl)-1-trimethylsilyl-1,3-butadiene with tert-butyllithium and subsequent treatment with phenyldibromobismuthine results in ring closure to form 1-phenyl-2-trimethylsilyl-1-benzobismepine [297]:

[6.150]

Without the trimethylsilyl group, the bismacycle ring becomes quite unstable. Similarly, trimethylsilyl-protected 3-benzobismepine and dibenzo[*b*,*d*]bismepine are prepared from (*Z*,*Z*)-1,2-bis(2-bromo-2-trimethylsilylvinyl)-benzene and *cis*-2,2′-dibromostilbene, respectively [297, 298]:

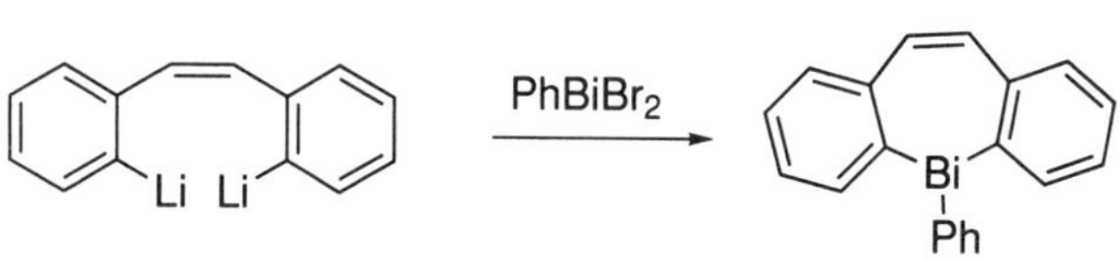

[6.151]

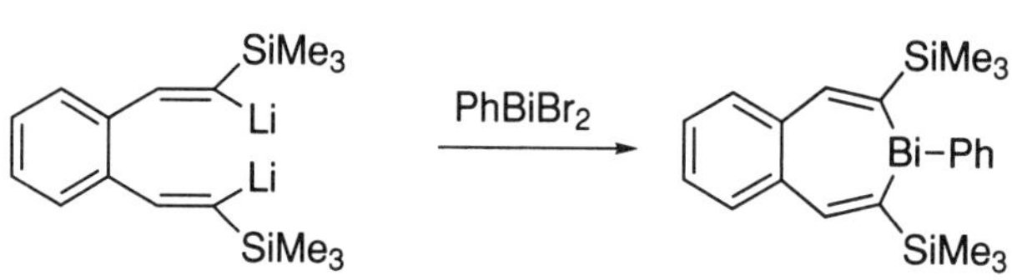

[6.152]

6.4.2 Heterobismacycles

The term 'heterobismacycle' is used here to refer to bismuth heterocyclic compounds containing an additional heteroatom or heteroatoms as a ring

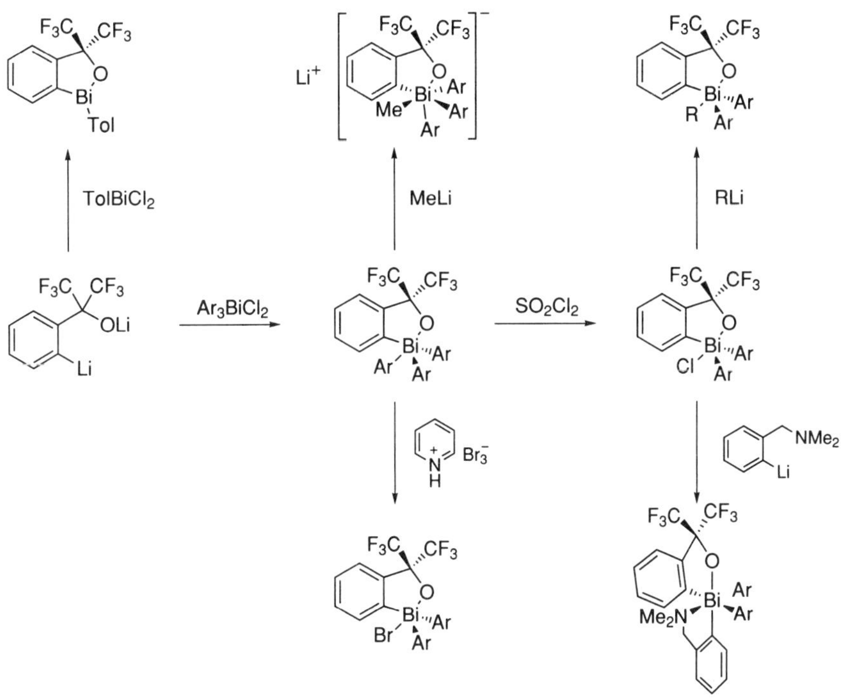

Scheme 6.8 Reaction modes of benzoxabismoles and their derivatives.

member. The most studied of this class of compounds are the benzoxabismoles which are readily obtained from the reaction of a dilithiated benzyl alcohol and a bismuth trihalide or an aryldihalobismuthine. Known benzoxabismoles are mostly derived from α,α-bis(trifluoromethyl)benzyl alcohol (Martin ligand) [299, 300]. Action of sulphuryl chloride on benzoxabismoles gives a pentacoordinated chlorobismuthorane almost quantitatively at room temperature [292]. A variety of benzoxabismoles and derivatives have been synthesized and their molecular structures examined by NMR spectroscopy and X-ray analysis [299, 300, 302, 303]. Permutational behaviours of these compounds are intensively studied in connection with the inversion mechanism of the pyramidal bismuth atom, and recent findings are consistent with an edge inversion mechanism. Benzoxabismoles and derivatives show the expected reaction modes of organobismuth(III) and organobismuth(V) compounds as summarized in Scheme 6.8 and which therefore will not be reviewed further.

2,2′-Spiro(benzoxabismole) is obtained by the reaction of bismuth trichloride with an excess of the dilithiated Martin ligand [304], and an X-ray analysis of this compound has revealed the special stabilization of the heterobismacycle system through the 3-centred-4-electron bond between the bismuth and two oxygen atoms [304]:

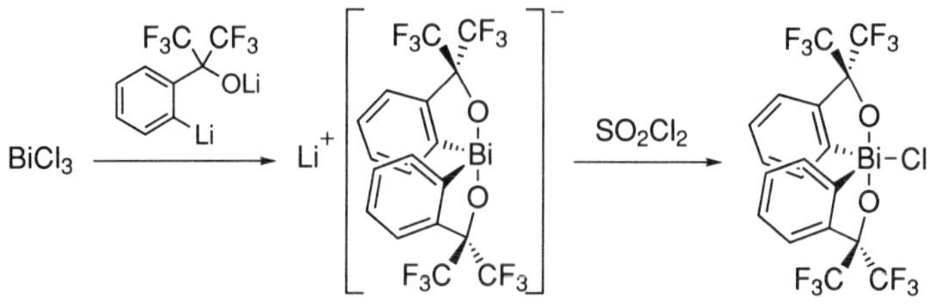

[6.153]

Action of carbon dioxide on pentaphenylbismuth at low temperatures leads to 2,1-benzoxabismole-3(1H)-one, thermolysis of which produces triphenylbismuthine, carbon dioxide and benzyne [305, 306]:

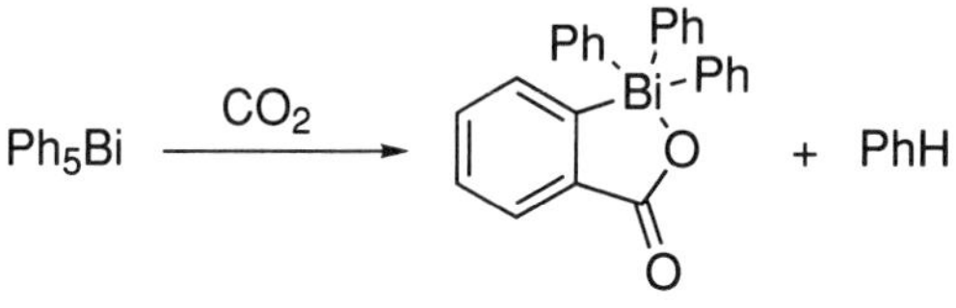

[6.154]

Acidolysis of triphenylbismuthine with oxalic, malonic, succinic or maleic acid gives rise to the corresponding five-, six- and seven-membered cyclic dicarboxylates, the high melting points of which suggest a polymeric nature for this class of compounds [307]:

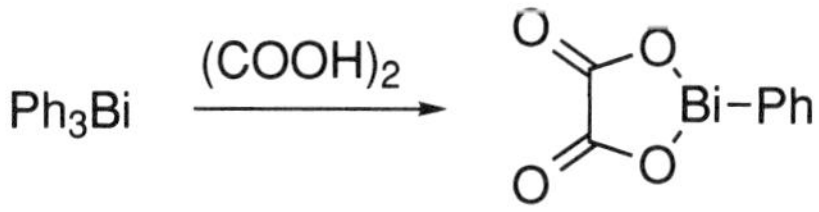

[6.155]

Transalkoxylation between dialkoxy(organyl)bismuthines and 1,2-alkanediols, 1,2-alkanedithiols or 1,5-alkanediols affords the corresponding five- or eight-membered heterobismacycles [308–310]:

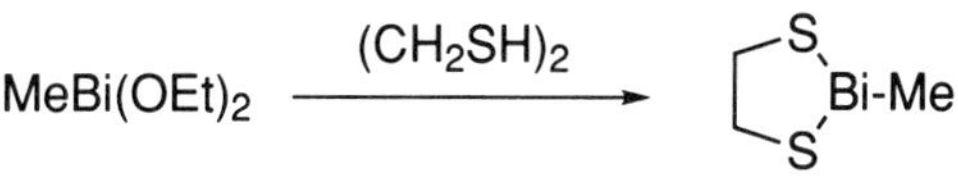

[6.156]

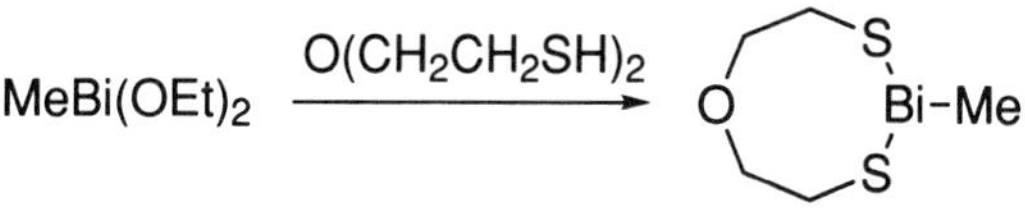

[6.157]

Reaction between dilithium salts of 1,2-benzenedithiol, 1,2-benzenediselenol or 2-aminobenzenethiol and organyl(dichloro)bismuthine yields the corresponding benzodithia-, benzodiselena- and benzothiazabismoles, respectively, as high-melting solids. Similarly, the reaction with thiocarboxylates gives the corresponding bismaheterocyclic ketones [311]:

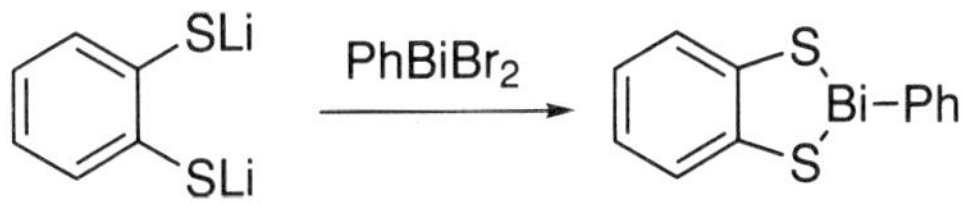

[6.158]

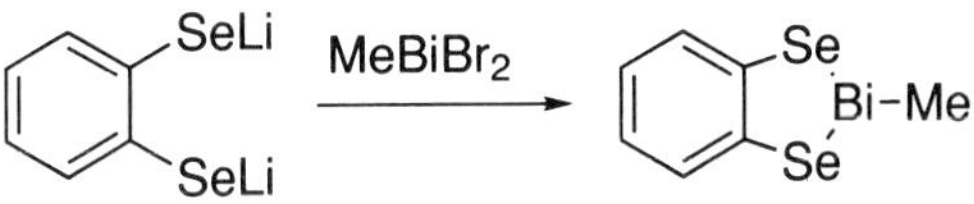

[6.159]

Reaction of 2,2′-dilithiated diphenyl ether or diphenyl sulphide with a bismuth trihalide or an aryldihalobismuthine affords the corresponding phenoxabismin and phenothiabismin as stable crystalline solids [290]:

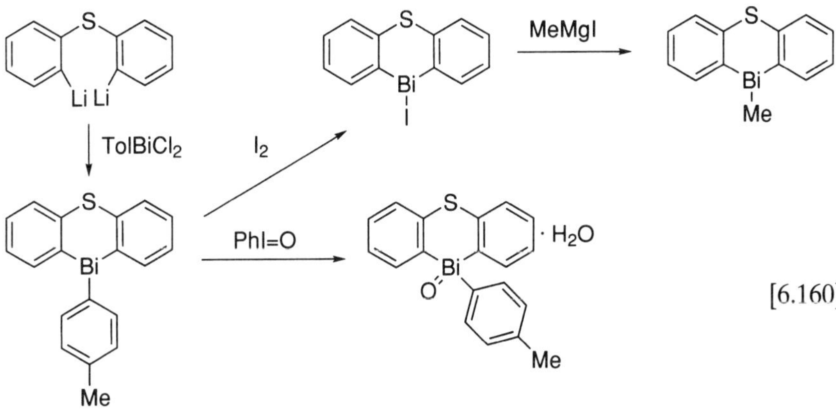

[6.160]

Reaction of 10-(4-methylphenyl)phenothiabismin with iodosylbenzene results in the preferential oxidation of the bismuth atom, giving the corresponding bismuthine oxide as an air-stable crystalline solid.

Treatment of the bis-Grignard reagent from bis(3-chloropropyl)methylamine with bismuth trichloride gives the corresponding 1,5-azabismocine derivative [312], and a similar reaction between 2,2′-dilithiated dibenzyl(methyl)amine and bismuth trichloride gives a 12-chloro-6-methyldibenzo[*b*,*g*][*1*,*5*]azabismocine, where a transannular interaction between the nitrogen and bismuth atoms has been suggested to operate for the stabilization of the Bi–Cl bond [313]:

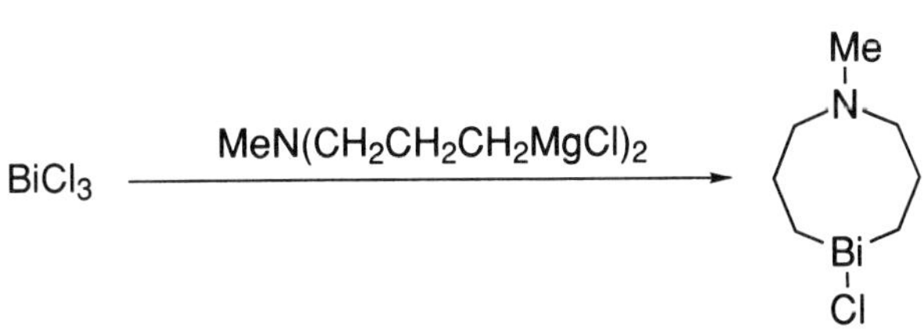

[6.161]

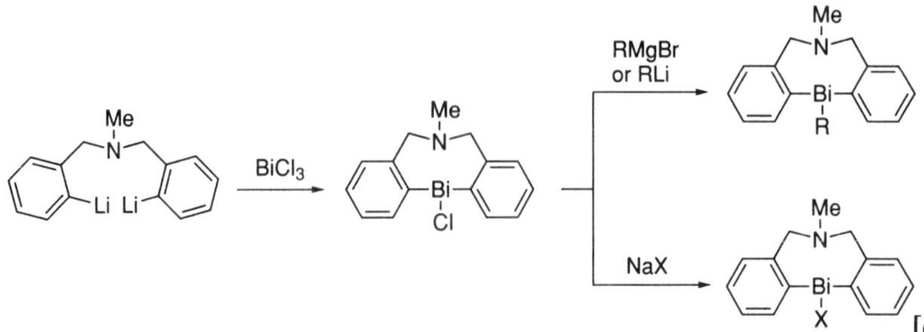

[6.162]

This chloride may be further reacted with organometallic reagents to afford 12-organyl azabismocine derivatives, and the anion exchange reaction with other halide anions leads to the corresponding 12-halogeno azabismocine derivatives. A similar ring closure of 2,2′-dilithiated dibenzyl ether by bismuth trichloride produces 12-chlorodibenzo[*b*,*g*][*1*,*5*]oxabismocine as an air-stable compound (Suzuki, H. and Ikegani, T., unpublished result):

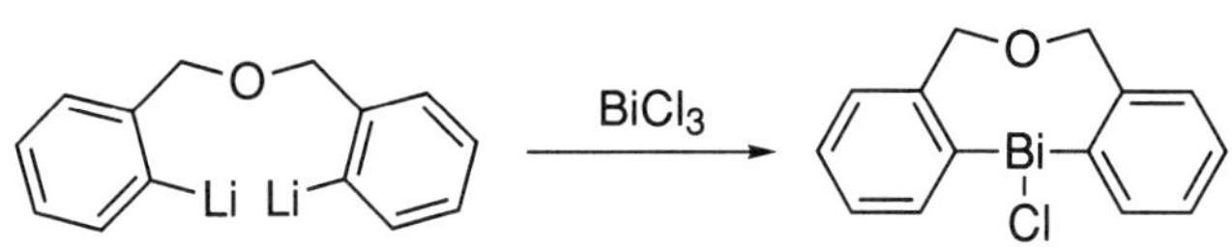

[6.163]

Reaction of 6-methyl-12-(4-methylphenyl)[*b*,*g*][*1*,*5*]dibenzoazabismocine with sulphuryl chloride or iodobenzene dichloride gives the corresponding dichloride in good yield. When this dichloride is treated with silver(I) oxide in dichloromethane, 6-methyl-12,12-bis(4-methylphenyl)-*b*,*g*][*1*,*5*]dibenzoazabismocine-12-onium chloride is formed, which readily undergoes anion exchange with silver(I) tetrafluoroborate to afford the corresponding tetrafluoroborate. On treatment with trifluoromethanesulphonamide in the presence of potassium tert-butoxide, the dichloride forms a stabilized bismuthine imide as a microcrystalline solid (Suzuki, H. and Ikegami, T., unpublished result):

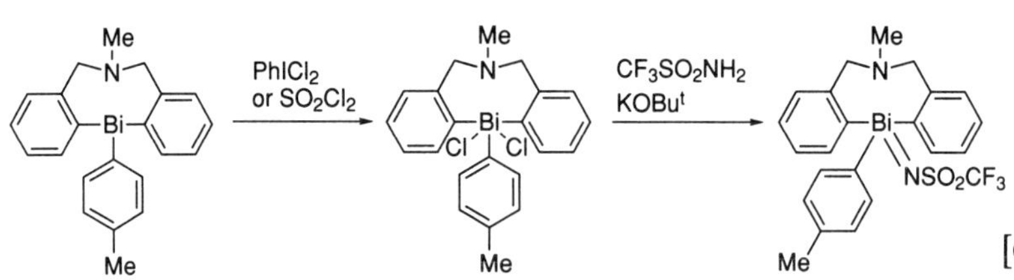

[6.164]

This is the first example of an air-stable bismuthine imide.

References

1. Löwig, C. and Schweizer, E. (1850) *Liebigs Ann. Chem.*, **75**, 315.
2. Gilman, H. and Yale, H. L. (1942) *Chem. Rev.*, **30**, 281.
3. Freedman, L. D. and Doak, G. O. (1982) *Chem. Rev.*, **82**, 15.
4. Doak, G. O. and Freedman, L. D. (1970) *Organometallic Compounds of Arsenic, Antimony, and Bismuth*, John Wiley, New York.
5. Freedman, L. D. and Doak, G. O. (1989) *The Chemistry of the Metal–Carbon Bond*, vol. 5 (ed. F. R. Hartley), John Wiley Interscience, Chichester, Sussex, p. 397.
6. Dub, M. (1970) *Organometallic Compounds*, Springer, New York.
7. Hellwinkel, D. (1983) *Top. Curr. Chem.*, **109**, 1.
8. Kitchin, J. P. (1986) *Organic Synthesis by Oxidation with Metal Compounds*, Plenum, New York, p. 817.
9. Barton, D. H. R. and Finet, J.-P. (1987) *Pure Appl. Chem.*, **59**, 937.
10. Abramovitch, R. A., Barton, D. H. R. and Finet, J.-P. (1988) *Tetrahedron*, **44**, 3039.
11. Barton, D. H. R. (1990) *Heteroatom Chemistry*, (ed. E. Block), VCH, New York, p. 95.
12. Finet, J.-P. (1989) *Chem. Rev.*, **89**, 1487.
13. Norman, N. C. (1988) *Chem. Soc. Rev.*, **17**, 269.
14. Compton, N. A., Errington R. J. and Norman, N. C. (1990) *Adv. Organometal. Chem.*, **31**, 91.
15. Wada, M. and Ohki, H. (1989) *Yuki Gosei Kagaku Kyokai Shi*, **47**, 425.
16. Seppelt, K. (1990) *Heteroatom Chemistry*, (ed. E. Block), VCH, New York, p. 335.
17. Seppelt, K. (1992) *Adv. Organomet. Chem.*, **34**, 207.
18. Ashe, A. J. III, (1990) *Adv. Organomet. Chem.*, **30**, 77.
19. Arduengo, A. J. III and Dixon, D. A. (1990) *Heteroatom Chemistry*, (ed. E. Block), VCH, New York, p. 47.
20. Akiba, K.-y. and Yamamoto, Y. (1994) *The Chemistry of Organic Arsenic, Antimony and Bismuth Compounds*, (ed. S. Patai), John Wiley, New York, p. 761.
21. Wardell, J. L. (1995) *Comprehensive Organometallic Chemistry*, vol. 2 (eds E. W. Abel, F. G. A. Stone and G. Wilkinson), Pergamon, New York, p. 321.
22. Huang, Y.-Z. and Zhou, Z.-L. (1995) *Comprehensive Organometallic Chemistry*, vol. 11 (eds E. W. Abel, F. G. A. Stone and G. Wilkinson), Pergamon, New York, p. 489.
23. Suzuki, H., Ikegami, T. and Matano, Y. (1997) *Synthesis*, 249.
24. Doak, G. O. and Freedman, L. D. (1995) *J. Organomet. Chem.*, **496**, 1.
25 Doak, G. O. and Freedman, L. D. (1995) *J. Organomet. Chem.*, **485**, 1.
26 Doak, G. O. and Freedman, L. D. (1994) *J. Organomet. Chem.*, **477**, 31.
27. Wardell, J. L. (1994) *Organomet. Chem.*, **23**, 138.
28 Wardell, J. L. (1992) *Organomet. Chem.*, **21**, 130.
29. Armitage, D. A. (1993) *Annual Reports Section A, The Royal Society of Chemistry*, 55.
30 Armitage, D. A. (1992) *Annual Reports Section A, The Royal Society of Chemistry*, 53.
31 Armitage, D. A. (1991) *Annual Reports Section A, The Royal Society of Chemistry*, 49.
32. Wieber, M. (1977) *Gmelin Handbuch der Anorganischen Chemie, Bisumut-Organische Verbindungen*, vol. 47, Springer, Berlin.
33. Samaan, S. (1978) *Methoden der Organischen Chemie. Metallorganische Verbindungen As, Sb, Bi*, vol. XIII, part 8, Georg Thieme Verlag, Stuttgart.
34. Banfi, A., Bartoletti, M., Bellora, E. *et al.* (1994) *Synthesis*, 775.
35. Pombrik, S. I., Kravtsov, D. N., Kvasov, B. A. and Fedin, E. I. (1977) *J. Organomet. Chem.*, **136**, 185.
36. Bras, P., van der Gen, A. and Wolters, J. (1983) *J. Organomet. Chem.*, **256**, C1.
37. Matano, Y., Miyamatsu T. and Suzuki, H. (1996) *Organometallics*, **15**, 1951.
38. Otero, A. and Royo, P. (1979) *J. Organomet. Chem.*, **171**, 333.
39. Kauffmann, T. and Steinseifer, F. (1985) *Chem. Ber.*, **118**, 1031.
40. Levason, W., Sheikh, B. and McCullough, F. P. (1982) *J. Coord. Chem.*, **12**, 53.
41. Murafuji, T., Mutoh, T. and Sugihara, Y. (1996) *J. Chem. Soc., Chem. Commun.*, 1693.
42. Rudolph, K. and Wieber, M. (1991) *Z. Naturforsch., B*, **46**, 1319.
43. Pasenok, S., Naumann D. and Tyrra, W. (1991) *J. Organomet. Chem.*, **417**, C47.
44. Naumann, D. and Tyrra, W. (1987) *J. Organomet. Chem.*, **334**, 323.

45. Nesmeyanov, A. N., Sazonova V. A. and Sedova, N. N. (1971) *Dokl. Akad. Nauk SSSR*, **198**, 590.
46. Nad, M. M., Kozminskaya, T. K. and Kocheshkov, K. A. (1946)*Zh. Obshch. Khim.*, **16**, 897.
47. Ptitsyna, O. A., Reutov O. A. and Ovodor, Y. S. (1962) *Izv. Akad. Nauk SSSR, Ser. Khim.*, 638.
48. Barton, D. H. R., Bhatnagar, N. Y., Finet, J.-P. and Motherwell, W. B. (1986) *Tetrahedron*, **42**, 3111.
49. Michaelis, A. and Polis, A. (1887) *Ber. Deut. Chem. Ges.*, **20**, 54.
50. Michaelis, A. and Marquardt, A. (1889) *Liebigs Ann. Chem.*, **251**, 323.
51. Gilmeister, A. (1897) *Ber. Deut. Chem. Ges.*, **30**, 2843.
52. Morrison, J. A. and Lagow, R. J. (1977) *Inorg. Chem.*, **16**, 1823.
53. Wieber, M. and Sauer, I. (1985) *Z. Naturforsch., B*, **40**, 1476.
54. Wieber, M. and Rudolph, K. (1988) *Z. Naturforsch., B*, **43**, 739.
55. Steinseifer, F. and Kauffmann, T. (1980) *Angew. Chem., Int. Ed. Engl.*, **19**, 723.
56. Kaufmann, T., Steinseifer, F. and Klas, N. (1985) *Chem. Ber.*, **118**, 1039.
57. Lorberth, J., Massa, W., Wocadlo, S. *et al.* (1995) *J. Organomet. Chem.*, **485**, 149.
58. Krommes, P. and Lorberth, J. (1975) *J. Organomet. Chem.*, **88**, 329.
59. Ziegler, K. and Steudel, O.-W. (1962) *Liebigs Ann. Chem.*, **652**, 1.
60. Chernykh I. N. and Tomilov, A. P. (1974) *Elektrokhimiya,*, **10**, 1424.
61. Deacon, G. B., Fallon, G. D. and Felder, P. W. (1971) *J. Organomet. Chem.*, **26**, C10.
62. Deacon, G. B. and Fallon, G. D. (1972) *Aust. J. Chem.*, **25**, 2107.
63. Bell, T. N., Pullman, B. J. and West, B. O. (1963) *Aust. J. Chem.*, **16**, 636.
64. Gilman, H. and Yablunky, H. L. (1940) *J. Am. Chem. Soc.*, **62**, 665.
65. Supniewski, J. V. and Adams R. (1926) *J. Am. Chem. Soc.*, **48**, 507.
66. Suzuki H. and Murafuji, T. (1992) *J. Chem. Soc., Chem. Commun.*, 1143.
67. Whitmire, K. H., Labahn, D., Roesky H. W. *et al.* (1991) *J. Organomet. Chem.*, **402**, 55.
68. Beagley, B. and McAloon, K. T. (1973) *J. Mol. Struct.*, **17**, 429.
69. Hawley, D. M. and Ferguson, G. (1968) *J. Chem. Soc. A*, 2059.
70. Murray, B., Hvoslef, J., Hope, H. and Power, P. P. (1983) *Inorg. Chem.*, **22**, 3421.
71. Sobolev, A. N., Romm, I. P., Bel'skii, V. K. and Gur'yanova, E. N. (1980) *Koord. Khim.*, **6**, 945.
72. Ogawa, T., Ikegami, T., Hikasa, T., Ono N. and Suzuki, H. (1994) *J. Chem. Soc., Perkin Trans. 1*, 3479.
73. Fukin, G. K., Pisarevskii, A. P., Yanovskii A. I. and Struchkov, Y. T. (1993) *Zh. Neorg. Khim.*, **38**, 1205.
74. Hey, D. H., Shingleton, D. A. and Williams, G. H. (1963) *J. Chem. Soc.*, 5612.
75. Suzuki, I. and Yamamoto, Y. (1993) *J. Org. Chem.*, **58**, 4783.
76. Barton, D. H. R., Motherwell W. B. and Stobie A. (1981) *J. Chem. Soc., Chem. Commun.*, 1232.
77. Barton, D. H. R., Finet, J.-P., Motherwell W. B. and Pichon, C. (1986) *Tetrahedron*, **42**, 5627.
78. Barton, D. H. R., Finet, J.-P. and Khamsi, J. (1987) *Tetrahedron Lett.*, **28**, 887.
79. Kawamura, T., Kikukawa, K., Takagi, M. and Matsuda, T. (1977) *Bull. Chem. Soc. Jpn.*, **50**, 2021.
80. Asano, R., Moritani, I., Fujiwara, Y. and Teranishi, S. (1973) *Bull. Chem. Soc. Jpn.*, **46**, 2910.
81. Barton, D. H. R., Ozbalik, N. and Ramesh, M. (1988) *Tetrahedron*, **44**, 5661.
82. Cho, C. S., Ohe, T., Itoh, O. and Uemura, S. (1992) *J. Chem. Soc., Chem. Commun.*, 453.
83. Cho, C. S., Yoshimori, Y. and Uemura, S. (1995) *Bull. Chem. Soc. Jpn.*, **68**, 950.
84. Ogawa, T., Hikasa, T., Ikegami, T. *et al.* (1993) *Chem. Lett.*, 815.
85. Ogawa, T., Hikasa, T. Ikegami, T. *et al.* (1994) *J. Chem. Soc., Perkin Trans. 1*, 3473.
86. Ogawa, T., Yoshikawa, A., Wada, H. *et al.* (1995) *J. Chem. Soc., Chem. Commun.*, 1407.
87. Huang, Y.- Z. and Liao, Y. (1991) *Heteroatom Chem.*, **2**, 297.
88. Barton, D. H. R., Ozbalik, N. and Ramesh, M. (1988) *Tetrahedron Lett.*, **29**, 857.
89. Matano, Y., Kinoshita, M. and Suzuki, H. (1992) *Bull. Chem. Soc. Jpn.*, **65**, 3504.
90. Ebert, K. H., Schulz, R. E., Breunig, H. J. *et al.* (1994) *J. Organomet. Chem.*, **470**, 93.
91. Blicke, F. F., Oakdale, U. O. and Smith, F. D. (1931) *J. Am. Chem. Soc.*, **53**, 1025.

92. Gilman, H. and Yablunky, H. L. (1941) *J. Am. Chem. Soc.*, **63**, 207.
93. Li, X.-W., Lorberth, J. Massa, W. and Wocadlo, S. (1995) *J. Organomet. Chem.*, **485**, 141.
94. Nakamura, E., Shimada, J. and Kuwajima, I. (1985) *Organometallics*, **4**, 641.
95. Frank, W. (1990) *J. Organomet. Chem.*, **386**, 177.
96. Bras, P., Herwijer, H. and Wolters, J. (1981) *J. Organomet. Chem.*, **212**, C7.
97. Jones, C., Engelhardt, L. M., Junk, P. C. *et al.* (1991) *J. Chem. Soc., Chem. Commun.*, 1560.
98. Atwood, P. A., Cowley, A. H. and Ruiz, J. (1992) *Inorg. Chim. Acta*, **198**, 271.
99. Suzuki, H., Murafuji, T. and Azuma, N. (1993) *J. Chem. Soc., Perkin Trans. 1*, 1169.
100. Murafuji, T., Mutoh, T., Satoh K. *et al.* (1995) *Organometallics*, **14**, 3848.
101. Suzuki, H., Murafuji, T., Matano, Y. and Azuma, N. (1993) *J. Chem. Soc., Perkin Trans. 1*, 2969.
102. Murafuji, T., Azuma N. and Suzuki, H. (1995) *Organometallics*, **14**, 1542.
103. Jutzi, P. and Schwartzen, K.-H. (1989) *Chem. Ber.*, **122**, 287.
104. Maier, L., Rochow, E. G. and Fernelius, W. C. (1961) *J. Inorg. Nucl. Chem.*, **16**, 213.
105. Wada, M. and Akiba, K.-y. (1985) *Tetrahedron Lett.*, **26**, 4211.
106. Wada, M., Ohki, H. and Akiba, K.-y. (1986) *Tetrahedron Lett.*, **27**, 4771.
107. Wada, M., Ohki, H. and Akiba, K.-y. (1987) *J. Chem. Soc., Chem. Commun.*, 708.
108. Wada, M., Ohki, H. and Akiba, K.-y. (1990) *Bull. Chem. Soc. Jpn.*, **63**, 1738.
109. Clegg, W., Errington, R. J., Fisher, G. A. *et al.* (1992) *J. Chem. Soc., Dalton Trans.*, 1967.
110. Clegg, W., Errington, R. J., Fisher, G. A. *et al.* (1993) *J. Chem. Soc., Dalton Trans.*, 637.
111. Clegg, W., Elsegood, M. R. J., Errington, R. J. *et al.* (1994) *J. Mater. Chem.*, **4**, 891.
112. Carmalt, C. J., Cowley, A. H., Decken, A. and Norman, N. C. (1995) *J. Organomet. Chem.*, **496**, 59.
113. Cassidy, J. M. and Whitmire, K. H. (1991) *Inorg. Chem.*, **30**, 2788.
114. Kaul, H., Greissinger, D., Luksza, M. and Malisch, M. (1982) *J. Organomet. Chem.*, **228**, C29.
115. Clegg, W., Compton, N. A., Errington, R. J. *et al.* (1990) *J. Organomet. Chem.*, **399**, C21.
116. Amberger, E. (1961) *Chem. Ber.*, **94**, 1447.
117. Faraglia, G. (1969) *J. Organomet. Chem.*, **20**, 99.
118. Allman, T., Goel, R. G. and Prasad, H. S. (1979) *J. Organomet. Chem.*, **166**, 365.
119. Carmalt, C. J., Norman, N. C., Orpen, A. G. and Stratford, S. E. (1993) *J. Organomet. Chem.*, **460**, C22.
120. Faleschini, S., Zanella, P., Doretti, L. and Faraglia, G. (1972) *J. Organomet. Chem.*, **44**, 317.
121. Okawara, R., Yasuda, K. and Inoue, M. (1966) *Bull. Chem. Soc. Jpn.*, **39**, 1823.
122. Minato, M. and Tsuji, J. (1988) *Chem. Lett.*, 2049.
123. Wieber, M. and Baudis, U. (1978) *Z. Anorg. Allg. Chem.*, **439**, 134.
124. Breunig, H. J., Ebert, K. H., Schulz, R. E. *et al.* (1995) *Z. Naturforsch., B*, **50**, 735.
125. Wieber, M. and Baudis, U. (1978) *Z. Anorg. Allg. Chem.*, **439**, 139.
126. Wieber, M. and Sauer, I. (1984) *Z. Naturforsch., B*, **39**, 1668.
127. Wieber, M. and Sauer, I. (1987) *Z. Naturforsch., B*, **42**, 695.
128. du Mont, W.-W., Severengiz, T., Breunig, H. J. and Müller, D. (1985) *Z. Naturforsch., B*, **40**, 848.
129. Calderazzo, F., Morvillo, A., Pelizzi, G. *et al.* (1988) *Inorg. Chem.*, **27**, 3730.
130. Wieber, M. and Sauer, I. (1984) *Z. Naturforsch., B*, **39**, 887.
131. Breunig, H. J. and Müller, D. (1986) *Z. Naturforsch., B*, **41**, 1129.
132. Calderazzo, F., Poli, R. and Pelizzi, G. (1984) *J. Chem. Soc., Dalton Trans.*, 2365.
133. Barton, D. H. R., Bridon, D. and Zard, S. Z. (1989) *Tetrahedron*, **45**, 2615.
134. Deacon, G. B., Jackson,W. R. and Pfeiffer, J. M. (1984) *Aust. J. Chem.*, **37**, 527.
135. Challenger, F., Smith, A. L. and Paton, F. J. (1923) *J. Chem. Soc.*, **123**, 1046.
136. Challenger, F., Peters, A. T. and Halevy, J. (1926) *J. Chem. Soc.*, 1648.
137. Whitmire, K. H., Hutchison, J. C., McKnight, A. L. and Jones, C. M. (1992) *J. Chem. Soc., Chem. Commun.*, 1021.
138. Burschka, C. and Wieber, M. (1979) *Z. Naturforsch., B*, **34**, 1037.
139. Burschka, C. (1982) *Z. Anorg. Allg. Chem.*, **485**, 217.
140. Ali, M., McWhinnie, W. R., West, A. A. and Hamor, T. A. (1990) *J. Chem. Soc., Dalton Trans.*, 899.

141. Curry, J. D. and Jandacek, R. J. (1972) *J. Chem. Soc., Dalton Trans.*, 1120.
142. Wieber, M., Rüdling, H. G. and Burschka, C. (1980) *Z. Anorg. Allg. Chem.*, **470**, 171.
143. Carmalt, C. J., Farrugia, L. J. and Norman, N. C. (1996) *J. Chem. Soc., Dalton Trans.*, 443.
144. Breeze, S. R. and Wang, S. (1993) *Angew. Chem., Int. Ed. Engl.*, **32**, 589.
145. Schere, O. J., Hornig, P. and Schmidt, M. (1966) *J. Organomet. Chem.*, **6**, 259.
146. Krommes, P. and Lorberth, J. (1975) *J. Organomet. Chem.*, **93**, 339.
147. Herberhold, M., Ehrenreich, W. and Guldner, K. (1984) *Chem. Ber.*, **177**, 1999.
148. Müller, J., Müller, U., Loss, A. *et al.* (1985) *Z. Naturforsch., B*, **40**, 1320.
149. Müller, J. (1971) *Z. Anorg. Allg. Chem.*, **381**, 103.
150. Ashe, A. J. III, Ludwig, E. G. Jr and Oleksyszyn, J. (1983) *Organometallics*, **2**, 1859.
151. Ashe, A. J. III, Kausch, C. M. and Eisenstein, O. (1987) *Organometallics*, **6**, 1185.
152. Calderazzo, F., Morvillo, A., Pelizzi, G. and Poli, R. (1983) *J. Chem. Soc., Chem. Commun.*, 507.
153. Ashe, A. J. III and Ludwig, E. G. Jr (1982) *Organometallics*, **1**, 1408.
154. Ashe, A. J. III and Drone, F. J. (1984) *Organometallics*, **3**, 495.
155. Ashe, A. J. III, Kampf, J. W., Puranik, D. B. and Al-Taweel, S. M. (1992) *Organometallics*, **11**, 2743.
156. Breunig, H. J. and Müller, D. (1982) *Angew. Chem. Int. Ed. Engl.*, **21**, 439.
157. Breunig, H. J. and Müller, D. (1983) *Z. Naturforsch., B*, **38**, 125.
158. Spence, R. E. v. H., Hsu, D. P. and Buchwald, S. L. (1992) *Organometallics*, **11**, 3492.
159. Becker, G. and Rössler, M. (1982) *Z. Naturforsch., B*, **37**, 91.
160. Mundt, V. O., Becker, G., Rössler, M. and Witthauer, C. (1983) *Z. Anorg. Allg. Chem.*, **506**, 42.
161. Mundt, O., Riffel, H., Becker, G. and Simon, A. (1988) *Z. Naturforsch., B*, **43**, 952.
162. Lohr, L. L. and Ashe, A. J. III, (1993) *Organometallics*, **12**, 343.
163. Ashe, A. J. III and Ludwig, E. G. Jr (1986) *J. Organomet. Chem.*, **303**, 197.
164. Ashe, A. J. III and Ludwig, E. G. Jr (1986) *J. Organomet. Chem.*, **308**, 289.
165. Calderazzo, F., Juris, A., Poli, R. and Ungari, F. (1991) *Inorg. Chem.*, **30**, 1274.
166. Nagase, S., Suzuki, S. and Kurakake, T. (1990) *J. Chem. Soc., Chem. Commun.*, 1724.
167. Gilman, H. and Yablunky, H. L. (1941) *J. Am. Chem. Soc.*, **63**, 212.
168. Davis, D. D. and Gray, C. E. (1970) *Organomet. Chem. Rev., Sect. A*, **6**, 283.
169. Doak, G. O. and Freedman, L. D. (1974) *Synthesis*, 328.
170. Breunig, H. J. and Müller, D. (1983) *J. Organomet. Chem.*, **253**, C21.
171. Bochkarev, M. N., Gur'ev, N. I. and Razuvaev, G. A. (1978) *J. Organomet. Chem.*, **162**, 289.
172. Razuvaev, G. A. (1980) *J. Organomet. Chem.*, **200**, 243.
173. Panster, P. and Malisch, W. (1977) *J. Organomet. Chem.*, **134**, C32.
174. Nesmeyanov, A. N., Anisimov, K. N., Kolobova N. E. and Khandozhko, V. N. (1964) *Dokl. Akad. Nauk SSSR*, **156**, 383.
175. Schrauzer, G. N. and Kratel, G. (1965) *Angew. Chem.*, **77**, 130.
176. Wallis, J. M., Müller, G. and Schmidbaur, H. (1987) *J. Organomet. Chem.*, **325**, 159.
177. Calderazzo, F., Poli, R. and Pelizzi, G. (1984) *J. Chem. Soc., Dalton Trans.*, 2535.
178. Challenger, F. (1916) *J. Chem. Soc.*, **109**, 250.
179. Glidewell, C. (1976) *J. Organomet. Chem.*, **116**, 199.
180. Keck, J.-M. and Klar, G. (1972) *Z. Naturforsch., B*, **27**, 596.
181. Schmuck, A. and Seppelt, K. (1989) *Chem. Ber.*, **122**, 803.
182. Tyrra, W. and Naumann, D. (1989) *Can. J. Chem.*, 67, 1949.
183. Frohn, H. J. and Maurer, H. (1986) *J. Fluorine Chem.*, **34**, 129.
184. Challenger, F. and Goddard, A. E. (1920) *J. Chem. Soc.*, **117**, 762.
185. Borisov, A. E., Osipova, M. A. and Nesmeyanov, A. N. (1963) *Izv. Akad. Nauk SSSR, Ser. Khim.*, 1507.
186. Wallenhauer, S. and Seppelt, K. (1994) *Angew. Chem., Int. Ed. Engl.*, **33**, 976.
187. Wittig, G. and Hellwinkel, D. (1964) *Chem. Ber.*, **97**, 789.
188. Challenger, F. and Wilson, V. K. (1927) *J. Chem. Soc.*, 209.
189. Wilkinson, J. F. and Challenger, F. (1924) *J. Chem. Soc.*, **125**, 854.
190. Suzuki, H., Ikegami, T., Matano, Y. and Azuma, N. (1993) *J. Chem. Soc., Perkin Trans. 1*, 2411.
191. Dahlmann, J. and Winsel, K. (1979) *J. Prakt. Chem.*, **321**, 370.

192. Goel, R. G. and Prasad, H. S. (1970) *Can. J. Chem.*, **48**, 2488.
193. Challenger, F. and Wilkinson, J. F. (1922) *J. Chem. Soc.*, **121**, 91.
194. Klemperer, W. G. and Liu, R.-S. (1980) *Inorg. Chem.*, **19**, 3863.
195. Ouchi, A., Uehiro, T. and Yoshino, Y. (1975) *J. Inorg. Nucl. Chem.*, **37**, 2347.
196. Ouchi, A., Honda, H. and Kitazima, S. (1975) *J. Inorg. Nucl. Chem.*, **37**, 2559.
197. Gilman, H. and Yale, H. L. (1951) *J. Am. Chem. Soc.*, **73**, 4470.
198. Barton, D. H. R., Bhatnagar, N. Y., Blazejewski, J.-C. *et al.* (1985) *J. Chem. Soc., Perkin Trans. 1*, 2657.
199. Rüther, R., Huber F. and Preut, H. (1986) *Z. Anorg. Allg. Chem.*, **539**, 110.
200. Challenger, F. and Richards, O. V. (1934) *J. Chem. Soc.*, 405.
201. Goel, R. G. and Prasad, H. S. (1973) *J. Organomet. Chem.*, **50**, 129.
202. Hawley, D. M. and Ferguson, G. (1968) *J. Chem. Soc. A*, 2539.
203. Ferguson, G., Kaitner, B., Glidewell, C. and Smith, S. (1991) *J. Organomet. Chem.*, **419**, 283.
204. Dittes, U., Keppler, B. K. and Nuber, B. (1996) *Angew. Chem., Int. Ed. Engl.*, **35**, 67.
205. Faraglia, G., Graziani, R., Volponi, L. and Casellato, U. (1983) *J. Organomet. Chem.*, **253**, 317.
206. Goel, R. G. and Prasad, H. S. (1976) *Spectrochim. Acta, Part A*, **32**, 569.
207. Dessy, R. E., Chivers, T. and Kitching, W. (1966) *J. Am. Chem. Soc.*, **88**, 467.
208. Challenger, F. and Rothstein, E. (1934) *J. Chem. Soc.*, 1258.
209. Barton, D. H. R., Lester, D. J., Motherwell, W. B. and Papoula, M. T. B. (1979) *J. Chem. Soc., Chem. Commun.*, 705.
210. Barton, D. H. R., Lester, D. J., Motherwell, W. B. and Papoula, M. T. B. (1980) *J. Chem. Soc., Chem. Commun.*, 246.
211. Barton, D. H. R., Blazejewski, J.-C., Charpiot, B. *et al.* (1985) *J. Chem. Soc., Perkin Trans. 1*, 2667.
212. Barton, D. H. R., Finet, J.-P., Khamsi, J. and Pichon, C. (1986) *Tetrahedron Lett.*, **27**, 3619.
213. Dodonov, V. A., Gushchin, A. V. and Brilkina, T. G. (1984) *Zh. Obshch. Khim.*, **54**, 2157.
214. Dodonov, V. A., Gushchin, A. V. and Brilkina, T. G. (1985) *Zh. Obshch. Khim.*, **55**, 466.
215. Dodonov, V. A., Gushchin, A. V. and Brilkina, T. G. (1985) *Zh. Obshch. Khim.*, **55**, 2514.
216. Gushchin, A. V., Brilkina, T. G. and Dodonov, V. A. (1985) *Zh. Obshch. Khim.*, **55**, 2630.
217. Dodonov, V. A., Gushchin, A. V., Brilkina, T. G. and Muratova, L. V. (1986) *Zh. Obshch. Khim.*, **56**, 2714.
218. Suzuki, H., Nakaya, C. and Matano, Y. (1992) *J. Chem. Res. (S)*, 34.
219. Suzuki, H., Nakaya, C. and Matano, Y. (1993) *Tetrahedron Lett.*, **34**, 1055.
220. Goel, R. G. and Prasad, H. S. (1972) *J. Organomet. Chem.*, **36**, 323.
221. Matano, Y., Azuma, N. and Suzuki, H. (1994) *J. Chem. Soc., Perkin Trans. 1*, 1739.
222. Rüther, R., Huber, F. and Preut, H. (1988) *J. Organomet. Chem.*, **342**, 185.
223. Beaumont, R. E., Goel, R. G. and Prasad, H. S. (1973) *Inorg. Chem.*, **12**, 944.
224. March, F. C. and Ferguson, G. (1975), *J. Chem. Soc., Dalton Trans.*, 1291.
225. Barton, D. H. R., Kitchin, J. P. and Motherwell, W. B. (1978), *J. Chem. Soc., Chem. Commun.*, 1099.
226. Ahond, A., Bowden, B. F., Coll, J. C. *et al.* (1979) *Aust. J. Chem.*, **32**, 1273.
227. Atwal, K. S., Sahoo, S. P., Tsai, T. Y. R. and Wiesner, K. (1982) *Heterocycles*, **19**, 641.
228. Wittig, G. and Clauss, K. (1952) *Liebigs Ann. Chem.*, **578**, 136.
229. Hellwinkel, D. and Bach, M. (1968) *Liebigs Ann. Chem.*, **720**, 198.
230. Ptitsyna, O. A., Gurskii, M. E., Mariorova, T. D. and Reutov, O. A. (1971) *Izv. Akad. Nauk SSSR, Ser. Khim.*, 2618.
231. Razuvaev, G. A., Osanova, N. A. and Sharutin, V. V. (1975) *Dokl. Akad. Nauk SSSR*, **225**, 581.
232. Barton, D. H. R., Charpiot, B., Dau, E. T. H. *et al.* (1984) *Helv. Chim. Acta*, **67**, 586.
233. Barton, D. H. R., Finet, J.-P., Motherwell, W. B. and Pichon, C. (1987) *J. Chem. Soc., Perkin Trans. 1*, 251.
234. Beaumont, R. E. and Goel, R. G. (1973) *J. Chem. Soc., Dalton Trans.*, 1394.

235. Goel, R. G. and Prasad, H. S. (1971) *J. Chem. Soc. A*, 562.
236. Matano, Y., Azuma, N. and Suzuki, H. (1993) *Tetrahedron Lett.*, **34**, 8457.
237. Matano, Y., Azuma, N. and Suzuki, H. (1995) *J. Chem. Soc., Perkin Trans. 1*, 2543.
238. Matano, Y., Yoshimune, M. and Suzuki, H. (1995) *Tetrahedron Lett.*, **36**, 7475.
239. Matano, Y., Yoshimune, M., Azuma, N. and Suzuki, H. (1996) *J. Chem. Soc., Perkin Trans.*, **1**, 1971.
240. Doak, G. O., Long, G. G., Kakar, S. K. and Freedman, L. D. (1966) *J. Am. Chem. Soc.*, **88**, 2342.
241. Amberger, E. and Gut, E. (1968) *Chem. Ber.*, **101**, 1200.
242. Suzuki, H., Ikegami, T. and Azuma, N. (1997) *J. Chem. Soc., Perkin Trans. 1*, 1609.
243. Matano, Y. and Suzuki, H. (1996) *Bull. Chem. Soc. Jpn.*, **69**, 2673.
244. Bordner, J. and Freedman, L. D. (1973) *Phosphorus*, **3**, 33.
245. Ptitsyna, O. A., Gurskii, M. E. and Reutov, O. A. (1973) *Izv. Akad. Nauk SSSR, Ser. Khim.*, 229.
246. Matano, Y. (1994) *J. Chem. Soc., Perkin Trans. 1*, 2703.
247. Schumuck, A., Buschmann, J., Fuchs, J. and Seppelt, K. (1987) *Angew. Chem., Int. Ed. Engl.*, **26**, 1180.
248. Schmuck, A., Pyykkö, P. and Seppelt, K. (1990) *Angew. Chem., Int. Ed. Engl.*, **29**, 213.
249. Schumuck, A., Leopold, D., Wallenhauer, S. and Seppelt, K. (1990) *Chem. Ber.*, **123**, 761.
250. Hellwinkel, D. and Kilthau, G. (1967) *Liebigs Ann. Chem.*, **705**, 66.
251. Wallenhauer, S., Leopold, D. and Seppelt, K. (1993) *Inorg. Chem.*, **32**, 3948.
252. Razuvaev, G. A., Osanova, N. A., Sharutin, V. V. *et al.* (1978) *Dokl. Akad. Nauk SSSR*, **238**, 361.
253. Barton, D. H. R., Donnelly, D. M. X., Finet, J.-P. and Stenson, P. H. (1988) *Tetrahedron*, **44**, 6387.
254. Barton, D. H. R., Blazejewski, J.-C., Charpiot, B. *et al.* (1980) *J. Chem. Soc., Chem. Commun.*, 827.
255. Lloyd, D. and Singer, M. I. C. (1967) *J. Chem. Soc., Chem. Commun.*, 1042.
256. Ferguson, G., Glidewell, C., Gosney, I. *et al.* (1988) *J. Chem. Soc., Perkin Trans. 2*, 1829.
257. Glidewell, C., Lloyd, D. and Metcalfe, S. (1988) *Synthesis*, 319.
258. Suzuki, H., Murafuji, T. and Ogawa, T. (1988) *Chem. Lett.*, 847.
259. Ogawa, T., Murafuji, T. and Suzuki, H. (1989) *J. Chem. Soc., Chem. Commun.*, 1749.
260. Kirij, N. V., Pasenok, S. V. Yagupolskii, Y. L. *et al.* (1994) *J. Fluorine Chem.*, **66**, 75.
261. Yasui, M., Kikuchi, T., Iwasaki, F. *et al.* (1990) *J. Chem. Soc., Perkin Trans. 1*, 3367.
262. Ogawa, T., Murafuji, T. and Suzuki, H. (1988) *Chem. Lett.*, 849.
263. Ogawa, T., Murafuji, T. Iwata, K. and Suzuki, H. (1989) *Chem. Lett.*, 325.
264. Suzuki, H. and Murafuji, T. (1990) *Bull. Chem. Soc. Jpn.*, **63**, 950.
265. Matano, Y., Yoshimune, M. and Suzuki, H. (1995) *J. Org. Chem.*, **60**, 4663.
266. Matano, Y. and Suzuki, H. (1996) *J. Chem. Soc., Chem. Commun.*, 2697.
267. Suzuki, H., Nakaya, C., Matano, Y. and Ogawa, T. (1991) *Chem. Lett.*, 105.
268. Suzuki, H. and Ikegami, T. (1996) *J. Chem. Res.*, 24.
269. Pasenok, S. V., Kirij, N. V., Yagupolskii, Y. L. *et al.* (1993) *J. Fluorine Chem.*, **63**, 179.
270. Suzuki, H., Ikegami, T. and Matano, Y. (1994) *Tetrahedron Lett.*, **35**, 8197.
271. Somasundaram, N. and Srinivasan, C. (1996) *J. Org. Chem.*, **61**, 2895.
272. Ogawa, T., Murafuji, T., Iwata, K. and Suzuki, H. (1988) *Chem. Lett.*, 2021.
273. Ogawa, T., Miyazaki, K. and Suzuki, H. (1990) *Chem. Lett.*, 1651.
274. Ashe, A. J. III, Kampf, J. W. and Al-Taweel, S. M. (1992) *Organometallics*, **11**, 1491.
275. Fagan, P. J. and Nugent, W. A. (1988) *J. Am. Chem. Soc.*, **110**, 2310.
276. Lohr, L. L. and Ashe, A. J. III (1993) *Organometallics* **12**, 343.
277. Grutzmacher, H. (1995) *Angew. Chem., Int. Ed. Engl.*, **34**, 295.
278. Mishra, B. K., Varghese, S. C., Roy, R. and Mishra, R. K. (1990) *Indian J. Chem. Sect. A*, **29**, 370.
279. Mishra, B. K., Varghese, S. C., Roy, R. and Mishra, R. K. (1990) *Chem. Abstr.*, **123**, 58162X.
280. Mishra, R. K. and Mishra, B. K. (1988) *Chem. Phys. Lett.*, **151**, 44.
281. Ashe, A. J. III, Kampf, J. W. and Al-Taweel, S. M. (1992) *J. Am. Chem. Soc.*, **114**, 372.
282. Ashe, A. J. III, Kampf, J. W. and Puranik, D. B. (1993) *J. Organomet. Chem.*, **447**, 197.

283. Kurita, J., Ishii, M., Yasuike, S. and Tsuchiya, T. (1993) *J. Chem. Soc., Chem. Commun.*, 1309.
284. Kurita, J., Ishii, M., Yasuike, S. and Tsuchiya, T. (1994) *Chem. Pharm. Bull.*, **42**, 1437.
285. Carmalt, C. J., Cowley, A. H., Decken, A. *et al.* (1996) *Organometallics*, **15**, 887.
286. Hellwinkel, D. and Bach, M. (1969) *J. Organomet. Chem.*, **17**, 389.
287. Ashe, A. J. III, Diephouse, T. R. and El-Sheikh, M. Y. (1982) *J. Am. Chem. Soc.*, **104**, 5693.
288. Ashe, A. J. III and Gordon, M. D. (1972) *J. Am. Chem. Soc.*, **94**, 7596.
289. Ashe, A. J. III (1976) *Tetrahedron Lett.*, 415.
290. Suzuki, H., Murafuji, T. and Azuma, N. (1992) *J. Chem. Soc., Perkin Trans. 1*, 1593.
291. Grüttner, G. and Wiernik, M. (1915) *Chem. Ber.*, **48**, 1473.
292. Al-Soudani, A. R. H. and Massey, A. G. (1988) *Appl. Organomet. Chem.*, **2**, 553.
293. Al-Soudani, A. R., Jones, J. B., Brown, D. S. *et al.* (1989) *Appl. Organomet. Chem.*, **3**, 459.
294. Al-Jabar, N. A. A. and Massey, A. G. (1985) *J. Organomet. Chem.*, **287**, 57.
295. Hassler, K. and Seidl, S. (1988) *J. Organomet. Chem.*, **347**, 27.
296. Kollegger, G. M., Seigl, H., Hassler K. and Gruber, K. (1996) *Organometallics*, **15**, 4337.
297. Yasuike, S., Ohta, H., Shiratori, S., Kurita J. and Tsuchiya, T. (1993) *J. Chem. Soc., Chem. Commun.*, 1817.
298. Yasuike, S., Kiharada, T., Kurita, J. and Tsuchiya, T. (1996) *J. Chem. Soc., Chem. Commun.*, 2183.
299. Chen, X., Ohdoi, K., Yamamoto, Y. and Akiba, K.-y. (1993*Organometallics*, **12**, 1857.
300. Yamamoto, Y., Chen, X. and Akiba, K.-y. (1992) *J. Am. Chem. Soc.*, **114**, 7906.
301. Akiba, K.-Y., Ohdoi, K. and Yamamoto, y. (1988) *Tetrahedron Lett.*, **29**, 3817.
302. Yamamoto, Y., Ohdoi, K., Chen, X. *et al.* (1993) *Organometallics*, **12**, 3297.
303. Akiba, K.-Y., Ohdoi, K. and Yamamoto, Y. (1989) *Tetrahedron Lett.*, **30**, 953.
304. Chen, X., Yamamoto, Y., Akiba, K.-y., Yoshida, S. *et al.* (1992) *Tetrahedron Lett.*, **33**, 6653.
305. Sharutin, V. V., Bychkov, V. T., Bolotova, O. P. and Kuzina, V. I. (1986) *Zh. Obshch. Khim.*, **56**, 330.
306. Sharutin, V. V. (1988) *Zh. Obshch. Khim.*, **58**, 2305.
307. Georgiades, A. and Latscha, H. P. (1980) *Z. Naturforsch., B,*, **35**, 1000.
308. Wieber, M. and Baudis, U. (1976) *Z. Anorg. Allg. Chem.*, **423**, 40.
309. Wieber, M. and Baudis, U. (1976) *Z. Anorg. Allg. Chem.*, **423**, 47.
310. Dräger, M. and Schmidt, B. M. (1985) *J. Organomet. Chem.*, **290**, 133.
311. Praeckel, U. and Huber, F. (1981) *Z. Naturforsch., B*, **36**, 70.
312. Bräu, E., Falke, R., Ellner, A. *et al.* (1994) *Polyhedron*, **13**, 365.
313. Ohkata, K., Takemoto, S., Ohnishi, M. and Akiba, K.-y. (1989) *Tetrahedron Lett.*, **30**, 4841.

7 Organotransition metal compounds with element to transition metal bonds

K.H. WHITMIRE

7.1 Introduction

Arsenic, antimony and bismuth readily form bonds to transition metals. There are many synthetic and structural relationships between these compounds and those of the other p-block elements in groups 13, 14, 15 and 16. In fact, a discussion and understanding of the structure and bonding of such compounds is incomplete without including those in the other groups. Because of space limitations in this volume, this chapter will not be able to discuss all of the related chemistry of other elements nor will it be 'comprehensive'; rather the most important aspects of synthetic strategies, bonding patterns and reactivities will be surveyed. The reader is referred to a number of review articles which have appeared concerning main group–transition metal complexes and transition metal cluster compounds in general [1–16].

Throughout this chapter, E will be used to designate p-block elements and M will be reserved to indicate transition metals. R will denote organic or inorganic functions, including halides and pseudo-halides, where either type of functionality is possible, and X will denote exclusively halide and pseudo-halide ligands. For ease of reference and brevity, counterions will be omitted since in many cases multiple salts of a compound are known but the chemistry is essentially unaffected.

7.2 Metal complexes with ER_3 ligands

Like amines and phosphines, the heavier group 15 congeners form ER_3 which are pyramidal and possess a lone pair of electrons which may coordinate to transition metals, $L_nM \leftarrow :ER_3$. Upon descending the periodic table there are three trends which are important in the chemistry observed.

- There is a tendency for these organo E complexes to become more reactive, partially as a result of weaker E–carbon bond energies. This weakening of the E–R bonds has implications for the formation of metallated organo-E complexes, and it is often observed that such bonds are very readily cleaved when E is attached to a transition metal.

Synthetic strategies must therefore accommodate these weaker E–C bonds, especially for the heavier elements. In fact, many M–BiR_3 compounds appear to have seriously limited thermal stability. Thermal or photochemical decomposition often leads to the formation of cluster complexes resulting from E–carbon bond breaking.

- The lone pairs of electrons on E become continually less basic as one descends the group 15 elements. This is a manifestation of the well-known inert pair effect which has its origins in the lowering of the energy of the s orbital with heavier elements owing to relativistic effects. Thus BiR_3 is considerably poorer as a donor ligand than is NR_3 or PR_3, and the list of simple coordination compounds with ER_3 as ligands becomes rapidly shorter on going from P to As to Sb to Bi. By the time one gets to Bi, for example, only the complexes $[M(CO)_5BiPh_3]$ (M = Cr, Mo, W) [17] and $[CpFe(CO)_2BiPh_3][BF_4]$ [18] have been structurally characterized.
- The third general observation is the increased tendency for the main group centre to function as a Lewis acid as the atomic number increases. This is the same phenomenon known as 'hypervalency' or 'octet expansion' which has been well documented in the case of conventional main group element complexes. The same general features found by valence shell electron pair repulsion theory (VSEPR theory) for those complexes also appear to operate when the substituents on E are transition metals.

The structure and bonding patterns of M–ER_3 complexes are otherwise conventional and often the ER_3 serve a similar function as other simple two-electron donor ligands. They will not be discussed in further detail.

7.3 EM_x complexes

7.3.1 Electron precise molecules

A number of metallated main group element compounds can be viewed as being obtained by replacement of R by an appropriate ML_n fragment, generating the series $ER_{3-n}\{ML_n\}_n$ where $n = 1$, 2 or 3 (structures 1, 2 and 3 in Scheme 7.1). These compounds may be viewed as having E in the +3 oxidation state and, like the parent ER_3, the products show a pyramidal E atom which retains a lone pair of electrons.

There are two general methods that have been used specifically to produce compounds of type 1 (shown in Scheme 7.1 ($n = 1$)). One is the deprotonation of $L_nM(ER_2H)$ with an alkyl lithium reagent as demonstrated for E = P. This methodology was employed for the related double deprotonation of $[(CO)_5Cr(AsPhH_2)]$ to give $Li_2[(CO)_5Cr(AsPh)]$ [19]:

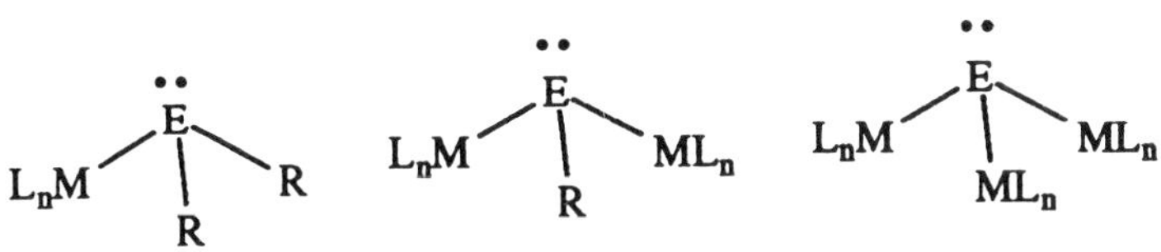

1 2 3

Scheme 7.1 The series $ER_{3-n}\{ML_n\}_n$: structure 1, $n = 1$; structure 2, $n = 2$; structure 3, $n = 3$.

$$L_nM-AsPhH_2 + 2LiR \rightarrow Li_2[L_nM-AsPh] + 2R\text{-}H \qquad [7.1]$$

The structure of the As compound has not been reported, but it is likely that the Li^+ are tightly associated with As. Another route is cross-exchange between R_2E–ER_2 and L_nM–ML_n ($ML_n = Cr(CO)_3Cp\{Cp = \eta\text{-}C_5H_5\}$; $ER_2 = AsPh_2$, $SbPh_2$; $ML_n = Co(CO)_4$; $ER_2 = BiPh_2$) [20–22]:

$$L_nM-ML_n + R_2E\text{-}ER_2 \rightarrow 2R_2E-ML_n \qquad [7.2]$$

These two methodologies have had limited application.

A third method, halide displacement from $EX_{3-n}R_n$ ($n = 0$, 1 or 2) by a metal carbonylate anion may be used to produce compounds with one or more metal centres attached to E:

$$R_2EX + [ML_n]^- \rightarrow R_2EML_n + X^- \qquad [7.3]$$

A variation is reaction of R_2ECl with $L_nM-SiMe_3$ complexes where elimination of Me_3SiCl is the reaction driving force [23]. A potential complication of these reactions is the tendency under certain circumstances for the bismuth compounds to undergo redistribution reactions of the following type [24]:

$$2[Cp(CO)_3MBi(Me)Br] \rightarrow [\{Cp(CO)_3M\}_2BiBr + BiMe_2Br] \qquad [7.4]$$

Complexes of types 1–3 (shown in Schemes 7.1) are compiled in Tables 7.1(a)–7.1(c), respectively. The structurally characterized examples show the expected pyramidal arrangement at E indicative of the presence of the lone pair of electrons. In the case of $[(CO)_4FeBiPh_2]^-$ the $BiPh_2$ group occupies an axial position at the trigonal bipyramidal metal centre, as is common for groups in trigonal bipyramidal complexes which have little π-bonding to the metal. This iron compound reacts further with $BiPh_2Cl$ to give $[Fe(CO)_4\{BiPh_2\}_2]$, which is the only example known of its type [25]. It is an octahedral complex with *cis* $BiPh_2$ groups which is of limited stability and decomposes with Bi–C bond cleavage.

The compound $[Bi_2\{Ge(C_6F_5)_2\}_2\{(C_6F_5)_2GePt(PPh_3)_2\}]$ is somewhat unique in this group of compounds, being synthesized from $Bi_2(\mu$-

Table 7.1 Examples of (a) L_nMER_2 Complexes; (b) pyramidal $\{L_nM\}_2$ ER complexes without M–M bonding; (c) pyramidal $E\{ML_n\}$ complexes without M–M bonding. L, M, E and R are defined in text, Section 7.1.

Compound	Reference
(a)	
$[Cp(CO)_3CrAsPh_2]$	[i]
$[(CO)_5MnAsCl_2]$	[ii]
$[Cp(CO)_2FeAsX_2]$, X = Cl, Br	[iii, iv]
$[Cp(CO)_3MoAs(CF_3)_2]$	[v]
$[Cp(CO)_2FeAs(CF_3)_2]$	[v]
$[Cp(CO)_2FeAs(C_6F_5)_2]$	[vi]
$[Cp(CO)_3MoAs(C_6F_5)_2]$	[vi]
$[Cp(CO)(L)FeAsMe_2]$, L = PMe_3, PEt_3, PMe_2Ph, $P(OMe)_3$	[vii]
$[(CO)_2(NO)(PPh_3)FeAsCl_2]$	[iv]
$[Cp(CO)_nMAsMe_2]$, M = Fe, $n = 2$; M = Cr, Mo, W, $n = 3$	[viii, ix]
$[Cp(CO)_3MSbMe_2]$, M = Cr, Mo, W	[x]
$[Cp(CO)_nMSb\{(N^tBu)_2SiMe_2\}]$, M = Fe, $n = 2$; M = Mo, W, $n = 3$	[v]
$[Cp(CO)_nMSb\{(N^tBu)_2SiMe_2\}Fe(CO)_4]$, M = Fe, $n = 2$; M = Mo, W, $n = 3$	[v]
$[Cp(CO)_3CrSbPh_2]$	[i]
$[Cp(CO)_2FeSbX_2]$, X = Cl, Br, I	[iii, iv]
$[Cp(CO)(L)FeSbMe_2]$, L = PMe_3, PEt_3, PMe_2Ph, $P(OMe)_3$	[vii]
$[Ph_2SbCo$(dimethylglyoximate)(pyridine)]	[xi]
$[Cp(CO)_nMBi\{(N^tBu)_2SiMe_2\}]$, M = Fe, $n = 2$; M = Mo, W, $n = 3$	[xii]
$[(CO)_5MnBiPh_2]$	[xiii]
$[(CO)_5ReBiPh_2]$	[xiv]
$[(CO)_5MnBiMe_2C_4H_2]$	[xv]
$[Cp(CO)_3CrBiPh_2]$	[i]
$[(CO)_4FeBiPh_2]^-$	[xi]
$[Cp(CO)_2FeBiX_2]$, X = Cl, Br, I	[iii, iv]
$[(CO)_2(NO)PPh_3FeBiCl_2]$	[iv]
$[Cp(CO)(L)FeBiMe_2]$, L = PMe_3, PEt_3, PMe_2Ph, $P(OMe)_3$	[vii]
$[Cp(CO)_3MBiMe_2]$, M = Cr, Mo, W	[xvi]
$[Cp(CO)_2FeBiX_2]$, X = $SC(S)NEt_2$, $SC(S)OMe$	[xvii]
$[(CO)_3(PPh_3)CoBiCl_2]$	[xviii]
$[(CO)_3LCoBiPh_2]$, L = CO, PPh_3	[xix]
[(dimethylglyoximate)(pyridine)$CoBiPh_2$]	[xi]
$[(CO)_4Fe\{BiPh_2\}_2]$	[xiii]
$[Bi_2\{Ge(C_6F_5)_2\}_2\{(C_6F_5)_2GePt(Ph_3P)_2\}]$	[xx]
(b)	
$[\{Cp(CO)_3Mo\}_2SbCl]$, M = Mo, W	[xxi]
$[\{Cp(CO)_2Fe\}_2SbX]$, X = Cl, Br, I	[xxi]
$[\{(CO)_3(PPh_3)Co\}_2BiCl]$	[xxii]
$[\{Cp(CO)_3M\}_2BiX]$, M = Cr, Mo, W; X = Me, halide	[xvi, xxiii–xxv]
$[\{Cp(CN^tBu)(CO)_2Mo\}_2BiCl]$	[xxv]
$[\{C_5H_4R(CO)_2Fe\}_2Bi(NO_3)]$, R = H, Me	[xxvi]
$[\{Cp(CO)_2Ru\}_2Bi(NO_3)]$	[xxvi]
$[\{Cp(CO)_2Ru\}_2BiCl]$	[xxvi]
$[\{C_5H_4R(CO)_3M\}_2Bi(NO_3)]$, M = Cr, Mo, W; R = H, Me	[xxvi]
$[(\{(C_5H_4R)(CO)_2Fe\}_2BiCl)_3]$, R = H, Me	[xxvii–xxix]
(c)	
$[As\{Mn(CO)_5\}_3]$	[ii]
$[Sb\{M(CO)_3Cp\}_3]$, M = Mo, W	[xxi]

Table 7.1 *Continued*

Compound	Reference
$[Sb\{Co(CO)_3PPh_3\}_3]$	[xviii]
$[Sb\{Mn(CO)_5\}_3]$	[ii]
$[Sb\{Fe(CO)_2Cp\}_3]$	[xxi]
$[Bi\{Cr(CO)_3Cp\}_3]$	[xxiv]
$[Bi\{Mo(CO)_3Cp\}_3]$	[xxv]
$[Bi\{Mn(CO)_5\}_3]$	[xxx]
$[Bi\{Fe(CO)_2C_5H_4R\}_3]$, R = H, Me	[xxvii]
$[Bi\{Fe(CO)_2(NO)PPh_3\}_3]$	[iv]
$[Bi\{Fe(CO)_3(NO)\}_3]$	[iv]
$[Bi\{Ru(CO)_2Cp\}_3]$	[xxvi]
$[Bi\{Co(CO)_4\}_3]$	[xxxi]
$[Bi\{Co(CO)_3SbPh_3\}_2\{Co(CO)_4\}]$	[xxxi]
$[Bi\{Co(CO)_3L\}_3]$, L = PPh_3, $AsPh_3$	[iv, xviii, xxxi]

References

i. F. Calderazzo, A. Juris, R. Poli, and F. Ungari, *Inorg. Chem.*, 1991, **30**, 1274.

ii. V. Grossbruchhaus and D. Rehder, *Inorg. Chim. Acta*, 1990, **172**, 141.

iii. W. R. Cullen, D. J. Patrnore, J. R. Sarns, M. J. Newlands, and L. K. Thompson, *Chem Commlm.*, 1971, 952.

iv. W. R. Cullen, D. J. Patmore, and J. R. Sams, *Inorg. Chem.*, 1973, **12**, 867.

v. W. R. Cullen and R. G. Hayter, J. *Am Chem Soc.*, 1964, **86**, 1030.

vi. M. Cooke, M. Green, and D. Kirkpatrick, J. *Chem Soc. (A)*, 1968, 1507.

vii. H.-A. Kaul, D. Greissinger, M. Luksza, and W. Malisch, J. *Organomet. Chem.*, 1982, **228**, C29.

viii. W. Malisch, M. Kuhn, W. Albert, and H Rößner, *Chem Ber.*, 1980, **113**, 3318.

ix. W. Malisch and M. Kuhn, *Angew. Chem Int. Ed. Eng.*, 1974, **13**, 84; *Angew.* ChenL, 1974, **86**, 51.

x. W. Malisch and P. Panster, J. *Organomet. Chem*, 1975, **99**, 421.

xi. G. N. Schrauzer and G. Kratel, *Angew. Chem Int. Ed. Eng.*, 1965, **4**, 146; *Angew. Chem*, 1965, **77**, 130.

xii. A.-M. Caminade, M. Veith, V. Huch and W. Malisch, *Organometallics*, 1990, **9**, 1798.

xiii. J. M. Cassidy and K. H. Whitmire, *Inorg. Chem.*, 1991, **30**, 2788.

xiv. A. N. Nesmeyanov, K. N. Anisimov, N. E. Kolobova, and V. N. Khandozhko, *Akad Nauk SSSR Proceedings Section Chemistry*, 1964, **156**, 383 (English translation, page 502).

xv. A. J. Ashe, III, J. W. Kampf, and D. B. Puranik, J. *Organomet. Chem.*, 1993, **447**, 197.

xvi. P. Panster and W. Malisch, J. *Organomet. Chem*, 1977, **134**, C32.

xvii. M Wieber, D. Wirth, and C. Burschka, Z. *Naturforsch.*, 1985, **40B**, 258

xviii. W. Clegg, N. A. Compton, R. J. Errington, D. C. R. Hockless, N. C. Norman, M. Ramshaw, and P. M. Webster, *J. Chem. Soc., Dalton Trans*, 1990, 2375.

xix. F. Calderazzo, R. Poli, and G. Pelizzi, *J. Chem. Soc., Dalton Trans.*, 1984, 2535.

xx. M. N. Bochkarev, G. A. Razuvaev, L. N. Zakharov, and Yu. T. Struchov, *J. Organomet. Chem.*, 1980, **199**, 205.

xxi. A. M. Barr, M. D. Kerlogue, N. C. Norman, P. M. Webster, and L. J. Farrugia, *Polyhedron*, 1989, 8, 2495.

xxii. U. Weber, G. Huttner, O. Scheidsteger, and L. Zsolnai, J. *Organomet. Chem.*, 1985, **289**, 357.

xxiii. W. Clegg, N. A. Compton, R. J. Errington, G. A. Fisher, D. C. R. Hockless, N. C. Norman, N. A. L. Williams, S. E. Stratford, S. J. Nichols, P. S. Jarrett, and A. G. Orpen, J. *Chem. Soc., Dalton Trans.*, 1992, 193.

xxiv. W. Clegg, N. A. Compton, R. J. Errington, G. A. Fisher, N. C. Norman, and T. B. Marder, *J. Chem. Soc., Dalton Trans.*, 1991, 2887.

Table 7.1 *Continued*

xxv. W. Clegg, N. A. Compton, R. J. Errington, and N. C. Norman, J. *Organomet.* Chem., 1987, **335**, C1.

xxvi R. J. Errington, L. J. Farrugia, G. A. Fisher, A. Niklaus, and N. C. Norman, J. *Chem Soc., Dalton Trans.*, 1993, 1201.

xxvii J. M. Wallis, G. Müller, and H. Schmidbaur, J. *Organomet. Chem.*, 1987, **325**, 159.

xxviii W. Clegg, N. A. Compton, R. J. Errington, and N. C. Norman, J. *Chem. Soc,* Dalton Trans. 1988, 1671

xxix W. Clegg, N. A. Compton, R. J. Errington, and N. C. Norman, *Polyhedron,* 1987, **6,** 2031.

xxx J. M. Wallis, G. Müller, and H. Schmidbaur, *Inorg. Chem.*, 1987, 26, 458.

xxxi G. Etzrodt, R. Boese, and G. Schmid, *Chem. Ber.*, 1979, **112**, 2574.

Ge $(C_6F_5)_2\}_3$ (no Bi–Bi bond) by addition of the Pt reagent across a Bi–Ge bond [26].

Note that the molecules of types 1–3 retain a lone pair of electrons which may also serve to bond to another metal substituent giving a tetra-coordinate E atom (4–6 in Scheme 7.2).

Vahrenkamp [27–49] and Malisch [50–52] have studied a large number of complexes of type 4 based upon the ER_2 group, especially $AsMe_2$, but complexes of Sb and Bi are also known. The synthesis of such compounds follows logically by using the L_nMAsMe_2 group as a complex ligand (Scheme 7.3). The lone pair on As is subject to electrophilic attack by organic or inorganic halides RX (R = $Me_3SiCH_2^+$, $PhC(O)CH_2^+$ or H^+) to give $[L_nM(AsMe_2R)]^+$ [51]. These molecules belong to the class of L_nMER_3 compounds described in Section 7.2 in which the metal fragment is a cationic 16-electron species.

The lone pair of structure 1 may coordinate to another metal fragment. Alternatively, the starting point may be L_nMAsMe_2X (structure 7) in reaction with a metal carbonylate anion. This chemistry has been extended to produce trimetallic (structure 8) and tetrametallic species containing two or three $AsMe_2$ bridging groups. For example, it is possible to prepare the disubstituted starting material $[(CO)_4M\{AsMe_2Cl\}_2]$ structure which can be reacted stepwise with one or two equivalents of metal carbonylate anions. The obvious advantage of this methodology is the ability to create a wide variety of homonuclear and heteronuclear complexes. In general, in Scheme 7.3 L_nM is a 16-electron fragment and L_nM' a 17-electron fragment. Another route to M_2ER_2 complexes is the reaction of $[M_2(CO)_{10}]^{2-}$ (M = Cr, Mo, W) with Ph_2BiCl which leads directly to the open complexes $[Ph_2Bi\{M(CO)_5\}_2]^-$ having no M–M bonding [53]. Photolysis of $[Cp(CO)_3MAsMe_2]$ leads to CO loss and formation of some oligomeric compounds of the type $[Cp(CO)_2\{Cp(CO)_3MAsMe_2\}MAsMe_2]$ [23]. When $HAsMe_2$ reacts with $HCo(CO)_4$, a compound formulated as $[Me_2AsCo(CO)_3]$ is obtained [54]. It is insoluble in most organic solvents and is assumed to be polymeric. Such an oligomerization–polymerization process is consistent with the chemistry documented in Scheme 7.3.

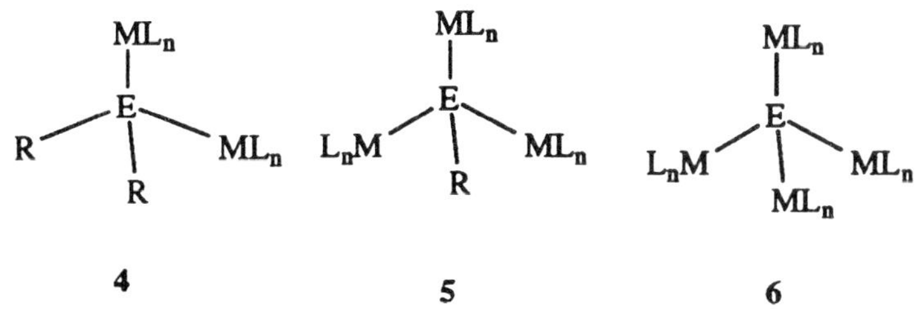

Scheme 7.2 The series $ER_{3-n}\{ML_n\}_{n+1}$: structure 4, $n = 1$; structure 5, $n = 2$; structure 6, $n = 3$.

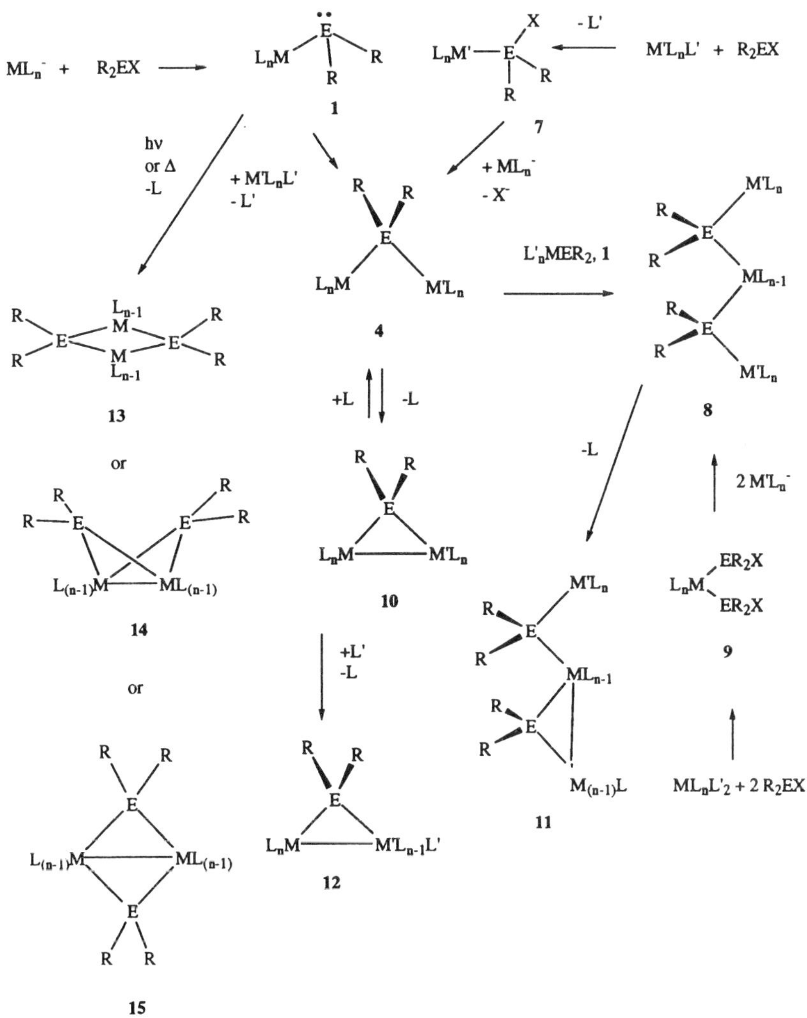

Scheme 7.3 Transformations of metallated ER_2 species.

It is common for the complexes of type 4 to undergo loss of one ligand to generate an M−M bond to give complex 10 or 11. This process may be induced thermally or photochemically. It may also occur spontaneously upon addition of the second metal group. Addition of L to complexes of type 4 may result in substitution (12) at one metal atom or the other (or both), but reaction of L with 10 may yield, besides substitution, addition products of type 4 in which the M−M bond has been broken. A number of molecules having both supported and unsupported ER_2 bridging units are known as a result of applying this strategy. Several of these binuclear $AsMe_2$-bridged complexes have been shown to catalyse the dimerization of norbornadiene, and the complex [{Fe(CO)$_4$}AsMe$_2${Co(CO)$_2$(norbornadiene)}] was isolated from one such reaction and structurally characterized [34]. Evidently, the ligation of the diene interrupts the Fe−Co bond. Photolysis or thermolysis of 1 can lead to dimerization giving molecules of types 13, 14 or 15.

Some molecules of type 4 have been obtained by addition of nucleophiles to the unsaturated 'inidene' complexes to be described in Section 7.3.3. Examples include [{(CO)$_5$Cr}$_2$As(R−O)] [55], [{(CO)$_5$Cr}$_2$As(Cl)X]$^-$ [56], and [{(C$_5$H$_4$Me)(CO)$_2$Mn}$_2$Bi(bipyridine)]$^+$ [57].

Alternatively, bismuth atoms have been shown to behave as Lewis acids and may add halides, as demonstrated by the addition of [NR$_4$]X to [{Cp(CO)$_2$Fe}$_2$BiX] to give [{Cp(CO)$_2$Fe}$_2$BiX$_2$]$^-$ [58, 59]. In contrast, the cationic [{Cp(CO)$_2$Fe}$_2$SbX$_2$]$^+$ compounds have been isolated from the reaction of [CpFe(CO)$_2$]$_2$ with $SbCl_3$ [60]. Unlike the hypervalent Bi molecules, these are electron precise, having Sb in the +5 oxidation state.

A couple of other reactions producing {L$_n$M}$_2$ER$_2$ compounds deserve brief comment. Photolysis of $Fe_3(CO)_{12}$ in the presence of $SbPh_3$ produces as the major product [(CO)$_4$FeSbPh$_2${Fe(CO)$_3$(Ph)(SbPh$_3$)}] (structure 16, Scheme 7.4) in addition to [Fe(CO)$_{5-x}$(SbPh$_3$)$_x$] ($x = 1, 2$) [61]. A bridging configuration has also been observed for As in [{CpFe(CO)}{CpFe(CO)$_2$}$_2$As$_4$O$_5$][BF$_4$] (structure 17, Scheme 7.5) [62]. This compound is produced in the reaction of [{CpFe(CO)$_2$}$_2$] with AsF_3 (reaction with the pyrex walls of the container is involved) and may be viewed as a deriv-

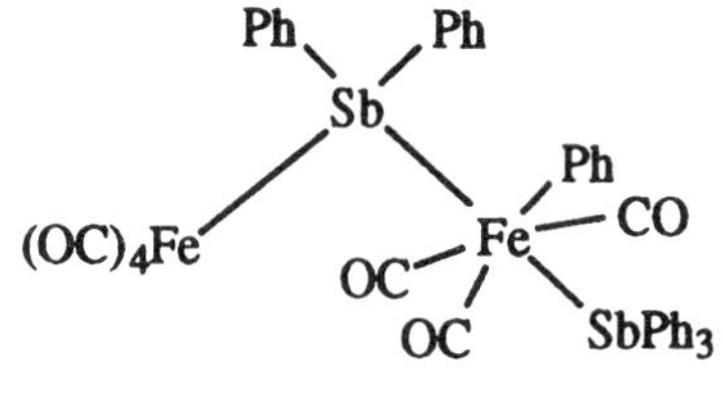

16

Scheme 7.4 The major product of photolysis of $Fe_3(CO)_{12}$ in the presence of $SbPh_3$.

Table 7.2 $L_nM_3ER_2$ complexes.

Compound	Reference
$[Fe_3(CO)_9(SC_6H_{11})(AsPh_2)]$	[i]
$[Fe_3(CO)_9(S^tBu)(AsR_2)]$, R = Me, Ph, $C_{12}H_9N$	[i]
$[HRu_3(CO)_9(As(Ph)CH_2AsPh_2)]$	[ii]
$K[Ru_3(CO)_9(As(Ph)CH_2AsPh_2)]$	[ii]
$[Ru_3(CO)_9\{As(Ph)CH_2AsPh_2\}(N = NPh)]$	[ii]
$[HRu_3(CO)_9\{As(Ph)CH_2AsPh_2\}(N = NC_6H_4)]$	[ii]
$[(\mu\text{-}H)Ru_2(CO)_6\{\mu_3\text{-}\eta\text{-}[(\eta^6\text{-}C_6H_4)Cr(CO)_3]\}(\mu\text{-}AsMe_2)]$	[iii]
$[(\mu\text{-}H)Ru_3(CO)_8\{\mu_3\text{-}\eta\text{-}[(\eta^6\text{-}C_6H_4)Cr(CO)_3]\}(\mu\text{-}AsMe_2)]$	[iii]
$[(\mu\text{-}H)Ru_3(CO)_7(PhAsCH_2AsPh_2)(Ph_2AsCH_2AsPh_2)]$	[iv]
$[(\mu\text{-}H)_2Ru_3(CO)_6(PhAsCH_2AsPh_2)_2$	[v]
$[PPN][Ru_3(CO)_9(AsPh_3)(AsPh_2)]$	[vi]
$[HOs_3(CO)_{11}(\mu\text{-}AsMe_2)]$	[vii]
$[HOs_3(CO)_{10}(\mu\text{-}AsMe_2)]$	[vii]
$[H_2Os_3(CO)_8(AsMe_2)_2$	[vii]
$[Os_3(CO)_7(C_6H_3R)(AsMe_2)_2]$, R = H, Pr^i	[viii]
$[HOs_3(CO)_9(C_6H_4)(AsMe_2)]$	[viii]
$[H_2Os_3(Me_3SiC_5H_4As)(CO)_9]$	[ix]
$[HOs_3\{(Me_3Si)_2C_{10}H_9As_2\}(CO)_9]$	[ix]
$[HOs_3(CO)_9\{\mu\text{-}As(C_6H_4Me)_2\}]$	[x]
$[HOs_3(CO)_{10}(SbPh_2)]$	[xi]

References

i. A. Winter, L. Zsolnai and G. Huttner, J *Organomet. Chem.*, 1983, **250**, 409.
ii. M. I. Bruce, E. Horn, M. R. Snow, and M. L. Williams, J. *Organomet. Chem.*, 1984, **276**, C53.
iii. W. R. Cullen, S. J. Rettig and H. Zhang, *Organometallics,* 1993, **12**, 1964.
iv. N. Lugan, J.-M Savariault, G. Lavigne and J.-J. Bonnet, J. *Cryst. Spect. Res.*, 1983, **13**, 389.
v. G. Lavigne, N. Lugan and J.-J. Bonnet, *Organometallics,* 1982, **1**, 1040.
vi. G. Gremaud, H. Jungbluth, H. Stoeckli-Evans, and G. Süss-Fink, J. *Organomet.* Chem, 1990, **388**, 351.
vii. K. Guldner, B. F. G. Johnson, J. Lewis, S. M. Owen, and P. R. Raithby, J. *Organomet. Chem.*, 1988, **341**, C45.
viii. A. J. Deeming, I. P. Rothwell, M B. Hursthouse, and J. D. J. Backer-Dirks, J. *Chem. Soc., Dalton Trans.*, 1981, 1879.
ix. A. J. Arce, A. J. Deeming, Y. De Sanctis, A. M. García, J. Manzur, and E. Spodine, *Organometallics,* 1994, **13**, 3381.
x. P. A. Jackson, B. F. G. Johnson, J. Lewis, A. D. Massey, D. Braga, C. Gradella, and F. Grepioni, J. *Organomet. Chem.*, 1990, **391**, 225.
xi. B. F. G. Johnson, J. Lewis, A. J. Whitton, and S. G. Bott, J. *Organomet. Chem.*, 1990, **389**, 129.

ative of As_4O_6 in which one of the oxide ligands has been replaced by a μ-CpFe(CO) group. The arsenic atoms bound to this μ-CpFe(CO) group use their lone pairs to ligate $CpFe(CO)_2$ moieties. The simple bridged complex $[Me_2As\{Fe(CO)_4\}_2]^-$ was produced via reaction of Me_2AsCl with $[Fe_3(CO)_{11}]^{2-}$ or $[Fe_4(CO)_{13}]^{2-}$ [31].

In reactions of trimetallic Ru and Os clusters with ER_2H and ER_3, the metal triangles are retained. Evidently, the EH and EC bonds are readily broken. Examples are given in Table 7.2. Trinuclear anions $[Fe_3(CO)_9(\mu_3\text{–}SR)]^-$ (R=Bu^t, Ph) react with R'_2EX (E = As, Sb) to give

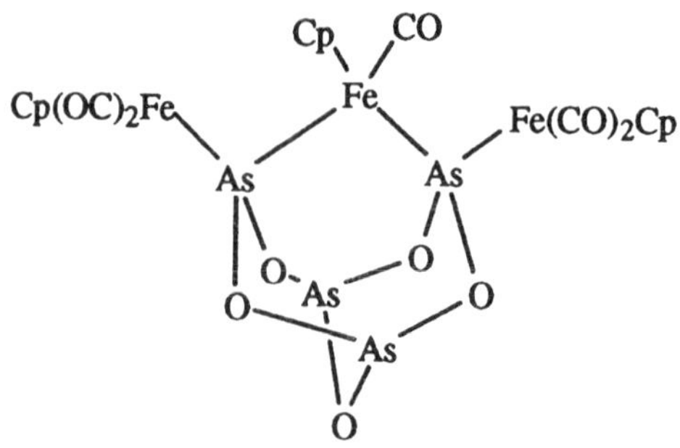

17

Scheme 7.5 The product of the reaction of $[\{CpFe(CO)_2\}_2]$ with AsF_3.

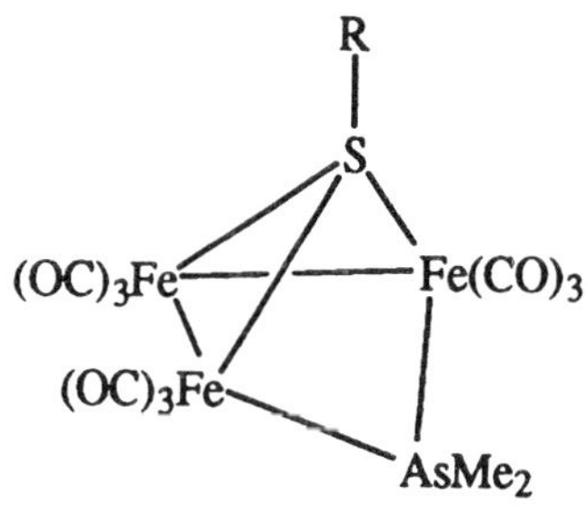

18

Scheme 7.6 The product of the reaction of $[Fe_3(CO)_9(\mu_3\text{-}SR)]^-$ (R = Bu^t, Ph) with R'_2EX (E = As, Sb).

$[Fe_3(CO)_9(\mu_3\text{-}SR)(\mu\text{-}ER_2)]$ (structure 18, Scheme 7.6). The Fe_3 unit remains intact but with one Fe–Fe bond broken, in particular the one bridged by the ER_2 group [63].

Structures which contain more than one AsR_2 unit bridging between two metals are also known. These include the complexes listed in Table 7.3. They generally fall into one of three general structure classifications (13–15, Scheme 7.3). Compounds which have 34 valence electrons should have an M–M bond, whereas those that have 36 should not, as predicted by the 18-electron rule. Whether the compound adopts a butterfly configuration (structure 14) or is planar (structure 15) is determined by the number and type of ligands on M. $M_2(CO)_6$ fragments prefer the butterfly arrangement whereas the $M_2(CO)_8$ compounds adopt a planar arrangement.

The routes to $(L_nM)_2(ER_2)_2$ compounds are similar to those found for L_nMER_2, and often the di-bridged compounds arise from dimerization of the latter species. For example, addition of the dimers $[ML_n]_2$

Table 7.3 Compounds $(L_nM)_2(ER_2)_2$ units.

Compound	Reference
$[Cr_2(CO)_8(AsMe_2)_2]$	[i]
$[CrFe(CO)_7(AsMe_2)_2]$	[i]
$[Cp_2(CO)_4Mo_2\{As(CF_3)_2\}_2]$	[ii]
$[Mn_2(CO)_8\{As(CF_3)_2\}_2]$	[ii]
$[\{M(CO)_4AsMe_2\}_2]$, M = Mn, Re	[iii, iv]
$[Fe_2(CO)_6\{Bu^tAsNSNAsBu^t\}]$	[v]
$[Fe_2(CO)_6(AsMe_2)_2]$	[i]
$[M_2(CO)_6\{As(C_6F_5)_2\}_2]$, M = Fe, Ru	[vi]
$[Fe_2(CO)_6\{AsMe\}_4]$	[vii]
$[Fe_2(NO)_4\{As(CF_3)_2\}_2]$	[viii]
$[Cp_2(CO)_2Fe_2\{As(CF_3)_2\}_2]$	[viii]
$[H(CO)_6Fe_2Bi_2Cl_2]^-_\infty$	[ix]
$[(C_5Me_4Et)_2Fe_2As_2S_2]$	[x]
$[(C_5Me_4Et)_2Fe_2As_2S_2\{Cr(CO)_5\}_2]$	[x]
$[(C_5Me_4Et)_2Fe_2As_2S_3]$	[x]
$[(C_5Me_4Et)_2Fe_2As_2S_3\{Cr(CO)_5\}_2]$	[x]

References

i. H. Vahrenkamp and E. Keller, *Chem. Ber.*, 1979, **112**, 1991.
ii. A.-M. Caminade, M Veith, V. Huch and W. Malisch, *Organometallics*, 1990, **9**, 1798.
iii. R. G. Hayter, J. *Am Chem. Soc.*, 1964, **86**, 823.
iv. W. Ehrl and H. Vahrenkamp, *Chem Ber.* 1971, **104**, 3261.
v. M. Herberhold, K. Schamel, G. Herrmann, A. Gieren, C. Ruiz-Pérez, and T. Hübner, Z. *Anorg. Allg. Chem.*, 1988, **562**, 49.
vi. M. Cooke, M. Green, and D. Kirkpatrick, J. *Chem Soc. (A)*, 1968, 1507.
vii. B. M. Gatehouse, *Chem. Commun.*, 1969, 948.
viii. W. R. Cullen and R. G. Hayter, J. *Am. Chem Soc.*, 1964, **86**, 1030.
ix. J. R. Eveland, J.-Y. Saillard and K H. Whitmire, submitted for publication.
x. H. Brunner, L. Poll, J. Wachter, and B. Nuber, J. *Organomet. Chem.*, 1994, **471**, 117.

$[ML_n = Mn(CO)_5$, $CpFe(CO)_2$ or $CpMo(CO)_3]$ to R_2AsAsR_2 leads to $(ML_{n-1})_2\{AsR_2\}_2$ (R = Me, CF_3) [21, 64]. The intermediacy of L_nMAsR_2 has been noted in some cases [64]. The corresponding reaction between the diarsines R_2AsAsR_2 and the mononuclear complexes $Fe(CO)_5$ (R = Me, Ph), $M(CO)_6$ (M = Cr, Mo, W; R = Me, Ph) or $Fe(CO)_2(NO)_2$ (R = CF_3) gave $[Fe_2(CO)_6(AsR_2)_2]$, $[M_2(CO)_8(AsR_2)_2]$ or $[Fe_2(NO)_4\{As(CF_3)_2\}_2]$, respectively [64, 65]. Similarly, $[(CO)_5CrAsMe_2AsMe_2]$ upon heating dimerizes to $[Cr_2(CO)_8(AsMe_2)_2]$ with loss of $AsMe_2$ fragments [44]. Thermolysis and photolysis of $[(CO)_5MAsMe_2M'(CO)_5]$ (M = Cr, Mo, W; M′ = Mn, Re) gave the dimers $[M'_2(CO)_8(AsMe_2)_2]$ [28]. Loss of CO from $[Cp(CO)_3MAsMe_2]$ (M = Mo, W) leads to $[\{Cp(CO)_2MAsMe_2\}_2]$ [23]. Reaction of As_2R_4 (R = C_6F_5) with $M_3(CO)_{12}$ (M = Fe, Ru) produces $[M_2(CO)_6(AsR_2)_2]$ [66]. The complex $[Fe_2(CO)_6(RAsNSNAsR)]$ (R = Bu^t) is similar to $[M_2(CO)_6(AsR_2)_2]$ but an R group on each As atom has been replaced by a bridging NSN function. It arises from reaction of $Bu^tAs(NSN)_2AsBu^t$ with $Fe_3(CO)_{12}$ accompanied by loss of one NSN group [67]. Treatment of $(AsMe)_5$ with $Fe(CO)_5$ leaves some of the As–As bonds intact, giving $[Fe_2(CO)_6(AsMe)_4]$ in which a chain of four arsenic atoms bridges the two iron atoms. It is similar to $[Fe_2(CO)_6(RAsNSNAsR)]$ with the bridging NSN unit replaced by a bridging As(Me)As(Me) fragment [68]. Ion metathesis also works, as exemplified by the production of $[FeM(CO)_7(AsMe_2)_2]$ upon addition of $[Fe(CO)_4]^{2-}$ to $[M(CO)_4(AsMe_2Cl)_2]$ (M = Cr, Mo, W) [43]. One product obtained upon reaction of $[Bi_4Fe_4(CO)_{13}]^{2-}$ with $MePCl_2$ is $[HFe_2(CO)_6Bi_2Cl_2]^-_\infty$ (structure 19, Scheme 7.7) which is unique in this list in that there is only one X group per Bi atom [69]. Formation of Cl bridges in the solid state effectively generates a structural type analogous to the $[Fe_2(CO)_6(ER_2)_2]$ complexes described above, but the bismuth atom adopts a hypervalent bonding configuration in order to achieve this. The solution spectroscopic data and solubility properties suggest that this molecule really exists as the symmetric monomer $[HFe_2(CO)_6(\mu\text{-}BiCl)_2]^-$ when dissolved [69].

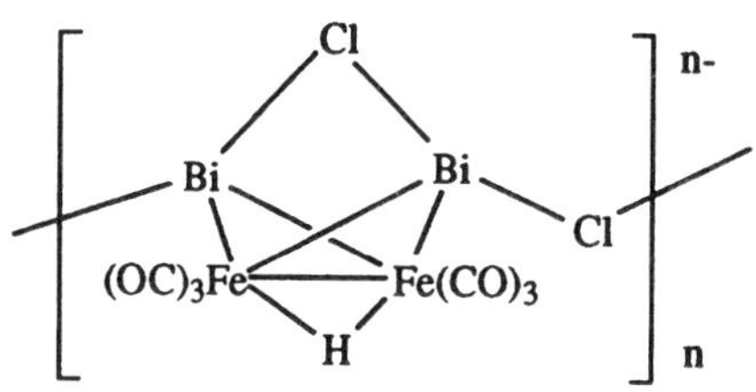

19

Scheme 7.7 A product of the reaction of $[Bi_4Fe_4(CO)_{13}]^{2-}$ with $MePCl_2$.

A series of related compounds arises from the reaction of $[\{(C_5Me_4Et)Fe(CO)_2\}_2]$ with realgar (As_4S_4). These molecules loosely fit the structural types discussed here but retain S atoms attached to As (structures 20 and 21, Scheme 7.8) [70]. The arsenic atoms retain their basicity and are found to coordinate to $Cr(CO)_5$ groups.

Higher degrees of metallation of E are observed as noted for the structural types 5 and 6 (Scheme 7.2). The addition of arsenic or antimony

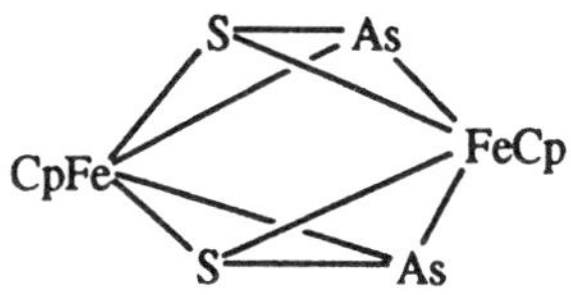

20

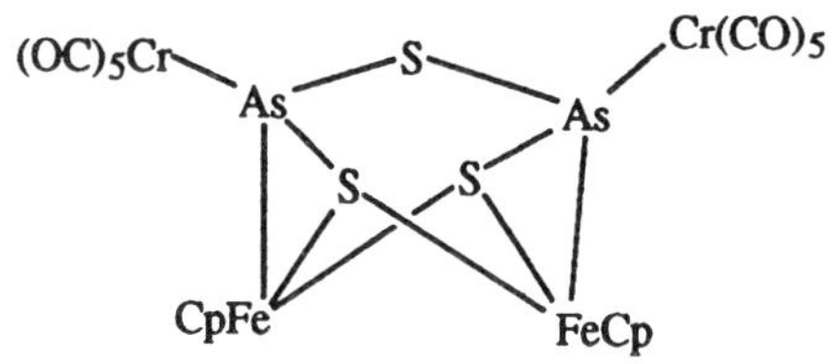

21

Scheme 7.8 Products of the reaction of $[\{(C_5Me_4Et)Fe(CO)_2\}_2]$ with realgar (As_4S_4).

reagents to a methanolic KOH solution of $Fe(CO)_5$ produces the anions $[HE\{Fe(CO)_4\}_3]^{2-}$ [71–73], but the same method applied to $NaBiO_3$ led instead to $[Bi\{Fe(CO)_4\}_4]^{3-}$, presumably because of the better accessibility of the carbonylate anions to the larger Bi atom [74, 75]. The analogous Sb complex, $[Sb\{Fe(CO)_4\}_4]^{3-}$, had to be obtained via an alternate route – the addition of $SbCl_3$ or $SbCl_5$ to $[Fe(CO)_4]^{2-}$ [76]. When complexes $[E\{Fe(CO)_4\}_4]^{3-}$ (E = Sb, Bi) are treated with ECl_3 or alkyl halides (for E = Bi) the resulting compounds are $[\{(CO)_4Fe\}_3ER]^{2-}$ [76–78]. Although this may appear at first glance to be a simple replacement of $[Fe(CO)_4]^{2-}$ with X^-, such a simple substitution does not occur for MX_n, which cannot also serve as potential oxidizing agents. The E = Sb, R = Cl compound was also obtained via reaction of $SbCl_3$ with $[PPN][HFe(CO)_4]$ [79]. The alkyl derivatives for E = Bi may be decomposed oxidatively to yield $[Fe(CO)_4BiR]_2$ (structure 22, Scheme 7.9) [77, 78].

In addition to the anionic compounds listed in the preceding paragraph several other tetrasubstituted compounds are known which are cationic. These include $[As\{Fe(CO)_2Cp\}_4]^+$ [80], $[HOAs\{Fe(CO)_2Cp\}_3]^+$ [80], $[[Sb\{Co(CO)_3PPh_3\}_4]^+$ [81], $[Cl_2Sb\{Fe(CO)_2Cp\}_2]^+$ [60, 81, 82] and $[ClSb\{Fe(CO)_2Cp\}_3]^+$ [83]. The tetrametallated arsenic compound is obtained via the reaction of $As(SiMe_3)_3$ with $[CpFe(CO)_2Cl]$. It undergoes hydrolysis to give the hydroxy species. The antimony compounds are obtained via reaction of $SbCl_3$ with $[Co(CO)_3PPh_3]^-$, $[CpFe(CO)_2]^-$ or $[\{CpFe(CO)_2\}_2]$. These compounds highlight the oxidation state assignment of E, which is here best represented as +5, whereas the oxidation states of E in the anionic compounds are more ambiguous. They can be viewed as +5 or +3 (or even −3) depending upon how one assigns charges to the metal fragments. A neutral example of an Sb^{5+} compound is $[Sb(O)(OH)\{Co(CO)_3L\}_2]$ (L = PPh_3, $P(o\text{-tolyl})_3$), a metallated derivative of stibinic acid, which arises from the oxidation of solutions containing $[Sb_2Co_2(CO)_4L_2]$ [84].

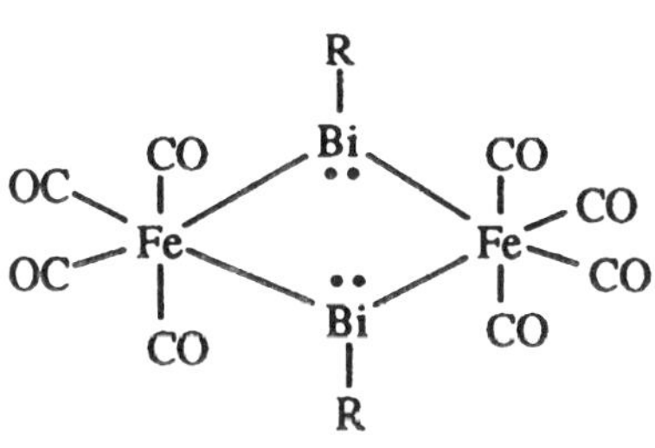

22

Scheme 7.9 Product of the oxidative decomposition of the alkyl derivatives of $[\{(CO)_4Fe\}_3BiR]^{2-}$.

An interesting structural difference between the pyramidal three-coordinate and tetrahedral four-coordinate complexes lies in the M−E−M angles. In general, it is observed that these angles are smaller for the three-coordinate complexes than they are for the four-coordinate complexes. One can explain this by using classical VSEPR theory which ascribes a repulsive effect to the presence of a lone pair of electrons which tends to repel the bonding electrons. In molecular orbital terms, the three-coordinate complexes have the lone pair of electrons residing in an orbital of primarily s character whereas the bonding orbitals are mostly p-like – another manifestation of the well-known inert pair effect. In the four-coordinate complexes, however, the s orbitals are required to be involved in the bonding. Thus a significant difference between the three- and four-coordinate complexes is the hybridization at E. It actually appears that this subtle effect in the bonding has a profound influence on the reactivity of the molecules. A number of three-coordinate complexes with a lone pair of electrons readily undergo simple CO loss leading to *closo*-clusters:

$$[Bi\{Co(CO)_4\}_3] \xrightarrow{\Delta} [BiCo_3(CO)_9] + 3\ CO \qquad [7.5]$$

On the other hand, four-coordinate complexes such as $[HAs\{Fe(CO)_4\}_3]^{2-}$ react via an apparently different process to give coupled products:

$$[HAs\{Fe(CO)_4\}_3]^{2-} \xrightarrow{\Delta} [Fe_3(CO)_9\{AsFe(CO)_4\}_2]^{2-} + \cdots \qquad [7.6]$$

This observation appears to be general for a number of E, not just As, Sb and Bi.

There is another class of compound having tri-metallated E atoms but no M−M bonds. These compounds adopt the well-known cubane geometry (structure 23, Scheme 7.10) which has been extensively examined for the metal chalcogenides, some of which are of biological importance. The distinguishing feature of these molecules is that they contain a distorted cube of atoms in which vertices alternate between the metal and non-metallic elements. The examples found here include $[Fe_4(CO)_{12}\{AsMe\}_4]$, [85, 86], $[Co_4(CO)_{12}Sb_4]$ [87] and $[Co_4(CO)_{12}Bi_4]$ [88]. The molecule $[\{CpMoAs\}_4]$ has also been mentioned but its crystallographic characterization was complicated by disorder problems [89]. These molecules were all obtained via thermolysis reactions – the arsenic compound by heating $[Fe(CO)_4AsMeH_2]$, the antimony compound from $Co(OAc)_2 \cdot 4\ H_2O$ and $SbCl_3$ under a pressure of CO, and the bismuth compound from pyrolysis of $[Bi\{Co(CO)_4\}_3]$. The electron counts of all three of these clusters are the same. The Co atoms have one more valence electron than Fe, but the difference is made up by the presence of the R groups in the Fe–As compound which provide one more electron each compared with the naked E atoms.

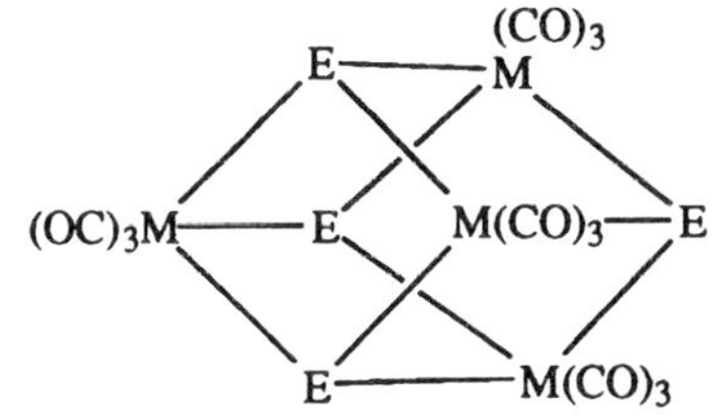

23

Scheme 7.10 An example of a structure with cubane geometry.

7.3.2 Hypervalent compounds

Complexes of the formula $[Cp(CO)_3MBiCl_3]^-$ (M = Mo, W) have been described and the molybdenum compound has been structurally characterized, comprising two *trans* related, basal edge-sharing square-based pyramids with the $CpMo(CO)_3$ fragments in the apical sites [58, 59]. In these molecules the Bi atom is formally hypervalent and acts as a Lewis acid to attract one or more ions to itself. $[Cp(CO)_3M]^-$ serves the same function as a halide ion (cf. complex halide ions such as $[Bi_2Cl_8]^{2-}$). The sense of the bonding interaction is opposite to what one normally expects – generally EX_3 would donate a lone pair of electrons to M whereas here the electron pair is donated from M to E. A related compound, $[(CO)_4Fe\{BiCl_3\}_2]^{2-}$ (structure 24, Scheme 7.11) is formed via the reaction

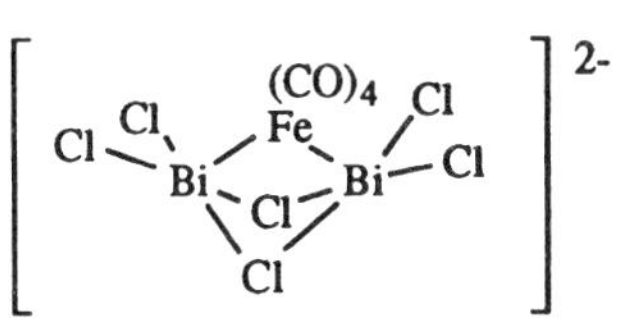

24

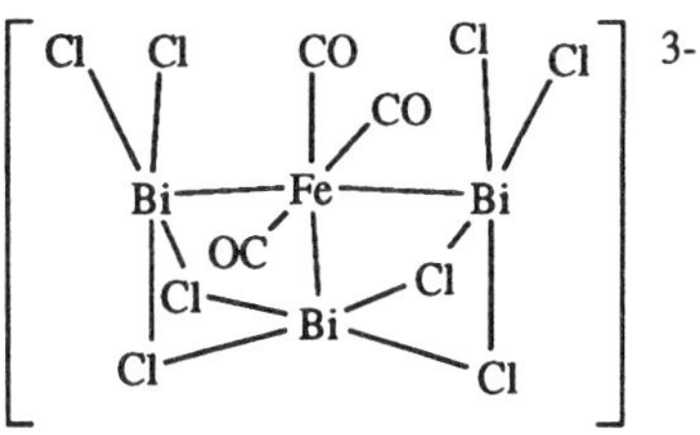

25

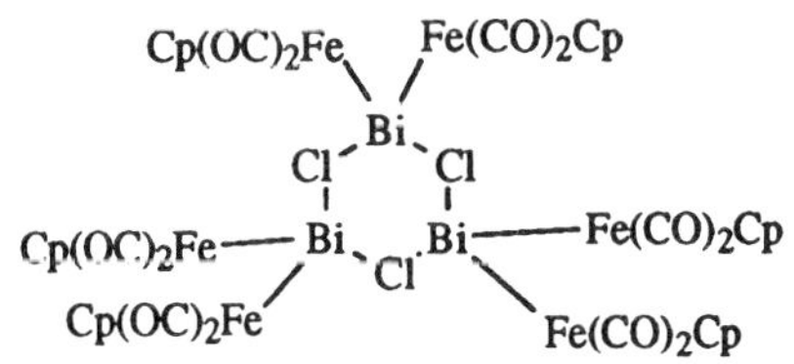

26

Scheme 7.11 Examples of hypervalent compounds.

of two equivalents of $BiCl_3$ with $[Fe(CO)_4]^{2-}$ or from the reaction of $[Bi_4Fe_4(CO)_{13}]^{2-}$ with $MePCl_2$ [90]. Extended Hückel molecular orbital calculations indicate that the Bi–Fe bonding in structure 24 is very covalent and the metal centre remains highly negatively charged. The hypervalent bonding pattern is one that is now well-established for metallated Bi complexes of a number of structural types (*vide infra*). Also arising from the reaction of $[Bi_4Fe_4(CO)_{13}]^{2-}$ with $MePCl_2$ is $[Fe(CO)_3Bi_3Cl_8]^{3-}$ (structure 25) which can be viewed as arising from structure 24 via replacement of a CO with a $BiCl_2^-$ fragment. There are no direct Bi–Bi bonds in these molecules. Another example of hypervalency is the molecule $[(\{(C_5H_4R)(CO)_2Fe\}_2BiCl)_3]$ (R = H, Me) which exists as a cyclic trimer (structure 26) formed via donation of the lone pair on a Cl^- ion of one molecule to the Bi atom of another [91–93].

The open molecule $[Bi\{Co(CO)_4\}_3]$ (structure 3) adds an additional equivalent of $[Co(CO)_4]^-$ to produce the regular tetrahedral complex $[Bi\{Co(CO)_4\}_4]^-$ (structure 6) [94, 95] which is isostructural but not isoelectronic with $[E\{Fe(CO)_4\}_4]^{3-}$ (E = Sb, Bi) [74–76]. The former differs in that it has a 10-electron Bi centre. This hypervalency at Bi is reflected by the long Bi–Co bond distances – 2.906 Å (average) compared with 2.766 Å in $[Bi\{Co(CO)_4\}_3]$. Hypervalency of this sort has also been seen in complexes of Te.

7.3.3 *'Electron deficient' species: arsinidene, stibinidene and bismuthinidene complexes*

It is possible for the lone pair of electrons present on ER_2, ER or E to form a π-interaction to M giving rise to a double or triple bond (structures 27–29, Scheme 7.12). Such interactions are more common for E = N and high oxidation state, early transition metals, but are rare for E = As, Sb or Bi coordinated to a single metal, especially those later in the transition series. Loss of L from complexes of type 1 would be one method of generating such a π-interaction as it would create an empty orbital into which the lone pair of electrons could move. To date, no examples of complex types 27 or 28 have been reported for E = As, Sb or Bi, but

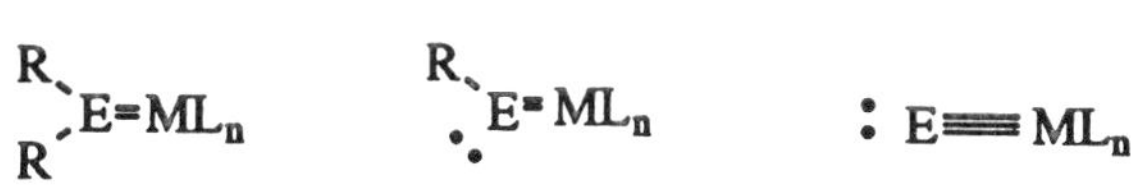

27 **28** **29**

Scheme 7.12 Example of compounds with E–M multiple bonds.

cumulene-type complexes $[\{(C_5H_4Me)(CO)_2Mn\}_2E]O_3SCF_3$ (E = As, Bi) are proposed to form when $[\{(C_5H_4Me)(CO)_2Mn\}_2ECl]$ is treated with $Me_3SiO_3SCF_3$ [96]. These unstable compounds have not been structurally characterized but they have been trapped as base adducts.

Table 7.4 'Inidene' complexes of As, Sb and Bi.

Compound	Reference
$[\{Cp^*(CO)_2Mn\}_2AsH]$	[i, ii]
$[\{(CO)_5Cr\}_2AsCl]$	[iii, iv]
$[\{(CO)_5Cr\}_2AsPh]$	[v, vi]
$[\{Cp(CO)_2Mn\}_2AsCl]$	[vii]
$[\{(CO)_5Mn\}_2As\{=Mn(CO)_4\}]$	[viii]
$[\{(CO)_5Cr\}_2AsN(SiMe_3)_2]$	[ix]
$[\{(CO)_4Fe\}_2AsN(SiMe_3)_2]$	[ix]
$[\{Cp(CO)_7CrMo\}As\{Cr(CO)_5\}]$	[x]
$[\{CpCO)_7CrW\}As\{Cr(CO)_5\}]$	[x]
$[\{Cp(C_5H_4Me)(CO)_4MoMn\}As\{Mn(CO)_2C_5H_4Me\}]$	[x]
$[\{L_nM\}_2SbR]$, M = $Cr(CO)_5$, R = Bu^t, Cl, EtS; ML_n = $Mn(CO)_2Cp$, R = Cl; ML_n = C_5H_4Me, R = Br	[xi, xii]
$[\{Cp(CO)_2Mn\}_2SbPh]$	[xiii]
$[\{(\eta^6\text{-arene})(CO)_2Cr\}_2SbX]$, X = Cl, Br, I; arene = benzene, mesitylene, hexamethylbenzene, $C_6H_5CF_3$	[xiv]
$[Bi\{Mn(CO)_2C_5H_4Me\}_2][O_3SCF_3]$	[xv]
$[\{(C_5H_4Me)(CO)_2Mn\}_2BiX]$, X = Cl, I	[xv]
$[(\{Cp(CO)_2Mn\}_2BiCl)_2]$	[xvi]

References

i. W. A. Herrmann, B. Koumbouris, T. Zahn and M. L. Ziegler, *Angew. Chem. Int.* Ed. Eng., 1984, 23, 812; *Angew. Chem.,* 1984, **96**, 802.

ii. W. A. Herrmann, B. Koubouris, A. Schafer, T. Zahn, and M. L. Ziegler, *Chem.* Ber., 1985, **118**, 2488.

iii. J. von Seyerl, B. Sigwarth, and G. Huttner, *Chem. Ber.,* 1981, **114**, 1407.

iv. J. von Seyerl, B. Sigwarth, and G. Huttner, *Chem Ber.,* 1981, **114**, 727.

v. G. Huttner and H.-G. Schmid, *Angew. Chem Int. Ed. Eng.,* 1975, **14**, 433; *Angew. Chem.*, 1975, **87**, 454.

vi. G. Huttner, J. von Seyerl, M. Marsili, and H.-G. Schmid, *Angew. Chem. Int. Ed. Eng.,* 1975, **14**, 434; *Angew. Chem,* 1975, **87**, 455.

vii. J. von Seyerl, U. Moering, A. Wagner, A. Frank, and G. Huttner, *Angew. Chem. Int. Ed. Eng.*, 1978, **17**, 844; *Angew. Chem.,* 1978, **90**, 912.

viii. V. Grossbruchhaus and D. Rehder, *Inorg. Chem. Acta,* 1990, **172**, 141.

ix. K M. Flynn, B. D. Murray, M. M. Olmstead, and P. P. Power, *J. Am. Chem. Soc.,* 1983, **105**, 7460.

x. G. Huttner, B. Sigwarth, J. von Seyerl, and L. Zsolnai, *Chem Ber.,* 1982, **115**, 2035.

xi. U. Weber, G. Huttner, O. Scheidsteger, and L. Zsolnai, *J. Organomet. Chem,* 1985, **289**, 357.

xii. U. Wever, L. Zsolnai, and G. Huttner, J. *Organomet. Chem.,* 1984, **260**, 281.

xiii. J. von Seyerl and G. Huttner, *Angew. Chem. Int. Ed. Eng.,* 1978, **17**, 843; *Angew.* Chem, 1978, **90**, 911.

xiv. F. Bringewski, G. Huttner, W. Imhof, and L. Zsolnai, J. *Organomet. Chem,* 1992, **439**, 33.

xv. S. J. Davies, N. A. Compton, G. Huttner, L. Zsolnai, and S. E. Garner, *Chem. Ber.,* 1991, **124**, 2731.

xvi. J. von Seyerl and G. Huttner, J. *Organomet. Chem.,* 1980, **193**, 207.

A mass spectroscopic investigation has led to the characterization of $[(CO)_{n-2}MSb]^-$ (M = Cr, Mo, W, $n = 6$; M = Fe, n = 5) which are believed to have M – Sb triple bonds [97]. Another mass spectroscopic study has identified the $CrBi^+$ and Cr_2Bi^+ ions [98]. $FeBi^+$ has also been observed but is much less stable while the Bi_2Fe^+ ion is more accessible [98]. The detailed bonding situation in these species remains unknown but presumably the E and M atoms are multiply bonded to each other.

In contrast a number of species where two or three metal fragments are attached to E show multiple bonding between E and M and now constitute a well-known class of so-called 'inidene' compounds. Owing to the multiple bonding these compounds should be planar and this has been confirmed in a number of single-crystal X-ray structure determinations. Table 7.4 contains a list of reported molecules. These compounds (structure 30, Scheme 7.13) have the multiple bond delocalized over both metal centres and show symmetrical, somewhat short, E–M bond distances. A fully metallated derivative has also been observed: $[As\{Mn(CO)_4\}\{Mn(CO)_5\}_2]$ (structure 31) [96]. Several EM_3 compounds with M–M bonds (structure 32) are represented in Table 7.4 as well. One compound, namely $[(\{Cp(CO)_2Mn\}_2BiCl)_2]$, differs from the others in that, to an extent, the unsaturation at Bi has been ameliorated via dimer formation with halide bridges (structure 33, Scheme 7.14) [99]. Compare this compound with $[(\{C_5H_4R(CO)_2Fe\}_2BiCl)_3]$ discussed above [91–93].

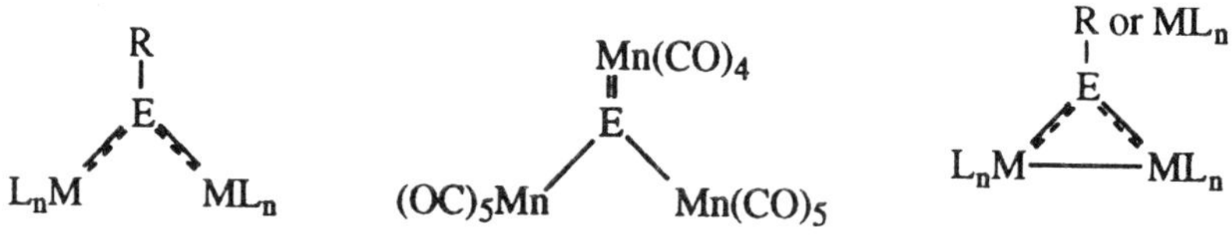

30 31 32

Scheme 7.13 Examples of 'inidenes' and their derivatives: structure 30, multiple bond delocalized over both metal centres; structure 31, fully metallized derivative; structure 32, compound with M–M bond.

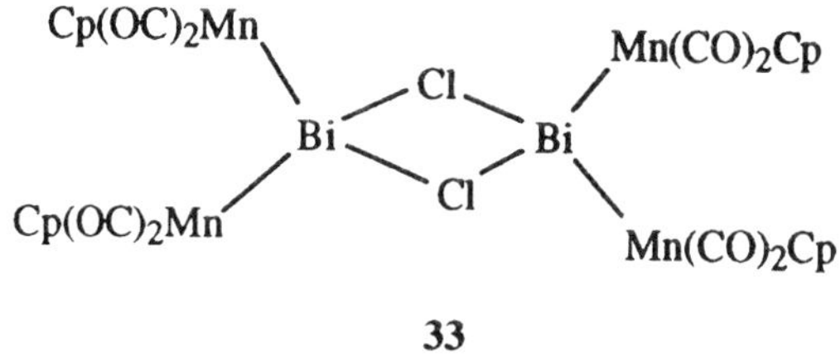

33

Scheme 7.14 Compound in which unsaturation is amleriorated by halide bridges.

Unsaturation at E has been established by the addition of X ($X = Cl^-$, Br^-, I^-, N_3^-, CN^-, SCN^-) to $[ClAs\{Cr(CO)_5\}_2]$ to form the tetrahedral anionic adducts $[Cl(X)As\{Cr(CO)_5\}_2]$ [56] and by displacement of the Cl^- with monoanionic chelating ligands in $[\{(CO)_5Cr\}_2ECl]$ (E = As, Bi) [55]. The ligands studied included 2,4-pentanedioate, 1,3-diphenyl-1,3-propanedioate, tropolonate and chinolinate. The cumulene complex $[\{Cp(CO)_2Mn\}_2Bi]^+$ is also unsaturated and adds bipyridine to give $[\{Cp(CO)_2Mn\}_2Bi(bipyridine)]^+$ in which the Bi atom is four-coordinate [57].

7.3.4 Electron precise EM_x complexes with one or more M–M bonds

Electron precise molecules may be generated in forms having M–M bonding. These structural types are shown in Scheme 7.15. Compounds of these types are listed in Table 7.5. Conceptually they may be thought of as arising via ligand loss from the corresponding open clusters. Loss of two electrons via oxidation or one two-electron-donating ligand results in

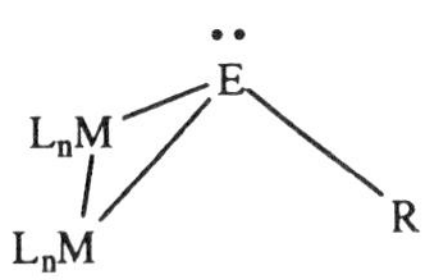

34

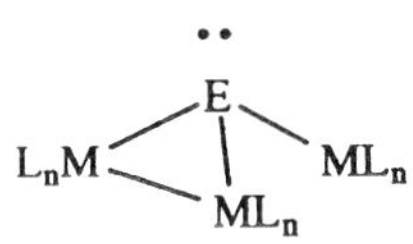

35

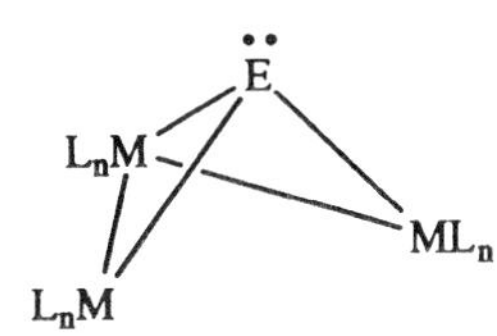

36

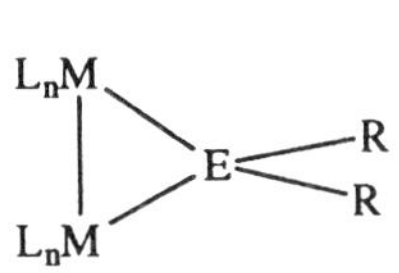

37

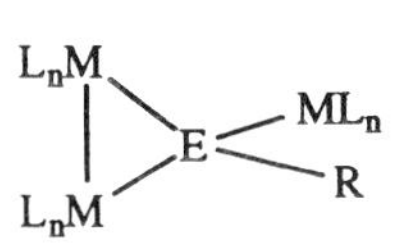

38

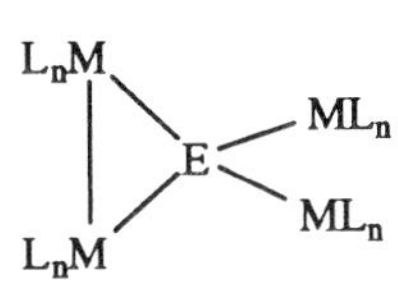

39

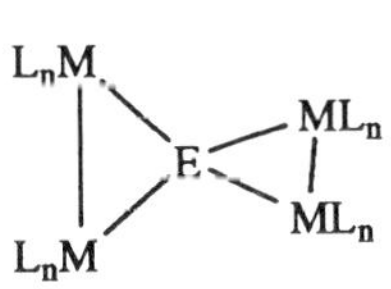

40

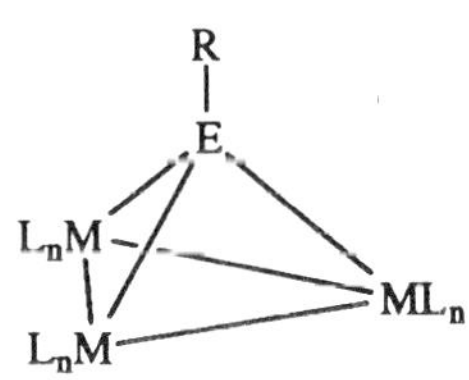

41

Scheme 7.15 Electron precise molecules with M–M bonding. R = lone pair of electrons; H – halide or pseudo-halide, alkyl group, aryl group or ML_n.

Table 7.5 Electron precise EM_x compounds with M–M bonds

Compound	Reference
M_xE and M_xER complexes with one M–M bond (type 20, Scheme 7.8):	
$[\{CpRu(CO)\}_2(\mu\text{-}CO)(\mu\text{-}AsR)]$, R = Me, Ph	[i]
$[FeCo(CO)_6(PMe_2Ph)AsMe\{Co(CO)_3(PMe_2Ph)\}]$, one Fe–Co bond	[ii]
EM_x complexes with one M–M bond (type 23), Scheme 7.10):	
$[Sb\{Fe_2(CO)_8\}\{Fe(CO)_4\}_2]^-$	[iii]
$[Sb\{Fe_2(CO)_8\}\{Fe(CO)_4\}\{Cr(CO)_5\}]^-$	[iv]
$[Bi\{Fe_2(CO)_8\}\{Fe(CO)_4\}\{Cr(CO)_5\}]^-$	[iv]
EM_x complexes with two M–M bonds (type 21, Scheme 7.8):	
$[Cp(CO)_4(PPh_3)_2Pt_2MoAs]$, two Mo–Pt bonds	[v]
Spirocyclic complexes (type 25, Scheme 7.11):	
$[(CO)_8Fe_2AsFe_2(CO)_6Cl]$	[vi]
$[\{Cp(CO)_2MoC(H)(PPh_3)As\}Mo_2(CO)_4Cp_2]$	[vii][a]
$[\{Fe_2(CO)_8(\mu_4\text{-}As)\}\{Fe_2(CO)_6\}]$	[viii]
$[\{Cp_2(CO)_2Co_2\}_2As][BF_4]$	[ix]
$[Fe_2(CO)_6Sb_2\{Fe_2(CO)_8\}_2]$	[x, xi]
$[Ru_2(CO)_8Bi\{HRu_3(CO)_{10}\}]$	[xii]
Tetrahedral M_3ER (type 26, Scheme 7.11):	
$[Cp_3(CO)_6M_3As]$, M = Mo, W	[xiii, xiv]
$[Cp_3Ni_2Mo(CO)_2AsMo_3Cp_3(CO)_5As]$	[v]
$[(\mu_3\text{-}HC)Fe_3(CO)_9As]$	[xv]
$[(\mu_3\text{-}HC)Fe_3(CO)_9Sb]$	[xv]
$[PPN][Fe_3(CO)_{10}Sb]$	[xvi]
$[Et_4N]_2[(OC)_3Mo(OMe)_3Mo_3(CO)_9Bi]$	[xvii]
$[Fe_3(CO)_{10}Bi]^-$	[xvi, xviii]
$[(\mu_3\text{-}MeOC)Fe_3(CO)_9Bi]$	[xix]
$[H_3Fe_3(CO)_9Bi]$	[xix]
$[(\mu_3\text{-}HC)Fe_3(CO)_9Bi]$	[xv]
$[(C_5H_4R)_3Fe_3(CO)_3Bi]$, R = H, Me	[xx, xxi]
$[H_3Ru_3(CO)_9Bi]$	[xxii, xxiii]
$[H_3Ru_3(CO)_{9-x}(L)_xBi]$, L = PPh_3, $As(tolyl)_3$, $SbPh_3$, $P(OMe)_3$, $x = 1, 2, 3$; L = PEt_3, $x = 1, 2$	[xxiv]
$[H_3Os_3(CO)_9Bi]$	[xxiii]
$[Co_3(CO)_9Bi]$	[xxv, xxvi]
$[Ir_3(CO)_9Bi]$	[xxvii]
$[FeCo_2(CO)_9AsMe]$	[xxviii]
$[CpMoFeCo(CO)_8AsMe]$	[xxviii]
$[H_2Ru_3(CO)_8(AsPh_3)(AsPh)]$	[xxix]
$[H_2Ru_3(CO)_7(AsPh_3)_2(AsPh)]$	[xxix]
$[(\mu\text{-}H)_2Ru_3(CO)_8(\mu_2\text{-}NSNAsBu^t_2)(AsBu^t)]$	[xxx]
$[HOs_3(CO)_8(NSNAsBu^t_2)(AsBu^t)]$	[xxxi, xxxii]
$[Os_3(CO)_9(\mu_3\text{-}C_6H_3Me)(AsC_6H_4Me)]$	[xxxiii, xxxiv]
$[Os_3(CO)_8(\mu_3\text{-}C_6H_3Me)(AsPh_3)(AsC_6H_4Me)]$	[xxxiii, xxxiv]
$[HFe_3(CO)_9SbBu^t]^-$	[xvi]
$[(HC)Fe_3(CO)_9AsML_n]$ $ML_n = M(CO)_5$, M = Cr, Mo, W	[xv]
$[\{(HC)Fe_3(CO)_9As\}_2ML_n]$ $ML_n = M(CO)_4$; M = Cr, Mo	[xv]
$[(HC)Fe_3(CO)_9EML_n]$, E = As, Sb; $ML_n = C_5H_4Me(CO)_2Mn$	[xv]
$[Et_4N]_2[(OC)_3Mo(OMe)_3Mo_3(CO)_9AsMo(CO)_5]$	[xxxv]
$[Mo_3Cp_3(CO)_6AsML_n]$, $ML_n = Cp(CO)_2Mn$, $Cr(CO)_5$, $Mo(CO)_5$, $W(CO)_5$)	[v, vii, xiii]
$[Cp_3Ni_2Mo(CO)_2AsMo_3Cp_3(CO)_5As]$	[v]
$[(\mu\text{-}H)Fe_2Mo(CO)_8CpAsMo(CO)_3Cp]$	[vii, xiii]
$[CpMo_2Fe(CO)_7AsMo(CO)_3Cp]$	[xiii]
$[Co_3(CO)_9AsCr(CO)_5]$	[xxxvi]

Table 7.5 *continued*

Compound	Reference
$[Co_3(CO)_9AsCo_4(CO)_{11}]$	[xxxvi]
$[\{Co_3(CO)_8As\}_3]$	[viii, xxxvi, xxxvii]
$[(PPh_3)_4Co_4(\mu_3\text{-}As)_3(\mu_3, \eta^3\text{-}As_3)]$	[xxxviii]
$[HFe_3(CO)_9\{SbFe(CO)_4\}]^{2-}$	[xxxix]
$[H_2Fe_3(CO)_9\{SbFe(CO)_4\}]^-$	[xxxix]
$[Fe_3(CO)_{10}\{SbFe(CO)_4\}]^-$	[xl]

[a]Considering the C atom bonded to arsenic as a vertex in the spirocyclic array.

References

i. A.-J. DiMaio, T. E. Bitterwolf, and A. L. Rheingold, *Organometallics,* 1990, **9**, 551.
ii. W.-R. Dietrich and H. Vahrenkamp, J. *Chem. Res. (S),* 1985, 200.
iii. S. Luo and K. H. Whitmire, J. *Organomet. Chem.,* 1989, **376**, 297.
iv. K. H. Whitmire, M. Shieh, and J. Cassidy, *Inorg. Chem.,* 1989, **28**, 3164.
v. M. Gorzellik, B. Nuber, T. Bohn, M. L. Ziegler, J. *Organomet. Chem.,* 1992, **429**, 173.
vi. G. Huttner, G. Mohr, B. Pritzlaff, J. von Seyerl, and L. Zsolnai, *Chem. Ber.,* 1982, **115**, 2044.
vii. K. Blechsmitt, T. Zahn, and M. L. Ziegler, *Angew. Chem. Int. Ed. Eng.,* 1985, **24**, 702; *Angew. Chem.,* 1985, 97, 686.
viii. L J. Arnold, K. M. Mackay, and B. K. Nicholson, J. *Organomet. Chem,* 1990, **387**, 197.
ix. C. F. Campana and L. F. Dahl, J. *Organomet. Chem.,* 1977, **127**, 209.
x. A. L. Rheingold, S. J. Geib, M. Shieh, and K. H. Whitmire, *Inorg. Chem.,* 1987, **26**, 463.
xi. A. M. Arif, A. H. Cowley, and M. Pakulski, J. *Chem. Soc., Chem. Commun.,* 1987, 622.
xii. B. F. G. Johnson, J. Lewis, P. R. Raithby, and A. J. Whitton, J. *Chem Soc., Chem Commun.,* 1988, 401.
xiii. M. L. Ziegler, K. Blechsmitt, B. Nuber and T. Zahn, *Chem. Ber.,* 1988, **121**, 159.
xiv. K. Blechsmitt, H. Pfisterer, T. Zahn, and M. L. Ziegler, *Angew. Chem. Int. Ed.* Eng., 1985, **24**, 66; *Angew. Chem.,* 1985, **97**, 73.
xv. C. Caballero, B. Nuber, and M. L. Ziegler, J. *Organomet. Chem.,* 1990, **386**, 209.
xvi. W. Deck and H. Vahrenkarnp, Z. *Anorg. Allg. Chena,* 1991, **598/599**, 83
xvii. M. Shieh, F.-D. Mia, S.-M. Peng, and G.-H Lee, *Inorg. Chem,* 1993, **32**, 2787.
xviii. J. S. Leigh and K H. Whitmire, *Angew. Chem Int. Ed. Eng.,* 1988, **27**, 396; *Angew. Chem.,* 1988, **100**, 399
xix. K. H. Whitmire, C B. Lagrone, and A. L. Rheingold, *Inorg. Chem.,* 1986, **25**, 2472.
xx. J. M. Wallis, G. Müller, and H. Schmidbaur, J. *Organomet. Chem.,* 1987, **325**, 159.
xxi. J. M. Wallis, G. Müller, J. Riede, and H. Schmidbaur, J. *Organomet. Chem.,* 1989, **369**, 165.
xxii. C. M. Hay, B. F. G. Johnson, J. Lewis, P. R. Raithby, and A. J. Whitton, J. *Chem. Soc., Dalton Trans.,* 1988, 2091.
xxiii. H. G. Ang, C. M. Hay, B. F. G. Johnson, J. Lewis, P. R. Raithby, and A. J. Whitton, J. *Organomet. Chem.,* 1987, **330**, C5.
xxiv. B. F. G. Johnson, J. Lewis, and A. J. Whitton, J. *Chem. Soc., Dalton Trans.,* 1990, 3129.
xxv. K. H. Whitmire, J. S. Leigh, and M. E. Gross, J. *Chem. Soc., Chem. Commun.,* 1987, 926.
xxvi. S. Martinengo and G. Ciani, J. *Chem. Soc., Chem. Commun.,* 1987, 1589.
xxvii. W. Kruppa, D. Blaser, R. Boese, and G. Schmid, Z *Naturforsch.,* 1982, **37B**, 209.
xxviii. J. von Seyerl, U. Moering, A. Wagner, A. Frank, and G. Huttner, *Angew. Chem. Int. Ed. Eng.,* 1978, **17**, 844; *Angew. Chem.,* 1978, **90**, 912.
xxix. G. Gremaud, H. Jungbluth, H. Stoeckli-Evans, and G. Süss-Fink, J. *Organomet. Chem.,* 1990, **388**, 351.

Table 7.5 *continued*

xxx. M. Herberhold, K. Schamel, G. Herrmann, A. Gieren, C. Ruiz-Pérez, and T. Hubner, *Z Anorg. Allg. Chem,* 1988, **562**, 49.
xxxi. H. Betz, A. Gieren and T. Hübner, *Acta Cryst. A (Fnd. Cryst.)* 1984, **40**, C265.
xxxii. G. Suss-Fink, K. Guldner, M. Herberhold, A. Gieren, and T. Hübner, J. *Organomet. Chem.,* 1985, **279**, 447
xxxiii. P. A. Jackson, B F G. Johnson, J. Lewis, A. D Massey, D. Braga, C Gradella, and F Grepioni, J *Organomet Chem,* 1990, **391**, 225 .
xxxiv. B. F. G. Johnson, J Lewis, A. D. Massey, D. Braga, and F. Grepioni, J. *Organomet. Chem.,* 1989, **369**, C43.
xxxv. J. W. van Hal, K. H. Whitmire, B. Zouchoune, J.-F. Halet, and J.-Y. Saillard, *Inorg. Chem.,* 1995, **34**, 5455.
xxxvi. H. Lang, G. Huttner, B. Sigwarth, I. Jibril, L. Zsolnai, and O. Orama, J. *Organomet. Chem,* 1986, **304**, 137.
xxxvii. A. Vizi-Orosz, V. Galamb, G. Palyi, L. Markó, G. Bor and G. Natile, J. *Organomet. Chem,* 1976, **107**, 235.
xxxviii. D. Fenske and J. Hachgenei, *Angew. Chem. Int. Ed. Eng.,* 1986, **25**, 175; *Angew. Chem.,* 1986, **98**, 165.
xxxix. K. H. Whitmire, C. B. Lagrone, M. R. Churchill, J. C. Fettinger, and L. V. Biondi, *Inorg. Chem.,* 1984, **23**, 4227.
xl. K. H. Whitmire, J. S. Leigh, S. Luo, M. Shieh, M. D. Fabiano, and A. L. Rheingold, *New J. Chem,* 1988, **12**, 397. l. K. H. Whitmire, J. *Cluster Science,* 1991, **2**, 231.

formation of one M–M bond. Thus it has been shown that two-electron oxidation of $[Sb\{Fe(CO)_4\}_4]^{3-}$ with $[Cu(NCMe)_4]BF_4$ generates $[Sb\{Fe_2(CO)_8\}\{Fe(CO)_4\}_2]^-$ of structure type (Scheme 7.16). Another example is the reversible loss of three CO ligands from $[Bi\{Co(CO)_4\}_3]$ to give $[BiCo_3(CO)_9]$ (reaction [7.5], Section 8.3.1). Likewise, the addition of L to a closed cluster can result in M–M bond breakage, as was reported for addition of PMe_2Ph to $[FeCo_2(CO)_9AsMe]$ to form $[\{FeCo(CO)_6(PMe_2Ph)\}\{Co(CO)_3(PMe_2Ph)\}AsMe]$ which has only one M–M bond [100]. The arsinidene complexes $[\{CpRu(CO)\}_2(\mu\text{-}CO)(\mu\text{-}AsR)]$ (R = Me, Ph) are obtained via reactions of $[Cp_2Ru_2(CO)_3(HCCPh)]$ with the polyarsines $(AsMe)_5$ and $(AsPh)_6$ [101].

The heterometallic complexes $[E\{Fe_2(CO)_8\}\{Fe(CO)_4\}\{Cr(CO)_5\}]^-$ (E = Sb, Bi) arise from addition of $[Cr(CO)_5(thf)]$ (thf = tetrahydrofuran) to

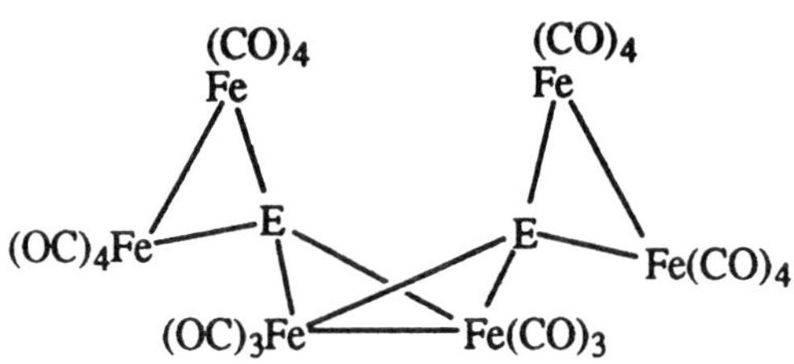

42

Scheme 7.16 Product of the two-electron oxidation of $[Sb\{Fe(CO)_4\}_4]^{3-}$ with $\{Cu(NCMe)_4]BF_4$.

solutions prepared from ECl_3 and $[Fe_2(CO)_8]^{2-}$ [102]. The intermediate cluster was identified as $[\{(CO)_4Fe\}_3ECl]^{2-}$. The reaction proceeds, however, as if the intermediate was $[E\{Fe(CO)_4\}_3]^-$ which could conceivably arise via dissociation of Cl^-. The hypothetical $[E\{Fe(CO)_4\}_3]^-$ could have one of two structures – a trigonal planar, 'electron deficient' structure as has been established for $[E\{Fe(CO)_4\}_3]^{2-}$ (E = Sn, Pb) or a structure like 34 containing an Fe–Fe bond. It is also possible that these two isomers could exist in equilibrium with each other. This type of exchange was also mentioned in connection with the stibinidene complexes $[\{(CO)_5\}W\}_2SbR]$ which were found to function as both Lewis acids and Lewis bases [103].

In many cases, M–M bond formation occurs spontaneously during synthesis and the corresponding open compounds are never seen. In fact, the closed tetrahedral fragment M_3E is the most often observed type of cluster arrangement for this class of compound. It was once thought that the size of the Bi atom would not allow formation of M–M bonds when attached to transition metals of the first row [104], but this idea of bond strain has been shown to be inaccurate. On the other hand, there does appear to be a significant tendency for the lighter elements to form *closo* as opposed to open geometries owing to the proximity of the metals brought about by the shorter E–M distances.

The compound $[Cp(CO)_4(PPh_3)_2Pt_2MoAs]$ arose directly when the *closo* cluster $[Cp_3(CO)_6Mo_3As]$ was photolysed with $[(Ph_3P)_2Pt(C_2H_4)]$ [105].

The spirocyclic complex $[Fe_2(CO)_8AsFe_2(CO)_6Cl]$ (structure 40) is obtained via reaction of $Fe_2(CO)_9$ with $AsCl_3$ [106]. A compound with two spirocyclic centres (structure 42) (E = As) forms upon heating AsH_3 with $Fe_2(CO)_9$ at 70°C [107]. If the temperature is higher, the resultant species is $[Fe_3(CO)_9As_2]$ and not the spirocyclic compound. The analogous compound to 42 for the E = Sb cluster was obtained by Cu^+ oxidation of the reaction of $SbCl_3$ with $[Fe_2(CO)_8]^{2-}$ or by reaction of $SbCl_3$ with a mixture of $[Fe(CO)_4]^{2-}$ and $Fe_2(CO)_9$ [108, 109]. Upon reaction of AsF_3 with $CpCo(CO)_2$ in benzene $[\{Cp_2(CO)_2Co_2\}_2As]^+$ is obtained, which also shows a simple spirocyclic core. The compound $[Ru_2(CO)_8BiRu_3(CO)_{10}H]$ (structure 43, Scheme 7.17) contains a spirocyclic Bi atom connecting a Ru_2 fragment and a Ru_3 triangular unit. It is produced via addition of bismuth nitrate to methanolic solutions of $[HRu_3(CO)_{11}]^-$ [110].

Complexes with the tetrahedral M_3E (structure 41) unit are common and this fragment appears to have particular stability, and a wide variety of synthetic routes have been observed. As noted above, an easy route to the *closo* geometry is via ligand loss from an open cluster as observed for $[Co_3(CO)_9Bi]$ [111, 112] and $[(C_5H_4R)_3(CO)_3Fe_3Bi]$ [91, 113]. It is interesting to note that an open compound was not observed upon addition of $[Ir(CO)_4]^-$ to $BiCl_3$, but the closed cluster formed directly [114]. Likewise, addition of $[Co(CO)_4]^-$ to AsX_3 [113] (or AsH_3 to $Co_2(CO)_8$] [107] led directly to $[\{Co_3(CO)_8As\}_3]$ which, in addition to showing the

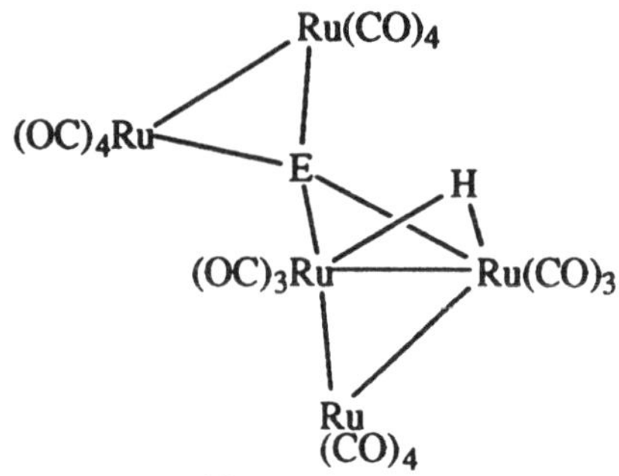

43

Scheme 7.17 Example of a compund with a spirocyclic E atom.

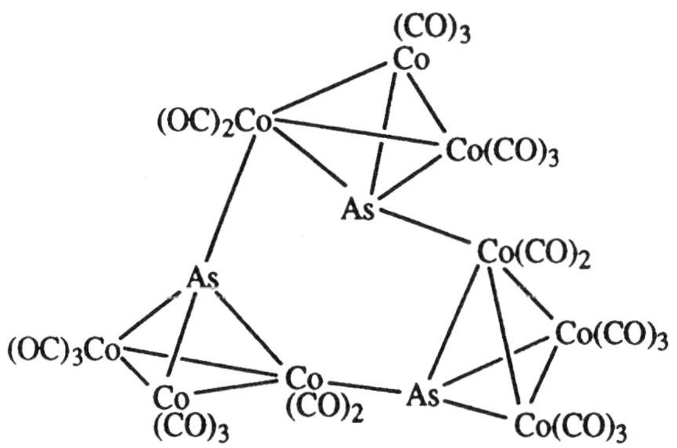

44

Scheme 7.18 Cyclic trimer of $[CO_3(CO)_9As]$.

closed framework, indicates such a high basicity of As that the parent $[Co_3(CO)_9As]$ is not isolated but undergoes cyclic trimerization having the As atom of one cluster unit acting as a ligand to another such unit (structure 44, Scheme 7.18). The $[Co_3(CO)_9As]$ unit may be trapped by complexation to other metal fragments such as $Cr(CO)_5$ or $Co_4(CO)_{11}$. These are prepared via addition of $[Co(CO)_4]^-$ to $[\{Cr(CO)_5\}_2AsCl]$ in the first case or addition of $Co_2(CO)_8$ to $[Cr(CO)_5AsPhH_2]$ in the latter [115].

Other compounds with tetrahedral frameworks appear to be formed directly without the observation of an open framework precursor. Elemental arsenic reacts with $[\{CpM(CO)_3\}_2]$ to produce $[Cp_3(CO)_6M_3As]$ (M = Mo, W) [115, 117]. Pentacarbonyl iron and $NaBiO_3$ react in refluxing MeOH to give $[Fe_3(CO)_{10}Bi]^-$ via elimination of CO_2 [94]. The compound

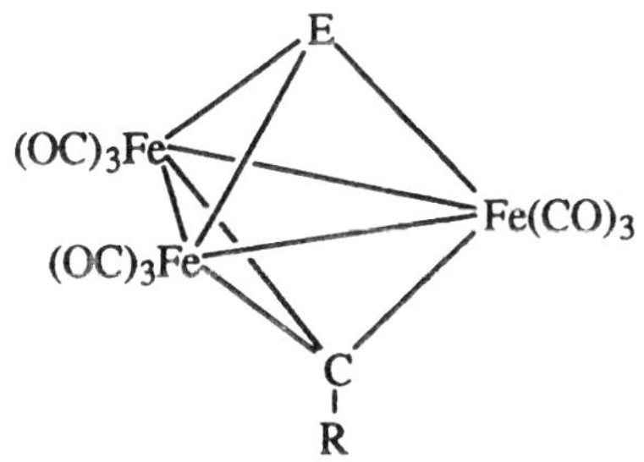

45

Scheme 7.19 A mixed alkylidene–E structure with a tetrahedral framework.

$[BiFe_3(CO)_{10}]^-$ could also be obtained via reaction of $[Fe_3(CO)_{11}]^{2-}$ or $[Fe_4(CO)_{13}]^{2-}$ with BiI_3 or $Bi(NO_3)_3{\cdot}5H_2O$ [31]. The analogous Sb compound was not obtained from sodium antimonate and $Fe(CO)_5$ but could be produced in low yields by the reaction of $[Fe_3(CO)_{11}]^{2-}$ with $SbCl_3$ followed by addition of $[Cu(NCMe)_4][BF_4]$ or by reaction of $SbCl_3$ with $[Fe_4(CO)_{13}]^{2-}$ [31]. Reaction of $SbCl_3$ with $[Fe_4(CO)_{13}]^{2-}$ was also observed to give $[Fe_3(CO)_{10}SbFe(CO)_4]^-$, which has the same structure as $[Fe_3(CO)_{10}Sb]^-$ but with an $Fe(CO)_4$ appended to the Sb atom [31]. The Bi compound reacts with MeO_3SCF_3, alkylating the μ_3-CO ligand, to give the mixed alkylidyne–Bi cluster $[(\mu_3\text{-}MeOC)Fe_3(CO)_9Bi]$ (structure 45, Scheme 7.19) (R = OMe) [118] An antimony analogue in which an R group is attached to Sb was reported but arose via use of Bu^tSbCl_2 in reaction with $[Fe_3(CO)_{11}]^{2-}$ [31]. It may be possible, given the more basic nature of Sb, to alkylate directly at the Sb atom of $[Fe_3(CO)_{10}Sb]^-$, but this has not yet been investigated. The addition of $NaBiO_3$ was extended to $Mo(CO)_6$/MeOH which formed $[(CO)_3Mo(\mu_3\text{-}OMe)_3Mo_3(CO)_9Bi]^{2-}$ (structure 46, Scheme 7.20) [118]. The similar reaction with $NaAsO_2$ gave $[(CO)_3Mo(\mu_3\text{-}OMe)_3Mo_3(CO)_9AsMo(CO)_5]^{2-}$, which shows the same core structure but with the As atom coordinated to an external metal fragment as might be expected from its higher basicity [119].

Protonation reactions may also promote cluster closing. For example, acidification of solutions of $[Bi\{Fe(CO)_4\}_4]^{3-}$ results in low yields of $[H_3Fe_3(CO)_9Bi]$ [108] while similar protonations of $[Sb\{Fe(CO)_4\}_4]^{3-}$, if carefully controlled, can be used to produce $[HFe_3(CO)_9SbFe(CO)_4]^{2-}$, which is further protonated to $[H_2Fe_3(CO)_9SbFe(CO)_4]$ [76]. For both molecules, an oxidative pathway appeared to compete with protonation. For bismuth, $[Bi_2Fe_3(CO)_9]$ is formed in higher yield than is the hydride, and for antimony the similar oxidative coupling product $[Fe_3(CO)_9\{SbFe(CO)_4\}_2]^{2-}$ results when an excess of acid is present [120]. Osmium and ruthenium analogues of the hydrido compound $[H_3M_3(CO)_9Bi]$ have

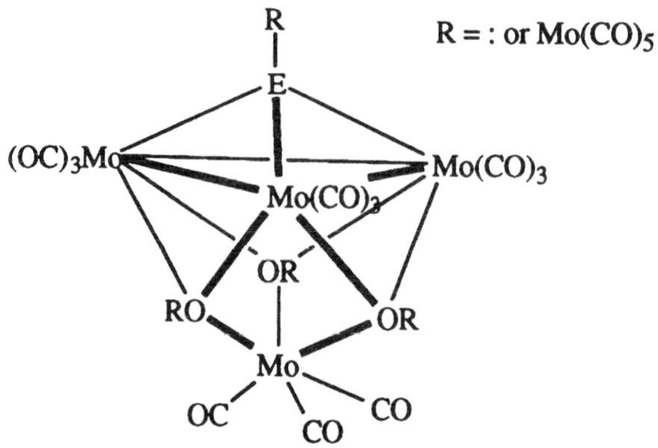

46

Scheme 7.20 A compound with a tetrahedral framework produced by an addition reaction between $NaBiO_3$ and $Mo(CO)_6$/MeOH. R = lone pair of electrons or $Mo(CO)_5$.

been reported [116, 117]. They are obtained from the reaction of $NaBiO_3$ with $M_3(CO)_{12}$. The ruthenium compound was shown to undergo substitution reactions with PR_3, $As(tolyl)_3$ and $SbPh_3$ producing mono-, di- and tri-substituted derivatives [111].

The methyne clusters (structure 45) (R = H; E = As, Sb and Bi) are obtained in low yield via reaction of As/H_2SO_4, Bi/H_2O or $SbCl_5$ with $Fe_5(CO)_{15}C$ [121]. The As atom was found to coordinate to transition metal fragments $M(CO)_5$ (M = Cr, Mo, W) or $Cp(CO)_2Mn$, and two of these clusters can be placed upon $M(CO)_4$ (M = Cr, Mo) fragments. The Sb compound was found to coordinate to $Cp(CO)_2Mn$. It would be interesting to examine the chemistry of this series of compounds further to see what influence the main group element has on the reactivity of the CH function.

As also observed for the formation of μ-ER_2 groups, a μ_3-AsR (R = Bu^t, aryl group) function bonded to a triangle of ruthenium or osmium atoms is readily obtained via cleavage of two As–C or As–N bonds, starting from $AsPh_3$ or $Bu^t_2AsNSNAsBu^t_2$, for example (see Table 7.5). The μ_3-AsMe function is readily introduced into tri-nuclear clusters via reaction of the appropriate mononuclear metal complexes with $MeAsH_2$ [83]. It has also been shown that new cluster complexes could be accessed by metal vertex substitution reactions starting with $[FeCo_2(CO)_9AsMe]$, a reaction type explored extensively by Vahrenkamp not only for this system but for cluster compounds in general [14, 15].

As mentioned earlier, the lone pairs of electrons are much more basic for the lighter elements and this can be seen in the tables. A high

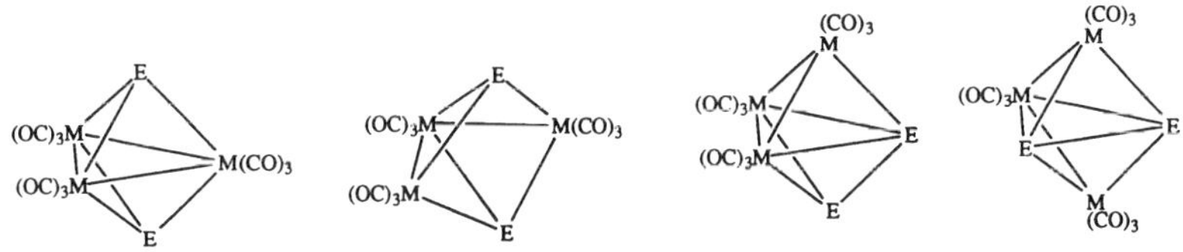

47 48 49 50

Scheme 7.21 Isomers of $M_3(CO)_9E_2$.

percentage of the As and Sb compounds retain a fourth group attached to E whereas the Bi-containing clusters commonly have a non-bonded lone pair of electrons.

To this point, discussion has concentrated on compounds with a single EM_x fragment but it is possible to have more than one such fragment in a single molecule, and M_3E_2 and M_4E_2 core structures are routinely encountered. The M_3E_2 compounds have two common geometries, the trigonal bipyramid (structure 47, Scheme 7.21) and the square pyramid (structure 48). There are two other isomers of M_3E_2 that are possible (structures 49 and 50), but structurally characterized examples to date have all had the 47 geometry. It is possible that $[Os_3(CO)_9Bi_2]$ is one of the other isomeric forms but its X-ray structure has not yet been determined. Regardless of which isomer is present, trigonal bipyramidal compounds $M_3(CO)_9E_2$ should (and here do) have 48 valence electrons in agreement with either the 18-electron rule (a localized bonding model) or Wade's rules for cluster electron counting which predicts six skeletal electron pairs for the *closo*-trigonal bipyramidal cluster (a delocalized bonding picture).

To date structures with 'naked' E have been observed only for the iron triad: $[As_2Fe_3(CO)_9]$ [122], $[Bi_2Fe_3(CO)_9]$ [123], $Bi_2Ru_3(CO)_9]$ [116] and $[Bi_2Os_3(CO)_9]$ [117]. It is surprising that the antimony analogues have not yet been reported but metallated versions $[Fe_3(CO)_9\{SbML_n\}_2]$ $[ML_n = CpMn(CO)_2, M(CO)_5$ (M = Cr, Mo, W)] are known as are the corresponding As derivatives [124]. Reaction of AsF_3 with $Fe(CO)_5$ gave $[As_2Fe_3(CO)_9]$ whereas $[Bi_2Fe_3(CO)_9]$ was obtained from protonation of solutions of $[Bi\{Fe(CO)_4\}_4]^{3-}$. In the cases of $[Bi_2Ru_3(CO)_9]$ and $[Bi_2Os_3(CO)_9]$ the compounds were obtained by reaction of $M_3(CO)_{12}$ with $NaBiO_3$. Metallated versions were obtained via reaction of the corresponding inidene complexes $[\{L_nM\}_2EX]$ with $Na_2[Fe(CO)_4]$.

Complexes of type 48 are related to 47 by the addition of two electrons in more than a formal sense, as has been demonstrated by the reduction of $[Bi_2Fe_3(CO)_9]$ to give $[Bi_2Fe_3(CO)_9]^{2-}$. This compound has also been obtained via the pyrolysis of $[Bi_4Fe_4(CO)_{13}]^{2-}$ [69]. A partially metallated version $[Fe_3(CO)_9(Bi)\{BiFe(CO)_4\}]^{2-}$ was obtained via the addition of excess $NaBiO_3$ to $[Bi\{Fe(CO)_4\}_4]^{3-}$ or by the addition of $[FE(CO)_4]^{2-}$ to $[Bi_2Fe_3(CO)_9]$ [125, 126] Similar dimetallated complexes of the formula $[Fe_3(CO)_9\{EFe(CO)_4\}_2]^{2-}$ (E = As, Sb) were obtained via the pyrolysis of $[\{(CO)_4Fe\}_3AsH]^{2-}$ [73] or the reaction of $NaSbO_3$ with $Fe(CO)_5$ in MeOH [120]. When $PhAsCl_2$ is treated with $[Fe(CO)_4]^{2-}$, the phenyl ring is retained and the presence of two of them in the compound $[Fe_3(CO)_9(AsPh)_2]$ supplies the added two electrons necessary to form the square pyramidal rather than trigonal bipyramidal structure [127, 128] All of these compounds, too, are based on iron with the exception of $[Co_2Cr(CO)_{10}\{AsCr(CO)_5\}_2]$ which is obtained from the addition of $Na[Co(CO)_4]$ to the arsinidene complex $[\{Cr(CO)_5\}_2AsCl]$ [129].

An additional structural type (51, Scheme 7.22) has been observed for M_3E_2 complexes which have one more skeletal pair of electrons than structure 48. As expected, another M–M bond is broken. As may be seen in Scheme 7.22, the unexpected feature is that an E–E interaction is apparently created that is not anticipated by the electron-counting formalisms. This phenomenon was first observed for addition of L (L = CO or phosphines) to $[Fe_3(CO)_9Te_2]$ to give $[Fe_3(CO)_9LTe_2]$. The same structural type arises when $[Bi_2Fe_3(CO)_9]$ is treated with $[Co(CO)_4]^-$. The product was found to be $[Fe_2(CO)_6Bi_2Co(CO)_4]^-$, resulting from a vertex exchange reaction [130]. Compare this with the addition of $[Fe(CO)_4]^{2-}$ to $[Bi_2Fe_3(CO)_9]$ which adds the metal fragment to give a cluster of type 48. The Bi–Bi distance in structure 51 is *c.* 3.1 Å, which is consistent with single-bond formation. A molecular orbital analysis of this class of compounds has shown that the electronegativity difference between E and M gives rise to a non-bonding molecular orbital which becomes the HOMO of the cluster [131]. This bonding situation is similar to that arising in other capped tetrahedral structures formed with M and E constituents. The isoelectronic molecule $[HFe_2(CO)_6Bi_2Fe(CO)_4]^-$ has recently been

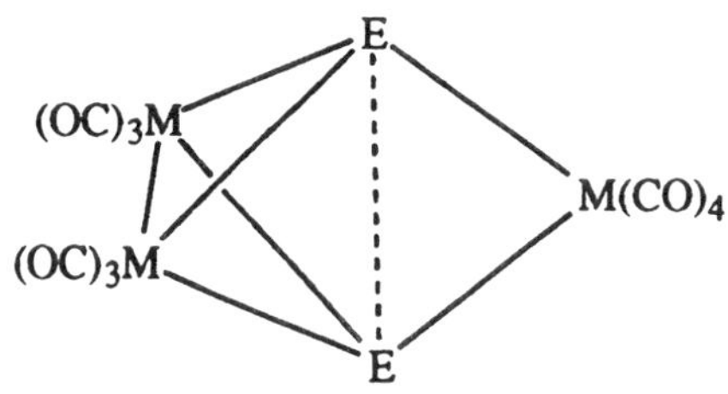

51

Scheme 7.22 Structure of type M_3E_2.

obtained from the reaction of $MePCl_2$ with $[Bi_4Fe_4(CO)_{13}]^{2-}$ and shows essentially the same core geometry.

It was noted above that compounds 48 can be viewed according to Wade's rules of cluster electron counting as a *nido*-octahedral structure. This suggests immediately that another metal vertex could be added to form the *closo*-octahedral structure 52 (Scheme 7.23) in which E bridges a square array of metal atoms. Such complexes have been reported for Sb and Bi (they are even more plentiful where E = PR or PML_n). The known examples include $[Fe_4(CO)_{12}\{SbFe(CO)_4\}_2]^{2-}$ [132] and $[Bi_2Ru_4(CO)_{12}]$ [116, 117]. The antimony compound is obtained upon photolysis of $[Sb\{Fe_2(CO)_8\}\{Fe(CO)_4\}_2]^-$ whereas the ruthenium cluster arises from the reaction of $Ru_3(CO)_{12}$/MeOH/KOH with $Bi(NO_3)_3{\cdot}5H_2O$. It is interesting that the similar reaction of $Os_3(CO)_{12}$ with $NaBiO_3$ in MeOH gives the isomeric octahedral compound 53 (Scheme 7.24) which has a Bi_2 bond [117]. The reasons for the alternative isomer being produced for the osmium case are not yet clear, although a similar phenomenon appears to occur in the M_3E_2 series of compounds.

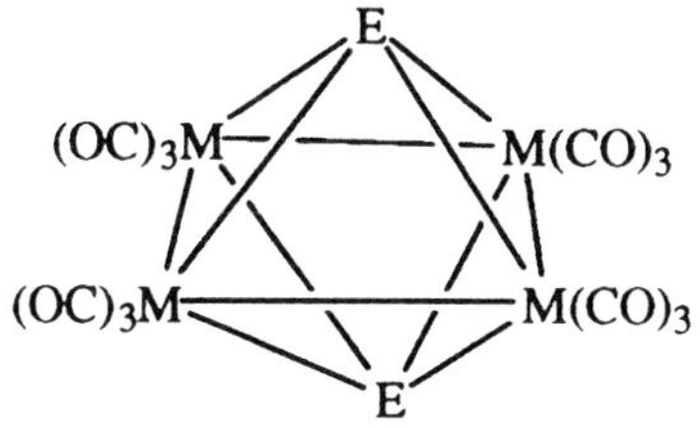

52

Scheme 7.23 Example of a *closo*-octahedral structure.

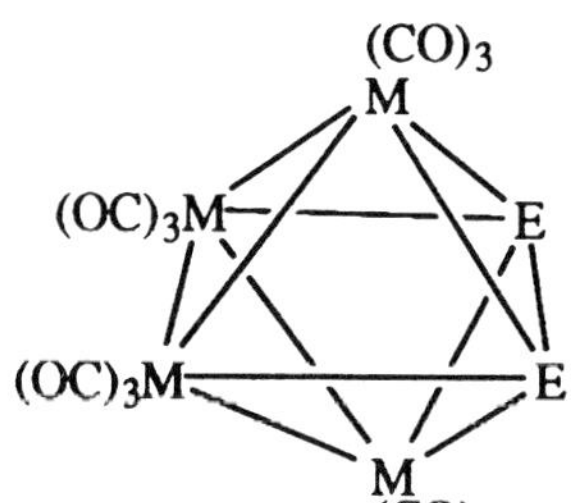

53

Scheme 7.24 The octahedral product of the reaction of $Os_3(CO)_{12}$ with $NaBiO_3$ in MeOH.

The ruthenium and osmium compounds possess seven skeletal electron pairs as predicted by the cluster-counting formalisms for a *closo*-octahedral molecule. This electron count, however, will not satisfy the 18-electron rule so the delocalized bonding model is clearly preferred for these molecules. The antimony complex has one excess pair of electrons for a total of eight skeletal electron pairs, as has been observed and discussed for a number of other clusters. This electron count cannot be rationalized by the 18-electron rule either. It appears that the extra electron pair occupies a b_u orbital of π^* nature localized on the transition metals. A structural comparison of the seven and eight electron pair systems shows the M_4 unit in the latter to be significantly puckered ([132] and references therein).

A general route to preparing numerous main group element–transition metal cluster compounds has been developed by Fenske and co-workers. This route involves addition of main group element silyl complexes to metal halides. Topical to this section is the production of $[Ni_9(\mu_4\text{-}As)_6(PPh_3)_5Cl_3]$ and $[Ni_9(\mu_4\text{-}As)_6(PPh_3)_6Cl_2]$ from $PhAs(SiMe_3)_2$ and $[NiCl_2(PPh_3)_2]$ [133]. In contrast to similar reactions with phosphorus reagents, the As–C bond is cleaved. The As-containing molecules consist of a Ni-centred cube of Ni atoms with an As atom capping each face of the cube (structure 54, Scheme 7.25). Each corner nickel atom is in turn coordinated either to a PPh_3 or chloride ligand. For the trichloroderivative the distances between As and the nickel atoms in the cube range from 2.298 Å to 2.350(5) Å whereas the $As{-}Ni_{interstitial}$ distances are longer, ranging from 2.610 Å to 2.641(6) Å. The electron count of the trichloro

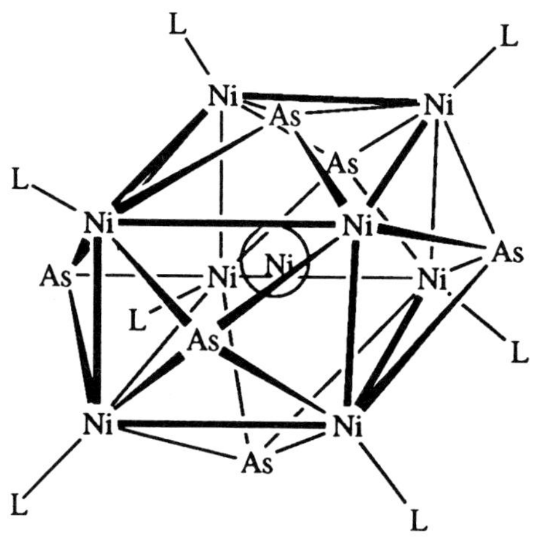

54

Scheme 7.25 Structure formed from $PhAs(SiMe_3)_2$ and $[NiCl_2(PPH_3)_2]$.

compound is 121 electrons, and that of the dichloro complex is 122. The 18-electron rule predicts a value of 122 electrons whereas an electron-counting scheme for close-packed metal clusters predicts values of 114 or 120 valence electrons.

Two related bismuth-containing clusters also exhibit the μ_4-E configuration [134]. These are $[Bi_4Co_9(CO)_{16}]^{2-}$ and $[Bi_8Co_{14}(CO)_{20}]^{2-}$ which arise from the slow oxidation of solutions of $Mo(CO)_3$(toluene) with $[Bi_2Co_4(CO)_{11}]^-$. The metal frameworks are shown as structures 55 and 56, respectively, in Scheme 7.26. Structure 56 can be viewed as two units of structure 55 fused about a Co_4 face. The parent compound $[Bi_2Co_4(CO)_{11}]^-$ can be viewed as a *nido*-pentagonal bipyramidal cluster. It was hoped initially that the $Mo(CO)_3$ unit would fill in the missing vertex but no Mo is incorporated into the cluster. Instead, close-packed arrays of Bi and Co are created which have a structural relationship to the $Ni_9(\mu_4\text{-}As)_6L_8$ clusters described above. As for those clusters, the basic building block is a cube of Co atoms containing an interstitial Co atom in the middle. Unlike the Ni compounds, however, the array is not cubic but rather tetragonal. Thus structure 55 may be viewed in layers, with a Co_4 square over a Bi_4Co cross over another Co_4 square. There are no inter-layer Co–Co bonds, although these connectivities are indicated in the diagram. The compound 56 is then composed of five layers, three of Co_4 and two of Bi_4Co which alternate. The close-packed nature of this arrangement is best seen in structure 56 where the two interstitial Co atoms and the central Co_4 layer constitute a Co_6 octahedron, each atom of which lies in the centre of a cubic array formed by the eight Bi atoms.

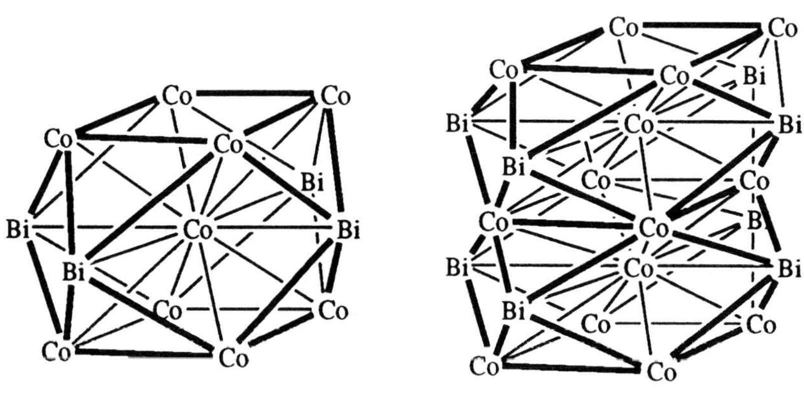

55 **56**

Scheme 7.26 Products of the slow oxidation of solutions of $Mo(CO)_3$(toluene) with $[Bi_2Co_4(CO)_{11}]^-$. Structure 56 may be viewed as two units of structure 55 fused about a Co_4 face.

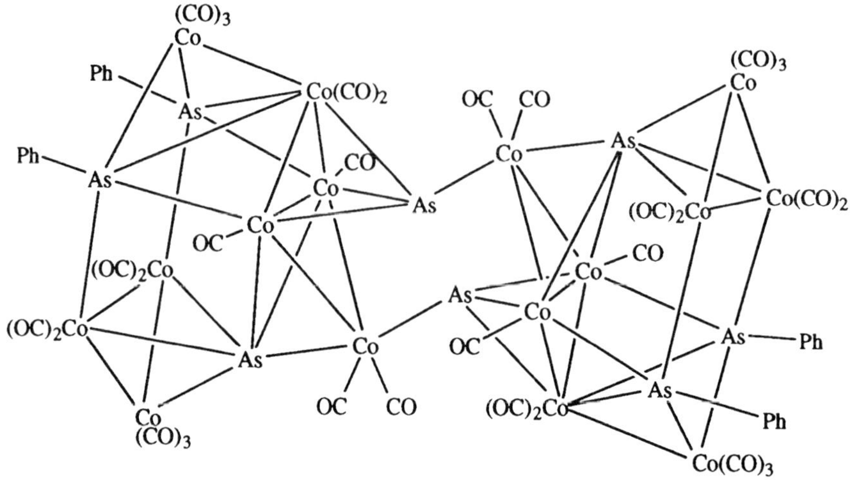

57

Scheme 7.27 Product of the reaction of $Co_2(CO)_8$ with $(AsPh)_6$.

In comparison with the Ni–As clusters, the Co–Bi compounds have a considerably higher valence electron count (127 for structure 55) which in part explains the lack of Co Co bonding between the layers. The actual bonding situation is much more complicated and a molecular orbital analysis is in progress (Saillard, J.-Y., personal communication).

A cluster with the formula $[Co_8(\mu_6\text{-}As)(\mu_4\text{-}As)(\mu_4\text{-}AsPh)_2(CO)_{16}]$, arising from the reaction of $Co_2(CO)_8$ with $(AsPh)_6$, contains AsPh groups bridging four metal centres (structure 57, Scheme 7.27) in addition to other naked As atoms [135].

Pentagonal arrays of metal atoms may also be capped by E (E = Sb or Bi) [136–139]. This geometry is observed in a series of icosahedral clusters of ten Ni atoms and two E atoms (metal framework structure 58, Scheme 7.28) in which the main group atoms occupy opposite vertices. When $MeAsCl_2$ or $PhSbCl_2$ is added to $[Ni_6(CO)_{12}]^{2-}$ a compound which has the R functionality still attached to E is observed, $[Ni_{10}(CO)_{18}(ER)_2]^{2-}$ [136]. The similar reaction employing (p-tolyl)$BiBr_2$ was unsuccessful [136]. When $PhAsCl_2$ was employed the cluster $[Ni_9(CO)_{15}(AsPh)_3]^{2-}$ was obtained [140]. The geometry of the Ni_9As_3 cluster is the same as that of structure 58 with one of the Ni vertices replaced by As. It has the same skeletal electron count (13 pairs of electrons) as $[Ni_{10}(CO)_{18}(ER)_2]^{2-}$, which is also the same as that of the well-known $[B_{12}H_{12}]^{2-}$ icosahedral cluster.

If $SbCl_3$ is used instead of $RSbCl_2$, the clusters $[Ni_{10}(CO)_{18}(\mu_{12}\text{-}Ni)\{SbNi(CO)_3\}_2]^{n-}$ (n = 2, 3 or 4) are obtained [138, 141]. Unlike the SbPh

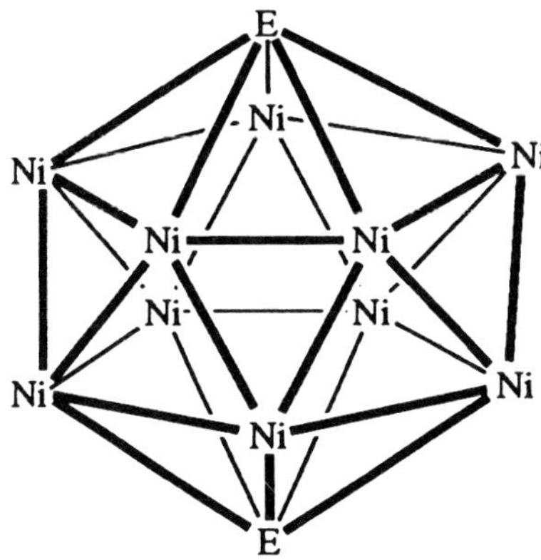

58

Scheme 7.28 Example of a pentagonal array of metal atoms capped by E (E = Sb, Bi).

compound these molecules contain an interstitial Ni atom at the centre of the cluster framework and each Sb atom is ligated to a tetrahedral $Ni(CO)_3$ group. Likewise, the interstitial clusters $[Ni_{10}(CO)_{18}(\mu_{12}\text{-}Ni)Bi_2]^{n-}$ ($n = 2$, 3 or 4) arise in the reaction of $BiCl_3$ with $[Ni_6(CO)_{12}]^{2-}$ [139]. The only difference between the Bi and Sb clusters is that the lone pairs on the Bi atoms are not coordinated to external ML_n fragments. Addition of MeI to $[Ni_{10}(CO)_{18}(\mu_{12}\text{-}Ni)\{SbNi(CO)_3\}_2]^{n-}$ results in formation of $[Ni_{10}(CO)_{18}(\mu_{12}\text{-}Ni)\{SbMe\}_2]^{2-}$ and $[Ni(CO)_3I]^-$, but attempted alkylation of the bismuth analogues resulted only in cluster oxidation [138]. These metal-centred clusters have anomalous electron counts according to the conventional Wade–Mingos counting formalism, and this was attributed to the involvement of the d orbitals on the interstitial atom [138].

7.3.5 Complexes with encapsulated E atoms

Complete encapsulation of the heaviest main group elements is not often seen because of the size requirements of E. With sufficiently large M, however, it is possible to place As or Sb inside a metal framework. Thus reaction of $Rh(CO)_2(acac)$ – (acac = acetylacetonate or 2,4-pentanedionate) with Ph_3As in the presence of alkali carboxylates and high CO pressure leads to $[Rh_{10}(CO)_{22}As]^{3-}$ [142]. The structure is best-described as a bi-capped square anti-prism with As in the centre (structure 59, Scheme 7.29). The distances of As to the Rh atoms in the square faces is *c.* 2.5 Å whereas the $Rh_{capping}$–As distances are on the order of 3.0 Å. This cluster could also be obtained via addition of $[Rh(CO)_4]^-$ with $[Rh_9(CO)_{21}As]^{2-}$ which was identified on the basis of the similarity of its

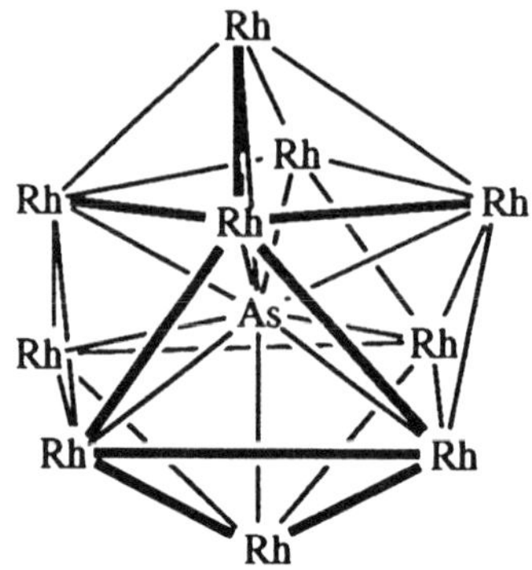

59

Scheme 7.29 Product of the reaction of $Rh(CO)_2(acac)$ with Ph_3As in the presence of alkali carbonates and under high CO pressure. acac = acetylacetonate or 2,4-pentanedionate.

infrared spectrum to the known $[Rh_9(CO)_{21}P]^{2-}$ cluster. The core structure of that compound is the same as structure 59, with one of the capping Rh atoms missing.

$Rh(CO)_2(acac)$ reacts with $SbPh_3$ under 400 atm of CO/H_2 (1:1) in the presence of caesium benzoate to give $[Rh_{12}(CO)_{27}Sb]^{3}$ for which the cluster core is structure 60, Scheme 7.30 [143]. This structure is best described as a distorted icosahedral array of Rh atoms with the Sb atom

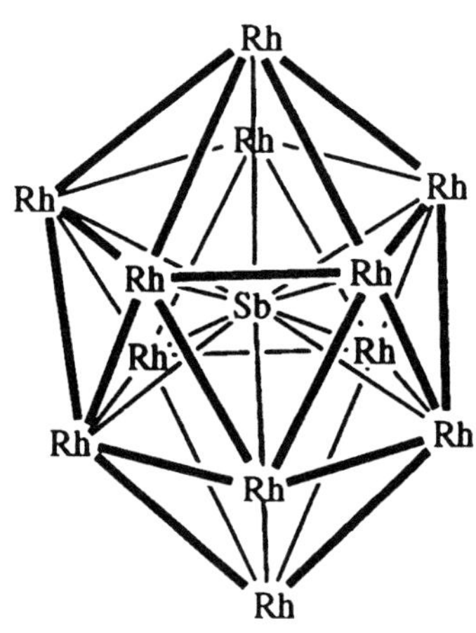

60

Scheme 7.30 Product of the reaction of $Rh(CO)_2(acac)$ with $SbPh_3$ under 400 atm of CO/H_2 (1:1 in the presence of caesium benzoate. acac = acetylacetonate or 2,4-pentanedionate.

at the centre. The Rh–Rh distances show a considerable variation (2.80–3.33 Å). The radius of a regular Rh_{12} icosahedron would be in the order of 2.085 Å, which is considerably larger than the radius of the Sb atom (1.41 Å). For comparison, the radius of the square anti-prismatic cavity in structure 59 is *c.* 0.90 Å, whereas the radius of the P and As atoms which have been found to reside there are 1.10 Å and 1.21 Å, respectively. The mismatch between the radius of Sb and the icosahedral cavity may in part explain some of the observed structural distortions in the direction of forming a cuboctahedral core. The cuboctahedron has been proposed as an intermediate geometry in the isomerization of icosahedral clusters. The cluster is stable under pressures of CO.

7.4 Metal complexes with fragments containing E–E bonds

The pnictogens are known to form catenated compounds, so it is not surprising that metallated versions of complexes with E–E bonds have also been observed. The E–E bond strengths decrease as one descends the periodic table, making the presence of E–E bonding much more important for the lighter elements. As the E fragments become larger there is a clear relationship to the Zintl ions E_x^{n-} – naked main group element anions created by reduction of the main group elements in some suitable solvent such as liquid ammonia or ethylenediamine. Synthetic strategies for metallation of the Zintl ions directly has also been developed (*vide infra*).

7.4.1 Complexes with E_2 fragments

The most commonly observed E_x fragment is that for which $x = 2$ (structures 61–64, Scheme 7.31). In order to rationalize the bonding, one can

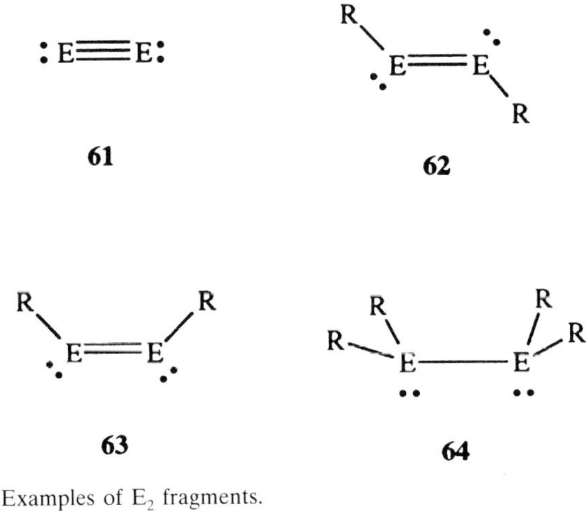

Scheme 7.31 Examples of E_2 fragments.

Table 7.6 Complexes with E_2 fragments.

Compound	Reference
$[Mn_2(CO)_7X(AsMe_2)(As_2Me_4)]$	[i]
$[\{(\mu\text{-Bu}^tS)Fe_2(CO)_6\}_2(\eta^2,\mu_4\text{-MeAsAsMe})]$	[ii]
$[Fe(CO)_4As_2(C_6F_5)_2]$	[iii]
$[\{MeNC_4O_2(PPh_2)_2\}NiAs_2Ph_2]$	[iv]
$[Fe(CO)_4Sb_2\{CH(SiMe_3)_2\}_2]$	[v, vi]
$[Ph_2As_2\{Cr(CO)_5\}_3]$	[vii]
$[R_2Sb_2\{W(CO)_5\}_3]$, R = Me, Ph, Bu^t	[viii, ix]
$[Bi_2\{SmCp^*2\}_2]$	[x]
$[\{Cp^*(CO)_2Mn\}_2As_2]$	[xi, xii]
$[\{Cp^*Co\}_3\{As_2\}\{As_4\}]$	[xiii]
$[\{Cp(CO)_2Mn\}_4As_2]$	[xi, xii, xiv]
$[As_2\{Mo(CO)_5\}_3]$	[xiv]
$[As_2\{W(CO)_5\}_3]$	[xv]
$[Sb_2\{W(CO)_5\}_3]$	[ix]
$[Bi_2\{W(CO)_5\}_3]$	[xvi, xvii]
$[Bi_2\{Mn(CO)_2C_5H_4Me\}_3]$	[xviii, xix]
$[W_2(CO)_7(I)(\mu\text{-I})As_2]$	[xiv]
$[Cp^*_2(CO)_4Mo_2As_2]$	[xx]
$[Cp_2(CO)_4M_2As_2]$, M = Mo, W	[xiv, xxi]
$[Cp_2(CO)_4M_2As\{AsCr(CO)_5\}]$, M = Mo, W	[xiv]
$[Cp_2(CO)_4M_2\{AsCr(CO)_5\}_2]$, M = Mo, W	[xiv]
$[Cp_2(CO)_4Mo_2\{AsMn(CO)_2Cp\}_2]$	[xxi]
$[Co_2(CO)_6As_2]$	[xxii, xxiii]
$[Co_2(CO)_{6-x}(PPh_3)_xAs_2]$, $x = 1, 2$	[xxii, xxiii]
$[Co_2(CO)_{6-x}(L)_xAs_2]$, $x = 1, 4$; L = PMe_3, $P(OMe)_3$	[xxiv]
$[Co_2(CO)_6As\{AsM(CO)_5\}]$, M = Cr, Mo, W	[xxiv]
$[Co_2(CO)_4\{P(OMe)_3\}_2\{AsM(CO)_5\}_2]$, M – Cr, Mo, W	[xxiv]
$[Co_2(CO)_{6-x}(PMe_3)x\{AsW(CO)_5\}_2]$, $x = 1, 2$	[xxiv]
$[Co_2(CO)_3\{P(OMe)_3\}_3\{AsCr(CO)_5\}_2]$	[xxiv]
$[Sb_2Co_2(CO)_4(PR_3)_2]$, R = Ph, *p*-tolyl	[xxv]
$[Sb_2\{Mo(CO)_2Cp\}_2]$	[xxvi]
$[Bi_2Co_2(CO)_4(PPh_3)_2]$	[xxv]
$[Bi_2M_2(CO)_4(C_5H_4R)_2]$, M = Mo, W; R = H, Me	[xxvii, xxviii]
$[Bi_2W_2(CO)_8\{Bi(Me)W(CO)_5\}]$	[xvii]
$[Fe_2(CO)_6Bi_2\{\mu\text{-}Co(CO)_4\}]^-$	[xxix, xxx]
$[HFe_2(CO)_6Bi_2\{\mu\text{-}Fe(CO)_4\}]^-$	[xxxi]
$[PPN]_x[Sb_2Co(CO)_{11}]$, $x = 1, 2$	[xxxii, xxxiii]
$[Bi_2Co_4(CO)_{11}]^-$	[xxxiii, xxxiv]
$[Bi_2Os_4(CO)_{12}]$	[xxxv]

References

i. E. Röttinger, A. Trenkle, R. Muller and H. Vahrenkamp, *Chem. Ber.*, 1980, **113**, 1280.
ii. L.-C. Song and Q.-M. Hu, J. *Organomet. Chem.*, 1991, **414**, 219
iii. P. S. Elmes, P. Leverett, and B. O West, J. *Chem. Soc., Chem. Commun.*, 1971, 747.
iv. D. Fenske and K. Merzweiler, *Angew. Chem. Int. Ed. Eng.*, 1984, **23**, 635; *Angew. Chem*, 1984, **96**, 600.
v. A. H. Cowley, N. C. Norman, M. Pakulsi, D. L. Bricker, and D. H. Russell. *J. Am. Chem Soc.*, 1985,**107**, *8211.*
vi. A. H. Cowley, N. C. Norman, and M. Pakulski, J. *Am. Chem. Soc.*, 1984, **106**, 6844.
vii. G. Huttner, H.-G. Schmid, A. Frank, and O. Orama, *Angew. Chem. Int. Ed. Eng.*, 1976, **15**, 234.
viii. U. Weber, G. Huttner, O. Scheidsteger, and L. Zsolnai, J. *Organomet. Chem.*, 1985, **289**, 357.

Table 7.6 *Continued*

ix. G. Huttner, U. Wever, B. Sigwarth, and O. Scheidsteger, *Angew. Chem. Int. Ed Eng.*, 1982, **21**, 215; Angew. Chem.., 1982, **94**, 210.
x. W. J. Evans, S. L. Gonzales, and J. W. Ziller, J. *Am. Chem. Soc.*, 1991, **113**, 9880.
xi. W. A. Herrmann, B. Koumbouris, T. Zahn and M. L. Ziegler, *Angew. Chem. Int. Ed. Eng.*, 1984, **23**, 812, *Angew. Chem.*, 1984, **96**, 802.
xii. W. A. Herrmann, B. Koumbouris, A. Schäfer, T. Zahn, and M. L. Ziegler, *Chem Ber.*, 1985, **118**, 2488.
xiii. M. Detzel, K. Pfeiffer, O. J. Scherer, and G. Wolmershäuser, *Angew. Chem. Int. Ed. Eng.*, 1993, **32**, 914; *Angew Chem.*, 1993, **105**, 936.
xiv. G. Huttner, B. Sigwarth, O Scheidsteger, L Zsolnai, O Orama, *Organometallics*, 1985, **4**, 326.
xv. B. Sigwarth, L Zsolnai, H. Berke and G. Huttner, J. *Organomet Chem.*, 1982, **226**, C5
xvi. G. Ciani, M. Moret, A. Fumagalli, and S. Martinengo, J. *Organomet. Chem.*, 1989, **362**, 291.
xvii. A. M. Arif, A. H. Cowley, N. C. Norrnan, and M. Pakulski, *Inorg. Chem.*, 1986, **25**, 4836.
xviii. S. J. Davies, N. A. Compton, G. Huttner, L. Zsolnai, and S. E. Garner, *Chem Ber.*, 1991, **124**, 2731.
xix. K. Plößl, G. Huttner, and L. Zsolnai, Angew. Chem. Int. Ed. Eng., 1989, **28**, 446; *Angew. Chem.*, 1989, **101**, 482.
xx. P. J. Sullivan and A. L. Rheingold, *Organometallics*, 1982, **1**, 1547.
xxi. M. L. Ziegler, K. Blechsmitt, B. Nuber and T. Zahn, *Chem. Ber.*, 1988, **121**, 159.
xxii. A. S. Foust, M. S. Foster and L. F. Dahl, *J. Am. Chem. Soc.*, 1969, **91**, 5633.
xxiii. A. S. Foust, C. F. Campana, J. D. Sinclair, and L. F. Dahl, *Inorg. Chem.*, 1979, **18**, 3047.
xxiv. M. Muller and H. Vahrenkamp, J. *Organomet. Chem.*, 1983, **252**, 95.
xxv. W. Clegg, N. A. Compton, R. J. Errington, D. C. R. Hockless, N. C. Norman, M. Ramshaw, and P. M. Webster, J. *Chem. Soc., Dalton Trans.*, 1990, 2375.
xxvi. J. R. Harper and A. L. Rheingold, J. *Organomet. Chem.*, 1990, **390**, C36.
xxvii. W. Clegg, N. A. Compton, R. J. Errington, G. A. Fisher, N. C. Norman, and T. B. Marder, J. *Chem. Soc., Dalton Trans.*, 1991, 2887.
xxviii. W. Clegg, N. A. Compton, R. J. Errington, and N. C. Norman, *Polyhedron*, 1988, **7**, 2239.
xxix. K. H. Whitmire, M. Shieh, C. B. Lagrone, B. H. Robinson, M. R. Churchill, J. C. Fettinger, and R. F. See, *Inorg. Chem.*, 1987, **26**, 2798.
xxx. K. H. Whitmire, K. S. Rhaguveer, M. R. Churchill, J. C. Fettinger, and R. F. See, *J. Am. Chem Soc*, 1986, **108**, 2778.
xxxi. J. R. Eveland, J.-Y. Saillard and K H. Whitmire, submitted for publication.
xxxii. J S. Leigh and K. H. Whitmire, J. *Am Chem. Soc*, 1989, **111**, 2726.
xxxiii. T. A. Albright, K. A. Yee, J.-Y. Saillard, S. Kahlal, J.-F. Halet, J. S. Leigh, and K. H. Whitmire, *Inorg. Chem.*, 1991, **30**, 1179
xxxiv. S. Martinengo and G. Ciani, J. *Chem. Soc., Chem. Commun.*, 1987, 1589.
xxxv. H. G. Ang, C. M. Hay, B. F. G. Johnson, J. Lewis, P. R. Raithby, and A. J. Whitton, J. *Organomet. Chem*, 1987, **330**, C5.

consider these fragments separate from the metal constituents. For example, E_2 as an isolated unit is isoelectronic and isostructural to N_2 (structure 61), having a triple bond, and E_2R_2 (structures 62 and 63) would possess a double bond and could exist either as *cis* (63) or as *trans* (62) isomers. Metallated derivatives can then be derived by using the lone pairs and/or the π electron system of the precursor fragment. Table 7.6 lists molecules having two E atoms bonded to each other.

The E_2R_4 molecules can function the same way as the simple ligands ER_3. It has already been noted that $L_nM{\leftarrow}ER_2ER_2$ can serve as precursors to $\{L_nM\}_2\{ER_2\}_2$ compounds [44]. A more complex molecule with each lone pair donated to a different metal has been observed for $[Mn_2(CO)_7X(AsMe_2)(As_2Me_4)]$ (structure 65, Scheme 7.32) (X = Cl, Br) [27]. It is obtained from the reaction of $K[Mn(CO)_5]$ with Me_2AsCl.

Compounds derived from structure 62 or 63 are more interesting as the multiple bond may also be involved in the bonding to metals. In general, only the more stable *trans* configuration 62 is observed. Two similar molecules from different synthetic strategies have been reported in which an E_2R_2 fragment is π coordinated to a 16-electron transition metal fragment (structure 66, Scheme 7.33). The known examples include $[Fe(CO)_4\{As_2(C_6F_5)_2\}]$ [144] from reaction of $Fe(CO)_5$ with $\{As(C_6F_5)\}_4$ and $[Fe(CO)_4\{SbCH(SiMe_3)_2\}_2]$ from treatment of $Na_2[Fe(CO)_4]$ with $Cl_2SbCH(SiMe_3)_2$ [145, 146]. The first example shows partial retention of the As–As bonding present in the starting cyclic tetramer, but the second case is of note for the coupling of two Sb units to give the distibine ligand.

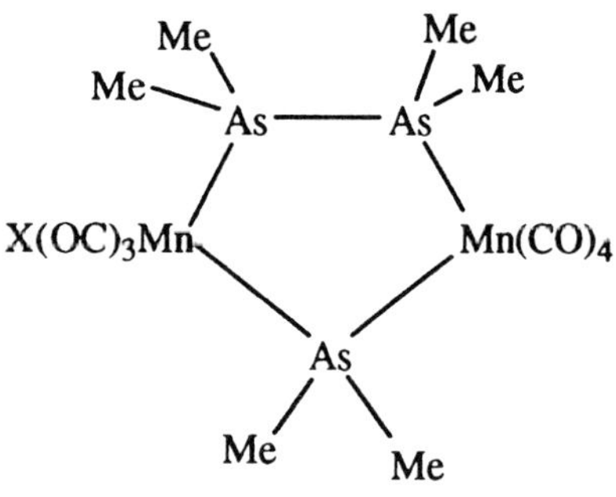

65

Scheme 7.32 Product of the reaction of $K[Mn(CO)_5]$ with Me_2AsCl.

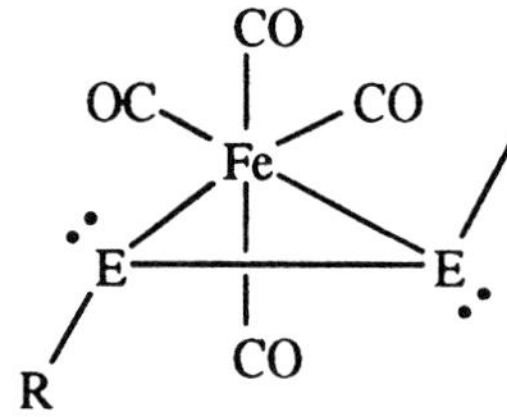

66

Scheme 7.33 Example of an E_2R_2 fragment π coordinated to a 16-electron transition metal fragment.

An As_2Ph_2 ligand is created in the reaction of $[\{MeNC_4O_2(PPh_2)_2\}NiCl_2]$ with $PhAs(SiMe_3)_2$ [147]. It has essentially the same structure as structure 66 but with the $Fe(CO)_4$ unit replaced by the $\{MeNC_4O_2(PPh_2)_2\}Ni$ group. The relationship of the Ph rings is also *trans* in this molecule.

The complexes $[R_2E_2\{M(CO)_5\}_3]$ (E = As, R = Ph, M = Cr; E = Sb, R = Me, Ph, Bu^t, M = W) have essentially the same geometry as structure 66 except that the lone pairs of electrons on E coordinated to an external metal fragment (structure 67, Scheme 7.34). These molecules were obtained by a coupling process in which $[M_2(CO)_{10}]^{2-}$ (M = Cr, Mo, W) are treated with $RECl_2$ [103, 144, 145]. The E_2 unit serves as a six-electron donor, two electrons being contributed to each of the metal centres either via a lone pair or via donation of the double bond. A different coordination mode is seen for the As_2Me_2 group in $[\{(\mu\text{-}Bu^tS)Fe_2(CO)_6\}_2\{As_2Me_4\}]$ (structure 68, Scheme 7.35) [148]. In this molecule, the overall contribution is again six electrons, but each As atom is coordinated to a different Fe_2 unit as a three-electron donor. It is prepared from $MeAsI_2$ and $[Fe_2(CO)_6)(\mu\text{-}CO)(\mu\text{-}SBu^t)]^-$.

Naked E_2 fragments, which are isostructural and isoelectronic to dinitrogen, have been observed to coordinate to two or three metal centres.

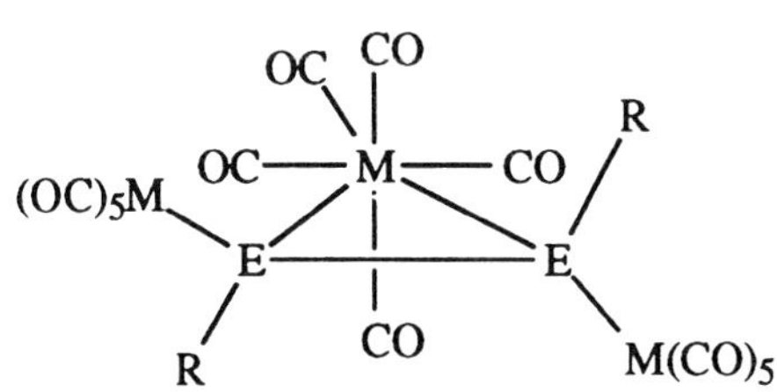

67

Scheme 7.34 Geometry of the product of the reaction between $[M_2(CO)_{10}]^{2-}$ (M = Cr, Mo, W) with $RECl_2$.

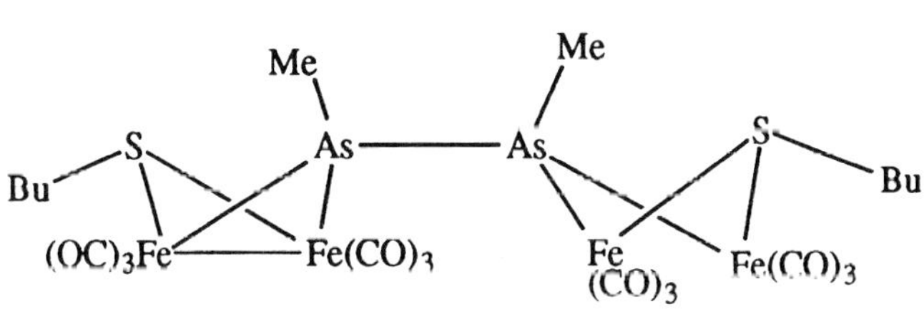

68

Scheme 7.35 Geometry of the product between $MeAsI_2$ and $[Fe_2(CO)_6(\mu\text{–}CO)(\mu\text{–}SBu^t)]^-$.

In the former case they are considered as four-electron donors whereas in the latter they donate six electrons. When they coordinate to two metal centres there are two basic geometries that are observed. These include the open structure 69 (Scheme 7.36) and the tetrahedral arrangement 70. The compound $[Cp^*_2SmBi_2SmCp^*_2]$ ($Cp^* = \eta$-C_5Me_5) of type 69 is formed along with biphenyl and $[Cp^*_2SmPh]$ when $[Cp^*_2Sm]$ reacts with $BiPh_3$ (the only example of a Bi compound of the lanthanides) [149]. This complex is planar whereas the compound $[\{Cp^*(CO)_2Mn\}_2As_2]$ created in the thermal decomposition of $[\{Cp^*(CO)_2Mn\}_2AsH]$ has a dihedral angle of 120.8° [150]. Compare this with thermolysis of $[\{Cp(CO)_2Mn\}_2AsH]$ or $[\{Cp(CO)_2Mn\}_2AsI]$ which yields a coupling product but with a different geometry (structure 71, Scheme 7.37) [151, 152]. The differing outcomes are attributable to the steric differences between the Cp* and Cp rings. Two As_2 units each bonded to two Co atoms are found in the molecule $[\{Cp^*Co\}_2\{As_2\}_2]$ which results when $[Cp^*CoAs_4Co(CO)Cp^*]$ is heated in xylene [153]. This thermolysis prompts the loss of CO which causes insertion of Co into one of the As–As bonds of the As_4 chain structure (Section 7.4.3). Likewise, two isolated As_2 units occur in $[R_2Mo_2(CO)_2(As_2)_2]$ ($R = C_5H_4Me$, Cp*) [154, 155] in which there are also Mo–Mo bonds. The As–As bonds are very short, again indicative of multiple bonding between these atoms. The C_5H_4Me derivative was prepared from $[\{(C_5H_4Me)Mo(CO)_3\}_2]$ and $(AsMe)_5$ whereas the Cp* complex was isolated from the reaction of $[\{Cp^*Mo(CO)_2\}_2]$ with As_4.

Coupling of two E fragments occurs when the trihalides are treated with $[W_2(CO)_{10}]^{2-}$, $[(C_5H_4Me)Mn(CO)_2(thf)]$ or $[(C_5H_4Me)Mn(CO)_2$

69 **70**

Scheme 7.36 Observed geometries of naked E_2 fragments coordinated to two metal centres: open structure (69) and tetrahedral arrangement (70).

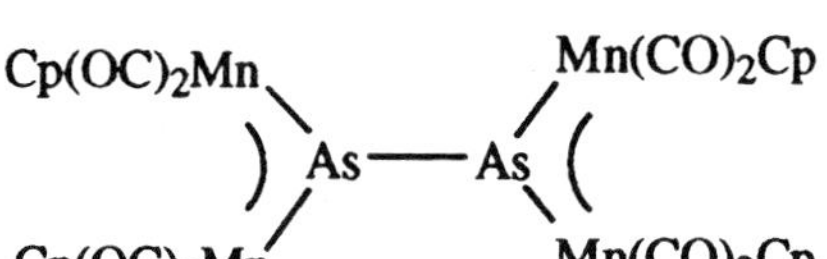

71

Scheme 7.37 Product of the thermolysis of $[\{Cp(CO)_2Mn\}_2AsH]$ or $[\{Cp(CO)_2Mn\}_2AsI]$.

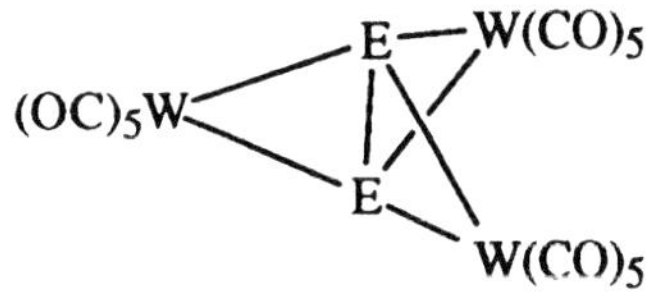

72

Scheme 7.38 Coupled E fragments.

$SiPh_3]^-$ [57, 156, 157–160]. The resultant structure type [72] is shown in Scheme 7.38. The entire series of compounds $[E_2\{W(CO)_5\}_3]$ (E = As, Sb, Bi) has been characterized while only the Bi compound has been reported for $ML_n = (C_5H_4Me)Mn(CO)_2$. The most intriguing feature is the E–E distance. In the $E_2\{W(CO)_5\}_3$ compounds these values (E = As, 2.279 Å; E = Sb, 2.663 Å; E = Bi, 2.818 Å) are indicative of substantial multiple bonding that is not required by conventional electron-counting formalisms. The distance in $[\{(C_5H_4Me)Mn(CO)_2\}_3Bi_2]$ (2.813 Å) is comparable to that of the tungsten compound. The compound $[As_2\{Mo(CO)_5\}_3]$ has also been prepared but has not been structurally characterized [152].

The most commonly observed structural type for the M_2E_2 series of compounds is the tetrahedron (structure 70). It has been encountered for a large number of molecules as seen in Table 7.6. These cluster cores possess six skeletal pairs of electrons and may be viewed in either a localized bonding picture (there are six edges in a tetrahedron corresponding to the six bonding pairs) or in a delocalized fashion using Wade's rules which treat the tetrahedron as a *nido*-trigonal bipyramid.

Compounds of this class may donate their lone pairs of electrons to external metal fragments, and ligand substitution at the metal centres may also be supported. The synthetic methods follow the same patterns as those given above for other types of EM cluster compounds. For example, reaction of $Co_2(CO)_8$ and $AsCl_3$ produces $[Co_2(CO)_6As_2]$ which undergoes substitution reactions or coordination of As to ML_n fragments [161, 162]. This can be compared to the isoelectronic $[Fe_2(CO)_6(AsS)]^-$ ion which has recently been prepared from $NaAsO_2$, $Na_2S_2O_4$, $Fe_3(CO)_{12}$ and $[PPh_4]Br$ in H_2O/MeOH [163]. Reaction of the arsinidene compound $\{CpMCr(CO)_7\}As\{Cr(CO)_5\}]$ (M = Mo, W) with PPh_3 gives $[\{Cp(CO)_2M\}_2As_2]$, $[\{Cp(CO)_2M\}_2As\{AsCr(CO)_5\}]$ or $[\{Cp(CO)_2M\}_2\{AsCr(CO)_5\}_2]$, depending upon the amount of PPh_3 added as a '$Cr(CO)_5$' scavenger [152]. Elemental arsenic also reacts with $[\{Cp(CO)_3M\}_2]$ (M = Mo, W) to give $[\{Cp(CO)_2M\}_2As_2]$ which can be metallated at As by addition of reagents such as $[Cr(CO)_5(thf)]$ or $[CpMn(CO)_2(thf)]$ [104, 152]. Iodination of $[As_2\{W(CO)_5\}_3]$ (structure 60) results in the tetrahedral cluster $[W_2(CO)_7(I)(\mu\text{-}I)As_2]$ in which one I atom bridges the W–W bond and the other is terminal [152]. Antimony metal reacts with [$\{CpMo(CO)_2\}_2$] similarly to arsenic to give [$\{CpMo(CO)_2\}_2Sb_2$] [164]. Reaction of $BiCl_3$ with K[$Co(CO)_3(PPh_3)$] yields, as one of several products, a compound formulated as [$Bi_2Co_2(CO)_4(PPh_3)_2$], but it did not prove possible to determine the structure of that compound [84]. Photolysis of [$Bi\{Mo(CO)_2(C_5H_4Me)\}_3$] afforded [$(C_5H_4Me)_2(CO)_4Mo_2Bi_2$] in *c.* 20% yield [74, 165]. Another cluster containing a Bi_2M_2 unit, [$Bi_2\{W(CO)_4\}_2(\mu\text{-}Bi(Me)\{W(CO)_5\})$], is obtained from the reaction of $(Me_3Si)_2HCSbCl_2$ with [$W_2(CO)_{10}$]$^{2-}$ [159]. [$Bi_2\{W(CO)_5\}_3$] is also produced in that reaction [159]. In [$Bi_2\{W(CO)_4\}_2(\mu\text{-}Bi(Me)\{W(CO)_5\})$], the W–W bond is bridged by a tetrahedral BiMe{$W(CO)_5$} group.

The E–E distances in [$Co_2(CO)_5(PPh_3)As_2$], [161] [$\{CpMo(CO)_2\}_2Sb_2$], [164] [$(C_5H_4Me)_2(CO)_4Mo_2Bi_2$] [166] and [$Bi_2\{W(CO)_4\}_2(\mu\text{-}Bi(Me)\{W(CO)_5\})$] [159] are short and this observation was interpreted to arise from the charge separation between the M and E fragments. This suggests an alternate picture to structure 70 in which the molecule is considered as an E_2 unit π coordinated to the dimetal fragment (structure 73, Scheme 7.39).

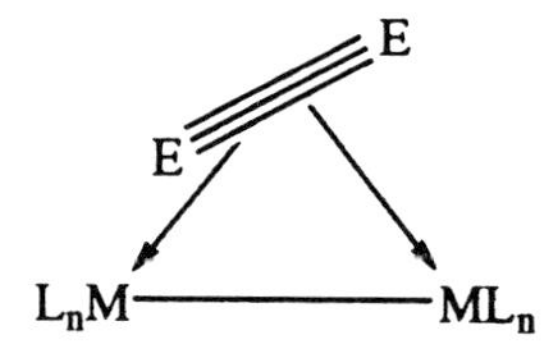

73

Scheme 7.39 An E_2 unit π coordinated to the dimetal fragment.

There are two other structural types of molecules containing E_2 fragments which are based upon the M_2E_2 tetrahedron. One of these, structure 51, was discussed above (Section 7.3.4) because of the structural and chemical relationship to the square pyramidal M_3E_2 clusters (structure 48). Addition of a two-electron donor ligand results in breakage of one M–M bond, as might be expected. What was unusual about this structural type and the reason it is mentioned again here is that the E–E distance is relatively short and, for Bi compounds of this type, consistent with a single bond distance. The two examples known to date include [$Fe_2(CO)_6Bi_2\{\mu\text{-}Co(CO)_4\}$]$^-$ formed in the reaction of [$Co(CO)_4$]$^-$ with [$Bi_2Fe_3(CO)_9$] [125, 126] and the isoelectronic and isostructural [$HFe_2(CO)_6Bi_2\{Fe(CO)_4\}$]$^-$ obtained from reaction of $MePCl_2$ with [$Bi_4Fe_4(CO)_{13}$]$^{2-}$ [69].

Another structural type based on the M_2E_2 tetrahedron is the bicapped derivative 74 (Scheme 7.40) which is known for both Sb and Bi. In the

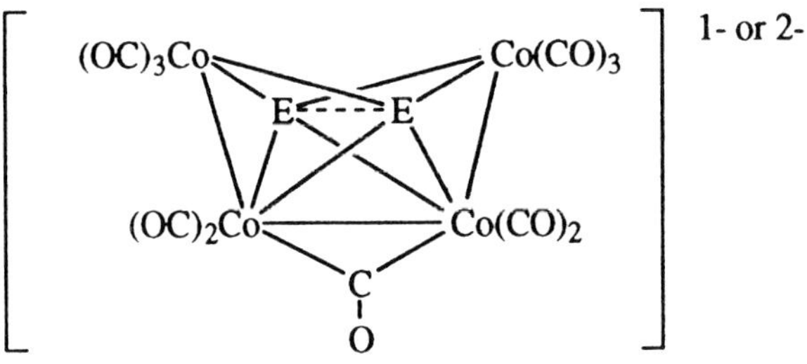

74

Scheme 7.40 Bi-capped derivative of the M_2E_2 tetrahedron (known for E = Sb, Bi).

case of antimony, the compound is obtained as a mixture of 1– and 2– ions formed upon the addition of antimony halides to $[Co(CO)_4]^-$ [167, 168]. The E = Bi monoanionic compound is obtained from the pyrolysis of $[BiCo_3(CO)_{12}]$ or via reduction of $[BiCo_3(CO)_9]$ [168, 169]. It may be reduced to a dianion as well but this compound is not as stable as the antimony derivative. These compounds are formally electron rich by the Wade–Mingos counting formalisms. A molecular orbital analysis indicated that the mismatch between orbitals on E and M creates nonbonding orbitals which accommodate the extra electrons.

7.4.2 Complexes with E_3 fragments

Compounds which contain an intact E_3 unit are well-represented although still less common than those with an E_2 unit. The prevalent structure for the E_3 group is the triangle which can be viewed as a fragment of the E_4 tetrahedron – the form adopted by white phosphorus and gaseous arsenic. The E_4 tetrahedron is an electron precise molecule with each E atom providing three electrons for localized edge bonding. It is reasonable to expect that each E of the unit could be replaced by a suitable three-electron donor metal fragment. Thus the entire series of Co and As compounds (structures 75–79, Scheme 7.41) are known in some form [note: the structure of $Co_4(CO)_{12}$ (structure 79) is actually more complicated, with bridging carbonyl ligands, but is drawn to emphasize the relationship to the other molecules in the series]. As noted above (Section 7.3.4), structure 78 spontaneously trimerizes (Scheme 7.18) but the basic structural Co_3As unit is retained in the $[\{Co_3(CO)_8As\}_3]$ molecule. Complex 76 is obtained from the reaction of $Co_2(CO)_8$ with $(AsMe)_5$ at 200°C under 100 atm CO [170].

When $[\{Cp^*Mo(CO)_2\}_2]$ is treated with As_4S_4, one of the products is $[Cp^*Mo(CO)_2As_3]$ [171]. Analogous Cr complexes arise from reaction of

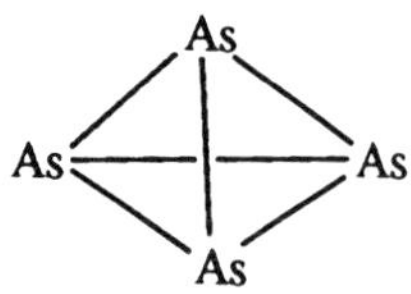

75

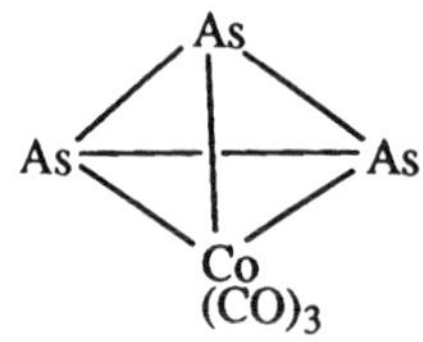

76

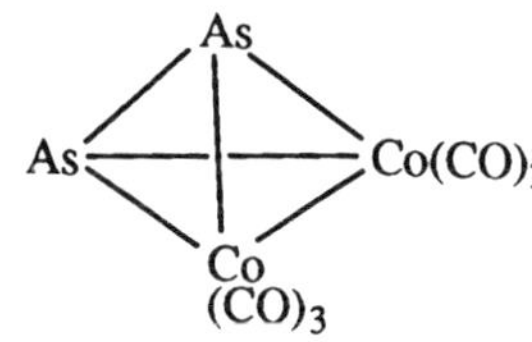

77

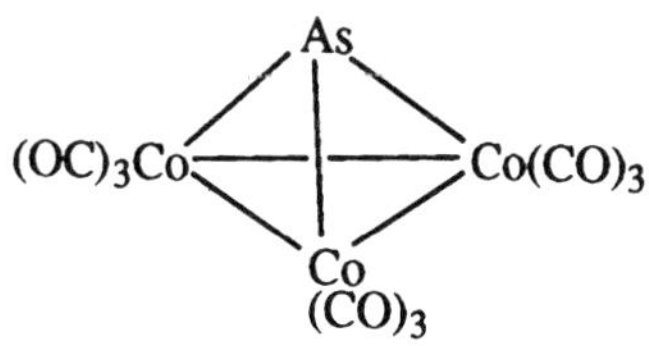

78

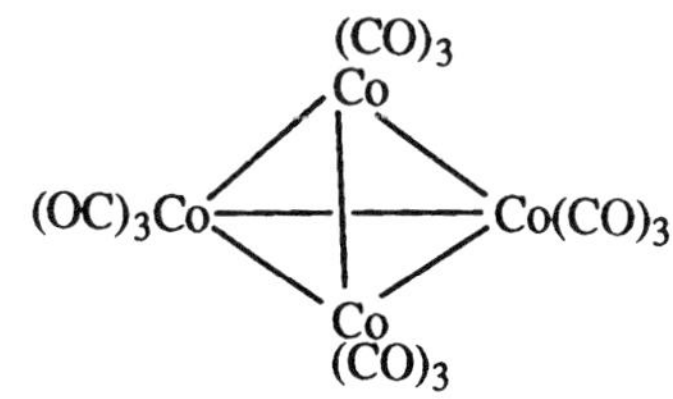

79

Scheme 7.41 The series of Co and As compounds with the E_4 tetrahedral structure.

$[\{(C_5Me_4R)Cr(CO)_2\}_2]$ (R = Me, Et) with yellow arsenic [172]. As with the Co complex, the As–As distances in the Mo and Cr complexes are essentially equal and shorter than that found in As_4.

In $[Co_4(\mu_3\text{-}As)_3(As_3)(PPh_3)_4]$ (structure 80, Scheme 7.42) an As_3 triangle is fused onto a face of a Co_4 tetrahedron in such a fashion to form a

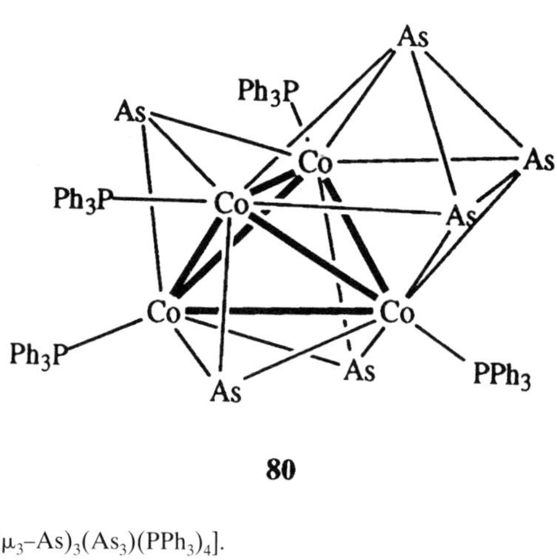

80

Scheme 7.42 $Co_4(\mu_3\text{–}As)_3(As_3)(PPh_3)_4]$.

Co_3As_3 octahedral unit [173]. The other three faces of the Co_4 tetrahedron are capped by μ_3-As atoms. Here the As–As distances are comparable to those found in As_4. Yellow arsenic reacts with $M[BF_4]_2$ (M = Co, Ni) in the presence of $MeC(CH_2CH_2PPh_2)_3$(triphos) to produce $[(triphos)MAs_3M(triphos)]^{n+}$ (n = 1, 2) [174, 175]. These molecules have a trigonal bipyramidal arrangement of three As and two metal atoms with the metals in apical positions.

Two other clusters have been reported in which an E_3 unit is coordinated to three metal centres. These include $[(C_5Me_4R)_3Ru_4(\mu_3,\eta^3\text{-}As_3)(\eta^3\text{-}As_3)(\mu_3\text{-}As)_3]$ (structure 81, Scheme 7.43) [162] and $[Bi_3\{Fe(CO)_3\}_3$

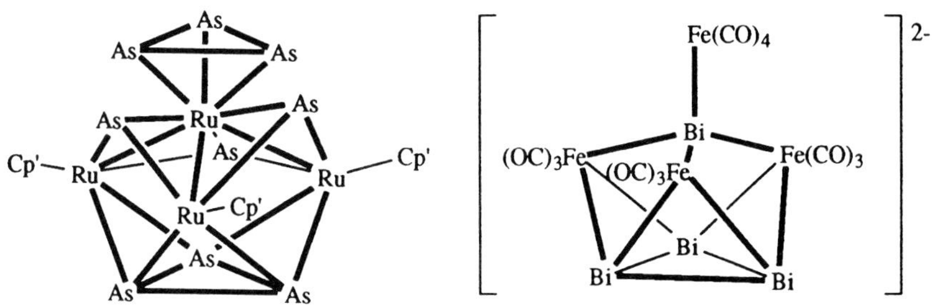

81

82

Scheme 7.43 Examples of clusters with E_3 units coordinated to three metal centres.

$\{BiFe(CO)_4\}]^{2-}$ (structure 82) [176, 177]. The ruthenium compound arises from the reaction of $[\{(C_5Me_4R)Ru(CO)_2\}_2]$ with As_4 whereas the bismuth compound is obtained when $[BiFe_3(CO)_{10}]^-$ is placed under pressures of CO in CH_2Cl_2. Compound 81 has two As_3 fragments, one of which is similar to the tetrahedral E_3M compounds already mentioned. The other fragment is attached to three $Ru(C_5Me_4R)$ groups such that each Ru is bridging between two of the As atoms of the As_3 group. It is this last arrangement which is very similar to the array found in structure 82 where the Bi_3 triangle is similarly attached to three $Fe(CO)_3$ groups. These structures bear some resemblance to the Zintl ions E_7^{3-} (E = P, As, Sb) which have structure 83 (shown in Scheme 7.44). This picture is easiest to see for structure 82 in which one replaces each of the μ_2-E atoms of structure 83 with an $Fe(CO)_3$ unit and then allows that group to twist to form the bridging structure.

One compound containing an As_3 group which is not a triangular array is known. It is $[(CpMo)_2As_3(AsS)]$ (structure 84, Scheme 7.45) arising from the reaction of $[\{CpMo(CO)_3\}_2]$ with $(MeAsS)_n$ (n = 3, 4) [178]. The As–S unit is statistically disordered in the crystal lattice. The As–As bond distance [2.374(1) Å] is short and has been described as allylic. It is

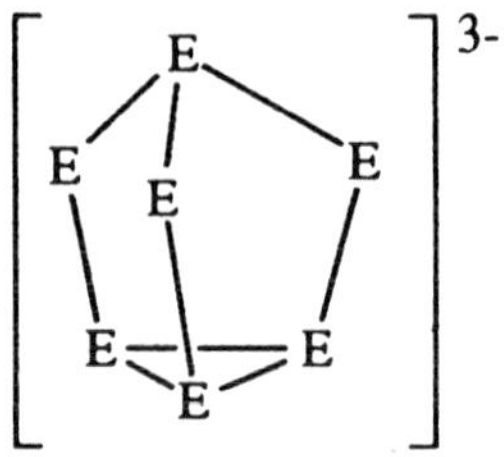

83

Scheme 7.44 A Zintl ion.

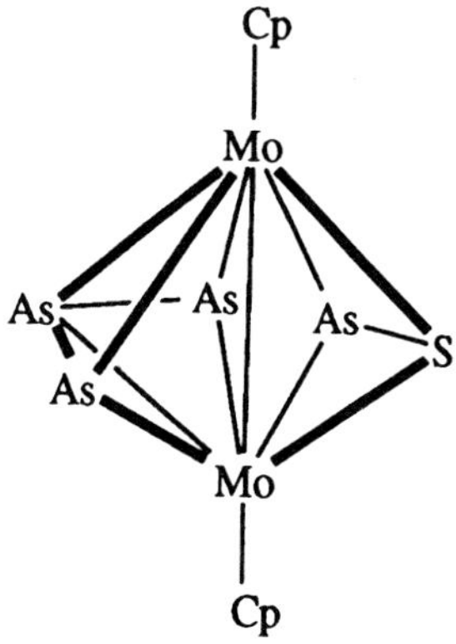

84

Scheme 7.45 Example of an As_3 group that is not in a triangular array.

comparable in length to the short As–As distances in the M_2E_2 tetrahedral compounds described earlier. Another compound with which it is similar is $[(CpMo)_2As_5]$ discussed below. This compound fits into a general classification known as triple-decker sandwich complexes. The three layers are composed of the two Cp rings and the As_3,AsS atoms in this case.

7.4.3 Complexes with E_x fragments ($x \geqslant 4$)

Two molecules with chains of E_4 or E_5 have been reported. From reaction of $Fe(CO)_5$ with $(AsMe)_5$ the compound $[Fe_2(CO)_6(AsMe)_4]$ is obtained (structure 85, Scheme 7.46) [68]. It is essentially the same as structural type 14 and possesses an Fe–Fe bond. The As–As distances are not uniform, with the central As–As bond being somewhat shorter than

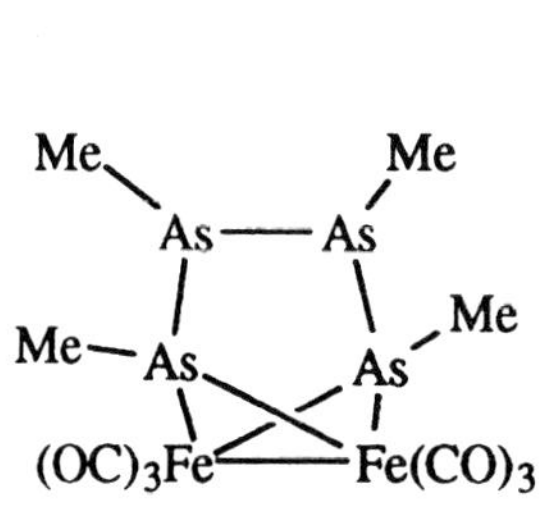

85

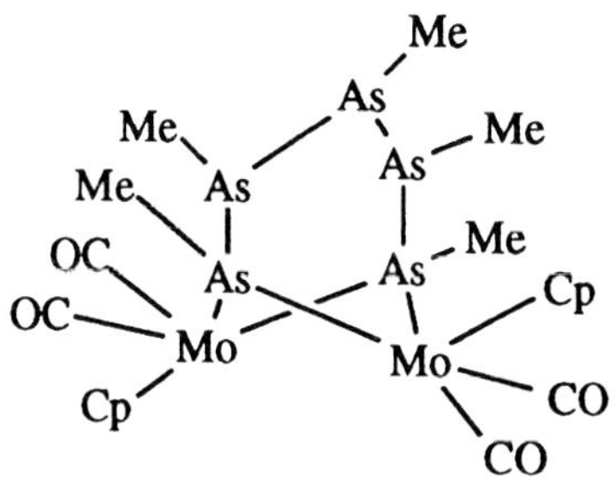

86

Scheme 7.46 Two molecules with chains of E_4 (85) and E_5 (86).

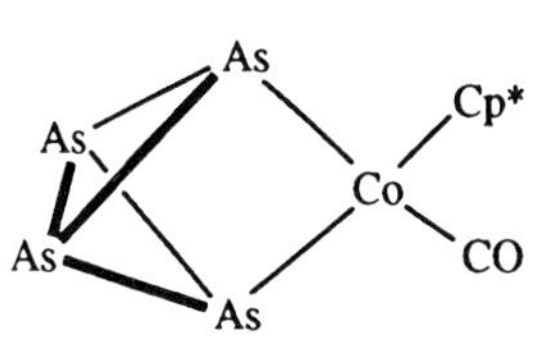

87

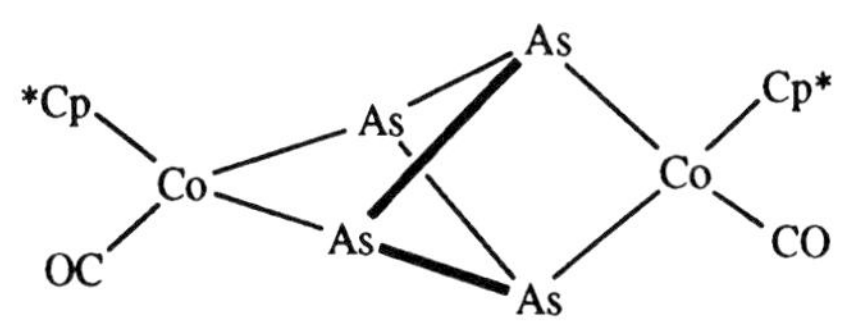

88

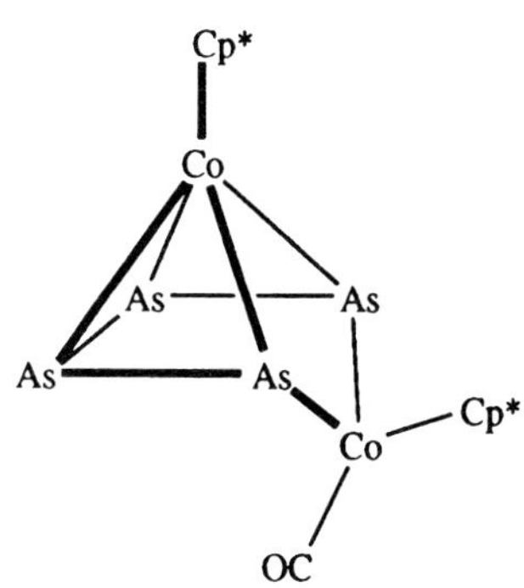

89

Scheme 7.47 Molecules with naked As_4 units.

those to the iron-bonded As atoms. A pentaarsenic chain is retained upon reaction of [{CpMo(CO)$_3$}$_2$] with (AsMe)$_5$ to give structure 86 in which there is no Mo–Mo bond [179].

Three molecules have been reported which have 'naked' As_4 units. They are produced in the reaction of [{Cp*Co(CO)$_2$}$_2$] with As_4 in refluxing

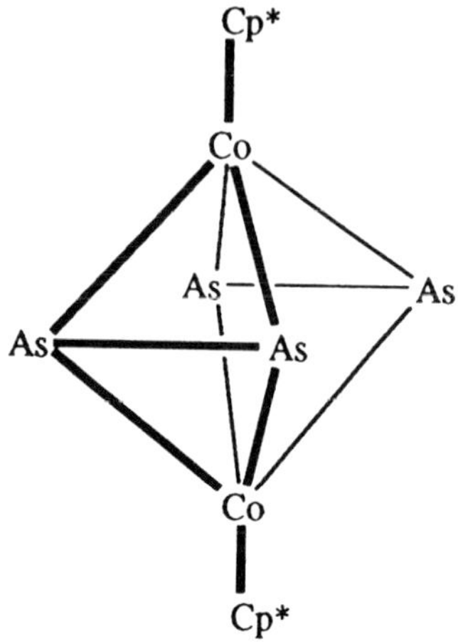

90

Scheme 7.48 Compound produced by the loss of two CO ligands from structure 88.

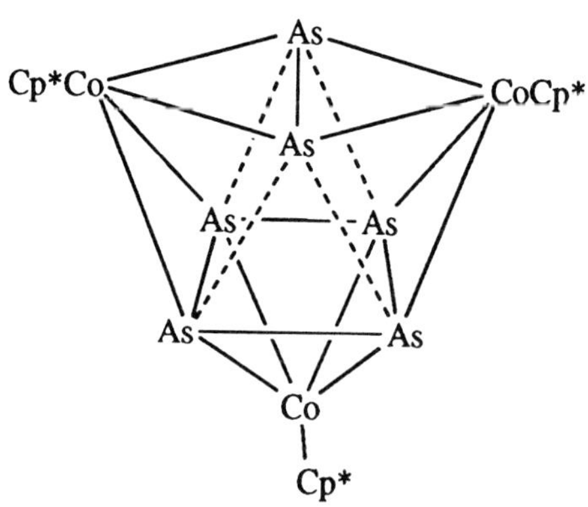

91

Scheme 7.49 Compound with a planar As_4 square and an As_2 unit.

toluene [71]. The first two (structures 87 and 88, Scheme 7.47) can be viewed as a metal fragment insertion into one or two of the As–As bonds of the original As_4 tetrahedron. The third (structure 89) is obtained via loss of CO from structure 88. Loss of one more CO ligand results in the compound [{Cp*Co}$_2${As_2}$_2$] (structure 90, Scheme 7.48) [180]. In the same system, the compound [{Cp*Co}$_3${As_4}{As_2}] (structure 91, Scheme 7.49)

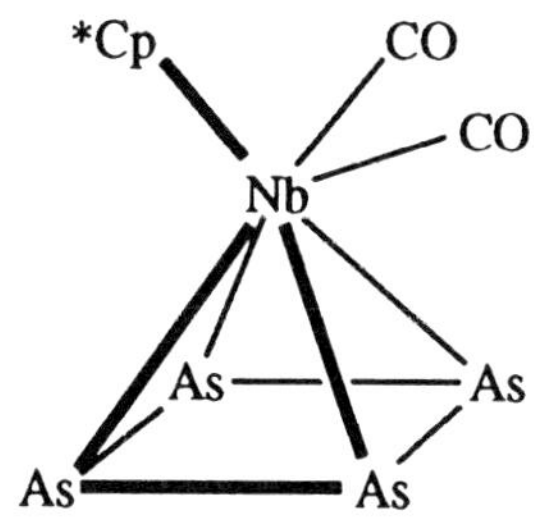

92

Scheme 7.50 Product of the reaction of $Cp^*Nb(CO)_4]$ with As_4.

contains both a planar As_4 square and an As_2 unit. There are weak interactions between the two As fragments so that the unit may be viewed as deriving from As_6, which has been found in other molecules of this same system.

Reaction of $[Cp^*Nb(CO)_4]$ with As_4 yields $[Cp^*(CO)_2NbAs_4]$ (structure 92, Scheme 7.50), a cluster with a square pyramidal arrangment of atoms having the niobium in the apical position [181]. The As–As distances (2.38 Å, average) are similar to those found in other sandwich-type complexes. The planar As_4 unit may be viewed as derived from cyclotetra-arsadiene and so the As–As bond distances may be expected to show some multiple bond character as seen for the M_2E_2 tetrahedral compounds described earlier.

Compounds which contain an As_5 unit are also known. The molecules to date show a relatively planar pentagonal As_5 ring system capped on one or both sides by ML_n fragments (structures 93 and 94, Scheme 7.51). Analogues in phosphorus chemistry are known but antimony and bismuth examples are still lacking. The reported compounds include $[(C_5Me_4R)FeAs_5]$ (R = Me, Et) [182], $[Cp^*FeAs_5FeCp]^+$ [182], $[(C_5Me_4R)RuAs_5]$ [183] and $[(Cp^*)MAs_5M'(CO)_3]$ (M = Fe, Ru′, M′ = Cr, Mo, W) [184]. This system is accessed by direct reaction of As_4 with $[\{(C_5Me_4R)M(CO)_2\}_2]$. Once formed, the L_nMAs_5 complexes will readily add the second metal fragment. For example, the $M'(CO)_3$ complexes are obtained by reaction of $[(C_5Me_4R)MAs_5]$ with $[M'(CO)_3(NCMe)_3]$, and $[Cp^*FeAs_5FeCp]$ was obtained from Cp*Fe and $[CpFe(C_6H_6)]PF_6$ [153]. Heating $[\{CpMo(CO)_3\}_2]$ with $(AsMe)_5$ and toluene in a sealed tube at 190°C affords $[CpMoAs_6MoCp]$ directly [87]. There is a Mo–Mo bonding interaction through the As_6 ring. The As_5 unit here is essentially planar but, unlike the other As_5 molecules, the As–As bond distances are very asymmetric suggesting the segregation of the As atoms into As_3 and As_2 fragments.

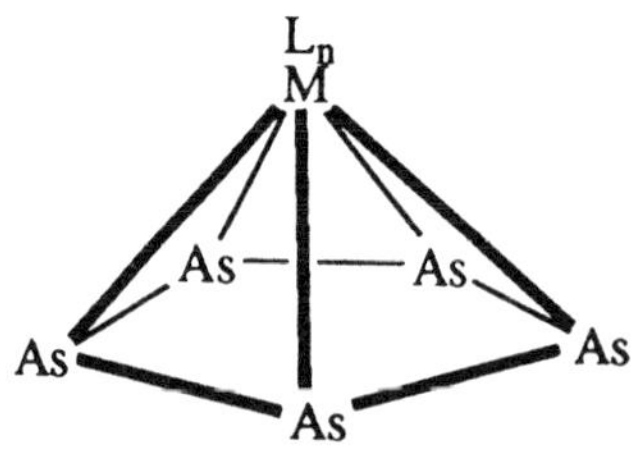

93

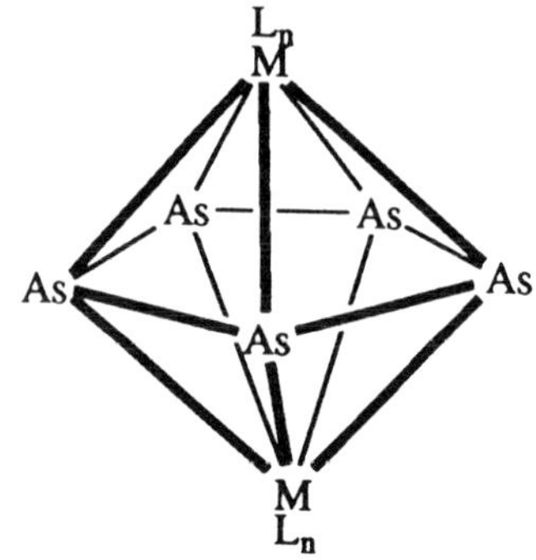

94

Scheme 7.51 Examples of compounds with an As_6 unit.

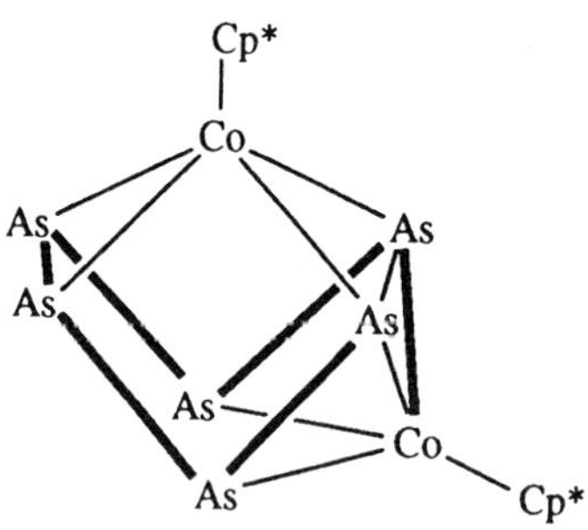

95

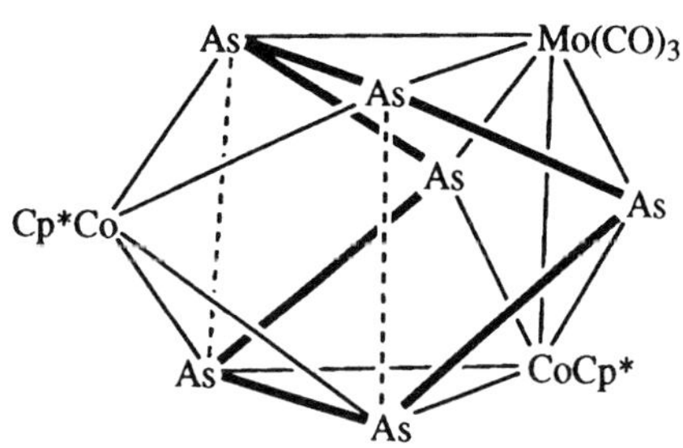

96

Scheme 7.52 Examples of compounds with an As_6 unit.

This is similar to the situation in [CpMo(As_3)(AsS)MoCp] described earlier [178].

The As_6 unit is represented in [{Cp*Co}$_2$$As_6$] (structure 95, Scheme 7.52) and [{Cp*Co}$_2${Mo(CO)$_3$}As_6] (structure 96) [185]. The six arsenic atoms lie in a folded array rather than in a plane as do the As_5 units discussed above. Structures 95 and 96 can be viewed as capped trigonal prisms having 11 skeletal electron pairs. The latter compound has been compared to Bi_9^{5+}, in which the structural distortions are similar. The ^1H NMR spectra of these two compounds indicate that they are fluxional, as evidenced by equivalent Cp* rings. This phenomenon is common for nonanuclear clusters. A structure intermediate between an As_5 and As_6

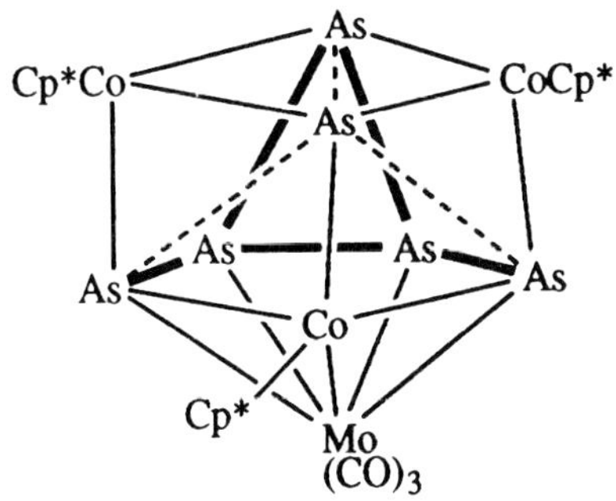

97

Scheme 7.53 Compound with the rare sphenocorona structure.

unit is seen in [$\{Cp^*Co\}_3\{Mo(CO)_3\}As_6$] (structure 97, Scheme 7.53) in which there is one arsenic atom whose distances to the other arsenic atoms are long. It has been analysed as having the rare sphenocorona structure.

Direct metallation of the Zintl ions E_7^{3-} (E = P, As, Sb) has been reported. The parent ion (structure 83) adds $M(CO)_3$ fragments (M = Cr, Mo, W) {from [$M(CO)_3$(arene)] or [$Mo(CO)_4$(bipyridine)]} in a μ_4-M fashion as shown in structure 98, Scheme 7.54 [186, 187]. One of the E–E bonds of the starting E_7^{3-} ion is broken upon $M(CO)_3$ addition. A molecular orbital analysis of these complexes indicates that the binding of the metal to the E_7 unit induces π interactions between the E atoms which are bound to the metal centre. Protonation of these anions occurs at the non-metallated μ-E group (structure 99). Other metal fragments can also be added to give [$\{L_2M'(CO)_3\}E_7M(CO)_3]^{3-}$ (L_2 = ethylenediamine or $(DMF)_2$] (structure 100) which shows the additional metal unit ligated at the same site which undergoes protonation.

A complex rearrangement takes place upon reaction of K_3Sb_7 with [$Ni(CO)_2(PPh_3)_2$] in ethylenediamine in the presence of 2,2,2-crypt (structure 101, Scheme 7.55) [188]. The molecule possesses 12 skeletal electron pairs which corresponds to a *nido*-hexadecahedral structure (102) according to conventional electron counting formalisms. The actual structure, however, is not derivable from structure 102 as it contains neither a six-connected vertex nor a hexagonal face, one of which would be expected to be present upon removing a single vertex from the hexadecahedron. An alternate view bases the cluster framework upon an electron precise '$[Sb_7Ni(CO)]^{3-}$' cuneane (structure 103) in which the pentagonal Sb_4Ni faces are capped with Ni(CO) fragments. Addition of a neutral Ni(CO) unit to the cluster does not change the skeletal electron pair count.

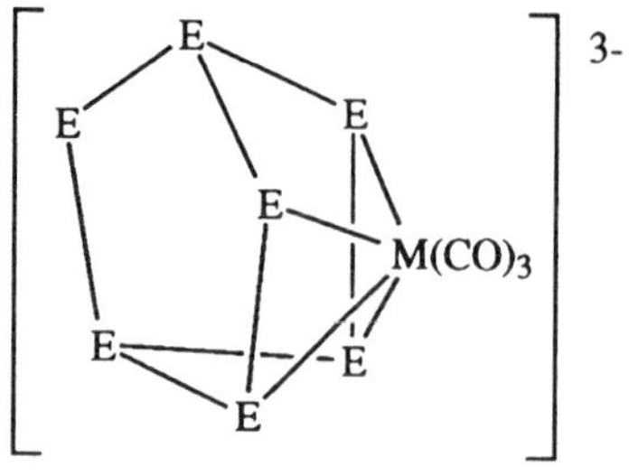

98

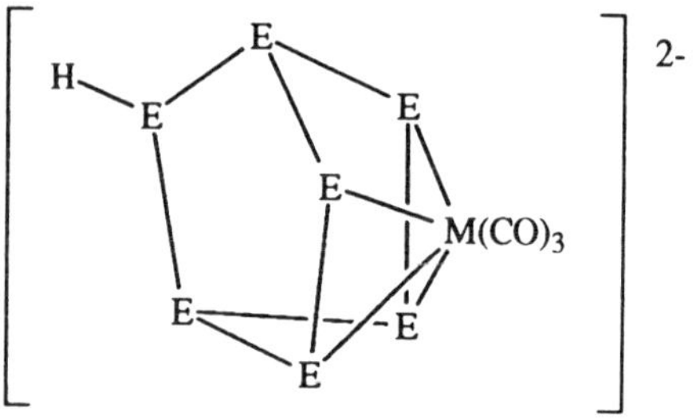

99

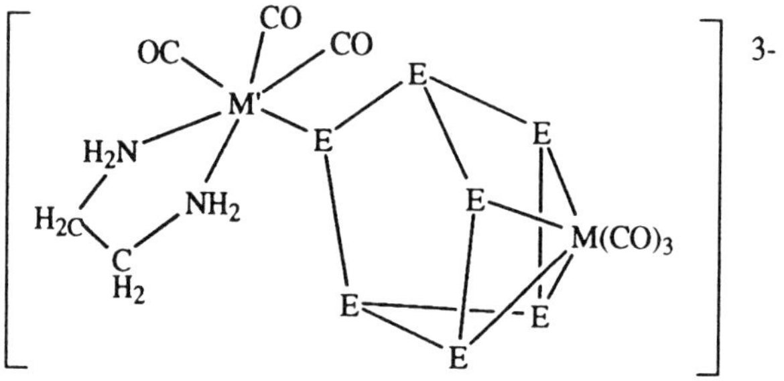

100

Scheme 7.54 A metallated Zintl ion (98) [the parent ion is structure 83 (Scheme 7.44)] and derivatives (99 and 100).

7.5 Concluding comments and future directions

It is clear that there is considerable flexibility in the formation of main group element–transition metal compounds. As molecules become more complicated current theories for rationalizing the structural patterns and bonding schemes become less adequate. Although nuclearities approaching those in the related metal–chalcogenide system have not yet been produced for E = As, Sb, Bi it is clear that results are pushing in that direction. At sufficiently high nuclearities, relationships (and perhaps even applications) to solid state materials can be envisioned. Many of the compounds discussed in this chapter have analogues in chalcogenide derivatives (and also those from groups 13 and 14), but numerous M–E combinations remain to be explored as well as variations in the appended

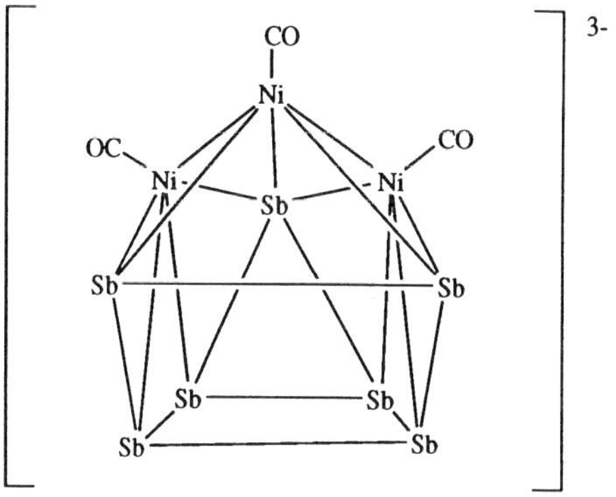

101

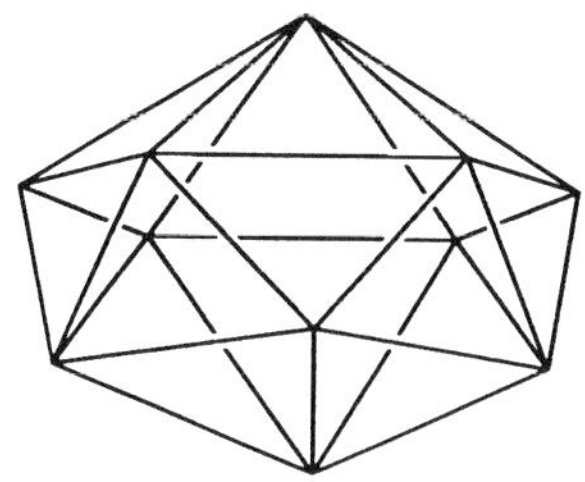

102

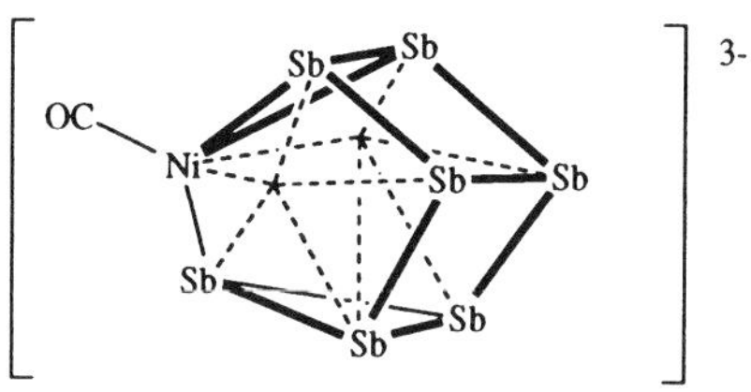

103

Scheme 7.55 The product of the reaction of K_3Sb_7 with $[Ni(CO)_2(PPH_3)_2]$ in ethylenediamine in the presence of 2,2,2-crypt (101) has 12 skeletal electron pairs which corresponds to a *nido*-hexadecahedral structure (102); however, it may be based on the electron precise '$[Sb_7Ni(CO)]3^-$' cuneane (103).

ligands. Given the progress in the field to date it is likely that surprises are still in store for those working in this field.

One obvious area that needs to be addressed is the reactivity of these complexes. There is some indication that the presence of main group elements in transition metal clusters helps to stabilize the metal framework. This is a considerable limitation for the homonuclear metal carbonyl clusters which often fragment into mononuclear complexes in reactions involving CO pressure. Of interest is catalysis under reducing conditions (i.e. H_2 and/or CO) for reactions such as hydrogenation, carbonylation (e.g. hydroformylation), water–gas shift chemistry, and so on. To date such reactions have not been extensively explored for EM compounds in general, nor in fact even the stoichiometric interactions between EM clusters and organic substrates. It will be particularly interesting to see what influence the presence of E has on the bonding modes of organic molecules to transition metals in cluster compounds. There is mounting evidence that E may in fact be very much involved in the cluster reactivity and thus constitute a cluster vertex as opposed to being simply a spectator ligand.

References

1. Whitmire, K. H. (1991) *J. Cluster Science*, **2,** 231.
2. Scheer, M. and Herrmann, E. (1990) *Z. Chem.*, **30,** 41.
3. Compton, N. A., Errington, J. and Norman, N. C. (1990) *Adv. Organomet. Chem.*, **31,** 91.
4. Scherer, O. J. (1990) *Angew. Chem. Int. Ed. Eng.*, **29,** 1104; (1990) *Angew. Chem.*, **102,** 1137.
5. Whitmire, K. H. (1989) *Clusters of Metals and Nonmetals*, in *Rings, Clusters and Polymers of Main Group and Transition Elements*, (ed. H. W. Roesky), Elsevier, New York, Amsterdam, p. 503.
6. Fehlner, T. P. (1988) *Comments Inorg. Chem.*, **7,** 307.
7. Whitmire, K. H. (1988) *J. Coord. Chem. B*, **17,** 95.
8. Norman, N. C. (1988) *Chem. Soc. Rev.*, **17,** 269.
9. Fenske, D., Ohmer, J., Hachgenei, J. and Merzweiler, K. (1988) *Angew. Chem. Int. Ed. Eng.*, **27,** 1277; (1988) *Angew. Chem.*, **100,** 1300.
10. Huttner, G. (1986) *Pure Appl. Chem.*, **58,** 585.
11. Herrmann, W. A. (1986) *Angew. Chem. Int. Ed. Eng.*, **25,** 56; (1986) *Angew. Chem.*, **98,** 57.
12. Scherer, O. J. (1985) *Angew. Chem. Int. Ed. Eng.*, **24,** 924; (1985) *Angew. Chem.*, **97,** 905.
13. Nicholls, J. N. (1984) *Polyhedron*, **3,** 1307.
14. Vahrenkamp, H. (1983) *Pure Appl. Chem.*, **22,** 169.
15. Vahrenkamp, H. (1983) *Adv. Organomet. Chem.*, **22,** 169.
16. Schmid, G. (1978) *Angew. Chem. Int. Ed. Eng.*, **17,** 392; (1978) *Angew. Chem.*, **17,** 392.
17. Carty, A. J., Taylor, N. J., Coleman, A. W., Lappert, M. F. (1979) *J. Chem. Soc., Chem. Commun.*, 639.
18. Schumann H. and Eguren, L. (1991) *J. Organomet. Chem.*, **403,** 183.
19. Huttner G. and Schmid, H.-G. (1975) *Angew. Chem. Int. Ed. Eng.*, **14,** 433; (1975) *Angew. Chem.*, **87,** 454.
20. Calderazzo, F., Juris, A., Poli, R. and Ungari, F. (1991) *Inorg. Chem.*, **30,** 1274.
21. Hayter, R. G. (1964) *J. Am. Chem. Soc.*, **86,** 823.
22. Calderazzo, F., Poli, R. and Pelizzi, G. (1984) *J. Chem. Soc., Dalton Trans., 2535.*
23. Malisch, W., Kuhn, M., Albert, W. and Rößner, H. (1980) *Chem. Ber.*, **113,** 3318.
24. Panster, P. and Malisch, M. (1977) *J. Organomet. Chem.*, **134,** C32.
25. Cassidy, J. M. and Whitmire, K. H. (1991) *Inorg. Chem.*, **30,** 2788.
26. Bochkarev, M. N., Razuvaev, G. A., Zakharov, L. N. and Struchov, Yu. T. (1980) *J. Organomet. Chem.*, **199,** 205.
27. Röttinger, E., Trenkle, A., Müller R. and Vahrenkamp, H. (1980) *Chem. Ber.*, **113,** 1280.
28. Ehrl, W. and Vahrenkamp, H. (1971) *Chem. Ber.*, **104,** 3261.
29. Ehrl, W. and Vahrenkamp, H. (1973) *Chem. Ber.*, **106,** 2550.
30. Ehrl, W. and Vahrenkamp, H. (1973) *Chem. Ber.*, **106,** 2556.
31. Deck, W. and Vahrenkamp, H. (1991) *Z. Anorg. Allg. Chem.*, **598/599,** 83.
32. Ehrl, W. and Vahrenkamp, H. (1973) *Chem. Ber.*, **106,** 2563.
33. Vahrenkamp, H. (1973) *Chem. Ber.*, **106,** 2570.
34. Langenbach, H. J., Keller, E. and Vahrenkamp, H. (1979) *J. Organomet. Chem.*, **171,** 259.

35. Mayr, A., Ehrl, W. and Vahrenkamp, H. (1974) *Chem. Ber.*, **107,** 3860.
36. Müller, R. and Vahrenkamp, H. (1977) *Chem. Ber.*, **110,** 3910.
37. Börner, M and Vahrenkamp, H. (1977) *J. Chem. Res.* (S), 74.
38. Börner, M. and Vahrenkamp, H. (1977) *J. Chem. Res.* (M), 801.
39. Börner, M. and Vahrenkamp, H. (1978) *Chem. Ber.*, **111,** 2190.
40. Langenbach, H.-J. and Vahrenkamp, H. (1977) *Chem. Ber.*, **110,** 1195.
41. Vahrenkamp, H. (1974) *Chem. Ber.*, **107,** 3867.
42. Röttinger, E. and Vahrenkamp, H. (1978) *Chem. Ber.*, **111,** 2199.
43. Trenkel, A. and Vahrenkamp, H. (1978) *J. Organomet. Chem.*, **155,** C51.
44. Vahrenkamp, H. and Keller, E. (1979) *Chem. Ber.*, **112,** 1991.
45. Keller, E., Trenkle, A. and Vahrenkamp, H. (1977) *Chem. Ber.*, **110,** 441.
46. Hirt, J., Börner M. and Vahrenkamp, H. (1977) *Chem. Ber.*, **110,** 1187.
47. Langenbach, H. J., Keller, E. and Vahrenkamp, H. (1980) *J. Organomet. Chem.*, **191,** 95.
48. Langenbach, H.-J. and Vahrenkamp, H. (1977) *Chem. Ber.*, **110,** 1206.
49. Keller, E. and Vahrenkamp, H. (1977) *Chem. Ber.* **110,** 430.
50. Kaul, H.-A., Greissinger, D., Luksza, M. and Malisch, W. (1982) *J. Organomet. Chem.*, **228,** C29.
51. Malisch, W. and Kuhn, M. (1974) Angew. *Chem. Int. Ed. Eng.*, **13,** 84; (1974) *Angew. Chem.*, **86,** 51.
52. Malisch, W. and Panster, P. (1975) *J. Organomet. Chem.*, **99,** 421.
53. Bachman, R. E. and Whitmire, K. H. (1995) *Inorg. Chem.*, **34,** 1542.
54. Baay, Y. L. and MacDiarmid, A. G. (1967) *Inorg. Nucl. Chem. Lett.*, **3,** 159.
55. von Seyerl, J., Sigwarth, B. and Huttner, G. (1981) *Chem. Ber.*, **114,** 1407.
56. von Seyerl, J., Sigwarth, B. and Huttner, G. (1981) *Chem. Ber.*, **114,** 727.
57. Davies, S. J., Compton, N. A., Huttner, G. *et al.* (1991) *Chem. Ber.*, **124,** 2731.
58. Clegg, W., Compton, N. A., Errington, R. J., *et al.* (1992) *J. Chem. Soc., Dalton Trans.*, 3515.
59. Errington, R. J. Fisher, G. A., Norman, N. C. *et al.* (1994) *Z. Anorg. Allg. Chem.*, **620,** 457.
60. Einstein., F. W. B. and Jones, R. D. G. (1973) *Inorg. Chem.*, **12,** 1690.
61. Cane, D. J., Forbes, E. J. and Hamor, T. (1976) *J. Organomet. Chem.*, **117,** C101.
62. Campana, C. F., Sinclair, J. D. and Dahl, L. F. (1977) *J. Organomet. Chem.*, **127,** 223.
63. Winter, A., Zsolnai L. and Huttner, G. (1983) *J. Organomet. Chem.*, **250,** 409.
64. Cullen, W. R. and Hayter, R. G. (1964) *J. Am. Chem. Soc.*, **86,** 1030.
65. Chatt J. and Thornton, D. A. (1964) *J. Chem. Soc.*, 1005.
66. Cooke, M., Green, M. and Kirkpatrick, D. (1968) *J. Chem. Soc. (A)*, 1507.
67. Herberhold, M., Schamel, K., Herrmann, G. *et al.* (1988) *Z. Anorg. Allg. Chem.*, **562,** 49.
68. Gatehouse, B. M. (1969) Chem. Commun., 948.
69. Eveland, J. R., Saillard J.-Y. and Whitmire, K. H. submitted for publication.
70. Brunner, H., Poll, L., Wachter, J. and Nuber, B. (1994) *J. Organomet. Chem.*, **471,** 117.
71. Henderson, P., Rossignoli, M., Burns, R. C. *et al.* (1994) *J. Chem. Soc., Dalton Trans.*, 1641.
72. Bachman, R. E., Miller S. K. and Whitmire, K. H. (1994) *Inorg. Chem.*, **33,** 2075.
73. Bachman, R. E., Miller S. K. and Whitmire, K. H. (1995) *Organometallics*, **14,** 796.
74. Whitmire, K. H., Lagrone, C. B., Churchill, M. R. *et al.* (1984) *Inorg. Chem.*, **23,** 4227.
75. Churchill, M. R., Fettinger, J. C., Whitmire, K. H. and Lagrone, C. B. (1986) *J. Organomet. Chem.*, **303,** 99.
76. Luo, S. and Whitmire, K. H. (1989) *Inorg. Chem.*, **28,** 1424.
77. Shieh, M., Liou, Y., Peng, S.-M. and Lee, G.-H. (1993) *Inorg. Chem.*, **32,** 2212.
78. Shieh, M., Liou, Y. and Jeng, B.-W. (1993) *Organometallics*, **12,** 4926.
79. Ferrer, M., Rossell, O., Seco, M. and Braunstein, P. (1989) *J. Organomet. Chem.*, **364,** C5.
80. Lorenz, L.-P., Effinger, G. and Hiller, H. (1990) *Chem. Ber.*, **123,** 251.
81. Cullen, W. R., Patmore, D. J. and Sams, J. R. (1973) *Inorg. Chem.*, **12,** 867.
82. Cullen, W. R., Patmore, D. J., Sams, J. R. *et al.* (1971) *J. Chem. Soc., Chem. Commun.*, 952.

83. Trinh-Toan, and Dahl, L. F. (1971) *J. Am. Chem. Soc.*, **93,** 2654.
84. Clegg, W., Compton, N. A., Errington, R. J. *et al.* (1990) *J. Chem. Soc., Dalton Trans.*, 2375.
85. Röttinger, E. and Vahrenkamp, H. (1978) *Angew. Chem. Int. Ed. Eng.*, **17,** 273; (1978) *Angew. Chem.* **90,** 294.
86. Röttinger E. and Vahrenkamp, H. (1981) *J. Organomet. Chem.*, **213,** 1.
87. Foust A. S. and Dahl, L. F. (1970) *J. Am. Chem. Soc.*, **92,** 7337.
88. Ciani, G., Moret, M., Fumagalli, A. and Martinengo, S. (1989) *J. Organomet. Chem.*, **362,** 291.
89. Rheingold, A. L., Foley, M. J. and Sullivan, P. J. (1982) *J. Am. Chem. Soc.*, **104,** 4727.
90. Johnson, B. F. G., Lewis, J., Whitton, A. J. and Bott, S. G. (1990) *J. Organomet. Chem.*, **389,** 129.
91. Wallis, J. M., Müller, G. and Schmidbaur, H. (1987) *J. Organomet. Chem.*, **325,** 159.
92. Clegg, W., Compton, N. A., Errington, R. J. and Norman, N. C. (1988) *J. Chem. Soc, Dalton Trans.*, 1671.
93. Clegg, W., Compton, N. A., Errington, R. J. and Norman, N. C. (1987) *Polyhedron*, **6,** 2031.
94. Leigh, J. S. and Whitmire, K. H. (1988) *Angew. Chem. Int. Ed. Eng.*, **27,** 396; (1988) *Angew. Chem.*, **100,** 399.
95. Martinengo, S., Fumagalli, A., Ciani, G. and Moret, M. (1988) *J. Organomet. Chem.*, **347,** 413.
96. Grossbruchhaus V. and Rehder, D. (1990) *Inorg. Chim. Acta*, **172,** 141.
97. Arnold, F. P. Jr., Ridge, D. P. and Rheingold, A. L. (1995) *J. Am. Chem. Soc.*, **117,** 4427.
98. Taylor, T. G., Willey, K. F., Bishop, M. B. and Duncan, M. A. (1990) *J. Phys. Chem.*, **94,** 8016.
99. von Seyerl, J. and Huttner, G. (1980) *J. Organomet. Chem.*, **193,** 207.
100. Dietrich, W.-R. and Vahrenkamp, H. (1985) *J. Chem. Res. (S)*, 200.
101. DiMaio, A.-J., Bitterwolf, T. E. and Rheingold, A. L. (1990) *Organometallics*, **9,** 551.
102. Whitmire, K. H., Shieh, M. and Cassidy, J. (1989) *Inorg. Chem.*, **28,** 3164.
103. Weber, U., Huttner, G., Scheidsteger, O. and Zsolnai, L. (1985) *J. Organomet. Chem.*, **289,** 357.
104. Ziegler, M. L., Blechsmitt, K., Nuber B. and Zahn, T. (1988) *Chem. Ber.*, **121,** 159.
105. Gorzellik, M., Nuber, B., Bohn, T. and Ziegler, M. L. (1992) *J. Organomet. Chem.*, **429,** 173.
106. Huttner, G., Mohr, G., Pritzlaff, B. *et al.* (1982) *Chem. Ber.*, **115,** 2044.
107. Arnold, L. J., Mackay, K. M. and Nicholson, B. K. (1990) *J. Organomet. Chem.*, **387,** 197. Campana, C. F. and Dahl, L. F. (1977) *J. Organomet. Chem.*, **127,** 209.
108. Rheingold, A. L., Geib, S. J., Shieh, M. and Whitmire, K. H. (1987) *Inorg. Chem.*, **26,** 463.
109. Arif, A.M., Cowley, A.H. and Pakulski, M. (1987) *J. Chem. Soc., Chem. Comm.*, 622.
110. Johnson, B. F. G., Lewis, J., Raithby, P. R. and Whitton, A. J. (1988) *J. Chem. Soc., Chem. Commun.*, 401.
111. Johnson, B. F. G., Lewis, J. and Whitton, A. J. (1990) *J. Chem. Soc., Dalton Trans.*, 3129.
112. Whitmire, K. H., Leigh, J. S. and Gross, M. E. (1987) *J. Chem. Soc., Chem. Commun.*, 926.
113. Whitmire, K. H., Lagrone, C. B. and Rheingold, A.L. (1986) *Inorg. Chem.*, **25,** 2472.
114. Kruppa, W., Bläser, D., Boese, R. and Schmid, G. (1982) *Z. Naturforsch.*, **37B,** 209.
115. Wallis, J. M., Müller, G., Riede, J. and Schmidbaur, H. (1989) *J. Organomet. Chem.*, **369,** 165.
116. Hay, C. M., Johnson, B. F. G., Lewis, J. *et al.* (1988) *J. Chem. Soc., Dalton Trans.*, 2091.
117. Ang, H. G., Hay, C. M., Johnson, B. F. G. *et al.* (1987) *J. Organomet. Chem.*, **330,** C5.
118. Shieh, M., Mia, F.-D., Peng, S.-M. and Lee, G.-H. (1993) *Inorg. Chem.*, **32,** 2787.
119. van Hal, J. W., Whitmire, J. H., Zouchoune, B. *et al.* (1995) *Inorg. Chem.*, **34,** 5455.
120. Whitmire, K. H., Leigh, J. S., Luo, S. *et al.* (1988) *New J. Chem.*, **12,** 397.
121. Caballero, C., Nuber, B. and Ziegler, M. L. (1990) *J. Organomet. Chem.*, **386,** 209.

122. Delbaere, L. T. J., Kruczynski, L. J. and McBride, D. W. (1973) *J. Chem. Soc., Dalton Trans.*, 307.
123. Churchill, M. R., Fettinger, J. C. and Whitmire, K. H. (1985) *J. Organomet. Chem.*, **1284,** 13.
124. Lang, H., Huttner, G., Zsolnai, L. *et al.* (1986) *J. Organomet. Chem.*, **304,** 157.
125. Whitmire, K. H., Shieh, M., Lagrone, C. B. *et al.* (1987) *Inorg. Chem.*, **26,** 2798.
126. Whitmire, K. H., Rhaguveer, K. S., Churchill, M. R. *et al.* (1986) *J. Am. Chem. Soc.*, **108,** 2778.
127. Jacob, M. and Weiss, E. (1977) *J. Organomet. Chem.*, **131,** 263.
128 Huttner, G., Mohr, G., Frank, A. and Schubert, U. (1976) *J. Organomet. Chem.*, **118,** C73.
129. Lang, H., Huttner, G., Sigwarth, B. *et al.* (1986) *Organomet. Chem.*, **304,** 137.
130. Whitmire, K. H., Rhaguveer, K. S., Churchill, M. R. *et al.* (1986) *J. Am. Chem. Soc.* **108,** 2778.
131. Kahlal, S., Halet, J.-F., Saillard, J.-Y. and Whitmire, K. H. (1994) *J. Organomet. Chem.* **478,** 1.
132. Luo, S. and Whitmire, K. H. (1989) *J. Organomet. Chem.*, **376,** 297.
133. Fenske, D., Merzweiler, K. and Ohmer, J. (1988) *Angew. Chem. Int. Ed. Eng.*, **27,** 1512; (1988) *Angew. Chem.*, **100,** 1572.
134. Whitmire K. H. and Eveland, J. R. (1994) *J. Chem. Soc., Chem. Commun.*, 1335.
135. Rheingold A. L. and Sullivan, P. J. (1983) *J. Chem. Soc., Chem. Commun.*, 39.
136. DesEnfants, R. E. II, Gavney, J. A. Jr., Hayashi, R. K. *et al.* (1990) *J. Organomet. Chem.*, **383,** 543.
137. Albano, V. G., Demartin, F., Iapalucci, M. C. *et al.* (1990) *J. Chem. Soc., Chem. Commun.*, 547.
138. Albano, V. G., Demartin, F., Iapalucci, M. C. *et al.* (1991) *J. Chem. Soc., Dalton Trans.*, 739.
139. Albano, V. G., Demartin, F., Iapalucci, M. C. *et al.* (1992) *J. Chem. Soc., Dalton Trans.*, 497.
140. Rieck, D. F., Montag, R. A., McKechnie, T. F. and Dahl, L. F. (1986) *J. Am. Chem. Soc.*, **108,** 1330.
141. Longoni, G. (1990) *Pure Appl. Chem.*, **62,** 1183.
142. Vidal, J. L. (1981) *Inorg. Chem.*, **20,** 243.
143. Vidal, J. and Troup, J. M. (1981) *J. Organomet. Chem.*, **213,** 351.
144. Elmes, P. S., Leverett, P. and West, B. O. (1971) *J. Chem. Soc., Chem. Commun.*, 747.
145. Cowley, A. H., Norman, N. C., Pakulsi, M., Bricker, D. L. and Russell, D. H. (1985) *J. Am. Chem. Soc.*, **107,** 8211.
146. Cowley, A. H., Norman, N. C. and Pakulski, M. (1984) *J. Am. Chem. Soc.*, **106,** 6844.
147. Fenske, D. and Merzweiler, K. (1984) *Angew. Chem. Int. Ed. Eng.*, **23,** 635; (1984) *Angew. Chem.*, **96,** 600.
148. Song, L.-C., and Hu, Q.-M. (1991) *J. Organomet. Chem.*, **414,** 219.
149. Evans, W. J., Gonzales, S. L. and Ziller, J. W. (1991) *J. Am. Chem. Soc.*, **113,** 9880.
150. Herrmann, W. A., Koubouris, B., Schäfer, A. *et al.* (1985) *Chem. Ber.*, **118,** 2488.
151. Herrmann, W. A., Koumbouris, B., Zahn T. and Ziegler, M. L. (1984) *Angew. Chem. Int. Ed. Eng.*, **23,** 812; (1984) *Angew. Chem.*, **96,** 802.
152. Huttner, G., Sigwarth, B., Scheidsteger, O. *et al.* (1985) *Organometallics*, **4,** 326.
153. Scherer, O. J., Pfeiffer, K. and Wolmershäuser, G. (1992) *Chem. Ber.*, **125,** 2367.
154. Scherer, O. J., Sitzmann, H. and Wolmershäuser, G. (1986) *J. Organomet. Chem.*, **309,** 77.
155. DiMaio, A.-J. and Rheingold, A. L. (1987) *J. Chem. Soc., Chem. Commun.*, 404.
156. Huttner, G., Wever, U., Sigwarth, B. and Scheidsteger, O. (1982) *Angew. Chem. Int. Ed Eng.*, **21,** 215; (1982) *Angew. Chem.*, **94,** 210.
157. Sigwarth, B., Zsolnai, L., Berke H. and Huttner, G. (1982) *J. Organomet. Chem.*, **226,** C5.
158. Huttner, G., Weber, U. and Zsolnai, L. (1982) *Z. Naturforsch.*, **37B,** 707.
159. Arif, A. M., Cowley, A. H., Norman, N. C. and Pakulski, M. (1986) *Inorg. Chem.*, **25,** 4836.
160. Plößl, K., Huttner, G. and Zsolnai, L. (1989) *Angew. Chem. Int. Ed. Eng.*, **28,** 446; (1989) *Angew. Chem.*, **101,** 482.

161. Foust, A. S., Foster M. S. and Dahl, L. F. (1969) *J. Am. Chem. Soc.*, **91,** 5633.
162. Foust, A. S., Campana, C. F., Sinclair, J. D. and Dahl, L. F. (1979) *Inorg. Chem.*, **18,** 3047.
163. Das, B. K. and Kanatzidis, M. G. (1995) *Inorg. Chem.*, **34,** 6505.
164. Harper, J. R. and Rheingold, A. L. (1990) *J. Organomet. Chem.*, **390,** C36.
165. Clegg, W., Compton, N. A., Errington, R. J. *et al.* (1991) *J. Chem. Soc., Dalton Trans.*, 2887.
166. Clegg, W., Compton, N. A., Errington, R. J. and Norman, N. C. (1988) *Polyhedron*, **7,** 2239.
167. Leigh J. S. and Whitmire, K. H. (1989) *J. Am. Chem. Soc.*, **111,** 2726.
168. Albright, T. A., Yee, K. A., Saillard, J.-Y. *et al.* (1991) *Inorg. Chem.*, **30,** 1179.
169. Martinengo S. and Ciani, G. (1987) *J. Chem. Soc., Chem. Commun.*, 1589.
170. Foust, A. S., Foster, M. S. and Dahl, L. F. (1969) *J. Am. Chem. Soc.*, **91,** 5631.
171. Bernal, I., Brunner, H., Meier, W. *et al.* (1984) *Angew. Chem. Int. Ed. Eng.*, **23,** 438; (1984) *Angew. Chem.*, **96,** 428.
172. Scherer, O. J., Wiedeman, W. and Wolmershäuser, G. (1990) *Chem. Ber.*, **123,** 3.
173. Fenske, D. and Hachgenei, J. (1986) *Angew. Chem. Int. Ed. Eng.*, **25,** 175; (1986) *Angew. Chem.*, **98,** 165.
174. Di Vaira, M., Midollini, S., Sacconi, L. and Zanobini, F. (1978) *Angew. Chem. Int. Ed. Eng.*, **17,** 676; (1978) *Angew. Chem.*, **90,** 720.
175. Di Vaira, M., Midollini, S. and Sacconi, L. (1979) *J. Am. Chem. Soc.*, **101,** 1757.
176. Whitmire, K. H., Churchill, M. R. and Fettinger, J. C. (1985) *J. Am. Chem. Soc.*, **107,** 1056.
177. Whitmire, K. H., Albright, T. A., Kang, A.-K. *et al.* (1986) *Inorg. Chem.*, **25,** 2799.
178. DiMaio, A.-J. and Rheingold, A. L. (1990) *Inorg. Chem.*, **29,** 798.
179. Rheingold, A. L. and Churchill, M. R. (1983) *J. Organomet. Chem.*, **243,** 165.
180. Scherer, O. J., Pfeiffer, K., Heckmann, G. and Wolmershäuser, G. (1992) *J. Organomet. Chem.*, **425,** 141.
181. Scherer, O. J., Vondung, J. and Wolmershäuser, G. (1989) *J. Organomet. Chem.*, **376,** C35.
182. Scherer, O. J., Blath, C. and Wolmershäuser, G. (1990) *J. Organomet. Chem.*, **387,** C21.
183. Scherer, O. J., Blath, C., Heckmann, G. and Wolmershäuser, G. (1991) *J. Organomet. Chem.*, **409,** C15.
184. Rink, B., Scherer, O. J., Heckman, G. and Wolmershäuser, G. (1992) *Chem. Ber.*, **125,** 1011.
185. Detzel, M., Pfeiffer, K., Scherer, O. J. and Wolmershäuser, G. (1993) *Angew. Chem. Int. Ed. Eng.*, **32,** 914; (1993) *Angew. Chem.*, **105,** 936.
186. Bolle, U. and Tremel, W. (1994) *J. Chem. Soc., Chem. Commun.*, 217.
187. Charles, S., Eichhorn, B. W., Rheingold, A. L. and Bott, S. G. (1994) *J. Am. Chem. Soc.*, **116,** 8077.
188. Charles, S., Eichhorn, B. W. and Bott, S. G. (1993) *J. Am. Chem. Soc.*, **115,** 5837.

8 Environmental and medicinal chemistry of arsenic, antimony, and bismuth

J. REGLINSKI

8.1 Introduction

The lower group 15 elements arsenic, antimony and bismuth are of low natural abundance in the Earth's crust (Table 8.1) being predominantly found as oxides and sulphides. As they are mainly locked into stable compounds of low concentration with respect to the surrounding mineralization they pose little hazard to the environment, except for a few sites around the world where they can be released by natural events such as volcanic activity [1] and water action [2]. Much of the presence of these elements in the biosphere is due to a result of the action of humans and is a direct consequence of the desire to win other elements of greater commercial importance (e.g. gold, copper and lead) from their ores, as a result of the combustion of fossil fuels [3–5] or their direct use as medicines [6–8], pesticides [9] or advanced materials [10].

8.2 Group 15 in the environment

These elements are introduced into the biosphere as a direct result of industrial activity across the wider spectrum of phases, that is as volatile species (e.g. AsH_3), solutes (e.g. $[SbO_2]^-$) and as solids (e.g. Bi_2O_3). Environmental concern rests mainly with the toxic, lighter, elements, antimony and arsenic rather than the more benign bismuth, which at levels of approximately ng g^{-1} rarely achieves concentrations of environmental concern [11]. Environmentally, bismuth has one unique feature. As a daughter nucleus of the radioactive decay of radon (reaction [8.1]), estimations of the concentration of ^{214}Bi (half-life, $t_{1/2} = 19.7$ mins, α emitter) can be used to assess the exposure of miners and households to airborne radioactivity [12]:

$$^{222}_{86}\text{Rn} \xrightarrow{3.8\ \text{days}} {}^{218}_{84}\text{Po} \xrightarrow{\alpha;\ 3.1\ \text{min}} {}^{214}_{82}\text{Pb} \xrightarrow{\beta;\ 2.7\ \text{min}} {}^{214}_{83}\text{Bi} \qquad [8.1]$$

As mentioned above, significant amounts [13] of these species are released into the atmosphere as a consequence of the isolation of the

Table 8.1 A simple environmental profile of the group 15 elements.

Element	Concentration in the Earth's crust (g tonne^{-1})	In order of abundance with respect to the periodic table	Common ores	Formula
Arsenic	1.8	51st	Realgar	As_4S_4
			Orpiment	As_2S_3
			Arsenolite	As_2O_3
			Arsenopyrite	FeAsS
			Enargite	Cu_3AsS_4
Antimony	0.2	62nd	Stibnite	Sb_2S_3
			Ullmanite	NiSbS
			Livingstonite	$HgSb_4S_8$
			Tetrahedrite	Cu_3SbS_3
			Jamesonite	$FePb_4Sb_6S_{14}$
Bismuth	0.008	69th	Bismite	α-Bi_2O_3
			Bismuthinite	Bi_2S_3
			Bismutite	$(BiO)_2CO_3$

elements, burning of fossil fuels [3–5] and the smelting of ores (i.e. combustion). The oxides generated thermally during these processes are particulate in nature, which owing to their small size (*c.* nanometre diameter) are held up in the exhaust gases and are easily vented into the local atmosphere. They become distributed widely by the prevailing air currents [14–16] and can be detected in the air [17] some distance from their source, being finally deposited via natural precipitation [18]. Once deposited they become available to the local flora and fauna [19]. Models for the transport of heavy metals by this mechanism [17] predict that currently approximately 4 tonnes of antimony and 285 tonnes of arsenic are deposited annually on the Arctic as a consequence of industrial activity in the northern hemisphere. Thus through this mechanism these elements can affect the environment some distance from their source. It can be argued that monitoring the impact of these species on the remote environment is poorly served by measuring the fluctuating levels of arsenic and antimony in the mobile phases (air and water) at any given time. It is probably more suitable to measure the problems of the long-term build-up of the heavy metals and the damage they do to the environment by measuring the elemental uptake in sedentary marker species such as simple plants (*Hylocomium splendens* and honey mesquite) [20, 21] which are known to accumulate heavy elements in their tissues. This approach has the further advantage of providing details of the long-term impact of heavy metals on the environment as a function of 'acceptable' practices rather than reporting the effects of accidental emissions which although dramatic may obscure a wider problem.

The discharge of soluble waste and suspended solid material into the rivers [22–24] is a more direct route by which these elements enter the immediate biosphere. However, industrial effluent deposits a cocktail of heavy metals and toxins into the biosphere, which the environment initially finds difficult to assimilate. Thus there is some interest in sites where there is a progressive introduction of low level concentrations of heavy metals into the biosphere from sources such as mine tailings [25] and from old industrial sites [26]. Here the spoil slowly begins to give up these trace elements once the minerals are removed from their anaerobic strata and are exposed to the air and the acidic action of rainwater. In a similar manner the presence of unique species such as ^{125}Sb ($t_{1/2} = 2.7$ years β emitter) in the environment can be used a signature of effluent from nuclear reactors (cf. ^{214}Bi) [16, 27]. The fate of trace antimony species from nuclear reactors has been followed through the soils [28] and plants [29] and into the small mammals [30, 31]. These studies suggest that the transport, accumulation and transformation of antimony in the natural environment is poor in comparison with arsenic and that the observed distributions of deposited particulate oxides occurs via independent inhalation and absorption by the various species rather than by transmission through the food chain [32].

In the aqueous environment the total burden of antimony and the antimony (III)/(V) oxidation state ratio has received some attention. However, one must remain aware that the redox nature of the ecosystems will play a major role in determining the species present. Many sites are oxidizing and the major species are thus observed to be the stibnates. Reducing systems (e.g. sediments) do exist which can generate appreciable amounts of the lower oxidation state. Antimony in the lower oxidation state has been shown to have an affinity for humic acids [33] which can bind up to *c.* 10 nmol Sb g^{-1} humic acid. Thus, great care must be taken when estimating the total antimony and antimony(III) content at a site, as through environmental reduction these elements can become adsorbed onto the sediments and particulate matter in the rivers. In contrast both arsenic(III) or arsenic(V) will bind to humic acids at roughly similar concentrations (*c.* 90 nmol g^{-1}) [34].

As stated above the incorporation of antimony into plants and animals is poor; however, some formative studies have identified the simple organospecies $MeSbO(OH)_2$ and $Me_2SbO(OH)$ [35, 36] in some ecosystems. These studies and the observation that antimony can be incorporated into the shells of molluscs further suggests that some capability for antimony transformation exists [36, 37]. In contrast to antimony the ability of arsenic to enter the food chain and become transformed into a variety of organic forms as it passes through the biological hierarchy has been known for some time. Simple arsenic oxides ($[AsO_2]^-$ or $[AsO_4]^{3-}$) are slowly organified to produced simple methylarsonic acids [37], more complex betaines, cholines, sugar derivatives and lipids (Scheme 8.1) [38–45].

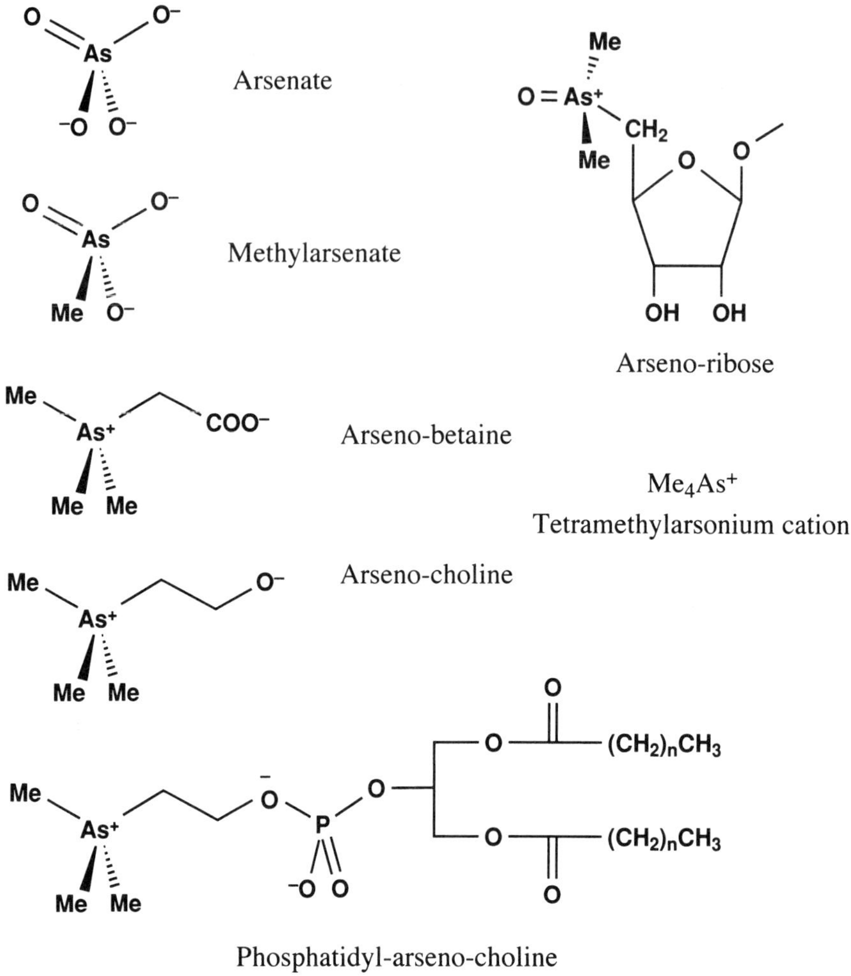

Scheme 8.1 The structures of some common arsenic-containing natural products.

The introduction of arsenic into the biosphere is not limited to its release through industrial activity. Significant amounts of this metalloid are used in herbicides and biocides. Chromated copper arsenate(V) and ammoniacial copper arsenite(III) are used in wood treatment as preservatives. Arsenic (e.g. lead arsenate) was a common pesticide used extensively in fruit production to control pH and eradicate the codling moth in orchards. Sodium arsenate is used to control fungi which cause measles on grapes. Sodium arsenite and monomethylarsonate are used as simple bulk weedkillers, and arsenicals such as sodium arsenate and copper acetoarsenite ($Cu(CH_3COO)_2Cu(AsO_2)_2$) can be found in commercial household insecticides.

Arsenicals are initially toxic to plants [46]. However, the lower organisms (bacteria, fungi, plants) by virtue of their reduced mobility can develop such that they can tolerate arsenicals to a greater extent than can higher species [47]. Where there is an appreciable concentration of arsenic in soil (45 mg kg^{-1}) many plants exhibit a uniform [48] uptake of the element, possibly through the phosphate pathway [49] accumulating it in their leaves (15 mg kg^{-1}), stems (20 mg kg^{-1}) and roots (200 mg kg^{-1}) [50]. Certain organisms (*Candida* sp., *Pseudomonas* sp.) have become tolerant to these agents and can mobilize and metabolize the arsenic, producing a range of methylated products and eventually the volatile organic compound trimethylarsine [51–53]. Similarly, some marine organisms may accumulate arsenic from anti-fouling treatments such as chromated copper arsenate [54]. It is thus likely that it is these species that begin the organification of arsenic by generating a range of methylated species [37]. In contrast, it was only recently confirmed that certain micro-organisms (e.g. *Pseudomonas* sp.) were also capable of methylating antimony [55].

Arsenic and antimony are common components in fire retardants. Recently it has been suggested that sudden infant death syndrome (SIDS), or cot death, can be linked to the release of poisonous stibine or arsine [56] from the arsenic and antimony present in the fire retardant material incorporated into the infant's mattress. The generation of these toxic gases in the cot requires a rather unique microenvironment in the mattress, that is, moisture, complete with nutrients (e.g. urine or sweat), coupled with a fungal infection (e.g. *Scopulariopsis brevicaulis*). These conditions are similar to those required for the bacterial release of volatile arsines from fungicides. It is the extension of these fungicide studies to the aetiology of SIDS which led to the initial hypothesis that the microbes could release these dense, volatile gases at levels capable of contributing to the death of the infant. However, a recent analysis of the heavy metal content of tissues obtained from infants succumbing to SIDS [57] in comparison with other types of fatality have shown that the levels of antimony deposited on the lung of the infants of these two groups is roughly similar at *c.* 6 ng g^{-1}. As such, this hypothesis has recently met with some criticism, with the debate continuing.

Although the higher life forms (e.g. mammals) have a capacity to methylate arsenic [58], it is likely that their main defence against inorganic arsenic ($[AsO_4]^{3-}$, $[AsO_2]^-$) is via a combination of methylation and excretion [58]. In contrast, the main defence against antimony relies on excretion only as there would seem to be little capacity to methylate antimony in mammals [55a]. Studies using human volunteers have shown that for a given, specific, controlled intake of arsenic (182 ng) comprising inorganic arsenic (6%), monomethylarsonic acid (19%), dimethylarsonic acid (27%) and organic arsenic (i.e. Me_4As^+-, 48%), 46 ng of the dose was excreted in the faeces and 148 ng in the urine. Of the latter 1.4%

remained in the inorganic form, and 3.5% as monomethylarsonic acid. *In situ* methylation raises the relative amounts of the dimethylarsonic acid and organic arsenic to 34% and 61%, respectively. The organification process in many species is rapid [59], and methylation would seem to be a parallel process (with excretion) for the detoxification of arsenic. Less than 1% of the arsenic remains in the tissues [60]. However, this is in a +3 oxidation form, which has become associated with the circulating erythrocytes, the liver and kidneys. Biological methylation is a natural intracellular process which occurs in the liver rather than by the microflora which inhabit the gut. As stated above, the major metabolite would seem to be soluble di-methylarsonate which can be excreted easily. However, some further methylation can occur to form trace amounts of trimethylarsine oxide.

In the absence of high arsenic backgrounds derived from an industrial or environmental source (e.g. well water) it is possible that the bulk of the arsenic uptake by mammals is derived from the ingestion of simple organisms (plants and fish) which contain predominantly organic species [61] and that they acquire a greater proportion of their heavy metal intake in a partially transformed, more encapsulated and less toxic form, such as arsenobetaine and arsenocholine. It has been shown that healthy adults can increase their dietary intake of arsenic (normally about 10 ng per day [62]) 3–5-fold simply by eating fish. Although a complete linear study of the behaviour of arsenic in the food chain has not been reported, the data compiled and presented in Table 8.2 can be used to indicate that the concentration of arsenic falls by roughly one order of magnitude as one progresses up the food chain (e.g. algae, 50 ng g^{-1}; shell fish, 10 ng g^{-1}; fish, *c.* 1 ng g^{-1} and animals fed on fish meal, *c.* 0.02 ng g^{-1}). Thus one observes a dilution in the total arsenic present and an increase in the degree of sophistication of the dominant molecules. This trend may be a result of the more efficient extrusion of the soluble forms of the heavy metal (e.g. $[AsO_4]^{3-}$ and $[Me_2AsO_2]^-$). However, it can be postulated that not only is there some transformation of ingested inorganic arsenic but also of the partially transformed arsenic. For example in fish as little as 3% of the arsenic burden will be in an inorganic form [63]. Although the arsenic is accumulated in less toxic, stable forms such as arsenobetaine which is 50 times less toxic than arsenite [64], it would seem that a wider biochemical cycle for arsenic exists. Some evidence exists that certain disparate species such as microbes and shark can break these more sophisticated compounds down further to produce trimethylarsine oxide or reform inorganic arsenic [65, 66].

Most of the products isolated thus far indicate that arsenic enters the biosynthetic pathways associated with nitrogen and lipid synthesis [67] in particular. That arsenic is deposited in lipid-type species suggests that acute toxicity can be ameliorated in this manner by depositing the element

Table 8.2 A simple food chain for arsenic, showing species and some of the complex arsenic species isolated

	Natural water	Algae	Molluscs and bivalves	Crustaceans	Fish	Mammals
Species	Arsenate	Arseno-sugars	Arsenobetaine and arseno-sugars	Arsenobetaine and arsenocholine	Arsenoribose and arsenobetaine	Lipid-arsenate
Total arsenic	Highly variable	50 μg g^{-1}	*c.* 0.4 μg ml^{-1}	10 ng g^{-1} wet weight	30 μg g^{-1} wet weight	0.02 ng g^{-1}

in a less reactive hydrophobic environment. For species such as shellfish a portion of the heavy metals are deposited in the exoskeletons, another method of removing heavy metals from the organism biochemistry. Owing to the mutagenic properties of arsenicals the progressive build up of toxins in this manner can be expected to create more long-term effects such as an increased chance of malignancy. Considering the similarities between arsenate and phosphate it is curious that there are only a few reports which indicate that arsenic follows the biosynthetic pathways of phosphorus. This is evidenced by the recovery of arsenic from chemical species such as adenosine-5′-arsenate [68].

Whereas biological methylation is usually effected by cobalamins by means of radical or methyl transfer, arsenic is somewhat unique in the sphere of inorganic biochemistry as methylation is effected via the transfer of a carbocation, a process familiarly known as oxidative methylation. The active agent for this step can vary from simple sulphonium salts (Me_3S^+),

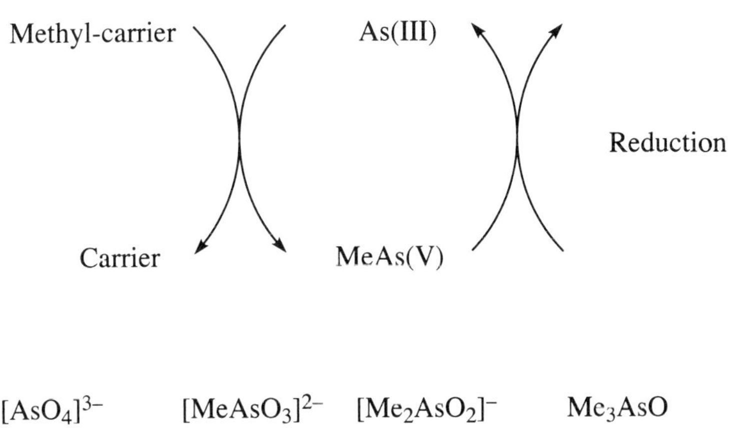

Scheme 8.2 A schematic representation of the biological cycles required for the methylation of arsenic.

for example S-adenosyl-methionine [69], to specific methyltransferase enzymes [70]. The cycle is completed (Scheme 8.2) through the re-reduction of the new methylated arsenical(V) by a range of non-specific reductants. Bacteria can make use of lipoic acid (6,8-dithio-octanoic acid) and mammals the ubiquitous thiolate glutathione (γ-glutamyl-cysteinyl-glycine), both of which have been reported to be capable of reducing these organoarsenicals *in vitro* [71, 72].

Studies of microbial systems have shown that the production of trimethylarsine from the various methylarsonates (e.g. Me_3AsO, $[Me_2AsO_2]^-$, $[MeAsO_3]^{2-}$) is more rapid than their respective thioarsinites [e.g. Me_2AsSR, $MeAs(SR)_2$] [73], thus suggesting that although small thioarsinite species can be methylated [74] they are further downstream in the methylation process than are their parent arsenic acid [75]. It is unlikely that they are primary substrates for oxidative methylation. If biological reduction and thiol ligation operate at different rates then the key species in the biological cycle and the substrate for oxidative methylation is likely to be the respective arsine oxides which will form first:

$$MeAsO(OH)_2 + 2RSH \xrightarrow{\text{fast}} MeAsO + RSSR + 2H_2O \qquad [8.2]$$

$$MeAsO + 2RSH \xrightarrow{\text{slow}} MeAs(SR)_2 + H_2O \qquad [8.3]$$

However, there is a possible role for the thioarsinite in arsenic methylation. Studies on the binding of arsenic to the methyltransferase enzyme suggest that prior to the transfer of the methyl group the arsine oxide becomes bound to a thiolate at the active site. By taking advantage of the hard–soft nature of arsenic(V)–arsenic(III), it is possible to assist the binding of the softer arsenic(III) substrate prior to methylation and facilitate the release of the harder arsenic(V) product from the soft metal recognition site [70] once the reaction is complete.

Cycles which progressively exchange oxygen ligands on arsenate for methyl groups have been constructed (Scheme 8.2) which explain the sequential production of the arsenic(V) species shown. These species are all known to occur in the environment.

8.3 Toxicology

As can be seen from Table 8.3, arsenic, infamous for its use as a poison, is the most toxic of the three lower group 15 elements. Even in the current literature there continues to be a remarkable number of reports of accidental death, suicide and murder in connection with arsenic. Its toxicity to humans has been known for many centuries and although its use today is considered somewhat dated and staid, even in recent times aspects of

Table 8.3 The relative toxicity of simple chlorides, oxides and oxyanions of arsenic, antimony and bismuth. Figures are in LD_{50} (50% of the dose) (mg of compound per kg body weight). ipr = intraperitoneally; ims = intramuscularly; oral = orally.

	Arsenic[a]		Antimony[b]		Bismuth(III)[c]
	III	V	III	V	
Chlorides					
rat, oral		not known	525	1100	
Oxy-chlorides					
rat, oral					22 000
Oxides					
rat, oral	14·6	8	20 000		5000
mouse, oral	31	55			10 000
rat, ipr	871		3250	4000	
mouse, ipr			172	1000	
Oxyanions					
rat, oral	41				
mouse, ipr	1.17				
mouse, ims	14	87			

[a]Minimum exposure limit 0.2 mg m^{-3}.
[b]Minimum exposure limit 0.5 mg m^{-3}.
[c]Minimum exposure limit not specified.

Table 8.4. The relative toxicity of some substituted arsenicals.

Species	Name	LD_{50} (mg kg^{-1})[a]	Use or nature
$[AsO_4]^{3-}$	Arsenate	20	Inorganic
$[p\text{-}NH_2PhAsO_3]^{2-}$	Atoxyl	75	Therapeutic
$[Me_2AsO_2]^-$	Dimethylarsonate	700	Herbicide
$[Me_3As^+CH_2COO^-]$	Arsenobetaine	*c.* 900	Natural product

[a]This is defined as 50% of a lethal dose (given here as mg compound per kg body weight).

its chemistry were explored in the development of Lewisite (β-chlorovinylarsine dichloride) as a nerve gas. In reality there is no need to be so sophisticated, most methylarsine chlorides (and methylphosphine or methylstibine chlorides for that matter) have a low vapour pressure and will provide a nasty skin-peeling burn. The LD_{50} (50% of a lethal dose) of AsH_3 which induces massive haemolysis is approximately 5 mg kg^{-1}; that is, about 0.3 g. This makes it 3–5 times more toxic than strychnine (with a LD_{50} of 16 mg kg^{-1}).

Table 8.3 clearly indicates that the toxicity of arsenicals is not uniform and many factors play a role.* Predictions are dangerous: for example, the table reconfirms the assertion that, in general, toxicity falls as one progresses down the group, to the extent that less concern exists when supplying minimum exposure limits and values for the LD_{50} for bismuth

* This is a classic example of the variability of the toxicity found for compounds of specific elements of notoriety where the difference between the toxicity of the starting materials and products (or *vice versa*) can be staggering.

complexes. It is clear that the lower oxidation states are in general slightly more toxic than are the higher oxidation states. However, in a redox active environment such as the human body, one can expect the two oxidation states to coexist in a dynamic cycle. Thus the immediate toxicity of these species, estimated by using acute animal models, can only be expected to obtain values of a roughly similar magnitude owing to the fact that the two oxidation states available to arsenic and antimony are probably linked through their biochemistry [75].

The nature of ligands can also have a profound effect on the overall toxicity of the compounds. A cursory glance at some substituted arsenicals in Table 8.4 indicates that the LD_{50} can be reduced quite simply by altering the ligands.

In general small inorganic species are more toxic than are organoarsenicals. Exchangeable ligands such as Cl^-, NH_2 and the presence of lone pairs (e.g. Me_2AsCl: $LD_{50} \simeq 5\ mg\ kg^{-1}$) would seem to increase toxicity. However, if the arsenic is sealed into a quaternary arsonium salt (e.g. $[Me_3A^+sCH_2COO^-)$ it can be almost inert and relatively non-toxic. The behaviour of the compounds, that is, their phase distribution in the body, is also important. Lipid solubility can lead to storage in tissue or fat and long-term problems, whereas ionic character, although allowing good solubility and rapid clearing from the body, can lead to more immediate problems during the period of time leading up to excretion.

Generally, arsenic(V) carries a long-term threat (cancer). Its immediate effect is unclear, but its toxicity is likely a result of its reduction to arsenic(III). Arsenic(III), especially with labile ligands, has a strong affinity for sulphur (as the soft thiolate ligand) and will affect key enzymes, for example acetylcholine esterases, lipoic acid and haemoglobin. Long-term contact with the organometallic forms cause hyperactivity, dizziness, speech and psychological problems, possibly linked to methylation (MeAs−) which facilitates its transport across the blood–brain barrier.

The acute effects of arsenic are common knowledge. Much of the implications of the long-term exposure to arsenic and antimony in humans can be gleaned from studies of the effects of prolonged therapy with anti-parasitic agents and the occupational exposure to arsenic, antimony and bismuth in industrial workers employed in heavy industry (e.g. at smelters, etc.). Exposure to intravenous arsenic was common in the treatment of parasitic diseases (e.g. syphilis) and would seem to correlate with specific forms of malignancies (Bowens disease, hepatic angiosarcomas [76, 77]). Within the industrial framework, arsenic and antimony have been clearly identified as a definite risk to the health of workers, and routine monitoring of their exposure to arsenous and antimonous environments is commonplace. Short term there is a possibility of an increased risk of dermatitis, pulmonary problems, pneumoconiosis (antimoniosis) and cardiovascular problems. In the long term many studies have suggested that a link can be established with cancer within the work force. Even when workers do not develop any visible medical problems it can be shown that substantial numbers (*c.* 25%) may be unknowingly sensitized to these elements [78].

Arsenic and antimony can be deposited on the lung as particulate metal oxides or inhaled as arsine and stibine gas, which are produced in reducing environments. Apart from the deposition of insoluble compounds (e.g. calcium arsenate) once inhaled the distribution of these elements in the body follows the argument discussed above. Furthermore, the rate of excretion of these elements via the kidneys would seem to be unaffected by the nature of the species (oxide or hydride) originally deposited on the lung [79]. Although care is taken to keep the exposure of workers to the individual elements down to their threshold limits, the analysis of smelter workers shows that there is generally a distinct and comparable presence of both elements in their bodies reaching values an order of magnitude greater than found normally. Overall, the conflicting reports (i.e. poor correlation) linking occupational exposure of arsenic and antimony to the incidence of cancer in the workers and the inhabitants in the local environment to these smelters is possibly a result of a number of competing factors which exist in these areas. However, it is curious that, although arsenic and antimony are impurities common to many ores and starting materials and the synergistic effects of many metals have been well reported, the synergistic effects of arsenic and antimony in the workplace have been ignored. Considering the chemical similarity between these two elements, a re-analysis of the data using instead a measure of the total amount of the two group 15 elements present may provide a stronger correlation with the increased incidence of cancer. This hypothesis is based on the similarity in the chemistry and biochemistry of arsenic and antimony and, if correct, would result in a lower threshold limit for the occupational exposure to these elements [80, 81] which in part would not be dependent on the element alone.

Elevated levels of these two elements can also be demonstrated in the human population living in the vicinity of these plants, with again a weak correlation with a higher incidence of health problems normally associated with these species. However, the distribution of these elements can be highly variable and depends heavily on the prevailing air currents. Some studies have shown that for emissions of light particles the concentration of arsenic in the local population can be lower than found in the populations some distance away. However, the link between arsenic and cancer can be substantiated further by studies on native populations exposed to arsenic in their natural drinking water, which show that DNA is another definite site of action [82] for the element *in vivo*. In contrast, a similar volume of evidence [83] does not exist to link antimony to cancer even though the studies on human populations exposed to antimony indicate that a link may exist. There would seem to be little evidence that bismuth is carcinogenic in its simple complexes. As will be discussed below it is available in over-the-counter medicines and has been consumed in large quantities for many years without any signs of carcinogenesis. Also, in contrast to arsenic and antimony, bismuth may actually have a beneficial role in cancer therapy by providing some protection against the nephrotoxic side effects which are associated with many common therapeutic compounds (e.g. *cis*-platin).

The clinical (field) studies on arsenic can, however, generate variable results [84]. This variation can be accounted for by the inability adequately to control these 'experiments', the difficulty in compensating for other contributing factors and the inability to differentiate between the mutagenic and co-mutagenic behaviour of the arsenic, that is, its ability to potentiate other naturally occurring toxins such as ultraviolet light and organic carcinogens [85]. Attempts to model accurately the behaviour of this metalloid at DNA are also difficult. The extrapolation of data derived from studies of the short-term (weeks), high-concentration ($mg\,kg^{-1}$) effects of arsenic on simple models (isolated DNA, bacteria, small animals) cannot be expected to predict accurately the behaviour of prolonged exposure (years) of humans to low concentrations ($ng\,kg^{-1}$) of arsenic. Thus although it is possible to substantiate the clinical link between arsenic and certain forms of cancer and the teratogenic (fetotoxic) effects of arsenic [86, 87], the exact mode (or major path) by which this overall result is achieved is not as yet clear.

Arsenic once present in the bloodstream will quickly transfer to the cell nucleus [88]. In the nucleus it exerts a biphasic response on DNA synthesis. At low concentrations it stimulates synthesis whereas at high concentrations it would seem to inhibit synthesis [88, 89]. In the various models employed, arsenic(III) is reported to be an order of magnitude more carcinogenic than is arsenic(V). However, at the concentrations normally present *in vivo*, and considering the redox cycles and biotransformations available to arsenic, it could be suggested that the speciation of the metalloid as introduced into the body will be less important than the serendipitous biochemistry it encounters *en route* to the nucleus.

At this stage a number of mechanisms behind the carcinogenic effects of arsenic have received some attention.

Arsenic has been reported to induce damage directly on DNA, causing gaps, strand breaks and aberrations in the biological information located in the chromosomes and chromatids [90–93]. The ability of arsenic(III) to form stable complexes with oxygen and sulphur donors has been implicated in the formation of DNA–protein cross-links [94, 95] and conformational changes in the DNA. The recent isolation of adenosine-5′-arsenate [68] suggests that arsenic could replace phosphate in the DNA backbone [96]. Reductive modification of the sugar-arsenate linkage could subsequently lead to strand breakage or conformational changes. Each of these interactions would compromise DNA transcription [97, 98] or the DNA repair mechanisms [99], an interaction necessary for chemical carcinogenesis. It is thus apparent that via any of these routes arsenic could interfere in the synthesis and repair of DNA in a protracted and sustained manner at very low, non-toxic levels [100].

As oxidizing agents arsenate and its bioproduct methylarsonate are reported directly to induce oxidative damage on DNA [101]. This hypothesis is supported by studies which report the protective effects of superoxide dismutase and catalase against arsenic(V) insult [102] at DNA. However, considering that arsenite is more carcinogenic than is arsenate [96] this mechanism would seem to be less important than those described above. It is of some interest, however, that one of the main products of arsenic detoxification in humans, dimethylarsonic acid, should still be carcinogenic whereas arsenobetaine, which is one of the major natural products in fish, is neither mutagenic nor co-mutagenic [103]. This is consistent with the view that the former is for excretion and the latter for storage.

8.4 The medicinal chemistry of group 15

The group 15 elements have a rich if somewhat dubious history in their use as medicines prior to the turn of the century. However, as with many toxic compounds, a belief existed that there was a therapeutic window for these species and that through studies some clinical benefit could be achieved in the face of extreme toxicity. These hazardous, serendipitous studies led to the adoption of many compounds by modern medicine but not necessarily for the purpose for which they were initially intended. The group 15 elements were actively researched for their potential to treat parasitic diseases such as syphilis and leprosy.

The simple tartrates (emetics) were common antiparasitic agents, each metal tartrate providing some benefit, with the main emphasis falling on arsenic. However, as a consequence of the stability of the metal–carbon bond in both oxidation states and the accessibility of both the lower and the higher oxidation states of arsenic, it was possible to synthesize and investigate a wider range of arsenic-containing compounds, a selection of which are shown in Table 8.5. These are all aryl complexes, which as shown above impart a slightly more acceptable LD_{50} than the simple compounds shown in Table 8.3 while maintaining the basic reactivity (redox, geometry, ligand preference) of the arsenic(III) or arsenic(V) centre. The three classes of compound presented are related through the underlying chemistry which is active *in vivo*. The functional groups present can be considered as modifications on a theme (as with many inorganic drugs the parent compounds are in reality pro-drugs which are modified *in situ*):

Table 8.5 Some common arsenicals in use in human and veterinary medicine.

Structure	Compound (trivial name)	Use	LD_{50} (mg kg^{-1})[a]
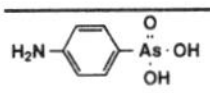	*p*-Arsanilic acid (Atoxyl)	Sleeping sickness	75
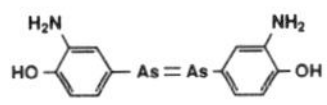	Diaminodihydroxyarsenobenzene (Arsephenamine)	Leprosy, syphilis	100
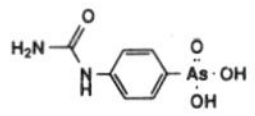	*N*-Carbamoylarsanilic acid (Carbasone)	Amoebiasis	—
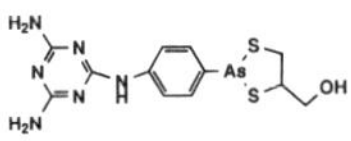	Diaminotriazinylaminothioarsolane-4- (Melarsoprol) methanol	Meningoencephalitis (trypanosomes)	250
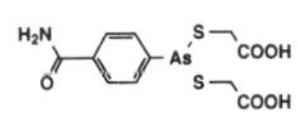	Carbamoyl phenylarsene-dithio-diacetic acid (Thiacetarsimide)	Canine filarisi	—

[a]Defined as 50% of a lethal dose (given here as mg compound per kg body weight).

$$PhAsO(OH)_2 + 4RSH \rightarrow PhAs(SR)_2 \quad [8.4]$$

$$PhAs{=}AsPh + 2RSSR \rightarrow PhAs(SR)_2 \quad [8.5]$$

Thioarsinites (e.g. melarsoprol, Table 8.5) are well established products of arsenic acid reduction by thiols [72, 73]. The interaction of arsphenamines with disulphides readily allows the formation of thioarsinites *in situ*. This interrelationship within the chemistry can be used to explain the mechanisms behind cross-resistance [104] between the various compounds.

However, as the medical fraternity became more aware of the toxicity of these elements, more efficient organic therapies were developed. Some interest remains in the use of arsenicals for non-human use in herbicides and animal husbandry. Thiacetarsimide can be used as a deworming powder in cats and dogs. Compounds such as atoxyl (*p*-arsanilic acid) and roxarsone (3-nitro-4-hydroxyphenylarsonic acid) are used to increase the weight gain of chickens and pigs. This is achieved by eradicating parasites thus preventing dysentery while also increasing food efficiency. However, within the third world some interest and use is still made of melarsoprol derivatives for sleeping sickness as well as for simple, less toxic, antimony preparations in the treatment of leishmaniasis.

As a consequence of the interest in the lower group 15 elements as antiparasitic agents it became accepted that many of these preparations were antibacterial and had certain antiseptic properties. The low toxicity of bismuth in mammals made it especially suitable for incorporation into simple medical preparations for internal use and into topical pastes for external use. Bismuth thus became a common ingredient in preparations for disorders of the gut such as in antacid tablets (bismuth carbonate), diarrhoeal treatments (e.g. bismuth subsalicylate – PEPTO-BISMOL®)

and haemorrhoidal creams (bismuth subgallate, bismuth oxide) where it provides symptomatic relief. Externally it has been incorporated into dressings for wounds (e.g. BIPP®, bismuth subnitrate iodoform paraffin paste), burns and skin grafts (e.g. xeroform, bismuth tribromophenate). Bismuth subgallate as a saline solution (containing adrenaline) is used during surgery for tonsillectomies and andenoidectomies. However, there remains little evidence that bismuth is itself directly and primarily antibacterial in these preparations [105, 106]. In preparations such as BIPP® and bismuth sub-gallate in surgery, its primary function would seem to be the activation of Hagemen clotting factor XII. By activating the coagulation cascade [107] the amount and duration of bleeding during surgery is reduced. Whereas problems have been reported with large volumes of these topical applications the dangers are small and reversible.

8.4.1 Sleeping sickness, melarsoprol and arsenic

Diaminotriazinylaminothioarsolane-4-methanol (melarsoprol, Table 8.5) was introduced in the 1940s and is still used to treat advanced African trypanosomiasis (sleeping sickness) caused by *Trypanosome brucei* which is transmitted to the human by the tsetse fly. In the African subcontinent this disease remains a significant problem with 50 million people at risk with approximately 25 000 new cases every year. Considering that the application of melarsoprol is a last line of defence, recovery rates are good (*c.* 85%–90%) with low rates of relapse (3%) within two years. This compound which has a relatively high LD_{50} (250 mg kg^{-1}) is administered at dosages of approximately 2×16 mg kg^{-1} day^{-1} for 10–20 days, that is, a total body burden of approximately 30–50 g. Although the therapeutic regime remains below the threshold limit for a healthy person, the general health problems associated with parasitic infections means that the effects on the patients can be variable and problematic and side effects such as arsenic reactive encephalopoly leading to fatalities do occur [6].

The compounds enter the bloodstream where there is an initial increase in the serum arsenic concentration (2–4 $\mu g\ l^{-1}$ at 24 h falling to 0.1 $\mu g\ l^{-1}$ after 120 h). Of key importance is the ability of the compounds to penetrate the cerebrospinal fluid (CSF) generating a concentration (*c.* 0.1 $\mu g\ l^{-1}$) sufficient to eradicate the key parasitic infestation in the CSF, but not induce encephalopoly. Studies on cysteamine thioarsinite derivatives of melarsoprol (Cymelarsen®) clearly indicate that the integrity of these compounds is not maintained in the bloodstream and they quickly disproportionate to form a wide variety of aryl arsine species [108]:

$$\underset{(33\%)}{\phi As(SR)_2} \rightleftharpoons \underset{(24\%)}{\phi As(SR)(OH)} \rightleftharpoons \underset{(33\%)}{\phi AsO} \qquad [8.6]$$

where ϕAsO is melarsine oxide and RSH is cysteamine ($HSCH_2CH_2NH_2$).

Exchange processes allow the arsenic to form a weak chelate with lipoic acid (6,8-dithio-octanoic acid) which is central to parasite electron transport [109], trypanothione [*N*,*N*-bis-(glutathionyl)-sperimidine] and trypanothione reductase. Trypanothione is the principal low molecular weight thiol responsible for antioxidant activity in these parasites which is in turn regulated by its respective reductase enzyme. Inactivation of the reductase by antimonyls [110] or melarsoprol is effected by the formation of a thioarsinite or thiostibnite at the active site through an interaction with two vicinal cysteinyl residues. However, for this important antiparasitic process to occur the species must be present in their lower +3 oxidation state. The modifications of the parent compound (e.g. melarsoprol) or reduction of the parent arsenic acid (e.g. arsanilic acid) in the blood-stream will be discussed below.

8.4.2 Leishmaniasis and antimony

Leishmaniasis is another parasitic infection still treated by compounds of the group 15 elements, with antimony being the element of choice. This disease remains one of the largest killers in the tropics with approximately 1.5 billion at risk and 20 million cases being reported annually (this is comparable to one third of the population of the United Kingdom!). Its reemergence is of note as it is increasingly resistant to standard organic therapies and has begun to be seen as an opportunistic infection associated with autoimmunodeficiency syndrome (AIDS) in the southern Mediterranean and the USA. Owing to the cost effectiveness and the need to treat large numbers of people, there is a resurgent interest in the use of antimonyls as antiparasitic compounds in this area. Leishmaniasis is a parasitic infection of the mononuclear phagocytic system. A successful outcome of therapy depends heavily on the immune cells and their activity. Consequently, where the immune system is activated, clearance of the parasite can be achieved with lower drug dosages [111, 112]. This has led to the suggestion that the inhibitory action of antimony compounds are mediated through the macrophage rather than through direct toxic action of the heavy metal on the parasite [113]. Thus although the parasites reside in the organs of the body, the immune cells represent the therapeutic battlefield.

There are a number of accepted therapies, including *N*-methyl-glucamine antimonate(V) (Glucantime®), stibo(V)gluconate (Pentosam®), sodium antimonyl(III) gluconate (Triostam®) and potassium antimony(III) tartrate (Scheme 8.3). In comparison with the arsenicals (shown in Table 8.5), these antileishmania compounds employ simple biocompatible ligands. Similar to the situation discussed above for Cymelarsen®, many of these compounds (e.g. Pentosam®) as delivered to the patient are not pure species but a dynamic mixture of compounds each with a varying degree of

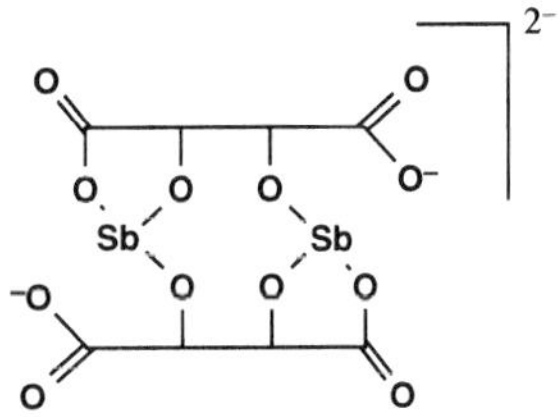

Antimony(III) tartrate
Tartar emetic

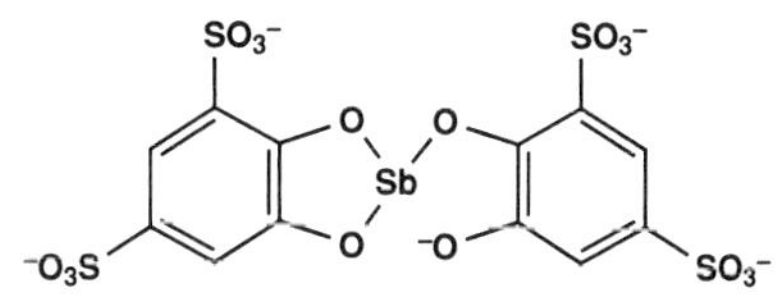

Stibophen
Antimony bis(4,5-dihydroxy-benzene-3,5-disulphonate

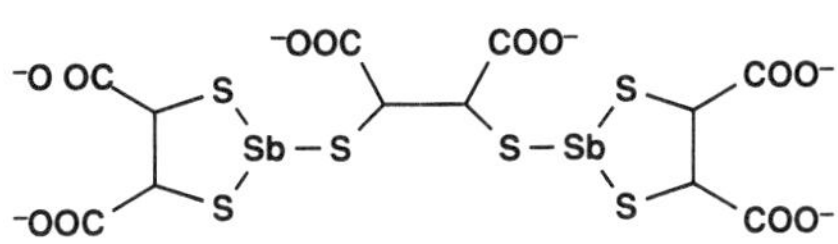

Astiban
Antimony-2,3-dimercaptosuccinate

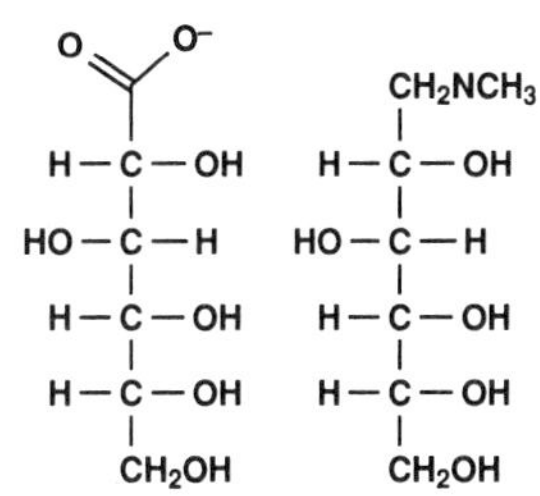

Gluconate
N-Methyl-glucamine

Scheme 8.3 Some antileishmanial compounds and ligands.

activity [114]. The overall dosage of each of these compounds is roughly the same and, depending on the amount of antimony, can reach levels of 20 mg kg^{-1} day^{-1} (up to a maximum dosage of 850 mg day^{-1}) for 20 days. Although not as toxic as arsenicals (Table 8.3), this amount of heavy metal still places a heavy burden on the parasite-weakened body and some fatalities do arise.

These drugs are introduced intramuscularly or directly into the lesions. The bloodstream is a mildly reducing environment and there is substantial evidence to support the formation of antimony(III) from high-valent antimony complexes in the circulation [115]. A large proportion of the antimony becomes bound to the cell populations in the blood causing a reduction in the numbers of red cells and an increase in the number of white cells [116]. The increase in the white cell population is beneficial to the clearance of the parasites. Similar to arsenic(III), antimony(III) will inhibit trypanothione reductase via an interaction with the sulphydryl function, but curiously it does not form a strong complex with the thiol substrate trypanothione itself [110].

In the bloodstream antimony can reach levels of 8 mg l^{-1} during the short period after the sample is introduced. However, 80% of the bolus is excreted within 6 h. In the absence of a methylation pathway a large proportion of the antimony is excreted unchanged. However, some studies have reported the excretion of thiol conjugates of glutathione [e.g. $Sb(SG)_3$, GSH = glutathione] [117] suggesting that the reductive step necessary for the biomethylation (Scheme 8.2) may still be active. The circulating blood will distribute the antimony around the body delivering it to the infected organs [115] and the skin [118].

As with many therapies the development of resistance by the infecting organism remains a problem. As will be discussed below, studies have shown that the parasite can become resistant to arsenicals and antimonyls by inducing the synthesis of specific proteins designed to pump drugs from the cell.

Unlike with arsenic there remains some interest in the further development of antimonyls as antiparasitic agents. Work has been reported on derivatives of quinolenes [119, 120] and mannan. The latter are designed to be recognized by the α-D-mannosyl receptors present on the parasite. Other workers are looking at methods of increasing the efficacy of existing compounds by means of combination therapy, that is, antimonyls plus interferon-γ or interleukins or by encapsulating the antimony-based drugs in liposomes [121, 122].

8.4.3 Helicobacter pylori, *bismuth and the eradication of ulcers*

Preparations which incorporate bismuth (e.g. bismuth carbonate) have been used to treat dyspepsia for more than a 100 years. Bismuth subsalicylate is a common ingredient in antidiarrhoeal treatments. This acceptance of bismuth by practitioners in simple medicinal preparations predates the current drug approval methods and procedures which are not generally sympathetic to the use of metal based compounds in therapy. As such, by the time *Helicobacter pylori* (*Campylobacter pyloridis* prior to 1990) was recognized as being central to the pathology of ulceration of the gut, the use of bismuth preparations in the maintenance of certain symptoms associated with this condition had already been accepted and the value of the inorganic ion appreciated.

Helicobacter pylori is a gram negative bacterium which in humans has its unique reservoir in the stomach, inhabiting the mucus which coats the stomach lining. Its presence is detected by IgG serology, a urea breath test, histology and culturing. In the stomach *Helicobacter* hydrolyses urea to ammonia and carbon dioxide, produces monochloroamines and secretes a variety of enzymes and cytotoxins. The bacterium requires a slightly lower environmental pH for survival, and its presence alters acid production and assists mucosal degradation. Thus infection by *Helicobacter*, although not the primary cause of the ulcer, generates the conditions necessary for ulcer formation. Furthermore, prolonged exposure to the toxins produced by *Helicobacter* in the stomach have been linked to the onset of gastric cancers [123, 124] and lymphoproliferative diseases [125, 126]. As a consequence, *Helicobacter* has recently been classed as a human pathogen, suggesting that certain forms of gastric ulcers and cancers result from bacterial infection. Complete eradication of this infection would seem to be important to the complete recovery of the patient from ulceration. Of some significance is the regression observed in lymphomatous tissue on treatment with a therapeutic regime which includes bismuth [125, 126]. Consequently, some workers have begun to suggest that vaccination against *Helicobacter* would be of some value.

Eradication of *Helicobacter* is achieved by means of combination therapy (usually referred to as triple therapy), which uses a bismuth compound, an antimicrobial (e.g. amoxicillin, metronidazole) and an anti-(H+)-secretory (omeprazole). Triple therapy leads to eradication rates of greater than 95% with low rates of relapse. Significantly, no regime which does not include a bismuth compound has exhibited an eradication rate greater than 90% [8]. It is clear that this element is playing a central role in therapy.

Bismuth in solid preparations can be administered in several forms. Some of these preparations remain very simple, such as bismuth carbonate, bismuth subnitrate, bismuth subgallate and bismuth subsalicylate. More complex formulations have appeared since the importance of bismuth was recognized, for example colloidal bismuth subcitrate in suspension (De-Nol®) or in tablet form (De-Noltab®) and ranitidene bismuth citrate. Many of the older compounds have a history of use in treatment of other ailments such as antacids (carbonate), diarrhoea (salicylate), and haemostatic agents (gallate) – symptoms which are common side effects of stomach ulcers, which may help explain their therapeutic durability. In effect these compounds all deliver simple, insoluble bismuth oxides into the patient's gut. The dosage can be between 100 mg day^{-1} for two weeks, to approximately 4 × 400 mg day^{-1} for two weeks, which means that the total loading of bismuth can be as high as 20 g. Three aspects to the action of bismuth as an antiulcer complex have been identified: its structural effects on the stomach wall, its biochemical effects on the stomach wall and its action on the bacterium.

Dosing with these preparations distributes insoluble bismuth particles around the stomach wall. Studies using ranitidene bismuth citrate [127] have shown bismuth particles distributed in the epithelia antral mucosa, and columns of bismuth could be observed down the lamina propia into positions surrounding the blood vessels. In the vicinity of the ulceration bismuth would seem to combine with 'gastric glycoproteins', coagulating to form a coating at the base of the ulcer. This coating forms a barrier [128] which prevents further irritation as a consequence of the natural acidic environment. It is possible that the deep penetration of particulate bismuth into the stomach wall will effectively seal microfissures throughout the stomach wall. This ability of bismuth complexes to heal ulcers independently of an antibacterial process is confirmed by the studies on the efficacy of bismuth in ethanol-induced ulceration [129] where it is observed to enhance the rate of tissue repair and prevent the establishment of persistent erosions [130]. This effect can be explained via the inhibition of alcohol dehydrogenase [129] and acetaldehyde production by bismuth. This interaction may be of some significance as enhanced cytosolic alcohol dehydrogenase activity has been demonstrated in *Helicobacter pylori* [131].

Deposited bismuth inhibits the adherence of *Helicobacter* to the gastric epithelial cells by interfering with the bacterial anchoring mechanism [132] and by allowing the regeneration of protective mucus, slightly increasing the barrier depth. The presence of bismuth on the stomach wall also stimulates prostaglandin (PEG-2) production [133] which in turn stimulates bicarbonate production. As a consequence, there is a change in localized stomach pH, behaviour which is not necessarily antacid in nature but which compromises the viability of *Helicobacter* on the stomach wall to a small extent. By altering the bacterial microenvironment in this manner it is possible that bismuth increases the efficacy of the other components of triple therapy which act directly on the bacterium.

Many workers have tried to establish a biochemical pathology for the action of bismuth compounds. *In vitro* studies have shown that bismuth does not act on replicating cells, only those in stationary phase [134]. The fact that it acts on only a subset of the bacterial population goes some way to explain why bismuth therapy alone has a high relapse rate. Its anti-(H+)-secretory or antacid properties [135] are poor and it is an order of magnitude less effective as an antibactericide [136] compared with its co-partners in triple therapy. Small amounts of the bismuth are endocytosed by enterocytes [137] and some would seem to enter the bacterial cell wall [138]. There is evidence that bismuth in this arena directly inhibits lipases [139], pepsin activity [140] and gastrin production [136, 141]. Bismuth is known to form insoluble precipitates with phosphates, a process which can be used to stain functional groups prior to electron microscopy. As such it would be expected to inhibit a number of enzymes such as alkaline phosphatase [142] and most ATPases [143, 144]. Bismuth has also been reported to operate via an interaction with the sulphydryl function (SH−) [143, 144] which are present on many of the membrane proteins. However, the dominance of this chemical process would seem to diminish as one progresses from arsenic to antimony, and it is thought that this process will have little impact on the antiulcer properties of bismuth. Many other factors are observed to change as a function of bismuth such as the fall in pepsinogen levels [145, 146]. However, in the complex system prevailing in the stomach it is difficult to identify if these are direct effects, effects arising from an interaction in the stomach or effects of bismuth on the bacterium,

It is apparent that bismuth in this system is multifunctional, affecting the stomach beneficially to a small extent and bacterium detrimentally to a small extent creating sufficient protective and inhibitory properties to assist stomach wall repair while allowing the other components in triple therapy to be therapeutically more efficacious.

Although bismuth penetrates deeply into the stomach lining, the transfer of bismuth to the bloodstream can be very low ($< 1\%$ of the dose) [147] with the bulk of the bismuth passing through the lower gut being excreted in the faeces as the sulphide (Bi_2S_3). Although transfer of bismuth into the circulating system is low, immediately after dosing a sharp increase in plasma bismuth (*c.* 20–100 ng l^{-1}) is observed. This non-toxic bolus of bismuth is rapidly cleared from the body by the kidneys. In reverse, however, owing to the slow transfer of bismuth between the tissues and the bloodstream, the excretion of the bismuth accumulated from the large dosages used (*c.* 20 g) can continue for some months after the cessation of therapy.

Considering the prolonged and sustained nature of the therapy (20 g in two weeks) it is not surprising that, with time, circulating bismuth is observed to transfer slowly to the tissues such as the kidneys, brain, bone, spleen and liver [148]. The slow build up of metal in the brain causes great concern within the medical community; once in the brain it can induce delirium, psychosis, ataxia, myolonus and seizures. Clinically there are few problems [149] except under unusual circumstances (e.g. kidney failure). However, it is more common for problems to arise through the excessive overadministration of commercially available preparations by individuals [150]. Many of the neurotoxic effects of bismuth are observed to be reversible rather than fatal [149].

The success of bismuth therapy in treatment of stomach ailments has led to the suggestion that it may play a role in the treatment of other bowel disorders, especially those in the lower gastrointestinal tract, that is, in diseases such as ulcerative colitis [151]. Certain organisms similar to *Helicobacter pylori* (e.g. *Helicobacter hepaticus*) have been shown to colonize the ceca and colon of mice, suggesting that erosive diseases of the lower gastrointestinal tract may also arise through infection. The ability of the standard bismuth preparations to stimulate prostaglandin production [133] and inhibit sulphatase activity (the ability to sulphate mucin) [152] suggests that bismuth may eventually have a role in the treatment of these illnesses [153].

8.4.4 *New therapies*

With some justification, this area still holds some fascination for many workers. There is a belief that there remains a place for group 15 compounds in therapy and that compounds of these elements can stimulate therapeutic effects which cannot be rivalled by standard organic therapies.

Early studies sought to develop these elements as anticancer compounds. With the current reports of the *in vitro* anticancer properties of diphenylantimony compounds, specifically derivatives which incorporate dithiophosphorus ligands ($Ph_2Sb(S_2PR_2)$, R = Ph or OPr^i), it would seem that the conviction that this is a viable goal remains [154]. However, these elements will always be perceived as being too toxic and it is doubtful that any of these compounds will ever reach the clinic. Only in fields where there is no viable therapy will compounds of these elements be accepted, and even then only as interim solutions. This is the case with viral infections and specifically AIDS. Simple nitrate and tartrates of bismuth(III) have shown some activity in cell models [155]. Attempts have been made to generate combination therapies, such as bismuth-*N*-acetylcysteine complexes and compounds of antimony and tungsten, namely ammonium 21-tungsto-9-antimonate which would seem to have some ability to inhibit viral reverse transcriptase [156].

8.5 Group 15 in the human circulatory system

Compounds of the group 15 elements enter the circulatory system either as part of therapy, naturally or by accident. The body's circulation is a highly active environment comprising many binding sites and redox active areas. However, the body is organized such that the intracellular environment is reducing in nature, with the extracellular environment being more redox tolerant. This is especially true for mild oxidants such as arsenic(V) and antimony(V) which require vicinal thiolates for efficient reduction *in vivo*. Many proteins in serum (e.g. albumin) contain a single protected thiol function which although capable of binding soft metals [As(III) or Au(I)] is independently incapable of reducing arsenic(V) and antimony(V). Thus although the redox nature of arsenic and antimony make them more susceptible to adopt the lower oxidation state *in vivo* and to form stable complexes with thiolates, the metals must first cross the cell wall to an appreciable extent to initiate the chemistry required to increase substantially their toxicity.

Concurrent with the slow transport of arsenic and antimony species across cell walls is their excretion from the serum by the kidneys. These two processes occur at comparable rates. However, a small proportion of the high-valency metals present in the serum will still be transported across the cell wall. In the bloodstream the major cell type is the erythrocyte which contains *c.* 2 mM glutathione and *c.* 5 mM haemoglobin, both of which can act as reducing agents for the higher oxidation state species and as ligands for the lower oxidation states species as shown in reactions [8.7]–[8.13] below which illustrate a selection of reactions available to arsenicals in the intracellular environment, indicating the possible mechanisms for glutathione depletion, the formation of protein cross-links and the formation of metal–thiol conjugates.

$$[HAsO_4]^{2-}{}_{(extracellular)} \rightleftharpoons [HAsO_4]^{2-}{}_{(intracellular)} \quad [8.7]$$

$$[HAsO_4]^{2-} + 2GSH \rightarrow [AsO_2]^- + GSSG \quad [8.8]$$

$$[HAsO_4]^{2-} + \text{protein-SH} + GSH \rightarrow [AsO_2]^- + \text{protein-S-SG} \quad [8.9]$$

$$[HAsO_4]^{2-} + 2\text{protein-SG} \rightarrow [AsO_2]^- + \text{protein-S-S-protein} \quad [8.10]$$

$$\text{protein-S-SG} + GSH \rightarrow \text{protein-S-SG} + \text{protein-SH} \quad [8.11]$$

$$[AsO_2]^- + 3GSH + H^+ \rightarrow As(SG)_3 + 2H_2O \quad [8.12]$$

$$[AsO_2]^- + \text{protein-SH} \rightarrow \text{As(III)-S-protein} \quad [8.13]$$

The entry of high oxidation state metals into the cytosolic environment initiates a sequence of redox reactions and ligand exchange process which effectively distributes soft metals across a wide range of ligand sites. Within the erythrocyte the bulk of the arsenic has been shown to complex to small molecules such as glutathione (*c.* 70%–75%) and the major protein, haemoglobin (*c.* 20%), which contains a free cysteinyl residue [157]. The depletion of intracellular glutathione through oxidation, metal binding and ligand exchange processes (reactions [8.7]–[8.13]) has a profound effect on organelles such as the mitochondria [158] (and therefore respiration) and many enzymes such as glutathione peroxidase [159] and glutathione reductase [160], both of which utilize glutathione as a co-factor. Thus, not only is arsenate an oxidant in its own right but it acts in a manner which allows it to impair the function of two key enzymes involved in the cell's natural antioxidant defence mechanisms. In this manner it can exacerbate the stress (e.g. lipid peroxidation) experience by the cell [161] from other sources. The reactions shown above are equally available to the antiparasitic agents of arsenic and antimony serving to demonstrate the dynamic behaviour of these species *in vivo*. Equally, it becomes almost impossible to identify a key antiparasitic reaction which is ultimately responsible for the efficacy of these species.

Where metal-induced stress prevails for a lengthy period of time the cells will begin to adjust to their presence by producing elevated levels of glutathione [159]. Considering the amount of haemoglobin present inside the erythrocyte, at low concentrations, binding of soft metals to the cysteinyl residues (apart from some haem degradation and the release of iron which induces further oxidation through Fenton chemistry) does not create a major problem for cell function. The production of thioarsinites such as $As(SG)_3$ [160], although potent inhibitors of certain enzymes such as glutathione reductase, are species which are easily recognized and either metabolized (methylated) or excreted though the hepatobiliary system.

Whereas reactions [8.7]–[8.13] are available to the organoarsenicals (monomethylarsonate and dimethylarsonate) [157], these organometallic species do not bind as efficiently to glutathione as does arsenite. Furthermore, although these ligand exchange reactions (reactions [8.11]–[8.13]) are equally available to the low oxidation state species of these three elements, when added to cellular systems or introduced into thc body they are more likely to bind to scavenger proteins in the serum (e.g. albumin) and on the cell walls prior to their entry into the cell, thus diluting their effect.

Of greater concern is the binding of arsenic and antimony to the active sites of low concentration soft donors (i.e. the remaining 5%–10%) at key functional sulphydryl (HS−) sites. The binding of many cell messengers, for example steroids [162] and insulin [163], which control a vast range of process, for example nutrient uptake and protein degradation [164], commonly involve the use of vicinal thiolates. Arsenite and antimonite form strong complexes with these sites, inducing conformational changes and dysfunction.

An important facet of these effects is the ability of low valent arsenic to interfere in glucose metabolism. This interaction is an early event in arsenic and antimony toxicity and creates serious problems for the production of the cellular energy required to offset the stress created elsewhere by these heavy metals. As will be discussed below, the effects of arsenate on cellular energy pathways is not restricted to glucose-derived energy production. It will also uncouple phosphorylation pathways and deplete ATP, that is, energy transduction. Thus in severe cases of heavy metal toxicity the assault on the cell comes from many directions, all of which are detrimental to viability.

Unlike the effects on phosphate pathways which are mainly caused by pentavalent metals, the effects on glucose metabolism is more pronounced for the lower valent forms (compared with higher valent forms by an order of magnitude). The effect on glucose metabolism is not a single-site effect, and arsenic compounds would seem to affect a number of key points in the production of energy from this source. For example, the enzymes involved in glucose catabolism and which effect pyruvate [165] and keto-glutaric acid oxidation [163] are both inhibited by arsenite.

In model studies arsenite and phenylarsine oxide have been shown to inhibit the cell's ability to take up glucose from the suspending media under stress conditions [163]. The mechanism behind this effect would seem to arise from an interaction of phenylarsine oxide either at the insulin binding site or by binding directly to the transporter at exposed sulphydryl sites thus altering (locking) the transport protein conformation. Two sulphydryl functions exist on the plasma membrane surface of the glucose transporter remote from the glucose binding site. These residues are known to be capable of binding other thiophylic species such as Ellman's reagent and organomercurials [166]. The rationale would seem to be that by binding to the protein the arsine oxide marshals the transporters to the plasma membrane surface preventing the translocation of its natural substrate. The outcome of this interaction has many similarities to the effects of insulin. Phenylarsine oxide is an order of magnitude more effective than arsenite at inhibiting glucose uptake. This is possibly because the former requires only two thiolate donors (which the protein can provide), generating a protein–metal complex which exhibits slow exchange kinetics. Arsinite, however, requires a further donor group to stabilize the protein–arsenic complex which if not apposite to its ligand preference and geometry will facilitate more rapid ligand exchange and thus allow some residual activity.

Not only is the transport of glucose, the cell's primary supply of nutrient, inhibited, but its back up system, glucogenesis, is also found to be impaired. These changes in turn lead to carbohydrate depletion, glucose intolerance and hyperglycaemia, which are the more visible side effects off arsenite poisoning [167]. It is not surprising therefore that there would seem to be a weak link between diabetes mellitus and the consumption of arsenic in drinking water [168].

As has been discussed above, the presence of heavy metals, especially soft species such as arsenic(III), antimony(III) and bismuth(III), can be highly detrimental to mammalian biochemistry. As such, compounds have been developed to lower the toxicity and accelerate the removal of such chemical species. One of the most efficient methods developed thus far is known as chelation therapy, which is based on the generation of highly stable complexes *in situ* susceptible to excretion.

A wide range of chelating species have been developed and a small selection is shown in Scheme 8.4. All of these species are designed to chelate the lower oxidation forms of arsenic, antimony and bismuth, which are more toxic especially when bound to functional protein sites. The successful compounds employ vicinal thiols such as 2,3-dimercaptopropanol (BAL, British anti-Lewisite or dimercaprol), 2,3-dimercaptosuccinic acid (DMSA) and 2,3-dimercaptopropane-1-sulphonic acid (DMPS). These species increase the LD_{50} of the arsenic approximately fourfold [169] and increase the rate of heavy metal excretion by a similar factor [170]. DMSA and DMPS are preferred at this time to BAL as they can be delivered orally and are relatively non-toxic. In comparison, BAL

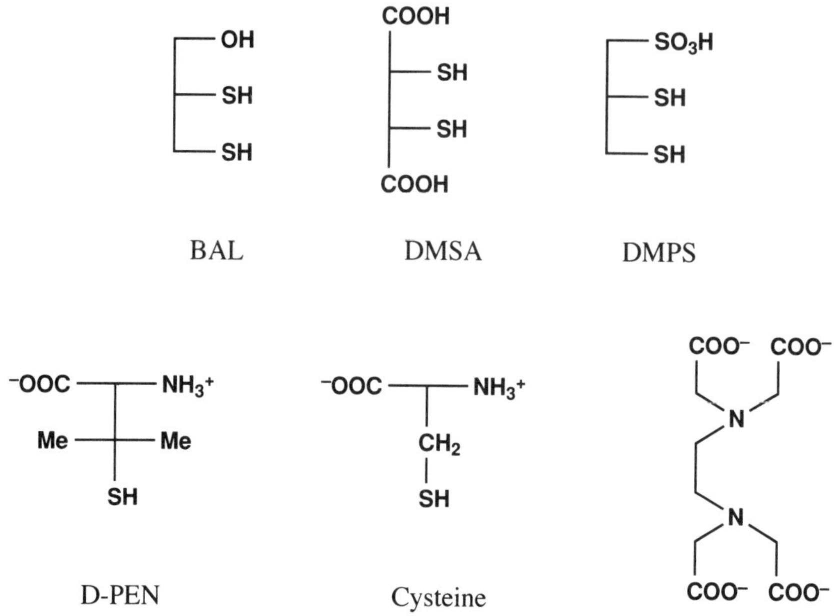

BAL DMSA DMPS

D-PEN Cysteine EDTA

Scheme 8.4 Some standard chelates for the heavy metals arsenic, antimony and bismuth. BAL = British anti-Lewisite (2,3-dimercapto-propanol); DMSA = 2,3-dimercapto-succinic acid; DMPS = 2,3-dimercaptopropane-1-sulphonic acid; D-Pen = D-penicillamine; EDTA = ethylenediaminetetraacetic acid.

is insoluble in water and is itself quite toxic, thus preventing the administration of large amounts in response to an unknown amount of heavy metal, which is a common occurrence in accidental poisoning. It could be implied that BAL is of little value compared with DMSA and DMPS. However, because of its hydrophobic nature, BAL will remove arsenic and bismuth from brain tissue in a more efficient manner than will DMSA and DMPS and as such still has a place in heavy metal detoxification [148].

Normally these elements are removed from the body via the kidneys. Treatment with chelation therapy somewhat alters the biochemistry of these elements allowing them to be excreted via the bile into the faeces also. The success of these agents in removing these elements, however, depends heavily on the timing of the intervention, the region of the body in which the metalloid resides and the nature of the intoxication.

D-penicillamine (D-PEN) is a renowned metal chelator and has a special importance in the treatment of copper overload (Wilson's disease). It has been used for the group 15 elements but would seem to be inferior to the others shown [148]. Cysteine can chelate arsenite [171] but, if used to treat arsenate poisoning, as a good reducing agent it will exacerbate the condition by facilitating the reduction of arsenate to the more toxic arsenite. In the process cysteine becomes oxidized, producing the ineffective chelator cystine. EDTA will chelate all of these metals but it may not assist their complete excretion. It is more likely to increase their distribution, thus lowering the metal concentration at the site of insult [172]. For example, bismuth–EDTA therapy has been shown to increase the amount of bismuth in the brain. Through treatment with chelation therapy it is possible to abstract heavy metals from strong binding sites. As a result the activity of many of the enzymes inhibited by low valency group 15 elements can be restored through chelation therapy.

As discussed above, the body acts to detoxify arsenic via methylation and, after prolonged intoxication, the concentration of the methylated arsenicals $Me_2AsO(OH)$ and $MeAsO(OH)_2$ increase in the bloodstream. Chelation therapy relies on the presence of arsenic in the lower oxidation state, which if not present initially can be produced via the reducing capacity of the detoxification agents themselves, producing Me_2AsSR, $MeAsSR_2$ and $[AsO_2]^-$. Whereas the monomethylthioarsinites continue to have two vacant coordination sites and are thus amenable to the formation of a chelate with DMSA, DMPS or BAL, the dimethylthioarsinte, with only one available coordination site, no longer has any capacity to chelate with these agents. As the latter is the major product in long-term arsenic intoxication it can be appreciated why 'chelation' therapy is ineffective for clearing arsenic if the treatment is not effected quickly.

The complexity of the behaviour of these species *in vivo* is demonstrated further by the formation of antimony(III)–lactic acid complexes which are soluble in non-polar solvents. This suggests that the antimonite- and arsenite-induced stress will produce elevated levels of lactic acid which, once combined with the small anions, will be capable of distributing these species further into the hydrophobic regions (lipids) of the cell [173].

8.6 Stress and inducable proteins

Within the tolerances available to a biological system the structure of the three tetrahedral anions, phosphate, arsenate and stibnate, pose a significant problem for their discriminatory transport and regulation by the cell. Had their structures or pK_as been substantially different it might have been possible for the biochemistry operating in species such as bacteria or parasites simply to ignore these anions. Great specificity for phosphate over arsenate has, however, been developed in specific places (e.g. the kidney), where it is very important to effect efficient clearance of these toxic anions and prevent metal-induced organ or organism failure. Even many of these proteins, although non-functional with arsenate as a substrate, are inhibited by the presence of the anion and it is likely that the anion is still recognized but not transported. This hypothesis is further supported by the studies which have found that monomethylarsonate and dimethylarsonate, which are substantially structurally different from phosphate, are relatively unreactive in phosphate biochemistry.

Where the tolerance for the tetrahedral anions is not so great it is possible to have the co-transport of phosphate and arsenate. This is the case for the human band III (anion transporter) and certain sodium pumps. Here arsenate is accepted along with phosphate [174]. Furthermore, there is some evidence to show that the entry of ingested arsenate into mammalian biochemistry occurs via a shared phosphate–arsenate transporter in the gastrointestinal tract [175].

Intracellularly, arsenate can substitute at phosphate chemistry. Many proteins, for example the Na/K/ATPase, which are routinely phosphorylated and dephosphorylated, can be arsenylated and dearsenylated. Arsenate uncouples ATP synthesis in bacteria [176], leading to a progressive depletion in cellular ATP and glucose-6-phosphate levels. In turn this leads to a build up of intracellular phosphate [177] and an increase in glucose consumption [178], the supply of which is seriously affected by arsenite. By uncoupling oxidative phosphorylation, arsenate induces a graded reduction in the transport of sodium and phosphate [179] and an increased calcium efflux [180] (via an interaction with (Ca-ATPase)). The reversible loss of flagellar action and loss of motility in *Escherichia coli* treated with arsenate can be ascribed to down-regulation of the phosphorylation process responsible for motion in these bacteria [181, 182]. The loss of motility of parasites in this manner may be yet another beneficial effect of these compounds in therapy [183].

Some synthetic arseno-compounds, for example arsenopyruvate [184], will behave as substrates for the enzymes normally reserved for the parent phosphorus species. This similarity between arsenate and phosphate complexes can be used in a subtle manner. For example, use of the isosteric analogue of glycerol-3-phosphate, 3,4-dihydroxybutylarsonic acid as a substrate for glycerol-3-phosphate dehydrogenase facilitates the enzymatic release of arsenite as a toxin [185]. Similarly, purine nucleoside phosphorylase catalyses the arsenolysis of inosine to form hypoxanthine and ribose-1-arsenate, which spontaneously hydrolyses to produce ribose and arsenate [186]. This is a further example of how arsenic can effectively drain biological energy by inducing a redundant cycle of ineffective chemistry.

As discussed above, a further concern exists for the biological system as both arsenate and stibnate are reasonable oxidizing agents within the context of biology. For this reason and those mentioned above (reactions [8.7]–[8.13]) it can be appreciated why these two anions must be efficiently excluded from the cell or, once admitted, removed:

$$H_3PO_4 + 2H^+ + 2e^- \rightarrow H_3PO_3 + H_2O, \quad E^0(H^+) = -0.28\ V \qquad [8.14]$$

$$H_3AsO_4 + 2H^+ + 2e^- \rightarrow HAsO_2 + 2H_2O, \quad E^0(H^+) = 0.56\ V \qquad [8.15]$$

$$H_3SbO_4 + 2H^+ + 2e^- \rightarrow HSbO_2 + 2H_2O, \quad E^0(H^+) = 0.75\ V \qquad [8.16]$$

Biomethylation-producing anions such as methylarsonate and volatile species such as trimethylarsine or stibine offer the cell one route by which it can expel these species. However, this type of process is relatively inefficient and consumes a reasonable amount of biological energy and has an attendant potential to produce toxic side effects. Many bacterial (e.g. *Escherichia coli*, *Staphylococcus aureus*) systems have thus developed a series of proteins specifically designed to expel arsenic and antimony from the cell. At this time there are two classes of protein known: the anion translocating ATPase and the less well defined *p*-glycoproteins [187, 188], both of which actively remove arsenic and antimony as the oxyanions $[AsO_2]^-$ and $[SbO_2]^-$. Significantly, deletion of the genes for these two transporting systems renders the cells hypersensitive to antimonyls and arsenicals [189]. Conversely, the presence of these proteins generates multidrug resistance on the host cell. As such, there is some concern that these transporting species are now being observed to occur in a number of parasites such as leishmaniasis. As stated above, these proteins are gene encoded, being induced by the presence of the trivalent anions of arsenic, antimony and, in some species, bismuth [190] in the intracellular space. These anions will be produced normally under stress conditions once arsenate or stibnate reach concentrations where they begin to react significantly with functional thiols *in situ*, becoming reduced in the process (reactions [8.7]–[8.13]).

The anion-translocating ATPase consists of three polypeptides, a (15 kDa) catalytic unit, the membrane channel (35 kDa) and a reductase (15 kDa). The reductase binds arsenate or stibnate via three cysteine residues in an allosteric interaction producing arsenite or stibnite, respectively; this process, by taking advantage of the favourable oxidation potentials of the two toxins, confers the necessary discrimination required to separate these ions from phosphate [191]. The protein dimerizes and brings the anion to the attention of the channel through which it is exported [192]. Anion translocation is a process which is effectively active efflux and is achieved at the expense of ATP. The binding site of the reductase involves three rather than two sulphydryls, indicative that reduction is achieved by the formation of a disulphide bond, with the metalloid being held on the reductase *post* reduction as a thioarsenite or thiostibnite, consistent with asymmetry in the cysteinyl residues. Subtle discrimination exists within this system. For the export of arsenite as opposed to arsenate the reductase is not required [193] and there would seem to be greater specificity for stibnate over arsenate [194].

There is weak evidence to suggest that arsenite can be expelled from a cell through an interaction with glutathione-S-transferase [195]. It is envisaged that the functional thiol residue acts as the transfer agent in this process. However, what is not so clear is the mechanism by which the metalloid is removed from the protein on the plasma surface. The importance of the transferase enzyme to arsenic toxicity is demonstrated by blocking the protein with ethacrynic acid, which significantly decreases the cell's resistance to arsenite.

The presence of group 15 metals would seem to induce the production of other protective proteins. The induction of metallothionein has been investigated as one mechanism by which cells develop arsenic tolerance [196]. The speciation of the circulating arsenical species is important to the degree of metallothionein induction. Arsenite is the more potent, inducing a 30-fold increase in metallothionein. In comparison and at similar concentrations the inducing capacity falls as the arsenic progresses up the biotransformation hierarchy, that is, in the order arsenate (30% of arsenite values), monomethylarsonate (2%) and dimethylarsonate ($< 1\%$) [197]. Although hepatic metallothionein concentrations rise, and arsenite toxicity is ameliorated, little binding of these soft metals to the polypeptide is observed. These studies would seem to indicate that although metallothionein concentrations are elevated, toxicity is ameliorated through more efficient clearance of arsenic from the host [198].

Of more value is the ability to lower significantly the nephrotoxicity of the anticancer compounds *cis*-platin [198, 199], 5-flurouracil and adrymycin [200] and reduce the nephrotoxicity and hepatotoxicity of cadmium [201–203] by pretreating animals with bismuth nitrate (5–50 nmol kg^{-1}). Studies using cell culture have shown that bismuth nitrate acts by inducing the intracellular accumulation of metallothionein in organs (e.g. kidney [199, 204] and liver, lung and bone marrow) [201–203] but not in the tumour [205]. As bismuth compounds are common treatments for intestinal disorders with trace amounts being cleared by the kidney, some workers have put forward the hypothesis that co-administration of bismuth with platinum may offer a regime which is less toxic. As bismuth is cleared by the kidneys there is a greater chance of interaction at this site in the body. Small molecular weight binding proteins (*c.* 4) of approximate molecular weight 7500 Da (bismuth content *c.* 0.1 $\mu g\ mg^{-1}$) which may be related to copper binding proteins have been detected [206, 207].

Not only do the low valency compounds of these elements induce metallothionein, but they also stimulate the production of the synthesis of haem oxygenases and heat-shock proteins. The induction of haem oxygenases would seem to occur via the desorption of these species into the mitochondria membranes [196] altering their structures, stimulating the production of proteins which [208] enhance the degradation of haem in the liver and kidney.

In contrast, the arrival of the trivalent ions in the nucleus and their complexation to thiol residues on the nuclear proteins [209] induce the production of low molecular weight heat-shock proteins (28–110 kDa). These small proteins are normally produced as a response to thermal stress and assist the organism in becoming more thermotolerant. Here these proteins are induced by the exposure of cells to heavy metal shock, that is, 50–100 nM arsenite for 1 h. However, there is evidence that the range of proteins is not exactly the same for thermal and arsenite, arsenite producing a unique 28 kDa protein [210].

8.7 Immunochemistry and radiochemistry

Both antimony and bismuth have, at this time, limited roles as therapeutic vehicles for biologically targeted imaging and radiotherapy. Antimony colloid is used as a vehicle for technetium in tomography. Antimony sulphide (Sb_2S_3) colloid is prepared by reducing potassium antimony tartrate with H_2S in the presence of polyvinylpyrolidene. The colloid is filtered generating a homogenous suspension of particles of approximately similar size (*c.* 3–30 nm). Sodium ^{99m}Tc-pertechnitate is added and is adsorbed onto the colloid by lowering the pH of the solution. The solutions are then autoclaved and diluted in phosphate-buffered saline. Typical use of colloid technology is to monitor the flow of fluids inside the body, that is, drainage in the lymphatic system and the imaging of the gut. As these species are cleared by the liver they can be used to image the function of this organ also. Although the functional image arises via the presence of the technetium nucleus, the information obtained really reflects the behaviour of the antimony sulphide colloid in the body.

Unlike antimony, bismuth, in common with many of the heavier elements, has a number of available radioactive isotopes: ^{206}Bi, ^{212}Bi and ^{213}Bi (Table 8.6) [211–213].

The α emitter ^{212}Bi is preferred as it offers the best compromise between the linear energy transfer of the particle (highest) and a reasonable half life. Although the α particle has a low penetration and is limited in distance to only several cell diameters it has a good destructive capacity, producing a good yield of DNA double-strand breaks which are, furthermore, more difficult for the tumour cell to repair than strand breaks caused by X-rays [212]. The activity of the bismuth nucleus also transiently arrests the activity of cells in a dose-dependent manner [214].

Owing to the relatively short half-life and the moderate penetrative capacity of the α-particles it is important that the bismuth is efficiently chaperoned to the site of importance. This is achieved in two ways through the nature of the ligand. The complex itself can be biospecific as is the case for ethylenediaminetetramethylene phosphonate (EDTMP,

Table 8.6 Isotopes of bismuth relevant to biological targeted radiotherapy.

Isotope	Daughter	Particle	Half-life	Particle energy (MeV)
$^{206}_{83}Bi$	$^{206}_{82}Pb$	β^+	6 days	0.98
$^{212}_{83}Bi$	$^{208}_{81}Tl$	α	1 hr	6.21
$^{213}_{83}Bi$	$^{209}_{81}Tl$	α	47 min	5.98

Sources: Kaspersen, F. M., Bos, E., Doornmalen, A. V. *et al.*, *Nucl. Med. Comm.*, **16,** 468, 1995; Kozak, R. W., Atcher, R. W. and Gansow, O. A., *Proc. Natl. Acad. Sci. USA*, **83,** 474, 1986; Shadley, J. D., Whitlock, J. L., Rotmensch, J. *et al.*, *Mut. Res.*, **248,** 73, 1991.

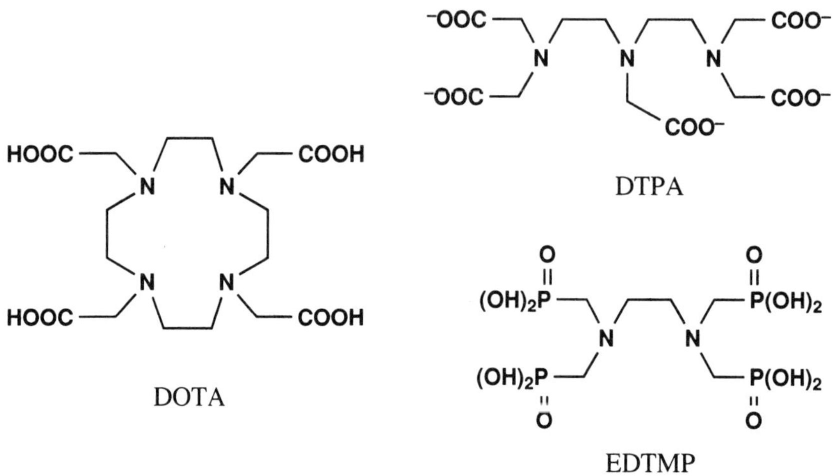

Scheme 8.5 Some of the common ligands currently being used to chelate bismuth isotopes for radiopharmaceuticals. EDTMP = ethylenediaminetetramethylene phosphonate; DOTA = 1,4,7,10-tetraazacyclododecane *N,N′,N″,N‴*-tetraacetic acid; DTPA = diethylenetriamine-pentaacetic acid.

Scheme 8.5) which generates a complex which preferentially accumulates in bone. Of some interest at this time is the development of tagged antibodies. By using an organic coupling agent the metal–ligand unit can be attached to an antibody raised against the tumour [211, 215, 216]. In this way the bismuth-tagged antibody, once introduced into the host, seeks out and attaches itself directly onto the malignant tissue. In this manner it is possible to arrange preferentially the release of high-energy particles in the vicinity of the malignancy. Keeping the release of the α particles in proximity to the tumour tissue in this manner prevents the α particles damaging the surrounding healthy tissue. Two ligands are currently being researched for this purpose: diethylenetriaminepentaacetic acid (DTPA) [211, 215] and 1,4,7,10-tetraazacyclododecane *N,N′,N″,N‴*-tetraacetic acid (DOTA) [216] (Scheme 8.5).

8.8 Staining

Antimony and bismuth are both used for the staining and histological examination of the subcellular distribution of functional groups in the cell ultrastructure. The contrast in the chemistry of these two group members is elegantly demonstrated here as antimony is used to stain cations and bismuth anions. Potassium pyroantimonite $[KSb(OH)_6]$ is used to precipitate calcium, magnesium and sodium in biological tissues thus generating an electron-dense precipitate suitable for electron microscopy or X-ray analysis. Calcium is by far the most sensitive cation with a detection limit in the range 10^{-6}–10^{-7} mol l^{-1} followed by magnesium (10^{-5} mol l^{-1}) and sodium (10^{-2} mol l^{-1}).

In contrast, bismuth oxynitrate ($BiO{\cdot}NO_3$) is used as a stain for phosphate groups, and bismuth periodate for polysaccharides. For the analysis of highly phosphorylated proteins and nuclear components the samples are usually fixed in gluteraldehyde solution to block exposed reactive amino residues. However, not all phosphates will be stained by bismuth.

References

1. Bichler, M., Platschka R. and Polijanc, K. (1992) *J. Radioanal. Nucl. Chem.*, **165**, 61.
2. Chen, S.L., Yeh S.J. and Lin, T.H. (1995) *Biol. Trace Elem. Res.*, **48**, 263.
3. Bettinelli, M., Baroni, U. and Pastorelli, N. (1988) *J. Anal. Atom. Spec.*, **3**, 1005.
4. Austin, L.S. and Millward, G.E. (1988) *Atmosph. Environ.*, **22**, 1395.
5. Cullen, G.V. and Siviour, N.G. (1982) *Water Res.*, **16**, 1357.
6. Pepin, J. and Milord, F. (1991) *Trans. Roy. Soc. Trop. Med. Hyg.*, **85**, 222.
7. Roberts, W.L. and Rainey, P.M. (1993) *Antimicrob. Agents Chemother.*, **37**, 1842.
8. Berstad, K. and Berstad, A. (1994) *Tidsskrift Norske Laegeforehing*, **114**, 932.
9. Abdelghani, A.A., Anderson, A.C., Jaghabir, M. and Mather, F. (1986) *Arch. Environ. Health,* **41**, 163.
10. Ungers, L.J., Jones, J.H., McIntyre, A.J. and McHenry, C.R. (1985) *Am. Ind. Hyg. Ass. J.,* **46**, 416.
11. Ni, Z.-M., Le X.-C and Han, H.B. (1986) *Anal. Chim. Acta*, **186**, 147.
12. Hartley, B.M. and Hartley, M.I. (1989) *J. Radiolog. Protect.*, **9**, 165.
13. Reddy, R.G. (1982) *Resources and Conservation,* **9**, 333.
14. Bloch, P., Vanderborght, B., Adams, F. and Van Landuyt, J. (1983) *Int. J. Environ. Anal. Chem.*, **14**, 257.
15. Landsberger, S. (1988) *Anal. Chem.*, **60**, 1842.
16. Papastefanou, C., Manolopoulou, M. and Charalambous, S. (1988) *Sci. Total Environ.*, **72**, 81.
17. Akeredolu, F.A., Barrie, L.A., Olson, M.P., Oikawa, *et al.* (1994) *Atmos. Environ.*, **28**, 1557.
18. Andreae, M.O. (1980) *J. Geophys. Res.*, **85**, 4512.
19. Siversten, T., Daae, H.L., Godal, A. and Sand, G. (1995) *Environ. Pollut.*, **90**, 75.
20. Steinnes, E. and Henriksen, A. (1993) *Water Air and Soil Pollut.*, **71**, 167.
21. Berg, T., Royset, O. and Steinnes, E. (1995) *Atmos. Environ.*, **29**, 353.
22. de la Calle-Guntinas, M.B., Madrid, Y. and Camara, C. (1991) *Analyst.*, **116**, 1029.
23. Abbasi, S.A. (1989) *Anal. Lett.*, **22**, 237.
24. Andreae, M.O., Byrd, J.T. and Froelich, P. (1983) *Environ. Sci. Tech.*, **17**, 731.
25. Azcue, J.M., Murdroch, A., Rosa, F. and Hall, G.E.M. (1994) *Environ. Tech.*, **15**, 669.
26. Aurilio, A.C., Durant, J.L., Hemond, H.F. and Knox, M.L. (1995) *Water Air Soil Pollut.*, **81**, 265.

27. McCartney, M., Kershaw, P.J., Woodhead, D.S. and Denoon, D.C. (1994) *Sci. Total Environ.*, **141**, 103.
28. Ghuman, G.S., Motes, B.G., Fernandez, S.J. *et al.* (1993) *J. Environ. Radioact.*, **21**, 161.
29. Eckl, P., Hofmann, W. and Turk, R. (1986) *Rad. Environ. Biophys.*, **25**, 43.
30. Ainsworth, N., Cooke, J.A. and Johnson, M.S. (1990) *Environ. Pollut.*, **65**, 65.
31. Ainsworth, N., Cooke, J.A. and Johnson, M.S. (1990) *Environ. Pollut.*, **65**, 79.
32. Ainsworth, N., Cooke, J.A. and Johnson, M.S. *Water Air Soil Pollut.*, **57-58**, 193.
33. Pilarski, J., Waller, P. and Pickering, W.F. (1995) *Water Air and Soil Pollut.*, **84**, 51.
34. Thanabalasingam, P. and Pickering, W.F. (1986) *Environ. Poll. Ser. B. Cem. Phys.*, **12**, 233.
35. Andreae, M.O., Asmode, J.F., Foster, P. and Vant Dack, L. (1981) *Anal. Chem.*, **53**, 1766.
36. Ndiokwere, C.L. (1983) *Radioisotopes,* **32**, 117.
37. Challenger, F. (1945) *Chem. Rev.*, **36**, 315.
38. Attar, K.M., El Faer, M.Z., Rawdah, T.N. and Tawabini, B.S. (1992) *Marine Pollut. Bull.*, **24**, 94.
39. Jorhem, L., Slorach, S., Sundstrom, B. and Ohlin, B. (1991) *Food Add. Contam.*, **8**, 201.
40. Nixon, D.E. and Moyer, T.P. (1992) *Clin. Chem.*, **38**, 2479.
41. Leah, R.T., Evans, S.J. and Johnson, M.S. (1992) *Marine Pollut. Bull.*, **24**, 544.
42. Le, S.X.C., Cullen, W.R., and Reimer, K.J. (1994) *Environ. Sci. Tech.*, **28**, 1598.
43. Edmonds, J.S., Shibata, Y., Francesconi, K.A. *et al.* (1992) *Sci. Total Environ.*, **122**, 321.
44. Francesconi, K.A., Stick, R.V. and Edmonds, J.S. (1990) *Experimentia,* **46**, 464.
45. Hanaoka, K., Koga, H., Tagawa, S. and Kaise, T. (1992) *Comp. Biochem. Physiol. B Biochem. Mol. Biol.*, **101**, 595.
46. Kapustka, L.A., Lipton, J., Galbraith, H. *et al.* (1995) *Environ. Toxicol. Chem.*, **14**, 1905.
47. Maeda, S., Nakashima, S., Takeshita, T. and Higashi, S. (1985) *Sep. Sci. Tech.*, **20**, 153.
48. Onken, B.M. and Hossner, L.R. (1995) *J. Environ. Qual.*, **24**, 373.
49. Meharg, A.A., Naylor, J. and McNair, M. (1994) *J. Environ. Qual.*, **23**, 234.
50. Weaver, R.W., Melton, J.R., Wang, D. and Duble, R.L. (1984) *Environ. Pollut. Ser. A Eco. Biol.*, **33**, 133.
51. Cullen, W.R., McBride, B.C., Pickett, A.W. and Reglinski, J. (1984) *Appl. Environ. Microbiol.*, **47**, 443.
52. Cheng, C.N. and Focht, D.D. (1979) *Appl. Environ. Microbiol.*, **38**, 494.
53. De Koe, T. (1994) *Sci. Tot. Environ.*, **145**, 103.
54. Weis, P., Weis, J. and Lores, E. (1993) *Marine Pollut. Bull.*, **26**, 428.
55. Gurleyuk, H., van Fleet-Stalder, V. and Chasteen, T.G. (1997) *Appl. Organomet. Chem.*, **11**, 471.
55a. Bailly, R., Lauwerys, L., Buchet, J.P. *et al.* (1991) *Brit. J. Indust. Med.*, **48**, 93.
56. Richardson, B.A. (1994) *J. Forensic Sci. Soc.*, **34**, 199.
57. Howatson, A.G., Patrick, W.J.A., Fell, G.S. *et al.* (1995) *Lancet,* **345**, 1044.
58. Mohri, T., Hisanaga, A. and Ishinishi, N. (1990) *Food Chem. Toxicol.*, **28**, 521.
59. Oladimeji, A.A., Qadri, S.U., Tam, G.K.H. and DeFreitas, A.S.W. (1979) *Ecotox. Environ.*, **3**, 394.
60. Bertolero, F., Marafante, E. and Edel Rade, J. (1981) *Toxicol.*, **20**, 35.
61. Vahter, M. and Lind, B. (1986) *Sci. Total. Environ.*, **54**, 1.
62. Buchet, J.P., Lauwerys, L., Vandevoorde, A. and Pycke, J.M. (1981) *Arch. Belg. Med.*, **39**, 465.
63. Buchet, J.P., Pauwels, J. and Lauwerys, R. (1994) *Environ. Res.*, **66**, 44.
64. Sabbioni, E., Fischbach, M., Pozzi, G. *et al.* (1991) *Carcinogenis,* **12**, 1287.
65. Hanaoka, K., Kogure, T., Miura, Y. *et al.* (1993) *Chemosphere,* **27**, 2163.
66. Kaise, T., Hanaoka, K. and Tagawa, S. (1987) *Chemosphere,* **16**, 2551.
67. Phillips, D.J.H. and Depledge, M.H. (1986) *Water Sci. Tech.*, **18**, 213.
68. Lagunas, R., Pestana, D. and Diez Masa, J.C. (1984) *Biochemistry,* **23**, 955.
69. Cullen, W.R., Li, H., Pergantis, S.A. *et al.* (1994) *Chemosphere,* **28**, 1009.
70. Tsao, D.H.H. and Maki, A.H. (1991) *Biochemistry,* **30**, 4565.

71. Cullen, W.R., McBride, B.C. and Reglinski, J. (1984) *J. Inorg Biochem.*, **21**, 45.
72. Cullen, W.R., McBride, B.C. and Reglinski, J. (1984) *J. Inorg Biochem.*, **21**, 179.
73. Cullen, W.R., McBride, B.C., Manji, H. *et al.* (1989) *Appl. Organomet. Chem.*, **3**, 71.
74. Lam, W.C., Tsao, D.H.H., Maegley, A.H. and Reich, N.O. (1992) *Biochemistry,* **31**, 10438.
75. Bertolero, F., Pozzi, G., Sabbioni, E. and Saffiotti, U. (1987) *Carcinogenisis,* **8**, 803.
76. Falk, H., Caldwell, C.G. and Ishak, K.G. (1981) *Am. J. Ind. Med.*, **2**, 43.
77. Ohyama, K., Sonoda, K. and Kuwahara, H. (1982) *Dermatologica,* **164**, 161.
78. Motolese, A., Truzzi, M., Giannini, A. and Seidenari, S. (1993) *Contact Dermatitis,* **28**, 59.
79. Kentner, M., Leinemann, M., Schaller, K.H. *et al.* (1995) *Int. Arch. Occ. Soc. Med. Environ. Health,* **67**, 119.
80. White, G.P., Mathias, C.G.T. and Davin, J.S. (1993) *J. Occ. Med.*, **35**, 392.
81. Brune, D., Nordberg, G., and Wester, P.O. (1980) *Sci. Tot. Envir.*, **16**, 13.
82. Lai, M.K. and Jones, P.A. (1994) *Carcinogenisis,* **15**, 1085.
83. Kuroda, K., Endo, G., Okamoto, A. *et al.* (1991) *Mut. Res.*, **264**, 163.
84. Ostrosky Wegman, P., Gonsebatt, M.E., Montero, R. *et al.* (1991) *Mut. Res.*, **250**, 477.
85. Dong, J.T. and Luo, X.M. (1994) *Mut. Res.*, **315**, 11.
86. Ferm, V.H. and Hanlon, D.P. (1985) *Environ. Res.*, **37**, 425.
87. Yamamoto, S., Konishi, Y., Matsuda, T. *et al.* (1995) *Cancer Res.*, **55**, 1271.
88. Meng, Z. (1993) *Arch. Environ. Contam. Toxicol.*, **25**, 525.
89. Meng, Z. and Meng, N. (1994) *Biol Trace Elem. Res.*, **42**, 201.
90. Tezuka, M., Hanioka, K.I., Yamanaka, K. and Okada, S. (1993) *Biochem. Biophys. Res. Comm.*, **191** 1178.
91. Wiencke, J.K. and Yager, J.W. (1992) *Environ. Mol. Mutagen.*, **19**, 195.
92. Christie, N.T. and Costa, M. (1983) *Biol. Trace Elem. Res.*, **5**, 55.
93. Nordenson, I., Sweins, A. and Beckman, L. (1981) *Scan. J. Work. Environ. Health,* **7**, 277.
94. Yamanaka, K., Tezuka, M., Kato, K. *et al.* (1993) *Biochem. Biphys. Res. Comm.*, **191**, 1184.
95. Dong, J.T. and Luo, X.M. (1993) *Mut. Res.*, **302**, 97.
96. Jacobson Kram, D. and Montalbano, D. (1985) *Environ. Mut.*, **7**, 787.
97. Kachinskas, D.J., Phillips, M.A., Qin, Q. *et al.* (1994) *Cell Growth Different.*, **5**, 1235.
98. McCabe, M., Maguire, D. and Nowak, M. (1983) *Environ. Res.*, **31**, 323.
99. Li, J.H. and Rossman, T.G. (1989) *Mol. Toxicol.*, **2**, 1.
100. Hartwig, A. (1995) *Biometals,* **8**, 3.
101. Yamanaka, K., Hoshino, O., Okamoto, M. *et al.* (1990) *Biochem. Biophys. Res. Comm.*, **168**, 58.
102. Nordenson, I. and Beckman, L. (1991) *Hum. Heredity,* **41**, 71.
103. Jongen, W.M.F., Cardinaals, J.M., Bos, P.M.J. and Hagel, P. (1985) *Food Chem. Toxicol.*, **23**, 669.
104. Fairlamb, A.H., Carter, N.S., Cunningham, M. and Smith, K. (1992) *Mol. Biochem. Parasitol.*, **53**, 213.
105. Nigam, A. and Allwood, M.C. (1990) *Clin. Otolaryngol. Allied Sci.*, **15**, 173.
106. Thorisdottir, H., Ratnoff, O.D. and Maniglia, A.J. (1988) *J. Lab. Clin. Med.*, **112**, 481.
107. Callanan, V., Curran, A.J., Smyth, D.A. and Gormley, P.K. (1995) *J. Laryngol. Otol.*, **109**, 206.
108. Berger, B.J. and Fairlamb, A.H. (1994) *Antimicrob. Agents and Chemother.*, **38**, 1298.
109. Fairlamb, A.H., Smith, K. and Hunter, K.J. (1992) *Mol. Biochem. Parasitol.*, **53**, 223.
110. Cunningham, M.L. and Fairlamb, A.H. (1995) *Eur. J. Biochem.*, **230**, 460.
111. Haidaris, C.G. and Bonventre, P.F. (1983) *Am. J. Trop. Med Hyg.*, **32**, 286.
112. Berger, B.J. and Fairlamb, A.H. (1993) *Parasitology,* **107**, S71.
113. Ibraham, M.E., Hag Ali, M., El-Hassam, A.M. *et al.* (1994) *Parasitol Res.*, **80**, 569.
114. Roberts,W.L. and Rainey, P.M. (1993) *Antimicrob. Agents Chemother.*, **37**, 1842.
115. Lugo de Yarbuh, A., Anez, N., Petit de Pena, Y. *et al.* (1994) *Ann. Trop. Med. Parasitol.*, **88**, 37.
116. Al-Khawajah, A., Larbi, E.B., Jain, S. *et al.* (1992) *Hum. Expt. Toxicol.*, **11**, 283.
117. Gyurasics, A., Koszorus, L., Varga, F. and Gregu, Z. (1992) *Biochem. Pharmacol.*, **44**, 1275.

118. Dorea, J.G., Merchan-Hamann, E., Ryan, D.E. and Holzbecher, J. (1990) *Clin. Chem.*, **36**, 680.
119. Cantos, G., Barbieri, C.L., Iacomini, M. *et al.* (1993) *Biochem. J.*, **289**, 155.
120. Shoeb, H.A., Korkor, M.I., Tammam, G.M. and El-Amin, S.M. (1980) *Can. J. Pharm. Sci.*, **15**, 66.
121. Murray, H.W., Granger, A.M. and Mohanty, S.K. (1991) *J. Infect. Diseas.*, **163**, 622.
122. El-Ridy, M., Akbarieh, M., Kassem, M., *et al.* (1989) *Int. J. Pharm.*, **56**, 23.
123. Thomson, A.B.R. and Williams, C.N. (1995) *Can. J. Gastroenterol.*, **9**, 91.
124. Ahmed, E.E.M., Nicholson. G. and Seddon, I. (1993) *Lancet,* **341**, 1666.
125. Roggero, E., Zucca, E., Pinotti, G. *et al.* (1995) *Annals Int. Med.*, **122**, 767.
126. Blecker, U., McKeithan, T.W., Hart, J. and Kirschner, B.S. (1995) *Gastroenterology,* **109**, 973.
127. Fraser, A.G., Lewin, J.F. and Pounder, R.E. (1995) *Alimentary Pharm. Ther.*, **9**, 447.
128. Koo, J., Ho, J. and Lam, S.K. (1982) *Gasteoenterology,* **82**, 864.
129. Salmela, K.S., Rhoine, R.P., HookNikanne, J. *et al.* (1994) *Scan. J. Gastroenterol.*, **29**, 528.
130. Hinsull, S.M. and Bellamy, D. (1990) *Gut,* **31**, 389.
131. Roine, R.P., Salmela, K.S., Hookinkanne, J. *et al.* (1992) *Life Sci.*, **51**, PL195.
132. Wagner, S., Beil, W., Mai, U.E.H. *et al.* (1994) *Pharmacology,* **49**, 226.
133. Konturek, S.J., Bilski, J., Kwiecien, N. *et al.* (1987) *Gut,* **28**, 1557.
134. Coudron, P.E. and Stratton, C.W. (1995) *Antimicrobial Agents Chemother.*, **39**, 66.
135. Wieriks, J., Hespe, W. and Jaitly, K.D. (1982) *Scand. J. Gastroenterol.*, **17**, 11.
136. Waldum, H.L., Qvigstad, G., Marvik, R. *et al.* (1994) *Alimentary Pharm. Ther.*, **8**, 425.
137. Coghill, S.B., Hopwood, D., McPherson, S. and Hislop, S. (1983) J. Pathol., 139, 105.
138. Armstrong, J.A., Wee, S.H., Goodwin, C.S. and Wilson, D.H. (1987) *J. Med. Microbiol.*, **24**, 343.
139. Ottlecz, A., Romero, J.J., Hazell, S.L. *et al.* (1993) *Digest. Dis. Sci.*, **38**, 2071.
140. Beil, W., Beibaum, S. and Sewing, K.F. (1993) *Pharmacol.*, **47**, 141.
141. Yasunaga, Y., Shinomura, Y., Kanayama, S. *et al.* (1994) *Gut,* **35**, 1571.
142. Komoda, T., Sonoda, M. and Ikeda, M. (1981) *Clin. Chim. Acta.*, **116**, 161.
143. Beil, W., Beibaum, S. and Sewing, K.F. (1993) *Pharmacol.*, **47**, 135.
144. Beil, W., Birkholz, C., Wagner, S. and Sewing, K.F. (1995) *Pharmacol.*, **50**, 333.
145. Fraser, A.G., Prewett, E.J., Pounder, R.E. and Samloff, I.M. (1992) *Alimen. Pharm. Ther.*, **6**, 389.
146. Hunter, F.M., Correa, P., Fontham, E. *et al.* (1993) *Dig. Dis. Sci.*, **38**, 2081.
147. Slikkerveer, A., Helmich, R.B., Van der Voet, G.B. and de Wolff, F.A. (1995) *J. Pharm. Sci.*, **84**, 512.
148. Slikkerveer, A., Jong, H.B., Helmich, R.B. and de Wolff, F.A. (1992) *J. Lab. Clin. Med.*, **119**, 529.
149. Gordon, M.F., Abrams, R.I., Rubin, D.B. *et al.* (1995) *Movement Disorders,* **10**, 220.
150. Vernace, M.A., Bellucci, A.G. and Wilkes, B.M. (1994) *Am. J. Med.*, **97**, 308.
151. Zahavi, I., Burg, Z., Marcus, H. *et al.* (1995) *Digestion,* **56**, 211.
152. Tsai, H.H., Dwarakanath, A.D., Hart, C.A. *et al.* (1995) *Gut,* **36**, 570.
153. Pullan, R.D., Ganesh, S., Mani, V. *et al.* (1993) *Gut,* **34**, 676.
154. Bara, A., Socaciu, C., Silvestru, C. and Haiduc, J. (1991) *Anti-Cancer Res.*, **11**, 1651.
155. Mahmmod, N., Burke, A., Hussain, S., Anner, R.M. and Anner, B.M. (1995) *Antiviral Chem. Chemother.*, **6**, 187.
156. Chandra, P., Chandra, A., Demirhan, I. and Gerber, T. (1987) *AIDS Forschung.*, **2**, 196.
157. Delnomdedieu, M., Styblo, M. and Thomas, D.J. (1995) *Chem. Biol. Interact.*, **98**, 69.
158. Costantini, P., Chernyak, B.V., Petronilli, V. and Bernardi, P. (1996) *J. Biol. Chem.*, **271**, 6746.
159. Lee, T.C. and Ho, I.C. (1995) *Arch. Toxicol.*, **69**, 498.
160. Styblo, M. and Thomas, D.J. (1995) *Biochem. Pharmacol.*, **49**, 971.
161. Lantz, R.C., Parliman, G., Guan, C.-J. and Carter, D.E. (1994) *Environ Res.*, **67**, 183.
162. Stancato, L.F., Hutchinson, K.A., Chakraborti, P.K. *et al.* (1993) *Biochemistry,* **32**, 3729.
163. Liebl, B., Muckter, H., Doklea, E. *et al.* (1995) *Analyst,* **120**, 771.

164. Yih, L.H. and Lee, T.C. (1994) *Biochem. Biophys. Res. Comm.*, **202**, 1015.
165. Aposhian, H.V. and Aposhian, M.M. (1989) *J. Amer. Coll. Toxicol.*, **8**, 1297.
166. Chilles, C., Mulheron, M., McCrae, F. *et al.* (1990) *Ann. Rheum. Dis.*, **49**, 668.
167. Ghafghazi, T., Ridington, J.W. and Fowler, B.A. (1980) *Toxicol. Appl. Pharmacol.*, **55**, 126.
168. Lai, M.S., Hsueh, Y.M., Chen, C.J. *et al.* (1994) *Amer. J. Epidemiol.*, **139**, 484.
169. Aposhian, H.V., Tadlock, C.H. and Moon, T.E. (1981) *Toxicol. Appl. Pharmacol.*, **61**, 385.
170. Reichl, F.X., Kreppel, H., Szinicz, L. *et al.* (1990) *Vet. Hum. Toxicol.*, **32**, 223.
171. Baker, D.H. and Czarnecki-Maulden, G.L. (1987) *J. Nutrit.*, **117**, 1003.
172. Allain, P., Krari, N., Chaleil, D. *et al.* (1991) *Clin. Pharmacol.*, **5**, 39.
173. de La Calle Guntinas, M.B., Madrid, Y. and Camara, C. (1991) *Anal. Chim. Acta,* **247**, 7.
174. Kenney, L.J. and Kaplan, J.H. (1988) *J. Biol. Chem.*, **263**, 7954.
175. Gonzalez, M.J., Aguilar, M.V. and Para, M.C.M. (1995) *Vet. Hum. Toxicol.*, **37**, 131.
176. Bock, A.K. and Schonheit, P. (1995) *J. Bacteriol.*, **177**, 2002.
177. Chen, B., Burt, C.T. and Goering, P.L. (1986) *Biochem. Biophys. Res. Comm.*, **139**, 228.
178. Ataullakhanov, F.I., Vitvitskii, V.M. and Zhabotinskii, A.M. (1978) *Biophysics,* **23**, 1044.
179. Brazy, P.C., Balaban, R.S. and Gullans, S.R. (1980) *J. Clin. Invest.*, **66**, 1211.
180. Alves, E.W. (1992) *Arch. Biochem. Biophys.*, **292**, 613.
181. Margolin, Y., Barak, R. and Eisenbach, M. (1994) *J. Bacteriol.*, **176**, 5547.
182. Welch, M., Margolin, Y., Calan, S.R. and Eisenbach, M. (1995) *Biochim. Biophys. Acta Mol. Cell Res.*, **1268**, 81.
183. Ercoli, N., Minelli, E.B. and Olivo, N. (1980) *Chemother (swis),* **26**, 254.
184. Chawla, S., Mutenda, E.K., Dixon, H.B.F. *et al.* (1995) *Biochem J.*, **308**, 931.
185. Mutenda, E.K., Sparkes, M.J. and Dixon, H.B.F. (1995) *Biochem J.*, **310**, 983.
186. Kline, P.C. and Schram, V.L. (1993) *Biochemistry,* **32**, 13212.
187. Legare, D., Hettema, E. and Ouellette, M. (1994) *Mol. Biochem. Parasitol.*, **68**, 81.
188. Borst, P. and Ouellett, M. (1995) *Ann. Rev. Microbiol.*, **49**, 427.
189. Zhou, T., Liu, S. and Rosen, B.P. (1995) *Biochemistry,* **34**, 13622.
190. Ji, G. and Silver, S. (1992) *J. Bacteriol.*, **174**, 3684.
191. Rosen, B.P., Bhattacharjee, H. and Shi, W. (1995) *J. Bioenerg. Biomemb.*, **27**, 85.
192. Bhattacharjee, H., Li, J., Yu-Ksenzenko, M. and Rosen, B.P. (1995) *J.Biol. Chem.*, **270**, 11245.
193. Ji, G. and Silver, S. (1992) *Proc. Nat. Acad. Sci. USA,* **89**, 9474.
194. Hanel, F., Krugel, H. and Fiedler, G. (1989) *J. Gen. Microbiol.*, **135**, 583.
195. Wang, H.F. and Lee, T.C. (1993) *Biochem. Biophys. Res. Comm.*, **192**, 1093.
196. Kreppel, H., Lui, J., Lui, Y. *et al.* (1994) *Fund. Appl. Toxicol.*, **23**, 32.
197. Kreppel, H., Bauman, J.W., Lui J. *et al.* (1993) *Fund. Appl. Toxicol.*, **20**, 184.
198. Naganuma, A., Satoh, M. and Imura, M. (1987) *Cancer Res.*, **47**, 983.
199. Boogaard, P.J., Slikkerveer, A., Nagelkerke, F. and Mulder, G.J. (1991) *Biochem. Pharmacol.*, **41**, 369.
200. Satoh, M., Tsuchiya, T., Kumada, Y. *et al.* (1993) *J. Trace Elem. Expt. Med.*, **6**, 41.
201. Naruse, I. and Hayashi, Y. (1989) *Teratology,* **40**, 459.
202. Satoh, M., Kondo, Y., Mita, M. *et al.* (1993) *Cancer Res.*, **53**, 4767.
203. Nakagawa, I., Nishi, E., Naganuma, A. and Imura, N. (1995) *Mut. Res. – Mut. Res. lett.*, **348**, 37.
204. Kaji, T., Suzuki, M., Yamamoto, C. *et al.* (1994) *Res. Comm. Mol. Path. Pharm.*, **86**, 25.
205. Kondo, Y., Satoh, M., Imura, N. and Akimoto, M. (1992) *Anticancer Res.*, **12**, 2303.
206. Cherian, M.G. and Nordberg, M. (1983) *Toxicology,* **28**, 1.
207. Szymanska, J.A. and Piotrowski, J.K. (1980) *Biochem. Pharmacol.*, **29**, 2913.
208. Oskarsson, A. and Fowler, B.A. (1987) *Ann. New York Acad. Sci.*, **514**, 268.
209. Koizumi, T., Negishi, M. and Ichikawa, A. (1993) *Biochem. Pharmacol.*, **45**, 2475.
210. Van Bergen en Henegouwen, P.M.P. and Linnemans, W.A.M. (1987) *Expt. Cell Res.*, **171**, 367.
211. Kozak, R.W., Atcher, R.W. and Gansow, O.A. (1986) *Proc. Natl. Acad. Sci. USA,* **83**, 474.

212. Shadley, J.D., Whitlock, J.L., Rotmensch, J. *et al.* (1991) *Mut. Res.*, **248**, 73.
213. Kaspersen, F.M., Bos, E., Doornmalen, A.V. *et al.* (1995) *Nucl. Med. Comm.*, **16**, 468.
214. Palayoor, S.T., Humm, J.L., Atcher, R.W. *et al.* (1993) *Nucl. Med. Biol.*, **20**, 795.
215. Tiffany, L.J., Dobbs, D., Garmestani, K. *et al.* (1994) *Antibody, Immunoconj. Radiopharmaceut.*, **7**, 99.
216. Junghams, R.P., Dobbs, D., Brechbiel, M.W. *et al.* (1993) *Cancer Res.*, **53**, 5683.

9 Analytical methods

H. ONISHI

This chapter presents an overview of classical methods of separation and determination (quantification or quantitation) and of recent advances in analytical science involving arsenic, antimony and bismuth; however, the literature search is not comprehensive. To keep abreast of developments and advances the reader is referred to *Analytical Chemistry*, *Fundamental Reviews* (the recent issues were published in 1996 and 1994) and *Application Reviews* (the recent issues were published in 1997 and 1995).

Analytically, arsenic and antimony, especially the former, show both metallic and non-metallic properties (semimetallic); bismuth shows metallic properties. Analytically important oxidation states of arsenic and antimony are +3, +5 and –3 (hydrides), but for bismuth they are +3 and –3 (hydride).

Atomic spectroscopic or spectrometric methods, typically atomic absorption spectrometry and atomic emission spectrometry, are particularly useful for the determination of low concentrations or traces of arsenic [1], antimony and bismuth in a variety of samples. Mass spectrometry is also gaining in importance. In order to use these methods, decomposition and solution (dissolution) of samples is required, acid digestion in a microwave oven (microwave digestion) being frequently used.

To bring the concentration of an analyte (substance to be determined) to a suitable determination range in a determination method, preconcentration is used. This is a field closely connected to separation. Sorption (adsorption, ion exchange and complex-forming sorption) [2] or solid phase techniques [3] are used for preconcentration. A definite oxidation state of arsenic (+3, +5 or –3) generally corresponds to the separation and determination of arsenic. Except for the case of speciation (see below), however, determination of arsenic means that of total arsenic. This is also true for antimony.

Speciation, that is, identification, examination or determination of chemical species or forms (oxidation states, ions, molecules, etc.) has been studied by many workers. The speciation of arsenic in environmental materials (samples) and biological materials by using combinations of chromatography and atomic spectrometry is noteworthy.

Control of blanks and prevention of loss of an analyte throughout the entire course of the analysis are always important.

9.1 Arsenic

Because most forms of arsenic are toxic or poisonous, safe practice is required throughout the analysis and with disposal of wastes.

The description in the introductory part of this chapter suggests that speciation deserves a separate section, but because separation and determination are involved in speciation, this section is placed after Methods of determination. Speciation is a matter of considerable relevance to the environmental chemistry of arsenic.

9.1.1 Decomposition or digestion of samples

In many cases, samples must be brought into solution by decomposition or digestion before separation and/or determination [4–7].

In wet and dry methods of decomposition losses of arsenic as a result of volatilization in the form of $AsCl_3$, AsF_3, or AsH_3, etc. must be prevented. In general, decomposition of samples should be carried out under oxidizing conditions (HNO_3 present); arsenic is lost from fuming $HClO_4$–HF solutions.

Procedures for decomposing a variety of samples prior to photometric (spectrophotometric) determination of arsenic are given elsewhere [8].

A microwave oven is a relatively new heating device, and microwave digestion with oxidizing acids has been applied to various samples including biological materials [9,10], silicate rocks [11], coal [11] and oily wastes [12].

Additional comments on the decomposition of specific materials are given below.

Instead of fusion with NaOH or with NaOH + Na_2O_2, silicate rocks can be decomposed with $HClO_4$, HNO_3, HF or $KMnO_4$ [13]. Soil is decomposed with HNO_3–HF–$HClO_4$ in a pressure vessel. Air-borne particles or atmospheric particulates are decomposed by microwave digestion with an acid mixture of HNO_3–$HClO_4$–HF in closed vessels [14].

Steels are dissolved by microwave digestion with HNO_3, HCl, HF and water [15]. Microwave bomb digestion (with HNO_3) of sulphur is effected in less than 10 min [16].

Reviews of atomic spectrometric analysis of foods for arsenic [17] and of hydride generation atomic absorption spectrometric determination of As, Sb, Bi, etc. in wines and beverages [18] contain sections on sample preparation.

A study [19] on wet digestion of marine organisms for the determination of total arsenic has shown that wet oxidation with HNO_3–$HC10_4$–H_2SO_4 is most satisfactory. Microwave digestion with HNO_3 or HNO_3–H_2SO_4 is applied to biological tissues [20]. Vegetable and herbage samples in a conical flask are decomposed by HNO_3 in the presence of

$Mg(NO_3)_2$ and $Ni(NO_3)_2$, the contents are then evaporated to dryness and the residue heated at 450°C in a muffle furnace [21]. This is a combination of wet and dry oxidation.

Dry oxidation (dry ashing) of tomato products, beer and seafood products [22] has been described. The sample is mixed with HNO_3 and an ashing-aid suspension containing $Mg(NO_3)_2$ and MgO, the mixture is evaporated to dryness and the residue carefully heated in a muffle furnace to a final temperature of 450°C. Various samples, including gasoline, are decomposed by using an oxyhydrogen flame.

Microwave oven digestion of sludge samples with HNO_3 and H_2SO_4 has given results comparable to those obtained by reflux digestion with HNO_3, H_2SO_4 and $HClO_4$ [23].

9.1.2 Separations

A large percentage of speciation studies (Section 9.1.4) use separations, especially chromatography; reviews of preconcentration methods have already been cited [2, 3]. Reductive precipitation with $Na[BH_4]$ may be used for preconcentration of arsenic from water [24]. Arsenic is collected on a nitrocellulose membrane filter as an ion associate of arsenomolybdate with tetrapentylammonium ion [25]. Solid-phase extraction of arsenic from aqueous samples using on-column formation of arsenic(III) pyrrolidine dithiocarbamate has also been described [26]. Zirconium-loaded or zirconium-treated activated carbon and a fibrous ion exchanger are used for preconcentration of arsenic(V). Online anion-exchange preconcentration of arsenic is followed by inductively coupled plasma atomic emission spectrometry. A detailed account of separation procedures is given elsewhere [8].

Arsenic has been separated from wastewater, Cu [27], Mo and Ni by co-precipitation of As(V) or As(III) with $La(OH)_3$. Other inorganic and organic co-precipitants have also been studied.

Arsenic is separated from copper by using an online flow-injection analysis system containing a cation-exchange resin [28]. Chloride is removed by anion exchange before determination of arsenic by inductively coupled plasma mass spectrometry.

Extraction of AsI_3 from a solution of HCl or H_2SO_4 with KI or NaI has been used in the analysis of zinc concentrates and lead concentrates (chloroform as an immiscible solvent) [29], food (carbon tetrachloride), silicate rocks (toluene), alloys and biological materials (xylene). Isobutyl methyl ketone (IBMK or 4-methyl-2-pentanone) also has been used as a solvent in the analysis of alloys and steels.

Arsenic in ores, concentrates and related materials is separated by extraction of ethylxanthate of As(III) into cyclohexane from about 9 M HCl [30, 31]. Ammonium pyrrolidine dithiocarbamate has been used for the extraction of arsenic(III) from seawater (chloroform as an immiscible solvent) and soils (IBMK).

IBMK extraction of molybdoarsenic acid is applied to steel, wastewater and river water. The same heteropoly acid is also extracted with trioctylamine in molten stearic acid.

The separation of arsenic by distillation of $AsCl_3$ from HCl solution is described elsewhere [8]. This distillation has been incorporated in the voltammetric determination of arsenic.

Evolution as AsH_3 or hydride generation has been applied to various samples; the hydride is formed by the action of zinc or $Na[BH_4]$ in HCl or H_2SO_4 solution.

Chromatography. Arsenate, not arsenite, in smelter flue dusts is determined by ion chromatography with conductivity detection (measurement) [32]. Arsenate and phosphate are determined by ion chromatography with conductivity detection or with post-column molybdenum blue detection at 840 nm [33]. Ion exclusion chromatography with amperometric detection of arsenic(III) has also been studied.

Separation and determination of As(III), Sb(III) and Bi(III) as diethyl dithiophosphinates are effected by reversed-phase liquid chromatography (detection at 280 nm) [34]. 12-Molybdoheteropoly acids of arsenate, phosphate and silicate are separated and determined similarly.

Separation and determination of Ge, As, Sb, etc. by hydride generation gas chromatography has been described by many workers [35, 36]. Alkylarsenic acids have also been determined by gas chromatography.

Arsenic(III), antimony(III), etc. are determined by chelation with bis(trifluoroethyl)dithiocarbamate and supercritical fluid (CO_2) chromatography [37].

9.1.3 Methods of determination

Earlier reviews of this subject are to be found in other works [38–40]; for a more up-to-date account, however, the reader is directed to *Vogel* [41].

Comparison of methods. Immediately after World War II two silicate rock samples, granite (G-1) and dibase (W-1), were prepared jointly by the US Geological Survey, the Massachusetts Institute of Technology and the Geophysical Laboratory of the Carnegie Institution of Washington. Thus, a cooperative investigation of precision and accuracy in chemical, spectrochemical and modal analysis of the granite and diabase samples began [42, 43]. Since then not only major and minor constituents but also trace constituents ($<$ 0.01% or 100 ppm) of G-1 and W-1 were determined by many workers. Early (1955–65) arsenic determinations in G-1 and W-1 were made mainly by photometric (spectrophotometric) methods and by radiochemical neutron activation analysis [44]. Geochemical reference sample programmes since G-1 and W-1 have been reviewed by Kane [45].

Comparison of methods may be based either on results obtained by a single laboratory or on results obtained by different laboratories, the latter being generally more significant. Examples of the former and the latter are given in Table 9.1. For details of the conclusions reached, the original reports should be consulted.

The work closely related to interlaboratory comparison is the preparation of certified reference materials (or standard samples). Preparation of water [46], geological materials [47], white clover [48] and milk [49] has been recently reported.

Gravimetric methods. Arsenic(III) can be precipitated with hydrogen sulphide from 9 M HCl at 15°C–20°C and weighed as As_2S_3 after drying at 105°C [38, 41].

Table 9.1 Comparison of analytical methods used for the determination of arsenic.

Sample	Method	Reference
Water	Polarography, photometry and HG AAS	[i]
Water	AAS, AES and XRF, all after sorption concentration	[ii]
Air-borne particles	ICP MS, INAA and XRF	[iii]
Seafood products	HG AAS, platform furnace Zeeman-effect AAS and ICP AES	[iv]
Seafood products	ICP AES and ICP MS	[v]
Acid digestates of soils	Interlaboratory comparison: ICP AES, ICP MS, flame AAS, electrothermal AAS, HG AAS and energy-dispersive XRF	[vi]
Timber treated with preservative	Titration with Ce(IV), AAS and XRF	[vii]
Kidney, liver and human hair	Nuclear-related analytical techniques	[viii]

Note: AAS = atomic absorption spectrometry; AES = atomic emission spectrometry; HG AAS = hydride generation atomic absorption spectrometry; ICP AES = inductively coupled plasma atomic emission spectrometry; ICP MS = inductively coupled plasma mass spectrometry; INAA = instrumental neutron activation analysis; XRF = X-ray fluorescence.

References

i Howe, L. H. (1977) *U.S. Environ. Prot. Agency, Office Res. Dev., Rep.*, EPA-600/7-77-036.
ii Andreeva, I. Yu. and Drogobuzhskaya, S. V. (1990) *Zh. Anal. Khim.*, **45,** 904.
iii Wang, C. F., Chang, E. E., Chiang, P. C. and Aras, N. K. (1995) *Analyst*, **120,** 2521.
iv Ybanez, N., Cervera, M. L. and Montoro, R. (1992) *Anal. Chim. Acta*, **258,** 61.
v Sheppard, B. S., Heitkemper, D. T. and Gaston, C. M. (1994) *Analyst*, **119,** 1683.
vi Kimbrough, D. E. and Wakakuwa, J. (1994) *Analyst*, **119,** 383.
vii Dawson, B. S. W., Parker, G. F., Cowan, F. J. and Hong, S. O. (1991) *Analyst*, **116,** 339.
viii Cortes Toro, E., De Goeij, J. J. M. *et al.* (1993) *J. Radioanal. Nucl. Chem.*, **167,** 413.

Table 9.2 Oxidation–reduction titrations for As(III) recommended or positively described (plus sign) by some representative books (Kolthoff and Belcher's book is also relevant [i]).

Oxidizing agent	Reference			
	[ii]	[iii]	[iv]	[v]
Potassium bromate		+	+	+
Iodine	+	+	+	+
Potassium iodate				+
Potassium permanganate		+		
Cerium(IV)			+	

References

i Kolthoff, I. M. and Belcher, R. (1957) *Volumetric Analysis*, vol. III, Wiley-Interscience, New York.

ii Hillebrand, W. F., Lundell, G. E. F., Bright, H. A. and Hoffman, J. I. (1953) *Applied Inorganic Analysis*, 2nd edn, John Wiley, New York.

iii Abel, G. J. Jr (1968) Arsenic, in *Encyclopedia of Industrial Chemical Analysis*, vol. 6 [eds F. D. Snell and C. L. Hilton), John Wiley, New York.

iv Skonieczny, R. F. and Hahn, R. B. (1978) Arsenic, in *Treatise on Analytical Chemistry*, part II, vol. 10 (eds I. M. Kolthoff and P. J. Elving), John Wiley, New York.

v Jeffery, G. H., Bassett, J., Mendham, J. and Denney, R. C. (1989) *Vogel's Textbook of Quantitative Chemical Analysis*, 5th edn, Longman Scientific and Technical, Harlow, Essex.

Volumetric (titrimetric) methods. Arsenic(III) or arsenic(V) can be conveniently determined by oxidation–reduction titration. The standard reduction potential (E^0) for the electrode reaction

$$H_3AsO_4 + 2H^+ + 2e^- \rightarrow HAsO_2 + 2H_2O \qquad [9.1]$$

is 0.559 V. Titrimetric methods for As(III) recommended or positively described by some representative books are listed in Table 9.2. In the titration with bromate, methyl orange and other dyestuffs are used as irreversible indicators.

Arsenic(V) can be determined in hydrochloric acid solution by reaction with iodide, and the iodine produced titrated with standard sodium thiosulphate solution using starch as the indicator [39, 41].

Electrochemical methods. Arsenic(III) can be titrated potentiometrically with permanganate or cerium(IV). Potentiometric determination or arsenate with silver nitrate has also been described [50]. Electrochemical sensors for gaseous arsenic oxides and arsine have been prepared.

Coulometric titration of arsenic(III) with electrogenerated iodine gives good results for GaAs analysis [51, 52].

The polarography [53] cited in Table 9.1 is the differential pulse polarography for the sensitive determination of arsenic(III) in 1 M HCl [54]. This method has been further applied to the analysis of drinking water and tuna.

Molybdoarsenate is used for differential pulse anodic voltammetric determination of arsenate and for indirect polarographic determination of arsenic.

Nanogramme levels of arsenic(III) have been determined by anodic stripping voltammetry in 7 M HCl [55], 0.1 M HCl [56] and 0.1 M $LiClO_4$ in acetonitrile [57]. Determination of arsenic(III) by cathodic stripping analysis has been described [58–61], Greulach and Henze [61] also analysing for arsenic(V).

Arsenic(III) and antimony(III) can be titrated amperometrically in hydrochloric acid solution containing potassium bromide with standard potassium bromate solution [41].

Photometric (spectrophotometric) methods. Two important photometric methods are the heteropoly molybdenum blue (or molybdenum blue) method and silver diethyldithiocarbamate method [8]. The former is based on the reduction of heteropoly molybdoarsenate and the latter on the reduction of silver(I) with AsH_3.

The heteropoly molybdenum blue method has been used after chloroform extraction of AsI_3 [29]. The method is also used after evolution as AsH_3 (hydride generation). Flow injection spectrophotometry has been described [62] and determination of arsenate and phosphate by kinetic spectrophotometry is also available [63].

The ion associate of molybdoarsenate with basic or cationic dyes is utilized for the determination of arsenic after liquid–liquid extraction, flotation, membrane filtration and in aqueous medium containing poly(vinyl alcohol).

The photometric method [53] cited in Table 9.1 is that with silver diethyldithiocarbamate. AsH_3 is usually passed into a pyridine solution of the reagent, although attempts are being made to avoid the use of this solvent, with both silver diethyldithiocarbamate and hexamethylenetetramine in chloroform having been used [64].

The use of ethanol solution containing silver nitrate and poly(vinyl alcohol) in place of silver diethyldithiocarbmate solution has been studied by Chinese workers [65] and formation of a formazan by reaction with arsine was reported by Russian workers [66].

Gas-phase molecular absorption spectrometry of AsH_3 and SbH_3 has been studied, and chemiluminescence methods have been proposed for determination of arsenic.

Atomic absorption spectrometry. Numerous papers on atomic spectrometric methods for determination of arsenic have appeared over the past 10–15 years.

Atomic absorption spectrometry (AAS) may be divided into flame AAS, hydride generation (HG) AAS and electrothermal (ET) AAS, with a wavelength of 193.7 nm being exclusively used. ET indicates ET atomization; the majority of ET AAS employs a graphite furnace (GF).

Determination of As, Bi, Sb and Se in steels and nickel alloys by GF AAS has been compared with that by flow injection HG AAS [67]. For comparisons, see also Table 9.1. A review of the atomic spectrometric analysis of foods for arsenic is mentioned above [17]. The paper by Ybanez, Cervera and Montoro [22] may also be referred to.

Flame AAS is occasionally used [68] notwithstanding a high detection limit. (Detection limit is usually defined as three times the standard deviation of the blank; 'high' means that a numerical value is large.) An argon–hydrogen or N_2O–acetylene flame is used.

In 1969 Holak [69] described HG AAS in which arsine, AsH_3, was generated in a then conventional arsine generator, collected in a cold trap (liquid nitrogen) and passed into the burner (air–acetylene flame) of an atomic absorption spectrometer. Determination of 0.5–1.5 μg of arsenic was possible, although much lower amounts of arsenic may now be determined.

The use of sodium tetrahydroborate instead of zinc for the generation of arsine in AAS was described by several workers in 1973–74 and has been commonly used since. In acid solution tetrahydroborate changes to boric acid, liberating hydrogen, but aqueous solutions of the reagent are fairly stable in the presence of small amounts of sodium hydroxide. Instead of the reagent in solution, the use of an anion-exchange resin of a tetrahydroborate form has also been proposed.

For atomization, arsine is passed into a flame or electrically heated quartz cell (silica tube) or graphite furnace.

Present HG AAS makes use of continuous hydride generation and flow injection analysis, and a new continuous hydride generator for the determination of As, Sb and Sn has been described [70]. An early paper on flow injection HG AAS is that of Yamamoto, Yasuda and Yamamoto (for As, Sb, Bi, Se and Te) [71].

Conditions for hydride generation include:

- the concentration of $NaBH_4$;
- HCl concentration or other media;
- the arsenic species forming AsH_3 and other arsines, namely, As(III), As(V) and organoarsenic compounds;
- pre-reduction of As(V);
- interference.

A hydrochloric acid medium is generally used, but buffer solutions may also be used for As(III) [71–73]. Pre-reduction of As(V) to As(III) is often carried out in hydrochloric acid with KI or KI plus ascorbic acid. L-Cysteine has also been used instead of KI. Moreover, in 0.3–0.7 M HCl containing 2% L-cysteine, hydride generation responses are the same from

As(III), As(V), methylarsonic (monomethylarsonic) acid and dimethylarsinic acid [74].

Effects of other hydride-forming elements [75–78] and of the other elements [79–82] have been studied and interference from Co, Ni and Cu has been reported.

HG AAS has been applied to a variety of samples, especially geological, environmental, and organic and biological materials, including foods.

In GF AAS a detection limit of about 10 pg of As may be obtained, which appears to be lower than that obtained in HG AAS. However, the actual range of determination seems to be approximately the same. Matrix modifiers such as Pd nitrate and a mixture of Pd and Mg nitrates are often added to the sample solution. Formation of $[Pd_xAs_yO_m]$ [83] or an intermetallic solid solution [84] has been suggested as a mechanism of Pd-induced stabilization of arsenic in ET atomization AAS. Simplex optimization of conditions has been described for the determination of arsenic in microwave-digested oyster tissue by using ET AAS [85]. The conditions considered are ashing temperature, atomization temperature, modifier (palladium) concentration and atomization ramping time.

Stabilized temperature platform furnace Zeeman AAS has been described, and, instead of sample solutions, slurries have been analysed directly by GF AAS.

ET and GF AAS have been applied to a variety of samples, including environmental samples, metals and alloys, and foods. The use of AAS in speciation is mentioned in Section 9.1.4.

Atomic fluorescence spectrometry (AFS) is not as widely used as AAS, but Chinese workers [86] have described procedures for the determination of arsenic by HG AFS (Stockwell and Corns' work [87] may also be referred to). ET AFS has also been studied.

Atomic emission spectrometry. Excitation with an inductively coupled plasma (ICP) is the principal excitation method of atomic emission spectrometry (AES). ICP was first described in 1964–65, and ICP AES instruments are now commercially available. For introduction for ICP analysis, the sample solution is passed through a nebulizer, hydride (arsine) is generated, or electrothermal vaporization is employed. A detection limit of 10 ng ml^{-1} As has been reported for the former at 193.696 nm, but the detection limit is much lower where arsine is generated.

Comparisons of analytical methods are given in Table 9.1. Quality control of HG ICP AES data (As and Sb) has been discussed [88].

Arsenic in water has been determined by ICP AES with direct sample nebulization (observing the arsenic 228.812 nm line) after preconcentration [25].

Continuous hydride generation is advantageous and flow injection analysis is often carried out. Interference of Co, Ni and Cu in the hydride generation may be eliminated or reduced by adding thiourea. Arsine has been formed after extraction of AsI_3 into xylene by mixing the organic phase with $NaBH_4$ in dimethylformamide and with acetic acid (a wavelength of 228.812 nm being used).

ICP AES has been used for the analysis of various samples, including environmental samples, metals and alloys, biological materials and foods. Preconcentration is carried out when necessary. The use of ICP AES in speciation is mentioned in Section 9.1.4.

Microwave-induced plasma (MIP) and direct current plasma (DCP) have both been used for excitation in AES where hydride generation is usually involved.

X-Ray fluorescence. In this method, a solid sample is irradiated or excited with X-rays from an X-ray tube or radiation from radioactive isotopes, and characteristic or fluorescent X-rays are measured. An X-ray fluorescence (XRF) spectrometer is either wavelength-dispersive or energy-dispersive. The As K_α X-ray line, 10.530 keV or 0.1177 nm, may be used. A suitable concentration or content of arsenic for XRF is not very low and preconcentration is made where necessary. Various samples including biological samples [21] have been analysed (Table 9.1).

Radioactive ^{109}Cd as an excitation source is used for energy-dispersive XRF determination of arsenic, and the use of synchrotron X-ray sources is also being studied. Determination of arsenic by electron-probe X-ray microanalysis has been described.

Mass spectrometry. ICP used for AES can also be used as an ion source for mass spectrometry (MS), although ICP MS is relatively new; Kane *et al.* [89] describe ICP MS chronology. As in the case of ICP AES, the sample solution is introduced into an ICP system through a nebulizer, or the hydride (arsine) is introduced. Detection limits of 1–10 pg ml^{-1} As (or higher) have been reported with the former and may be lower with the latter. A comparison of detection limits of As, Sb and Bi with atomic spectrometry and ICP MS has been made by Powell, Boomer and McVicars [90]. Laser ablation and electrothermal vaporization are also studied.

Mass spectrometric measurements are made at a mass-to-charge (m/z) ratio of 75, but precautions have to be taken against interference from the $ArCl^+$ ion (m/z 75), Ar being used as the carrier gas for ICP. Measures include corrections, conditions of sample introduction, addition of nitrogen to the Ar carrier gas and removal of Cl^- by anion exchange.

In HG ICP MS, the hydride generated may be concentrated or trapped in a graphite furnace and then vaporized electrothermally.

Nebulization ICP MS and to a lesser extent laser ablation ICP MS have been applied to the analysis of SiH_4, air-borne particulate matter, geological materials (including seawater), steels, Cu and biological materials, including urine. Comparison of ICP MS and instrumental neutron activation analysis for the determination of trace elements including As and Sb in petroleum exploration samples has also been made [91]. Comparisons of analytical methods are given in Table 9.1.

HG ICP MS (including continuous HG) has been applied to the analysis of natural waters, environmental samples and biological samples, including urine.

High-performance liquid chromatography coupled with ICP MS is useful for the separation and determination of arsenite, arsenate and organoarsenic compounds (Section 9.1.4) [92].

MIP MS has been investigated, and secondary-ion mass spectrometry (SIMS) is a tool of surface analysis.

Neutron activation analysis. Original papers by Smales and Pate on the determination of traces of arsenic in GeO_2, seawater and biological material by neutron activation appeared in 1952 [93]. In neutron activation analysis (NAA) $^{75}As(n,\gamma)^{76}As$ is utilized, the half-life of ^{76}As being 26.3 h and its γ-ray (559.1 keV) being measured. Early NAA work entailed radiochemical separation(s) (RNAA) and RNAA is still used for the analysis of geological or geochemical materials (e.g. determination of As(III) and As(V) in seawater) and of biological materials. The recent trend of NAA is, however, instrumental NAA (INAA), which does not entail separation(s) and permits non-destructive, multi-element determination.

A variety of samples have been analysed by INAA, including geological or geochemical materials, environmental materials, high-purity titanium, hair, food and other biological materials. A comparative study has been made of particle-induced X-ray emission (PIXE) and INAA for the determination of As and other metals in estuarine sediment (see Table 9.1) [94].

Photon activation analysis has been used for the analysis of coal fly ash.

9.1.4 Speciation

At this stage, it will be useful to consider the background of speciation, the definition of which was given at the beginning of this chapter. Two factors apparently have contributed to speciation, namely environment and analytical methodology. With regard to the former, environmental pollution or contamination has caused interest in and concern about the environment and also about environmental protection. With regard to the latter, HG AAS and the use of various chromatographic methods – especially high-performance liquid chromatography – appear to be important developments. Speciation work on arsenic began to appear in about 1975.

Table 9.3 Arsenic species that have been studied for speciation.

Name	Formula
arsenic(III), arsenite	$[AsO_3]^{3-}$ [a]
arsenic(V), arsenate	$[AsO_4]^{3-}$ [a]
methylarsonic acid, monomethylarsonic acid, methanearsonic acid (MMAA)	$CH_3AsO_3H_2$ or $CH_3AsO(OH)_2$
dimethylarsinic acid, cacodylic acid (DMAA)	$(CH_3)_2AsO_2H$ or $(CH_3)_2AsO(OH)$
trimethylarsine oxide (TMAO)	$(CH_3)_3AsO$
tetramethylarsonium ion	$[(CH_3)_4As]^+$
arsenobetaine (AB)	$(CH_3)_3As^+CH_2COO$
arsenocholine (AC)	$(CH_3)_3As^+CH_2CH_2OH$
dimethylarsinoylethanol	$(CH_3)_2As(O)CH_2CH_2OH$
dimethylarsinoylribosides (dimethylarsinylribosides, arsenosugars)	
trimethylarsonioriboside	
arsenolipids	

[a]In fully dissociated form.

One may cite, however, examples of speciation in a broad sense before this date. One such example are the Standard Samples (now Standard Reference Materials) of glasses issued by the US National Bureau of Standards (now the National Institute of Standards and Technology) before and after World War II. These samples contain 0.07%–0.4% As_2O_5 and 0.03%–0.1% As_2O_3. Presumably, As(III) was volatilized as AsF_3 or $AsCl_3$ and the remaining As(V) was determined. Another portion of the sample was used for the determination of total As, and As(III) was determined from the difference. Another example is the determination of As(V) and As(III) in seawater [95]; separation by liquid–liquid extraction and spectrophotometric determination were employed.

Reviews of speciation include:

- the determination of arsenic species in environmental and biological samples [96];
- selenium and arsenic in biology, their chemical forms and biological functions [97];
- arsenic and organoarsenicals [98];
- metal speciation in solid matrices [99].

Table 9.3 shows the main arsenic species that have been studied for speciation. They are practically the same as the arsenic compounds reported in living organisms. In addition, trivalent methylarsenicals, namely, monomethylarsonous acid [MMAA(III)] and dimethylarsinous acid [DMAA(III)], have been reported in natural waters [100].

Many papers have appeared since the publication of Morita and Edmonds' [96] and the samples studied include waters (mostly natural waters), soil and sediments, urine and marine organisms. The methods

Table 9.4 Speciation methods for (a) air-borne particulates, waters, soil and sediments; (b) urine and marine biological materials

Sample	Species	Method	Reference
(a)			
air-borne particulates	methylarsenic compounds	HG GC–AAS	[i]
natural waters	As^{III}, As^{V}, MMAA, DMAA	HG AAS	[ii]
freshwaters	As^{III}, As^{V}, methylarsenic compounds	HG, cold trapping, chromatographic separation, AES	[iii]
wastewater and sediments	As^{III}, As^{V}	HG spectrophotometry, HG AAS	[iv]
seawater	As^{III}, As^{V}	thionalide loaded on silica gel, spectrophotometry	[v]
natural waters	As^{III}, As^{V}, MMAA(III), DMAA(III), MMAA, DMAA	diethyldithiocarbamate extraction, HG, cold trapping, chromatographic separation, AAS	[vi]
wastewater	As^{III}, As^{V}	extraction chromatography–GF AAS	[vii]
river water and urine	As^{III}, As^{V}	bis(trifluoroethyl)dithiocarbamate extraction, GC	[viii]
river water	As^{III}, As^{V}	ion chromatography	[ix]
wastewater	As^{III}, As^{V}	ion chromatography	[x]
river water	As^{III}, As^{V}, MMAA, DMAA, inert, unidentified species	cation exchange, GF AAS, HG GF AAS	[xi]
soil water and commercial bottled waters	As^{III}, As^{V}, MMAA, DMAA	HPLC (ion-exchange)–HG AAS	[xii]
water and sediment extracts	As^{III}, As^{V}, MMAA, DMAA, AB, AC	HPLC (ion-exchange)–HG AAS	[xiii]
mineral water	As^{III}, As^{V}, MMAA, DMAA, AB, AC	HPLC (ion-exchange)–HG ICP AES	[xiv]
natural waters	As^{III}, As^{V}, MMAA, DMAA	HPLC (ion-pair)–HG ICP MS	[xv]
natural waters	As^{III}, As^{V}, DMAA, Se compounds	CZE	[xvi]
(not specified)	As^{III}, As^{V}, MMAA, DMAA	CZE	[xvii]
contaminated soil	As^{III}, As^{V}	$AsCl_3$ extraction, HG AAS	[xviii]
soil	As^{III}, As^{V}, MMAA, DMAA, AB, AC, tetramethylarsonium	HPLC (ion-exchange)–flame AAS and ICP MS	[xix]
marine sediment	As^{III} and total As	extraction of As^{III} diethyldithiocarbamate, GF AAS	[xx]

Table 9.4 *Continued*

Sample	Species	Method	Reference
(b)			
human urine	As^{III}, As^{V}, MMAA, DMAA, AB, AC, tetramethylarsonium	HG AAS and HG GC–AAS	[xxi]
rat urine	As^{III}, As^{V}, MMAA, DMAA, TMAO, AB	ion chromatography–ICP MS	[xxii]
human urine	inorganic As, MMAA, DMAA	ion-exchange chromatography–HG AAS	[xxiii]
human urine and seafoods	AB, AC, tetramethylarsonium	HPLC (ion-pair)–HG AAS	[xxiv]
human urine	15 As compounds, including AB and DMAA	HPLC (ion-pair)–ICP MS	[xxv]
human urine	As^{III}, As^{V}, MMAA, DMAA, AB	HPLC (ion-exchange or ion-pair)–HG AAS and ICP MS	[xxvi]
algae	As^{III}, As^{V}, MMAA, DMAA	substoichiometric isotope dilution	[xxvii]
marine animals, including dogfish muscle	inorganic As, MMAA, DMAA, AB	ion-exchange chromatography–INAA	[xxviii]
dogfish muscle	10 As compounds, including AB and MMAA	HPLC–AAS	[xxix]
marine animals	12 As compounds, including AB, arsenosugars and tetramethylarsonium	HPLC (gel permeation, cation-exchange, anion-exchange)–ICP AES	[xxx]
marine biological materials, including dogfish muscle	As^{V}, MMAA, DMAA, AB, AC	HPLC (ion-exchange)–HG ICP AES	[xxxi]
dogfish muscle	15 As compounds, including AB	HPLC (ion-pair)–ICP MS	[xxxii]
shrimp extracts	AC	HPLC (ion-exchange)–ICP MS	[xxxiii]
shellfish	arsenosugars	HPLC (ion-exchange)–ICP MS	[xxxiv]

Note: AAS = atomic absorption spectrometry; AB = arsenobetaine; AC = arsenocholine; AES = atomic emission spectrometry; CZE = capillary zone electrophoresis; DMAA = dimethylarsinic acid; GC = gas chromatography; GF = graphite furnace; HG = hydride generation; HPLC = high-performance liquid chromatography; ICP = inductively coupled plasma; INAA = instrumental neutron activation analysis; MMAA = monomethylarsonic acid; MS = mass spectrometry; TMAO = trimethylarsine oxide.

References

i Mukai, H. and Ambe, Y. (1987) *Anal. Chim. Acta*, **193,** 219.
ii Anderson, R. K., Thompson, M. and Culbard, E. (1986) *Analyst*, **111,** 1143, 1153.
iii Braman, R. S., Johnson, D. L., Foreback, C. C. *et al.* (1977) *Anal. Chem.*, **49,** 621.
iv Lopez, A., Torralba, R., Palacios, M. A. and Camara, C. (1992) *Talanta*, **39,** 1343.
v Terada, K., Matsumoto, K. and Inaba, T. (1984) *Anal. Chim. Acta*, **158,** 207.
vi Hasegawa, H., Sohrin, Y., Matsui, M. *et al.* (1994) *Anal. Chem.*, **66**, 3247.
vii Russeva, E., Havezov, I. and Detcheva, A. (1993) *Fresenius' J. Anal. Chem.*, **347,** 320.
viii Yu, J.-J. and Wai, C. M. (1991) *Anal. Chem.*, **63,** 842.
ix Fung, Y. S. and Dao, K. L. (1995) *Anal. Chim. Acta*, **300,** 207.
x Li, Z.-L., Mou, S.-F., Ni, Z.-M. and Riviello, J. M. (1995) *Anal. Chim. Acta*, **307,** 79.
xi Sturgeon, R. E., Siu, K. W. M., Willie, S. N. and Berman, S. S. (1989) *Analyst*, **114,** 1393.
xii Tye, C. T., Haswell, S. J., O'Neill, P. and Banckroft, K. C. C. (1985) *Anal. Chim. Acta*, **169,** 195.
xiii Martin, I., Lopez Gonzalvez, M. A., Gomez, M. *et al.* (1995) *J. Chromatogr. B, Biomed. Appl.*, **666,** 101.
xiv Rubio, R., Padro, A., Alberti, J. and Rauret, G. (1993) *Anal. Chim. Acta*, **283,** 160.
xv Huang, C.-J. and Jiang, S.-J. (1994) *Anal. Chim. Acta*, **289,** 205.
xvi Li, K. and Li, S. F. Y. (1995) *Analyst*, **120,** 361.
xvii Morin, P., Amran, M. B., Favier, S. *et al.* (1992) *Fresenius' J. Anal. Chem.*, **342,** 357.
xviii Chappell, J., Chiswell, B. and Olszowy, H. (1995) *Talanta*, **42,** 323.
xix Hansen, S. H., Larsen, E. H., Pritzl, G. and Cornett, C. (1992) *J. Anal. At. Spectrom.*, **7,** 629.
xx Bermejo Barrera, P., Barciela Alonso, M. C., Ferron Novais, M. and Bermejo Barrera, A. (1995) *J. Anal. At. Spectrom.*, **10,** 247.
xxi Le, X.-C., Cullen, W. R. and Reimer, K. J. (1994) *Anal. Chim. Acta*, **285,** 277.
xxii Inoue, Y., Kawabata, K., Takahashi, H. and Endo, G. (1994) *J. Chromatogr. A*, **675,** 149.
xxiii Jimenez de Blas, O., Gonzalez, S. V., Rodriguez, R. S. and Mendez, J. H. (1994) *J. AOAC Int.*, **77,** 441.
xxiv Momplaisir, G. M., Blais, J.-S., Quinteiro, M. and Marshall, W. D. (1991) *J. Agri. Food Chem.*, **39,** 1448.
xxv Shibata, Y. and Morita, M. (1989) *Anal. Sci.*, **5,** 107.
xxvi Le, X.-C., Cullen, W. R. and Reimer, K. (1994) *Talanta*, **41,** 495.
xxvii Jitoh, F., Imura, H. and Suzuki, N. (1990) *Anal. Chim. Acta*, **228,** 85.
xxviii Slejkovec, Z., Byrne, A. R. and Dermelj, M. (1993) *J. Radioanal. Nucl. Chem.*, **173,** 357.
xxix Momplaisir, G.-M., Lei, T. and Marshall, W. D. (1994) *Anal. Chem.*, **66,** 3533.
xxx Morita, M. and Shibata, Y. (1987) *Anal. Sci.*, **3,** 575.
xxxi Alberti, J., Rubio, R. and Rauret, G. (1995) *Fresenius' J. Anal. Chem.*, **351,** 415.
xxxii Shibata, Y. and Morita, M. (1989) *Anal. Chem.*, **61,** 2116.
xxxiii Larsen, E. H. and Stürp, S. (1994) *J. Anal. At. Spectrom.*, **9,** 1099.
xxxiv Larsen, E. H. (1995) *Fresenius' J. Anal. Chem.*, **352,** 582.

studied are diverse and depend considerably on the sample and the species sought.

Stability of species in the sample during storage and analysis is a prerequisite to speciation. Preservation of As(III) and As(V) in freshwater and seawater samples has been studied [101], and the effect of storage of oxic and anoxic seawater samples on As(III) and As(V) and Sb(III) and Sb(V) has been examined [102].

Table 9.4 shows examples of speciation methods, although the selection is somewhat arbitrary. Speciation of arsenic in a dogfish muscle reference material (DORM-1) has attracted the attention of many workers [Table 9.4(b)]. Generally speaking, high-performance liquid chromatography (several types) coupled with atomic spectrometry (AAS and ICP AES) or ICP MS is very useful for speciation, but efficient extraction of arsenic-containing compounds from a solid sample is necessary before chromatographing.

9.1.5 *Analysis of arsenic, gallium arsenide and arsine*

Arsenic. Old methods for the determination of impurities in arsenic can be found in Abel's work [39]. Methods that have been described recently include:

- stripping voltammetry for Cu, Pb and Bi;
- spectrophotometric determination of Si;
- GF AAS for Co, Cr, Cu, Fe and Ni;
- AES, spark-source mass spectrometry and INAA for many impurity elements.

Gallium arsenide. Analysis of gallium arsenide is treated elsewhere [103]. Coulometric titrations afford precise determination of both Ga and As (Section 9.1.3). PIXE analysis of GaAs has been described. AAS has been used to determine Cr and In and INAA for the determination of impurities in GaAs.

Characterization of GaAs by SIMS, sputtered neutral mass spectrometry [104], Auger electron spectroscopy and atomic force microscopy [105] has been described, and several techniques have been examined for measurement of particle size of nanocrystalline GaAs [106].

Arsine. Determination of arsine has been effected by means of a tape monitor, sensors, gas chromatography (GC), AAS, Mössbauer spectrometry and an ultrasonic time-of-flight method [107].

Impurities in arsine have been determined by GC, AES and photometric methods (H_2S in AsH_3).

9.2 Antimony

Some forms of antimony are toxic or poisonous and safe practice is necessary. At present the main part of the speciation of antimony is the separation and/or determination of Sb(III) and Sb(V) which is treated in Section 9.2.4.

9.2.1 Decomposition or digestion of samples

Some references applicable to arsenic are also relevant to antimony [4–8]. In the decomposition or dissolution of samples, possible losses of antimony owing to volatilization of $SbCl_3$ and SbH_3 must be considered. In addition, formation of insoluble oxide (Sb_2O_4) by evaporation with nitric acid should be avoided. Addition of tartaric acid stabilizes solutions of antimony(III).

Microwave digestion with oxidizing acids has been applied to air-borne particles [14], silicate rocks [11], coal [11], oil [20] and oily wastes [12]. Silicate rocks can be decomposed with H_2SO_4 and HF. Decomposition of silicate rocks with $HClO_4$, HNO_3, HF and $KMnO_4$, which has been used for the determination of arsenic [13], is also applicable to the determination of antimony [108].

The review of analysis of wines and beverages [18] contains a section of sample preparation (dry ashing and wet digestion).

9.2.2 Separations

A detailed account of separation procedures is given elsewhere [8] and the extraction of SbI_3 and $SbBr_3$ into benzene (or toluene) is particularly useful.

Extraction of $HSbCl_6$ with butyl acetate has been used in the analysis of Pb and Zn concentrates [109], and extraction of Sb(V) and Sb(III) from HCl solution with crown ethers has also been studied [110].

Prior to its photometric determination, antimony was separated as the hydride (stibane or stibine, SbH_3) that is formed by the reaction with $Na[BH_4]$.

High-performance liquid chromatography (HPLC) of benzyl-(3-nitrilopropyl)dithiocarbamates of Cd, Pb, Zn, Sb and Tl has been described (detection at 260 nm) [111] (also see Section 9.1.2), and determination of some antibilharzial antimony compounds has been effected by HPLC [112].

Antimony in ores and alloys has been determined by chlorination and gas chromatography (GC) [113]. Hydride generation GC is described in Section 9.1.2, which also mentions supercritical fluid chromatography.

9.2.3 Methods of determination

Earlier reviews of this subject are to be found in other works [38, 114, 115], but for a more up-to-date account, the reader is directed to *Vogel* [41].

In terms of a comparison of methods, the antimony values obtained by photometry or colorimetry and by radiochemical neutron activation analysis are in good agreement (about 1 ppm or $\mu g\ g^{-1}$ Sb) for W-1 diabase (cf. Section 9.1.3), but agreement is not so good for G-1 granite (< 1 ppm Sb), details of which are given elsewhere [116].

Analysis of air-borne particles for antimony by X-ray fluorescence, instrumental neutron activation analysis and inductively coupled plasma mass spectrometry has been evaluated [117].

Gravimetric methods. Antimony(III) can be precipitated with hydrogen sulphide from dilute HCl solution and finally weighed as Sb_2S_3 [38, 41, 115], or complexation with pyrogallol may be used [41].

Volumetric (titrimetric) methods. Antimony(III) or antimony(V) can be determined by oxidation–reduction titration for which the standard reduction potential (E^0) for the electrode reaction

$$Sb_2O_5 + 6H^+ + 4e^- \rightarrow 2SbO^+ + 3H_2O \qquad [9.2]$$

is 0.581 V or 0.605 V, depending on the sources. If a table is prepared for antimony in the same way as arsenic (Table 9.2), it can be seen that the following titrations are more usefully adopted: potassium bromate, iodine and potassium permanganate. Titration of antimony(III) in hydrochloric acid solution with cerium(IV) using methyl orange or methyl red indicator is also useful [118, 119]. Compleximetric titrations of antimony(III) have also been described (actually back-titrations).

Electrochemical methods. Antimony(III) can be determined by potentiometric titration with hexaamminecobalt(III) tricarbonatocobaltate(III), $[Co(NH_3)_6][Co(CO_3)_3]$ [120], the reaction involved being as follows:

$$Sb(III) + 2Co(III) \rightarrow Sb(V) + 2Co(II) \qquad [9.3]$$

Alternatively, potentiometric titration of Sb(V) as $SbCl_6^-$ with 1,2,4,6-tetraphenylpyridinium cation (a precipitation titration) can be used [121].

Traces of antimony in leaves have been determined by stripping potentiometry (potentiometric stripping analysis) [122].

The accuracy of the coulometric titration of antimony(III) in 3 M HCl by electrogenerated iodine has been studied [123], and a purity of $99.79\% \pm 0.02\%$ m m^{-1} was obtained for antimony(III) oxide.

Dilute hydrochloric acid, for example 1 M, is a suitable supporting electrolyte for the polarographic determination of antimony(III), and differential pulse polarography of antimony(III) in 0.01 M oxalic acid has also been studied [124]. Adsorptive voltammetric determination of antimony in hair has been described [125].

Traces of antimony and bismuth in lithium iodate have been determined by anodic stripping voltammetry in a dilute hydrochloric acid medium [126], whereas differential pulse anodic stripping voltammetry has been applied to the determination of Pb, Cu, Sb and Bi in seawater [127] and river water [128].

Amperometric titration of antimony(III) is covered in Section 9.1.3.

Photometric (spectrophotometric) methods. Antimony is usually determined either by extracting an ion pair or ion associate of a cationic (basic) dye with $SbCl_6^-$ into an immiscible organic solvent, or by forming a yellow Sb(III)–I^- compex (or complexes) [8]. In the former method Rhodamine B and Brilliant Green are representative dyes, and benzene and toluene representative solvents; the formation of $SbCl_6^-$ is a critical step.

Antimony has been determined after extraction of Brilliant Green hexachloroantimonate(V) with microcrystalline 1,4-dichlorobenzene [129] and thermal lens spectrometry has been used after benzene extraction of Malachite Green (Basic Green 4) hexachloroantimonate(V) [130].

The Sb(III)–I^- method described above is simpler to carry out, the determination usually being made in dilute H_2SO_4 solution with KI plus ascorbic acid; the I^- concentration must be kept constant.

Determination of antimony with use of ion associates of cationic dyes with anionic Sb(III) complexes has also been studied, the anionic Sb(III) complexes being formed with I^-, mandelic acid, etc. Derivatives of phenylfluorone give sensitive colour reactions with antimony(III) in the presence of surfactants (mostly cationic).

Kinetic spectrophotometric methods are available for the determination of antimony(III), and kinetic fluorimetric determination of both Sb(III) and Sb(V) has been described.

Atomic absorption spectrometry. Like arsenic, antimony can be determined by AAS by using flame, hydride generation (HG) and electrothermal (ET) atomization where a wavelength of 217.6 nm is generally used, and a wavelength of 231.1 nm occasionally used.

In AAS using an air–acetylene flame, the detection limit of antimony is higher than that of the other AAS method. Nevertheless, flame AAS has been applied to the analysis of geological materials and cobalt salts after liquid–liquid extraction.

The outline of HG AAS for antimony is much the same as that for arsenic; additional comments are given below. Determination of Sb(III), Sb(V), methylstibonic acid, $CH_3SbO(OH)_2$, and dimethylstibinic acid, $(CH_3)_2SbO(OH)$, by HG AAS has been described by Andreae *et al.* [131], this being a speciation method [cf. Table 9.4(a), ref. iii]. Sb(III) and Sb(V) are reduced together to SbH_3 with $NaBH_4$ at a pH of about 1 in the presence of KI. Sb(III) is selectively reduced at near-neutral pH, where no reduction of Sb(V) takes place and Sb(V) is then determined by subtraction. The methylantimony acids are reduced to CH_3SbH_2 and $(CH_3)_2SbH$ at acid pH without addition of I^-. SbH_3, methyl- and dimethylstibines are then collected with a liquid nitrogen trap and separated chromatographically after heating the trap, where they elute into a quartz cuvette burner or a graphite furnace. L-Cysteine can also be used instead of KI as a pre-reductant for Sb(V) [132].

For the determination of antimony in seawater, the SbH_3 generated is trapped on the activated carbon surface in a graphite furnace at 187°C and atomized at 2000°C [133].

For the effects of other elements, see refs 76–78, 81 and 134; the interference from Co, Ni and Cu has been noted.

HG AAS has been applied to the analysis of various samples including natural waters (river, sea), geochemical reference samples, ores, steel, copper and food. With regard to ET AAS or graphite furnace (GF) AAS, it appears that the detection limits of HG AAS and GF AAS are much the same if a suitable volume (e.g. 10 μl) of sample is assumed in the former. As in the case of arsenic, matrix modifiers such as Pd nitrate or a mixture of palladium and magnesium nitrates are often added to the sample solution and the role of matrix modifiers has also been studied [135, 136]. Zeeman-effect background correction has been made and the use of a stabilized temperature platform furnace has been described.

GF AAS has also found wide applications, for example waters, geological materials, non-ferrous metals, plant materials and urine, and determination of antimony by atomic fluorescence spectrometry, mostly with hydride generation, has been studied mainly by Chinese workers [137].

Atomic emission spectrometry. Hydride generation (HG) inductively coupled plasma (ICP) AES is currently typical of AES methods, with continuous hydride generation being preferred where the antimony lines 217.581 nm, 231.147 nm and 206.833 nm are used. The detection limit of HG ICP AES for antimony is higher than that of HG AAS.

Co, Ni and Cu all interfere in HG ICP AES and the use of thiourea as a masking agent has been proposed. The sample solution may also be introduced into an ICP system through a nebulizer.

ICP AES has been used for the analysis of environmental samples, ferrous and non-ferrous metals and biological materials and hydride generation microwave-induced plasma AES has also been used for the determination of antimony. Hydride generation direct current plasma AES has also been studied [138].

X-ray fluorescence. XRF is not suitable for the determination of trace constituents (< 100 ppm or 0.01%), but traces of antimony have been determined after preconcentration where the Sb K_α X-ray line at 26.271 keV or 0.0472 nm may be used. The use of synchrotron X-ray sources is also being studied, and total reflection X-ray fluorescence has been described.

Mass spectrometry. Very low concentrations of antimony as well as arsenic can be determined by ICP MS, and sample solutions may be introduced into an ICP system through a nebulizer or via hydride generation; laser ablation ICP MS also has been described. Detection limits of 1–10 pg ml^{-1} Sb (or higher) have been reported with nebulization and they may be lower with the hydride. Mass spectrometric measurements are made at $m/z = 121$ (rarely at $m/z = 123$).

ICP MS has been applied to the determination of antimony in natural waters (fresh and sea), wastewater, air-borne particulates, silicate rocks, soils, sediments, steels, copper, petroleum exploration samples, etc., and isotope dilution analysis combined with ICP MS has been applied to the determination of more than 5 ng g^{-1} of antimony in copper [139].

Spark source mass spectrometry has been used to determine antimony in silicate rock reference materials [140], and secondary-ion mass spectrometry (SIMS) has been used for measurement of Sb in Si.

Neutron activation analysis. A look at the literature on NAA reveals the following papers: determination of antimony by Hudgens, and Cali (1952) [141]; determination of antimony in G-1 and W-1 (sections 9.1.3 and 9.2.3) by Hamaguchi *et al.* (1961) [142]; and determination of antimony in biological materials by Howie *et la.* (1965) [143]. All of these are concerned with radiochemical NAA (RNAA), that is, NAA with separation(s). In NAA, the nuclear reactions $^{121}Sb(n,\gamma)$ ^{122}Sb and ^{123}Sb ^{124}Sb are utilized. The half-life of ^{122}Sb is 2.70 day and its γ-ray at 564.3 keV is measured. The half-life of ^{124}Sb is 60.20 days and its γ-rays at 602.7 keV and 1691 keV are measured. RNAA is still used for the analysis of silicate rocks and biological materials, but recent applications of instrumental NAA (INAA), that is, NAA without separations, include the analysis of volcanic gases, oil shale, environmental materials, high-purity titanium, C_{60}, food and biological materials.

Photon activation analysis has been used for the analysis of coal fly ash and substoichiometric radiochemical determination of antimony has also been described.

9.2.4 *Speciation*

Data on the speciation of antimony are contained in Das *et al.*'s review [99], and the effect of storage of seawater on Sb(III) and Sb(V) has been examined by Cutter, Cutter and San Diego-McGlone [102].

Determination of Sb(III) and Sb(V) by differential pulse anodic stripping voltammetry has been described [144], and determination of Sb(V) and total Sb in water by extraction spectrophotometry has been proposed [145].

Following the work of Andrcac *et al.* for natural waters [131], determination of Sb(III) and Sb(V) by selective hydride generation AAS has been studied by many workers, including Yamamoto, Yasuda and Yamamoto [71]. Generally speaking, Sb(III) is reduced with $NaBH_4$ at a pH of about 6 (citrate), and Sb(III) and Sb(V) are reduced in dilute HCl solution containing KI. This approach has been extended to the analysis of marine sediments by slurry formation–HG AAS [146]; SbH_3 from total Sb is formed in 4 M HCl solution containing 2% (mass/volume) KI and Sb(III) in 6% (mass/volume) citric acid solution. An attempt has been made to determine Sb(III) and Sb(V) in liver tissue [147].

Many papers have appeared on the determination of Sb(III) and Sb(V) in water samples by NAA, AAS or ICP AES after separation by precipitation, on columns or by extraction. Extractive separation of Sb(III) from Sb(V) with 1-pyrrolidinecarbodithioic acid, ammonium salt, appears to be especially useful [148–150].

Antimony(V) and antimony(III) in copper electroplating solutions have been separated and determined by cation-exchange chromatography–ICP AES [151].

9.2.5 *Analysis of antimony and antimony(III) oxide*

Early accounts of the analysis of antimony and antimony compounds may be found elsewhere [114, 115]. Recent papers describe the determination of As, Bi, Cu, Fe, Hg, Pb and Se in Sb by ICP AES and photometric determination of Fe and Pb.

Arsenic in Sb_2O_3 has recently been determined by stripping potentiometry, Pb by AAS, and U and Th by ICP MS.

9.3 Bismuth

A comparison of the analytical methods for bismuth with those for arsenic and antimony reveals that oxidation–reduction titrimetric methods and neutron activation methods have rarely been employed for bismuth, although compleximetric titrations or titrations with EDTA (ethylenediaminetetraacetic acid) have been used for bismuth.

Bismuth(V) sodium trioxide or sodium bismuthate, $NaBiO_3$, is a powerful oxidizing agent in acid solution, and bismuth(V) in the superconductors containing the elements Bi, Sr, Ca, Cu and O has been determined spectrophotometrically by its reaction with manganese(II) to form permanganate [152].

A comparison of methods is presented in Section 9.3.3.

9.3.1 *Decomposition or digestion of samples*

Works cited for arsenic and antimony are also useful for bismuth [4–8]. Nitric acid is suitable for dissolution of metals and alloys and silicate rocks can be decomposed with $HClO_4$, HNO_3 and HF. The destruction of organic matter is effected by digestion with H_2SO_4 and HNO_3, often together with $HClO_4$ and an organobismuth sample has been decomposed by oxygen flask combustion.

Oxidation of urine with $KBrO_3$–KBr–HCl (bromination) is suitable for online microwave digestion and hydride generation atomic absorption spectrometric determination of bismuth [153]. A microwave oven has also been used for decomposition of silicate rocks [154] and steels [15].

9.3.2 *Separations*

Reviews of preconcentration already cited also include a discussion of bismuth [2, 3]. Thionalide-loaded silica gel has been used for preconcentration of bismuth in river water and a more detailed account of separations is given elsewhere [8].

Co-precipitation of bismuth with iron(III) hydroxide is suitable for separation from copper and silver, and bismuth is co-precipitated with magnesium hydroxide by simply adding sodium hydroxide to seawater [155]. Ion flotation and precipitate flotation of bismuth have also been described.

Extraction with dithizone and cupferron is mentioned elsewhere [8] and Sb, Bi, Cd, Cu and In in ores and concentrates are extracted from 2 M H_2SO_4–0.1 M KI into isobutyl methyl ketone (IBMK) [156].

Anion-exchange separation of Bi from large amounts of Fe [157] and Zn [158] has been described and traces of bismuth in silicate rocks and biological materials have been separated or preconcentrated by anion exchange [159].

Bismuth can be separated from gramme amounts of Tl(I) and Ag(I) by cation-exchange chromatography in nitric acid [160]. The sample solution, 0.2 M in HNO_3, is passed through a column of AG 50W-X4 and the column is washed with 0.3 M HNO_3. Tl(I) and Ag(I) are eluted with 0.6 M HNO_3, and Bi(III) with 0.2 M HBr containing 20% vol./vol. acetone. Cation-exchange separation of Bi from Tl(III) and Hg(II) on a macro-

Table 9.5 Results of determination of bismuth in seawater.

Location	Bi (ng l^{-1})[a]	Method	Reference
North Atlantic:		Anion-exchange separation, photometric determination with dithizone	[i]
surface water	33		
from 2000 m	15		
North Pacific:		Anodic stripping voltammetry	[ii]
from 1–2 m	15–94[b,c]		
from 2550 m	40[b,c]		
Inland Sea, Japan	10	Precipitate flotation, HG AAS	[iii]
North Sea, from 3 m	200–680	Anodic stripping voltammetry	[iv]
North Pacific, coastal surface water	25–32[d]	Diethyldithiocarbamate extraction, GF AAS	[v]
North Pacific:		Precipitation (co-precipitation) by adding NaOH, HG AAS	[vi]
surface water	0.053[e]		
from 2500 m	<0.003		
Tokyo Bay, surface water	0.6–2.1[c]	Almost the same as above.	[vii]
Sea of Japan, coastal surface water	3.6 ± 1.3	Anion-exchange preconcentration, anion-exchange spectrophotometry as iodo complexes	[viii]
North Pacific, coastal:		Extraction preconcentration, ICP MS	[ix]
from 320 m	0.161[c]		
from 1 m	0.453[c]		
from 99 m	0.471[c]		
North Pacific, coastal surface water	1.6	Diethyldithiocarbamate extraction, GF AAS	[x]

Note: GF AAS = graphite furnace atomic absorption spectrometry; HG AAS = hydride generation atomic absorption spectrometry; ICP MS = inductively coupled plasma mass spectrometry.

[a] Unless otherwise noted, the sample was filtered before analysis.

[b] ng kg^{-1}.

[c] Seawater was not filtered before analysis.

[d] Seawater was adjusted to pH 0.7 with HNO_3, boiled for 1 h and filtered.

[e] The total Bi in the surface water was estimated as 0.15 ng l^{-1}.

References

i Portmann, J. E. and Riley, J. P. (1966) *Anal. Chim. Acta*, **34,** 201.
ii Gilbert, T. R. and Hume, D. N. (1973) *Anal. Chim. Acta*, **65,** 451.
iii Nakashima, S. (1980) *Fresenius' Z. Anal. Chem.*, **303,** 10.
iv Gillain, G., Duyckaerts, G. and Disteche, A. (1979) *Anal. Chim. Acta*, **106,** 23.
v Shimizu, T., Kamata, Y., Kimura, Y. *et al.* (1982) *Bunseki Kagaku*, **31,** 299.
vi Lee, D. S. (1982) *Anal. Chem.*, **54,** 1682.
vii Ohyama, J., Maruyama, F. and Dokiya, Y. (1987) *Anal. Sci.*, **3,** 413.
viii Yoshimura, K. (1987) *Bunseki Kagaku*, **36,** 656.
ix Shabani, M. B., Sahoo, S. K. and Masuda, A. (1992) *Analyst*, **117,** 1477.
x Shijo, Y., Mitsuhashi, M., Shimizu, T. and Sakurai, S. (1992) *Analyst*, **117,** 1929.

porous resin, AG MP-50, is also available [161]. A chelating resin, Chelex 100®, has been used in the determination of Bi in sulphide ores, concentrates, Cu, Ag and Ag–Cd alloys [162].

Bismuth and other metals have been separated and determined by high-performance liquid chromatography of their complexes with the following reagents: picolinaldehyde 4-phenyl-3-thiosemicarbazone [163], diethyl phosphorodithioate [34], sodium diethyldithiocarbamate [164] and 4-(2-thiazolylazo)resorcinol [165].

9.3.3 Methods of determination

Early accounts of this subject are given in refs 38, 166 and 167 whereas ref. 41 provides a more up-to-date account. A series of reviews of the chemical analysis of bismuth [168, 169] are also useful.

Results of seawater and rock analyses. Table 9.5 summarizes results of seawater analysis for bismuth. The earlier bismuth values are much higher than the recent values, and Table 9.5 shows the difficulty, or difficulties, in determining extremely low concentrations (ultra-trace) of bismuth. The reason for the difference (or discrepancy) is not clear; sampling and storage of seawater and correct blank determination might be factors.

In 1973, a report [170] on values for international geochemical reference samples was presented. Bismuth values for many US Geological Survey (USGS) samples are included, but recommended values are given for only four samples, namely, AGV-1 (0.057 ppm Bi), BCR-1 (0.050 ppm), DTS-1 (0.010 ppm) and G-2 (0.043 ppm). Results obtained later by several workers are concordant with the recommended values.

Reports [171, 172] on 1994 values for 17 Geological Survey of Japan (GSJ) igneous rock reference samples have also been published. Only four samples are given mean bismuth content with standard deviation: JB-1 (0.033 ± 0.0038 ppm Bi), JG-1 (0.50 ± 0.087 ppm), JR-1 (0.56 ± 0.090 ppm) and JR-2 (0.62 ± 0.12 ppm). Four papers that were published during 1990–92, and not collected in the 1994 compilation [172], deal with bismuth determination by atomic absorption spectrometry; three out of four papers give the results that are concordant with the above-mentioned values.

Gravimetric methods. Bismuth can be precipitated from hot, dilute nitric acid solution with $(NH_4)_2HPO_4$, after which the precipitate is filtered, heated at 800°C, cooled and weighed as $BiPO_4$ [38, 41, 167]. Other gravimetric methods involve weighing as BiOCl [38, 167], BiOI [41] or Bi_2O_3 [38, 167].

Volumetric (titrimetric) methods. The stability constant (as $\log_{10}K$) of the 1:1 complex formed by bismuth(III) with the ethylenediaminetetraacetate anion (EDTA) is 27.9. Few metal ions have such a large value, and bismuth can be determined by direct titration with EDTA (disodium salt) in nitric acid solution of pH $\geqslant 1$. Many metal indicators have been proposed for the titration of bismuth. Xylenol Orange appears to be especially useful, with colour changes from red to yellow at the end point. Procedure is discussed by, for example, Jeffery *et al.* [41].

Compleximetric determination of Bi in the presence of Ti and Al using Semi-Xylenol Orange as indicator has been described [173], and determination of Bi and Cu in the Bi–Cu–Ca–Sr oxide system by spectrophotometric titration with EDTA and Pyrocatechol (Catechol) Violet indicator has also been reported [174].

Electrochemical methods. Comparison of potentiometric stripping analysis and square-wave voltammetry has been made with respect to the influence of Triton X–100 [175]. Coulometric compleximetric back-titration with electrolytically generated mercury(II) permits the determination of bismuth in $Bi_{12}GeO_{20}$ [176].

Bismuth(III) in a supporting electrolyte of 1 M HCl shows a polarographic half-wave potential of –0.09 V compared with SCE and complexes of bismuth with some organic reagents show polarographic adsorptive complex waves that can be used for bismuth determination. One such reagent is glyoxal bis(2-hydroxyanil) [177] with which traces of bismuth in copper-base alloys have been determined. Voltammetric analysis for bismuth with chemically modified electrodes containing organic reagents, for example 1-(2-pyridylazo)-2-naphthol [178], has also been studied.

Anodic stripping voltammetry, including differential pulse anodic stripping voltammetry, is a useful method of trace analysis for bismuth. Samples analysed include water [128], seawater [127] (Table 9.5), minerals [179], aluminium [180], arsenic [181], $LiIO_3$ [126], soil, grapes and wine [182], and blood [183].

Photometric (spectrophotometric) methods. Four important photometric methods are the dithizone method (extraction), the iodide method (aqueous medium or extraction), the thiourea method (aqueous medium) and the diethyldithiocarbamate method (extraction) [8]. The iodide method [8, 41] in aqueous medium is simple to carry out and its sensitivity index (Sandell sensitivity) corresponding to an absorbance of 0.001 is 0.015–0.02 μg Bi cm^{-2} at 460–465 nm.

Thermal lens spectrometry has been applied to the iodide method [184], and ion-exchange spectrophotometry has been described (Table 9.5, ref. viii). Several cationic reagents have been proposed to extract tetraiodobismuthate(III); for example, Burns, Chimpalee and Harriott describe flow injection extraction–spectrophotometric determination of bismuth [185].

The thiourea method has been used after separations by extraction and back-extraction and a cyclic thiourea is used for extraction–photometric determination of bismuth [186].

Porphyrins show very sensitive colour reactions with bismuth and bisazo derivatives of chromotropic acid (1,8-dihydroxy-3,6-naphthalenedisulphonic acid) have been proposed as sensitive reagents for bismuth.

Gas-phase molecular absorption spectrometry of BiH_3 has been studied [187], the hydride being generated in dilute hydrochloric acid solution with $NaBH_4$.

Atomic absorption spectrometry. Like arsenic and antimony, bismuth can be determined by AAS by using flame, hydride generation (HG) and electrothermal (ET) atomization; bismuth lines at 223.1 nm and 306.8 nm are used.

In comparing these methods, flame AAS, graphite furnace AAS and ICP AES have been evaluated for the determination of trace elements, including bismuth in high-purity iron [188], and determination of bismuth in steels and nickel alloys by flow injection HG AAS has been compared with that by graphite furnace (GF) AAS [67, 189]. The range of bismuth determination in the Bi–Sr–Ca–Cu–O system by flame AAS is much higher than that by ET AAS [190].

Detection limits of 50 ng ml^{-1}, 0.2 ng ml^{-1} and 4 pg of Bi have been reported for flame AAS, HG AAS and GF AAS, respectively.

An air–acetylene flame is used in flame AAS. To determine lower concentrations of bismuth, liquid–liquid extraction is often practised where the aqueous phase after back-extraction is sprayed or the organic phase is directly sprayed. Various samples including ores and concentrates [156], geological materials [191], Co salts, Cu-based alloys, Zn, NaCl solutions, superconductors and biological materials have been analysed by this method.

HG AAS is reviewed by Baluja-Santos and Gonzalez-Portal [18]. The hydride (bismuthane, bismuthine, BiH_3) is usually generated in dilute HCl solution with $NaBH_4$. For atomization BiH_3 is passed into a flame (e.g. argon–hydrogen), heated quartz cell or a graphite furnace. Continuous hydride generation and flow injection HG AAS are in use but there are reports of problems in collecting BiH_3 in a liquid nitrogen trap. Concentration of BiH_3 in a graphite furnace coated with Pd has been described and the effects of other elements on HG AAS have been studied by many workers [71, 81, 192]. Generation of BiH_3 in non-aqueous media has also been investigated.

HG AAS has been applied to a variety of samples including river water, seawater (Table 9.5), geological materials, copper and biological materials.

Both ET AAS and GF AAS are widely used and many matrix modifiers have been studied. Use of Pd nitrate or a mixture of Pd and Mg nitrates is notable, and Zeeman-effect background correction is often made. A stabilized temperature platform furnace has been used, and the introduction of solid samples into furnaces has been described.

Samples that have been analysed by GF AAS cover seawater (Table 9.5), geological [154] and environmental [159] materials, metals and alloys, and biological materials.

Atomic fluorescence spectrometry (AFS) has been employed for bismuth as for arsenic and antimony. Laser-excited AFS has also been studied, with very low detection limits being claimed.

Atomic emission spectrometry. Inductively coupled plasma (ICP) AES is widely used. Either the hydride (BiH_3) is generated or the sample solution is passed through a nebulizer. A detection limit of 5 ng ml^{-1} Bi has been reported for the latter at 223.061 nm. The detection limit for the former is much lower [90]. The bismuth line 306.772 nm is also used.

Continuous hydride generation is advantageously used and flow injection analysis is carried out. Interference studies in HG ICP AES show that Co, Ni and Cu interfere, but thiourea will mask them; copper is likely to interfere with the 223.061 nm bismuth line.

ICP AES has been used for the determination of bismuth in waters, geological material, metals and alloys, superconductors, and biological and environmental samples. HG direct current plasma AES has been described, and excitation with arc discharge (solid samples) is still used.

X-Ray fluorescence. The Bi $L_{\alpha1}$ line at 10.837 keV or 0.1144 nm and the $L_{\beta1}$ line at 13.021 keV or 0.09520 nm may both be used. Bismuth in superconductors has been determined by this method as has bismuth in ores, concentrates and alloys after extraction preconcentration. Traces of Bi in Cu and Ag can be determined after co-precipitation of Bi with Fe(III) hydroxide.

Copper, Sr and Bi in thin sputtered films have been determined by radioisotope (^{241}Am)-excited XRF.

Traces of impurities including bismuth in high-purity iron have been determined by total reflection XRF after removal of iron by isobutyl methyl ketone extraction.

Mass spectrometry. Like As and Sb, very low concentrations of Bi can be determined by ICP MS. Sample solutions are introduced into the ICP system through a nebulizer or via hydride generation, and the hydride also can be collected in a graphite furnace before ICP MS. Laser ablation ICP MS has also been described, and a detection limit of 0.4 pg ml^{-1} Bi (or higher) has been reported with nebulization, although it may be lower with the hydride [90]. Measurements are made at a m/z of 209.

ICP MS has been used for the determination of bismuth in seawater (Table 9.5), minerals and rocks, steels, Cu, Ni-base alloys, Zn, biological materials and lead shotgun pellets. Spark source MS has been used to determine bismuth in silicate rock reference samples, the results of which have been compared with ICP MS data [140].

Radiochemical methods. ^{207}Bi (half-life 32.2 years, γ-rays at 569 keV and 1063 keV) has been used as a radioactive tracer to determine recoveries of separations.

^{210}Bi (half-life 5.01 days, β-rays) is a nuclide that belongs to the uranium series, and procedures are available for the determination of ^{210}Bi in environmental samples; separations are involved.

9.3.4 *Analysis of bismuth and bismuth(III) oxide*

Mostly classical wet methods for the determination of impurities in bismuth and bismuth compounds are found elsewhere [167–169], but more recent papers describe the analysis of Bi for Li by flame AAS, Te by spectrophotometry and many impurities by radiochemical neutron activation analysis.

AAS, both flame and electrothermal, is used for the determination of many impurities in Bi_2O_3 after separations. Use of ICP AES for the determination of impurities in Bi compounds has also been described.

The methods for the analysis of Bi should be applicable to Bi_2O_3, and vice versa.

References

1. Sanz-Medel, A. (1995) *Analyst*, **120,** 799.
2. Terada, K. (1991) *Anal. Sci.*, **7,** 187.
3. Nickson, R. A., Hill, S. J. and Worsfold, P. J. (1995) *Anal. Proc.*, **32,** 387.
4. Sandell, E. B. and Onishi, H. (1978) *Photometric Determination of Traces of Metals, Part I: General Aspects*, 4th edn, John Wiley, New York, Ch. 3.
5. Bock, R. (1979) *A Handbook of Decomposition Methods in Analytical Chemistry*, International Textbook, Glasgow.
6. Sulcek, Z. and Povondra, P. (1989) *Methods of Decomposition in Inorganic Analysis*, CRC Press, Boca Raton, FL.
7. Gorsuch, T. T. (1970) *The Destruction of Organic Matter*, Pergamon Press, Oxford.
8. Onishi, H. (1986) *Photometric Determination of Traces of Metals, Part IIA: Individual Metals, Aluminum to Lithium*, 4th edn, John Wiley, New York.
9. Abu-Samra, A., Morris, J. S. and Koirtyohan, S. R. (1975) *Anal. Chem*, **47,** 1475.
10. Mayer, D., Haubenwallner, S., Kosmus, W. and Beyer, W. (1992) *Anal. Chim. Acta*, **268,** 315.
11. Bettinelli, M., Baroni, U. and Pastorelli, N. (1989) *Anal. Chim. Acta*, **225,** 159.
12. Campbell, M. B. and Kanert, G. A. (1992) *Analyst*, **117,** 121.
13. Terashima, S. (1976) *Anal. Chim. Acta*, **86,** 43.
14. Wang, C. F., Chen, W. H., Yang, M. H. and Chiang, P. C. (1995) *Analyst*, **120,** 1681.
15. Coedo, A. G. and Dorado, M. T. (1994) *J. Anal. At. Spectrom.*, **9,** 1111.
16. Matusiewicz, H. and Kurzawa, Z. (1991) *Acta Chim. Hung.*, **128,** 401.

17. Cervera, M. L. and Montoro, R. (1994) *Fresenius' J. Anal. Chem.*, **348,** 331.
18. Baluja-Santos, C. and Gonzalez-Portal, A. (1992) *Talanta*, **39,** 329.
19. Jin, K., Ogawa, H. and Taga, M. (1983) *Bunseki Kagaku*, **32,** E171, E259.
20. Liu, J., Sturgeon, R. E. and Willie, S. N. (1995) *Analyst*, **120,** 1905.
21. Horler, B. A. T. (1989) *Analyst*, **114,** 919.
22. Ybanez, N., Cervera, M. L. and Montoro, R. (1992) *Anal. Chim. Acta*, **258,** 61.
23. Saraswati, R., Vetter, T. W. and Watters, R. L. Jr (1995) *Analyst*, **120,** 95.
24. Mentasti, E., Nicolotti, A., Porta, V. and Sarzanini, C. (1989) *Analyst*, **114,** 1113.
25. Hata, N., Kasahara, I., Taguchi, S. and Goto, K. (1989) *Analyst*, **114,** 1255.
26. Van Elteren, J. T., Gruter, G. J. M., Das, H. A. and Brinkman, U. A. T. (1991) *Int. J. Environ. Anal. Chem.*, **43,** 41.
27. British Standards Institution (1990) British Standard 7317, part 3.
28. Wang, H. and Chen, Y. (1994) *Fenxi Huaxue*, **22,** 816.
29. Rao, V. S. S., Rajan, S. C. S. and Rao, N. V. (1993) *Talanta*, **40,** 653.
30. Donaldson, E. M. (1988) *Talanta*, **35,** 47.
31. Donaldson, E. M. and Leaver, M. E. (1988) *Talanta*, **35,** 297.
32. Hansen, L. D., Richter, B. E., Rollins, D. K. *et al.* (1979) *Anal. Chem.*, **51,** 633.
33. Jones, P., Stanley, R. and Barnett, N. (1991) *Anal. Chim. Acta*, **249,** 539.
34. Irth, H., Brouwer, E., De Jong, G. J. *et al.* (1988) *J. Chromatogr.*, **439,** 63.
35. Kadeg, R. D. and Christian, G. D. (1977) *Anal. Chim. Acta*, **88,** 117.
36. Skogerboe, R. K. and Bejmuk, A. P. (1979) *Anal. Chim. Acta*, **94,** 297.
37. Laintz, K. E., Yu, J.-J. and Wai, C. M. (1992) *Anal. Chem.*, **64,** 311.
38. Hillebrand, W. F., Lundell, G. E. F., Bright, H. A. and Hoffman, J. I. (1953) *Applied Inorganic Analysis*, 2nd edn, John Wiley, New York.
39. Abel, G. J. Jr (1968) Arsenic in *Encyclopedia of Industrial Chemical Analysis*, vol. 6 (eds F. D. Snell and C. L. Hilton), John Wiley, New York.
40. Skonieczny, R. F. and Hahn, R. B. (1978) Arsenic, in *Treatise on Analytical Chemistry*, part II, vol. 10 (eds I. M. Kolthoff and P. J. Elving), John Wiley, New York.
41. Jeffery, G. H., Bassett, J., Mendham J. and Denney, R. C. (1989) *Vogel's Textbook of Quantitative Chemical Analysis*, 5th edn, Longman Scientific and Technical, Harlow, Essex.
42. Fairbairn, H. W. *et al.* (1951) *U.S. Geological Survey Bulletin*, 980.
43. Stevens, R. E. *et al.* (1960) *U.S. Geological Survey Bulletin*, 1113.
44. Onishi, H. (1969) Arsenic, in *Handbook of Geochemistry*, vol. II (ed. K. H. Wedepohl), Springer, Berlin.
45. Kane, J. S. (1991) *Spectrochim. Acta*, **46B,** 1623.
46. Gupta, P. K., Kaw, C., Ganju, S. *et al.* (1993) *Fresenius' J. Anal. Chem.*, **345,** 278.
47. Richardson, J. M. (1995) *Analyst*, **120**, 1513.
48. Quevauviller, P., Vercoutere, K. and Griepink, B. (1992) *Anal. Chim. Acta*, **259**, 281.
49. Nagaraja, K. V. Vibhakar, S., Rao, S. N. R. and Rani, U. (1995) *Fresenius' J. Anal. Chem.*, **352**, 33.
50. Pietrogrande, A. and Zancato, M. (1989) *Anal. Lett.*, **22**, 2521.
51. Tanaka, T., Watakabe, K., Kurooka, K. and Yoshimori, T. (1989) *Bunseki Kagaku*, **38**, 724.
52. Cullen, M. and Mearns, J. M. P. (1993) *Anal. Chim. Acta*, **277**, 113
53. Howe, L. H. (1977) *U.S. Environ. Prot. Agency, Office Res. Dev., Rep.*, EPA-600/7–77–036.
54. Myers, D. J. and Osteryoung, J. (1973) *Anal. Chem.*, **45,** 267.
55. Davis, P. H., Dulude, G. R., Griffin, R. M. *et al.* (1978) *Anal. Chem.*, **50,** 137.
56. Svintsova, L. D., Kaplin, A. A. and Vartan'yan, S. V. (1991) *Zh. Anal. Khim.*, **46,** 896.
57. Ramadan, A. A. and Mandil, H. (1989) *Indian J. Chem., Sect. A*, **28A,** 984.
58. Kotoucek, M., Vasicova, J. and Ruzicka, J. (1993) *Mikrochim. Acta*, **111,** 55.
59. Mrzljak, R. I., Bond, A. M., Cardwell, T. J. *et al.* (1994) *Analyst*, **119,** 1051.
60. Zima, J. and van den Berg, C. M. G. (1994) *Anal. Chim. Acta*, **289,** 291.
61. Greulach, U. and Henze, G. (1995) *Anal. Chim. Acta*, **306,** 217.
62. Frenzel, W., Titzenthaler, F. and Elbel, S. (1994) *Talanta*, **41,** 1965.
63. Lopez Carreto, M., Sicilia, D., Rubio, S. and Perez-Bendito, D. (1993) *Anal. Chim. Acta*, **283,** 481.
64. Nowicka, J. (1994) *Chem. Anal. (Warsaw)*, **39,** 741.

65. Du, H. and Wang, B. (1992) *Fenxi Huaxue*, **20,** 623.
66. Kolesnikova, A. M. and Lazarev, A. I. (1989) *Zavod. Lab.*, **55**(5), 7.
67. Bettinelli, M., Spezia, S., Baroni, U. and Bizzarri, G. (1994) *At. Spectrosc.*, **15,** 115.
68. British Standards Institution (1991) British Standard 5666, part 3.
69. Holak, W. (1969) *Anal. Chem.*, **41,** 1712.
70. Le, X.-C., Cullen, W. R., Reimer, K. J. and Brindle, I. D. (1992) *Anal. Chim. Acta*, **258,** 307.
71. Yamamoto, M., Yasuda, M. and Yamamoto, Y. (1985) *Anal. Chem.*, **57,** 1382.
72. Anderson, R. K., Thompson, M. and Culbard, E. (1986) *Analyst*, **111,** 1143, 1153.
73. Torralba, R., Bonilla, M., Palacios, A. and Camara, C. (1994) *Analysis*, **22,** 478.
74. Le, X.-C., Cullen, W. R. and Reimer, K. J. (1994) *Anal. Chim. Acta*, **285,** 277.
75. De la Calle-Guntinas, M. B., Torralba, R., Madrid, Y. *et al.* (1992) *Spectrochim. Acta, Part B*, **47B,** 1165.
76. Barth, P., Krivan, V. and Hausbeck, R. (1992) *Anal. Chim. Acta*, **263,** 111.
77. Henrion, G., Henrion, R., Hebisch, R. and Boeden, B. (1992) *Anal. Chim. Acta*, **268,** 115.
78. Walcerz, M., Bulska, E. and Hulanicki, A. (1993) *Fresenius' J. Anal. Chem.*, **346,** 622.
79. Aggett, J. and Boyes, G. (1989) *Analyst*, **114,** 1159.
80. Schmidt, C. and Bahadir, M. (1993) *Fresenius' J. Anal. Chem.*, **346,** 683.
81. Alexandrov, S., Gafur, I., Mandzhukov, P. and Nedeltchev, O. (1993) *Fresenius' J. Anal. Chem.*, **347,** 303.
82. Wickstrøm, T., Lund, W. and Bye, R. (1995) *Analyst*, **120,** 2695.
83. Styris, D. L., Prell, L. J. and Redfield, D. A. (1991) *Anal. Chem.*, **63,** 503.
84. Yang, P., Ni, Z., Zhuang, Z. *et al.* (1992) *J. Anal. At. Spectrom.*, **7,** 515.
85. Pergantis, S. A., Cullen, W. R. and Wade, A. P. (1994) *Talanta*, **41,** 205.
86. Cheng, X. and Ma, Z. (1991) *Fenxi Shiyanshi*, **10**(3), 69.
87. Stockwell, P. B. and Corns, W. T. (1994) *Analyst*, **119,** 1641.
88. Howarth, R. J. (1995) *Analyst*, **120,** 1851.
89. Kane, J. S., Beary, E. S., Murphy, K. E. and Paulsen, P. J. (1995) *Analyst*, **120,** 1505.
90. Powell, M. J., Boomer, D. W. and McVicars, R. J. (1986) *Anal. Chem.*, **58,** 2864.
91. Olsen, S. D., Filby, R. H., Brekke, T. and Isaksen, G. H. (1995) *Analyst*, **120,** 1379.
92. Pergantis, S. A., Heithmar, E. M. and Hinners, T. A. (1995) *Anal. Chem.*, **67,** 4530.
93a. Smales, A. A. and Pate, B. D. (1952) *Anal. Chem.*, **24,** 717.
93b. Smales, A. A. and Pate, B. D. (1952) *Analyst*, **77,** 188.
93c. Smales, A. A. and Pate, B. D. (1952) *Analyst*, **77,** 196.
94. Al-Jundi, J., Mamas, C., Earwaker, L. G. *et al.* (1993) *Anal. Proc.*, **30,** 153.
95. Sugawara, K., Terada, K., Kanamori, S. *et al.* (1962) *J. Earth Sci., Nagoya Univ.*, **10,** 34.
96. Morita, M. and Edmonds, J. S. (1992) *Pure Appl. Chem.*, **64,** 575.
97. Shibata, Y., Morita, M. and Fuwa, K. (1992) *Adv. Biophys.*, **28,** 31.
98. Shibata, Y. and Morita, M. (1994) Arsenic and organoarsenicals, in *Analysis of Contaminants in Edible Aquatic Resources* (eds J. W. Kiceniuk and S. Ray), VCH, Cambridge.
99. Das, A. K., Chakraborty, R., Cervera, M. L. and de la Guardia, M. (1995) *Talanta*, **42,** 1007.
100. Hasegawa, H., Sohrin, Y., Matsui, M. *et al.* (1994) *Anal. Chem.*, **66,** 3247.
101. van Elteren, J. T., Hoegee, J., van der Hoek, E. E. *et al.* (1991) *J. Radioanal. Nucl. Chem.*, **154,** 343.
102. Cutter, L. S., Cutter, G. A. and San Diego-McGlone, M. L. C. (1991) *Anal. Chem.*, **63,** 1138.
103. Downs, A. J. (ed.) (1993) *Chemistry of Aluminium, Gallium, Indium and Thallium*, Blackie Academic & Professional, Glasgow, pp. 307–309, 311, 499.
104. Wise, M. L., Emerson, A. B. and Downey, S. W. (1995) *Anal. Chem.*, **67,** 4033.
105. Prohaska, T., Friedbacher, G., Grasserbauer, M. *et al.* (1995) *Anal. Chem.*, **67,** 1530.
106. Hagan, C. R. S., Kher, S. S., Halaoui, L. I. *et al.* (1995) *Anal. Chem.*, **67,** 528.
107 Valdes, J. L. and Cadet, G. (1991) *Anal. Chem.*, **63,** 366.
108. Terashima, S. (1984) *Bunseki Kagaku*, **33,** 561.
109. Murti, S. S., Rajan, S. C. S. and Subrahmanyam, J. (1988) *Talanta*, **35,** 443.
110. Koshima, H. and Onishi, H. (1986) *Analyst*, **111,** 1261.

111. Park, Y. J. and Hardy, J. K. (1989) *J. Chromatogr.*, **481,** 287.
112. El-Rabbat, N. A., Farag, H. H., El-Kommos, M. E. and Refaat, I. H. (1990) *Talanta*, **37,** 951.
113. Iatridis, B. and Parissakis, G. (1977) *Anal. Chim. Acta*, **89,** 347.
114. Abel, G. J. Jr (1968) Antimony, in *Encyclopedia of Industrial Chemical Analysis*, vol. 6 (eds F. D. Snell and C. L. Hilton), John Wiley, New York.
115. Hahn, R. B. (1978) Antimony, in *Treatise on Analytical Chemistry*, part II, vol. 10 (eds I. M. Kolthoff and P. J. Elving), John Wiley, New York.
116. Onishi, H. (1969) Antimony, in *Handbook of Geochemistry*, vol. II (ed. K. H. Wedepohl), Springer, Berlin.
117. Wang, C. F., Chang, E. E., Chiang, P. C. and Aras, N. K. (1995) *Analyst*, **120,** 2521.
118. British Standards Institution (1989) British Standards, 3908, part 10.
119. Rajan, S. C. S. and Rao, I. V. S. (1989) *Analyst*, **114,** 987.
120. Walash, M. I., Belat, F. and Aly, F. A. (1988) *Analyst*, **113,** 955.
121. Sanchez-Pedreno, C., Ortuno, J. A. and Alvarez, J. (1991) *Anal. Chem.*, **63,** 764.
122. Adeloju, A. and Young, T. M. (1995) *Anal. Chim. Acta*, **302,** 225.
123. Hioki, A., Kubota, M. and Kawase, A. (1992) *Analyst*, **117,** 997.
124. Hasebe, K., Kakizaki, T., Tanaka, S. and Yoshida, H. (1983) *Nippon Kagaku Kaishi*, 1137.
125. Zhang, X., Ma, C., Wang, L. and Zhang, J. (1995) *Talanta*, **42,** 897.
126. Kaplin, A. A., Gubina, N. G., Pikula, N. P. *et al.* (1990) *Zavod. Lab.*, **56** (6), 26.
127. Gillain, G., Duyckaerts, G. and Disteche, A. (1979) *Anal. Chim. Acta*, **106,** 23.
128. Wang, F., Li, S., Liu, S. *et al.* (1994) *Anal. Lett.*, **27,** 1779.
129. Burns, D. T., Chimpalee, D. and Bullick, H. J. (1993) *Anal. Chim. Acta*, **284,** 195.
130. Du, J. X., Chen, H. T. and Li, H. C. (1993) *Fenxi Huaxue*, **21,** 572.
131. Andreae, M. O., Asmode, J.-F., Foster, P. and Van't dack, L. (1981) *Anal. Chem.*, **53,** 1766.
132. Welz, B. and Sucmanova, M. (1993) *Analyst*, **118,** 1417, 1425.
133. Sturgeon, R. E., Willie, S. N. and Berman, S. S. (1985) *Anal. Chem.*, **57,** 2311.
134. Narsito, Agterdenbos, J. and Bax, D. (1991) *Anal. Chim Acta*, **244,** 129.
135. Morishige, Y., Hirokawa, K. and Yasuda, K. (1994) *Fresenius' J. Anal. Chem.*, **350,** 410.
136. Havezov, I., Detcheva, A. and Rendl, J. (1995) *Mikrochim. Acta*, **119,** 147.
137. Dai, J. (1989) *Fenxi Shiyanshi*, **8** (2), 61.
138. Chen, H., Brindle, I. D. and Zheng, S. (1992) *Analyst*, **117,** 1603.
139. Chiba, K., Inamoto, I. and Saeki, M. (1992) *J. Anal. At. Spectrom.*, **7,** 115.
140. Jochum, K. P. and Jenner, G. (1994) *Fresenius' J. Anal. Chem.*, **350,** 310.
141. Hudgens, J. E. Jr and Cali, P. J. (1952) *Anal. Chem.*, **24,** 171.
142. Hamaguchi, H., Kuroda, R., Tomura, K. *et al.* (1961) *Geochim. Cosmochim. Acta*, **23,** 296.
143. Howie, R. A., Molokhia, M. M. and Smith, H. (1965) *Anal. Chem.*, **37,** 1059.
144. Waller, P. A. and Pickering, W. F. (1995) *Talanta*, **42,** 197.
145. Sharma, M. and Patel, K. S. (1993) *Int. J. Environ. Anal. Chem.*, **50,** 63.
146. de la Calle Guntinas, M. B., Madrid, Y. and Camara, C. (1991) *Analyst*, **116,** 1029.
147. Rondon, C., Burguera, J. L., Burguera, M. *et al.* (1995) *Fresenius' J. Anal. Chem.*, **353,** 133.
148. Iwamoto, E., Inoike, Y., Yamamoto, Y. and Hayashi, Y. (1986) *Analyst*, **111,** 295.
149. Xu, Z. G., Wu, Y. L. and Xu, D. R. (1992) *Fenxi Huaxue*, **20,** 1321.
150. Menendez Garcia, A., Perez Rodriguez, M. C., Sanchez Uria, J. E. and Sanz-Medel, A. (1995) *Fresenius' J. Anal. Chem.*, **353,** 128.
151. Sayama, Y., Fukaya, T. and Kuno, Y. (1995) *Bunseki Kagaku*, **44,** 569.
152. Watanabe, K., Takamoto, N. and Fueki, K. (1993) *Bunseki Kagaku*, **42,** 33.
153. Tsalev, D. L., Sperling, M. and Welz, B. (1992) *Analyst*, **117,** 1729, 1735.
154. Sen Gupta, J. G. and Bouvier, J. L. (1995) *Talanta*, **42,** 269.
155. Lee, D. S. (1982) *Anal. Chem.*, **54,** 1682.
156. Donaldson, E. M. and Wang, M. (1986) *Talanta*, **33,** 233.
157. Grossman, A. M., Ciba, J. and Jurczyk, J. (1988) *Chem. Anal. (Warsaw)*, **33,** 933.
158. Sayama, Y., Hayashibe, Y. and Fukuda, M. (1995) *Fresenius' J. Anal. Chem.*, **353,** 162.
159. Kuroda, R., Oguma, K., Mori, Y. *et al.* (1991) *Chromatographia*, **32,** 583.

160. Meintjies, E., Strelow, F. W. E. and Victor, A. H. (1987) *Talanta*, **34,** 401.
161. Strelow, F. W. E. (1991) *Talanta*, **38,** 923.
162. Adsul, J. S., Dias, C. C., Iyer, S. G. and Venkateswarlu, Ch. (1987) *Talanta*, **34,** 503.
163. Uehara, N., Morimoto, K. and Shijo, Y. (1992) *Analyst*, **117,** 977.
164. Steenkamp, P. A. and Coetzee, P. P. (1993) *Fresenius' J. Anal. Chem.*, **346,** 1017.
165. Saraswati, R. and Rao, T. H. (1993) *Talanta*, **40,** 565.
166. Fritz, J. S. (1963) Bismuth, in *Treatise on Analytical Chemistry*, part II, vol. 8 (eds I. M. Kolthoff and P. J. Elving), John Wiley, New York.
167. Wynne, E. A. (1968) Bismuth, in *Encyclopedia of Industrial Chemical Analysis*, vol. 7 (eds F. D. Snell and C. L. Hilton), John Wiley, New York.
168. Young, R. S. (1978) *Bull. Bismuth Inst.*, Supplement 20.
169. Young, R. S. (1979) *Bull. Bismuth Inst.*, Supplements 21, 22.
170. Flanagan, F. J. (1973) *Geochim. Cosmochim. Acta*, **37,** 1189.
171. Imai, N., Terashima, S., Itoh, S. and Ando, A. (1995) *Geochem. J.*, **29,** 91.
172. Imai, N., Terashima, S., Itoh, S. and Ando, A. (1995) *Geostand. Newslett.*, **19,** 135.
173. Hafez, M. A. E.-H. (1992) *Talanta*, **39,** 1189.
174. Vanchikova, E. V., Rychkova, L. V. and Belykh, L. I. (1993) *Zavod. Lab.*, **59** (1), 10.
175. Komorsky-Lovric, S. and Branica, M. (1993) *Anal. Chim. Acta*, **276,** 361.
176. Kravchenko, V. B., Lebedeva, Z. M. and Sobolev, A. T. (1991) *Zavod. Lab.*, **57** (11), 15.
177. Jiang, M., Meng, F., Gong, C. and Zhao, Z. (1990) *Analyst*, **115,** 49.
178. Dong, K. L., Kryger, L., Christensen, J. K. and Thomsen, K. N. (1991) *Talanta*, **38,** 101.
179. Liu, K. and Wu, Q. (1992) *Electroanalysis (N.Y.)*, **4,** 569.
180. Kawai, H., Harada, S., Arai, K. and Inoue, S. (1980) *Bunseki Kagaku*, **29,** 99.
181. Asatiani, N. G., Vasil'eva, M. G. and Korostyleva, L. D. (1990) *Zh. Anal. Khim.*, **45,** 1006.
182. Bauer, K.-H., Neeb, R. and Eschnauer, H. (1989) *Fresenius' Z. Anal. Chem.*, **335,** 808.
183. Tian, Q., Zhao, W., Xie, Y. and Qin, Y. (1994) *Fenxi Huaxue*, **22,** 642.
184. Proskurnin, M. A., Abroskin, A. G., Golovko, I. V. and Ivanova, E. K. (1994) *Zh. Anal. Khim.*, **49,** 263.
185. Burns, D. T., Chimpalee, N. and Harriott, M. (1989) *Anal. Chim. Acta*, **225,** 449.
186. Presnyak, I. S., Shelikhina, E. I., Antonovich, V. P. and Nazarenko, V. A. (1990) *Zh. Anal. Khim.*, **45,** 1548.
187. Sanz, J., De Marcos, S., Galban, J. and Gallarta, F. (1993) *Analusis*, **21,** 27.
188. Chen, J.-S., Berndt, H. and Tölg, G. (1992) *Fresenius' J. Anal. Chem.*, **344,** 526.
189. Bettinelli, M., Spezia, S. and Bizzarri, G. (1992) *At. Spectrosc.*, **13,** 75.
190. Bel'skii, N. K., Ochertyanova, L. I. and Krasilov, Yu. I. (1990) *Zh. Anal. Khim.*, **45,** 809.
191. Chattopadhyay, P. and Nathan, S. S. (1991) *Analyst*, **116,** 1145.
192. Xuan, W. and Chen, M. (1989) *Fenxi Huaxue*, **17,** 835.

Index

For the sake of convenience, the index has been compiled with most headings under bold main headings for each of three elements. Each of the element main headings is divided into a general category followed by headings according to oxidation state. For the transition metal derivatives where the oxidation state formalism is not always unambiguous, a separate main heading has been used; in these sections, R = organo group and X = halide.